A Textbook of

Integral Calculus

for

JEE Main & Advanced

Leading Edge Texts

A Textbook of Integral Calculus for JEE Main & Advanced

The Only text book which Starts from fundamentals and gradually builds your concepts upto the level required for Engineering Entrances and finally will place you among toppers.

Amit M Agarwal

arihant
Inspiring Minds.Inspiring Lives

Arihant Prakashan, Meerut

Arihant Prakashan, Meerut

Administrative & Production Offices

Corporate Office: 4577/15, Agarwal Road, Darya Ganj, New Delhi -110002
Tele: 011- 47630600, 23280316; Fax: 011- 23280316

Head Office: Kalindi, TP Nagar, Meerut (UP) - 250002
Tele: 0121-2401479, 2512970, 4004199; Fax: 0121-2401648

All disputes subject to Meerut (UP) jurisdiction only.

Sales & Support Offices

Agra, Ahmedabad, Bengaluru, Bhubaneswar, Chennai, Delhi(I&II), Guwahati, Haldwani, Hyderabad, Jaipur, Kolkata, Kota, Lucknow, Nagpur, Meerut, Patna & Pune

ISBN

978-93-5094-396-0

Price : ₹ 340

Typeset by Arihant DTP Unit at Meerut

Printed & Bound in India
by Arihant Print Solution

For further information about the products from Arihant,
log on to www.arihantbooks.com or email to info@arihantbooks.com

Preface

It is a matter of great pride and honour for me to have received such an overwhelming response to the previous editions of this book from the readers. In a way, this has inspired me to revise this book thoroughly as per the changed pattern of JEE Main & Advanced other Engineering Entrances. I have tried to make the contents more relevant to the needs of students, many topics have been re-written, a lot of new problems of new types have been added in this book now. I have made all possible efforts to remove all the printing errors that had crept in previous editions. The book is now in such a shape that the students would feel at ease while going through the problems which will in turn clear their concepts too.

Some of the Sailent Features of the New Edition are:

- Thoroughly revised theory covering latest changes in the subject & the exam pattern.
- Solved Examples of new types has been added.
- Exercise have been enriched with Objective Questions of latest types of questions; Single Option Correct, Multiple Option Correct, Linked Comprehension Based, Matching Type, Assertion-Reason, Single Integer Answer Type etc.
- All Exercises have been fully solved.

Valuable suggestions from students and teachers are welcome, and these will find due places in the ensuing editions.

Amit M Agarwal

Contents

1

Indefinite Integrals

Chapter in a Snapshot

- Definition
- Fundamental Integration Formulae
- Methods of Integration
- Integration by Parts
- Integration of the Type $\int(\sin^m x \cdot \cos^n x)\, dx$
- Some Special Integrals
- Some Important Substitutions
- Trigonometric Integrals
- Integration of Rational Functions Using Partial Fraction
- Indirect and Derived Substitutions
- Integration of Some Special Irrational Algebraic Functions
- Integrals of the Form $\int x^m (a+bx^n)^P dx$
- Integration Using Euler's Substitutions
- Introduction of Reduction Formulae Over Indefinite Integrals
- Integration Using Differentiation
- Some Integrals which Cannot be Found
- Review of Methods of Integration

Do You Know

It was in this aspect that the process of integration was treated by Leibnitz, the symbol of $\int$ being regarded as the initial letter of the word Sum, in the same way as the symbol of differentiation d is the initial letter in the word difference.

Definition

If f and g are functions of x such that $g'(x) = f(x)$, then the function g is called a anti-derivative (or primitive function or simply integral) of f w.r.t. x.

It is written symbolically,

$$\int f(x)\,dx = g(x), \text{ where } \frac{d}{dx} g(x) = f(x)$$

Points to Consider

1. In other words,

$$\int f(x)\,dx = g(x) \text{ iff } g'(x) = f(x)$$

2. $\int f(x)\,dx = g(x) + c$, where c is constant, $[\because\ (g(x)+c)' = g'(x) = f(x)]$ and c is called constant of integration.

Illustration 1 If $\frac{d}{dx}[x^{n+1} + c] = (n+1)x^n$, then find $\int x^n\,dx$.

Solution. As, $\frac{d}{dx}[x^{n+1} + c] = (n+1)x^n$

$\Rightarrow$ $(x^{n+1} + c)$ is anti-derivative or integral of $(n+1)x^n$

$\therefore$ $$\int x^n\,dx = \frac{x^{n+1}}{n+1} + C$$

Illustration 2 Evaluate $\int e^x(\cos x - \sin x)\,dx$.

Solution. Here, $e^x(\cos x - \sin x)$ is the derivative of $e^x \cos x$

$\Rightarrow$ $$\int e^x(\cos x - \sin x)\,dx = e^x \cos x + C$$

Fundamental Integration Formulae

Since, $$\frac{d}{dx}\{g(x)\} = f(x)$$

$\Leftrightarrow$ $$\int f(x)\,dx = g(x) + C$$

Therefore, based upon this definition and various standard differentiation formulas, we obtain the following integration formulae

(i) $\frac{d}{dx}\left(\frac{x^{n+1}}{n+1}\right) = x^n,\ n \neq -1 \quad \Rightarrow \quad \int x^n\, dx = \frac{x^{n+1}}{n+1} + C,\ n \neq -1$

(ii) $\frac{d}{dx}(\log|x|) = \frac{1}{x} \quad \Rightarrow \quad \int \frac{1}{x}\, dx = \log|x| + C$, when $x \neq 0$

(iii) $\frac{d}{dx}(e^x) = e^x \quad \Rightarrow \quad \int e^x\, dx = e^x + C$

(iv) $\frac{d}{dx}\left(\frac{a^x}{\log_e a}\right) = a^x,\ a > 0,\ a \neq 1 \quad \Rightarrow \quad \int a^x\, dx = \frac{a^x}{\log a} + C$

(v) $\frac{d}{dx}(-\cos x) = \sin x \quad \Rightarrow \quad \int \sin x\, dx = -\cos x + C$

(vi) $\frac{d}{dx}(\sin x) = \cos x \quad \Rightarrow \quad \int \cos x\, dx = \sin x + C$

(vii) $\frac{d}{dx}(\tan x) = \sec^2 x \quad \Rightarrow \quad \int \sec^2 x\, dx = \tan x + C$

(viii) $\frac{d}{dx}(-\cot x) = \text{cosec}^2 x \quad \Rightarrow \quad \int \text{cosec}^2 x\, dx = -\cot x + C$

(ix) $\frac{d}{dx}(\sec x) = \sec x \tan x \quad \Rightarrow \quad \int \sec x \tan x\, dx = \sec x + C$

(x) $\frac{d}{dx}(-\text{cosec}\, x) = \text{cosec}\, x \cot x \quad \Rightarrow \quad \int \text{cosec}\, x \cot x\, dx = -\text{cosec}\, x + C$

(xi) $\frac{d}{dx}(\log|\sin x|) = \cot x \quad \Rightarrow \quad \int \cot x\, dx = \log|\sin x| + C$

(xii) $\frac{d}{dx}(-\log|\cos x|) = \tan x \quad \Rightarrow \quad \int \tan x\, dx = -\log|\cos x| + C$

(xiii) $\frac{d}{dx}(\log|\sec x + \tan x|) = \sec x \quad \Rightarrow \quad \int \sec x\, dx = \log|\sec x + \tan x| + C$

(xiv) $\frac{d}{dx}(\log|\text{cosec}\, x - \cot x|) = \text{cosec}\, x$

$\Rightarrow \quad \int \text{cosec}\, x\, dx = \log|\text{cosec}\, x - \cot x| + C$

(xv) $\frac{d}{dx}\left(\sin^{-1}\frac{x}{a}\right) = \frac{1}{\sqrt{a^2 - x^2}} \quad \Rightarrow \quad \int \frac{dx}{\sqrt{a^2 - x^2}} = \sin^{-1}\left(\frac{x}{a}\right) + C$

(xvi) $\frac{d}{dx}\left(\cos^{-1}\frac{x}{a}\right) = \frac{-1}{\sqrt{a^2 - x^2}} \quad \Rightarrow \quad \int \frac{-1}{\sqrt{a^2 - x^2}}\, dx = \cos^{-1}\left(\frac{x}{a}\right) + C$

(xvii) $\frac{d}{dx}\left(\frac{1}{a}\tan^{-1}\frac{x}{a}\right) = \frac{1}{a^2 + x^2} \quad \Rightarrow \quad \int \frac{dx}{a^2 + x^2} = \frac{1}{a}\tan^{-1}\left(\frac{x}{a}\right) + C$

(xviii) $\frac{d}{dx}\left(\frac{1}{a}\cot^{-1}\frac{x}{a}\right) = \frac{-1}{a^2 + x^2} \quad \Rightarrow \quad \int \frac{-1}{a^2 + x^2}\, dx = \frac{1}{a}\cot^{-1}\left(\frac{x}{a}\right) + C$

(xix) $\frac{d}{dx}\left(\frac{1}{a}\sec^{-1}\frac{x}{a}\right) = \frac{1}{x\sqrt{x^2 - a^2}} \quad \Rightarrow \quad \int \frac{dx}{x\sqrt{x^2 - a^2}} = \frac{1}{a}\sec^{-1}\left(\frac{x}{a}\right) + C$

(xx) $\frac{d}{dx}\left(\frac{1}{a}\text{cosec}^{-1}\frac{x}{a}\right) = \frac{-1}{x\sqrt{x^2 - a^2}} \Rightarrow \int \frac{-dx}{x\sqrt{x^2 - a^2}} = \frac{1}{a}\text{cosec}^{-1}\left(\frac{x}{a}\right) + C$

Points to Consider

1. $\int k f(x)\, dx = k \int f(x)\, dx$, where k is constant.

   ie, The integral of the product of a constant and a function = the constant × integral of the function.

2. $\int \{f_1(x) \pm f_2(x) \pm \ldots \pm f_n(x)\}\, dx = \int f_1(x)\, dx \pm \int f_2(x)\, dx \pm \ldots \pm \int f_n(x)\, dx.$

   ie, The integral of the sum or difference of a finite number of functions is equal to the sum or difference of the integrals of the various functions.

3. **Geometrical interpretation**

   eg, $$y = \int x\, dx = \frac{x^2}{2} + C$$

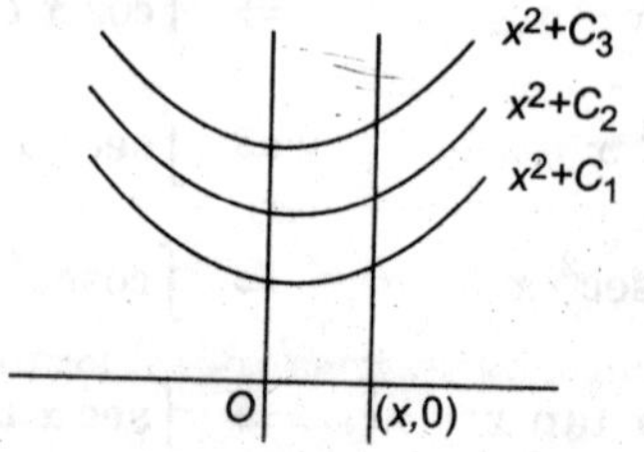

Fig. 1.1

$$\therefore \quad y = \int f(x)\, dx = F(x) + C$$

$$\Rightarrow \quad F'(x) = f(x);\ F'(x_1) = f(x_1)$$

Hence, $y = \int f(x) dx$ denotes a family of curves such that the slope of the tangent at $x = x_1$ on every member is same *ie*, $F'(x_1) = f(x)$ [when x_1 lies in the domain of $f(x)$]

Hence, anti-derivative of a function is not unique. If $g_1(x)$ and $g_2(x)$ are two anti-derivatives of a function $f(x)$ on $[a, b]$, then they differ only by a constant.

ie, $$g_1(x) - g_2(x) = C$$

4. Anti-derivative of a continuous function is differentiable.

   ie, If $f(x)$ is continuous, then

$$\int f(x)\, dx = F(x) + C$$

$$\Rightarrow \quad F'(x) = f(x)$$

$\Rightarrow$ $F'(x)$

$\Rightarrow$ always exists and is continuous.

5. If integrand is discontinuous at $x = x_1$, then its anti-derivative at $x = x_1$ need not be discontinuous.

   eg, $\int x^{-1/3} dx$. Here, $x^{-1/3}$ is discontinuous at $x = 0$.

   But $\int x^{-1/3} dx = \frac{3}{2} x^{2/3} + C$ is continuous at $x = 0$.

6. Anti-derivative of a periodic function need not be a periodic function.

   eg, $f(x) = \cos x + 1$ is periodic but $\int (\cos x + 1)\, dx = \sin x + x + C$ is aperiodic.

Illustration 3 Evaluate

(i) $\int \frac{x^2+5x-1}{\sqrt{x}}\,dx$ (ii) $\int (x^2+5)^3\,dx$

Solution. (i) $I = \int \frac{x^2+5x-1}{\sqrt{x}}\,dx$

$$= \int \left(\frac{x^2}{x^{1/2}} + \frac{5x^1}{x^{1/2}} - \frac{1}{x^{1/2}}\right) dx$$

$$= \int (x^{3/2} + 5x^{1/2} - x^{-1/2})\,dx$$

$$= \frac{x^{3/2+1}}{3/2+1} + \frac{5x^{1/2+1}}{1/2+1} - \frac{x^{-1/2+1}}{-1/2+1} + C$$

$$\left[\text{using } \int x^n\,dx = \frac{x^{n+1}}{n+1} + C\right]$$

$$\Rightarrow \quad I = \frac{2}{5}x^{5/2} + \frac{2}{3}\cdot 5x^{3/2} - 2x^{1/2} + C$$

(ii) $I = \int (x^2+5)^3\,dx$

Expanding the integrand by the binomial formula

$$I = \int (x^6 + 15x^4 + 75x^2 + 125)\,dx$$

$$I = \frac{x^7}{7} + \frac{15x^5}{5} + \frac{75x^3}{3} + 125x + C$$

$$I = \frac{x^7}{7} + 3x^5 + 25x^3 + 125x + C$$

Illustration 4 Evaluate

(i) $\int \tan^2 x\,dx$ (ii) $\int \frac{dx}{\sin^2 x \cos^2 x}$

(iii) $\int \frac{\sin^6 x + \cos^6 x}{\sin^2 x \cos^2 x}\,dx$ (iv) $\int \frac{\cos x - \cos 2x}{1-\cos x}\,dx$

Solution. (i) $I = \int \tan^2 x\,dx$

$$I = \int (\sec^2 x - 1)\,dx$$

$$I = \int \sec^2 x\,dx - \int 1\,dx$$

$$\Rightarrow \quad I = \tan x - x + c$$

(ii) $I = \int \frac{1}{\sin^2 x \cos^2 x}\,dx$

$$I = \int \frac{\sin^2 x + \cos^2 x}{\sin^2 x \cos^2 x}\,dx \qquad (\text{as } \sin^2 x + \cos^2 x = 1)$$

$$I = \int \frac{\sin^2 x}{\sin^2 x \cos^2 x}\,dx + \int \frac{\cos^2 x}{\sin^2 x \cos^2 x}\,dx$$

$$I = \int \sec^2 x\,dx + \int \operatorname{cosec}^2 x\,dx$$

$$I = \tan x - \cot x + C$$

(iii) $$I = \int \frac{\sin^6 x + \cos^6 x}{\sin^2 x \cos^2 x} dx$$

$$I = \int \frac{(\sin^2 x + \cos^2 x)^3 - 3\sin^2 x \cos^2 x (\sin^2 x + \cos^2 x)}{\sin^2 x \cos^2 x} dx$$

[using $(a+b)^3 = a^3 + b^3 + 3ab(a+b)$]

$$I = \int \frac{1 - 3\sin^2 x \cos^2 x}{\sin^2 x \cos^2 x} dx$$

$$= \int \frac{1}{\sin^2 x \cos^2 x} dx - \int 3\, dx$$

$$= \int \frac{(\sin^2 x + \cos^2 x)}{\sin^2 x \cos^2 x} dx - 3x + C$$

$$= \int \sec^2 x\, dx + \int \operatorname{cosec}^2 x\, dx - 3x + C$$

$$= \tan x - \cot x - 3x + C$$

(iv) $$I = \int \frac{\cos x - \cos 2x}{1 - \cos x} dx$$

$$I = \int \frac{\cos x - (2\cos^2 x - 1)}{1 - \cos x} dx$$

$$I = \int \frac{-(2\cos x + 1)(\cos x - 1)}{-(\cos x - 1)} dx$$

$$I = \int (2\cos x + 1)\, dx$$

$$I = 2\sin x + x + C$$

Illustration 5 Evaluate

(i) $\int \frac{x^3}{x+2} dx$ (ii) $\int \frac{x^2}{x^2+5} dx$

Solution. (i) $$I = \int \frac{x^3}{x+2} dx$$

$$I = \int \left(x^2 - 2x + 4 - \frac{8}{x+2} \right) dx$$

$$I = \frac{x^3}{3} - x^2 + 4x - 8 \log |x+2| + C$$

(ii) $$I = \int \frac{x^2}{x^2+5} dx$$

$$= \int \left(1 - \frac{5}{x^2+5} \right) dx = x - 5 \int \frac{dx}{x^2 + (\sqrt{5})^2}$$

$$= x - \frac{5}{\sqrt{5}} \tan^{-1}\left(\frac{x}{\sqrt{5}} \right) + C$$

$\therefore$ $$I = x - \sqrt{5} \tan^{-1}\left(\frac{x}{\sqrt{5}} \right) + C$$

Points to Consider

In rational algebraic functions if the degree of numerator is greater than or equal to degree of denominator, then always divide the numerator by denominator and use the result of integration.

Illustration 6 Solve

(i) $\int 5^{\log_e x}\, dx$ (ii) $\int 2^{\log_4 x}\, dx$

Solution. (i)

$$I = \int 5^{\log_e x}\, dx$$

$$= \int x^{\log_e 5}\, dx \quad (\text{as } a^{\log_e b} = b^{\log_e a})$$

$$= \frac{x^{\log_e 5 + 1}}{(\log_e 5 + 1)} + C$$

$$\therefore \quad \int 5^{\log_e x}\, dx = \frac{x^{\log_e 5+1}}{\log_e 5 + 1} + C$$

(ii)

$$I = \int 2^{\log_4 x}\, dx$$

$$= \int 2^{\log_{2^2} x}\, dx$$

$$= \int 2^{1/2 \log_2 x}\, dx \quad \left(\text{using } \log_{b^n} x = \frac{1}{n}\log_b x\right)$$

$$= \int 2^{\log_2 \sqrt{x}}\, dx$$

$$= \int \sqrt{x}\, dx \quad (\text{using } a^{\log_a b} = b)$$

$$= \frac{x^{3/2}}{3/2} + C$$

$$\therefore \quad \int 2^{\log_4 x}\, dx = \frac{2}{3}x^{3/2} + C$$

Methods of Integration

If the integrand is not a derivative of a simple function, then the corresponding integrals cannot be found directly. In order to find the integral of complex problems, we follow the following rules :

Integration by Substitution (or by change of the independent variable)

If $g(x)$ is a continuously differentiable function, then to evaluate integrals of the form,

$$I = \int f(g(x)) \,.\, g'(x)\, dx,$$

we substitute $g(x) = t$ and $g'(x)\, dx = dt$

The substitution reduces the integral to $\int f(t)\, dt$.

After evaluating this integral we substitute back the value of t.

Illustration 7 Evaluate

(i) $\int \frac{\sin(\log x)}{x}\, dx$ (ii) $\int \left(\frac{3\sin x + 4\cos x}{4\sin x - 3\cos x} \right) dx$

(iii) $\int \frac{e^{m\tan^{-1} x}}{1+x^2}\, dx$ (iv) $\int x \sin(4x^2 + 7)\, dx$

Solution. (i) $I = \int \frac{\sin(\log x)}{x}\, dx$

We know, $\frac{d}{dx}(\log x) = \frac{1}{x}$

Thus, let $\log x = t \Rightarrow \frac{1}{x}\, dx = dt$ ...(i)

$\therefore$ $I = \int \sin(t)\, dt = -\cos(t) + C$

$= -\cos(\log x) + C$ [using Eq. (i)]

(ii) $I = \int \frac{3\sin x + 4\cos x}{4\sin x - 3\cos x}\, dx$

We know, $\frac{d}{dx}(4\sin x - 3\cos x) = (4\cos x + 3\sin x)$

Thus, let $4\sin x - 3\cos x = t$ ...(i)

$\Rightarrow$ $(4\cos x + 3\sin x)\, dx = dt$

$\therefore$ $I = \int \frac{dt}{t} = \log|t| + C$

$\Rightarrow$ $I = \int \frac{3\sin x + 4\cos x}{4\sin x - 3\cos x}\, dx$

$= \log|4\sin x - 3\cos x| + C$ [using Eq. (i)]

(iii) $I = \int \frac{e^{m\tan^{-1} x}}{1+x^2}\, dx$ let $m\tan^{-1} x = t$

$I = \int e^t \cdot \frac{dt}{m}$ $\Rightarrow \frac{m}{1+x^2}\, dx = dt$

$I = \frac{1}{m} \int e^t\, dt$ $\Rightarrow \frac{1}{1+x^2}\, dx = \frac{1}{m}\, dt$

$I = \frac{1}{m} e^t + C = \frac{1}{m} e^{m\tan^{-1} x} + C$

(iv) $I = \int x \sin(4x^2 + 7)\, dx$

Let $4x^2 + 7 = t$

$\Rightarrow$ $8x\, dx = dt,\ x\, dx = \frac{1}{8}\, dt$

$\therefore$ $I = \int \sin(t) \frac{dt}{8}$

$\Rightarrow$ $I = -\frac{1}{8}\cos(t) + C$

$I = -\frac{1}{8}\cos(4x^2 + 7) + C$

Points to Consider

1. If $\int f(x)\,dx = g(x) + C$, then $\int f(ax+b)\,dx = \frac{1}{a}\,g(ax+b) + C$

2. If $\int \frac{1}{x}\,dx = \log|x| + C$, then $\int \frac{1}{ax+b}\,dx = \frac{1}{a}\log|ax+b| + C$

Thus, in any fundamental integral formulae given in article fundamental integration formulae if in place of x we have $(ax+b)$, then same formula is applicable but we must divide by coefficient of x or derivative of $(ax+b)$ *ie*, a.

Here is the list of some of frequently used formulae

(i) $\int (ax+b)^n\,dx = \frac{(ax+b)^{n+1}}{a(n+1)} + C,\ n \neq -1$

(ii) $\int \frac{1}{ax+b}\,dx = \frac{1}{a}\log|ax+b| + C$

(iii) $\int e^{ax+b}\,dx = \frac{1}{a}\,e^{ax+b} + C$

(iv) $\int a^{bx+c}\,dx = \frac{1}{b}\cdot\frac{a^{bx+c}}{\log a} + C$

(v) $\int \sin(ax+b)\,dx = -\frac{1}{a}\cos(ax+b) + C$

(vi) $\int \cos(ax+b)\,dx = \frac{1}{a}\sin(ax+b) + C$

(vii) $\int \sec^2(ax+b)\,dx = \frac{1}{a}\tan(ax+b) + C$

(viii) $\int \operatorname{cosec}^2(ax+b)\,dx = -\frac{1}{a}\cot(ax+b) + C$

(ix) $\int \sec(ax+b)\tan(ax+b)\,dx = \frac{1}{a}\sec(ax+b) + C$

(x) $\int \operatorname{cosec}(ax+b)\cot(ax+b)\,dx = -\frac{1}{a}\operatorname{cosec}(ax+b) + C$

(xi) $\int \tan(ax+b)\,dx = -\frac{1}{a}\log|\cos(ax+b)| + C$

(xii) $\int \cot(ax+b)\,dx = \frac{1}{a}\log|\sin(ax+b)| + C$

(xiii) $\int \sec(ax+b)\,dx = \frac{1}{a}\log|\sec(ax+b) + \tan(ax+b)| + C$

(xiv) $\int \operatorname{cosec}(ax+b)\,dx = \frac{1}{a}\log|\operatorname{cosec}(ax+b) - \cot(ax+b)| + C$

Illustration 8 Evaluate

(i) $\int \cos 4x \cos 7x \, dx$ (ii) $\int \cos x \cos 2x \cos 5x \, dx$

Solution. When calculating such integrals it is advisable to use the trigonometric product formulae.

(i) $\int \cos 4x \cos 7x \, dx$

Here, $\cos 4x \cos 7x = \frac{1}{2}(\cos 3x + \cos 11x)$

$$\therefore \quad I = \int \cos 4x \cos 7x \, dx$$

$$= \frac{1}{2}\int (\cos 3x + \cos 11x)\, dx$$

$$= \frac{1}{2}\int \cos 3x \, dx + \frac{1}{2}\int \cos 11x \, dx$$

$$= \frac{\sin 3x}{6} + \frac{\sin 11x}{22} + C$$

(ii) $\int \cos x \cos 2x \cos 5x \, dx$

We have, $(\cos x \cos 2x)\cos 5x = \frac{1}{2}(\cos x + \cos 3x)\cos 5x$

$$= \frac{1}{4}\{2\cos x \cos 5x + 2\cos 3x \cos 5x\}$$

$$= \frac{1}{4}\{(\cos 4x + \cos 6x) + (\cos 2x + \cos 8x)\}$$

$$\therefore \quad \cos x \cos 2x \cos 5x = \frac{1}{4}\{\cos 2x + \cos 4x + \cos 6x + \cos 8x\}$$

$$\therefore \quad I = \int (\cos x \cos 2x \cos 5x)\, dx$$

$$= \frac{1}{4}\int (\cos 2x + \cos 4x + \cos 6x + \cos 8x)\, dx$$

$$= \frac{\sin 2x}{8} + \frac{\sin 4x}{16} + \frac{\sin 6x}{24} + \frac{\sin 8x}{32} + C$$

Points to Consider

While solving such problems it is expedient to use the following trigonometric identities :

1. $\sin mx \cos nx = \frac{1}{2}\{\sin(m-n)x + \sin(m+n)x\}$
2. $\cos mx \sin nx = \frac{1}{2}\{\sin(m+n)x - \sin(m-n)x\}$
3. $\sin mx \sin nx = \frac{1}{2}\{\cos(m-n)x - \cos(m+n)x\}$
4. $\cos mx \cos nx = \frac{1}{2}\{\cos(m-n)x + \cos(m+n)x\}$

Illustration 9 Evaluate

(i) $\int x\sqrt{x-5}\,dx$ (ii) $\int \frac{8x+13}{\sqrt{4x+7}}\,dx$

(iii) $\int \frac{x^3}{\sqrt{1+2x}}\,dx$ (iv) $\int \frac{x^2+1}{(x^2-1)^2}\,dx$

(v) $\int \frac{x^2}{(a+bx)^2}\,dx$ (vi) $\int \frac{dx}{\sqrt{x+1}-\sqrt{x}}$

Solution. (i) $I = \int x\sqrt{x-5}\,dx$

Let $x-5 = t^2 \Rightarrow dx = 2t\,dt$

$\therefore$ $I = \int (t^2+5)\cdot t\cdot 2t\,dt = 2\int (t^4+5t^2)\,dt$

$$= 2\left\{\frac{t^5}{5} + \frac{5t^3}{3}\right\} + C$$

$$= 2\left[\frac{(x-5)^{5/2}}{5} + \frac{5(x-5)^{3/2}}{3}\right] + C$$

(ii) $I = \int \frac{8x+13}{\sqrt{4x+7}}\,dx$

Let $4x+7 = t^2$

$\therefore$ $4dx = 2t\,dt,\quad dx = \frac{1}{2}t\,dt$

$\therefore$ $I = \int \frac{2(t^2-7)+13}{t}\cdot\frac{1}{2}t\,dt$

$\Rightarrow$ $I = \frac{1}{2}\int (2t^2-14+13)\,dt$

$$= \frac{1}{2}\int (2t^2-1)\,dt$$

$$I = \frac{1}{3}t^3 - \frac{t}{2} + C$$

$\therefore$ $I = \frac{1}{3}(4x+7)^{3/2} - \frac{1}{2}(4x+7)^{1/2} + C$

(iii) $I = \int \frac{x^3}{\sqrt{1+2x}}\,dx$, let $1+2x = t^2$

$2dx = 2t\,dt,\ dx = t\,dt$

$\therefore$ $I = \int \frac{\left(\frac{t^2-1}{2}\right)^3\cdot t\,dt}{t}$

$$I = \frac{1}{8}\int (t^6-3t^4+3t^2-1)\,dt$$

$$= \frac{1}{8}\left(\frac{t^7}{7} - \frac{3}{5}t^5 + t^3 - t\right) + C$$

$\therefore$ $I = \frac{1}{8}\left\{\frac{(2x+1)^{7/2}}{7} - \frac{3}{5}(1+2x)^{5/2} + (2x+1)^{3/2} - (2x+1)^{1/2}\right\} + C$

(iv) $$I = \int \frac{x^2+1}{(x^2-1)^2}\,dx$$

$$I = \int \frac{1+1/x^2}{\left(x-\frac{1}{x}\right)^2}\,dx \qquad \text{(dividing } N^r \text{ and } D^r \text{ by } x^2)$$

$\therefore$ $$I = \int \frac{dt}{t^2} \quad \text{let } x-\frac{1}{x}=t \Rightarrow \left(1+\frac{1}{x^2}\right)dx = dt$$

$$= -\frac{1}{t}+C$$

$$I = -\frac{1}{\left(x-\frac{1}{x}\right)}+C = \frac{x}{1-x^2}+C$$

(v) $$I = \int \frac{x^2}{(a+bx)^2}\,dx = \int \frac{1}{b^2} + \frac{\left(-\frac{2a}{b}x-\frac{a^2}{b^2}\right)}{(bx+a)^2}\,dx$$

$$= \frac{1}{b^2}\int 1\,dx - \frac{a}{b^2}\int \frac{2bx+a}{(bx+a)^2}\,dx$$

$$I = \frac{1}{b^2}x - \frac{a}{b^2}\left\{\int \frac{2bx+2a-a}{(bx+a)^2}\,dx\right\}$$

$$= \frac{1}{b^2}x - \frac{a}{b^2}\left\{2\int \frac{1}{(bx+a)}\,dx - a\int (bx+a)^{-2}\,dx\right\}$$

$$= \frac{1}{b^2}x - \frac{a}{b^2}\left\{\frac{2}{b}\log|bx+a| + \frac{a}{b(bx+a)}\right\}+C$$

$$= \frac{1}{b^3}\left\{bx - 2a\log|bx+a| - \frac{a^2}{(a+bx)}\right\}+C$$

(vi) $$I = \int \frac{dx}{\sqrt{x+1}-\sqrt{x}} \qquad \text{(rationalising } D^r)$$

$$I = \int \frac{\sqrt{x+1}+\sqrt{x}}{(x+1)-(x)}\,dx$$

$$I = \int (\sqrt{x+1}+\sqrt{x})\,dx$$

$$I = \frac{2}{3}(x+1)^{3/2} + \frac{2}{3}(x)^{3/2} + C$$

Integral of the Form

1. $\int f\left(x+\frac{1}{x}\right)\left(1-\frac{1}{x^2}\right)dx$

Put $\quad x+\frac{1}{x}=t \Rightarrow \left(1-\frac{1}{x^2}\right)dx = dt$

2. $\int f\left(x-\frac{1}{x}\right)\left(1+\frac{1}{x^2}\right)dx \quad$ Put $x-\frac{1}{x}=t \Rightarrow \left(1+\frac{1}{x^2}\right)dx = dt$

3. $\int \frac{x^2+1}{x^4+kx^2+1}dx$

Divide numerator and denominator by x^2

4. $\int \frac{x^2-1}{x^4+kx^2+1}dx$

Divide numerator and denominator by x^2

Illustration 10 $\int \frac{x^x(x^{2x}+1)(\ln x+1)}{x^{4x}+1}dx$

Solution. $I = \int \frac{x^x(x^{2x}+1)(\ln x+1)}{x^{4x}+1}dx$

Put $x^x = y \Rightarrow x^x(\ln x+1)\,dx = dy$

$$I = \int \frac{y^2+1}{y^4+1}dy = \int \frac{1+\frac{1}{y^2}}{y^2+\frac{1}{y^2}}dy = \int \frac{1+\frac{1}{y^2}}{\left(y-\frac{1}{y}\right)^2+2}dy$$

Let $y - \frac{1}{y} = t \Rightarrow \left(1+\frac{1}{y^2}\right)dy = dt$

$$I = \int \frac{dt}{t^2+2} = \frac{1}{\sqrt{2}}\left(\tan^{-1}\frac{t}{\sqrt{2}}\right)+C$$

$$= \frac{1}{\sqrt{2}}\tan^{-1}\left(\frac{y-\frac{1}{y}}{\sqrt{2}}\right)+C = \frac{1}{\sqrt{2}}\tan^{-1}\left(\frac{x^x-\frac{1}{x^x}}{\sqrt{2}}\right)+C$$

Illustration 11 Evaluate $\int \frac{(x^2-1)\,dx}{(x^4+3x^2+1)\tan^{-1}\left(x+\frac{1}{x}\right)}$.

Solution. The given integral can be written as

$$I = \int \frac{(1-1/x^2)\,dx}{(x^2+3+1/x^2)\tan^{-1}\left(x+\frac{1}{x}\right)} \quad \text{(dividing } N^r \text{ and } D^r \text{ by } x^2\text{)}$$

$$I = \int \frac{(1-1/x^2)\,dx}{\{(x+1/x)^2+1\}\tan^{-1}\left(x+\frac{1}{x}\right)}$$

Let $x+\frac{1}{x} = t \Rightarrow \left(1-\frac{1}{x^2}\right)dx = dt$

$\therefore$ $I = \int \frac{dt}{(t^2+1)\cdot\tan^{-1}(t)}$ ...(i)

Now, make one more substitution

$\tan^{-1}t = u$. Then, $\frac{dt}{t^2+1} = du$

$\therefore$ Eq. (i) becomes, $I = \int \frac{du}{u} = \log|u| + C$

$\Rightarrow$ $I = \log|\tan^{-1} t| + C = \log|\tan^{-1}(x + 1/x)| + C$

Illustration 12 $\int \frac{(x^{-7/6} - x^{5/6})\,dx}{x^{1/3}(x^2 + x + 1)^{1/2} - x^{1/2}(x^2 + x + 1)^{1/3}}$.

Solution. $I = \int \frac{x^{7/6}(x^{-7/6} - x^{5/6})\,dx}{x^{7/6} \cdot x^{1/3}(x^2 + x + 1)^{1/2} - x^{1/2} \cdot x^{7/6}(x^2 + x + 1)^{1/3}}$

$$= \int \frac{(1 - x^2)\,dx}{x^{3/2}(x^2 + x + 1)^{1/2} - x^{5/3}(x^2 + x + 1)^{1/3}}$$

$$= \int \frac{-\left(1 - \frac{1}{x^2}\right)dx}{\left(x + \frac{1}{x} + 1\right)^{1/2} - \left(x + \frac{1}{x} + 1\right)^{1/3}}$$

$$= -\int \frac{dt}{(t + 1)^{1/2} - (t + 1)^{1/3}} \qquad \left\{ \begin{array}{l} \text{Putting } \quad x + \frac{1}{x} = t \\ \Rightarrow \quad \left(1 - \frac{1}{x^2}\right) dx = dt \end{array} \right.$$

Substitute, $(t + 1) = u^6$

$$= -\int \frac{6u^5\,du}{u^3 - u^2}$$

$$= -6 \int \frac{u^3}{u - 1}\,du, \text{ put } u - 1 = z$$

$$= -6 \int \frac{(z + 1)^3}{z}\,dz$$

$$= -6 \int \frac{z^3 + 3z^2 + 3z + 1}{z}\,dz$$

$$= -6 \int \left(z^2 + 3z + 3 + \frac{1}{z}\right) dz$$

$$= -6 \left\{ \frac{z^3}{3} + \frac{3z^2}{2} + 3z + \log|z| \right\} + C$$

where, $z = \left(x + \frac{1}{x} + 1\right)^{1/6} - 1$

Illustration 13 The value of $\int \{\{[x]\}\}\,dx$, where {.} and [.] denotes fractional part of x and greatest integer function) is equal to

(a) 0 (b) 1 (c) 2 (d) –1

Solution. Let $I = \int \{\{[x]\}\}\,dx$

where, $[x] =$ Integer and we know $\{n\} = 0$; $n \in$ Integer.

$\therefore$ $I = \int 0\,dx = 0$

Hence, (a) is the correct answer.

Illustration 14 The value of $\int [\{x\}]\, dx$; (where [.] and {.} denotes greatest integer and fractional part of x is equal to

(a) 0 (b) 1 (c) 2 (d) –1

Solution. As, we know $y = \{x\}$ could be shown as

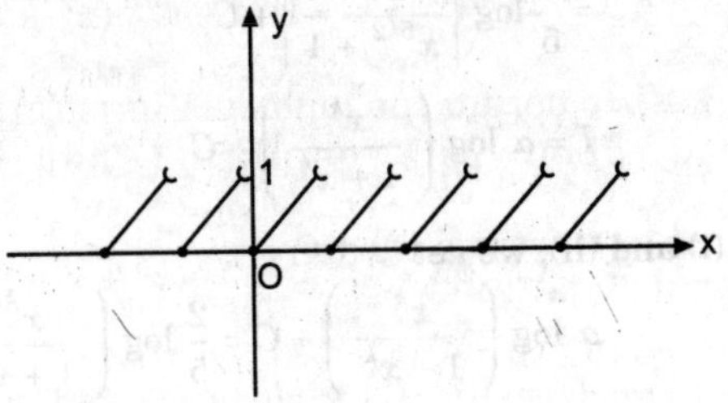

$\therefore \quad [\{x\}] = 0$

Thus, $\quad I = \int [\{x\}]\, dx = \int 0\, dx = 0$

Hence, (a) is the correct answer.

Illustration 15 The value of $\int \frac{d(x^2+1)}{\sqrt{x^2+2}}$, is

(a) $2\sqrt{x^2+2}+C$ (b) $\sqrt{x^2+2}+C$ (c) $x\sqrt{x^2+2}+C$ (d) None of these

Solution. Here, $I = \int \frac{d(x^2+1)}{\sqrt{x^2+2}}$

We know, $d(x^2+1) = 2x\, dx$

$$\therefore \quad I = \int \frac{2x\, dx}{\sqrt{x^2+2}}$$

Put, $x^2 + 2 = t^2$

$\therefore \quad 2x\, dx = 2t\, dt$

$$I = \int \frac{2t\, dt}{t} = 2t + C$$

$$\Rightarrow \quad I = 2\sqrt{x^2+2} + C$$

Hence, (a) is the correct answer.

Illustration 16 If $\int \frac{(\sqrt{x})^5}{(\sqrt{x})^7 + x^6}\, dx = a \log\left(\frac{x^k}{1+x^k}\right) + c$, then a and k are

(a) 2/5, 5/2 (b) 1/5, 2/5 (c) 5/2, 1/2 (d) 2/5, 1/2

Solution. Here, $I = \int \frac{(\sqrt{x})^5}{(\sqrt{x})^7 + x^6}\, dx$

$$= \int \frac{dx}{(\sqrt{x})^2 + (\sqrt{x})^7}$$

$$= \int \frac{dx}{x^{7/2}\left\{1 + \frac{1}{x^{5/2}}\right\}}, \text{ put } \frac{1}{x^{5/2}} = y \Rightarrow -\frac{5}{2x^{7/2}}\, dx = dy$$

$$= -\frac{2}{5}\int \frac{dy}{1+y} = -\frac{2}{5}\log|1+y| + C$$

$$= \frac{2}{5}\log\left|\frac{1}{1+y}\right| + C$$

$$= \frac{2}{5}\log\left|\frac{x^{5/2}}{x^{5/2}+1}\right| + C \quad \text{...(i)}$$

where, $$I = a\log\left(\frac{x^k}{1+x^k}\right) + C \quad \text{(given)} \quad \text{...(ii)}$$

∴ From Eqs. (i) and (ii), we get

$$a\log\left(\frac{x^k}{1+x^k}\right) + C = \frac{2}{5}\log\left(\frac{x^{5/2}}{1+x^{5/2}}\right) + C$$

⇒ $$a = 2/5 \quad \text{and} \quad k = 5/2$$

Hence, (a) is the correct answer.

Illustration 17 The value of $\int \frac{\cos 5x + \cos 4x}{1 - 2\cos 3x}\,dx$, is equal to

(a) $\sin x + \sin 2x + C$

(b) $\sin x - \frac{\sin 2x}{2} + C$

(c) $-\sin x - \frac{\sin 2x}{2} + C$

(d) None of these

Solution. Here, $I = \int \frac{\cos 5x + \cos 4x}{1 - 2\cos 3x}\,dx$

$$= \int \frac{2\cos\frac{9x}{2}\cdot\cos\frac{x}{2}}{1 - 2\left(2\cos^2\frac{3x}{2} - 1\right)}\,dx$$

$$= \int \frac{2\cos\frac{x}{2}\cos\frac{9x}{2}}{3 - 4\cos^2\frac{3x}{2}}\,dx$$

Multiplying and dividing it by $\cos\frac{3x}{2}$, we get

$$= \int \frac{2\cos\frac{x}{2}\cdot\cos\frac{3x}{2}\cdot\cos\frac{9x}{2}}{3\cos\frac{3x}{2} - 4\cos^3\frac{3x}{2}}\,dx$$

$$= \int \frac{2\cos x/2\cdot\cos 3x/2\cdot\cos 9x/2}{-\cos 9x/2}\,dx$$

(using $\cos 3\theta = 4\cos^3\theta - 3\cos\theta$)

$$= -\int 2\cos\frac{x}{2}\cos\frac{3x}{2}\,dx = -\int(\cos 2x + \cos x)\,dx$$

$$= -\left(\frac{\sin 2x}{2} + \sin x\right) + C$$

Hence, (c) is the correct answer.

Illustration 18 The value of $\int \frac{\cos 7x - \cos 8x}{1 + 2\cos 5x}\,dx$, is equal to

(a) $\frac{\sin 2x}{2} + \frac{\cos 3x}{3} + C$ (b) $\sin x - \cos x + C$

(c) $\frac{\sin 2x}{2} - \frac{\cos 3x}{3} + C$ (d) None of these

Solution. Here, $I = \int \frac{\cos 7x - \cos 8x}{1 + 2\cos 5x}\,dx$

$$= \int \frac{2\sin \frac{15x}{2} \cdot \sin \frac{x}{2}}{1 + 2\cos 5x}\,dx$$

Multiplying and dividing by $\sin \frac{5x}{2}$,

$$= \int \frac{2\sin \frac{15x}{2} \cdot \sin \frac{x}{2} \cdot \sin \frac{5x}{2}}{\sin \frac{5x}{2} + 2\sin \frac{5x}{2} \cdot \cos 5x}\,dx$$

$$= \int \frac{2\sin \frac{15x}{2} \cdot \sin \frac{x}{2} \cdot \sin \frac{5x}{2}}{\sin \frac{5x}{2} + \sin \frac{15}{2}x - \sin \frac{5x}{2}}\,dx$$

$$= \int 2\sin \frac{x}{2} \cdot \sin \frac{5x}{2}\,dx$$

$$= \int (\cos 2x - \cos 3x)\,dx$$

$$I = \frac{\sin 2x}{2} - \frac{\sin 3x}{3} + C$$

Hence, (c) is the correct answer.

Illustration 19 If $\int f(x)\cos x\,dx = \frac{1}{2} f^2(x) + c$, then $f(x)$ can be

(a) x (b) 1 (c) $\cos x$ (d) $\sin x$

Solution. Here, $\int f(x)\cos x\,dx = \frac{1}{2} f^2(x) + C$

Differentiating both the sides, we get

$$f(x)\cos x = f(x) \cdot f'(x)$$

ie, $$\cos x = \frac{d}{dx}(f(x))$$

$\Rightarrow$ $$f(x) = \int \cos x\,dx$$

$\Rightarrow$ $$f(x) = \sin x + C$$

Hence, (d) is the correct answer.

Target Exercise 1.1

Solved the following integration :

1. $\int \frac{2+3x^2}{x^2(1+x^2)}\,dx$

2. $\int \frac{dx}{\sqrt{x+1}-\sqrt{x}}$

3. $\int \frac{x^2+3}{x^6(x^2+1)}\,dx$

4. $\int \frac{1+2x^2}{x^2(1+x^2)}\,dx$

5. $\int \frac{(1+x)^2}{x(1+x^2)}\,dx$

6. $\int \frac{x^4}{1+x^2}\,dx$

7. $\int \frac{x^6-1}{x^2+1}\,dx$

8. $\int \frac{x^4+x^2+1}{2(1+x^2)}\,dx$

9. $\int \frac{(\sqrt{x}+1)(x^2-\sqrt{x})}{x\sqrt{x}+x+\sqrt{x}}\,dx$

10. $\int \left(\frac{1-x^{-2}}{x^{1/2}-x^{-1/2}} - \frac{2}{x^{3/2}} + \frac{x^{-2}-x}{x^{1/2}-x^{-1/2}} \right) dx$

11. $\int \left(\frac{x^{-6}-64}{4+2x^{-1}+x^{-2}} \cdot \frac{x^2}{4-4x^{-1}+x^{-2}} - \frac{4x^2(2x+1)}{1-2x} \right) dx$

12. $\int \frac{(x^2+\sin^2 x)\sec^2 x}{1+x^2}\,dx$

13. $\int 2^x \cdot e^x\,dx$

14. $\int \frac{e^{3x}+e^{5x}}{e^x+e^{-x}}\,dx$

15. $\int (e^{a \ln x} + e^{x \ln a})\,dx$

16. $\int \frac{dx}{1+\sin x}$

17. $\int \sin x \cos x \cos 2x \cos 4x\,dx$

18. $\int \frac{1+\cos^2 x}{1+\cos 2x}\,dx$

19. $\int \frac{1-\tan^2 x}{1+\tan^2 x}\,dx$

20. $\int \frac{\cos 2x}{\cos^2 x \sin^2 x}\,dx$

21. $\int 4 \cos \frac{x}{2} \cdot \cos x \cdot \sin \frac{21}{2} x\,dx$

22. $\int \frac{\cos x - \sin x}{\cos x + \sin x}(2+2\sin 2x)\,dx$

23. $\int (3 \sin x \cos^2 x - \sin^3 x)\,dx$

24. $\int \cos x^\circ\,dx$

25. $\int \frac{\sec 2x - 1}{\sec 2x + 1}\,dx$

26. $\int \frac{\cos x - \cos 2x}{1-\cos x}\,dx$

27. $\int \frac{\sin x + \cos x}{\sqrt{1+\sin 2x}}\,dx \ (\cos x + \sin x > 0)$

28. $\int \frac{\cos 2x - \cos 2\alpha}{\cos x - \cos \alpha}\,dx$

29. $\int \frac{\sin^3 x + \cos^3 x}{\sin^2 x \cos^2 x}\,dx$

30. $\int \sec^2 x \operatorname{cosec}^2 x\,dx$

31. $\int \sqrt{1-\sin 2x}\,dx$

32. $\int \frac{\sin^6 x + \cos^6 x}{\sin^2 x \cdot \cos^2 x}\,dx$

33. $\int \left[\sin^2\left(\frac{9\pi}{8}+\frac{x}{4}\right) - \sin^2\left(\frac{7\pi}{8}+\frac{x}{4}\right) \right] dx$

34. $\int \frac{\cos 4x + 1}{\cot x - \tan x}\,dx$

35. $\int \left[\sin \alpha \sin (x - \alpha) + \sin^2\left(\frac{x}{2} - \alpha\right)\right] dx$ **36.** $\int \frac{\sin 2x + \sin 5x - \sin 3x}{\cos x + 1 - 2\sin^2 2x} dx$

37. $\int \left[\frac{\cot^2 2x - 1}{2\cot 2x} - \cos 8x \cot 4x\right] dx$ **38.** $\int \frac{\cos^4 x - \sin^4 x}{\sqrt{1 + \cos 4x}} dx \ (\cos 2x > 0)$

Integration by Parts

Theorem If u and v are two functions of x, then

$$\int uv\, dx = u \int v\, dx - \int \left\{\frac{du}{dx} \int v\, dx\right\} dx$$

ie, The integral of product of two functions = (first function) × (integral of second function) – integral of (differential of first function × integral of second function).

Proof For any two functions $f(x)$ and $g(x)$, we have

$$\frac{d}{dx}\{f(x) \cdot g(x)\} = f(x) \cdot \frac{d}{dx}\{g(x)\} + g(x) \cdot \frac{d}{dx}\{f(x)\}$$

$$\therefore \quad \int \left(f(x) \cdot \frac{d}{dx}\{g(x)\} + g(x) \cdot \frac{d}{dx}\{f(x)\}\right) dx = \int f(x) \cdot g(x)\, dx$$

$$\text{or} \quad \int \left(f(x) \cdot \frac{d}{dx}\{g(x)\}\right) dx + \int \left(g(x) \cdot \frac{d}{dx}\{f(x)\}\right) dx = \int f(x) \cdot g(x)\, dx$$

$$\text{or} \quad \int \left(f(x) \cdot \frac{d}{dx}\{g(x)\}\right) dx = \int f(x) \cdot g(x)\, dx - \int \left(g(x) \cdot \frac{d}{dx}\{f(x)\}\right) dx$$

Let $f(x) = u$ and $\frac{d}{dx}\{g(x)\} = v$

So that $g(x) = \int v\, dx$

$$\therefore \quad \int uv\, dx = u \cdot \int v\, dx - \int \left\{\frac{du}{dx} \cdot \int v\, dx\right\} \cdot dx$$

Points to Consider

While applying the above rule, care has to be taken in the selection of first function (u) and selection of second function (v). Normally we use the following methods :

1. If in the product of the two functions, one of the functions is not directly integrable (*eg*, $\log |x|$, $\sin^{-1} x$, $\cos^{-1} x$, $\tan^{-1} x, \ldots$, etc.) Then, we take it as the first function and the remaining function is taken as the second function. *eg*, In the integration of $\int x \tan^{-1} x\, dx$, $\tan^{-1} x$ is taken as the first function and x as the second function.
2. If there is no other function, then unity is taken as the second function. *eg*, In the integration of $\int \tan^{-1} x\, dx$, $\tan^{-1} x$ is taken as first function and 1 as the second function.
3. If both of the function are directly integrable, then the first function is chosen in such a way that the derivative of the function thus obtained under integral sign is easily integrable.

Usually we use the following preference order for selecting the first function. (Inverse, Logarithmic, Algebraic, Trigonometric, Exponent).

In above stated order, the function on the left is always chosen as the first function. This rule is called as **ILATE**.

Illustration 20 Evaluate

(i) $\int \sin^{-1} x \, dx$ (ii) $\int \log_e |x| \, dx$

Solution. (i) $I = \int \sin^{-1} x \, dx = \int \underset{\text{I}}{\sin^{-1} x} \cdot \underset{\text{II}}{1} \, dx$

Here, we know by definition of integration by parts that order of preference is taken according to ILATE. So, '$\sin^{-1} x$' should be taken as first and '1' as the second function to apply by parts.

Applying integration by parts, we get

$$I = \sin^{-1} x \cdot (x) - \int \frac{1}{\sqrt{1-x^2}} \cdot x \, dx$$

$$= x \cdot \sin^{-1} x + \frac{1}{2} \int \frac{dt}{t^{1/2}} \qquad \text{let } 1 - x^2 = t$$

$$-2x \, dx = dt$$

$$= x \sin^{-1} x + \frac{1}{2} \cdot \frac{t^{1/2}}{1/2} + C \qquad x \, dx = -\frac{1}{2} dt$$

$$I = x \sin^{-1} x + \sqrt{1-x^2} + C$$

$$\therefore \quad \int \sin^{-1} x \, dx = x \sin^{-1} x + \sqrt{1-x^2} + C$$

(ii) $I = \int \log_e |x| \, dx = \int \underset{\text{I}}{\log_e |x|} \cdot \underset{\text{II}}{1} \, dx$

Applying integration by parts, we get

$$= \log |x| \cdot x - \int \frac{1}{x} \cdot x \, dx$$

$$= x \log |x| - \int 1 \, dx$$

$$I = x \log |x| - x + C$$

Illustration 21 Evaluate

(i) $\int x \cos x \, dx$ (ii) $\int x^2 \cos x \, dx$

Solution. (i) $\int x \cos x \, dx$

$$I = \int \underset{\text{I}}{x} \, \underset{\text{II}}{\cos x} \, dx$$

Applying integration by parts,

$$I = x \left(\int \cos x \, dx\right) - \int \left\{\frac{d}{dx}(x)\right\}\left\{\int (\cos x) \, dx\right\} dx$$

$$I = x \sin x - \int 1 \cdot \sin x \, dx$$

$$I = x \sin x + \cos x + C$$

(ii) $I = \int \underset{\text{I}}{x^2} \underset{\text{II}}{\cos x}\, dx$

Applying integration by parts,

$$I = x^2 \left(\int \cos x\, dx\right) - \int \left\{\frac{d}{dx}(x^2)\right\} \cdot \left\{\int \cos x\right\} dx$$

$$= x^2 \sin x - \int 2x \cdot (\sin x)\, dx = x^2 \sin x - 2\int x\, (\sin x)\, dx$$

We again have to integrate $\int x \sin x\, dx$ using integration by parts,

$$= x^2 \cdot \sin x - 2 \int \underset{\text{I}}{x} \cdot \underset{\text{II}}{\sin x}\, dx$$

$$= x^2 \sin x - 2\left\{x\left(\int \sin x\, dx\right) - \int \left(\frac{dx}{dx}\right)\left(\int \sin x\, dx\right) dx\right\}$$

$$= x^2 \sin x - 2\{-x \cos x - \int 1 \cdot (-\cos x)\, dx\}$$

$$I = x^2 \sin x + 2x \cos x - 2 \sin x + C$$

Illustration 22 Evaluate $\int \frac{\sin^{-1}\sqrt{x} - \cos^{-1}\sqrt{x}}{\sin^{-1}\sqrt{x} + \cos^{-1}\sqrt{x}}\, dx.$

Solution. $\int \frac{\sin^{-1}\sqrt{x} - \cos^{-1}\sqrt{x}}{\sin^{-1}\sqrt{x} + \cos^{-1}\sqrt{x}}\, dx = \int \frac{\sin^{-1}\sqrt{x} - (\pi/2 - \sin^{-1}\sqrt{x})}{\pi/2}\, dx$

$(\because \sin^{-1}\theta + \cos^{-1}\theta = \pi/2)$

$\Rightarrow$ $$I = \frac{2}{\pi}\int (2\sin^{-1}\sqrt{x} - \pi/2)\, dx$$

$$I = \frac{4}{\pi}\int \sin^{-1}\sqrt{x}\, dx - \int 1\, dx$$

$$I = \frac{4}{\pi}\int \sin^{-1}\sqrt{x}\, dx - x + C \qquad \text{...(i)}$$

Let $x = \sin^2\theta$, then $dx = 2\sin\theta\cos\theta\, d\theta = \sin 2\theta\, d\theta$

$\therefore$ $\int \sin^{-1}\sqrt{x}\, dx = \int \underset{\text{I}}{\theta} \cdot \underset{\text{II}}{\sin 2\theta}\, d\theta$

Applying integration by parts

$$\int \sin^{-1}\sqrt{x}\, dx = -\theta \cdot \frac{\cos 2\theta}{2} + \int \frac{1}{2}\cos 2\theta\, d\theta = \frac{-\theta}{2}\cdot\cos 2\theta + \frac{1}{4}\sin 2\theta$$

$$= \frac{-1\cdot\theta}{2}\cdot(1 - 2\sin^2\theta) + \frac{1}{2}\cdot\sin\theta\cdot\sqrt{1-\sin^2\theta}$$

$$= \frac{-1}{2}\sin^{-1}\sqrt{x}\,(1-2x) + \frac{1}{2}\cdot\sqrt{x}\,\sqrt{1-x} \qquad \text{...(ii)}$$

From Eqs. (i) and (ii), we get

$$I = \frac{4}{\pi}\left\{\frac{-1}{2}(\sin^{-1}\sqrt{x})(1-2x) + \frac{1}{2}\sqrt{x}\,\sqrt{1-x}\right\} - x + C$$

$$= \frac{2}{\pi}\{\sqrt{x - x^2} - (1-2x)\sin^{-1}\sqrt{x}\} - x + C$$

Integral of the Form $\int e^x \{f(x) + f'(x)\}\, dx$

Theorem Prove that $\int e^x \{f(x) + f'(x)\}\, dx = e^x f(x) + C$

Proof We have, $\int e^x \{f(x) + f'(x)\}\, dx$

$$= \int \underset{\text{II}}{e^x} \cdot \underset{\text{I}}{f(x)}\, dx + \int e^x \cdot f'(x)\, dx$$

$$= f(x) \cdot e^x - \int f'(x) \cdot e^x\, dx + \int e^x \cdot f'(x)\, dx + C$$

$$= f(x) \cdot e^x + C$$

Thus, to evaluate the integrals of the type $\int e^x \{f(x) + f'(x)\}\, dx$, we first express the integral as the sum of two integrals $\int e^x f(x)\, dx$ and $\int e^x f'(x)\, dx$ and then integrate the integral involving $e^x f(x)$ as integrand by parts taking e^x as second function.

Points to Consider

The above theorem is also true, if we have e^{kx} in place of e^x. *ie*,

$$\int e^{kx} \{f(kx) + f'(kx)\}\, dx = e^{kx} f(kx) + C$$

General Concept

$\int e^{g(x)} \{f(x)g'(x) + f'(x)\}\, dx$

Proof $I = \int \underbrace{e^{g(x)}}_{\text{II}} \underbrace{f(x)}_{\text{I}} \underbrace{g'(x)}_{\text{II}}\, dx + \int e^{g(x)} f'(x)\, dx$

$$= f(x) \cdot e^{g(x)} - \int f'(x) \cdot e^{g(x)}\, dx + \int e^{g(x)} \cdot f'(x)\, dx = f(x) \cdot e^{g(x)}$$

eg, $\int e^{(x \sin x + \cos x)} \left(\frac{x^2 \cos^2 x - (x \sin x + \cos x)}{x^2} \right) dx$

$\Rightarrow \int e^{(x \sin x + \cos x)} \left(\cos^2 x - \frac{x \sin x + \cos x}{x^2} \right) dx$

$\Rightarrow \int e^{(x \sin x + \cos x)} \left(x \cos x \left(\frac{\cos x}{x} \right) + \left(\frac{\cos x}{x} \right)' \right) dx$

$\Rightarrow e^{(x \sin x + \cos x)} \cdot \frac{\cos x}{x} + C$

eg, $\int e^{\tan x} (\sin x - \sec x)\, dx = \int e^{\tan x} \sin x\, dx - \int e^{\tan x} \sec x\, dx$

$\Rightarrow -e^{\tan x} \cdot \cos x + \int e^{\tan x} \sec^2 x \cos x\, dx - \int e^{\tan x} \sec x\, dx$

$\Rightarrow -e^{\tan x} \cdot \cos x$

Illustration 23 Evaluate

(i) $\int e^x \left(\frac{1 + \sin x \cos x}{\cos^2 x} \right) dx$ (ii) $\int e^{2x} \left(\frac{1 + \sin 2x}{1 + \cos 2x} \right) dx$

Solution. (i) $I = \int e^x \left(\frac{1 + \sin x \cos x}{\cos^2 x} \right) dx$

$$I = \int e^x \left\{ \frac{1}{\cos^2 x} + \frac{\sin x \cos x}{\cos^2 x} \right\} dx$$

$$I = \int e^x \{\tan x + \sec^2 x\} \, dx$$

$$I = \int \underset{\text{II}}{e^x} \cdot \underset{\text{I}}{\tan x} \, dx + \int e^x (\sec^2 x) \, dx$$

$$I = \tan x \cdot e^x - \int \sec^2 x \cdot e^x \, dx + \int e^x \cdot \sec^2 x \, dx + C$$

$$I = e^x \tan x + C$$

(ii) $I = \int e^{2x} \left\{ \frac{1 + \sin 2x}{1 + \cos 2x} \right\} dx = \int e^{2x} \left\{ \frac{1 + 2 \sin x \cos x}{2 \cos^2 x} \right\} dx$

$$= \int e^{2x} \left\{ \frac{1}{2 \cos^2 x} + \frac{2 \sin x \cos x}{2 \cos^2 x} \right\} dx = \int e^{2x} \left\{ \frac{1}{2} \sec^2 x + \tan x \right\} dx$$

$$= \int \underset{\text{II}}{e^{2x}} \cdot \underset{\text{I}}{\tan x} \, dx + \frac{1}{2} \int e^{2x} \cdot \sec^2 x \, dx$$

$$= \tan x \cdot \frac{e^{2x}}{2} - \int \sec^2 x \cdot \frac{e^{2x}}{2} \, dx + \frac{1}{2} \int e^{2x} \cdot \sec^2 x \, dx$$

$$I = \frac{1}{2} e^{2x} \cdot \tan x + C$$

Illustration 24 Evaluate $\int e^x \left(\frac{1 - x}{1 + x^2} \right)^2 dx.$

Solution. $I = \int e^x \left(\frac{1 - x}{1 + x^2} \right)^2 dx = \int e^x \frac{(1 - 2x + x^2)}{(1 + x^2)^2} dx$

$$I = \int e^x \left\{ \frac{1 + x^2}{(1 + x^2)^2} - \frac{2x}{(1 + x^2)^2} \right\} dx$$

$$= \int e^x \left\{ \frac{1}{1 + x^2} - \frac{2x}{(1 + x^2)^2} \right\} dx \qquad \left[\text{as } \frac{d}{dx} \left(\frac{1}{1 + x^2} \right) = - \frac{2x}{(1 + x^2)^2} \right]$$

$$= \frac{e^x}{1 + x^2} + C$$

$\therefore \quad I = \frac{e^x}{1 + x^2} + C$

Integrals of the Form $\int e^{ax} \sin bx \, dx, \int e^{ax} \cos bx \, dx$

Let $\quad I = \int e^{ax} (\sin bx) \, dx$

Then, $\quad I = \int \underset{\text{I}}{\sin bx} \cdot \underset{\text{II}}{e^{ax}} \, dx$

$$I = \sin bx \cdot \left(\frac{e^{ax}}{a} \right) - \int b \cos bx \cdot \frac{e^{ax}}{a} \, dx$$

$$I = \frac{1}{a}\sin bx \cdot e^{ax} - \frac{b}{a}\left\{\cos bx \cdot \frac{e^{ax}}{a} - \int (-b\sin bx)\cdot \frac{e^{ax}}{a}\,dx\right\}$$

$$= \frac{1}{a}\sin bx \cdot e^{ax} - \frac{b}{a^2}\cos bx \cdot e^{ax} - \frac{b^2}{a^2}\int \sin bx \cdot e^{ax}\,dx$$

$$I = \frac{1}{a}\sin bx \cdot e^{ax} - \frac{b}{a^2}\cos bx \cdot e^{ax} - \frac{b^2}{a^2} I$$

$\therefore$ $$I + \frac{b^2}{a^2} I = \frac{1\cdot e^{ax}}{a^2}\cdot(a\sin bx - b\cos bx)$$

$\Rightarrow$ $$I\left(\frac{a^2+b^2}{a^2}\right) = \frac{e^{ax}}{a^2}(a\sin bx - b\cos bx)$$

or $$I = \frac{e^{ax}}{a^2+b^2}(a\sin bx - b\cos bx) + C$$

Thus, $$\int e^{ax}\sin bx\,dx = \frac{e^{ax}}{a^2+b^2}(a\sin bx - b\cos bx) + C$$

Similarly, $$\int e^{ax}\cos bx\,dx = \frac{e^{ax}}{a^2+b^2}(a\cos bx + b\sin bx) + C$$

Aliter Use Euler's equation

Let $$P = \int e^{ax}\cos bx\,dx \quad \text{and} \quad Q = \int e^{ax}\sin bx\,dx$$

Hence, $$P + iQ = \int e^{ax}\cdot e^{ibx}\,dx = \int e^{(a+ib)x}\,dx$$

$$P + iQ = \frac{1}{a+ib}e^{(a+ib)x} = \frac{a-ib}{a^2+b^2}e^{ax}(\cos bx + i\sin bx)$$

$$= \frac{(ae^{ax}\cos bx + be^{ax}\sin bx) - i(ae^{ax}\sin bx - be^{ax}\cos bx)}{a^2+b^2}$$

$\therefore$ $$P = \frac{e^{ax}(a\cos bx + b\sin bx)}{a^2+b^2}$$

$$Q = \frac{e^{ax}(a\sin bx - b\cos bx)}{a^2+b^2}$$

Illustration 25 Evaluate

(i) $\int e^x \cos^2 x\,dx$ (ii) $\int \sin(\log x)\,dx$

Solution. (i) $I = \int e^x \cdot \cos^2 x\,dx = \int e^x \cdot \left\{\frac{1+\cos 2x}{2}\right\}dx$

$$I = \frac{1}{2}\int e^x\,dx + \frac{1}{2}\int \cos 2x \cdot e^x\,dx$$

$$I = \frac{1}{2}e^x + \frac{1}{2}I_1 \qquad \text{...(i)}$$

where $$I_1 = \int \cos 2x \cdot e^x\,dx$$

$$I_1 = \int \underset{\text{I}}{\cos 2x} \cdot \underset{\text{II}}{e^x}\, dx = \cos 2x \cdot e^x - \int -2 \sin 2x \cdot e^x\, dx$$

$$= e^x \cdot \cos 2x + 2 \int \underset{\text{I}}{\sin 2x} \cdot \underset{\text{II}}{e^x}\, dx$$

$$= e^x \cdot \cos 2x + 2\left\{\sin 2x \cdot e^x - \int 2 \cos 2x \cdot e^x\, dx\right\}$$

$$= e^x \cdot \cos 2x + 2 \sin 2x \cdot e^x - 4 I_1$$

$$\therefore \quad I_1 = \frac{1}{5}\{e^x \cos 2x + 2 \sin 2x\, e^x\} \qquad \ldots\text{(ii)}$$

From Eqs. (i) and (ii), we get

$$I = \frac{1}{2} e^x + \frac{1}{2} \cdot \frac{1}{5}\{e^x \cos 2x + 2 \sin 2x \cdot e^x\}$$

$$I = \frac{1}{2} e^x + \frac{1}{10} e^x \{\cos 2x + 2 \sin 2x\} + C$$

(ii) $I = \int \sin(\log x)\, dx$

Let $\log x = t \Rightarrow x = e^t$ or $dx = e^t\, dt$

$$\therefore \quad I = \int \underset{\text{I}}{(\sin t)} \cdot \underset{\text{II}}{e^t}\, dt = \sin t \cdot e^t - \int \underset{\text{I}}{\cos t} \cdot \underset{\text{II}}{e^t}\, dt$$

$$I = \sin t \cdot e^t - \left\{\cos t \cdot e^t - \int -\sin t \cdot e^t\, dt\right\}$$

$$I = e^t \cdot \sin t - e^t \cdot \cos t - I$$

$$\therefore \quad I = \frac{1}{2} e^t (\sin t - \cos t) + C$$

$$I = \frac{x}{2}\{\sin(\log x) - \cos(\log x)\} + C$$

Illustration 26 Evaluate $\int \frac{x^2 dx}{(x \sin x + \cos x)^2}$.

Solution. Let $I = \int \frac{x^2}{(x \sin x + \cos x)^2}\, dx$

Multiplying and dividing it by $(x \cos x)$, we get

$$I = \int \underset{\text{I}}{(x \sec x)} \cdot \underset{\text{II}}{\frac{(x \cos x)}{(x \sin x + \cos x)^2}}\, dx$$

$$I = x \sec x \cdot \int \frac{x \cos x}{(x \sin x + \cos x)^2}\, dx - \int \left\{\frac{d}{dx}(x \sec x)\right\}\left\{\int \frac{x \cos x}{(x \sin x + \cos x)^2}\, dx\right\} dx$$

$$= x \sec x \cdot \frac{-1}{(x \sin x + \cos x)} - \int (x \sec x \cdot \tan x + \sec x) \cdot \frac{-1}{(x \sin x + \cos x)}\, dx$$

$$= \frac{-x \sec x}{(x \sin x + \cos x)} + \int \frac{(x \sin x + \cos x)}{\cos^2 x \cdot (x \sin x + \cos x)}\, dx$$

$$= \frac{-x \sec x}{(x \sin x + \cos x)} + \int \sec^2 x \, dx$$

$$I = \frac{-x \sec x}{(x \sin x + \cos x)} + \tan x + C$$

Integration of the Type $\int (\sin^m x \cdot \cos^n x) dx$

(i) Where m, n belongs to natural number.

(ii) If one of them is odd, then substitute for term of even power.

(iii) If both are odd, substitute either of them.

(iv) If both are even, use trigonometric identities only.

(v) If m and n are rational numbers and $\left(\frac{m+n-2}{2}\right)$ is a negative integer, then substitute $\cot x = p$ or $\tan x = p$ which so ever is found suitable.

Illustration 27 Evaluate $\int \sin^3 x \cdot \cos^5 x \, dx$.

Solution. $I = \int \sin^3 x \cdot \cos^5 x \, dx$

Let $\cos x = t \quad \Rightarrow \quad -\sin x \, dx = dt$

$$I = -\int (1 - t^2) \cdot t^{5} \, dt$$

$$I = \int t^7 \, dt - \int t^5 \, dt = \frac{t^8}{8} - \frac{t^6}{6} + C$$

$$I = \frac{\cos^8 x}{8} - \frac{\cos^6 x}{6} + C$$

Aliter $I = \int R^3 (1 - R^2)^2 \, dR$, if $\sin x = R$, $\cos x \, dx = dR$

$$I = \int R^3 \, dR - \int 2R^5 \, dR + \int R^7 dR$$

$$I = \frac{\sin^4 x}{4} - \frac{2 \sin^6 x}{6} + \frac{\sin^8 x}{8} + C$$

Points to Consider

This problem can also be handled by successive reduction or by trigonometrical identities. Answers will be in different form but identical with modified constant of integration.

Illustration 28 Evaluate $\int \sin^{-11/3} x \cdot \cos^{-1/3} x \, dx$.

Solution. Here, $\int \sin^{-11/3} x \cdot \cos^{-1/3} x \, dx$ *ie,* $\left(\frac{-\frac{11}{3} - \frac{1}{3} - 2}{2}\right) = -3$

$$\therefore \quad I = \int \frac{\cos^{-1/3} x}{\sin^{-1/3} x \cdot \sin^4 x} dx = \int (\cot^{-1/3} x)(\operatorname{cosec}^2 x)^2 dx$$

$$I = \int (\cot^{-1/3} x)(1+\cot^2 x)\operatorname{cosec}^2 x\, dx \ (\text{let } \cot x = t, -\operatorname{cosec}^2 x\, dx = dt)$$
$$= -\int t^{-1/3}(1+t^2)\, dt = -\int (t^{-1/3} + t^{5/3})\, dt$$
$$= -\left\{\frac{3}{2} t^{2/3} + \frac{3}{8} t^{8/3}\right\} + C = -\left\{\frac{3}{2}(\cot^{2/3} x) + \frac{3}{8}(\cot^{8/3} x)\right\} + C$$

Illustration 29 Evaluate

(i) $\int \frac{1}{\sin(x-a)\cos(x-b)}\, dx$ (ii) $\int \frac{1}{\cos(x-a)\cos(x-b)}\, dx$

Solution. (i) $I = \int \frac{1}{\sin(x-a)\cos(x-b)}\, dx$

$$I = \frac{\cos(a-b)}{\cos(a-b)} \cdot \int \frac{dx}{\sin(x-a)\cos(x-b)}$$
$$= \frac{1}{\cos(a-b)} \cdot \int \frac{\cos\{(x-b)-(x-a)\}}{\sin(x-a)\cos(x-b)}\, dx$$
$$= \frac{1}{\cos(a-b)} \cdot \int \left\{\frac{\cos(x-b)\cdot\cos(x-a)}{\sin(x-a)\cos(x-b)} + \frac{\sin(x-b)\cdot\sin(x-a)}{\sin(x-a)\cos(x-b)}\right\} dx$$
$$= \frac{1}{\cos(a-b)} \int \{\cot(x-a) + \tan(x-b)\}\, dx$$
$$= \frac{1}{\cos(a-b)} \{\log|\sin(x-a)| - \log|\cos(x-b)|\} + C$$
$$= \frac{1}{\cos(a-b)} \log_e \left|\frac{\sin(x-a)}{\cos(x-b)}\right| + C$$

(ii) $$I = \int \frac{1}{\cos(x-a)\cos(x-b)}\, dx$$
$$= \frac{1}{\sin(a-b)} \int \frac{\sin(a-b)}{\cos(x-a)\cos(x-b)}\, dx$$
$$= \frac{1}{\sin(a-b)} \int \frac{\sin\{(x-b)-(x-a)\}}{\cos(x-a)\cos(x-b)}\, dx$$
$$= \frac{1}{\sin(a-b)} \int \left\{\frac{\sin(x-b)\cos(x-a)}{\cos(x-a)\cos(x-b)} - \frac{\cos(x-b)\sin(x-a)}{\cos(x-a)\cos(x-b)}\right\} dx$$
$$= \frac{1}{\sin(a-b)} \int \{\tan(x-b) - \tan(x-a)\}\, dx$$
$$= \frac{1}{\sin(a-b)} [-\log|\cos(x-b)| + \log|\cos(x-a)|] + C$$
$$= \frac{1}{\sin(a-b)} \left[\log\left|\frac{\cos(x-a)}{\cos(x-b)}\right|\right] + C$$

Illustration 30 Evaluate $\int \frac{\sin(x+a)}{\sin(x+b)}\, dx.$

Solution. Let $I = \int \frac{\sin(x+a)}{\sin(x+b)}\, dx$

Put $x + b = t \Rightarrow dx = dt$

$$\therefore \quad I = \int \frac{\sin (t - b + a)}{\sin t}\, dt = \int \left\{ \frac{\sin t \cos (a-b)}{\sin t} + \frac{\cos t \sin (a-b)}{\sin t} \right\} dt$$

$$= \cos (a-b) \int 1\, dt + \sin (a-b) \int \cot (t)\, dt$$

$$= t \cos (a-b) + \sin (a-b) \log |\sin t| + C$$

$$= (x+b) \cos (a-b) + \sin (a-b) \log |\sin (x+b)| + C$$

Some Special Integrals

(i) $\int \frac{dx}{x^2 + a^2} = \frac{1}{a} \tan^{-1}\left(\frac{x}{a}\right) + C$

(ii) $\int \frac{dx}{x^2 - a^2} = \frac{1}{2a} \log \left| \frac{x-a}{x+a} \right| + C$

(iii) $\int \frac{dx}{a^2 - x^2} = \frac{1}{2a} \log \left| \frac{a+x}{a-x} \right| + C$

(iv) $\int \frac{dx}{\sqrt{a^2 - x^2}} = \sin^{-1}\left(\frac{x}{a}\right) + C$

(v) $\int \frac{dx}{\sqrt{a^2 + x^2}} = \log |x + \sqrt{x^2 + a^2}| + C$

(vi) $\int \frac{dx}{\sqrt{x^2 - a^2}} = \log |x + \sqrt{x^2 - a^2}| + C$

(vii) $\int \sqrt{a^2 - x^2}\, dx = \frac{1}{2} x \sqrt{a^2 - x^2} + \frac{1}{2} a^2 \sin^{-1}\left(\frac{x}{a}\right) + C$

(viii) $\int \sqrt{a^2 + x^2}\, dx = \frac{1}{2} x \sqrt{a^2 + x^2} + \frac{1}{2} a^2 \log |x + \sqrt{a^2 + x^2}| + C$

(ix) $\int \sqrt{x^2 - a^2}\, dx = \frac{1}{2} x \sqrt{x^2 - a^2} - \frac{1}{2} a^2 \log |x + \sqrt{x^2 - a^2}| + C$

Some Important Substitutions

Expression	Substitution
$a^2 + x^2$	$x = a \tan \theta$ or $a \cot \theta$
$a^2 - x^2$	$x = a \sin \theta$ or $a \cos \theta$
$x^2 - a^2$	$x = a \sec \theta$ or $a \operatorname{cosec} \theta$
$\sqrt{\frac{a-x}{a+x}}$ or $\sqrt{\frac{a+x}{a-x}}$	$x = a \cos 2\theta$
$\sqrt{\frac{x-\alpha}{\beta - x}}$ or $\sqrt{(x-\alpha)(x-\beta)}$	$x = \alpha \cos^2 \theta + \beta \sin^2 \theta$

Application of These Formulae

The above standard integrals are very important. We Given below integrals which are applications of these.

Type A

(i) $\int \frac{dx}{ax^2 + bx + c}$ **(ii)** $\int \frac{dx}{\sqrt{ax^2 + bx + c}}$ **(iii)** $\int \sqrt{ax^2 + bx + c}\, dx$

If $ax^2 + bx + c$ can be factorized, then the integration is easily done by the method of partial fractions (explained later).

If the denominator cannot be factorized, then express it as the sum or difference of two squares by the method of completing the square :

$$ax^2 + bx + c = a\left(x^2 + \frac{b}{a}x + \frac{c}{a}\right)$$

$$= a\left\{\left(x + \frac{b}{2a}\right)^2 + \left(\frac{c}{a}\right) - \frac{b^2}{4a^2}\right\}$$

make the substitution $x + \frac{b}{2a} = t$.

Type B

(i) $\int \frac{(px + q)\, dx}{ax^2 + bx + c}$ **(ii)** $\int \frac{(px + q)}{\sqrt{ax^2 + bx + c}}\, dx$

(iii) $\int (px + q)\sqrt{ax^2 + bx + c}\, dx$

The linear factor $(px + q)$ is expressed in terms of the derivative of the quadratic factor $ax^2 + bx + c$ together with a constant as :

$$px + q = \frac{\lambda d}{dx}(ax^2 + bx + c) + \mu$$

$$\Rightarrow \qquad px + q = \lambda(2ax + b) + \mu$$

Here, we have to find λ and μ and replace $(px + q)$ by $\{\lambda(2ax + b) + \mu\}$ in (i), (ii) and (iii).

Illustration 31 Evaluate

(i) $\int \frac{1}{x^2 - x + 1}\, dx$ (ii) $\int \frac{1}{2x^2 + x - 1}\, dx$

(iii) $\int \frac{1}{\sqrt{x^2 - 2x + 3}}\, dx$ (iv) $\int \sqrt{2x^2 - 3x + 1}\, dx$

Solution. (i) $I = \int \frac{dx}{x^2 - x + 1}$

$$= \int \frac{dx}{x^2 - x + 1/4 - 1/4 + 1} = \int \frac{dx}{(x - 1/2)^2 + 3/4}$$

$$\therefore \quad I=\int\frac{dx}{(x-1/2)^2+(\sqrt{3}/2)^2}=\frac{1}{\sqrt{3}/2}\tan^{-1}\left(\frac{x-1/2}{\sqrt{3}/2}\right)+C$$

$$\therefore \quad I=\frac{2}{\sqrt{3}}\tan^{-1}\left(\frac{2x-1}{\sqrt{3}}\right)+C \quad \left[\text{using}\int\frac{1}{x^2+a^2}dx=\frac{1}{a}\tan^{-1}\left(\frac{x}{a}\right)+C\right]$$

(ii) $$I=\int\frac{1}{2x^2+x-1}dx=\frac{1}{2}\int\frac{1}{x^2+x/2-1/2}dx$$

$$=\frac{1}{2}\int\frac{1}{x^2+x/2+1/16-1/16-1/2}dx$$

$$=\frac{1}{2}\int\frac{dx}{(x+1/4)^2-9/16}=\frac{1}{2}\int\frac{dx}{(x+1/4)^2-(3/4)^2}$$

$$=\frac{1}{2}\cdot\frac{1}{2(3/4)}\log\left|\frac{x+1/4-3/4}{x+1/4+3/4}\right|+C$$

$$\left[\text{using}\int\frac{dx}{x^2-a^2}=\frac{1}{2a}\log\left|\frac{x-a}{x+a}\right|+C\right]$$

$$=\frac{1}{3}\log\left|\frac{x-1/2}{x+1}\right|+C$$

$$\therefore \quad I=\frac{1}{3}\log\left|\frac{2x-1}{2(x+1)}\right|+C$$

(iii) $$I=\int\frac{dx}{\sqrt{x^2-2x+3}}=\int\frac{dx}{\sqrt{x^2-2x+1-1+3}}$$

$$=\int\frac{dx}{\sqrt{(x-1)^2+(\sqrt{2})^2}}=\log|(x-1)+\sqrt{(x-1)^2+(\sqrt{2})^2}|$$

$$\left[\text{using}\int\frac{dx}{\sqrt{x^2+a^2}}=\log|x+\sqrt{x^2+a^2}|+C\right]$$

$$\therefore \quad I=\log|(x-1)+\sqrt{x^2-2x+3}|+C$$

(iv) $$I=\int(\sqrt{2x^2-3x+1})\,dx$$

$$=\sqrt{2}\int(\sqrt{x^2-3x/2+1/2})\,dx$$

$$=\sqrt{2}\int(\sqrt{x^2-3x/2+9/16-9/16+1/2})\,dx$$

$$=\sqrt{2}\int(\sqrt{(x-3/4)^2-1/16})\,dx$$

$$=\sqrt{2}\left\{\begin{array}{l}\frac{1}{2}(x-3/4)\sqrt{(x-3/4)^2-1/16}\\ -\frac{1}{16\times 2}\log|(x-3/4)+\sqrt{(x-3/4)^2-1/16}|\end{array}\right\}+C$$

$$=\sqrt{2}\left\{\begin{array}{l}\frac{1}{8}(4x-3)\sqrt{x^2-3x/2+1/2}\\ -\frac{1}{32}\log|(x-3/4)+\sqrt{x^2-3x/2+1/2}|\end{array}\right\}+C$$

Illustration 32 Evaluate

(i) $\int \frac{1}{\sqrt{1-e^{2x}}}\,dx$ (ii) $\int \frac{2x}{\sqrt{1-x^2-x^4}}\,dx$

(iii) $\int \frac{a^x}{\sqrt{1-a^{2x}}}\,dx$ (iv) $\int \frac{e^x}{\sqrt{5-4e^x-e^{2x}}}\,dx$

(v) $\int \sqrt{\frac{x}{a^3-x^3}}\,dx$

Solution. (i) Here, $I = \int \frac{dx}{\sqrt{1-e^{2x}}}$

$\therefore$ $I = \int \frac{1}{\sqrt{t^2}} \cdot \frac{t}{t^2-1}\,dt$ let $1 - e^{2x} = t^2$

then, $-2e^{2x}\,dx = 2t\,dt$

$\Rightarrow dx = -\frac{-t\,dt}{e^{2x}},\ dx = \frac{t\,dt}{t^2-1}$

$$I = \int \frac{dt}{t^2-1} = \frac{1}{2\cdot 1}\log\left|\frac{t-1}{t+1}\right| + C$$

$$= \frac{1}{2}\log\left|\frac{\sqrt{1-e^{2x}}-1}{\sqrt{1-e^{2x}}+1}\right| + C$$

(ii) Here, $I = \int \frac{2x}{\sqrt{1-x^2-x^4}}\,dx$ let $x^2 = t,\ 2x\,dx = dt$

$\therefore$ $I = \int \frac{dt}{\sqrt{1-t-t^2}} = \int \frac{dt}{\sqrt{1+\frac{1}{4}-\frac{1}{4}-t-t^2}}$

$$I = \int \frac{dt}{\sqrt{5/4-(t+1/2)^2}} = \int \frac{dt}{\sqrt{(\sqrt{5}/2)^2-(t+1/2)^2}}$$

$$= \sin^{-1}\left(\frac{t+1/2}{\sqrt{5}/2}\right) + C = \sin^{-1}\left(\frac{2x^2+1}{\sqrt{5}}\right) + C$$

(iii) $I = \int \frac{a^x}{\sqrt{1-a^{2x}}}\,dx$ let $a^x = t$ $\therefore a^x \log a\,dx = dt,\ a^x\,dx = \frac{dt}{\log a}$

$\therefore$ $I = \int \frac{1}{\sqrt{1-t^2}} \cdot \frac{dt}{\log a} = \frac{1}{\log a} \cdot \sin^{-1}(t) + C$

$$I = \frac{1}{\log a}\sin^{-1}(a^x) + C$$

(iv) Here, $I = \int \frac{e^x\,dx}{\sqrt{5-4e^x-e^{2x}}}$ let $e^x = t \Rightarrow e^x\,dx = dt$

$\therefore$ $I = \int \frac{dt}{\sqrt{5-4t-t^2}}$

$$I = \int \frac{dt}{\sqrt{5+4-4-4t-t^2}} = \int \frac{dt}{\sqrt{(3)^2-(t+2)^2}} = \sin^{-1}\left(\frac{t+2}{3}\right) + C$$

$$\therefore \quad I = \sin^{-1}\left(\frac{e^x + 2}{3}\right) + C$$

(v) $I = \int \sqrt{\frac{x}{a^3 - x^3}}\, dx$ let $x^{3/2} = t, \frac{3}{2} x^{1/2}\, dx = dt, x^{1/2}\, dx = \frac{2}{3}\, dt$

$$= \int \frac{x^{1/2}\, dx}{\sqrt{(a^{3/2})^2 - (x^{3/2})^2}}$$

$$= \int \frac{\frac{2}{3}\, dt}{\sqrt{(a^{3/2})^2 - (t)^2}} = \frac{2}{3} \cdot \sin^{-1}\left(\frac{t}{a^{3/2}}\right) + C = \frac{2}{3} \sin^{-1}\left(\frac{x^{3/2}}{a^{3/2}}\right) + C$$

Illustration 33 Evaluate

(i) $\int \frac{\cos x}{\sqrt{\sin^2 x - 2 \sin x - 3}}\, dx$ (ii) $\int \sqrt{\frac{\sin (x - \alpha)}{\sin (x + \alpha)}}\, dx$

(iii) $\int \frac{2 \sin 2x - \cos x}{6 - \cos^2 x - 4 \sin x}\, dx$ (iv) $\int \sqrt{\frac{\cos x - \cos^3 x}{(1 - \cos^3 x)}}\, dx$

Solution. (i) Let $I = \int \frac{\cos x\, dx}{\sqrt{\sin^2 x - 2 \sin x - 3}}$

Put $\sin x = t \quad \therefore \cos x\, dx = dt$

$$\Rightarrow \quad I = \int \frac{dt}{\sqrt{t^2 - 2t - 3}} = \int \frac{dt}{\sqrt{t^2 - 2t + 1 - 1 - 3}}$$

$$I = \int \frac{dt}{\sqrt{(t-1)^2 - (2)^2}} = \log |(t - 1) + \sqrt{(t-1)^2 - (2)^2}| + C$$

$$\therefore \quad I = \log |(\sin x - 1) + \sqrt{\sin^2 x - 2 \sin x - 3}| + C$$

(ii) Let $I = \int \sqrt{\frac{\sin (x - \alpha)}{\sin (x + \alpha)}}\, dx$

$$I = \int \sqrt{\frac{\sin (x - \alpha)}{\sin (x + \alpha)} \times \frac{\sin (x - \alpha)}{\sin (x - \alpha)}}\, dx = \int \frac{\sin (x - \alpha)}{\sqrt{\sin^2 x - \sin^2 \alpha}}\, dx$$

$$I = \int \frac{\sin x \cos \alpha - \cos x \sin \alpha}{\sqrt{\sin^2 x - \sin^2 \alpha}}\, dx$$

$$= \cos \alpha \int \frac{\sin x\, dx}{\sqrt{\sin^2 x - \sin^2 \alpha}} - \sin \alpha \int \frac{\cos x\, dx}{\sqrt{\sin^2 x - \sin^2 \alpha}}$$

$$= \cos \alpha \int \frac{\sin x\, dx}{\sqrt{1 - \cos^2 x - \sin^2 \alpha}} - \sin \alpha \int \frac{\cos x\, dx}{\sqrt{\sin^2 x - \sin^2 \alpha}}$$

$$= \cos \alpha \int \frac{\sin x\, dx}{\sqrt{\cos^2 \alpha - \cos^2 x}} - \sin \alpha \int \frac{\cos x\, dx}{\sqrt{\sin^2 x - \sin^2 \alpha}}$$

In the first part put $\cos x = t$, so that $-\sin x\, dx = dt$ and in second part put $\sin x = u$, so that $\cos x\, dx = du$

$$\therefore \quad I = -\cos \alpha \int \frac{dt}{\sqrt{\cos^2 \alpha - t^2}} - \sin \alpha \int \frac{du}{\sqrt{u^2 - \sin^2 \alpha}}$$

$$= -\cos\alpha \cdot \sin^{-1}\left(\frac{t}{\cos\alpha}\right) - \sin\alpha \cdot \log|u - \sqrt{(u^2 - \sin^2\alpha)}| + C$$

$$= -\cos\alpha \cdot \sin^{-1}\left(\frac{\cos x}{\cos\alpha}\right) - \sin\alpha \cdot \log|\sin x - \sqrt{\sin^2 x - \sin^2\alpha}\ | + C$$

(iii) $I = \int \frac{2\sin 2x - \cos x}{6 - \cos^2 x - 4\sin x}\, dx$

$$= \int \frac{(4\sin x - 1)\cos x}{6 - (1 - \sin^2 x) - 4\sin x}\, dx = \int \frac{(4\sin x - 1)\cos x}{\sin^2 x - 4\sin x + 5}\, dx$$

Put $\sin x = t$, so that $\cos x\, dx = dt$

$\therefore$ $I = \int \frac{(4t - 1)\, dt}{(t^2 - 4t + 5)}$ ...(i)

Now, let $(4t - 1) = \lambda(2t - 4) + \mu$

Comparing coefficients of like powers of t, we get

$2\lambda = 4, -4\lambda + \mu = -1 \Rightarrow \lambda = 2, \mu = 7$ ...(ii)

$\therefore$ $I = \int \frac{2(2t - 4) + 7}{t^2 - 4t + 5}\, dt$ [using Eqs. (i) and (ii)]

$$= 2\int \frac{2t - 4}{t^2 - 4t + 5}\, dt + 7\int \frac{dt}{t^2 - 4t + 5}$$

$$= 2\log|t^2 - 4t + 5| + 7\int \frac{dt}{t^2 - 4t + 4 - 4 + 5}$$

$$= 2\log|t^2 - 4t + 5| + 7\int \frac{dt}{(t - 2)^2 + (1)^2}$$

$$= 2\log|t^2 - 4t + 5| + 7\tan^{-1}(t - 2) + C$$

$$= 2\log|\sin^2 x - 4\sin x + 5| + 7\tan^{-1}(\sin x - 2) + C$$

(iv) Let $I = \int \sqrt{\frac{\cos x - \cos^3 x}{(1 - \cos^3 x)}}\, dx$

$$= \int \frac{\sqrt{\cos x}\,(\sqrt{1 - \cos^2 x})}{\sqrt{1 - (\cos^{3/2} x)^2}}\, dx = \int \frac{\sqrt{\cos x} \cdot \sin x\, dx}{\sqrt{1 - (\cos^{3/2} x)^2}}$$

$\left[\text{If } \cos^{3/2} x = t \Rightarrow \frac{3}{2}\cos^{1/2} x \cdot (-\sin x)\, dx = dt\right]$

$\therefore$ $I = -\frac{2}{3}\int \frac{dt}{\sqrt{1 - t^2}} = -\frac{2}{3}\sin^{-1}(t) + C$

$$I = -\frac{2}{3}\sin^{-1}(\cos^{3/2} x) + C$$

Illustration 34 Evaluate

(i) $\int \sqrt{\frac{a - x}{a + x}}\, dx$ (ii) $\int x\sqrt{\frac{a^2 - x^2}{a^2 + x^2}}\, dx$

(iii) $\int x\sqrt{1 + x - x^2}\, dx$ (iv) $\int (x + 1)\sqrt{1 - x - x^2}\, dx$

Solution. (i) Let $I = \int \sqrt{\dfrac{a-x}{a+x}}\, dx$

$$I = \int \sqrt{\frac{a-x}{a+x} \times \frac{a-x}{a-x}}\, dx = \int \frac{a-x}{\sqrt{a^2-x^2}}\, dx$$

$$I = \int \frac{a}{\sqrt{a^2-x^2}}\, dx - \int \frac{x}{\sqrt{a^2-x^2}}\, dx$$

$$= a \cdot \sin^{-1}\left(\frac{x}{a}\right) + \int \frac{t\, dt}{t}$$

Put $a^2 - x^2 = t^2,\ -2x\, dx = 2t\, dt,\ x\, dx = -t\, dt$

$$= a \cdot \sin^{-1}\left(\frac{x}{a}\right) + t + C$$

$$= a \cdot \sin^{-1}\left(\frac{x}{a}\right) + \sqrt{a^2-x^2} + C$$

(ii) Let $I = \int x \sqrt{\dfrac{a^2-x^2}{a^2+x^2}}\, dx$ put $x^2 = t$, $2x\, dx = dt$

$$\therefore \quad I = \int \sqrt{\frac{a^2-t}{a^2+t}} \cdot \frac{dt}{2} = \frac{1}{2}\int \sqrt{\frac{a^2-t}{a^2+t} \cdot \frac{a^2-t}{a^2-t}}\, dt$$

$$= \frac{1}{2}\int \frac{a^2-t}{\sqrt{a^4-t^2}}\, dt = \frac{1}{2}a^2 \int \frac{dt}{\sqrt{(a^2)^2-(t)^2}} - \frac{1}{2}\int \frac{t\, dt}{\sqrt{a^4-t^2}}$$

$$= \frac{1}{2}a^2 \cdot \sin^{-1}\left(\frac{t}{a^2}\right) + \frac{1}{4}\int \frac{du}{\sqrt{u}} \quad \text{where } a^4 - t^2 = u \Rightarrow -2t\, dt = du$$

$$= \frac{1}{2}a^2 \cdot \sin^{-1}\left(\frac{t}{a^2}\right) + \frac{1}{4} \cdot \left(\frac{u^{1/2}}{1/2}\right) + C$$

where $t = x^2$ and $u = a^4 - x^4$

$$= \frac{1}{2}a^2 \cdot \sin^{-1}\left(\frac{x^2}{a^2}\right) + \frac{1}{2} \cdot \sqrt{a^4-x^4} + C$$

(iii) Let $I = \int x\sqrt{1+x-x^2}\, dx$

Put $x = \lambda \left\{\dfrac{d}{dx}(1+x-x^2)\right\} + \mu$

Then, comparing the coefficients of like powers of x, we get

$$1 = -2\lambda \text{ and } \lambda + \mu = 0 \Rightarrow \lambda = -1/2, \mu = 1/2$$

$$\therefore \quad I = \int x\sqrt{1+x-x^2}\, dx$$

$$= \int \{-\frac{1}{2}(1-2x) + \frac{1}{2}\}\sqrt{1+x-x^2}\, dx$$

$$= -\frac{1}{2}\int (1-2x)\sqrt{1+x-x^2}\, dx + \frac{1}{2}\int \sqrt{1+x-x^2}\, dx$$

$$= -\frac{1}{2}\int \sqrt{t}\, dt + \frac{1}{2}\int \sqrt{-(x^2 - x + \frac{1}{4} - \frac{1}{4} - 1)}\, dx \quad (\text{where } t = 1 + x - x^2)$$

$$= -\frac{1}{2}\left(\frac{t^{3/2}}{3/2}\right) + \frac{1}{2}\int \sqrt{\left(\frac{\sqrt{5}}{2}\right)^2 - \left(x - \frac{1}{2}\right)^2}\, dx$$

$$= -\frac{1}{3}t^{3/2} + \frac{1}{2}$$

$$\left[\frac{(x-1/2)}{2}\sqrt{\left(\frac{\sqrt{5}}{2}\right)^2 - \left(x-\frac{1}{2}\right)^2} + \frac{1}{2}\left(\frac{\sqrt{5}}{2}\right)^2 \sin^{-1}\left(\frac{x-1/2}{\sqrt{5}/2}\right)\right] + C$$

$$= -\frac{1}{3}(1+x-x^2)^{3/2} + \frac{1}{2}\left[\frac{(x-1/2)}{2}\sqrt{1+x-x^2} + \frac{5}{8}\sin^{-1}\left(\frac{2x-1}{\sqrt{5}}\right)\right] + C$$

(iv) Let $\quad I = \int (x+1)\sqrt{1-x-x^2}\,dx$

Put, $\quad (x+1) = \lambda \cdot \left(\frac{d}{dx}(1-x-x^2)\right) + \mu$

Then $(x+1) = \lambda(-1-2x) + \mu$ comparing the coefficients of like powers of x, we get $-2\lambda = 1$ and $\mu - \lambda = 1 \Rightarrow \lambda = -1/2$ and $\mu = 1/2$.

$\therefore \quad (x+1) = -\frac{1}{2}(-1-2x) + \frac{1}{2}$

So, $\int (x+1)\sqrt{1-x-x^2}\,dx = \int \left\{-\frac{1}{2}(-1-2x) + \frac{1}{2}\right\}\sqrt{1-x-x^2}\,dx$

$$= -\frac{1}{2}\int(-1-2x)\sqrt{1-x-x^2}\,dx + \frac{1}{2}\int\sqrt{1-x-x^2}\,dx$$

$$= -\frac{1}{2}\int(-1-2x)\sqrt{1-x-x^2}\,dx + \frac{1}{2}\int\sqrt{\left\{1-\left(x^2+x+\frac{1}{4}-\frac{1}{4}\right)\right\}}\,dx$$

$$= -\frac{1}{2}\int\sqrt{t}\,dt + \frac{1}{2}\int\sqrt{(\sqrt{5}/2)^2-(x+1/2)^2}\,dx \qquad \text{(where } t = 1+x-x^2\text{)}$$

$$= -\frac{1}{2}\left(\frac{t^{3/2}}{3/2}\right) + \frac{1}{2}\left\{\frac{1}{2}\left(x+\frac{1}{2}\right)\sqrt{1-x-x^2} + \frac{1}{2}\cdot\frac{5}{4}\sin^{-1}\left(\frac{x+1/2}{\sqrt{5}/2}\right)\right\} + C$$

$$= -\frac{1}{3}(1-x-x^2)^{3/2} + \frac{1}{8}(2x+1)\sqrt{1-x-x^2} + \frac{5}{16}\sin^{-1}\left(\frac{2x+1}{\sqrt{5}}\right) + C$$

Integrals of the Form $\int \frac{k(x)}{ax^2+bx+c}\,dx$, where $k(x)$ is a polynomial of degree greater than 2

To evaluate this type of integrals we divide the numerator by denominator and express the integral as :

$$Q(x) + \frac{R(x)}{ax^2+bx+c}, \text{ where } R(x) \text{ is a linear function of } x.$$

Illustration 35 Evaluate $\int \frac{x^2+x+3}{x^2-x-2}\,dx$.

Solution. Let $I = \int \frac{x^2+x+3}{x^2-x-2}\,dx$

$\therefore \quad I = \int\left(1 + \frac{2x+5}{x^2-x-2}\right)dx$

$$I = \int 1\,dx + \int \frac{2x+5}{x^2 - x - 2}\,dx$$

$$I = x + \int \frac{2x+5}{x^2 - x - 2}\,dx$$

Put, $2x + 5 = \lambda \left\{\frac{d}{dx}(x^2 - x - 2)\right\} + \mu$. Then, obtaining $2x + 5 = \lambda(2x - 1) + \mu$, comparing the coefficients of like terms. We get, $2 = 2\lambda$ and $5 = \mu - \lambda$

$\therefore \quad \lambda = 1$ and $\mu = 6$

$$\therefore \quad I = x + \int \frac{\lambda(2x-1) + \mu}{x^2 - x - 2}\,dx$$

$$= x + \int \frac{2x-1}{x^2 - x - 2}\,dx + 6\int \frac{1}{x^2 - x - 2}\,dx$$

$$= x + \int \frac{1}{t}\,dt + 6\int \frac{1}{x^2 - x + \frac{1}{4} - \frac{1}{4} - 2} \quad (\text{where } t = x^2 - x - 2)$$

$$= x + \log|t| + 6\int \frac{dx}{(x - 1/2)^2 - (3/2)^2}$$

$$= x + \log|x^2 - x - 2| + 6\cdot\frac{1}{2(3/2)}\log\left|\frac{x - \frac{1}{2} - \frac{3}{2}}{x - \frac{1}{2} + \frac{3}{2}}\right| + C$$

$$= x + \log|x^2 - x - 2| + 2\cdot\log\left|\frac{x-2}{x+1}\right| + C$$

Integrals of the Type

(i) $\int \frac{ax^2 + bx + c}{(px^2 + qx + r)}\,dx$ (ii) $\int \frac{(ax^2 + bx + c)}{\sqrt{px^2 + qx + r}}\,dx$

(iii) $\int (ax^2 + bx + c)\sqrt{px^2 + qx + r}\,dx$

In above cases substitute

$$ax^2 + bx + c = \lambda\,(px^2 + qx + r) + \mu\left\{\frac{d}{dx}(px^2 + qx + r)\right\} + \gamma.$$

Find λ, μ and γ. These integrations reduces to integration of three independent functions.

Illustration 36 Evaluate $\int \frac{2x^2 + 5x + 4}{\sqrt{x^2 + x + 1}}\,dx.$

Solution. Let $I = \int \frac{2x^2 + 5x + 4}{\sqrt{x^2 + x + 1}}\,dx$

Put $\quad 2x^2 + 5x + 4 = \lambda(x^2 + x + 1) + \mu\left\{\frac{d}{dx}(x^2 + x + 1)\right\} + \gamma$

or $$2x^2 + 5x + 4 = \lambda (x^2 + x + 1) + \mu (2x + 1) + \gamma$$

Comparing the coefficients of like terms, we get

$$2 = \lambda, \quad 5 = \lambda + 2\mu, \quad 4 = \lambda + \mu + \gamma$$

$$\therefore \quad \lambda = 2, \quad \mu = 3/2, \quad \gamma = 1/2$$

Hence, the above integral reduces to

$$I = \int \frac{2x^2 + 5x + 4}{\sqrt{x^2 + x + 1}} dx$$

$$= \int \left(2 \cdot \frac{(x^2 + x + 1)}{\sqrt{x^2 + x + 1}} + \frac{3}{2} \cdot \frac{(2x + 1)}{\sqrt{x^2 + x + 1}} + \frac{1}{2} \cdot \frac{1}{\sqrt{x^2 + x + 1}} \right) dx$$

$$= 2 \int \sqrt{x^2 + x + 1}\, dx + \frac{3}{2} \int \frac{dt}{\sqrt{t}} + \frac{1}{2} \int \frac{dx}{\sqrt{x^2 + x + 1}} \quad \text{(where } t = x^2 + x + 1\text{)}$$

$$= 2 \int \sqrt{(x + 1/2)^2 + (\sqrt{3}/2)^2}\, dx + \frac{3}{2} \cdot \frac{t^{1/2}}{1/2} + \frac{1}{2} \int \frac{dx}{\sqrt{(x + 1/2)^2 + (\sqrt{3}/2)^2}}$$

$$= 2 \left[\frac{1}{2} \left(x + \frac{1}{2} \right) \sqrt{x^2 + x + 1} + \frac{1}{2} \cdot \frac{3}{4} \cdot \log \left| \left(x + \frac{1}{2} \right) + \sqrt{x^2 + x + 1} \right| \right]$$

$$+ 3\sqrt{x^2 + x + 1} + \frac{1}{2} \log \left| \left(x + \frac{1}{2} \right) + \sqrt{x^2 + x + 1} \right| + C$$

$$\therefore \quad I = \left(x + \frac{1}{2} \right) \sqrt{x^2 + x + 1} + \frac{3}{4} \log \left\{ \left(x + \frac{1}{2} \right) + \sqrt{x^2 + x + 1} \right\}$$

$$+ 3\sqrt{x^2 + x + 1} + \frac{1}{2} \log \left\{ \left(x + \frac{1}{2} \right) + \sqrt{x^2 + x + 1} \right\} + C$$

$$\Rightarrow \quad I = \left(x + \frac{7}{2} \right) \sqrt{x^2 + x + 1} + \frac{5}{4} \log \left\{ \left(x + \frac{1}{2} \right) + \sqrt{x^2 + x + 1} \right\} + C$$

Trigonometric Integrals

(a) Integrals of the Form

$$\int \frac{1}{a \cos^2 x + b \sin^2 x} dx, \int \frac{1}{a + b \sin^2 x} dx, \int \frac{1}{a + b \cos^2 x} dx,$$

$$\int \frac{1}{(a \sin x + b \cos x)^2} dx, \int \frac{1}{a + b \sin^2 x + c \cos^2 x} dx$$

To evaluate this type of integrals, divide numerator and denominator both by $\cos^2 x$, replace $\sec^2 x$, if any, in denominator by $(1 + \tan^2 x)$ and put $\tan x = t$. So that $\sec^2 x\, dx = dt$.

Illustration 37 Evaluate

(i) $\int \frac{1}{4 \sin^2 x + 9 \cos^2 x} dx$ (ii) $\int \frac{\sin x}{\sin 3x} dx$

Solution. (i) $I = \int \frac{dx}{4 \sin^2 x + 9 \cos^2 x}$

Here, dividing N^r and D^r by $\cos^2 x$. We get

$$I = \int \frac{\sec^2 x}{4 \tan^2 x + 9} dx$$

Put $\qquad \tan x = t \quad \Rightarrow \quad \sec^2 x \, dx = dt$

$$\therefore \quad I = \int \frac{dt}{4t^2 + 9} = \frac{1}{4} \int \frac{dt}{t^2 + (3/2)^2} = \frac{1}{4} \cdot \frac{1}{3/2} \tan^{-1}\left(\frac{t}{3/2}\right) + C$$

$$I = \frac{1}{6} \tan^{-1}\left(\frac{2 \tan x}{3}\right) + C$$

(ii) Let $\qquad I = \int \frac{\sin x}{\sin 3x} dx = \int \frac{\sin x}{3 \sin x - 4 \sin^3 x} dx$

$$I = \int \frac{dx}{3 - 4 \sin^2 x}$$

Dividing N^r and D^r by $\cos^2 x$, we get

$$I = \int \frac{\sec^2 x \, dx}{3 \sec^2 x - 4 \tan^2 x} = \int \frac{\sec^2 x \, dx}{3(1 + \tan^2 x) - 4 \tan^2 x}$$

$$I = \int \frac{\sec^2 x \, dx}{3 - \tan^2 x}$$

Let $\tan x = t \Rightarrow \sec^2 x \, dx = dt$

$$\therefore \quad I = \int \frac{dt}{(\sqrt{3})^2 - (t)^2} = \frac{1}{2\sqrt{3}} \log \left| \frac{\sqrt{3} + t}{\sqrt{3} - t} \right| + C$$

$$I = \frac{1}{2\sqrt{3}} \log \left| \frac{\sqrt{3} + \tan x}{\sqrt{3} - \tan x} \right| + C$$

(b) Integrals of the Form $\int \frac{1}{a \sin x + b \cos x} dx$, $\int \frac{1}{a + b \sin x} dx$, $\int \frac{1}{a + b \cos x} dx$, $\int \frac{1}{a \sin x + b \cos x + c} dx$

To evaluate this type of integrals we put $\sin x = \frac{2 \tan x/2}{1 + \tan^2 x/2}$ and $\cos x = \frac{1 - \tan^2 x/2}{1 + \tan^2 x/2}$ and replace $\tan x/2 = t$, by performing these steps the integral reduces to the form $\int \frac{1}{at^2 + bt + c} dt$ which can be evaluated by methods discussed earlier.

Illustration 38 Evaluate

(i) $\int \frac{dx}{2 + \sin x + \cos x}$ $\qquad$ (ii) $\int \frac{dx}{\sqrt{3} \sin x + \cos x}$

Solution. For this type we use $\sin x = \frac{2 \tan x/2}{1 + \tan^2 x/2}$, $\cos x = \frac{1 - \tan^2 x/2}{1 + \tan^2 x/2}$.

(i) $\int \frac{dx}{2 + \sin x + \cos x}$

Let $$I = \int \frac{dx}{2 + \frac{2\tan x/2}{1+\tan^2 x/2} + \frac{1-\tan^2 x/2}{1+\tan^2 x/2}}$$

$$= \int \frac{\sec^2 \frac{x}{2}\, dx}{2 + 2\tan^2 \frac{x}{2} + 2\tan \frac{x}{2} + 1 - \tan^2 \frac{x}{2}}$$

$$I = \int \frac{\sec^2 \frac{x}{2}\, dx}{\tan^2 \frac{x}{2} + 2\tan \frac{x}{2} + 3}$$

Put $\tan \frac{x}{2} = t \Rightarrow \frac{1}{2}\sec^2 \frac{x}{2}\, dx = dt$

$$= \int \frac{2\, dt}{t^2 + 2t + 3} = 2\int \frac{dt}{t^2 + 2t + 1 + 2}$$

$$= 2\int \frac{dt}{(t+1)^2 + (\sqrt{2})^2} = 2 \cdot \frac{1}{\sqrt{2}} \tan^{-1}\left(\frac{t+1}{\sqrt{2}}\right) + C$$

$$I = \sqrt{2}\tan^{-1}\left(\frac{\tan x/2 + 1}{\sqrt{2}}\right) + C$$

(ii) Let $$I = \int \frac{dx}{\sqrt{3}\sin x + \cos x} = \int \frac{dx}{\frac{\sqrt{3}\cdot 2\tan x/2}{1+\tan^2 x/2} + \frac{1-\tan^2 x/2}{1+\tan^2 x/2}}$$

$$= \int \frac{\sec^2 \frac{x}{2}\, dx}{2\sqrt{3}\tan \frac{x}{2} + 1 - \tan^2 \frac{x}{2}}$$

Put $\tan \frac{x}{2} = t \Rightarrow \frac{1}{2}\sec^2 \frac{x}{2}\, dx = dt$

$$\therefore \quad I = \int \frac{2dt}{-t^2 + 2\sqrt{3}t + 1} = 2\int \frac{dt}{-t^2 + 2\sqrt{3}t - 3 + 3 + 1} = 2\int \frac{dt}{4 - (t - \sqrt{3})^2}$$

$$= 2\int \frac{dt}{(2)^2 - (t-\sqrt{3})^2} = 2 \cdot \frac{1}{2(2)} \log\left|\frac{2 + t - \sqrt{3}}{2 - t + \sqrt{3}}\right| + C$$

$$I = \frac{1}{2}\log\left|\frac{2 - \sqrt{3} + \tan x/2}{2 + \sqrt{3} - \tan x/2}\right| + C$$

(c) Alternative Method to Evaluate the Integrals of the Form $\int \frac{1}{a\sin x + b\cos x}\, dx$

To evaluate this type of integrals we substitute $a = r\cos\theta$, $b = r\sin\theta$ and so

$$r = \sqrt{a^2 + b^2}, \theta = \tan^{-1}\left(\frac{b}{a}\right)$$

$\therefore \quad a \sin x + b \cos x = r \sin (x + \theta)$

So, $\displaystyle\int \frac{1}{a \sin x + b \cos x} dx = \frac{1}{r} \int \frac{1}{\sin (x + \theta)} dx$

$$= \frac{1}{r} \int \operatorname{cosec} (x + \theta)\, dx = \frac{1}{r} \log \left| \tan \left(\frac{x}{2} + \frac{\theta}{2} \right) \right| + C$$

$$\therefore \quad \int \frac{1}{a \sin x + b \cos x} dx = \frac{1}{\sqrt{a^2 + b^2}} \log \left| \tan \left(\frac{x}{2} + \frac{1}{2} \tan^{-1} \frac{b}{a} \right) \right| + C$$

Illustration 39 Evaluate $\displaystyle\int \frac{1}{\sqrt{3} \sin x + \cos x} dx.$

Solution. Let $\sqrt{3} = r \sin \theta$ and $1 = r \cos \theta$. Then

$$r = \sqrt{(\sqrt{3})^2 + (1)^2} = 2 \quad \text{and} \quad \tan \theta = \frac{\sqrt{3}}{1} \quad \Rightarrow \quad \theta = \frac{\pi}{3}$$

$$\therefore \quad I = \int \frac{1}{\sqrt{3} \sin x + \cos x} dx$$

$$= \int \frac{1}{r \sin \theta \sin x + r \cos \theta \cos x} dx$$

$$= \frac{1}{r} \int \frac{dx}{\cos (x - \theta)}$$

$$= \frac{1}{r} \int \sec (x - \theta)\, dx = \frac{1}{r} \log \left| \tan \left(\frac{\pi}{4} + \frac{x}{2} - \frac{\theta}{2} \right) \right| + C$$

$$= \frac{1}{2} \log \left| \tan \left(\frac{\pi}{4} + \frac{x}{2} - \frac{\pi}{6} \right) \right| + C = \frac{1}{2} \log \left| \tan \left(\frac{x}{2} + \frac{\pi}{12} \right) \right| + C$$

(d) Integrals of the Form

$$\int \frac{p \cos x + q \sin x + r}{a \cos x + b \sin x + c} dx, \int \frac{p \cos x + q \sin x}{a \cos x + b \sin x} dx$$

Rule for (i) In this integral express numerator as,

λ (denominator) + μ (diffn. of denominator) + γ.

Find λ, μ and γ by comparing coefficients of sin x, cos x and constant term and split the integral into sum of three integrals.

$$\lambda \int dx + \mu \int \frac{\text{d.c. of (denominator)}}{\text{denominator}} dx + n \int \frac{dx}{a \sin x + b \cos x + c}$$

Rule for (ii) Express numerator as λ (denominator) + μ (diffn. of denominator) and find λ and μ as above.

Illustration 40 Evaluate $\displaystyle\int \frac{(2 + 3 \cos x)}{\sin x + 2 \cos x + 3} dx.$

Solution. Write the numerator = λ (denominator)

+ μ (d.c. of denominator) + γ

$$\Rightarrow \quad 2 + 3\cos x = \lambda(\sin x + 2\cos x + 3) + \mu(\cos x - 2\sin x) + \gamma$$

Comparing the coefficients of $\sin x$, $\cos x$ and constant terms, we get

$$0 = \lambda - 2\mu,\ 3 = 2\lambda + \mu,\ 2 = 3\lambda + \gamma$$

$$\Rightarrow \quad \lambda = 6/5,\ \mu = 3/5,\ \gamma = -8/5$$

Hence, $$I = \frac{6}{5}\int 1\,dx + \frac{3}{5}\int \frac{\cos x - 2\sin x}{\sin x + 2\cos x + 3}\,dx - \frac{8}{5}\int \frac{dx}{\sin x + 2\cos x + 3}$$

$$= \frac{6}{5}\cdot x + \frac{3}{5}\log|\sin x + 2\cos x + 3| - \frac{8}{5}I_3 \quad \text{...(i)}$$

where $$I_3 = \int \frac{dx}{\sin x + 2\cos x + 3} = \int \frac{dx}{\dfrac{2\tan x/2}{1+\tan^2 x/2} + \dfrac{2(1-\tan^2 x/2)}{1+\tan^2 x/2} + 3}$$

$$= \int \frac{\sec^2 \frac{x}{2}\,dx}{2\tan\frac{x}{2} + 2 - 2\tan^2\frac{x}{2} + 3 + 3\tan^2\frac{x}{2}}$$

$$= \int \frac{\sec^2\frac{x}{2}\,dx}{\tan^2\frac{x}{2} + 2\tan\frac{x}{2} + 5} \quad \text{let } \tan\frac{x}{2} = t \Rightarrow \frac{1}{2}\sec^2\frac{x}{2}\,dx = dt$$

$$= \int \frac{2dt}{t^2 + 2t + 5} = 2\int \frac{dt}{(t+1)^2 + 2^2}$$

$$I_3 = 2\cdot\frac{1}{2}\cdot\tan^{-1}\left(\frac{t+1}{2}\right) = \tan^{-1}\left(\frac{\tan\frac{x}{2} + 1}{2}\right) \quad \text{...(ii)}$$

From Eqs. (i) and (ii)

$$I = \frac{6}{5}x + \frac{3}{5}\log|\sin x + 2\cos x + 3| - \frac{8}{5}\tan^{-1}\left|\left(\frac{\tan\frac{x}{2} + 1}{2}\right)\right| + C$$

Illustration 41 Evaluate $\int \dfrac{3\sin x + 2\cos x}{3\cos x + 2\sin x}\,dx.$

Solution. Write the numerator = λ (denominator) + μ (d.c. of denominator)

$$\therefore \quad 3\sin x + 2\cos x = \lambda(3\cos x + 2\sin x) + \mu(-3\sin x + 2\cos x)$$

Comparing the coefficients of $\sin x$ and $\cos x$ on both the sides, we get

$$3 = 2\lambda - 3\mu \quad \text{and} \quad 2 = 3\lambda + 2\mu$$

$$\Rightarrow \quad \lambda = \frac{12}{13} \quad \text{and} \quad \mu = -\frac{5}{13}$$

Hence, $$I = \lambda\int dx + \mu\int \frac{-3\sin x + 2\cos x}{3\cos x + 2\sin x}\,dx$$

$$= \frac{12}{13}(x) - \frac{5}{13}\log|3\cos x + 2\sin x| + C$$

Illustration 42 The value of $\int \sqrt{\frac{3-x}{3+x}} \cdot \sin^{-1}\left(\frac{1}{\sqrt{6}}\sqrt{3-x}\right) dx$, is equal to

(a) $\frac{1}{4}\left\{-3\left(\cos^{-1}\left(\frac{x}{3}\right)\right)^2 + 2\sqrt{9-x^2} \cdot \cos^{-1}\left(\frac{x}{3}\right) + 2x\right\} + C$

(b) $\frac{1}{4}\left\{-3\left(\cos^{-1}\left(\frac{x}{3}\right)\right)^2 + 2\sqrt{9-x^2}\sin^{-1}\left(\frac{x}{3}\right) + 2x\right\} + C$

(c) $\frac{1}{4}\left\{-3\left(\sin^{-1}\left(\frac{x}{3}\right)\right)^2 + 2\sqrt{9-x^2}\sin^{-1}\left(\frac{x}{3}\right) + 2x\right\} + C$

(d) None of the above

Solution. Here, $I = \int \sqrt{\frac{3-x}{3+x}} \cdot \sin^{-1}\left(\frac{1}{\sqrt{6}}\sqrt{3-x}\right) dx$

Put $x = 3\cos 2\theta \Rightarrow dx = -6\sin 2\theta\, d\theta$

$$\therefore \quad I = \int \sqrt{\frac{3-3\cos 2\theta}{3+3\cos 2\theta}} \cdot \sin^{-1}\left(\frac{1}{\sqrt{6}}\sqrt{3-3\cos 2\theta}\right)(-6\sin 2\theta)\, d\theta$$

$$= \int \frac{\sin\theta}{\cos\theta} \cdot \sin^{-1}(\sin\theta) \cdot (-6\sin 2\theta)\, d\theta$$

$$= -6\int \theta \cdot (2\sin^2\theta)\, d\theta = -6\int \theta(1-\cos 2\theta)\, d\theta$$

$$= -6\left\{\frac{\theta^2}{2} - \int \theta\cos 2\theta\, d\theta\right\}$$

$$= -6\left\{\frac{\theta^2}{2} - \left(\theta\frac{\sin 2\theta}{2} - \int 1 \cdot \left(\frac{\sin 2\theta}{2}\right) d\theta\right)\right\}$$

$$= -3\theta^2 + 6\left\{\frac{\theta\sin 2\theta}{2} + \frac{\cos 2\theta}{4}\right\} + C$$

$$= \frac{1}{4}\left\{-3\left(\cos^{-1}\left(\frac{x}{3}\right)\right)^2 + 2\sqrt{9-x^2} \cdot \cos^{-1}\left(\frac{x}{3}\right) + 2x\right\} + C$$

Hence, (a) is the correct answer.

Illustration 43 The value of $\int \frac{\sec x\,(2+\sec x)}{(1+2\sec x)^2} dx$, is equal to

(a) $\frac{\sin x}{2+\cos x} + C$ (b) $\frac{\cos x}{2+\cos x} + C$ (c) $\frac{-\sin x}{2+\sin x} + C$ (d) $\frac{\cos x}{2+\sin x} + C$

Solution. Let $I = \int \frac{\sec x\,(2+\sec x)}{(1+2\sec x)^2} dx = \int \frac{2\cos x + 1}{(\cos x + 2)^2} dx$

$$= \int \frac{\cos x(\cos x + 2) + \sin^2 x}{(2+\cos x)^2} dx$$

$$= \int \frac{\cos x}{2+\cos x} dx + \int \frac{\sin^2 x}{(2+\cos x)^2} dx$$

$$= \int \underset{\text{II}}{\cos x} \cdot \underset{\text{I}}{\frac{1}{(2+\cos x)}} dx + \int \frac{\sin^2 x}{(2+\cos x)^2} dx$$

Applying Integration by parts to first by taking $\cos x$ as second function, keeping $\int \frac{\sin^2 x}{(2+\cos x)^2}\,dx$ as it is.

$$\therefore \qquad I = \frac{1}{2+\cos x}\cdot(\sin x) - \int \sin x \cdot \frac{\sin x}{(2+\cos x)^2}\,dx + \int \frac{\sin^2 x}{(2+\cos x)^2}\,dx$$

$$\therefore \qquad I = \frac{\sin x}{2+\cos x} + C$$

Hence, (a) is the correct answer.

Illustration 44 The value of $\int \log(\sqrt{1-x}+\sqrt{1+x})\,dx$, is equal to

(a) $x\log(\sqrt{1-x}+\sqrt{1+x}) + \frac{1}{2}x - \frac{1}{2}\sin^{-1}(x) + C$

(b) $x\log(\sqrt{1-x}+\sqrt{1+x}) + \frac{1}{2}x + \frac{1}{2}\sin^{-1}(x) + C$

(c) $x\log(\sqrt{1-x}+\sqrt{1+x}) - \frac{1}{2}x + \frac{1}{2}\sin^{-1}(x) + C$

(d) None of the above

Solution. Here, we have only one function. This can be solved easily by applying Integration by parts taking unity as second function. If we take $u = \log(\sqrt{1-x}+\sqrt{1+x})$ as the first function and $v = 1$ as the second function. Then

$$I = \int 1\cdot\log(\sqrt{1-x}+\sqrt{1+x})\,dx$$

$$= \{\log(\sqrt{1-x}+\sqrt{1+x})\}\cdot x - \int \frac{1}{\sqrt{1-x}+\sqrt{1+x}}\cdot\left(-\frac{1}{2\sqrt{1-x}}+\frac{1}{2\sqrt{1+x}}\right)\cdot x\,dx$$

$$= x\log(\sqrt{1-x}+\sqrt{1+x}) - \frac{1}{2}\int \frac{\sqrt{1-x}-\sqrt{1+x}}{\sqrt{1-x}+\sqrt{1+x}}\cdot\frac{1}{\sqrt{1-x^2}}\cdot x\,dx$$

$$= x\log(\sqrt{1-x}+\sqrt{1+x}) - \frac{1}{2}\int \frac{(1-x)+(1+x)-2\sqrt{1-x^2}}{(1-x)-(1+x)}\cdot\frac{1}{\sqrt{1-x^2}}\cdot 2x\,dx$$

$$= x\log(\sqrt{1-x}+\sqrt{1+x}) - \frac{1}{2}\int \frac{\sqrt{1-x^2}-1}{\sqrt{1-x^2}}\,dx$$

$$= x\log(\sqrt{1-x}+\sqrt{1+x}) - \frac{1}{2}\int 1\,dx + \frac{1}{2}\int \frac{1}{\sqrt{1-x^2}}\,dx$$

$$= x\log(\sqrt{1-x}+\sqrt{1+x}) - \frac{1}{2}x + \frac{1}{2}\sin^{-1}x + C$$

Hence, (c) is the correct answer.

Illustration 45 The value of $\int e^x\left(\frac{x^4+2}{(1+x^2)^{5/2}}\right)dx$, is equal to

(a) $\frac{e^x(x+1)}{(1+x^2)^{3/2}}$ (b) $\frac{e^x(1-x+x^2)}{(1+x^2)^{3/2}}$ (c) $\frac{e^x(1+x)}{(1+x^2)^{3/2}}$ (d) None of these

Solution. Let $I = \int e^x \left(\frac{x^4 + 2}{(1 + x^2)^{5/2}} \right) dx$

$$= \int e^x \left(\frac{1}{(1 + x^2)^{1/2}} + \frac{1 - 2x^2}{(1 + x^2)^{5/2}} \right) dx$$

$$= \int e^x \left(\frac{1}{(1 + x^2)^{1/2}} - \frac{x}{(1 + x^2)^{3/2}} + \frac{x}{(1 + x^2)^{3/2}} + \frac{1 - 2x^2}{(1 + x^2)^{5/2}} \right) dx$$

$$= \frac{e^x}{(1 + x^2)^{1/2}} + \frac{xe^x}{(1 + x^2)^{3/2}} + C = \frac{e^x \{1 + x^2 + x\}}{(1 + x^2)^{3/2}} + C$$

Hence, (d) is the correct answer.

Illustration 46 If $\int (\sin 3\theta + \sin \theta)\, e^{\sin \theta} \cos \theta \, d\theta$

$$= (A \sin^3 \theta + B \cos^2 \theta + C \sin \theta + D \cos \theta + E)\, e^{\sin \theta} + F, \text{ then}$$

(a) $A = -4, B = 12$ (b) $A = -4, B = -12$

(c) $A = 4, B = 12$ (d) $A = 4, B = -12$

Solution. Let $I = \int (\sin 3\theta + \sin \theta)\, e^{\sin \theta} \cdot \cos \theta \, d\theta$

$= \int (3 \sin \theta - 4 \sin^3 \theta) \cdot e^{\sin \theta} \cdot \cos \theta \, d\theta$; put $\sin \theta = t$, $\cos \theta \, d\theta = dt$

$= \int (3t - 4t^3)\, e^t \, dt$ ...(i)

As, $I = (A \sin^3 \theta + B \cos^2 \theta + C \sin \theta + D \cos \theta + E)\, e^{\sin \theta} + F$

$= (A \sin^3 \theta - B \sin^2 \theta + C \sin \theta + D \cos \theta + B + E)\, e^{\sin \theta} + F$

when, $\sin \theta = t$

$I = (At^3 - Bt^2 + Ct + B + E)\, e^t + F$ as by (i) $D = 0$ ...(ii)

From Eqs. (i) and (ii),

$\int (3t - 4t^3)\, e^t \, dt = \underbrace{(At^3 - Bt^2 + Ct + B + E)}_{f(t)}\, e^t + F$, differentiating both sides

$$(3t - 4t^3)\, e^t = (At^3 - Bt^2 + Ct + B + E)\, e^t + (3At^2 - 2Bt + C)\, e^t$$

$\Rightarrow$ $A = -4$ and $3A = B \Rightarrow B = -12$

Hence, (b) is the correct answer.

Illustration 47 The value of $\int \{1 + \tan x \cdot \tan (x + A)\}\, dx$ is equal to

(a) $\cot A \cdot \log \left| \frac{\sec x}{\sec (x + A)} \right| + C$ (b) $\tan A \cdot \log |\sec (x + A)| + C$

(c) $\cot A \cdot \log \left| \frac{\sec (x + A)}{\sec (x)} \right| + C$ (d) None of these

Solution. Let $I = \int \{1 + \tan x \cdot \tan (x + A)\}\, dx$

$$= \int \left\{ 1 + \frac{\sin x \cdot \sin (x + A)}{\cos x \cdot \cos (x + A)} \right\} dx$$

$$= \int \frac{\cos x \cdot \cos (x + A) + \sin x \cdot \sin (x + A)}{\cos x \cdot \cos (x + A)} dx$$

$$= \int \frac{\cos (x + A - x)\, dx}{\cos x \cdot \cos (x + A)} = \cos A \cdot \int \frac{dx}{\cos x \cdot \cos (x + A)}$$

Multiplying and dividing by $\sin A$, we get

$$= \cot A \cdot \int \frac{\sin A\, dx}{\cos x \cdot \cos (x + A)} = \cot A \cdot \int \frac{\sin (x + A - x)\, dx}{\cos x \cdot \cos (x + A)}$$

$$= \cot A \cdot \int \left\{ \frac{\sin (x + A) \cdot \cos x}{\cos x \cos (x + A)} - \frac{\cos (x + A) \cdot \sin x}{\cos x \cdot \cos (x + A)} \right\} dx$$

$$= \cot A \cdot \{ \int \tan (x + A)\, dx - \int \tan x\, dx \}$$

$$= \cot A \cdot \{ \log |\sec (x + A)| - \log |\sec x| \} + C$$

Hence, (c) is the correct answer.

Illustration 48 The value of $\int \frac{\sqrt{\cos 2x}}{\sin x} dx$, is equal to

(a) $\log |\cot x + \sqrt{\cot^2 x - 1}| + \sqrt{2} \log |\cos x + \sqrt{\cos^2 x - 1/2}| + C$

(b) $- \log |\cot x + \sqrt{\cot^2 x - 1}| + \sqrt{2} \log |\cos x + \sqrt{\cos^2 x - 1/2}| + C$

(c) $\log |\cot x + \sqrt{\cot^2 x - 1}| + 2 \log |\cos x + \sqrt{\cos^2 x - 1/2}| + C$

(d) $- \log |\cot x + \sqrt{\cot^2 x - 1}| + 2 \log |\cos x + \sqrt{\cos^2 x - 1/2}| + C$

Solution. Let $I = \int \frac{\sqrt{\cos 2x}}{\sin x} dx = \int \frac{\cos 2x}{\sin x \sqrt{\cos 2x}} dx = \int \frac{1 - 2\sin^2 x}{\sin x \sqrt{\cos 2x}} dx$

$$= \int \frac{1}{\sin x \sqrt{\cos^2 x - \sin^2 x}} dx - 2 \int \frac{\sin x}{\sqrt{2\cos^2 x - 1}} dx$$

$$= \int \frac{\operatorname{cosec}^2 x}{\sqrt{\cot^2 x - 1}} dx - \frac{2}{\sqrt{2}} \int \frac{\sin x}{\sqrt{\cos^2 x - 1/2}} dx$$

$$= - \int \frac{dt}{\sqrt{t^2 - 1}} - \sqrt{2} \int \frac{-ds}{\sqrt{s^2 - (1/\sqrt{2})^2}};$$

where $t = \cot x$ and $s = \cos x$

$$= - \log |t + \sqrt{t^2 - 1}| + \sqrt{2} \log |s + \sqrt{s^2 - 1/2}| + C$$

$$= - \log |\cot x + \sqrt{\cot^2 x - 1}| + \sqrt{2} \log |\cos x + \sqrt{\cos^2 x - 1/2}| + C$$

Hence, (b) is the correct answer.

Illustration 49 The value of

$\int e^{(x \sin x + \cos x)} \cdot \left(\frac{x^4 \cos^3 x - x \sin x + \cos x}{x^2 \cos^2 x} \right) dx$, is equal to

(a) $e^{(x \sin x + \cos x)} \cdot \left(x + \frac{1}{x \cos x} \right) + C$ (b) $e^{(x \sin x + \cos x)} \cdot \left(x \cos x + \frac{1}{x} \right) + C$

(c) $e^{(x \sin x + \cos x)} \cdot \left(x - \frac{1}{x \cos x} \right) + C$ (d) None of these

Solution. Let $I = \int e^{(x \sin x + \cos x)} \cdot \left(\dfrac{x^4 \cos^3 x - x \sin x + \cos x}{x^2 \cos^2 x} \right) dx$

$$= \int (x \cdot e^{(x \sin x + \cos x)} \cdot x \cos x \,) \, dx - \int e^{(x \sin x + \cos x)} \cdot \left(\frac{x \sin x - \cos x}{(x \cos x)^2} \right) dx$$

I II

Applying Integration by parts

$$= \{x \cdot e^{(x \sin x + \cos x)} - \int e^{(x \sin x + \cos x)} \, dx\}$$

$$- \left\{ e^{(x \sin x + \cos x)} \cdot \frac{1}{x \cos x} - \int e^{(x \sin x + \cos x)} \, dx \right\}$$

$$= e^{(x \sin x + \cos x)} \left(x - \frac{1}{x \cos x} \right) + C$$

Hence, (c) is the correct answer.

Illustration 50 The value of $\int \dfrac{(ax^2 - b)\, dx}{x \sqrt{c^2 x^2 - (ax^2 + b)^2}}$, is equal to

(a) $\dfrac{1}{c} \sin^{-1} \left(ax + \dfrac{b}{x} \right) + K$

(b) $c \sin^{-1} \left(ax + \dfrac{b}{x} \right) + K$

(c) $\sin^{-1} \left(\dfrac{ax + \frac{b}{x}}{c} \right) + K$

(d) None of these

Solution. Let $I = \int \dfrac{(ax^2 - b)\, dx}{x \sqrt{c^2 x^2 - (ax^2 + b)^2}} = \int \dfrac{\left(a - \frac{b}{x^2} \right) dx}{\sqrt{c^2 - \left(ax + \frac{b}{x} \right)^2}}$

Put $ax + \dfrac{b}{x} = t \quad \therefore \left(a - \dfrac{b}{x^2} \right) dx = dt$

$$\therefore \qquad I = \int \frac{dt}{\sqrt{c^2 - t^2}} = \sin^{-1} \left(\frac{t}{c} \right) + K = \sin^{-1} \left(\frac{ax + \frac{b}{x}}{c} \right) + K$$

Hence, (c) is the correct answer.

Target Exercise 1.2

1. The value of $\int \sec^3 \theta \, d\theta$, is equal to

(a) $\dfrac{1}{2} \{\sec \theta \tan \theta + \log |\sec \theta + \tan \theta|\} + C$

(b) $\{\sec \theta \tan \theta + \log |\sec \theta + \tan \theta|\} + C$

(c) $\dfrac{1}{3} \{\sec \theta \tan \theta + \log |\sec \theta + \tan \theta|\} + C$

(d) None of the above

2. The value of $\int \frac{dx}{(x-\beta)\sqrt{(x-\alpha)(\beta-x)}}$, is equal to

(a) $\frac{-1}{\beta-\alpha}\sqrt{\frac{x-\alpha}{\beta-x}}+C$ (b) $\frac{1}{\beta-\alpha}\sqrt{\frac{x-\alpha}{\beta-x}}+C$

(c) $\frac{2}{\alpha-\beta}\sqrt{\frac{x-\alpha}{\beta-x}}+C$ (d) None of these

3. The value of $\int \frac{\sqrt{x^2+1}\,\{\log_e(x^2+1)-2\log_e x\}}{x^4}\,dx$ is equal to

(a) $\frac{2}{3}\left(1+\frac{1}{x^2}\right)^{3/2}\cdot\left\{\log\left(1+\frac{1}{x^2}\right)-\frac{2}{3}\right\}+C$

(b) $-\frac{1}{3}\left(1+\frac{1}{x^2}\right)^{3/2}\cdot\left\{\log\left(1+\frac{1}{x^2}\right)-\frac{2}{3}\right\}+C$

(c) $\left(1+\frac{1}{x^2}\right)^{3/2}\cdot\left\{\log\left(1+\frac{1}{x^2}\right)+\frac{2}{3}\right\}+C$

(d) None of the above

4. The value of $\int\left(\frac{1}{\sqrt[3]{x}+\sqrt[4]{x}}+\frac{\log(1+\sqrt[6]{x})}{\sqrt[3]{x}+\sqrt{x}}\right)dx$, is equal to

(a) $12\left\{\sum_{r=1}^{8}{}^8C_r(-1)^r\frac{t^r}{r}+\log(1+t)\right\}$
$$+6\left\{\left(\frac{\theta}{3}-\frac{1}{9}\right)e^{3\theta}-\frac{3}{2}\left(\theta-\frac{1}{2}\right)e^{2\theta}+3(\theta-1)e^{\theta}-\frac{\theta^2}{2}\right\}+C$$

(b) $6\left\{\sum_{r=1}^{8}{}^8C_r(-1)^r\frac{t^r}{r}+\log(1+t)\right\}$
$$+12\left\{\left(\frac{\theta}{3}-\frac{1}{9}\right)e^{3\theta}-\frac{3}{2}(\theta-1/2)+3(\theta-1)e^{\theta}-\frac{\theta^2}{2}\right\}+C$$

(c) $\left\{\sum_{r=1}^{8}{}^8C_r(-1)^r\frac{t^r}{r}+\log(1+t)\right\}$
$$+6\left\{\left(\frac{\theta}{3}-\frac{1}{9}\right)e^{3\theta}-\frac{3\theta}{2}\left(\theta-\frac{1}{2}\right)+3(\theta-1)e^{\theta}-\theta^2/2\right\}+C$$

(d) None of the above

where $t=x^{1/12}$ and $\theta=\log(1+x^{1/6})$

5. The value of $\int \frac{dx}{\sec x+\operatorname{cosec} x}$, is equal to

(a) $\left\{(\sin x+\cos x)+\frac{1}{\sqrt{2}}\log\left|\frac{\tan x/2-1-\sqrt{2}}{\tan x/2-1+\sqrt{2}}\right|\right\}+C$

(b) $2\left\{(\sin x+\cos x)+\frac{1}{\sqrt{2}}\log\left|\frac{\tan x/2-1-\sqrt{2}}{\tan x/2-1+\sqrt{2}}\right|\right\}+C$

(c) $\frac{1}{2}\left\{(\sin x-\cos x)+\frac{1}{\sqrt{2}}\log\left|\frac{\tan x/2-1-\sqrt{2}}{\tan x-1+\sqrt{2}}\right|\right\}+C$

(d) None of the above

Integration of Rational Functions Using Partial Fraction

By now, you must have understood that in integration, there are few standard forms. By expressing the given integrand in the standard forms, we can reduce the given integrand into integrable forms. Here, we are interested in integration of a 'rational functions'.

Some Basic Definitions

(a) **Polynomial of degree n** An expression of the type $P(x) = a_0x^n + a_1x^{n-1} + \dots + a_{n-1}x + a_n$ where $a_0, a_1, a_2, \dots, a_n$ are real numbers, $a_0 \neq 0$ and n is positive integer is called a polynomial of degree n.

A function of the form P/Q where P and Q are polynomials is called **Rational function.** Consider the rational function

$$\frac{x+7}{(2x-3)(3x+4)} = \frac{1}{2x-3} - \frac{1}{3x+4}.$$

The two fractions on the RHS are called the partial fractions.

(b) **Proper and Improper functions** Any rational algebraic function is called a proper fraction, if the degree of numerator is less than that of its denominator, otherwise it is called an improper fraction.

For example $\dfrac{x^2+x+2}{x^3+4x^2-7x+1}$ is a proper fraction.

whereas $\dfrac{x^4-9x^2-10x+7}{x^2+4x+5} = \left\{(x^2-4x+2) + \dfrac{2x-3}{x^2+4x+5}\right\}$ is improper fraction.

To integrate the rational function on the LHS, it is enough to integrate the two fractions on the RHS, which is easy. This is known as the method of partial fractions. Here, we assume that the denominator can be fractionalised into linear or quadratic factors.

Points to Consider

In using the method of partial fractions, we must have the degree of polynomial in numerator ($P(x)$) always less than that of denominator ($Q(x)$). If it is not so, we carry out the division of $P(x)$ by $Q(x)$ and reduce the degree of the numerator to make it less than that of the denominator.

ie, $\dfrac{P(x)}{Q(x)} = P_1(x) + \dfrac{P_2(x)}{Q(x)}$, where the degree of $P_2(x) <$ degree of $Q(x)$.

Then to integrate, we apply the method of partial fractions to $\dfrac{P_2(x)}{Q(x)}$.

The partial fractions depend on the nature of the factors of $Q(x)$. We have to deal with the following different types when the factors of $Q(x)$ are

(i) Linear and non-repeated
(ii) Linear and repeated
(iii) Quadratic and non-repeated
(iv) Quadratic and repeated

Case I When denominator is expressible as the product of non-repeated linear factors :

Let $Q(x) = (x - a_1)(x - a_2)(x - a_3)\ldots(x - a_n)$.

Then we assume that

$$\frac{P(x)}{Q(x)} = \frac{A_1}{(x - a_1)} + \frac{A_2}{(x - a_2)} + \frac{A_3}{(x - a_3)} + \ldots + \frac{A_n}{(x - a_n)}$$

where $A_1, A_2, \ldots, A_n$ are constants and can be determined by equating the numerator on RHS to numerator on LHS and then substituting $x = a_1, a_2, \ldots, a_n$.

Illustration 51 Resolve $\frac{1}{(x-1)(x+2)(2x+3)}$ into partial fractions.

Solution. Let $\frac{1}{(x-1)(x+2)(2x+3)} = \frac{A}{x-1} + \frac{B}{x+2} + \frac{C}{2x+3}$, where A, B, C are constants.

$\Rightarrow \quad 1 = A(x+2)(2x+3) + B(x-1)(2x+3) + C(x-1)(x+2)$ ...(i)

For finding A, let $x - 1 = 0$ or $x = 1$ in Eq. (i), we get

$$1 = A(1+2)(2+3) + B(0) + C(0)$$

$\therefore \quad A = \frac{1}{15}$

Similarly, for getting B, let $x + 2 = 0$ or $x = -2$ in Eq. (i), we get

$$1 = A(0) + B(-2-1)(-4+3) + C(0) \quad \Rightarrow B = \frac{1}{3}$$

For getting C, let $2x + 3 = 0$ or $x = -3/2$ in Eq. (i), we get

$$1 = A(0) + B(0) + C\left(-\frac{3}{2} - 1\right)\left(-\frac{3}{2} + 2\right)$$

$\Rightarrow \quad C = -\frac{4}{5}$

Hence, $$\frac{1}{(x-1)(x+2)(2x+3)} = \frac{1}{15(x-1)} + \frac{1}{3(x+2)} - \frac{4}{5(2x+3)}$$

Illustration 52 Resolve $\frac{3x^3 + 2x^2 + x + 1}{(x+1)(x+2)}$ into partial fractions.

Solution. This is not a proper fraction. Hence, by division process it is to be expressed as the sum of an integral polynomial and a fraction.

Now, $3x^3 + 2x^2 + x + 1 = 3x(x^2 + 3x + 2) - 7(x^2 + 3x + 2) + (16x + 15)$

So, the given polynomial $\frac{3x^3 + 2x^2 + x + 1}{(x+1)(x+2)} = (3x - 7) + \frac{(16x + 15)}{(x+1)(x+2)}$ ...(i)

Now, the second term is proper fraction hence it can be expressed as a sum of partial fractions.

$$\frac{16x+15}{(x+1)(x+2)}=\frac{A}{x+1}+\frac{B}{x+2}$$

To find A put $x+1=0$, *ie*, $x=-1$ in the fraction except in the factor $(x+1)$.

$$\therefore \quad \frac{16(-1)+15}{(-1+2)}=A \quad \Rightarrow \quad A=-1 \qquad \text{...(ii)}$$

To find B put $x+2=0$, *ie*, $x=-2$ in the fraction except in the factor $(x+2)$.

$$\therefore \quad \frac{16(-2)+15}{(-2+1)}=B \quad \Rightarrow \quad B=17 \qquad \text{...(iii)}$$

$\Rightarrow$ The given expression $=(3x-7)-\dfrac{1}{x+1}+\dfrac{17}{x+2}$ [using Eqs. (i), (ii) and (iii)]

Case II When the denominator $g(x)$ is expressible as the product of the linear factors such that some of them are repeating. (Linear and Repeated)

Let $Q(x)=(x-a)^k(x-a_1)(x-a_2)\ldots(x-a_r)$. Then, we assume that

$$\frac{P(x)}{Q(x)}=\frac{A_1}{(x-a)}+\frac{A_2}{(x-a)^2}+\ldots+\frac{A_k}{(x-a)^k}+\frac{B_1}{(x-a_1)}+\frac{B_2}{(x-a_2)}+\ldots+\frac{B_r}{(x-a_r)}$$

Illustration 53 Expression $\dfrac{x+5}{(x-2)^2}$ has repeated (twice) linear factors in denominator, so find partial fractions.

Solution. Let $\quad \dfrac{x+5}{(x-2)^2}=\dfrac{A}{(x-2)}+\dfrac{B}{(x-2)^2}$

$$\therefore \quad (x+5)=A(x-2)+B$$

Comparing the like terms, $A=1$, $-2A+B=5$ or $B=7$

$$\therefore \quad \frac{x+5}{(x-2)^2}=\frac{1}{(x-2)}+\frac{7}{(x-2)^2}$$

Illustration 54 Resolve $\dfrac{3x-2}{(x-1)^2(x+1)(x+2)}$ into partial fractions.

Solution. Let $\dfrac{3x-2}{(x-1)^2(x+1)(x+2)}=\dfrac{A}{(x-1)}+\dfrac{B}{(x-1)^2}+\dfrac{C}{(x+1)}+\dfrac{D}{(x+2)}$

$$\therefore \quad 3x-2=A(x-1)(x+1)(x+2)+B(x+1)(x+2)+C(x-1)^2(x+2)+D(x-1)^2(x+1)$$

Putting $(x-1)=0$, we get $B=1/6$

Putting $(x+1)=0$, we get $C=-5/4$

Putting $(x+2)=0$, we get $D=8/9$

Now, equating the coefficient of x^3 on both the sides, we get

$$0=A+C+D \quad \Rightarrow A=\frac{5}{4}-\frac{8}{9}=\frac{13}{36}$$

$$\therefore \quad \frac{3x-2}{(x-1)^2(x+1)(x+2)}=\frac{13}{36(x-1)}+\frac{1}{6(x-1)^2}-\frac{5}{4(x+1)}+\frac{8}{9(x+2)}$$

Case III When some of the factors in denominator are quadratic but non-repeating.

Corresponding to each quadratic factor $ax^2 + bx + c$, we assume the partial fraction of the type $\dfrac{Ax + B}{ax^2 + bx + c}$, where A and B are constants to be determined by comparing coefficients of similar powers of x in numerator of both the sides.

Illustration 55 Resolve $\dfrac{2x + 7}{(x + 1)(x^2 + 4)}$ into partial fractions.

Solution. Let $\dfrac{2x + 7}{(x + 1)(x^2 + 4)} = \dfrac{A}{x + 1} + \dfrac{Bx + C}{x^2 + 4}$

$\therefore \quad 2x + 7 = A(x^2 + 4) + (Bx + C)(x + 1)$

Put $\quad x = -1 \quad \therefore \quad 5 = 5A$ or $A = 1$

Comparing the terms, $0 = A + B \quad \Rightarrow B = -1$

$7 = 4A + C \quad \Rightarrow C = 3$

$$\therefore \quad \frac{2x + 7}{(x + 1)(x^2 + 4)} = \frac{1}{x + 1} + \frac{(-x + 3)}{x^2 + 4}$$

Aliter To obtain values of A, B and C from

$$2x + 7 = A(x^2 + 4) + (Bx + C)(x + 1)$$

ie, $\quad 2x + 7 = (A + B)x^2 + (B + C)x + 4A + C$

Equating the coefficients of identical powers of x, we get $A + B = 0, B + C = 2$ and $4A + C = 7$.

Solving, we get $\quad A = 1, B = -1, C = 3$

Illustration 56 Find the partial fraction $\dfrac{2x + 1}{(3x + 2)(4x^2 + 5x + 6)}$.

Solution. Let $\dfrac{2x + 1}{(3x + 2)(4x^2 + 5x + 6)} = \dfrac{A}{(3x + 2)} + \dfrac{Bx + C}{(4x^2 + 5x + 6)}$, then

$2x + 1 = A(4x^2 + 5x + 6) + (Bx + C)(3x + 2)$ where A, B, C are constants.

For A, let $3x + 2 = 0$, *ie,* $x = -2/3$

$$2\left(-\frac{2}{3}\right) + 1 = A\left\{4 \cdot \frac{4}{9} - \frac{10}{3} + 6\right\} + \left(B\left(-\frac{2}{3}\right) + C\right)(0)$$

$$-\frac{1}{3} = A\left(\frac{40}{9}\right) \quad \Rightarrow \quad A = -\frac{3}{40}$$

Comparing coefficients of x^2 and constant term on both the sides for B and C, we get

$$4A + 3B = 0, \quad \therefore \quad B = -\frac{4}{3}A \quad \Rightarrow \quad B = \frac{1}{10}$$

and $\quad 6A + 2C = 1, \quad \therefore \quad C = \dfrac{1 - 6A}{2} \quad \Rightarrow \quad C = \dfrac{29}{40}$

$$\therefore \quad \frac{2x + 1}{(3x + 2)(4x^2 + 5x + 6)} = \frac{-3}{40(3x + 2)} + \frac{\left(x + \frac{29}{4}\right)}{10(4x^2 + 5x + 6)}$$

Case IV When some of the factors of the denominator are quadratic and repeating. For every quadratic repeating factor of the type $(ax^2+bx+c)^k$, we assume :

$$\frac{A_1x+A_2}{ax^2+bx+c}+\frac{A_3\,x+A_4}{(ax^2+bx+c)^2}+\ldots+\frac{A_{2k-1}\,x+A_{2k}}{(ax^2+bx+c)^k}$$

Illustration 57 Resolve $\dfrac{2x^4+2x^2+x+1}{x(x^2+1)^2}$ into partial fractions.

Solution. Let $\dfrac{2x^4+2x^2+x+1}{x(x^2+1)^2}=\dfrac{A}{x}+\dfrac{Bx+C}{x^2+1}+\dfrac{Dx+E}{(x^2+1)^2}$

or $\quad 2x^4+2x^2+x+1=A(x^2+1)^2+(Bx+C)\,x(x^2+1)+(Dx+E)\,x$

Comparing coefficients of x^4, x^3, x^2, x and constant term

$\therefore \quad A+B=2, C=0, 2A+D+B=2, E=1, A=1$

$\therefore$ we get $\quad A=1, B=1, C=0, D=-1, E=1$

Hence, the partial fraction,

$$\frac{2x^4+2x^2+x+1}{x(x^2+1)^2}=\frac{1}{x}+\frac{x}{1+x^2}+\frac{1-x}{(1+x^2)^2}$$

Illustration 58 Evaluate $\displaystyle\int\frac{1}{\sin x-\sin 2x}\,dx.$

Solution. Let $I=\displaystyle\int\frac{1}{\sin x-\sin 2x}\,dx=\int\frac{1}{\sin x-2\sin x\cos x}\,dx$

$$=\int\frac{1}{\sin x(1-2\cos x)}\,dx=\int\frac{\sin x}{\sin^2 x(1-2\cos x)}\,dx$$

$$=\int\frac{\sin x}{(1-\cos^2 x)(1-2\cos x)}\,dx$$

(put $\cos x=t \Rightarrow -\sin x\,dx=dt$)

$$\therefore \quad I=\int\frac{-dt}{(1-t^2)(1-2t)}=\int\frac{-1}{(1-t)(1+t)(1-2t)}\,dt \quad \ldots\text{(i)}$$

Here, in Eq. (i) we have linear and non-repeated factors thus we use partial fractions for;

$$\frac{-1}{(1-t)(1+t)(1-2t)}=\frac{A}{(1-t)}+\frac{B}{(1+t)}+\frac{C}{(1-2t)}$$

or $\quad -1=A(1+t)(1-2t)+B(1-t)(1-2t)+C(1-t)(1+t) \quad \ldots\text{(ii)}$

Putting $\quad (t+1)=0$ or $t=-1$, we get

$$-1=B(2)(1+2) \quad\Rightarrow\quad B=-\frac{1}{6}$$

Putting $\quad (1-t)=0$ or $t=1$, we get

$$-1=A(2)(-1) \quad\Rightarrow\quad A=\frac{1}{2}$$

Putting $\quad (1-2t)=0$ or $t=1/2$, we get

$$-1=C\left(1-\frac{1}{2}\right)\left(1+\frac{1}{2}\right) \quad\Rightarrow\quad C=-\frac{4}{3}$$

$$\therefore \quad \frac{-1}{(1-t)(1+t)(1-2t)} = \frac{1}{2(1-t)} - \frac{1}{6(1+t)} - \frac{4}{3(1-2t)}$$

So, Eq. (i) reduces to

$$I = \frac{1}{2}\int \frac{1}{1-t}\,dt - \frac{1}{6}\int \frac{1}{1+t}\,dt - \frac{4}{3}\int \frac{1}{1-2t}\,dt$$

$$= -\frac{1}{2}\log|1-t| - \frac{1}{6}\log|1+t| - \frac{4}{3}\times -\frac{1}{2}\log|1-2t| + C$$

$$= -\frac{1}{2}\log|1-\cos x| - \frac{1}{6}\log|1+\cos x| + \frac{2}{3}\log|1-2\cos x| + C$$

Illustration 59 Evaluate $\int \frac{(1-x\sin x)\,dx}{x(1-x^3e^{3\cos x})}$.

Solution. Here, $I = \int \frac{(1-x\sin x)\,dx}{x(1-(xe^{\cos x})^3)}$

Put $xe^{\cos x} = t$

So that $(xe^{\cos x}\cdot(-\sin x) + e^{\cos x})\,dx = dt$

$$\therefore \quad I = \int \frac{dt}{t(1-t^3)} = \int \frac{dt}{t(1-t)(1+t+t^2)}$$

$$= \int \left(\frac{A}{t} + \frac{B}{1-t} + \frac{Ct+D}{1+t+t^2}\right) dt$$

Comparing coefficients, we get

$$A = 1, B = \frac{1}{3}, C = -\frac{2}{3}, D = -\frac{1}{3}$$

$$\therefore \quad I = \int \frac{dt}{t} + \frac{1}{3}\int \frac{dt}{1-t} + \int \frac{\left(-\frac{2}{3}t - \frac{1}{3}\right)}{1+t+t^2}\,dt$$

$$= \log|t| - \frac{1}{3}\log|1-t| - \frac{1}{3}\log|1+t+t^2| \quad \text{where, } t = xe^{\cos x}$$

Illustration 60 Evaluate $\int \sin 4x \cdot e^{\tan^2 x}\,dx$.

Solution. The given integral could be written as,

$$I = \int 4\sin x\cdot\cos x\cdot\cos 2x\cdot e^{\tan^2 x}\,dx$$

$$= 4\int \tan x\cdot\cos^2 x\,(\cos^2 x - \sin^2 x)\cdot e^{\tan^2 x}\,dx$$

$$= 4\int \tan x\cdot\cos^4 x\,(1-\tan^2 x)\cdot e^{\tan^2 x}\,dx$$

$$= 4\int \frac{\tan x}{(\sec^2 x)^2}\cdot(1-\tan^2 x)\cdot e^{\tan^2 x}\,dx$$

Put $\tan^2 x = t \Rightarrow 2\tan x\cdot\sec^2 x\,dx = dt$

$$\Rightarrow \quad I = 4\int \frac{(1-t)e^t}{(1+t)^3}\cdot\frac{dt}{2} = 2\int \frac{(1-t)e^t}{(1+t)^3}\,dt$$

$$= 2\left\{\int \frac{2e^t - (1+t)e^t}{(1+t)^3}\,dt\right\} = 2\left\{\int e^t\,\frac{2}{(1+t)^3}, -\frac{1}{(1+t)^2}\right\}dt$$

$$= -2\frac{1}{(1+t)^2}e^t + C \quad [\text{using} \int e^x(f(x)+f'(x))\,dx = e^x \cdot f(x) + C]$$

$$= -\frac{2e^{\tan^2 x}}{(1+\tan^2 x)^2} + C$$

$$I = -2\cos^4 x \cdot e^{\tan^2 x} + C$$

Illustration 61 Solve $\int \frac{1 + x\cos x}{x(1 - x^2 e^{2\sin x})}\,dx.$

Solution. Put $(x\, e^{\sin x}) = t$

Differentiating both the sides, we get

$$(x\, e^{\sin x} \cdot \cos x + e^{\sin x})\,dx = dt \Rightarrow e^{\sin x}(x\cos x + 1)\,dx = dt$$

$$\Rightarrow \qquad I = \int \frac{1 + x\cos x}{x(1 - x^2 e^{2\sin x})}\,dx = \int \frac{dt}{t(1-t^2)}$$

$$= \int \frac{dt}{t(1-t)(1+t)} \text{ using partial fraction, we get}$$

$$= \int \left\{ \frac{1}{t} + \frac{1}{2(1-t)} - \frac{1}{2(1+t)} \right\} dt$$

$$= \log|t| - \frac{1}{2}\log|1-t| - \frac{1}{2}\log|1+t| + C$$

$$= \log|x\, e^{\sin x}| - \frac{1}{2}\log|1 - x^2 e^{2\sin x}| + C$$

Indirect and Derived Substitutions

(i) Indirect Substitution

If the integrand is of the form $f(x) \cdot g(x)$, where $g(x)$ is a function of the integral of $f(x)$, then put integral of $f(x) = t$.

Illustration 62 Evaluate $\int \frac{\sqrt{x}}{\sqrt{x^3 + a^3}}\,dx.$

Solution. Integral of numerator is $\frac{x^{3/2}}{3/2}$.

Put $\qquad x^{3/2} = t \Rightarrow \frac{3}{2}x^{1/2}\,dx = dt$, we get

$$I = \int \frac{x^{1/2}\,dx}{\sqrt{(x^{3/2})^2 + (a^{3/2})^2}} = \int \frac{\frac{2}{3}dt}{\sqrt{t^2 + (a^{3/2})^2}} + C$$

$$I = \frac{2}{3}\int \frac{dt}{\sqrt{t^2 + (a^{3/2})^2}} = \frac{2}{3} \cdot \log|t + \sqrt{t^2 + (a^{3/2})^2}| + C$$

$$I = \frac{2}{3}\log|x^{3/2} + \sqrt{x^3 + a^3}| + C$$

Illustration 63 Evaluate $\int \frac{\sin x + \cos x}{9 + 16 \sin 2x} dx$

Solution. Let $I = \int \frac{\sin x + \cos x}{9 + 16 \sin 2x} dx = \int \frac{\sin x + \cos x}{25 - 16(\sin x - \cos x)^2} dx$

[where $\sin 2x = 1 - (\sin x - \cos x)^2$]

$$I = \int \frac{\sin x + \cos x}{25 - 16(\sin x - \cos x)^2} dx$$

Let $(\sin x - \cos x) = t \Rightarrow (\cos x + \sin x)\, dx = dt$

$$= \int \frac{dt}{25 - 16t^2} = \frac{1}{16} \int \frac{dt}{(5/4)^2 - (t)^2}$$

$$= \frac{1}{16} \cdot \frac{1}{2(5/4)} \cdot \log \left| \frac{5/4 + t}{5/4 - t} \right| + C$$

$$= \frac{1}{40} \log \left| \frac{5 + 4t}{5 - 4t} \right| + C$$

$$\therefore \quad I = \frac{1}{40} \log \left| \frac{5 + 4(\sin x - \cos x)}{5 - 4(\sin x - \cos x)} \right| + C$$

(ii) Derived Substitutions

Some times it is useful to write the integral as a sum of two related integrals which can be evaluated by making suitable substitutions.

Examples of such integrals are

(a) Algebraic Twins

$$\int \frac{2x^2}{x^4 + 1} dx = \int \frac{x^2 + 1}{x^4 + 1} dx + \int \frac{x^2 - 1}{x^4 + 1} dx$$

$$\int \frac{2}{x^4 + 1} dx = \int \frac{x^2 + 1}{x^4 + 1} dx - \int \frac{x^2 - 1}{x^4 + 1} dx$$

$$\int \frac{2x^2}{x^4 + 1 + kx^2} dx, \int \frac{2}{(x^4 + 1 + kx^2)} dx$$

(b) Trigonometric Twins

$$\int \sqrt{\tan x}\, dx, \int \sqrt{\cot x}\, dx,$$

$$\int \frac{1}{(\sin^4 x + \cos^4 x)} dx, \int \frac{1}{\sin^6 x + \cos^6 x} dx,$$

$$\int \frac{\pm \sin x \pm \cos x}{a + b \sin x \cos x} dx$$

Method of evaluating these integral are illustrated by mean of the following examples :

Illustration 64 Evaluate $\int \frac{5}{1+x^4}\,dx$.

Solution. Let $I = \int \frac{5}{1+x^4}\,dx = \frac{5}{2}\int \frac{2}{1+x^4}\,dx$

$$= \frac{5}{2}\left[\int \frac{1+x^2}{1+x^4}\,dx + \int \frac{1-x^2}{1+x^4}\,dx\right]$$

$$= \frac{5}{2}\left[\int \frac{x^2+1}{x^4+1}\,dx - \int \frac{x^2-1}{x^4+1}\,dx\right]$$

Points to Consider

Here, dividing N^r and D^r by x^2 and converting D^r into perfect square so as to get differential in N^r.

ie, $$I = \frac{5}{2}\left[\int \frac{1+1/x^2}{x^2+1/x^2}\,dx - \int \frac{1-1/x^2}{x^2+1/x^2}\,dx\right]$$

$$= \frac{5}{2}\left[\int \frac{1+1/x^2}{(x-1/x)^2+2}\,dx - \int \frac{1-1/x^2}{(x+1/x)^2-2}\,dx\right]$$

$$= \frac{5}{2}\left[\int \frac{dt}{t^2+(\sqrt{2})^2} - \int \frac{du}{u^2-(\sqrt{2})^2}\right], \quad \text{where } t = x - \frac{1}{x} \text{ and } u = x + \frac{1}{x}$$

$$= \frac{5}{2}\left[\frac{1}{\sqrt{2}}\tan^{-1}\left(\frac{t}{\sqrt{2}}\right) - \frac{1}{2\sqrt{2}}\log\left|\frac{u-\sqrt{2}}{u+\sqrt{2}}\right|\right] + C$$

$$\therefore \quad I = \frac{5}{2}\left[\frac{1}{\sqrt{2}}\tan^{-1}\left(\frac{x-1/x}{\sqrt{2}}\right) - \frac{1}{2\sqrt{2}}\log\left|\frac{x+\frac{1}{x}-\sqrt{2}}{x+\frac{1}{x}+\sqrt{2}}\right|\right] + C$$

Illustration 65 Evaluate $\int \frac{1}{x^4+5x^2+1}\,dx$.

Solution. Let $I = \frac{1}{2}\int \frac{2}{x^4+5x^2+1}\,dx$

$$\Rightarrow \quad I = \frac{1}{2}\int \frac{1+x^2}{x^4+5x^2+1}\,dx + \frac{1}{2}\int \frac{1-x^2}{x^4+5x^2+1}\,dx$$

$$= \frac{1}{2}\int \frac{1+1/x^2}{x^2+5+1/x^2}\,dx - \frac{1}{2}\int \frac{1-1/x^2}{x^2+5+1/x^2}\,dx$$

(dividing N^r and D^r by x^2)

$$= \frac{1}{2}\int \frac{(1+1/x^2)}{(x-1/x)^2+7}\,dx - \frac{1}{2}\int \frac{(1-1/x^2)}{(x+1/x)^2+3}\,dx$$

$$= \frac{1}{2}\int \frac{dt}{t^2+(\sqrt{7})^2} - \frac{1}{2}\int \frac{du}{u^2+(\sqrt{3})^2}$$

where $t = x - \frac{1}{x}$ and $u = x + \frac{1}{x}$

$$\therefore \quad I = \frac{1}{2}\cdot\frac{1}{\sqrt{7}}\left(\tan^{-1}\frac{t}{\sqrt{7}}\right) - \frac{1}{2}\cdot\frac{1}{\sqrt{3}}\left(\tan^{-1}\frac{u}{\sqrt{3}}\right) + C$$

$$= \frac{1}{2}\left[\frac{1}{\sqrt{7}}\tan^{-1}\left(\frac{x - 1/x}{\sqrt{7}}\right) - \frac{1}{\sqrt{3}}\tan^{-1}\left(\frac{x + 1/x}{\sqrt{3}}\right)\right] + C$$

Illustration 66 Evaluate $\int \sqrt{\tan x}\, dx$.

Solution. Put $\tan x = t^2 \Rightarrow \sec^2 x\, dx = 2t\, dt$

$$\Rightarrow \quad dx = \frac{2t\, dt}{1 + t^4}$$

$$\therefore \quad I = \int t \cdot \frac{2t}{1 + t^4}\, dt = \int \frac{2t^2}{t^4 + 1}\, dt = \int \frac{t^2 + 1}{t^4 + 1}\, dt + \int \frac{t^2 - 1}{t^4 + 1}\, dt$$

$$= \int \frac{1 + 1/t^2}{t^2 + 1/t^2}\, dt + \int \frac{1 - 1/t^2}{t^2 + 1/t^2}\, dt$$

$$= \int \frac{1 + 1/t^2}{(t - 1/t)^2 + 2}\, dt + \int \frac{1 - 1/t^2}{(t + 1/t)^2 - 2}\, dt$$

$$I = \int \frac{ds}{s^2 + (\sqrt{2})^2} + \int \frac{dr}{r^2 - (\sqrt{2})^2}, \quad \text{where } s = t - \frac{1}{t} \text{ and } r = t + \frac{1}{t}$$

$$= \frac{1}{\sqrt{2}}\tan^{-1}\left(\frac{s}{\sqrt{2}}\right) + \frac{1}{2\sqrt{2}}\log\left|\frac{r - \sqrt{2}}{r + \sqrt{2}}\right| + C$$

$$= \frac{1}{\sqrt{2}}\left[\tan^{-1}\left(\frac{t - 1/t}{\sqrt{2}}\right) + \frac{1}{2}\log\left(\frac{t + \frac{1}{t} - \sqrt{2}}{t + \frac{1}{t} + \sqrt{2}}\right)\right] + C$$

(where $t = \sqrt{\tan x}$)

Illustration 67 Evaluate $\int \frac{4}{\sin^4 x + \cos^4 x}\, dx$.

Solution. Let $I = 4\int \frac{dx}{\sin^4 x + \cos^4 x}$

Dividing N^r and D^r by $\cos^4 x$, we get

$$I = 4\int \frac{\sec^4 x}{\tan^4 x + 1}\, dx$$

$$I = 4\int \frac{\sec^2 x\,(1 + \tan^2 x)}{1 + \tan^4 x}\, dx$$

Put $\quad \tan x = t \quad \Rightarrow \quad \sec^2 x\, dx = dt$

$$\therefore \quad I = 4\int \frac{t^2 + 1}{t^4 + 1}\, dt = 4\int \frac{1 + 1/t^2}{t^2 + 1/t^2}\, dt = 4\int \frac{1 + 1/t^2}{(1 - 1/t)^2 + 2}\, dt$$

Again, put $\quad z = t - \frac{1}{t}$

$$\therefore \quad I = 4\int \frac{dz}{z^2 + 2} = \frac{4}{\sqrt{2}}\tan^{-1}\left(\frac{z}{\sqrt{2}}\right) + C$$

$$I = 2\sqrt{2}\tan^{-1}\left(\frac{\tan x - 1/\tan x}{\sqrt{2}}\right) + C$$

Illustration 68 Evaluate $\int \frac{dx}{2 \sin x + \sec x}$.

Solution. Let $I = \int \frac{dx}{2 \sin x + \sec x} = \int \frac{\cos x \, dx}{\sin 2x + 1}$

$$= \frac{1}{2} \int \frac{2 \cos x \, dx}{1 + \sin 2x} = \frac{1}{2} \int \frac{(\cos x + \sin x) + (\cos x - \sin x)}{(\sin^2 + \cos^2 x + 2 \sin x \cos x)} dx$$

$$= \frac{1}{2} \int \frac{\cos x + \sin x}{(\sin x + \cos x)^2} dx + \frac{1}{2} \int \frac{(\cos x - \sin x)}{(\sin x + \cos x)^2} dx$$

$$= \frac{1}{2} \int \frac{dx}{\sin x + \cos x} + \frac{1}{2} \int \frac{dv}{v^2}, \quad \text{where } v = \sin x + \cos x$$

$$= \frac{1}{2\sqrt{2}} \int \frac{dx}{\frac{1}{\sqrt{2}} \sin x + \frac{1}{\sqrt{2}} \cos x} - \frac{1}{2v} + C$$

$$= \frac{1}{2\sqrt{2}} \int \frac{dx}{\sin\left(x + \frac{\pi}{4}\right)} - \frac{1}{2(\sin x + \cos x)} + C$$

$$= \frac{1}{2\sqrt{2}} \log \left| \operatorname{cosec}\left(x + \frac{\pi}{4}\right) - \cot\left(x + \frac{\pi}{4}\right) \right| - \frac{1}{2(\sin x + \cos x)} + C$$

Integration of Some Special Irrational Algebraic Functions

In this case we shall discuss four integrals of the form $\int \frac{\phi(x)}{P\sqrt{Q}} dx$, where P and Q are polynomial functions of x.

(a) Integrals of the Form $\int \frac{\phi(x)}{P\sqrt{Q}} dx$, where P and Q are both linear of x

To evaluate this type of integrals we put $Q = t^2$, *ie*, to evaluate integrals of the form $\int \frac{1}{(ax + b)\sqrt{cx + d}} dx$, put $cx + d = t^2$.

The following examples illustrate the procedure :

Illustration 69 Evaluate $\int \frac{1}{(x + 1)\sqrt{x - 2}} dx$.

Solution. Let $I = \int \frac{1}{(x + 1)\sqrt{x - 2}} dx$

Here, P and Q both are linear, so we put $Q = t^2$ *ie*, $x - 2 = t^2$ so that $dx = 2t \, dt$.

$$\therefore \quad I = \int \frac{1}{(t^2 + 2 + 1)\sqrt{t^2}} \cdot 2t \, dt = 2 \int \frac{dt}{t^2 + 3} = 2 \times \frac{1}{\sqrt{3}} \tan^{-1}\left(\frac{t}{\sqrt{3}}\right) + C$$

$$\therefore \quad I = \frac{2}{\sqrt{3}} \tan^{-1}\left(\frac{\sqrt{x - 2}}{\sqrt{3}}\right) + C$$

(b) Integrals of the Form $\int \frac{\phi(x)}{P\sqrt{Q}}\,dx$, where P is a quadratic expression and Q is a linear expression

To evaluate this type of integrals we put $Q = t^2$ *ie*, to evaluate the integrals of the form $\int \frac{1}{(ax^2 + bx + c)\sqrt{px + q}}\,dx$, put $px + q = t^2$.

Illustration 70 Evaluate $\int \frac{x+2}{(x^2+3x+3)\sqrt{x+1}}\,dx$.

Solution. Let $I = \int \frac{x+2}{(x^2+3x+3)\sqrt{x+1}}\,dx$ put $x + 1 = t^2 \Rightarrow dx = 2t\,dt$

$$\therefore \quad I = \int \frac{(t^2-1)+2}{\{(t^2-1)^2 + 3(t^2-1)+3\}\sqrt{t^2}} \cdot (2t)\,dt$$

$$= 2\int \frac{t^2+1}{t^4+t^2+1}\,dt = 2\int \frac{1+1/t^2}{t^2+1+1/t^2}\,dt$$

$$= 2\int \frac{1+1/t^2}{(t-1/t)^2 + (\sqrt{3})^2}\,dt = 2\int \frac{du}{u^2+(\sqrt{3})^2} \qquad \left(\text{where } u = t - \frac{1}{t}\right)$$

$$= \frac{2}{\sqrt{3}}\tan^{-1}\left(\frac{u}{\sqrt{3}}\right) + C$$

$$\therefore \quad I = \frac{2}{\sqrt{3}}\tan^{-1}\left(\frac{t^2-1}{\sqrt{3}t}\right) + C = \frac{2}{\sqrt{3}}\tan^{-1}\left(\frac{x}{\sqrt{3(x+1)}}\right) + C$$

(c) Integral of the Form $\int \frac{1}{(ax+b)} \cdot \frac{dx}{\sqrt{px^2+qx+r}}$, where in $\int \frac{dx}{P\sqrt{Q}}$ P is linear and Q is a quadratic put, $ax + b = \frac{1}{t}$.

Illustration 71 Evaluate $\int \frac{dx}{(x-1)\sqrt{x^2+x+1}}$.

Solution. Let $I = \int \frac{dx}{(x-1)\sqrt{x^2+x+1}}$ put $x - 1 = \frac{1}{t} \Rightarrow dx = -\frac{1}{t^2}\,dt$

$$\therefore \quad I = \int \frac{-1/t^2\,dt}{1/t\sqrt{\left(\frac{1}{t}+1\right)^2 + \left(\frac{1}{t}+1\right)+1}} = -\int \frac{dt}{\sqrt{3t^2+3t+1}}$$

$$= -\frac{1}{\sqrt{3}}\int \frac{dt}{\sqrt{\left(t+\frac{1}{2}\right)^2 + \frac{1}{12}}}$$

$$= -\frac{1}{\sqrt{3}} \log |(t + 1/2) + \sqrt{(t + 1/2)^2 + 1/12}| + C$$

$$= -\frac{1}{\sqrt{3}} \log \left| \left(\frac{1}{x-1} + \frac{1}{2}\right) + \sqrt{\frac{12\left(\frac{1}{x-1} + \frac{1}{2}\right)^2 + 1}{12}} \right| + C$$

(d) Integrals of the Form $\int \frac{dx}{P\sqrt{Q}}$, where P and Q both are pure quadratic expression in x, *ie*, $P = ax^2 + b$ and $Q = cx^2 + d$, *ie*, $\int \frac{dx}{(ax^2 + b)\sqrt{cx^2 + d}}$.

To evaluate this type of integrals of the form we put $x = \frac{1}{t}$.

Illustration 72 Evaluate $\int \frac{dx}{(1 + x^2)\sqrt{1 - x^2}}$.

Solution. Let $I = \int \frac{dx}{(1 + x^2)\sqrt{1 - x^2}}$

Put $x = \frac{1}{t}$, so that $dx = -\frac{1}{t^2} dt$

$\therefore$ $$I = \int \frac{-1/t^2 \, dt}{(1 + 1/t^2)\sqrt{1 - 1/t^2}} = -\int \frac{t \, dt}{(t^2 + 1)\sqrt{t^2 - 1}}$$

Again, let $t^2 = u$. So that $2t \, dt = du$.

$= -\frac{1}{2}\int \frac{du}{(u + 1)\sqrt{u - 1}}$ which reduces to the form $\int \frac{dx}{P\sqrt{Q}}$ where both P and Q are linear so that we put $u - 1 = z^2$ so that $du = 2z \, dz$

$\therefore$ $$I = -\frac{1}{2}\int \frac{2z \, dz}{(z^2 + 1 + 1)\sqrt{z^2}} = -\int \frac{dz}{(z^2 + 2)}$$

$$I = -\frac{1}{\sqrt{2}} \tan^{-1}\left(\frac{z}{\sqrt{2}}\right) + C$$

$$I = -\frac{1}{\sqrt{2}} \tan^{-1}\left(\frac{\sqrt{u - 1}}{\sqrt{2}}\right) + C = -\frac{1}{\sqrt{2}} \tan^{-1}\left(\frac{\sqrt{t^2 - 1}}{\sqrt{2}}\right) + C$$

$$= -\frac{1}{\sqrt{2}} \tan^{-1}\left(\frac{\sqrt{1 - x^2}}{\sqrt{2}x}\right) + C$$

Aliter Put $x = \cos\theta$, $dx = -\sin\theta \, d\theta$

$\therefore$ $$I = -\int \frac{\sin\theta \, d\theta}{(1 + \cos^2\theta)\sin\theta} = -\int \frac{d\theta}{1 + \cos^2\theta} = -\int \frac{\sec^2\theta \, d\theta}{\sec^2\theta + 1} = -\int \frac{\sec^2\theta \, d\theta}{\tan^2\theta + 2}$$

Put $\tan\theta = t$. So that $\sec^2\theta\, d\theta = dt$

$$\therefore \quad I = -\int \frac{dt}{t^2+2} = -\frac{1}{\sqrt{2}}\tan^{-1}\left(\frac{t}{\sqrt{2}}\right) + C = -\frac{1}{\sqrt{2}}\tan^{-1}\left(\frac{\tan\theta}{\sqrt{2}}\right) + C$$

$$= -\frac{1}{\sqrt{2}}\tan^{-1}\left(\frac{\sqrt{1-x^2}}{\sqrt{2}\,x}\right) + C \quad \left\{\begin{array}{l} \text{where} \\ \cos\theta = x \\ \sin\theta = \sqrt{1-x^2} \\ \therefore \tan\theta = \dfrac{\sqrt{1-x^2}}{x} \end{array}\right.$$

Illustration 73 Evaluate $I = \int \frac{(x-1)\sqrt{x^4+2x^3-x^2+2x+1}}{x^2(x+1)}\,dx.$

Solution. Here,

$$I = \int \frac{(x^2-1)\sqrt{x^4+2x^3-x^2+2x+1}}{x^2(x+1)^2}\,dx$$

$$= \int \frac{\left(1-\frac{1}{x^2}\right)\sqrt{x^2\left(x^2+2x-1+\frac{2}{x}+\frac{1}{x^2}\right)}}{\frac{x^2(x^2+2x+1)}{x^2}}\,dx$$

$$= \int \frac{\left(1-\frac{1}{x^2}\right)\sqrt{\left(x^2+\frac{1}{x^2}\right)+2\left(x+\frac{1}{x}\right)-1}}{\left(x+\frac{1}{x}+2\right)}\,dx$$

Put $\quad x+\frac{1}{x} = t$, *ie*, $\left(1-\frac{1}{x^2}\right)dx = dt$

$$= \int \frac{\sqrt{(t^2-2)+2t-1}}{(t+2)}\,dt = \int \frac{\sqrt{t^2+2t-3}}{(t+2)}\,dt$$

$$= \int \frac{t^2+2t-3}{(t+2)\sqrt{t^2+2t-3}}\,dt$$

$$= \int \frac{t(t+2)}{(t+2)\sqrt{t^2+2t-3}}\,dt - 3\int \frac{dt}{(t+2)\sqrt{t^2+2t-3}}$$

$$I = I_1 - 3I_2 \qquad \ldots\text{(i)}$$

where, $\quad I_1 = \int \frac{t\,dt}{\sqrt{t^2+2t-3}}$ and $I_2 = \int \frac{dt}{(t+2)\sqrt{t^2+2t-3}}$

$\therefore \quad I_1 = \int \frac{t\,dt}{\sqrt{(t+1)^2-4}}$ $\quad$ put, $t+1 = z$

$$= \int \frac{(z-1)\,dz}{\sqrt{z^2-2^2}}$$

$$= \int \frac{z\,dz}{\sqrt{z^2-2^2}} - \int \frac{dz}{\sqrt{z^2-2^2}} = \sqrt{z^2-2^2} - \log|z+\sqrt{z^2-4}|$$

$$= \sqrt{t^2+2t-3} - \log|(t+1)+\sqrt{t^2+2t+3}| \qquad \ldots\text{(ii)}$$

Also, $I_2 = \int \frac{dy}{y^2 \cdot \frac{1}{y}\sqrt{\left(\frac{1}{y}-2\right)^2 + 2\left(\frac{1}{y}-2\right) - 3}}$, put $t + 2 = \frac{1}{y}$

$$= \int \frac{dy}{\sqrt{1-2y-3y^2}} = \frac{1}{\sqrt{3}} \int \frac{dy}{\sqrt{\left(\frac{2}{3}\right)^2 - \left(y+\frac{1}{3}\right)^2}}$$

$$= \frac{1}{\sqrt{3}} \sin^{-1}\left(\frac{y+\frac{1}{3}}{\frac{2}{3}}\right) = \frac{1}{\sqrt{3}} \sin^{-1}\left(\frac{5+t}{2+t}\right) \quad \text{...(iii)}$$

$\therefore$ $I = \sqrt{t^2+2t-3} - \log\left(t+1+\sqrt{t^2+2t-3}\right) - \sqrt{3}\sin^{-1}\left(\frac{t+5}{t+2}\right)$

where $t = x + \frac{1}{x}$

(e) Integrals of the Form $\int \frac{dx}{(x-k)^r \sqrt{ax^2+bx+c}}$, where $r \geq 2$ and $r \in I$

Here, we substitute, $x - k = \frac{1}{t}$

Illustration 74 Evaluate $\int \frac{dx}{(x-3)^3\sqrt{x^2-6x+10}}$.

Solution. Substitute $(x-3) = \frac{1}{t} \Rightarrow dx = -\frac{1}{t^2}dt$

We get, $\int \frac{dx}{(x-3)^3\sqrt{x^2-6x+10}} = \int \frac{-1/t^2\,dt}{1/t^3\sqrt{(1/t+3)^2 - 6(1/t+3)+10}}$

$$= -\int \frac{t^2\,dt}{\sqrt{1+t^2}} = \int \frac{dt}{\sqrt{1+t^2}} - \int \sqrt{1+t^2}\,dt$$

$$= \log|t+\sqrt{1+t^2}| - \frac{t}{2}\sqrt{1+t^2} - \frac{1}{2}\log|t+\sqrt{1+t^2}| + C$$

$$= \frac{1}{2}\log|t+\sqrt{1+t^2}| - \frac{t}{2}\sqrt{1+t^2} + C$$

$$= \frac{1}{2}\left[\log\left|\frac{1+\sqrt{x^2-6x+10}}{|x-3|}\right| - \frac{\sqrt{x^2-6x+10}}{|x-3|^2}\right] + C$$

(f) Integrals of the Form $\int \frac{ax^2+bx+c}{(dx+e)\sqrt{fx^2+gx+h}}\,dx$

Here, we write $ax^2 + bx + c = A_1(dx+e)(2fx+g) + B_1(dx+e) + C_1$

Where A_1, B_1 and C_1 are constants which can be obtained by comparing the coefficients of like terms on both the sides.

Illustration 75 Evaluate $\int \frac{2x^2+5x+9}{(x+1)\sqrt{x^2+x+1}}\,dx$.

Solution. Let $2x^2+5x+9 = A(x+1)(2x+1)+B(x+1)+C$

or $2x^2+5x+9 = x^2(2A)+x(3A+B)+(A+B+C)$

$\Rightarrow$ $A=1, B=2, C=6$

Thus, $\int \frac{2x^2+5x+9}{(x+1)\sqrt{x^2+x+1}}\,dx$

$$= \int \frac{(x+1)(2x+1)}{(x+1)\sqrt{x^2+x+1}}\,dx + 2\int \frac{x+1}{(x+1)\sqrt{x^2+x+1}}\,dx + 6\int \frac{dx}{(x+1)\sqrt{x^2+x+1}}$$

$$= \int \frac{2x+1}{\sqrt{x^2+x+1}}\,dx + 2\int \frac{dx}{\sqrt{x^2+x+1}} + 6\int \frac{dx}{(x+1)\sqrt{x^2+x+1}}$$

$$= \int \frac{du}{\sqrt{u}} + 2\int \frac{dx}{\sqrt{(x+1/2)^2+(3/4)}} + 6\int \frac{-dt}{\sqrt{t^2-t+1}}$$

where $u = x^2+x+1$ and $\frac{1}{t} = x+1$

$$= 2\sqrt{x^2+x+1} + 2\cdot 1 \log|(x+1/2)+\sqrt{x^2+x+1}| - 6\int \frac{dt}{\sqrt{(t-1/2)^2+3/4}}$$

$$= 2\sqrt{x^2+x+1} + 2\log\left|\left(x+\frac{1}{2}\right)+\sqrt{x^2+x+1}\right| - 6\log\left|\left(t-\frac{1}{2}\right)+\sqrt{t^2-t+1}\right| + C$$

$$= 2\sqrt{x^2+x+1} + 2\log\left|\left(x+\frac{1}{2}\right)+\sqrt{x^2+x+1}\right| - 6\log\left|\frac{1-x+\sqrt{x^2+x+1}}{2(x+1)}\right| + C$$

Integrals of the Form $\int x^m (a+bx^n)^P\,dx$

Case I If $P \in N$. We expand using binomial and integrate.

Case II If $P \in I^-$ (*ie*, negative integer), write $x = t^k$, where k is the LCM of m and n.

Case III If $\frac{m+1}{n}$ is an integer and $P \leftrightarrow$ fraction put $(a+bx^n) = t^k$, where k is denominator of the fraction P.

Case IV If $\left(\frac{m+1}{n}+P\right)$ is an integer and $P \in$ fraction. We put $(a+bx^n) = t^k x^n$, where k is denominator of the fraction P.

Illustration 76 Evaluate $\int x^{1/3}(2+x^{1/2})^2\,dx$.

Solution. $I = \int x^{1/3}(2+x^{1/2})^2\,dx$

Since, P is natural number.

$\therefore$ $I = \int x^{1/3}(4+x+4x^{1/2})\,dx$

$$= \int (4x^{1/3} + x^{4/3} + 4x^{5/6})\, dx$$

$$= \frac{4x^{4/3}}{4/3} + \frac{x^{7/3}}{7/3} + \frac{4x^{11/6}}{11/6} + C$$

$$= 3x^{4/3} + \frac{3}{7}x^{7/3} + \frac{24}{11}x^{11/6} + C$$

Illustration 77 Evaluate $\int x^{-2/3}(1 + x^{2/3})^{-1}\, dx$.

Solution. If we substitute $x = t^3$ (as we know $P \in$ negative integer)

$\therefore$ Let $x = t^k$, where k is the LCM of m and n.

$\therefore$ $x = t^3 \Rightarrow dx = 3t^2\, dt$

or $I = \int \frac{3t^2}{t^2(1+t^2)}\, dt = 3\int \frac{dt}{t^2+1} = 3\tan^{-1}(t) + C$

$\Rightarrow$ $I = 3\tan^{-1}(x^{1/3}) + C$

Illustration 78 Evaluate $\int x^{-2/3}(1 + x^{1/3})^{1/2}\, dx$.

Solution. If we substitute $1 + x^{1/3} = t^2$, then $\frac{1}{3x^{2/3}}\, dx = 2t\, dt$

$\therefore$ $I = \int \frac{t \cdot 6t\, dt}{1} = 6\int t^2\, dt = 2t^3 + C$

or $I = 2(1 + x^{1/3})^{3/2} + C$

Illustration 79 Evaluate $\int \sqrt{x}\,(1 + x^{1/3})^4\, dx$.

Solution. Here, $m = \frac{1}{2}$ and $n = \frac{1}{3}$

Put $x = t^6 \Rightarrow dx = 6t^5\, dt$

$\Rightarrow$ $I = \int t^3(1+t^2)^4\, 6t^5\, dt$

$\Rightarrow$ $I = 6\int t^8(1 + 4t^2 + 6t^4 + 4t^6 + t^8)\, dt$

$$= 6\int (t^8 + 4t^{10} + 6t^{12} + 4t^{14} + t^{16})\, dt$$

$$= 6\left\{\frac{t^9}{9} + \frac{4t^{11}}{11} + \frac{6t^{13}}{13} + \frac{4t^{15}}{15} + \frac{t^{17}}{17}\right\} + C$$

$$I = 6\left\{x^{2/3} + \frac{4}{11}x^{11/6} + \frac{6}{13}x^{13/6} + \frac{4}{15}x^{5/2} + \frac{1}{17}x^{17/6}\right\} + C$$

Illustration 80 Evaluate $\int x^5(1 + x^3)^{2/3}\, dx$.

Solution. Here, $\int x^5(1+x^3)^{2/3}\, dx$ have $m = 5$, $n = 3$ and $p = \frac{2}{3}$

$\therefore$ $\frac{m+1}{n} = \frac{6}{3} = 2$ (an integer)

So, we substitute $1 + x^3 = t^2$ and $3x^2\, dx = 2t\, dt$

$\therefore$ $\int x^5(1+x^3)^{2/3}\, dx = \int x^3(1+x^3)^{2/3}\, x^2\, dx = \int (t^2 - 1)(t^2)^{2/3}\, \frac{2}{3}t\, dt$

$$= \frac{2}{3}\int (t^2-1)\, t^{7/3}\, dt = \frac{2}{3}\int (t^{13/3} - t^{7/3})\, dt$$

$$= \frac{2}{3}\left\{\frac{3}{16} t^{16/3} - \frac{3}{10} t^{10/3}\right\} + C$$

$$\frac{1}{8}(1+x^3)^{8/3} - \frac{1}{5}(1+x^3)^{5/3} + C$$

Illustration 81 Evaluate $\int x^{-11}(1+x^4)^{-1/2}\, dx.$

Solution. Here, $\left(\frac{m+1}{n}+p\right) = \left[\frac{-11+1}{4}+\frac{1}{2}\right] = -3$

If we substitute $(1+x^4) = t^2 x^4$, then $1+\frac{1}{x^4} = t^2$ and $\frac{-4}{x^5}\, dx = 2t\, dt$

$$\therefore \quad I = \int \frac{dx}{x^{11}(1+x^4)^{1/2}} = \int \frac{dx}{x^{11}\cdot x^2 (1+1/x^4)^{1/2}}$$

$$= \int \frac{dx}{x^{13}(1+1/x^4)^{1/2}} = -\frac{1}{4}\int \frac{2t\, dt}{x^8\, t}$$

$$= -\frac{1}{2}\int (t^2-1)^2\, dt = -\frac{1}{2}\int (t^4 - 2t^2 + 1)\, dt$$

$$= -\frac{1}{2}\left[\frac{t^5}{5} - \frac{2t^3}{3} + t\right] + C$$

where $\quad t = \sqrt{1+\frac{1}{x^4}}$

Illustration 82 Evaluate $\int \frac{1}{\sqrt[3]{x}+\sqrt[4]{x}}\, dx.$

Solution. Let $\quad I = \int \frac{1}{\sqrt[3]{x}+\sqrt[4]{x}}\, dx$

Put $\quad x^{1/12} = t, \quad \therefore\ x = t^{12}$ and $dx = 12t^{11}\, dt$

$$\therefore \quad I = \int \frac{1}{t^4+t^3}\cdot 12\, t^{11}\, dt = 12\int \frac{t^8}{t+1}\, dt \qquad \text{again put } (t+1) = y \quad \therefore\ dt = dy$$

$$= 12\int \frac{(y-1)^8}{y}\, dy$$

$$= 12\int \frac{(y^8 - 8y^7 + 28y^6 - 56y^5 + 70y^4 - 56y^3 + 28y^2 - 8y + 1)}{y}\, dy$$

(using binomial)

$$= 12\int (y^7 - 8y^6 + 28y^5 - 56y^4 + 70y^3 - 56y^2 + 28y - 8 + 1/y)\, dy$$

$$= 12\left(\frac{y^8}{8} - \frac{8y^7}{7} + \frac{28y^6}{6} - \frac{56y^5}{5} + \frac{70y^4}{4} - \frac{56y^3}{3} + \frac{28y^2}{2} - 8y + \log|y|\right) + C_1$$

where $y = x^{1/12} + 1$

Illustration 83 Evaluate $\int \frac{dx}{(x-1)^{3/4}(x+2)^{5/4}}$.

Solution. Let $I = \int \frac{dx}{(x-1)^{3/4}(x+2)^{5/4}} = \int \frac{dx}{(x+2)^2\left(\frac{(x-1)}{(x+2)}\right)^{3/4}}$

Let $\frac{x-1}{x+2} = t$. So that, $\frac{3}{(x+2)^2}\,dx = dt$

$\therefore$ $$I = \int \frac{dt}{3\,t^{3/4}} = \frac{1}{3}\int t^{-3/4}\,dt = \frac{1}{3}\cdot\frac{t^{1/4}}{1/4} + C = \frac{4}{3}\left(\frac{x-1}{x+2}\right)^{1/4} + C$$

Illustration 84 The value of $\int \frac{(1+x)}{x\,(1+xe^x)^2}\,dx$, is equal to

(a) $\log\left|\frac{x}{1+xe^x}\right| + \frac{1}{(1+xe^x)} + C$ (b) $\log\left|\frac{xe^x}{1+xe^x}\right| + \frac{1}{1+xe^x} + C$

(c) $\log\left|\frac{xe^x}{1+e^x}\right| + \frac{1}{1+xe^x} + C$ (d) None of these

Solution. Let $I = \int \frac{(1+x)}{x\,(1+xe^x)^2}\,dx = \int \frac{(1+x)\,e^x}{(xe^x)(1+xe^x)^2}\,dx$, put $1 + xe^x = t$

$\therefore (1+x)\,e^x\,dx = dt$

$= \int \frac{dt}{(t-1)\cdot t^2}$, applying partial fraction, we get

$$\frac{1}{(t-1)\,t^2} = \frac{A}{t-1} + \frac{B}{t} + \frac{C}{t^2}$$

$\Rightarrow$ $1 = A\,(t^2) + Bt\,(t-1) + C\,(t-1)$

For $t = 1 \Rightarrow A = 1$

For $t = 0 \Rightarrow C = -1$ and $B = -1$

$\therefore$ $$I = \int\left\{\frac{1}{t-1} - \frac{1}{t} - \frac{1}{t^2}\right\}dt = \log|t-1| - \log|t| + \frac{1}{t} + C$$

$$= \log|xe^x| - \log|1+xe^x| + \frac{1}{1+xe^x} + C = \log\left|\frac{xe^x}{1+xe^x}\right| + \frac{1}{1+xe^x} + C$$

Hence, (b) is the correct answer.

Illustration 85 The value of $\int \frac{dx}{x + \sqrt{a^2 - x^2}}$, is equal to

(a) $\frac{1}{2}\sin^{-1}\left(\frac{x}{a}\right) + \frac{1}{2}\log|x + \sqrt{a^2 - x^2}| + C_1$

(b) $\frac{1}{2}\sin^{-1}\left(\frac{x}{a}\right) - \frac{1}{2}\log|x + \sqrt{a^2 - x^2}| + C_1$

(c) $\frac{1}{2}\sin^{-1}\left(\frac{x}{a}\right) - \log|x + \sqrt{a^2 - x^2}| + C_1$

(d) $\frac{1}{2}\cos^{-1}\left(\frac{x}{a}\right) + \frac{1}{2}\log|x + \sqrt{a^2 - x^2}| + C_1$

Solution. Let $I = \int \frac{dx}{x + \sqrt{a^2 - x^2}}$, put $x = a \sin\theta \therefore dx = a\cos\theta\, d\theta$

$$= \int \frac{a\cos\theta\, d\theta}{a\sin\theta + \sqrt{a^2 - a^2\sin^2\theta}}$$

$$= \int \frac{\cos\theta\, d\theta}{\sin\theta + \cos\theta} = \frac{1}{2}\int \frac{\cos\theta + \sin\theta + \cos\theta - \sin\theta}{\sin\theta + \cos\theta}\, d\theta$$

$$= \frac{1}{2}\int 1\, d\theta + \frac{1}{2}\int \frac{\cos\theta - \sin\theta}{\sin\theta + \cos\theta}\, d\theta$$

$$= \frac{1}{2}\cdot\theta + \frac{1}{2}\log(\sin\theta + \cos\theta) + C$$

$$= \frac{1}{2}\sin^{-1}\left(\frac{x}{a}\right) + \frac{1}{2}\log\left|\frac{x}{a} + \sqrt{1 - \frac{x^2}{a^2}}\right| + C$$

$$= \frac{1}{2}\sin^{-1}\left(\frac{x}{a}\right) + \frac{1}{2}\{\log|x + \sqrt{a^2 - x^2}|\} - \frac{1}{2}\log a + C$$

$$= \frac{1}{2}\sin^{-1}\left(\frac{x}{a}\right) + \frac{1}{2}\log|x + \sqrt{a^2 - x^2}| + C_1 \qquad \left(\text{where } c_1 = c - \frac{1}{2}\log a\right)$$

Hence, (a) is the correct answer.

Illustration 86 The value of $\int \frac{x^2 - 1}{(x^2 + 1)\sqrt{x^4 + 1}}\, dx$, is equal to

(a) $\frac{1}{\sqrt{2}}\sec^{-1}\left(\frac{x^2 + 1}{\sqrt{2}x}\right) + C$ (b) $\sqrt{2}\sec^{-1}\left(\frac{x^2 + 1}{\sqrt{2}x}\right) + C$

(c) $\frac{1}{\sqrt{2}}\operatorname{cosec}^{-1}\left(\frac{x^2 + 1}{\sqrt{2}x}\right) + C$ (d) $\sqrt{2}\operatorname{cosec}^{-1}\left(\frac{x^2 + 1}{\sqrt{2}x}\right) + C$

Solution. Let $I = \int \frac{(x^2 - 1)}{(x^2 + 1)\sqrt{x^4 + 1}}\, dx$

$$= \int \frac{x^2(1 - 1/x^2)\, dx}{x^2\left(x + \frac{1}{x}\right)\sqrt{x^2 + \frac{1}{x^2}}} = \int \frac{(1 - 1/x^2)\, dx}{\left(x + \frac{1}{x}\right)\sqrt{\left(x + \frac{1}{x}\right)^2 - 2}}$$

Put $x + \frac{1}{x} = t \Rightarrow \left(1 - \frac{1}{x^2}\right) dx = dt$

$$= \int \frac{dt}{t\sqrt{t^2 - 2}} = \frac{1}{\sqrt{2}}\sec^{-1}\left(\frac{t}{\sqrt{2}}\right) + C = \frac{1}{\sqrt{2}}\sec^{-1}\left(\frac{x^2 + 1}{\sqrt{2}x}\right) + C$$

Hence, (a) is the correct answer.

Illustration 87 The value of $\int \frac{dx}{\cos^6 x + \sin^6 x}$, is equal to

(a) $\tan^{-1}(2\cot 2x) + C$ (b) $\tan^{-1}(\cot 2x) + C$

(c) $\tan^{-1}\left(\frac{1}{2}\cot 2x\right) + C$ (d) $\tan^{-1}(-2\cot 2x) + C$

Solution. Let $I = \int \frac{dx}{\cos^6 x + \sin^6 x} = \int \frac{\sec^6 x \, dx}{1 + \tan^6 x}$

$$= \int \frac{(1+\tan^2 x)^2 \cdot \sec^2 x}{1+\tan^6 x} dx$$

Put $\tan x = t \Rightarrow \sec^2 x \, dx = dt$

$$= \int \frac{(1+t^2)^2}{1+t^6} dt = \int \frac{(1+t^2)^2}{(1+t^2)(1-t^2+t^4)} dt$$

$$= \int \frac{1+t^2}{1-t^2+t^4} dt = \int \frac{(1+1/t^2)\, dt}{(1/t^2 - 1 + t^2)}$$

$$= \int \frac{\left(1+\frac{1}{t^2}\right) dt}{(t-1/t)^2+1}, \quad \text{put } t - \frac{1}{t} = z \quad \therefore \left(1+\frac{1}{t^2}\right) dt = dz$$

$$= \int \frac{dz}{z^2+1} = \tan^{-1}(z) + C$$

$$= \tan^{-1}\left(\frac{t^2-1}{t}\right) + C$$

$$= \tan^{-1}\left(\frac{\tan^2 x - 1}{\tan x}\right) + C = \tan^{-1}(-2\cot 2x) + C$$

Hence, (d) is the correct answer.

Target Exercise 1.3

1. The value of $\int \frac{dx}{(x+a)^{8/7}(x-b)^{6/7}}$, is equal to

(a) $\frac{3}{2(a+b)}\left(\frac{x+a}{x-b}\right)^{2/3} + C$ (b) $\frac{3}{(a+b)}\left(\frac{x-b}{x+a}\right)^{1/3} + C$

(c) $\frac{7}{(a+b)}\left(\frac{x-b}{x+a}\right)^{1/7} + C$ (d) None of these

2. The value of $\int \frac{dx}{(x+1)^{1/3}+(x+1)^{1/2}}$, is equal to

(a) $t^3 - 3t^2 + t - 6\log|1+t| + C$; where $t = (x+1)^{1/6}$

(b) $2t^3 - 3t^2 + 6t - 6\log|1+t| + C$; where $t = (x+1)^{1/6}$

(c) $t^3 - 3t^2 + 6t - 6\log|1+t| + C$; where $t = (x^2-1)^{1/6}$

(d) None of the above

3. The value of $\int \frac{(x+2)\, dx}{(x^2+3x+3)\sqrt{x+1}}$, is equal to

(a) $\frac{1}{\sqrt{3}}\tan^{-1}\left(\frac{x}{\sqrt{3}\sqrt{x+1}}\right) + C$ (b) $\frac{2}{\sqrt{3}}\tan^{-1}\left(\frac{x}{\sqrt{3}\sqrt{x+1}}\right) + C$

(c) $\frac{1}{2\sqrt{3}}\log\left|\frac{x+\sqrt{x+1}}{x-\sqrt{x+1}}\right| + C$ (d) None of these

4. The value of $\int \frac{x^4-1}{x^2(x^4+x^2+1)^{1/2}}\,dx$, is equal to

(a) $\sqrt{\frac{x^4+x^2+1}{x}}+C$ (b) $\sqrt{\frac{x^4+x^2+1}{x^3}}+C$

(c) $\sqrt{\frac{x^4+x^2+1}{x^2}}+C$ (d) None of these

5. The value of $\int \frac{\sec x\,dx}{\sqrt{\sin(2x+A)+\sin A}}$, is equal to

(a) $\sec A(\sqrt{\tan x+\sin A\cos A})+C$ (b) $\sqrt{2}\sec A(\sqrt{\tan x\sin A+\cos A})+C$

(c) $\frac{1}{\sqrt{2}}\sec A(\tan x\sin A+\cos A)^{3/2}+C$ (d) None of these

Integration Using Euler's Substitutions

Integrals of the form $\int f(x), \sqrt{ax^2+bx+c}\,dx$ are calculated with the aid of one of the three Euler's substitutions

(i) $\sqrt{ax^2+bx+c}=t\pm x\sqrt{a}$, if $a>0$.

(ii) $\sqrt{ax^2+bx+c}=tx+\sqrt{c}$, if $c>0$.

(iii) $\sqrt{ax^2+bx+c}=(x-\alpha)\,t$, if $ax^2+bx+c=a(x-\alpha)(x-\beta)$, *ie*, If α is real root of (ax^2+bx+c).

Points to Consider

The Euler's substitutions often lead to rather cumbersome calculations, therefore they should be applied only when it is difficult to find another method for calculating a given integral.

Illustration 88 Evaluate $I=\int \frac{x\,dx}{(\sqrt{7x-10-x^2})^3}$.

Solution. In this case $a<0$ and $c<0$. Therefore, neither (I) nor (II) Euler's Substitution is applicable. But the quadratic $7x-10-x^2$ has real roots $\alpha=2, \beta=5$.

$\therefore$ We use the (III)

ie, $\sqrt{7x-10-x^2}=\sqrt{(x-2)(5-x)}=(x-2)\,t$

where $(5-x)=(x-2)\,t^2$

or $5+2t^2=x(1+t^2)$ $\therefore$ $x=\frac{5+2t^2}{1+t^2}$

$$(x-2)\,t=\left(\frac{5+2t^2}{1+t^2}-2\right)t=\frac{3t}{1+t^2}$$

$$\therefore \quad dx=\frac{-6t}{(1+t^2)^2}\,dt$$

Hence, $$I=\int \frac{x\,dx}{(\sqrt{7x-10-x^2})^3}=\int \frac{\left(\frac{5+2t^2}{1+t^2}\right)\cdot\frac{-6t}{(1+t^2)^2}\,dt}{\left(\frac{3t}{1+t^2}\right)^3}$$

$$=\frac{-6}{27}\int \frac{5+2t^2}{t^2}\,dt=\frac{-2}{9}\int\left(\frac{5}{t^2}+2\right)dt=\frac{-2}{9}\left[\frac{-5}{t}+2t\right]+C$$

$$\therefore \int \frac{x\,dx}{(\sqrt{7x-10-x^2})^3}=\frac{-2}{9}\left(\frac{-5}{t}+2t\right)+C, \text{ where } t=\frac{\sqrt{7x-10-x^2}}{x-2}$$

Illustration 89 Evaluate $\int \frac{dx}{x+\sqrt{x^2-x+1}}$.

Solution. Since, here $c=1$, we can apply the second Euler's Substitution.

$$\sqrt{x^2-x+1}=tx-1$$

Therefore, $$(2t-1)\,x=(t^2-1)\,x^2$$

$$x=\frac{2t-1}{t^2-1}$$

$$\therefore \quad dx=-\frac{2\,(t^2-t+1)\,dt}{(t^2-1)^2}$$

and $$x+\sqrt{x^2-x+1}=\frac{t}{t-1}$$

$$\therefore \quad I=\int\frac{dx}{x+\sqrt{x^2-x+1}}=\int\frac{-2t^2+2t-2}{t\,(t-1)(t+1)^2}\,dt$$

Using partial fractions, we have

$$\frac{-2t^2+2t-2}{t\,(t-1)(t+1)^2}=\frac{A}{t}+\frac{B}{t-1}+\frac{C}{(t+1)}+\frac{D}{(t+1)^2}$$

or $$(-2t^2+2t-2)=A\,(t-1)(t+1)^2+Bt\,(t+1)^2+C\,(t-1)(t+1)\,t+Dt$$

we get $$A=2,\quad B=-1/2,\quad C=-3/2,\quad D=-3$$

Hence, $$I=2\int\frac{dt}{t}-\frac{1}{2}\int\frac{dt}{t-1}-\frac{3}{2}\int\frac{dt}{(t+1)}-3\int\frac{dt}{(t+1)^2}$$

$$=2\log_e|t|-\frac{1}{2}\log_e|t-1|-\frac{3}{2}\log_e|t+1|+\frac{3}{(t+1)}+C$$

$$\left(\text{where } t=\frac{\sqrt{x^2-x+1}+1}{x}\right)$$

Introduction of Reduction Formulae Over Indefinite Integrals

Reduction formulae makes it possible to reduce an integral depending on the index $n>0$, called the order of the integral, to an integral of the same type with smaller index.

(*ie*, To reduce the integrals into similar integrals of order less than or greater than given integral).

Application of reduction formula is given with the help of some examples.

Reduction Formula for $\int \sin^n x\, dx$

Let $$I_n = \int \sin^n x\, dx = \int \underset{\text{I}}{\sin^{n-1} x}\ \underset{\text{II}}{\sin x}\, dx$$

$$= -\sin^{n-1} x \cos x + \int (n-1) \sin^{n-2} x \cos^2 x\, dx$$

$$= -\sin^{n-1} x \cos x + (n-1) \int \sin^{n-2} x\, (1 - \sin^2 x)\, dx$$

$$= -\sin^{n-1} x \cos x + (n-1) \int (\sin^{n-2} x - \sin^n x)\, dx$$

$$= -\sin^{n-1} x \cos x + (n-1)\, I_{n-2} - (n-1)\, I_n$$

$$\therefore \quad n\, I_n = -\sin^{n-1} x \cos x + (n-1)\, I_{n-2}$$

$$\Rightarrow \quad I_n = -\frac{\sin^{n-1} x \cos x}{n} + \frac{n-1}{n} I_{n-2}$$

Thus, $$\int \sin^n x\, dx = \frac{-\sin^{n-1} x \cos x}{n} + \frac{n-1}{n} \int \sin^{n-2} x\, dx$$

Reduction Formula for $\int \cos^n x\, dx$

Let $$I_n = \int \cos^n x\, dx = \int \underset{\text{I}}{\cos^{n-1} x}\ \underset{\text{II}}{\cos x}\, dx$$

$$= \cos^{n-1} x \sin x + \int (n-1) \cos^{n-2} x \sin^2 x\, dx$$

$$= \cos^{n-1} x \sin x + (n-1) \int \cos^{n-2} x\, (1 - \cos^2 x)\, dx$$

$$= \cos^{n-1} x \sin x + (n-1)\, I_{n-2} - (n-1)\, I_n$$

$$\therefore \quad nI_n = \cos^{n-1} x \sin x + (n-1)\, I_{n-2}$$

or $$\int \cos^n x\, dx = \frac{\cos^{n-1} x \sin x}{n} + \frac{n-1}{n} \int \cos^{n-2} x\, dx$$

Reduction Formula for $\int \tan^n x\, dx$

Let $$I_n = \int \tan^n x\, dx$$

$$\Rightarrow \quad I_n = \int \tan^{n-2} x \tan^2 x\, dx = \int \tan^{n-2} x\, (\sec^2 x - 1)\, dx$$

$$= \int \tan^{n-2} x \sec^2 x - I_{n-2}$$

$$= \int t^{n-2}\, dt - I_{n-2} \quad \text{where, } \tan x = t \quad \Rightarrow \quad \sec^2 x\, dx = dt$$

$$I_n = \frac{t^{n-1}}{n-1} - I_{n-2}$$

$$\therefore \quad I_n = \frac{\tan^{n-1} x}{n-1} - I_{n-2}$$

$$\Rightarrow \quad \int \tan^n x \, dx = \frac{\tan^{n-1} x}{n-1} - \int \tan^{n-2} x \, dx$$

Reduction Formula for $\int \text{cosec}^n x \, dx$

Let $\quad I_n = \int \text{cosec}^n x \, dx = \int \underset{\text{I}}{\text{cosec}^{n-2} x} \, \underset{\text{II}}{\text{cosec}^2 x} \, dx$

$$= \text{cosec}^{n-2} x \, (-\cot x) - \int (n-2) \, \text{cosec}^{n-2} x \, (\text{cosec}^2 x - 1) \, dx$$

$$= -\text{cosec}^{n-2} x \cot x - (n-2) \int (\text{cosec}^n x - \text{cosec}^{n-2} x) \, dx$$

$$= -\text{cosec}^{n-2} x \cot x - (n-2) I_n + (n-2) I_{n-2}$$

$$\therefore \quad (n-1) I_n = -\text{cosec}^{n-2} x \cot x + (n-2) I_{n-2}$$

or $$I_n = -\frac{\text{cosec}^{n-2} x \cot x}{n-1} + \frac{n-2}{n-1} I_{n-2}$$

$$\therefore \quad \int \text{cosec}^n x \, dx = -\frac{\text{cosec}^{n-2} x \cot x}{n-1} + \frac{n-2}{n-1} \int \text{cosec}^{n-2} x \, dx$$

Reduction Formula for $\int \sec^n x \, dx$

Let $\quad I_n = \int \sec^n x \, dx = \int \underset{\text{I}}{\sec^{n-2} x} \, \underset{\text{II}}{\sec^2 x} \, dx$

$$= \sec^{n-2} x \tan x - \int (n-2) \sec^{n-3} x \sec x \tan x \cdot \tan x \, dx$$

$$= \sec^{n-2} x \tan x - (n-2) \int \sec^{n-2} x \, (\sec^2 x - 1) \, dx$$

$$= \sec^{n-2} x \tan x - (n-2) I_n + (n-2) I_{n-2}$$

$$\Rightarrow \quad (n-1) I_n = \sec^{n-2} x \tan x + (n-2) I_{n-2}$$

or $$I_n = \frac{\sec^{n-2} x \tan x}{(n-1)} + \frac{(n-2)}{(n-1)} I_{n-2}$$

$$\therefore \quad \int \sec^n x \, dx = \frac{\sec^{n-2} x \tan x}{(n-1)} + \frac{(n-2)}{(n-1)} \int \sec^{n-2} x \, dx$$

Reduction Formula for $\int \cot^n x \, dx$

Let $\quad I_n = \int \cot^n x \, dx$

$$= \int \cot^{n-2} x \cot^2 x \, dx$$

$$= \int \cot^{n-2} x \, (\text{cosec}^2 x - 1) \, dx$$

$$= \int \cot^{n-2} x \operatorname{cosec}^2 x \, dx - \int \cot^{n-2} x \, dx$$

$$= \int t^{n-2} \, dt - I_{n-2}, \qquad \text{where } t = \cot x$$

$$I_n = -\frac{\cot^{n-1} x}{n-1} - I_{n-2}$$

$$\therefore \quad \int \cot^n x \, dx = -\frac{\cot^{n-1} x}{n-1} - \int \cot^{n-2} x \, dx$$

Reduction Formula for $\int \sin^m x \cos^n x \, dx$

Let $\quad A = \sin^{m-1} x \cos^{n+1} x$

$$\therefore \quad \frac{dA}{dx} = (m-1) \sin^{m-2} x \cos^{n+2} x - (n+1) \sin^m x \cos^n x$$

$$= (m-1) \sin^{m-2} x \cos^n x (1 - \sin^2 x) - (n+1) \sin^m x \cos^n x$$

$$= (m-1) \sin^{m-2} x \cos^n x - (m - 1 + n + 1) \sin^m x \cos^n x$$

$$\Rightarrow \quad \frac{dA}{dx} = (m-1) \sin^{m-2} x \cos^n x - (m+n) \sin^m x \cos^n x$$

Integrating with respect to x on both the sides, we get

$$A = (m-1) \int \sin^{m-2} x \cos^n x \, dx - (m+n) \int \sin^m x \cos^n x \, dx$$

$$\Rightarrow \quad (m+n) \int \sin^m x \cos^n x \, dx = (m-1) \int \sin^{m-2} x \cos^n x \, dx - P$$

$$\Rightarrow \quad \int \sin^m x \cos^n x \, dx = \frac{(m-1)}{(m+n)} \int \sin^{m-2} x \cos^n x \, dx - \frac{\sin^{m-1} x \cos^{n+1} x}{m+n}$$

or

$$I_{m,n} = \frac{(m-1)}{(m+n)} I_{m-2,n} - \frac{\sin^{m-1} x \cos^{n+1} x}{(m+n)}$$

Points to Cosider

Similarly, we can show

1. $\int \sin^m x \cos^n x \, dx = \frac{\sin^{m+1} x \cos^{n+1} x}{m+n} + \frac{n-1}{m+n} \int \sin^m x \cos^{n-2} x \, dx$
2. $\int \sin^m x \cos^n x \, dx = \frac{\sin^{m+1} x \cos^{n+1} x}{m+1} + \frac{m+n+2}{m+1} \int \sin^{m+2} x \cos^n x \, dx$
3. $\int \sin^m x \cos^n x \, dx = \frac{\sin^{m+1} x \cos^{n+1} x}{n+1} + \frac{m+n+2}{n+1} \int \sin^m x \cos^{n+2} x \, dx$
4. $\int \sin^m x \cos^n x \, dx = -\frac{\sin^{m-1} x \cos^{n+1} x}{n+1} + \frac{m-1}{n+1} \int \sin^{m-2} x \cos^{n+2} x \, dx$
5. $\int \sin^m x \cos^n x \, dx = \frac{\sin^{m+1} x \cos^{n-1} x}{m+1} + \frac{n-1}{m+1} \int \sin^{m+2} x \cos^{n-2} x \, dx$

Reduction Formula for $\int \cos^m x \sin nx\, dx$

Let $I_{m,n} = \int \underset{\text{I}}{\cos^m x} \underset{\text{II}}{\sin nx}\, dx$

$$= -\frac{\cos^m x \cos nx}{n} - \frac{m}{n}\int \cos^{m-1} x \sin x \cos nx\, dx$$

$$= -\frac{\cos^m x \cos nx}{n} - \frac{m}{n}\int \cos^{m-1} x \{\sin nx \cos x - \sin (n-1)x\}\, dx$$

[using $\sin (n-1)x = \sin nx \cos x - \cos nx \sin x$

$\Rightarrow \sin x \cos nx = \sin nx \cos x - \sin (n-1)x$]

$$= -\frac{\cos^m x \cos nx}{n} - \frac{m}{n}\int \cos^m x \sin nx\, dx + \frac{m}{n}\int \cos^{m-1} x \sin (n-1)x\, dx$$

$$I_{m,n} = -\frac{\cos^m x \cos nx}{n} - \frac{m}{n} I_{m,n} + \frac{m}{n} I_{m-1,n-1}$$

$$\Rightarrow \quad \frac{m+n}{n} I_{m,n} = -\frac{\cos^m x \cos nx}{n} + \frac{m}{n} I_{m-1,n-1}$$

or $$I_{m,n} = -\frac{\cos^m x \cos nx}{m+n} + \frac{m}{m+n} I_{m-1,n-1}$$

Points to Consider

Similarly, we can show

1. $\int \cos^m x \cos nx\, dx = \dfrac{\cos^m x \sin nx}{m+n} + \dfrac{m}{m+n}\int \cos^{m-1} x \cos (n-1)x\, dx$

2. $\int \sin^m x \sin nx\, dx = \dfrac{n \sin^m x \cos nx}{m^2-n^2} - \dfrac{m \sin^{m-1} x \cos x \cos nx}{m^2-n^2} + \dfrac{m(m-1)}{m^2-n^2}\int \sin^{m-2} x \sin nx\, dx$

3. $\int \sin^m x \cos nx\, dx = \dfrac{n \sin^m x \sin nx}{m^2-n^2} - \dfrac{m \sin^{m-1} x \cos x \cos nx}{m^2-n^2} + \dfrac{m(m-1)}{m^2-n^2}\int \sin^{m-2} x \cos nx\, dx$

Illustration 90 Evaluate $I_n = \int \dfrac{dx}{(x^2+a^2)^n}$.

Solution. Here, $\quad I_n = \int \dfrac{dx}{(x^2+a^2)^n} = \int \dfrac{1}{(x^2+a^2)^n}\cdot 1\, dx$

Applying Integration by parts, we get

$$= \frac{1}{(x^2+a^2)^n}\cdot x - \int \frac{(2x)}{(x^2+a^2)^{n+1}}\cdot(-n)\cdot(x)\, dx$$

$$= \frac{x}{(x^2 + a^2)^n} + 2n \int \frac{x^2}{(x^2 + a^2)^{n+1}} dx$$

$$= \frac{x}{(x^2 + a^2)^n} + 2n \int \frac{x^2 + a^2 - a^2}{(x^2 + a^2)^{n+1}} dx$$

$$\therefore \quad I_n = \frac{x}{(x^2 + a^2)^n} + 2n \int \frac{1}{(x^2 + a^2)^n} dx - 2a^2 n \int \frac{dx}{(x^2 + a^2)^{n+1}}$$

$$I_n = \frac{x}{(x^2 + a^2)^n} + 2n I_n - 2n a^2 I_{n+1}$$

$$\therefore \quad 2n a^2 I_{n+1} = \frac{x}{(x^2 + a^2)^n} + (2n - 1) I_n$$

or
$$I_{n+1} = \frac{1}{2n a^2} \cdot \frac{x}{(x^2 + a^2)^n} + \frac{(2n - 1)}{2n} \frac{1}{a^2} I_n$$

Points to Consider

Above obtained formula reduces the calculations of the integral I_{n+1} to the calculations of the integral I_n and consequently, allows us to calculate completely an integral with natural index, as

$$I_1 = \int \frac{dx}{x^2 + a^2} = \frac{1}{a} \tan^{-1}\left(\frac{x}{a}\right) + C$$

$\therefore$ From above formula

Let $n = 1$

$$I_2 = \int \frac{dx}{(x^2 + a^2)^2} = \frac{1}{2a^2} \cdot \frac{x}{x^2 + a^2} + \frac{1}{2a^2} \cdot I_1$$

$$= \frac{1}{2a^2} \cdot \frac{x}{x^2 + a^2} + \frac{1}{2a^2} \cdot \frac{1}{a} \cdot \tan^{-1}\left(\frac{x}{a}\right) + C$$

$$= \frac{1}{2a^2} \cdot \frac{x}{x^2 + a^2} + \frac{1}{2a^3} \cdot \tan^{-1}\left(\frac{x}{a}\right) + C$$

Let $n = 2$

$$I_3 = \int \frac{dx}{(x^2 + a^2)^3} = \frac{1}{4a^2} \cdot \frac{x}{(x^2 + a^2)^2} + \frac{3}{4a^2} I_2$$

$$= \frac{1}{4a^2} \cdot \frac{x}{(x^2 + a^2)^2} + \frac{3}{8a^4} \cdot \frac{x}{x^2 + a^2} + \frac{3}{8a^5} \tan^{-1}\left(\frac{x}{a}\right) + C$$ ...and so on.

Illustration 91 Derive reduction formulae for $I_{(n, m)} = \int \frac{\sin^n x}{\cos^m x} dx.$

Solution. Using Integration by parts for $I_{(n, m)}$, we get

$$I_{(n, m)} = \int \underset{\text{I}}{\sin^{n-1} x} \; \underset{\text{II}}{\frac{\sin x}{\cos^m x}} dx$$

$$= \sin^{n-1} x \cdot \frac{(\cos x)^{-m+1}}{(m - 1)} - \int (n - 1) \sin^{n-2} x \cdot \cos x \cdot \frac{(\cos x)^{-m+1}}{(m - 1)} dx$$

$$= \frac{1}{m-1} \cdot \frac{\sin^{n-1} x}{\cos^{m-1} x} - \frac{(n-1)}{(m-1)} \cdot \int \frac{\sin^{n-2} x}{\cos^{m-2} x} dx$$

$$I_{n, m} = \frac{1}{(m-1)} \cdot \frac{\sin^{n-1} x}{\cos^{m-1} x} - \frac{(n-1)}{(m-1)} \cdot I_{(n-2, m-2)}$$

is required reduction formula.

Integration Using Differentiation

In $\int \frac{dx}{(a + b\cos x)^2}$, $\int \frac{dx}{(a + b\sin x)^2}$, $\int \frac{dx}{(\sin x + a\sec x)^2}$, $\int \frac{a + b\sin x}{(b + a\sin x)^2} dx, \ldots$ we follow the following method.

1. Let $A = \frac{\sin x}{a + b\cos x}$ or $A = \frac{\cos x}{a + b\sin x}$ according as the integral to evaluated is of the form $\int \frac{dx}{(a + b\cos x)^2}$ or $\int \frac{dx}{(a + b\sin x)^2}$
2. Find $\frac{dA}{dx}$ and express it in terms of $\frac{1}{a + b\cos x}$ or $\frac{1}{a + b\sin x}$ as the case may be.
3. Integrate both the sides of the expression obtained in step 2 to obtain the value of the required integral.

Illustration 92 Evaluate $\int \frac{dx}{(5 + 4\cos x)^2}$.

Solution. Here, $A = \frac{\sin x}{5 + 4\cos x}$, then

$$\frac{dA}{dx} = \frac{(5 + 4\cos x)(\cos x) - \sin x(-4\sin x)}{(5 + 4\cos x)^2}$$

$$\Rightarrow \quad \frac{dA}{dx} = \frac{5\cos x + 4}{(5 + 4\cos x)^2} = \frac{\frac{5}{4}(4\cos x + 5) + 4 - \frac{25}{4}}{(5 + 4\cos x)^2}$$

$$\Rightarrow \quad \frac{dA}{dx} = \frac{5}{4} \cdot \frac{1}{(5 + 4\cos x)} - \frac{9}{4} \cdot \frac{1}{(5 + 4\cos x)^2}$$

Integrating both the sides w.r.t. 'x', we get

$$A = \frac{5}{4} \int \frac{dx}{5 + 4\cos x} - \frac{9}{4} \int \frac{dx}{(5 + 4\cos x)^2}$$

$$\Rightarrow \quad \frac{9}{4} \int \frac{dx}{(5 + 4\cos x)^2} = \frac{5}{4} \int \frac{dx}{5 + 4\cos x} - A$$

$$= \frac{5}{4} \int \frac{dx}{5 + 4\frac{(1 - \tan^2 x/2)}{(1 + \tan^2 x/2)}} - \frac{\sin x}{(5 + 4\cos x)}$$

$$\Rightarrow \int \frac{dx}{(5+4\cos x)^2} = \frac{5}{9}\int \frac{1+\tan^2 x/2}{9+\tan^2 x/2}\, dx - \frac{4}{9}\cdot\frac{\sin x}{5+4\cos x}$$

$$\Rightarrow \int \frac{dx}{(5+4\cos x)^2} = \frac{5}{9}\int \frac{2\, dt}{9+t^2} - \frac{4}{9}\cdot\frac{\sin x}{5+4\cos x}$$

(where $\tan x/2 = t$)

$$\Rightarrow \int \frac{dx}{(5+4\cos x)^2} = \frac{10}{9}\cdot\frac{1}{3}\tan^{-1}\left(\frac{t}{3}\right) - \frac{4}{9}\cdot\frac{\sin x}{5+4\cos x}$$

$$\Rightarrow \int \frac{dx}{(5+4\cos x)^2} = \frac{10}{27}\tan^{-1}\left(\frac{\tan x/2}{3}\right) - \frac{4}{9}\left(\frac{\sin x}{5+4\cos x}\right) + C$$

Illustration 93 Evaluate $\int \frac{dx}{(16+9\sin x)^2}$.

Solution. Let $A = \frac{\cos x}{16+9\sin x}$ ...(i)

$$\Rightarrow \frac{dA}{dx} = \frac{(16+9\sin x)(-\sin x) - \cos x\,(9\cos x)}{(16+9\sin x)^2}$$

$$\Rightarrow \frac{dA}{dx} = \frac{-16\sin x - 9}{(16+9\sin x)^2}$$

$$\Rightarrow \frac{dA}{dx} = \frac{-\frac{16}{9}(9\sin x + 16) + \frac{256}{9} - 9}{(16+9\sin x)^2}$$

$$\Rightarrow \frac{dA}{dx} = -\frac{16}{9}\cdot\frac{1}{(16+9\sin x)} + \frac{175}{9(16+9\sin x)^2} \qquad \text{...(ii)}$$

Integrating both the sides Eq. (ii) w.r.t. 'x', we get

$$A = -\frac{16}{9}\int \frac{dx}{16+9\sin x} + \frac{175}{9}\int \frac{dx}{(16+9\sin x)^2}$$

$$\Rightarrow \frac{175}{9}\int \frac{dx}{(16+9\sin x)^2} = A + \frac{16}{9}\int \frac{(1+\tan^2 x/2)\, dx}{16+16\tan^2 x/2 + 18\tan x/2}$$

$$\Rightarrow \frac{175}{9}\int \frac{dx}{(16+9\sin x)^2} = A + \frac{16}{9}\int \frac{2\, dt}{16t^2+18t+16} \qquad \text{(where } \tan x/2 = t\text{)}$$

$$\Rightarrow \frac{175}{9}\int \frac{dx}{(16+9\sin x)^2} = A + \frac{2}{9}\int \frac{dt}{t^2+\frac{9}{8}t+1}$$

$$= A + \frac{2}{9}\int \frac{dt}{\left(t+\frac{9}{16}\right)^2 + \left(\frac{\sqrt{175}}{16}\right)^2}$$

$$= A + \frac{2}{9}\times\frac{16}{\sqrt{175}}\tan^{-1}\left(\frac{16t+9}{\sqrt{175}}\right)$$

$$\Rightarrow \int \frac{dx}{(16+9\sin x)^2} = \frac{9}{175}\cdot\frac{\cos x}{(16+9\sin x)}$$

$$+ \frac{2}{(175)^{3/2}}\tan^{-1}\left(\frac{16\tan x/2 + 9}{\sqrt{175}}\right) + C$$

Illustration 94 Evaluate $\int \frac{dx}{(\sin x + a \sec x)^2}$ when $|a| > 1/2$.

Solution. Here, $I = \int \frac{dx}{(\sin x + a \sec x)^2}$

or $$I = \int \frac{\cos^2 x \, dx}{(\sin x \cos x + a)^2}$$

$$I = \int \frac{\cos^2 x \, dx}{a^2 + 2a \sin x \cos x + \sin^2 x \cos^2 x}$$

$$= \int \frac{\cos^2 x \, dx}{a^2 + a \sin 2x + \frac{1}{4} \sin^2 2x}$$

$$= \int \frac{4 \cos^2 x \, dx}{(4a^2 + 4a \sin 2x + \sin^2 2x)} = 2 \int \frac{(1 + \cos 2x) \, dx}{(2a + \sin 2x)^2}$$

$$= 2 \int \frac{dx}{(2a + \sin 2x)^2} + 2 \int \frac{\cos 2x \, dx}{(2a + \sin 2x)^2}$$

$\Rightarrow$ $I = 2I_1 + \int \frac{dt}{t^2}$ [where $(2a + \sin 2x) = t, (2 \cos 2x) \, dx = dt$]

$\Rightarrow$ $I = 2I_1 - \frac{1}{(2a + \sin 2x)} + C$ where $I_1 = \int \frac{dx}{(2a + \sin 2x)^2}$ ...(i)

Put $A = \frac{\cos 2x}{2a + \sin 2x}$

$\Rightarrow$ $\frac{dA}{dx} = \frac{(2a + \sin 2x)(-2 \sin 2x) - \cos 2x \, (2 \cos 2x)}{(2a + \sin 2x)^2}$

$\Rightarrow$ $\frac{dA}{dx} = \frac{-4a \sin 2x - 2}{(2a + \sin 2x)^2}$

$\Rightarrow$ $\frac{dA}{dx} = \frac{-4a (\sin 2x + 2a) - 2 + 8a^2}{(2a + \sin 2x)^2}$

$\Rightarrow$ $\frac{dA}{dx} = -\frac{4a}{(2a + \sin 2x)} + \frac{(8a^2 - 2)}{(2a + \sin 2x)^2}$

Integrating both the sides w.r.t. 'x', we get

$\Rightarrow$ $A = -4a \int \frac{dx}{(2a + \sin 2x)} + (8a^2 - 2) I_1$

$\Rightarrow$ $(8a^2 - 2) I_1 = A + 4a \int \frac{\sec^2 x \, dx}{2a + 2 \tan x + 2a \tan^2 x}$

$$= A + \frac{4a}{2a} \int \frac{dt}{t^2 + \frac{t}{a} + 1} = A + 2 \int \frac{dt}{\left(t + \frac{1}{2a}\right)^2 + \left(1 - \frac{1}{4a^2}\right)}$$

$$= A + 2 \frac{(2a)}{\sqrt{4a^2 - 1}} \tan^{-1} \left(\frac{(2at + 1)}{\sqrt{4a^2 - 1}} \right)$$

$\Rightarrow$ $(8a^2 - 2) I_1 = \frac{\cos 2x}{2a + \sin 2x} + \frac{4a}{\sqrt{4a^2 - 1}} \tan^{-1} \left(\frac{(2a \tan x + 1)}{\sqrt{4a^2 - 1}} \right)$ ...(ii)

From Eqs. (i) and (ii)

$$I = \frac{1}{(4a^2-1)} \cdot \frac{\cos 2x}{(2a+\sin 2x)} + \frac{4a}{(4a^2-1)^{3/2}} \tan^{-1}\left(\frac{2a \tan x + 1}{\sqrt{4a^2-1}}\right) - \frac{1}{(2a+\sin 2x)} + C$$

Some Integrals which Cannot be Found

Any function continuous on an interval (a, b) has an anti-derivative in that interval. In other words, there exists a function $F(x)$ such that $F'(x) = f(x)$.

However, not every anti-derivative $F(x)$, even when it exists, is expressible in closed form in terms of elementary functions such as polynomials, trigonometric, logarithmic, exponential functions etc. Then, we say that such anti-derivatives or integrals "cannot be found".

Some typical examples are

(i) $\int \frac{\sin x}{x}\, dx$ **(ii)** $\int \frac{\cos x}{x}\, dx$ **(iii)** $\int \frac{1}{\log x}\, dx$

(iv) $\int \sqrt{1-k^2 \sin^2 x}\, dx$ **(v)** $\int \sqrt{\sin x}\, dx$ **(vi)** $\int \sin(x^2)\, dx$

(vii) $\int \cos(x^2)\, dx$ **(viii)** $\int x \tan x\, dx$ **(ix)** $\int e^{-x^2}\, dx$

(x) $\int e^{x^2}\, dx$ **(xi)** $\int \frac{x^2}{1+x^5}\, dx$ **(xii)** $\int \sqrt[3]{1+x^2}\, dx$

(xiii) $\int \sqrt{1+x^3}\, dx$ etc.

Review of Methods of Integration

1. $\int f\left(x+\frac{1}{x}\right)\cdot\left(1-\frac{1}{x^2}\right) dx$, put $x+\frac{1}{x} = t \Rightarrow \left(1-\frac{1}{x^2}\right) dx = dt$

2. $\int f\left(x-\frac{1}{x}\right)\cdot\left(1+\frac{1}{x^2}\right) dx$, put $x-\frac{1}{x} = t \Rightarrow \left(1+\frac{1}{x^2}\right) dx = dt$

3. $\int f\left(x^2+\frac{1}{x^2}\right)\cdot\left(x-\frac{1}{x^3}\right) dx$, put $x^2+\frac{1}{x^2} = t \Rightarrow \left(x-\frac{1}{x^3}\right) dx = \frac{dt}{2}$

4. $\int f\left(x^2-\frac{1}{x^2}\right)\cdot\left(x+\frac{1}{x^3}\right) dx$, put $x^2-\frac{1}{x^2} = t \Rightarrow \left(x+\frac{1}{x^3}\right) dx = \frac{dt}{2}$

5. $\int f(\phi(x)) \cdot \phi'(x)\, dx$, put $\phi(x) = t \Rightarrow \phi'(x)\, dx = dt$

6. $\int e^{\alpha x} f(x)\, dx$, where $f(x)$ is a polynomial of degree n, use integration by parts

ie, $\int e^{\alpha x} f(x)\, dx = e^{\alpha x}\left\{\frac{f(x)}{\alpha} - \frac{f'(x)}{\alpha^2} + \frac{f''(x)}{\alpha^3} - \ldots + (-1)^n \frac{f^n(x)}{\alpha^{n+1}}\right\} + C$

Points to Consider

Above definition is the direct result obtained by integration by parts.

7. $\int e^x \{f(x) + f'(x)\}\, dx = e^x f(x) + c$, obtained using integration by parts to first $\int e^x f(x)\, dx$, keeping $\int e^x f'(x)\, dx$ as it is.

8. $\int (\sin^m x \cdot \cos^n x)\, dx$

Case I When m, n belongs to natural number.

(a) If one of them is odd, then substitute for term of even power.
(b) If both are odd, substitute either of them.
(c) If m, n are both even, use trigonometric identities only.

Case II If m and n are rational numbers and $\left(\frac{m+n-2}{2}\right)$ is a negative integer, then substitute $\cot x = p$ or $\tan x = p$ whichever is found suitable.

9. $\int f(\sin x, \cos x)\, dx$, put $\tan \frac{x}{2} = t$

10. $\int \sin ax \sin bx\, dx$, $\int \sin ax \cos bx\, dx$, $\int \cos ax \cos bx\, dx$

Put
$$2 \sin ax \sin bx = \cos (a-b) x - \cos (a+b) x$$
$$2 \sin ax \cos bx = \sin (a+b) x + \sin (a-b) x$$
$$2 \cos ax \cos bx = \cos (a+b) x + \cos (a-b) x$$

11. $\int \frac{px+q}{\sqrt{ax^2+bx+c}}\, dx$, $\int (px+q)\sqrt{ax^2+bx+c}\, dx, \ldots$

Put $px + q = A\left\{\frac{d}{dx}(ax^2+bx+c)\right\} + B$

12. $\int \frac{dx}{a\cos^2 x + b\sin^2 x}, \int \frac{dx}{a + b\sin^2 x}, \int \frac{dx}{a+b\cos^2 x}, \ldots$

Divide numerator and denominator by $\cos^2 x$ and put, $\tan x = t$

13. $\int f(e^{ax})\, dx$, put $e^{ax} = t$

14. $\int f(x, x^{m_1/n_1}, x^{m_2/n_2}, \ldots, x^{m_a/n_a})\, dx$, put $x = t^k$ where k is a LCM of denominator of the fractions, $\frac{m_1}{n_1}, \frac{m_2}{n_2}, \ldots, \frac{m_a}{n_a}$

ie, $k = \text{LCM}\{n_1, n_2, \ldots, n_a\}$

15. $\int \frac{dx}{(x-k)^r \sqrt{ax^2+bx+c}}$, put $x - k = \frac{1}{t}$ where $r \geq 2$ and $r \in I$.

16. $\int \frac{\phi(x)}{P\sqrt{Q}}\, dx$, put $Q = t^2$ when P, Q are both linear.

Put $P = \frac{1}{t}$ when P is linear and Q is quadratic.

Put $x = \frac{1}{t}$ when P and Q are quadratic.

17. Algebraic Twins $\int \frac{2x^2}{x^4+1}\,dx = \int \frac{x^2+1}{x^4+1}\,dx + \int \frac{x^2-1}{x^4+1}\,dx,$

$$\int \sqrt{\tan x}\,dx, \int \sqrt{\cot x}\,dx, ...$$

18. $\int \frac{\phi_n(x)}{\sqrt{ax^2+bx+c}}\,dx$, where $\phi_n(x)$ is a polynomial of degree n.

Write,

$$\int \frac{\phi_n(x)}{\sqrt{ax^2+bx+c}}\,dx = Q_{n-1}(x)\sqrt{ax^2+bx+c} + k\int \frac{dx}{\sqrt{ax^2+bx+c}}$$

where $Q_{n-1}(x)$ is a polynomial of degree $(n-1)$ differentiating both the parts of this equality and multiplying by $\sqrt{ax^2+bx+c}$, we get

$$\phi_n(x) = Q_{n-1}(x)\sqrt{ax^2+bx+c} + \frac{1}{2}Q_{n-1}(x)(2ax+b) + k.$$

Now, compare the coefficients to get $Q_{n-1}(x)$ and k.

19. $I_n = \int \frac{dx}{(x^2+1)^n}$, reduction formula is applied

$$I_n = \frac{x}{(2n-2)(x^2+1)^{n-1}} + \frac{2n-3}{2n-2}I_{n-1}$$

20. $\int x^m (a+bx^n)^P\,dx$

Case I If $P \in N$ we expand using binomial and integrate.

Case II If $P \in I^-$ (*ie*, negative integer), write $x = t^k$, where k is the LCM of m and n.

Case III If $\frac{m+1}{n} \in$ Integer and $P \in$ fraction put $a + bx^n = t^k$, where k is denominator of fraction P.

Case IV If $\left(\frac{m+1}{n} + P\right)$ is an integer and $P \in$ fraction we put $(a+bx^n) = t^k x^n$, where k is denominator of the fraction P.

21. $\int f(x), \sqrt{ax^2+bx+c}\,dx$, then put

(i) $\sqrt{ax^2+bx+c} = t \pm x\sqrt{a}$, if $a > 0$

(ii) $\sqrt{ax^2+bx+c} = tx \pm \sqrt{c}$, if $c > 0$

(iii) $\sqrt{ax^2+bx+c} = (x-\alpha)\,t$, if $ax^2+bx+c = a(x-\alpha)(x-\beta)$
ie, when α is real root of ax^2+bx+c.

Worked Examples

Type 1 : Subjective Type Questions

Example 1 Evaluate $\int \frac{5x^4 + 4x^5}{(x^5 + x + 1)^2} dx.$

Solution. Here, $I = \int \frac{5x^4 + 4x^5}{(x^5 + x + 1)^2} dx = \int \frac{x^4 (5 + 4x)}{x^{10}\left(1 + \frac{1}{x^4} + \frac{1}{x^5}\right)^2} dx$

$$= \int \frac{5/x^6 + 4/x^5}{\left(1 + \frac{1}{x^4} + \frac{1}{x^5}\right)^2} dx$$

Put $1 + \frac{1}{x^4} + \frac{1}{x^5} = t \Rightarrow \left(-\frac{4}{x^5} - \frac{5}{x^6}\right) dx = dt$

$$= \int -\frac{dt}{t^2} = \frac{1}{t} + C = \frac{1}{1 + \frac{1}{x^4} + \frac{1}{x^5}} + C = \frac{x^5}{x^5 + x + 1} + C$$

Example 2 Evaluate $\int \sin^{-1} \left\{\frac{2x + 2}{\sqrt{4x^2 + 8x + 13}}\right\} dx.$ **[IIT JEE 2001]**

Solution. Here, $I = \int \sin^{-1} \left(\frac{2x + 2}{\sqrt{4x^2 + 8x + 13}}\right) dx$

$$= \int \sin^{-1} \left(\frac{2x + 2}{\sqrt{(2x + 2)^2 + 3^2}}\right) dx$$

Put $2x + 2 = 3 \tan \theta \Rightarrow 2\, dx = 3 \sec^2 \theta\, d\theta$

$$= \int \sin^{-1} \left(\frac{3 \tan \theta}{3 \sec \theta}\right) \frac{3}{2} \sec^2 \theta\, d\theta$$

$$= \frac{3}{2} \int \underset{\text{I}}{\theta}\, \underset{\text{II}}{\sec^2 \theta}\, d\theta$$

$$= \frac{3}{2} \{\theta \tan \theta - \int \tan \theta\, d\theta\}$$

$$= \frac{3}{2} \{\theta \tan \theta - \log |\sec \theta|\} + C$$

$$\Rightarrow \quad I = \frac{3}{2} \left\{\frac{2x + 2}{3} \tan^{-1} \left(\frac{2x + 2}{3}\right) - \log \left(\sqrt{1 + \left(\frac{2x + 2}{3}\right)^2}\right)\right\} + C$$

$$= \frac{3}{2} \left\{\frac{2}{3} (x + 1) \tan^{-1} \left(\frac{2}{3} (x + 1)\right) - \log \sqrt{4x^2 + 8x + 13}\right\} + C$$

$$\Rightarrow \quad I = (x + 1) \tan^{-1} \left(\frac{2}{3} (x + 1)\right) - \frac{3}{4} \log (4x^2 + 8x + 13) + C$$

Example 3 For any natural number m, evaluate

$$\int (x^{3m} + x^{2m} + x^m)(2x^{2m} + 3x^m + 6)^{1/m}\, dx,\ x > 0$$ **[IIT JEE 2002]**

Solution. Here,

$$I = \int (x^{3m} + x^{2m} + x^m)(2x^{2m} + 3x^m + 6)^{1/m}\, dx$$

$$I = \int (x^{3m} + x^{2m} + x^m)\frac{(2x^{3m} + 3x^{2m} + 6x^m)^{1/m}}{x}\, dx$$

$$I = \int (x^{3m-1} + x^{2m-1} + x^{m-1})(2x^{3m} + 3x^{2m} + 6x^m)^{1/m}\, dx \quad \text{...(i)}$$

Put $\quad 2x^{3m} + 3x^{2m} + 6x^m = t$

$\Rightarrow \quad 6m\,(x^{3m-1} + x^{2m-1} + x^{m-1})\, dx = dt$

$\therefore$ Eq. (i) becomes,

$$I = \int t^{1/m}\frac{dt}{6m} = \frac{1}{6m}\cdot\frac{t^{(1/m)+1}}{(1/m)+1} + C$$

$$I = \frac{1}{6(m+1)}\{2x^{3m} + 3x^{2m} + 6x^m\}^{\frac{m+1}{m}} + C$$

Example 4 Evaluate $\int \frac{dx}{\sqrt{(x-a)(b-x)}}$.

Solution. Let $x = a\cos^2\theta + b\sin^2\theta$ in the given integral.

So that, $\quad dx = a\,(2\cos\theta)(-\sin\theta) + b\,(2\sin\theta)(\cos\theta)\, d\theta$

$$dx = 2(b-a)\sin\theta\cos\theta\, d\theta$$

$$\therefore \quad I = \int \frac{2(b-a)\sin\theta\cos\theta\, d\theta}{\sqrt{(a\cos^2\theta + b\sin^2\theta - a)\,(b - a\cos^2\theta - b\sin^2\theta)}}$$

$$= 2(b-a)\int \frac{\sin\theta\cos\theta\, d\theta}{\sqrt{(b\sin^2\theta - a\sin^2\theta)(b\cos^2\theta - a\cos^2\theta)}}$$

$$= 2(b-a)\int \frac{\sin\theta\cos\theta\, d\theta}{(b-a)\sin\theta\cos\theta} = 2\int 1\, d\theta$$

$$= 2\theta + C = 2\sin^{-1}\sqrt{\frac{x-a}{(b-a)}} + C$$

Example 5 Prove that $\int (\cos x)^{2n+1}\, dx$

$$= \left\{\sin x - {}^nC_1\cdot\frac{\sin^3 x}{3} + {}^nC_2\cdot\frac{\sin^5 x}{5} + \ldots + (-1)^n\cdot\frac{\sin^{2n+1} x}{2n+1}\right\} + C$$

Solution. Here, $I = \int (\cos x)^{2n+1}\, dx$ could be written as

$$= \int (\cos x)^{2n}\cdot\cos x\, dx = \int (\cos^2 x)^n\cdot\cos x\, dx$$

$$= \int (1 - \sin^2 x)^n\cdot\cos x\, dx \qquad \left\{\begin{array}{l}\text{put } \sin x = t\\ \Rightarrow \cos x\, dx = dt\end{array}\right.$$

$$= \int (1 - t^2)^n\, dt$$

Using binomial expansion for $(1 - t^2)^n$, we get

$$= \int \{1 - {}^nC_1 t^2 + {}^nC_2 t^4 + \ldots + (-1)^n\, t^{2n}\}\, dt$$

$$= t - {}^nC_1 \frac{t^3}{3} + {}^nC_2 \frac{t^5}{5} + \ldots + (-1)^n \frac{t^{2n+1}}{(2n+1)} + C$$

$$\Rightarrow \int (\cos x)^{2n+1} dx = \sin x - {}^nC_1 \frac{\sin^3 x}{3} + {}^nC_2 \frac{\sin^5 x}{5} + \ldots + \frac{(-1)^n \sin^{2n+1} x}{(2n+1)} + C$$

Example 6 Evaluate $\int \log(1 + 2m\cos x + m^2)\, dx$, when $m > 0$.

Solution. As we know, $1 + 2m\cos x + m^2 = (1 + me^{ix})(1 + me^{-ix})$,

where $e^{ix} = \cos x + i\sin x$ and $e^{-ix} = \cos x - i\sin x$

$$\Rightarrow \log(1 + 2m\cos x + m^2) = \begin{cases} \log(1 + me^{ix})(1 + me^{-ix}); & 0 < m < 1 \\ \log m^2 \left(1 + \dfrac{e^{ix}}{m}\right)\left(1 + \dfrac{e^{-ix}}{m}\right); & m > 1 \end{cases}$$

Case I When $0 < m < 1$,

$$\log(1 + 2m\cos x + m^2) = \log(1 + me^{ix}) + \log(1 + me^{-ix})$$

$$= m(e^{ix} + e^{-ix}) - \frac{m^2}{2}(e^{2ix} + e^{-2ix}) + \ldots$$

$$= 2\left\{m\cos x - \frac{m^2}{2}\cos 2x + \frac{m^3}{3}\cos 3x - \ldots\right\}$$

$$\Rightarrow \int \log(1 + 2m\cos x + m^2)\, dx = 2\left\{m\sin x - \frac{m^2}{2^2}\sin 2x + \frac{m^3}{3^2}\sin 3x + \ldots\right\}$$

Case II When $m > 1$

$$\log(1 + 2m\cos x + m^2) = 2\log m + \log\left(1 + \frac{e^{ix}}{m}\right) + \log\left(1 + \frac{e^{-ix}}{m}\right)$$

$$= 2\log m + 2\left(\frac{\cos x}{m} - \frac{\cos 2x}{2m^2} + \frac{\cos 3x}{3m^3} - \ldots\right)$$

$$\Rightarrow \int \log(1 + 2m\cos x + m^2)\, dx = 2x\log m + 2\left(\frac{\sin x}{m} - \frac{\sin 2x}{2^2 m^2} + \frac{\sin 3x}{3^2 m^3} - \ldots\right)$$

Example 7 Evaluate $\int \dfrac{dx}{\cos x + \operatorname{cosec} x}$.

Solution. Let $I = \int \dfrac{dx}{\cos x + \dfrac{1}{\sin x}} = \int \dfrac{\sin x}{\cos x \cdot \sin x + 1}\, dx$

$$= \int \frac{2\sin x}{2 + 2\sin x\cos x}\, dx = \int \frac{2\sin x}{2 + \sin 2x}\, dx$$

$$= \int \frac{\{(\sin x + \cos x) + (\sin x - \cos x)\}}{2 + \sin 2x}\, dx$$

$$= \int \frac{\sin x + \cos x}{2 + \sin 2x}\, dx + \int \frac{\sin x - \cos x}{2 + \sin 2x}\, dx$$

$$= \int \frac{\sin x + \cos x}{3 - (1 - \sin 2x)}\, dx + \int \frac{\sin x - \cos x}{1 + (1 + \sin 2x)}\, dx$$

$$= \int \frac{\sin x + \cos x}{3 - (\sin x - \cos x)^2}\, dx + \int \frac{\sin x - \cos x}{1 + (\sin x + \cos x)^2}\, dx$$

Put $\sin x - \cos x = s$ and $\sin x + \cos x = t$

$\Rightarrow$ $(\cos x + \sin x)\, dx = ds$ and $(\cos x - \sin x)\, dx = dt$

$\therefore$
$$I = \int \frac{ds}{3 - s^2} - \int \frac{dt}{1 + t^2} = \int \frac{ds}{(\sqrt{3})^2 - (s)^2} - \int \frac{dt}{1 + t^2}$$

$$= \frac{1}{2\sqrt{3}} \log \left| \frac{\sqrt{3} + s}{\sqrt{3} - s} \right| - \tan^{-1} t + C$$

$$= \frac{1}{2\sqrt{3}} \log \left| \frac{\sqrt{3} + \sin x - \cos x}{\sqrt{3} - \sin x + \cos x} \right| - \tan^{-1} (\sin x + \cos x) + C$$

Example 8 Evaluate $\int \frac{x^2 (x \sec^2 x + \tan x)\, dx}{(x \tan x + 1)^2}$.

Solution. Here, $I = \int \underset{\text{I}}{x^2} \cdot \underset{\text{II}}{\frac{x \sec^2 x + \tan x}{(x \tan x + 1)^2}}\, dx$

$$= x^2 \left(- \frac{1}{(x \tan x + 1)} \right) - \int 2x \cdot \left(- \frac{1}{(x \tan x + 1)} \right) dx \qquad \ldots\text{(i)}$$

$$\left[\text{using} \int \frac{x \sec^2 x + \tan x}{(x \tan x + 1)^2}\, dx = \int \frac{dt}{t^2} = - \frac{1}{t} = - \frac{1}{(x \tan x + 1)} \right]$$

$\Rightarrow$
$$I = - \left(\frac{x^2}{x \tan x + 1} \right) + \int \frac{2x (\cos x)}{x \sin x + \cos x}\, dx$$

[put, $x \sin x + \cos x = u \Rightarrow (x \cos x + \sin x - \cos x)\, dx = du$]

$\Rightarrow$
$$I = - \frac{x^2}{(x \tan x + 1)} + 2 \int \frac{du}{u} = - \frac{x^2}{x \tan x + 1} + 2 \log |u| + C$$

$$= - \frac{x^2}{x \tan x + 1} + 2 \log |x \sin x + \cos x| + C$$

Example 9 Evaluate $\int \frac{x^2}{(x \sin x + \cos x)^2}\, dx.$

Solution. Let $I = \int \frac{x^2}{(x \sin x + \cos x)^2} \cdot dx$

We know,
$$\frac{d}{dx} \left(\frac{1}{x \sin x + \cos x} \right) = \frac{- \{\sin x + x \cos x - \sin x\}}{(x \sin x + \cos x)^2} = \frac{- x \cos x}{(x \sin x + \cos x)^2}$$

$\therefore$
$$\int \frac{- x \cdot \cos x}{(x \sin x + \cos x)^2}\, dx = \frac{1}{(x \sin x + \cos x)} \qquad \ldots\text{(i)}$$

$\Rightarrow$
$$I = \int \underset{\text{II}}{\frac{- x \cdot \cos x}{(x \sin x + \cos x)^2}} \cdot \underset{\text{I}}{\frac{- x}{\cos x}}\, dx$$

$$= \left(\frac{- x}{\cos x} \right) \cdot \frac{1}{(x \sin x + \cos x)} - \int \frac{- \cos x - x \sin x}{\cos^2 x} \cdot \frac{1}{x \sin x + \cos x}\, dx$$

$$= \frac{-x}{\cos x\,(x \sin x + \cos x)} + \int \sec^2 x\, dx$$

$$I = \frac{-x}{\cos x\,(x \sin x + \cos x)} + \tan x + C$$

Example 10 Evaluate $\int \frac{(x-1)}{(x+1)\sqrt{x^3+x^2+x}}\,dx.$

Solution. Let $I = \int \frac{(x-1)}{(x+1)\sqrt{x^3+x^2+x}}\,dx = \int \frac{(x^2-1)}{(x+1)^2\sqrt{x^3+x^2+x}}\,dx$

$$= \int \frac{x^2(1-1/x^2)}{(x^2+2x+1)\sqrt{x^3+x^2+x}}\,dx$$

$$= \int \frac{x^2(1-1/x^2)}{x(x+2+1/x)\cdot x\sqrt{x+1+1/x}}\,dx$$

put $x + \frac{1}{x} = t$, $(1-1/x^2)\,dx = dt$

$$= \int \frac{dt}{(t+2)\sqrt{t+1}},$$ which reduces to $\int \frac{dx}{P\sqrt{Q}}$.

Let $t + 1 = z^2 \therefore\ dt = 2z dz$

$$= \int \frac{2z\,dz}{(z^2+1)\sqrt{z^2}} = 2\int \frac{dz}{z^2+1} = 2\tan^{-1}(z) + C$$

$$= 2\tan^{-1}(\sqrt{t+1}) + C = 2\tan^{-1}\sqrt{\frac{x^2+x+1}{x}} + C$$

Example 11 Evaluate $\int \frac{(1+x^2)\,dx}{(1-x^2)\sqrt{1+x^2+x^4}}$.

Solution. Let $I = \int \frac{(1+x^2)\,dx}{(1-x^2)\sqrt{1+x^2+x^4}} = \int \frac{x^2\left(1+\frac{1}{x^2}\right)dx}{x\left(\frac{1}{x}-x\right)x\sqrt{\frac{1}{x^2}+1+x^2}}$

$$= -\int \frac{(1+1/x^2)\,dx}{(1-1/x)\sqrt{(x-1/x)^2+3}}$$

Put $x - \frac{1}{x} = t \Rightarrow -\int \frac{dt}{t\sqrt{t^2+3}}$ $\left(1+\frac{1}{x^2}\right)dx = dt$

Again, put $t^2 + 3 = s^2 \Rightarrow 2t\,dt = 2s\,ds$

$$= -\int \frac{s\,ds}{s(s^2-3)} = -\int \frac{ds}{s^2-(\sqrt{3})^2} = -\frac{1}{2\sqrt{3}}\log\left|\frac{s-\sqrt{3}}{s+\sqrt{3}}\right| + C$$

$$= -\frac{1}{2\sqrt{3}}\log\left|\frac{\sqrt{t^2+3}-\sqrt{3}}{\sqrt{t^2+3}+\sqrt{3}}\right| + C = -\frac{1}{2\sqrt{3}}\log\left|\frac{\sqrt{(x-1/x)^2+3}-\sqrt{3}}{\sqrt{(x-1/x)^2+3}+\sqrt{3}}\right| + C$$

$$= -\frac{1}{2\sqrt{3}}\log\left|\frac{\sqrt{x^2+\frac{1}{x^2}+1}-\sqrt{3}}{\sqrt{x^2+\frac{1}{x^2}+1}+\sqrt{3}}\right| + C$$

Example 12 Evaluate $I = \int \frac{dx}{(a + bx^2)\sqrt{b - ax^2}}$.

Solution. Substituting $ax^2 = b\sin^2\theta$

$\Rightarrow$ $$dx = \sqrt{\frac{b}{a}}\cos\theta\, d\theta$$

$\therefore$ $$I = \int \frac{\sqrt{\frac{b}{a}}\cos\theta\, d\theta}{\left(a + \frac{b^2}{a}\sin^2\theta\right)\sqrt{b - b\sin^2\theta}}$$

$$= \sqrt{a}\int \frac{\cos\theta\, d\theta}{(a^2 + b^2\sin^2\theta)\cdot\cos\theta}$$

$$= \sqrt{a}\int \frac{d\theta}{a^2 + b^2\sin^2\theta},$$ dividing N^r and D^r by $\cos^2\theta$, we get

$$= \sqrt{a}\int \frac{\sec^2\theta\, d\theta}{a^2\sec^2\theta + b^2\tan^2\theta},$$ put $\tan\theta = t$

$$= \sqrt{a}\int \frac{dt}{a^2(1 + t^2) + b^2t^2} = \sqrt{a}\int \frac{dt}{(a^2 + b^2)t^2 + a^2}$$

$$= \frac{\sqrt{a}}{a^2 + b^2}\int \frac{dt}{t^2 + \frac{a^2}{a^2 + b^2}}$$

$$= \frac{\sqrt{a}}{a^2 + b^2}\cdot\left(\frac{\sqrt{a^2 + b^2}}{a}\right)\tan^{-1}\left(\frac{t\sqrt{a^2 + b^2}}{a}\right) + C$$

$$= \frac{1}{\sqrt{a(a^2 + b^2)}}\cdot\tan^{-1}\left(\frac{x\sqrt{a^2 + b^2}}{a\sqrt{b - ax^2}}\right) + C$$

$$\left(\text{since } t = \tan\theta = \frac{x}{\sqrt{b - ax^2}}\right)$$

Example 13 Evaluate $I = \int \frac{dx}{2x\sqrt{1 - x}\sqrt{(2 - x) + \sqrt{1 - x}}}$.

Solution. Here, $I = \int \frac{dx}{2x\sqrt{1 - x}\sqrt{(2 - x) + \sqrt{1 - x}}}$, put $(1 - x) = t^2$

$$= -\int \frac{2t\, dt}{2(1 - t^2)\cdot t\sqrt{1 + t^2 + t}} = -\int \frac{dt}{(1 - t^2)\sqrt{t^2 + t + 1}}$$

$$= \int \frac{dt}{(t - 1)(t + 1)\sqrt{t^2 + t + 1}} = \frac{1}{2}\int\left(\frac{1}{t - 1} - \frac{1}{t + 1}\right)\cdot\frac{dt}{\sqrt{t^2 + t + 1}}$$

$$\because \frac{1}{(t - 1)(t + 1)} = \frac{1}{2}\left(\frac{1}{t - 1} - \frac{1}{t + 1}\right)$$

$$= \frac{1}{2}\int \frac{1}{(t - 1)\sqrt{t^2 + t + 1}}\, dt - \frac{1}{2}\int \frac{1}{(t + 1)\sqrt{t^2 + t + 1}}\, dt$$

Let $$I=\frac{1}{2}I_1-\frac{1}{2}I_2 \qquad \text{...(i)}$$

where, $$I_1=\int\frac{dt}{(t-1)\sqrt{t^2+t+1}}$$

and $$I_2=\int\frac{dt}{(t+1)\sqrt{t^2+t+1}}$$

For I_1, put $(t-1)=\dfrac{1}{z}$

$$I_1=\int\frac{-1/z^2\,dz}{\frac{1}{z}\sqrt{\left(1+\frac{1}{z}\right)^2+\left(1+\frac{1}{z}\right)+1}}=-\int\frac{dz}{\sqrt{\left(z+\frac{3}{2}\right)^2+\left(\frac{\sqrt{3}}{2}\right)^2}}$$

$$=-\log\left|\left(z+\frac{3}{2}\right)+\sqrt{z^2+3z+3}\right| \qquad \text{...(ii)}$$

For I_2, put $(t+1)=\dfrac{1}{S}$

$$I_2=-\int\frac{dS}{\sqrt{\left(S-\frac{1}{2}\right)^2+\frac{3}{4}}}=-\log\left|\left(S-\frac{1}{2}\right)+\sqrt{S^2-S+1}\right| \qquad \text{...(iii)}$$

$$\therefore \qquad I=-\frac{1}{2}\left\{\log\left(z+\frac{3}{2}+\sqrt{z^2+3z+3}\right)\right\}+\frac{1}{2}\log\left|\left(S-\frac{1}{2}\right)+\sqrt{S^2-S+1}\right|+C$$

where, $z=\dfrac{1}{\sqrt{1-x}-1}$ and $S=\dfrac{1}{\sqrt{1-x}+1}$

Example 14 Evaluate $\displaystyle\int\frac{dx}{(x^2+a^2)^2}$.

Solution. Here, we know

$$\int\frac{dx}{x^2+a^2}=\frac{1}{a}\tan^{-1}\frac{x}{a} \qquad \text{...(i)}$$

Also, $$\int\underset{\text{I}}{\frac{1}{x^2+a^2}}\cdot\underset{\text{II}}{1}\,dx=\frac{1}{x^2+a^2}\,x-\int\frac{-2x}{(x^2+a^2)^2}\,x\,dx$$

$$=\frac{x}{x^2+a^2}+2\int\frac{x^2+a^2-a^2}{(x^2+a^2)^2}\,dx$$

$$\int\frac{dx}{x^2+a^2}=\frac{x}{x^2+a^2}+2\int\frac{dx}{x^2+a^2}-2a^2\int\frac{dx}{(x^2+a^2)^2} \qquad \text{...(ii)}$$

From Eqs. (i) and (ii), we get

$$\frac{1}{a}\tan^{-1}\frac{x}{a}=\frac{x}{x^2+a^2}+2\frac{1}{a}\tan^{-1}\frac{x}{a}-2a^2\int\frac{dx}{(x^2+a^2)^2}$$

$$\Rightarrow \qquad 2a^2\int\frac{dx}{(x^2+a^2)^2}=\frac{x}{x^2+a^2}+\frac{1}{a}\tan^{-1}\frac{x}{a}$$

or $$\int\frac{dx}{(x^2+a^2)^2}=\frac{1}{2a^2}\left\{\frac{x}{x^2+a^2}+\frac{1}{a}\tan^{-1}\frac{x}{a}\right\}+C$$

Example 15 Evaluate $\int \frac{dx}{(x^2+a^2)^3}$.

Solution. Let $$I = \int \frac{dx}{(x^2+a^2)^3} \quad \text{...(i)}$$

and $$I_1 = \int \frac{1}{(x^2+a^2)^2}\, dx \quad \text{...(ii)}$$

$$= \int \underset{\text{I}}{\frac{1}{(x^2+a^2)^2}} \cdot \underset{\text{II}}{1}\, dx$$

$$= \frac{1}{(x^2+a^2)^2} \cdot x - \int \frac{-2(2x)}{(x^2+a^2)^3}\, x\, dx$$

$$= \frac{x}{(x^2+a^2)^2} + 4\int \frac{x^2+a^2-a^2}{(x^2+a^2)^3}\, dx$$

$$= \frac{x}{(x^2+a^2)^2} + 4\int \frac{1}{(x^2+a^2)^2}\, dx - 4a^2 \int \frac{dx}{(x^2+a^2)^3}$$

$$\Rightarrow \quad I_1 = \frac{x}{(x^2+a^2)^2} + 4I_1 - 4a^2 \cdot I \quad \text{[using Eqs. (i) and (ii)]}$$

$$\Rightarrow \quad 4a^2 I = \frac{x}{(x^2+a^2)^2} + 3I_1$$

$$\Rightarrow \quad I = \frac{x}{4a^2(x^2+a^2)^2} + \frac{3}{4a^2} I_1 \quad \text{...(iii)}$$

$$\left[\text{using previous example, } I_1 = \int \frac{dx}{(x^2+a^2)^2} = \frac{x}{2a^2(x^2+a^2)} + \frac{1}{2a^3}\tan^{-1}\left(\frac{x}{a}\right) + C\right]$$

$$\Rightarrow \quad I = \frac{x}{4a^2(x^2+a^2)^2} + \frac{3}{4a^2}\left\{\frac{x}{2a^2(x^2+a^2)} + \frac{1}{2a^3}\tan^{-1}\left(\frac{x}{a}\right)\right\} + C$$

Example 16 If $y(x-y)^2 = x$, then show that $\int \frac{dx}{(x-3y)} = \frac{1}{2}\ln[(x-y)^2-1]$.

Solution. Let $P = \int \frac{dx}{(x-3y)} = \frac{1}{2}\ln[(x-y)^2-1]$

$$\therefore \quad \frac{dP}{dx} = \frac{1}{x-3y} = \frac{(x-y)\left\{1-\frac{dy}{dx}\right\}}{\{(x-y)^2-1\}} \quad \text{...(i)}$$

Given, $y(x-y)^2 = x$, differentiating both the sides, we get

$$\therefore \quad \frac{dy}{dx} = \frac{1-2y(x-y)}{(x-y)(x-3y)} \quad \text{...(ii)}$$

$$\therefore \quad \frac{dP}{dx} = \frac{(x-y)\left\{1-\frac{1-2y(x-y)}{(x-y)(x-3y)}\right\}}{\{(x-y)^2-1\}}$$

$$= \frac{(x-y)(x-3y)-1+2y(x-y)}{(x-3y)\{(x-y)^2-1\}} = \frac{(x-y)^2-1}{(x-3y)\{(x-y)^2-1\}}$$

$$\therefore \qquad \frac{dP}{dx} = \frac{1}{(x-3y)} \qquad \text{...(iii)}$$

Which is true as given

$$\therefore \qquad \int \frac{dx}{(x-3y)} = \frac{1}{2} \log \{(x-y)^2 - 1\}$$

Example 17 Evaluate $\int (x + \sqrt{1+x^2})^n \, dx$.

Solution. Let $I = \int (x + \sqrt{1+x^2})^n \, dx$

Put $\quad x + \sqrt{1+x^2} = t \qquad \text{...(i)}$

$$\Rightarrow \qquad \left(1 + \frac{1}{2\sqrt{1+x^2}} \cdot 2x\right) dx = dt$$

$$\Rightarrow \qquad \left(\frac{\sqrt{1+x^2} + x}{\sqrt{1+x^2}}\right) dx = dt \qquad \text{...(ii)}$$

We know,

$$t = x + \sqrt{1+x^2} = x + \sqrt{1+x^2} \times \frac{x - \sqrt{1+x^2}}{x - \sqrt{1+x^2}}$$

$$t = \frac{-1}{x - \sqrt{1+x^2}} \qquad \therefore \quad t = x + \sqrt{1+x^2}$$

$$-\frac{1}{t} = x - \sqrt{1+x^2}$$

Subtracting, we get

$$2\sqrt{1+x^2} = t + \frac{1}{t} \quad \text{or} \quad \frac{1}{\sqrt{1+x^2}} = \frac{2t}{t^2+1} \qquad \text{...(iii)}$$

From Eqs. (i), (ii) and (iii), we get

$$dx = \frac{t^2+1}{2t^2} dt$$

$$\therefore \qquad I = \int t^n \cdot \frac{t^2+1}{(2t^2)} dt = \frac{1}{2} \int (t^n + t^{n-2}) \, dt = \frac{1}{2}\left[\frac{t^{n+1}}{n+1} + \frac{t^{n-1}}{n-1}\right] + C$$

$$\Rightarrow \qquad I = \frac{1}{2(n+1)} [x + \sqrt{(1+x^2)}]^{n+1} + \frac{1}{2(n-1)} (x + \sqrt{(1+x^2)})^{n-1} + C$$

Example 18 Evaluate $\int \frac{f(x)}{x^3 - 1} dx$, where $f(x)$ is a polynomial of degree 2 in x such that $f(0) = f(1) = 3f(2) = -3$.

Solution. Let $\quad f(x) = ax^2 + bx + c$

Given, $\quad f(0) = f(1) = 3f(2) = -3$

$$\therefore \qquad f(0) = c = -3$$

$$f(1) = a + b + c = -3$$

$$3f(2) = 3(4a + 2b + c) = -3$$

On solving, we get $\quad a = 1, b = -1, c = -3$

$\therefore \qquad f(x) = x^2 - x - 3$

$\Rightarrow \qquad I = \int \frac{f(x)}{x^3 - 1}\,dx = \int \frac{x^2 - x - 3}{(x-1)(x^2+x+1)}\,dx$

Using partial fractions, we get

$$\frac{(x^2 - x - 3)}{(x-1)(x^2+x+1)} = \frac{A}{(x-1)} + \frac{Bx + C}{(x^2+x+1)}$$

we get, $\quad A = -1, \quad B = 2, \quad C = 2$

$$\therefore \quad I = \int -\frac{1}{x-1}\,dx + \int \frac{(2x+2)}{(x^2+x+1)}\,dx$$

$$= -\log|x-1| + \int \frac{(2x+1)\,dx}{x^2+x+1} + \int \frac{1\,dx}{x^2+x+1}$$

$$= -\log|x-1| + \log|x^2+x+1| + \int \frac{dx}{(x+1/2)^2 + (\sqrt{3}/2)^2}$$

$$= -\log|x-1| + \log|x^2+x+1| + \frac{2}{\sqrt{3}}\tan^{-1}\left(\frac{2x+1}{\sqrt{3}}\right) + C$$

Example 19 If I_n denotes $\int z^n e^{1/z}\,dz$, then show that

$$(n+1)!\,I_n = I_0 + e^{1/z}(1!\,z^2 + 2!\,z^3 + \ldots + n!\,z^{n+1}).$$

Solution. $I_n = \int z^n e^{1/z}\,dz$, applying integration by parts taking $e^{1/z}$ as first function and z^n as second function. We get,

$$I_n = \frac{e^{1/z}\cdot z^{n+1}}{(n+1)} - \int e^{1/z}\left(-\frac{1}{z^2}\right)\cdot\frac{z^{n+1}}{n+1}\,dz$$

$$= \frac{e^{1/z}\cdot z^{n+1}}{(n+1)} + \frac{1}{(n+1)}\int e^{1/z}\cdot z^{n-1}\,dz$$

or $\quad I_n = \frac{e^{1/z}\cdot z^{n+1}}{(n+1)} + \frac{I_{n-1}}{(n+1)}$

$$= \frac{e^{1/z}\cdot z^{n+1}}{(n+1)} + \frac{1}{(n+1)}\left[\frac{e^{1/z}\cdot z^n}{n} + \frac{1}{n}I_{n-2}\right]$$

$$= \frac{e^{1/z}(z)^{n+1}}{(n+1)} + \frac{e^{1/z}\cdot(z)^n}{(n+1)\,n} + \frac{1}{(n+1)\,n}I_{n-2}$$

$$= \frac{e^{1/z}\cdot(z)^{n+1}}{(n+1)} + \frac{e^{1/z}\cdot(z)^n}{(n+1)\,n} + \frac{e^{1/z}\cdot(z)^{n-1}}{(n+1)\,n\cdot(n-1)} + \frac{1}{(n+1)\,n\,(n-1)}I_{n-3}$$

... ...

... ...

$$= \frac{e^{1/z}\cdot(z)^{n+1}}{n+1} + \frac{e^{1/z}\cdot(z)^n}{(n+1)\,n} + \ldots + \frac{e^{1/z}\cdot(z)^1}{(n+1)\,n\ldots 3\cdot 2} + \frac{1}{(n+1)\,n\,(n-1)\ldots 3\cdot 2}I_0$$

Multiplying both the sides by $(n+1)!$. We get,

$$(n+1)!\,I_n = (e^{1/z}\cdot z^{n+1}\cdot n! + e^{1/z}\cdot z^n\,(n-1)! + \ldots$$
$$+ \ldots + e^{1/z}\cdot z^3\cdot(2)! + e^{1/z}\cdot z^2\cdot 1!) + I_0$$

$\Rightarrow \qquad I_n\,(n+1)! = I_0 + e^{1/z}(1!\,z^2 + 2!\,z^3 + \ldots + n!\,z^{n+1})$

Example 20 If $I_n = \int x^n \sqrt{a^2 - x^2}\, dx$, prove that

$$I_n = -\frac{x^{n-1}(a^2 - x^2)^{3/2}}{(n+2)} + \frac{(n-1)}{(n+2)} a^2 I_{n-2}.$$

Solution. $I_n = \int x^n \sqrt{a^2 - x^2}\, dx = \int \underset{\text{I}}{x^{n-1}} \cdot \underset{\text{II}}{\{x\sqrt{a^2 - x^2}\}}\, dx$

Applying integration by parts, we get

$$= x^{n-1} \cdot \left\{ -\frac{(a^2 - x^2)^{3/2}}{3} \right\} + \int (n-1) x^{n-2} \cdot \left\{ -\frac{(a^2 - x^2)^{3/2}}{3} \right\} dx$$

$$= -\frac{x^{n-1}(a^2 - x^2)^{3/2}}{3} + \frac{(n-1)}{3} \int x^{n-2} \cdot (a^2 - x^2)\sqrt{a^2 - x^2}\, dx$$

$$\Rightarrow \quad I_n = -\frac{x^{n-1}(a^2 - x^2)^{3/2}}{3} + \frac{(n-1)a^2}{3} I_{n-2} - \frac{(n-1)}{3} I_n$$

$$\therefore \quad I_n + \frac{(n-1)}{3} I_n = -\frac{x^{n-1}(a^2 - x^2)^{3/2}}{3} + \frac{(n-1)a^2}{3} I_{n-2}$$

$$\left(\frac{n+2}{3}\right) I_n = -\frac{x^{n-1}(a^2 - x^2)^{3/2}}{3} + \frac{(n-1)a^2}{3} I_{n-2}$$

$$I_n = -\frac{x^{n-1}(a^2 - x^2)^{3/2}}{(n+2)} + \frac{(n-1)a^2}{(n+2)} I_{n-2}$$

Example 21 If $I_m = \int (\sin x + \cos x)^m\, dx$, then show that

$$m I_m = (\sin x + \cos x)^{m-1} \cdot (\sin x - \cos x) + 2(m-1) I_{m-2}$$

Solution. $\because I_m = \int (\sin x + \cos x)^m\, dx$

$$= \int (\sin x + \cos x)^{m-1} \cdot (\sin x + \cos x)\, dx,$$

applying integration by parts

$$= (\sin x + \cos x)^{m-1} (\cos x + \sin x) - \int (m-1)(\sin x + \cos x)^{m-2} \cdot (\cos x - \sin x) \cdot (\sin x - \cos x)\, dx$$

$$= (\sin x + \cos x)^{m-1} (\sin x - \cos x) + (m-1) \int (\sin x + \cos x)^{m-2} (\sin x - \cos x)^2\, dx$$

As we know, $(\sin x + \cos x)^2 + (\sin x - \cos x)^2 = 2$,

$$\therefore \quad I_m = (\sin x + \cos x)^{m-1} (\sin x - \cos x) + (m-1) \int (\sin x + \cos x)^{m-2} \cdot \{2 - (\sin x + \cos x)^2\}\, dx$$

$$= (\sin x + \cos x)^{m-1} (\sin x - \cos x) + (m-1) \int 2 (\sin x + \cos x)^{m-2}\, dx - (m-1) \int (\sin x + \cos x)^m\, dx$$

$$I_m = (\sin x + \cos x)^{m-1} (\sin x - \cos x) + 2(m-1) I_{m-2} - (m-1) I_m$$

or $\quad (m-1) I_m + I_m = (\sin x + \cos x)^{m-1} (\sin x - \cos x) + 2(m-1) I_{m-2}$

or $\quad m I_m = (\sin x + \cos x)^{m-1} (\sin x - \cos x) + 2(m-1) I_{m-2}$

Example 22 If $I_{m,n} = \int \cos^m x \cdot \cos nx\, dx$, show that

$$(m+n) I_{m,n} = \cos^m x \cdot \sin nx + m I_{(m-1,\, n-1)}$$

Solution. We have, $I_{m,n} = \int \underset{\text{I}}{\cos^m x} \cdot \underset{\text{II}}{\cos nx}\, dx$

$$= (\cos^m x)\left[\frac{\sin nx}{n}\right] - \int m \cos^{m-1} x(-\sin x) \cdot \frac{\sin nx}{n}\, dx$$

$$= \frac{1}{n}\cos^m x \cdot \sin nx + \frac{m}{n}\int \cos^{m-1} x \{\sin x \cdot \sin nx\}\, dx$$

As we have, $\cos(n-1)x = \cos nx \cos x + \sin nx \sin x$

$$\therefore \quad I_{m,n} = \frac{1}{n}\cos^m x \cdot \sin x + \frac{m}{n}\int \cos^{m-1} x \{\cos(n-1)x - \cos nx \cdot \cos x\}\, dx$$

$$= \frac{1}{n}\cos^m x \cdot \sin x + \frac{m}{n}\int \cos^{m-1} x \cdot \cos(n-1)x\, dx - \frac{m}{n}\int \cos^m x \cdot \cos nx\, dx$$

$$= \frac{1}{n}\cos^m x \cdot \sin nx + \frac{m}{n} I_{m-1,n-1} - \frac{m}{n} I_{m,n}$$

$$I_{m,n} + \frac{m}{n} I_{m,n} = \frac{1}{n}[\cos^m x \cdot \sin nx + m\, I_{m-1,n-1}]$$

$$\left(\frac{m+n}{n}\right) I_{m,n} = \frac{1}{n}[\cos^m x \cdot \sin nx + m\, I_{m-1,n-1}]$$

$$(m+n) I_{m,n} = \cos^m x \cdot \sin nx + m\, I_{m-1,n-1}$$

Example 23 Evaluate $\int \cot^{-1}(x^2 + x + 1)\, dx$.

Solution. Let $I = \int \cot^{-1}(x^2 + x + 1)\, dx$

$$= \int \tan^{-1}\left(\frac{1}{x^2 + x + 1}\right) dx = \int \tan^{-1}\left(\frac{(x+1)-(x)}{1 + x(x+1)}\right) dx$$

$$= \int \{\tan^{-1}(x+1) - \tan^{-1}(x)\}\, dx$$

$$= \int \tan^{-1}(x+1)\, dx - \int \tan^{-1}(x)\, dx$$

Applying integration by parts, we get

$$= \{\tan^{-1}(x+1)\} x - \int x \cdot \frac{1}{1+(x+1)^2}\, dx - \left\{(\tan^{-1} x) x - \int x \cdot \frac{1}{1+x^2}\, dx\right\}$$

$$= x\{\tan^{-1}(x+1) - \tan^{-1} x\} + \int \frac{x}{1+x^2}\, dx - \int \frac{x}{1+(1+x)^2}\, dx$$

$$I = x\{\tan^{-1}(x+1) - \tan^{-1}(x)\} + \frac{1}{2}\log(1+x^2) - I_1 \quad \text{...(i)}$$

Let $\quad I_1 = \int \frac{x}{1+(1+x)^2}\, dx$ put $1 + x = t$, $dx = dt$

$$= \int \frac{t-1}{1+t^2}\, dt = \int \frac{t}{1+t^2}\, dt - \int \frac{1}{1+t^2}\, dt = \frac{1}{2}\log|1+t^2| - \tan^{-1}(t)$$

$$I_1 = \frac{1}{2}\log|1+(1+x)^2| - \tan^{-1}(1+x) \quad \text{...(ii)}$$

From Eqs. (i) and (ii), we get

$$I = x\{\tan^{-1}(x+1) - \tan^{-1}(x)\} + \frac{1}{2}\log|1+x^2| - \frac{1}{2}\log|1+(1+x)^2| + \tan^{-1}(1+x) + C$$

Example 24 Evaluate $\int \dfrac{\tan\left(\dfrac{\pi}{4}-x\right)}{\cos^2 x\sqrt{\tan^3 x+\tan^2 x+\tan x}}\,dx.$

Solution. $I=\int \dfrac{\tan\left(\dfrac{\pi}{4}-x\right)dx}{\cos^2 x\sqrt{\tan^3 x+\tan^2 x+\tan x}}$

$$=\int \frac{(1-\tan^2 x)\,dx}{(1+\tan x)^2\cos^2 x\sqrt{\tan^3 x+\tan^2 x+\tan x}}$$

$$I=\int \frac{-\left(1-\dfrac{1}{\tan^2 x}\right)\sec^2 x\,dx}{\left(\tan x+2+\dfrac{1}{\tan x}\right)\sqrt{\tan x+1+\dfrac{1}{\tan x}}}$$

$$\left[\begin{array}{l}\text{let} \quad y=\sqrt{\tan x+1+\dfrac{1}{\tan x}}\\ \Rightarrow \quad 2y\,dy=\left(\sec^2 x-\dfrac{1}{\tan^2 x}\cdot\sec^2 x\right)dx\end{array}\right.$$

$\therefore$ $$I=\int\frac{-2y\,dy}{(y^2+1)\cdot y}=-2\int\frac{dy}{1+y^2}=-2\tan^{-1}y+C$$

$$=-2\tan^{-1}\left(\sqrt{\tan x+1+\frac{1}{\tan x}}\right)+C$$

Example 25 Evaluate $\int \dfrac{x^2+n(n-1)}{(x\sin x+n\cos x)^2}\,dx.$

Solution. Here, $I=\int \dfrac{x^2+n(n-1)}{(x\sin x+n\cos x)^2}\,dx$

Multiplying and dividing by x^{2n-2}, we get $I=\int \dfrac{\{x^2+n(n-1)\}\,x^{2n-2}}{(x^n\sin x+n\,x^{n-1}\cos x)^2}\,dx$

We know $x^n\sin x+n\,x^{n-1}\cos x=t$

$\Rightarrow$ $\left\{(n\,x^{n-1}\sin x)+(x^n\cos x)+n(n-1)\,x^{n-2}\cos x-(n\,x^{n-1}\sin x)\,dx=dt\right\}$

$\Rightarrow$ $x^{n-2}\cos x\cdot\{x^2+n(n-1)\}\,dx=dt$

Keeping this in mind, we put

$$I=\int\underset{\text{II}}{\frac{\{x^2+n(n-1)\}\cdot x^{n-2}\cdot\cos x}{(x^n\sin x+n\,x^{n-1}\cos x)^2}}\cdot\underset{\text{I}}{x^n\cdot\sec x}\,dx$$

and applying integration by parts, we get

$$=x^n\sec x\cdot\left(-\frac{1}{(x^n\sin x+n\,x^{n-1}\cos x)}\right)$$
$$+\int\frac{(x^n\sec x\tan x+n\,x^{n-1}\cdot\sec x)}{(x^n\sin x+n\,x^{n-1}\cos x)}\,dx$$

$$= -\frac{(x \sec x)}{(x \sin x + n \cos x)} + \int \sec^2 x \, dx$$

$$= -\frac{(x \sec x)}{(x \sin x + n \cos x)} + \tan x + C$$

Example 26 If $\cos\theta > \sin\theta > 0$, then evaluate

$$\int \left\{ \log \left(\frac{1 + \sin 2\theta}{1 - \sin 2\theta} \right)^{\cos^2 \theta} + \log \left(\frac{\cos 2\theta}{1 + \sin 2\theta} \right) \right\} d\theta.$$

Solution. Here, $I = \int \left\{ \log \left(\frac{1 + \sin 2\theta}{1 - \sin 2\theta} \right)^{\cos^2 \theta} + \log \left(\frac{\cos 2\theta}{1 + \sin 2\theta} \right) \right\} d\theta$

$$= \int \left\{ 2\cos^2\theta \log \left(\frac{\cos\theta + \sin\theta}{\cos\theta - \sin\theta} \right) - \log \left(\frac{\cos\theta + \sin\theta}{\cos\theta - \sin\theta} \right) \right\} d\theta$$

$$= \int (2\cos^2\theta - 1) \log \left(\frac{\cos\theta + \sin\theta}{\cos\theta - \sin\theta} \right) d\theta$$

$$= \int \underset{\text{II}}{\cos 2\theta} \cdot \underset{\text{I}}{\log \left(\frac{\cos\theta + \sin\theta}{\cos\theta - \sin\theta} \right)} d\theta, \text{ applying integration by parts}$$

$$= \log \left(\frac{\cos\theta + \sin\theta}{\cos\theta - \sin\theta} \right) \cdot \frac{\sin 2\theta}{2} - \int \frac{2}{\cos 2\theta} \cdot \frac{\sin 2\theta}{2} \, d\theta$$

$$= \frac{\sin 2\theta}{2} \log \left(\frac{\cos\theta + \sin\theta}{\cos\theta - \sin\theta} \right) + \frac{1}{2} \log |\cos 2\theta| + C$$

Example 27 Evaluate $\int \frac{\tan^{-1} x}{x^4} \, dx.$

Solution. $I = \int \frac{\tan^{-1} x}{x^4} \, dx = \int \underset{\text{I}}{\tan^{-1} x} \cdot \underset{\text{II}}{\frac{1}{x^4}} \, dx$

$$= (\tan^{-1} x) \left(-\frac{1}{3x^3} \right) - \int \frac{1}{1 + x^2} \cdot \frac{1}{(-3x^3)} \, dx$$

$$= -\frac{\tan^{-1} x}{3x^3} + \frac{1}{3} \int \frac{dx}{x^3 (1 + x^2)}, \qquad \text{put } 1 + x^2 = t, \quad 2x \, dx = dt$$

$$= -\frac{\tan^{-1} x}{3x^3} + \frac{1}{6} \int \frac{dt}{(t-1)^2 \cdot t}$$

$$I = -\frac{\tan^{-1} x}{3x^3} + \frac{1}{6} I_1 \qquad \ldots \text{(i)}$$

where, $I_1 = \int \frac{1}{(t-1)^2 \cdot t} \, dt = \int \left\{ \frac{A}{t-1} + \frac{B}{(t-1)^2} + \frac{C}{t} \right\} dt$

Comparing coefficients, we get

$$A = -1, B = 1, C = 1$$

$$\therefore \qquad I_1 = \int \left\{ -\frac{1}{(t-1)} + \frac{1}{(t-1)^2} + \frac{1}{t} \right\} dt$$

$$= -\log|t-1| - \frac{1}{(t-1)} + \log|t| \qquad \text{...(ii)}$$

$\therefore$ From Eqs. (i) and (ii), we get

$$I = -\frac{\tan^{-1} x}{3x^3} + \frac{1}{6}\left\{ -\log|x^2| - \frac{1}{x^2} + \log|1+x^2| \right\} + C$$

$$= -\frac{\tan^{-1} x}{3x^3} - \frac{1}{6}\log\left|\frac{x^2+1}{x^2}\right| - \frac{1}{6x^2} + C$$

Example 28 Evaluate $\int \frac{dx}{\tan x + \cot x + \sec x + \operatorname{cosec} x}$.

Solution. Here, $I = \int \frac{dx}{\tan x + \cot x + \sec x + \operatorname{cosec} x}$

$$= \int \frac{(\sin x \cos x)\, dx}{1 + \sin x + \cos x} = \int \frac{\sin x\, dx}{\sec x + \tan x + 1}$$

Multiplying and dividing by $(1 + \tan x - \sec x)$, we get

$$= \int \frac{\sin x\,(1 + \tan x - \sec x)}{(1+\tan x)^2 - \sec^2 x}\, dx$$

$$= \int \frac{\sin x\,(1 + \tan x - \sec x)}{2\tan x}\, dx$$

$$= \frac{1}{2}\int \cos x\,(1 + \tan x - \sec x)\, dx$$

$$= \frac{1}{2}\int (\cos x + \sin x - 1)\, dx$$

$$= \frac{1}{2}\{\sin x - \cos x - x\} + C$$

$$\therefore \qquad \int \frac{dx}{\tan x + \cot x + \sec x + \operatorname{cosec} x} = \frac{1}{2}(\sin x - \cos x - x) + C$$

Example 29 Evaluate $\int x^2 \log(1 - x^2)\, dx$, and hence prove that

$$\frac{1}{1\cdot 5} + \frac{1}{2\cdot 7} + \frac{1}{3\cdot 9} + \ldots = \frac{8}{9} - \frac{2}{3}\log 2.$$

Solution. We know,

$$\log(1-x) = -\left\{ x + \frac{x^2}{2} + \frac{x^3}{3} + \frac{x^4}{4} + \ldots \infty \right\}$$

Put x^2 instead of x in the above identity,

$$\Rightarrow \qquad \log(1-x^2) = -\left\{ x^2 + \frac{x^4}{2} + \frac{x^6}{3} + \frac{x^8}{4} + \ldots \infty \right\}$$

$$\Rightarrow \qquad x^2\log(1-x^2) = -\left\{ x^4 + \frac{x^6}{2} + \frac{x^8}{3} + \frac{x^{10}}{4} + \ldots \infty \right\}$$

Integrating both the sides, we get

$$\int x^2 \log(1-x^2)\,dx = -\left\{\frac{x^5}{1\cdot 5}+\frac{x^7}{2\cdot 7}+\frac{x^9}{3\cdot 9}+\dots\infty\right\}+C$$

Now, to find constant of integration, put $x = 0$

$\Rightarrow \quad 0 = 0 + C \quad \Rightarrow \quad C = 0$

$$\therefore \quad \int \underset{\text{II}}{x^2}\, \underset{\text{I}}{\log(1-x^2)}\,dx = -\left\{\frac{x^5}{1\cdot 5}+\frac{x^7}{2\cdot 7}+\frac{x^9}{3\cdot 9}+\dots\infty\right\}$$

Applying integration by parts, and taking limits 0 to 1 for LHS

$$\Rightarrow \quad \left(\frac{x^3}{3}\log(1-x^2)\right)_0^1 - \int_0^1 \frac{x^3}{3}\cdot\frac{1(-2x)}{1-x^2}\,dx$$

$$\Rightarrow \quad \left(\frac{x^3}{3}\log(1-x^2)\right)_0^1 + \frac{2}{3}\left(-\frac{x^3}{3}-x+\frac{1}{2}\log\left|\frac{1+x}{1-x}\right|\right)_0^1$$

Taking $\quad \log(1-x^2) = \log(1+x) + \log(1-x)$

and $\quad \log\left(\frac{1+x}{1-x}\right) = \log(1+x) - \log(1-x)$

$$\Rightarrow \quad \frac{1}{3}\log 2 + \frac{1}{3}\log 2 - \frac{2}{3} - \frac{2}{9} + \lim_{x\to 1}\left(\frac{x^3}{3}-\frac{1}{3}\right)\log(1-x)$$

$$\left[\because \lim_{x\to 1}(x^3-1)\log(1-x) = 0\right]$$

$$\Rightarrow \quad \frac{2}{3}\log 2 - \frac{8}{9} = \text{RHS}$$

$$\therefore \quad \frac{1}{1\cdot 5}+\frac{1}{2\cdot 7}+\frac{1}{3\cdot 9}+\dots = \frac{2}{3}\log(2) - \frac{8}{9}$$

Example 30 Evaluate $\int \frac{a+b\sin x}{(b+a\sin x)^2}\,dx$

Solution. Here, $I = \int \frac{a+b\sin x}{(b+a\sin x)^2}\,dx = \frac{b}{a}\int \frac{\frac{a^2}{b} - b + (b+a\sin x)}{(b+a\sin x)^2}\,dx$

$$I = \frac{a^2-b^2}{a}\int\frac{dx}{(b+a\sin x)^2} + \frac{b}{a}\int\frac{dx}{(b+a\sin x)} \qquad \text{...(i)}$$

Now, let $\quad A = \frac{\cos x}{b+a\sin x} \Rightarrow \frac{dA}{dx} = \frac{-b\sin x - a}{(b+a\sin x)^2}$

$$\Rightarrow \quad \frac{dA}{dx} = -\frac{b}{a}\left\{\frac{a\sin x + b + \frac{a^2}{b} - b}{(b+a\sin x)^2}\right\}$$

$$\Rightarrow \quad \frac{dA}{dx} = -\frac{b}{a}\left\{\frac{1}{b+a\sin x} + \frac{a^2-b^2}{b(b+a\sin x)^2}\right\}$$

Integrating both the sides w.r.t. 'x', we get

$$A = -\frac{b}{a}\int \frac{dx}{b + a\sin x} - \frac{(a^2 - b^2)}{a}\int \frac{dx}{(b + a\sin x)^2}$$

$$\Rightarrow \quad \frac{a^2 - b^2}{a}\int \frac{dx}{(b + a\sin x)^2} = -\frac{b}{a}\int \frac{dx}{b + a\sin x} - A \qquad \text{...(ii)}$$

From Eqs. (i) and (ii), we get

$$I = -\frac{b}{a}\int \frac{dx}{(b + a\sin x)} - A + \frac{b}{a}\int \frac{dx}{(b + a\sin x)}$$

$$\Rightarrow \quad I = -A + C \Rightarrow I = -\left(\frac{\cos x}{b + a\sin x}\right) + C$$

Type 2 : Only One Correct Option

Example 31 $\int \frac{x\,dx}{\sqrt{1 + x^2 + \sqrt{(1 + x^2)^3}}}$ is equal to

(a) $\frac{1}{2}\ln(1 + \sqrt{1 + x^2}) + C$

(b) $2\sqrt{1 + \sqrt{1 + x^2}} + C$

(c) $2(1 + \sqrt{1 + x^2}) + C$

(d) None of these

Solution. $\int \frac{x\,dx}{\sqrt{(1 + x^2)}\sqrt{1 + \sqrt{1 + x^2}}}$

Put $1 + \sqrt{1 + x^2} = t^2$

$$\Rightarrow \quad \frac{2x}{2\sqrt{1 + x^2}}\,dx = 2t\,dt \quad \therefore \quad \frac{x\,dx}{\sqrt{1 + x^2}} = 2t\,dt$$

$$\therefore \quad I = \int \frac{2t\,dt}{t} = 2t + C = 2\sqrt{1 + \sqrt{1 + x^2}} + C$$

Hence, (b) is the correct answer.

Example 32 $\int \frac{(2x + 1)}{(x^2 + 4x + 1)^{3/2}}\,dx$

(a) $\frac{x^3}{(x^2 + 4x + 1)^{1/2}} + C$

(b) $\frac{x}{(x^2 + 4x + 1)^{1/2}} + C$

(c) $\frac{x^2}{(x^2 + 4x + 1)^{1/2}} + C$

(d) $\frac{1}{(x^2 + 4x + 1)^{1/2}} + C$

Solution. $\int \frac{2x + 1}{(x^2 + 4x + 1)^{3/2}}\,dx = \int \frac{2x + 1}{x^3\left(1 + \frac{4}{x} + \frac{1}{x^2}\right)^{3/2}}\,dx = \int \frac{2x^{-2} + x^{-3}}{\left(1 + \frac{4}{x} + \frac{1}{x^2}\right)^{3/2}}\,dx$

Now, put $\frac{1}{x^2} + \frac{4}{x} + 1 = t^2 \Rightarrow \left(-\frac{2}{x^3} - \frac{4}{x^2}\right)dx = 2t\,dt$

$$\therefore \quad I = \int \frac{-t\,dt}{t^3} = \frac{1}{t} + C = \frac{x}{\sqrt{x^2 + 4x + 1}} + C$$

Hence, (b) is the correct answer.

Example 33 The evaluation of $\int \frac{p\,x^{p+2q-1} - qx^{q-1}}{x^{2p+2q} + 2x^{p+q} + 1}\,dx$ is

(a) $-\frac{x^p}{x^{p+q}+1} + C$ (b) $\frac{x^q}{x^{p+q}+1} + C$

(c) $-\frac{x^q}{x^{p+q}+1} + C$ (d) $\frac{x^p}{x^{p+q}+1} + C$

Solution. $\int \frac{px^{p+2q-1} - qx^{q-1}}{(x^{p+q}+1)^2}\,dx = \int \frac{px^{p-1} - qx^{-q-1}}{(x^p + x^{-q})^2}\,dx$

Taking x^q as x^{2q} common from denominator and take it in numerator.

Put $x^p + x^{-q} = t \Rightarrow (px^{p-1} - qx^{-q-1})\,dx = dt$

$$\therefore \quad I = \int \frac{dt}{t^2} = -\frac{1}{t} + C = -\left(\frac{x^q}{x^{p+q}+1}\right) + C$$

Hence, (c) is the correct answer.

Example 34 $\int \frac{x^2(1-\ln x)}{\ln^4 x - x^4}\,dx$ is equal to

(a) $\frac{1}{2}\ln\left(\frac{x}{\ln x}\right) - \frac{1}{4}\ln(\ln^2 x - x^2) + C$ (b) $\frac{1}{4}\ln\left(\frac{\ln x - x}{\ln x + x}\right) - \frac{1}{2}\tan^{-1}\left(\frac{\ln x}{x}\right) + C$

(c) $\frac{1}{4}\ln\left(\frac{\ln x + x}{\ln x - x}\right) + \frac{1}{2}\tan^{-1}\left(\frac{\ln x}{x}\right) + C$ (d) $\frac{1}{4}\left(\ln\left(\frac{\ln x - x}{\ln x + x}\right) + \tan^{-1}\left(\frac{\ln x}{x}\right)\right) + C$

Solution. $I = \int \frac{x^2(1-\ln x)}{x^4\left(\left(\frac{\ln x}{x}\right)^4 - 1\right)}\,dx = \int \frac{1-\ln x}{x^2\left(\left(\frac{\ln x}{x}\right)^4 - 1\right)}\,dx$

Put $\frac{\ln x}{x} = t \Rightarrow \frac{1-\ln x}{x^2} = dt$

$$I = \int \frac{dt}{(t^4-1)} = \int \frac{dt}{(t^2+1)(t^2-1)} = \frac{1}{2}\int \frac{(t^2+1)-(t^2-1)}{(t^2+1)(t^2-1)}\,dt$$

$$I = \frac{1}{2}\left(\int \frac{dt}{t^2-1} - \int \frac{dt}{t^2+1}\right) = \frac{1}{2}\left(\frac{1}{2}\ln\frac{t-1}{t+1} - \tan^{-1} t\right)$$

$$= \frac{1}{4}\ln\left(\frac{\ln x - x}{\ln x + x}\right) - \frac{1}{2}\tan^{-1}\left(\frac{\ln x}{x}\right) + C$$

Hence, (b) is the correct answer.

Example 35 $\int \frac{x^2-1}{x^3\sqrt{2x^4 - 2x^2 + 1}}\,dx$ is equal to **[IIT JEE 2006]**

(a) $\frac{\sqrt{2x^4 - 2x^2 + 1}}{x^2} + C$ (b) $\frac{\sqrt{2x^4 - 2x^2 + 1}}{x^3} + C$

(c) $\frac{\sqrt{2x^4 - 2x^2 + 1}}{x} + C$ (d) $\frac{\sqrt{2x^4 - 2x^2 + 1}}{2x^2} + C$

Solution. Let $I = \int \frac{x^2 - 1}{x^3 \sqrt{2x^4 - 2x^2 + 1}}\, dx = \int \frac{x^2 - 1}{x^5} \cdot \frac{dx}{\sqrt{2 - \frac{2}{x^2} + \frac{1}{x^4}}}$

$$= \frac{1}{4} \int \frac{\frac{4}{x^3} - \frac{4}{x^5}}{\sqrt{2 - \frac{2}{x^2} + \frac{1}{x^4}}}\, dx = \frac{1}{4} \cdot 2\sqrt{2 - \frac{2}{x^2} + \frac{1}{x^4}} \left(\because \int \frac{f'(x)}{\sqrt{f(x)}}\, dx = 2\sqrt{f(x)} + C \right)$$

$$= \frac{\sqrt{2x^4 - 2x^2 + 1}}{2x^2} + C$$

Hence, (d) is the correct answer.

Example 36 Let $f(x) = \frac{x}{(1 + x^n)^{1/n}}$ for $n \geq 2$ and $g(x) = \underbrace{fofo \ldots of}_{n \text{ times}}(x)$, then $\int x^{n-2} g(x)\, dx$ equals to **[IIT JEE 2007]**

(a) $\frac{1}{n(n-1)} (1 + nx^n)^{1 - \frac{1}{n}} + C$ (b) $\frac{1}{n-1} (1 + nx^n)^{1 - \frac{1}{n}} + C$

(c) $\frac{1}{n(n+1)} (1 + nx^n)^{1 + \frac{1}{n}} + C$ (d) $\frac{1}{n+1} (1 + nx^n)^{1 + \frac{1}{n}} + C$

Solution. $\because f(x) = \frac{x}{(1 + x^n)^{1/n}}$

$$\therefore \quad f(f(x)) = \frac{y}{(1 + x^n)^{1/n}}, \text{ where } y = \frac{x}{(1 + x^n)^{1/n}} = \frac{x}{(1 + 2x^n)^{1/n}}$$

Similarly, $f(f(f(x))) = \frac{x}{(1 + 3x^n)^{1/n}}$ and $\underbrace{(fofofo \ldots of)}_{n \text{ times}}(x) = g(x) = \frac{x}{(1 + nx^n)^{1/n}}$

$$\therefore \quad \int x^{n-2} g(x)\, dx = \int \frac{x^{n-1}}{(1 + nx^n)^{1/n}}\, dx = \frac{1}{n^2} \int n^2 \cdot x^{n-1} \cdot (1 + nx^n)^{-1/n}\, dx$$

$$= \frac{1}{n^2} \cdot \frac{(1 + nx^n)^{1 - \frac{1}{n}}}{1 - \frac{1}{n}} + C = \frac{(1 + nx^n)^{1 - \frac{1}{n}}}{n(n-1)} + C$$

Hence, (a) is the correct answer.

Example 37 $\int \frac{e^{\tan^{-1} x}}{(1 + x^2)} \left[(\sec^{-1} \sqrt{1 + x^2})^2 + \cos^{-1} \left(\frac{1 - x^2}{1 + x^2} \right) \right] dx, \ (x > 0)$

(a) $e^{\tan^{-1} x} \cdot \tan^{-1} x + C$ (b) $\frac{e^{\tan^{-1} x} \cdot (\tan^{-1} x)^2}{2} + C$

(c) $e^{\tan^{-1} x} \cdot (\sec^{-1}(\sqrt{1 + x^2}))^2 + C$ (d) $e^{\tan^{-1} x} \cdot (\text{cosec}^{-1} (\sqrt{1 + x^2}))^2 + C$

Solution. Note that $\sec^{-1} \sqrt{1 + x^2} = \tan^{-1} x$; $\cos^{-1} \left(\frac{1 - x^2}{1 + x^2} \right) = 2 \tan^{-1} x$, for $x > 0$

$$I = \int \frac{e^{\tan^{-1} x}}{1 + x^2} \{(\tan^{-1} x)^2 + 2 \tan^{-1} x\}\, dx, \text{ put } \tan^{-1} x = t$$

$$= \int e^t (t^2 + 2t)\, dt = e^t \cdot t^2 = e^{\tan^{-1} x} (\tan^{-1} x)^2 + C$$

Hence, (c) is the correct answer.

Example 38 Let $I = \int \frac{e^x}{e^{4x} + e^{2x} + 1} dx$, $J = \int \frac{e^{-x}}{e^{-4x} + e^{-2x} + 1} dx$. Then, for an arbitrary constant c, the value of $J - I$ equals to **[IIT JEE 2008]**

(a) $\frac{1}{2} \log \left(\frac{e^{4x} - e^{2x} + 1}{e^{4x} + e^{2x} + 1} \right) + C$ (b) $\frac{1}{2} \log \left(\frac{e^{2x} + e^x + 1}{e^{2x} - e^x + 1} \right) + C$

(c) $\frac{1}{2} \log \left(\frac{e^{2x} - e^x + 1}{e^{2x} + e^x + 1} \right) + C$ (d) $\frac{1}{2} \log \left(\frac{e^{4x} + e^{2x} + 1}{e^{4x} - e^{2x} + 1} \right) + C$

Solution. $J = \int \frac{e^{3x}}{1 + e^{2x} + e^{4x}} dx$

$$J - I = \int \frac{(e^{3x} - e^x)}{1 + e^{2x} + e^{4x}} dx = \int \frac{(u^2 - 1)}{1 + u^2 + u^4} du \qquad (u = e^x)$$

$$= \int \frac{\left(1 - \frac{1}{u^2}\right) du}{1 + \frac{1}{u^2} + u^2} = \int \frac{\left(1 - \frac{1}{u^2}\right) du}{\left(u + \frac{1}{u}\right)^2 - 1} = \int \frac{dt}{t^2 - 1} \qquad \left(t = u + \frac{1}{u}\right)$$

$$= \frac{1}{2} \log \left| \frac{t-1}{t+1} \right| + C = \frac{1}{2} \log \left| \frac{u^2 - u + 1}{u^2 + u + 1} \right| + C = \frac{1}{2} \log \left| \frac{e^{2x} - e^x + 1}{e^{2x} + e^x + 1} \right| + C$$

Hence, (c) is the correct answer.

Example 39 Integral of $\sqrt{1 + 2 \cot x (\cot x + \operatorname{cosec} x)}$ w.r.t. x, is

(a) $2 \ln \cos \frac{x}{2} + C$ (b) $2 \ln \sin \frac{x}{2} + C$

(c) $\frac{1}{2} \ln \cos \frac{x}{2} + C$ (d) $\ln \sin x - \ln (\operatorname{cosec} x - \cot x) + C$

Solution. $I = \int \sqrt{1 + 2 \operatorname{cosec} x \cot x + 2 \cot^2 x}\, dx$

$$= \int \sqrt{\operatorname{cosec}^2 x + 2 \operatorname{cosec} x \cot x + \cot^2 x}\, dx = \int (\operatorname{cosec} x + \cot x)\, dx$$

$$= \int \frac{1 + \cos x}{\sin x} dx = \int \cot\left(\frac{x}{2}\right) dx = 2 \log \left| \sin \frac{x}{2} \right| + C$$

Hence, (b) is the correct answer.

Example 40 If $I_n = \int \cot^n x\, dx$, then $I_0 + I_1 + 2(I_2 + I_3 + \ldots + I_8) + I_9 + I_{10}$ equals to (where $u = \cot x$)

(a) $u + \frac{u^2}{2} + \ldots + \frac{u^9}{9}$ (b) $-\left(u + \frac{u^2}{2} + \ldots + \frac{u^9}{9}\right)$

(c) $-\left(u + \frac{u^2}{2!} + \ldots + \frac{u^9}{9!}\right)$ (d) $\frac{u}{2} + \frac{2u^2}{3} + \ldots + \frac{9u^9}{10}$

Solution. $I_n = \int \cot^n x\, dx = \int \cot^{n-2} x \cdot (\operatorname{cosec}^2 x - 1)\, dx$

$$I_n = -\frac{u^{n-1}}{n-1} - I_{n-2} \quad \text{or} \quad I_n + I_{n-2} = -\frac{u^{n-1}}{n-1} \quad (\text{put } n = 2, 3, 4, \ldots, 10)$$

$$I_2 + I_0 = -\frac{u}{1}$$

$$I_3 + I_1 = -\frac{u^2}{2}$$

$$I_4 + I_2 = -\frac{u^3}{3}$$

$$\dots\dots\dots\dots\dots\dots$$

$$\dots\dots\dots\dots\dots\dots$$

$$I_{10} + I_9 = -\frac{u^9}{9}$$

Adding, $I_0 + I_1 + 2(I_2 + I_3 + \dots + I_8) + I_9 + I_{10} = -\left(u + \frac{u^2}{2} + \dots + \frac{u^9}{9}\right)$

Hence, (b) is the correct answer.

Example 41 Let $f(x) = x + \sin x$. Suppose g denotes the inverse function of f. The value of $g'\left(\frac{\pi}{4} + \frac{1}{\sqrt{2}}\right)$ has the value equal to

(a) $\sqrt{2} - 1$ (b) $\frac{\sqrt{2}+1}{\sqrt{2}}$ (c) $2 - \sqrt{2}$ (d) $\sqrt{2} + 1$

Solution. $f(x) = y = x + \sin x$

$$\frac{dy}{dx} = 1 + \cos x$$

$$g'(y) = \frac{dx}{dy} = \frac{1}{1 + \cos x}$$

where $y = \frac{\pi}{4} + \frac{1}{\sqrt{2}} = x + \sin x \Rightarrow x = \frac{\pi}{4}$

$$\therefore \quad g'\left(\frac{\pi}{4} + \frac{1}{\sqrt{2}}\right) = \frac{1}{1 + (1/\sqrt{2})} = \frac{\sqrt{2}}{\sqrt{2}+1} = \sqrt{2}(\sqrt{2} - 1) = 2 - \sqrt{2}$$

Hence, (c) is the correct answer.

Type 3 : More Than One Correct Options

Example 42 $\int \frac{\sqrt{4 + x^2}}{x^6}\, dx = \frac{A(4 + x^2)^{3/2}(Bx^2 - 6)}{x^5} + C$

(a) $A = \frac{1}{120}$ (b) $B = 1$ (c) $A = -\frac{1}{120}$ (d) $B = -1$

Solution. $I = \int \frac{\sqrt{4 - x^2}}{x^6}\, dx$

$$I = \int \frac{\sqrt{1 + \frac{4}{x^2}}}{x^5}\, dx = \int \frac{\sqrt{1 + \frac{4}{x^2}}}{x^2 \cdot x^3}\, dx$$

Put $t = \sqrt{1 + \frac{4}{x^2}} \Rightarrow t^2 = 1 + \frac{4}{x^2} \quad \therefore \quad 2t\, dt = -\frac{8}{x^3}\, dx$

$$\Rightarrow \quad I = \frac{1}{16}\int (t^2 - t^4)\, dx = \frac{1}{16}\left\{\frac{t^3}{3} - \frac{t^5}{5}\right\} + C = \frac{1}{120} \cdot \frac{(4 + x^2)^{3/2}}{x^5}(x^2 - 6) + C$$

$$A = \frac{1}{120}, B = 1$$

Hence, (a) and (b) are the correct answers.

Example 43 The value of the integral $\int e^{\sin^2 x}(\cos x + \cos^3 x)\sin x\,dx$ is

(a) $\frac{1}{2}e^{\sin^2 x}(3 - \sin^2 x) + C$ (b) $e^{\sin^2 x}\left(1 + \frac{1}{2}\cos^2 x\right) + C$

(c) $e^{\sin^2 x}(3\cos^2 x + 2\sin^2 x) + C$ (d) $e^{\sin^2 x}(2\cos^2 x + 3\sin^2 x) + C$

Solution. Put $t = \sin^2 x$

The integral reduces to $I = \frac{1}{2}\int e^t(2 - t)\,dt = \frac{3}{2}e^t - \frac{te^t}{2} + C$

$$= \frac{1}{2}e^{\sin^2 x}(3 - \sin^2 x) + C \qquad \text{[option (a)]}$$

$$= e^{\sin^2 x}\left(1 + \frac{1}{2}\cos^2 x\right) + C \qquad \text{[option (b)]}$$

Hence, (a) and (b) are the correct answers.

Example 44 If $I = \int(\sqrt{\tan x} + \sqrt{\cot x})\,dx = f(x) + c$, then $f(x)$ is equal to

(a) $\sqrt{2}\sin^{-1}(\sin x - \cos x)$ (b) $\frac{\pi}{\sqrt{2}} - \sqrt{2}\cos^{-1}(\sin x - \cos x)$

(c) $\sqrt{2}\tan^{-1}\left(\frac{\tan x - 1}{\sqrt{2}\sqrt{\tan x}}\right)$ (d) None of these

Solution. $I = \int(\sqrt{\tan x} + \sqrt{\cot x})\,dx = \int \sqrt{2}\cdot\frac{\sin x + \cos x}{\sqrt{2\sin x\cos x}}\,dx$

If $\sin x - \cos x = p$, then $(\cos x + \sin x)\,dx = dp$

$$I = \sqrt{2}\int\frac{dp}{\sqrt{1 - p^2}} = \sqrt{2}\sin^{-1} p + c = \sqrt{2}\sin^{-1}(\sin x - \cos x) + c$$

$$= \frac{\pi}{\sqrt{2}} - \sqrt{2}\cos^{-1}(\sin x - \cos x) = \sqrt{2}\tan^{-1}\frac{\sin x - \cos x}{\sqrt{1 - (\sin x - \cos x)^2}}$$

$$= \sqrt{2}\tan^{-1}\frac{\sin x - \cos x}{\sqrt{2\sin x\cos x}} = \sqrt{2}\tan^{-1}\left(\frac{\tan x - 1}{\sqrt{2\tan x}}\right)$$

Hence, (a), (b) and (c) are the correct answers.

Type 4 : Assertion and Reason

Directions
(Q. Nos. 45 to 48)

For the following questions, choose the correct answers from the codes (a), (b), (c) and (d) defined as follows:

(a) Statement I is true, Statement II is also true; Statement II is the correct explanation of Statement I.

(b) Statement I is true, Statement II is also true; Statement II is not the correct explanation of Statement I.
(c) Statement I is true, Statement II is false.
(d) Statement I is false, Statement II is true.

45 Statement I $\int \frac{d(x^2+1)}{\sqrt{x^2+2}}$ is equal to $2\sqrt{x^2+2}+C$

Statement II $\int \frac{x^{9/2}}{\sqrt{1+x^{11}}}dx$ is $\frac{2}{11}\ln|x+\sqrt{1+x^{11}}|+C$

Solution. $I = \int \frac{dt}{\sqrt{1+t}}$, where $x^2+1=t$

$$= 2\sqrt{1+t}+C = 2\sqrt{x^2+2}+C$$

$$I = \int \frac{x^{9/2}}{\sqrt{1+x^{11}}}dx$$

Put $x^{11/2} = t \Rightarrow I = \frac{2}{11}\int \frac{dt}{\sqrt{1+t^2}}$

$$I = \frac{2}{11}\ln(x^{11/2}+\sqrt{1+x^{11}})+C$$

Hence, (c) is the correct answer.

46 Statement I If $x > 0, x \neq 0$, then

$$\int [\log_x e - (\log_x e)^2]\,dx = x\log_x e + C$$

Statement II $\int e^x[f(x)+f'(x)]\,dx = e^x f(x)+C$ and $e^t = x$ iff $t = \ln x$

Solution. $\int [(\log_x e) - (\log_x e)^2]dx = \int \left(\frac{1}{\ln x} - \frac{1}{(\ln x)^2}\right)dx = \int \left(\frac{1}{t} - \frac{1}{t^2}\right)e^t\,dt$

(where $t = \ln x$)

$$= \frac{e^t}{t} + C = \frac{x}{\ln x} + C = x\log_x e + C$$

Hence, (a) is the correct answer.

47 Let $F(x)$ be an indefinite integral of $\sin^2 x$.

Statement I The function $F(x)$ satisfies $F(x+\pi) = F(x)$ for all real x.

Statement II $\sin^2(x+\pi) = \sin^2 x$ for all real x. **[IIT JEE 2007]**

Solution. $F(x) = \int \sin^2 x\,dx = \int \frac{1-\cos 2x}{2}dx$

$\Rightarrow$ $F(x) = \frac{1}{4}(2x - \sin 2x) + C$

Since, $F(x+\pi) \neq F(x)$

Therefore, Statement I is false.

But Statement II is true as $\sin^2 x$ is periodic with period π.

Hence, (d) is the correct answer.

48 **Statement I** If $y = \sin^{-1} x$, then $\int \sin^{-1} x\, dx = \int y \cos y\, dy + C$

Statement II If $y = f^{-1}(x)$, then $\int f^{-1}(x)\, dx = \int y f'(y)\, dy + C$

Solution. Since, $y = f^{-1}(x)$, therefore $x = f(y)$

$\therefore \quad dx = f'(y)\, dy$. Thus, $\int f^{-1}(x)\, dx = \int y f'(y)\, dy + C$

$\therefore$ Statement I is true for $f^{-1}(x) = \sin^{-1}(x)$ in Statement II.

Hence, (a) is the correct answer.

Type 5 : Linked–Comprehension Based Questions

Passage
(Q. Nos. 49 to 51)

For integral $\int f\left(x - \frac{a}{x}\right) \cdot \left(1 + \frac{a}{x^2}\right) dx$, *put* $x - \frac{a}{x} = t$

For integral $\int f\left(x + \frac{a}{x}\right) \cdot \left(1 - \frac{a}{x^2}\right) dx$, *put* $x + \frac{a}{x} = t$

For integral $\int f\left(x^2 - \frac{a}{x^2}\right) \cdot \left(x + \frac{a}{x^3}\right) dx$, *put* $x^2 - \frac{a}{x^2} = t$

For integral $\int f\left(x^2 + \frac{a}{x^2}\right) \cdot \left(x - \frac{a}{x^3}\right) dx$, *put* $x^2 + \frac{a}{x^2} = t$

many integrands can be brought into above forms by suitable reductions or transformations.

49 $\int \frac{x^4 - 2}{x^2 \sqrt{x^4 + x^2 + 2}}\, dx$

(a) $\sqrt{x^2 + 1 + \frac{1}{x^2}} + C$ (b) $\sqrt{x^2 + 1 + \frac{2}{x^2}} + C$

(c) $\sqrt{x^2 + \frac{1}{x^2}} + C$ (d) $\sqrt{x^2 + \frac{2}{x^2}} + C$

Solution. $\int \frac{x - \frac{2}{x^3}}{\sqrt{x^2 + 1 + \frac{2}{x^2}}}\, dx$

Put $x^2 + \frac{2}{x^2} + 1 = t \Rightarrow 2\left(x - \frac{2}{x^3}\right) dx = dt$

$$= \frac{1}{2} \int \frac{dt}{\sqrt{t}} = \frac{1}{2} \cdot \frac{t^{1/2}}{1/2} + C = \sqrt{t} + C = \sqrt{x^2 + \frac{2}{x^2} + 1} + C$$

Hence, (b) is the correct answer.

50 $\int \frac{(x - 1)}{(x + 1)\sqrt{x^3 + x^2 + x}}\, dx$

(a) $\tan^{-1}\left(x + \frac{1}{x} + 1\right) + C$ (b) $\tan^{-1} \sqrt{x + \frac{1}{x} + 1} + C$

(c) $2 \tan^{-1} \sqrt{x + \frac{1}{x} + 1} + C$ (d) None of these

Solution. $\int \frac{x^2-1}{(x+1)^2\sqrt{x^3+x^2+x}}\,dx = \int \frac{\left(1-\frac{1}{x^2}\right)}{\left(x+\frac{1}{x}+2\right)\sqrt{x+\frac{1}{x}+1}}\,dx$

Put $x+\frac{1}{x}+1=t^2 \Rightarrow \left(1-\frac{1}{x^2}\right)dx = 2t\,dt$

$$= \int \frac{2t\,dt}{(t^2+1)t} = 2\int \frac{1}{(t^2+1)}\,dt$$

$$= 2\cdot\tan^{-1}(t) + C = 2\tan^{-1}\left(\sqrt{x+\frac{1}{x}+1}\right)+C$$

Hence, (c) is the correct answer.

51 $\int \frac{5x^4+4x^5}{(x^5+x+1)^2}\,dx$

(a) $x^5+x+1+C$ (b) $\frac{x^5}{x^5+x+1}+C$ (c) $x^{-4}+x^{-5}+C$ (d) $\frac{x^5}{x^5+x+1}+C$

Solution. Divide numerator and denominator by x^{10}, we get

$$I = \int \frac{5x^{-6}+4x^{-5}}{(1+x^{-4}+x^{-5})^2}\,dx$$

Put $1+x^{-4}+x^{-5}=t \Rightarrow (-4x^{-5}-5x^{-6})\,dx = dt$

$\therefore$ $$I = -\int \frac{dt}{t^2} = \frac{1}{t}+C = \frac{1}{1+x^{-4}+x^{-5}}+C = \frac{x^5}{x^5+x+1}+C$$

Hence, (d) is the correct answer.

Type 6 : Match the Columns

Example 52 If $x \in (0, 1)$ then match the entries of Column I with Column II considering 'c' as an arbitrary constant of integration.

Column I	Column II
(A) $\int \tan\left(2\tan^{-1}\sqrt{\frac{\sqrt{1+\sqrt{x}}-1}{\sqrt{1+\sqrt{x}}+1}}\right)dx$	(p) $\frac{4}{3}x^{3/4}+C$
(B) $\int \cot\left(2\tan^{-1}\sqrt{\frac{\sqrt{1+\sqrt{x}}-\sqrt[4]{x}}{\sqrt{1+\sqrt{x}}+\sqrt[4]{x}}}\right)dx$	(q) $\frac{4}{5}x^{5/4}+C$
(C) $\int \left(\frac{1-\tan\left(\frac{1}{2}\sin^{-1}\left(\frac{1-\sqrt{x}}{1+\sqrt{x}}\right)\right)}{1+\tan\left(\frac{1}{2}\sin^{-1}\left(\frac{1-\sqrt{x}}{1+\sqrt{x}}\right)\right)}\right)dx$	(r) $\frac{2}{3}x^{3/4}+C$
(D) $\int \sqrt{x}\tan\left(2\tan^{-1}\left(\frac{\sqrt{\sqrt{1+\sqrt{x}}+1}-\sqrt{\sqrt{1+\sqrt{x}}-1}}{\sqrt{\sqrt{1+\sqrt{x}}+1}+\sqrt{\sqrt{1+\sqrt{x}}-1}}\right)\right)dx$	(s) $\frac{2}{5}x^{5/4}+C$

Solution. (A) → (q); (B) → (q); (C) → (q); (D) → (q)

Let $\sqrt{x} = \tan^2\theta$

$x = \tan^4\theta \quad \Rightarrow \quad \tan\theta = x^{1/4}$

$\left[\begin{array}{l} \because x \in (0, 1) \\ \therefore \theta \in (0, \pi/4) \end{array}\right]$

$\therefore \quad dx = 4\tan^3\theta\sec^2\theta\, d\theta$

$\therefore \quad \sqrt{1+\sqrt{x}} = \sec\theta$

(A) $I = \int \tan\left(2\tan^{-1}\sqrt{\dfrac{\sqrt{1+\sqrt{x}}-1}{\sqrt{1+\sqrt{x}}+1}}\right)dx$

$$\Rightarrow \quad \tan\left(2\tan^{-1}\sqrt{\frac{\sqrt{1+\sqrt{x}}-1}{\sqrt{1+\sqrt{x}}+1}}\right) = \tan\left(2\tan^{-1}\sqrt{\frac{\sec\theta-1}{\sec\theta+1}}\right)$$

$$= \tan\left(2\tan^{-1}\left(\tan\frac{\theta}{2}\right)\right) = \tan\theta$$

$$\therefore \quad I = \int \tan\theta\cdot 4\tan^3\theta\sec^2\theta\, d\theta = \frac{4}{5}\tan^5\theta + C = \frac{4}{5}(x^{5/4}) + C$$

(B) $\int \cot\left(2\tan^{-1}\sqrt{\dfrac{\sqrt{1+\sqrt{x}}-\sqrt[4]{x}}{\sqrt{1+\sqrt{x}}+\sqrt[4]{x}}}\right)dx$

$$\therefore \quad \cot\left(2\tan^{-1}\sqrt{\frac{\sqrt{1+\sqrt{x}}-\sqrt[4]{x}}{\sqrt{1+\sqrt{x}}+\sqrt[4]{x}}}\right) = \cot\left(2\tan^{-1}\sqrt{\frac{\sec\theta-\tan\theta}{\sec\theta+\tan\theta}}\right)$$

$$= \cot\left(2\tan^{-1}\sqrt{(\sec\theta-\tan\theta)^2}\right)$$

$$= \cot\left(2\tan^{-1}\frac{1-\sin\theta}{\cos\theta}\right)$$

If $\theta \in (0, \pi/4)$, then $\sec\theta - \tan\theta > 0$

$$= \cot\left(2\tan^{-1}\tan\left(\frac{\pi}{4}-\frac{\theta}{2}\right)\right) = \cot\left(\frac{\pi}{2}-\theta\right) = \tan\theta$$

$$\therefore \quad I = \int \tan\theta\cdot 4\tan^3\theta\sec^2\theta\, d\theta = \frac{4}{5}\tan^5\theta + C = \frac{4}{5}(x^{5/4}) + C$$

(C) $\dfrac{1}{2}\sin^{-1}\left(\dfrac{1-\sqrt{x}}{1+\sqrt{x}}\right) = \dfrac{1}{2}\sin^{-1}\left(\dfrac{1-\tan^2\theta}{1+\tan^2\theta}\right) = \dfrac{1}{2}\sin^{-1}(\cos 2\theta)$

$$= \frac{1}{2}\sin^{-1}\sin\left(\frac{\pi}{2}-2\theta\right) = \frac{\pi}{4}-\theta$$

$$\therefore \quad \int\left(\frac{1-\tan\left(\frac{1}{2}\sin^{-1}\left(\frac{1-\sqrt{x}}{1+\sqrt{x}}\right)\right)}{1+\tan\left(\frac{1}{2}\sin^{-1}\left(\frac{1-\sqrt{x}}{1+\sqrt{x}}\right)\right)}\right)dx = \int \frac{1-\tan\left(\frac{\pi}{4}-\theta\right)}{1+\tan\left(\frac{\pi}{4}-\theta\right)}\,4\tan^3\theta\sec^2\theta\, d\theta$$

$$= \int \tan\left(\frac{\pi}{4}-\left(\frac{\pi}{4}-\theta\right)\right)4\tan^3\theta\sec^2\theta\, d\theta = \int 4\tan^4\theta\sec^2\theta\, d\theta$$

$$= \frac{4}{5}\tan^5\theta + C = \frac{4}{5}(x^{5/4}) + C$$

(D) Let $\sqrt{x} = \tan^2\theta \Rightarrow x = \tan^4\theta$

$\therefore \quad dx = 4\tan^3\theta\sec^2\theta\, d\theta \qquad \begin{bmatrix} \because x \in (0, 1) \\ \therefore \theta \in (0, \pi/4) \end{bmatrix}$

$\therefore \quad \sqrt{1+\sqrt{x}} = \sec\theta$

$$I = \int \sqrt{x}\tan\left(2\tan^{-1}\left(\frac{\sqrt{\sqrt{1+\sqrt{x}}+1} - \sqrt{\sqrt{1+\sqrt{x}}-1}}{\sqrt{\sqrt{1+\sqrt{x}}+1} + \sqrt{\sqrt{1+\sqrt{x}}-1}}\right)\right)dx$$

$$\therefore \quad \sqrt{x}\tan\left(2\tan^{-1}\left(\frac{\sqrt{\sqrt{1+\sqrt{x}}+1} - \sqrt{\sqrt{1+\sqrt{x}}-1}}{\sqrt{\sqrt{1+\sqrt{x}}+1} + \sqrt{\sqrt{1+\sqrt{x}}-1}}\right)\right)$$

$$= \tan^2\theta\cdot\tan\left(2\tan^{-1}\left(\frac{\sqrt{\sec\theta+1} - \sqrt{\sec\theta-1}}{\sqrt{\sec\theta+1} + \sqrt{\sec\theta-1}}\right)\right)$$

$$= \tan^2\theta\cdot\tan\left(2\tan^{-1}\left(\frac{\cos\frac{\theta}{2} - \sin\frac{\theta}{2}}{\cos\frac{\theta}{2} + \sin\frac{\theta}{2}}\right)\right)$$

$$= \tan^2\theta\cdot\tan\left(2\tan^{-1}\tan\left(\frac{\pi}{4} - \frac{\theta}{2}\right)\right)$$

$$= \tan^2\theta\cdot\tan\left(\frac{\pi}{2} - \theta\right) = \tan^2\theta\cdot\cot\theta = \tan\theta$$

$$\therefore \quad I = \int \tan\theta\cdot 4\tan^3\theta\sec^2\theta\, d\theta = \frac{4\tan^5\theta}{5} + C = \frac{4}{5}(x^{5/4}) + C$$

Type 7 : Integer Answer Type Questions

Example 53 If the primitive of the function $f(x) = \dfrac{x^{2009}}{(1+x^2)^{1006}}$ w.r.t. x is equal to $\dfrac{1}{n}\left(\dfrac{x^2}{1+x^2}\right)^m + C$, then $\dfrac{n}{m}$ is equal to ....

Solution. (2) $f(x) = \displaystyle\int \frac{x^{2009}}{(1+x^2)^{1006}}\,dx$

Put $\quad 1 + x^2 = t \Rightarrow 2x\,dx = dt$

$$\therefore \quad I = \frac{1}{2}\int \frac{(t-1)^{1004}\,dt}{t^{1006}} = \frac{1}{2}\int\left(1 - \frac{1}{t}\right)^{1004}\cdot\frac{1}{t^2}\,dt$$

Put $1 - \dfrac{1}{t} = y \Rightarrow \dfrac{1}{t^2}\,dt = dy$

$$\therefore \quad I = \frac{1}{2}\int y^{1004}dy = \frac{1}{2}\cdot\frac{y^{1005}}{1005} + C$$

$$= \frac{1}{2010}\cdot\left(\frac{t-1}{t}\right)^{1005} + C = \frac{1}{2010}\cdot\left(\frac{x^2}{1+x^2}\right)^{1005} + C$$

$$\Rightarrow \quad m = 1005, n = 2010 \Rightarrow \frac{n}{m} = \frac{2010}{1005} = 2$$

Example 54 Suppose $\begin{vmatrix} f'(x) & f(x) \\ f''(x) & f'(x) \end{vmatrix} = 0$ where $f(x)$ is continuous differentiable function with $f'(x) \neq 0$ and satisfies $f(0) = 1$ and $f'(0) = 2$, then $f(x) = e^{\lambda x} + k$, then $\lambda + k$ is equal to ....

Solution. (2) $f'(x) \cdot f'(x) - f(x) \cdot f''(x) = 0$

or $$\frac{[f'(x)]^2 - f(x) f''(x)}{[f'(x)]^2} = 0$$

$$\frac{d}{dx}\left[\frac{f(x)}{f'(x)}\right] = 0$$

Integrating, $$\frac{f(x)}{f'(x)} + C \quad \text{...(i)}$$

Put $x = 0$, $$\frac{f(0)}{f'(0)} = C$$

$\Rightarrow \quad C = \frac{1}{2}$. Hence, $$\frac{f(x)}{f'(x)} = \frac{1}{2} \quad \text{...(ii)}$$

From Eq. (i), $$2f(x) = f'(x)$$

$\therefore$ $$\frac{f'(x)}{f(x)} = 2$$

Again, integrating, $\ln [f(x)] = 2x + k$

Put $x = 0$ to get, $k = 0$

$$f(x) = e^{2x} \Rightarrow \lambda + k = 2 + 0 = 2$$

Example 55 $\int \{\sin(101x) \cdot \sin^{99} x\}\, dx = \frac{\sin(100x)(\sin x)^{\lambda}}{\mu}$, then $\frac{\lambda}{\mu}$ is equal to ......

Solution. (1) $I = \int \{\sin(100x + x) \cdot (\sin x)^{99}\} dx$

$$= \int \{\sin(100x) \cos x + \cos 100x \sin x\} (\sin x)^{99}\, dx$$

$$= \int \underbrace{\sin(100x)}_{\text{I}} \underbrace{\cos x \cdot (\sin x)^{99}}_{\text{II}}\, dx + \int \cos(100x) \cdot (\sin x)^{100}\, dx$$

$$= \frac{\sin(100x)(\sin x)^{100}}{100} - \frac{100}{100} \int \cos(100x)(\sin x)^{100}\, dx + \int \cos(100x)(\sin x)^{100}\, dx$$

$$= \frac{\sin(100x)(\sin x)^{100}}{100} + C$$

$\Rightarrow \quad \lambda = 100, \mu = 100$

$\Rightarrow \quad \frac{\lambda}{\mu} = \frac{100}{100} = 1$

Proficiency Exercise 1

Type 1 : Only One Correct Option

1. Let $f(x)=\int \dfrac{x^2}{(1+x^2)(1+\sqrt{1+x^2})}\,dx$ and $f(0)=0$. Then $f(1)$ is equal to

(a) $\log_e(1+\sqrt{2})$ (b) $\log_e(1+\sqrt{2})-\dfrac{\pi}{4}$

(c) $\log_e(1+\sqrt{2})+\dfrac{\pi}{4}$ (d) None of these

2. If $\int f(x)\,dx=f(x)$, then $\int \{f(x)\}^2\,dx$ is equal to

(a) $\dfrac{1}{2}\{f(x)\}^2$ (b) $\{f(x)\}^3$ (c) $\dfrac{\{f(x)\}^3}{3}$ (d) $\{f(x)\}^2$

3. If $\int f(x)\,dx=F(x)$, then $\int x^3 f(x^2)\,dx$ is equal to

(a) $\dfrac{1}{2}[x^2\{F(x)\}^2-\int \{F(x)\}^2\,dx]$ (b) $\dfrac{1}{2}[x^2F(x^2)-\int F(x^2)\,d(x^2)]$

(c) $\dfrac{1}{2}[x^2F(x)-\dfrac{1}{2}\int \{F(x)\}^2\,dx]$ (d) None of these

4. If n is an odd positive integer, then $\int |x^n|\,dx$ is equal to

(a) $\left|\dfrac{x^{n+1}}{n+1}\right|+C$ (b) $\dfrac{x^{n+1}}{n+1}+C$

(c) $\dfrac{|x^n|\,x}{n+1}+C$ (d) None of these

5. Let $F(x)$ be the primitive of $\dfrac{3x+2}{\sqrt{x-9}}$ w.r.t. x. If $F(10)=60$, then the value of $F(13)$ is

(a) 66 (b) 132 (c) 248 (d) 264

6. $\int (x^x)^x\,(2x\log_e x+x)\,dx$ is equal to

(a) $x^{(x^x)}+C$ (b) $(x^x)^x+C$

(c) $x^2\cdot\log_e x+C$ (d) None of these

7. The value of $\int x\log x(\log x-1)\,dx$ is equal to

(a) $2(x\log x-x)^2+C$ (b) $\dfrac{1}{2}(x\log x-x)^2+C$

(c) $(x\log x)^2+C$ (d) $\dfrac{1}{2}(x\log x)^3+C$

8. $\int \frac{x^2-1}{x^3\sqrt{2x^4-2x^2+1}}\,dx$ is equal to

(a) $\frac{\sqrt{2x^4-2x^5+1}}{x^2}+C$ (b) $\frac{\sqrt{2x^4-2x^2+1}}{x^3}+C$

(c) $\frac{\sqrt{2x^4+2x^2+1}}{x}+C$ (d) $\frac{\sqrt{2x^4-2x^2+1}}{2x^2}+C$

9. Let $f(x)$ be a polynomial satisfying $f(0)=2$, $f'(0)=3$ and $f''(x)=f(x)$. Then $f(4)$ is equal to

(a) $\frac{5(e^8+1)}{2e^4}$ (b) $\frac{5(e^8-1)}{2e^4}$ (c) $\frac{2e^4}{5(e^8-1)}$ (d) $\frac{2e^4}{5(e^8+1)}$

10. $\int \frac{e^{(x^2+4\ln x)}-x^3e^{x^2}}{x-1}\,dx$ is equal to

(a) $\left(\frac{e^{3\ln x}-e^{\ln x}}{2x}\right)e^{x^2}+C$ (b) $\frac{(x-1)xe^{x^2}}{2}+C$

(c) $\frac{(x^2-1)}{2x}-e^{x^2}+C$ (d) None of these

11. $\int \tan^4 x\,dx = A\tan^3 x + B\tan x + f(x)$, then

(a) $A=\frac{1}{3}, B=-1, f(x)=x+C$ (b) $A=\frac{2}{3}, B=-1, f(x)=x+C$

(c) $A=\frac{1}{3}, B=1, f(x)=x+C$ (d) $A=\frac{2}{3}, B=1, f(x)=-x+C$

12. If the anti-derivative of $\int \frac{\sin^4 x}{x}\,dx$ is $f(x)$, then $\int \frac{\sin^4\{(p+q)x\}}{x}\,dx$ in terms of $f(x)$ is

(a) $f\{(p+q)x\}$ (b) $\frac{f\{(p+q)x\}}{p+q}$

(c) $f\{(p+q)x\}(p+q)$ (d) None of these

13. $\int\left(\frac{\sin\theta}{\cos 3\theta}+\frac{\sin 3\theta}{\cos 9\theta}+\frac{\sin 9\theta}{\cos 27\theta}\right)d\theta$ is equal to

(a) $\frac{1}{2}\log\left|\frac{\sec 27\theta}{\sec\theta}\right|+C$ (b) $\frac{1}{2}\log\left|\frac{\sec\theta}{\sec 27\theta}\right|+C$

(c) $\frac{1}{2}\log\left|\frac{\sqrt[27]{\sec 27\theta}}{\sec\theta}\right|+C$ (d) None of these

14. Let $x^2 \neq n\pi - 1$, $n \in N$. Then, the value of $\int x\sqrt{\frac{2\sin(x^2+1)-\sin 2(x^2+1)}{2\sin(x^2+1)+\sin 2(x^2+1)}}\,dx$ is equal to

(a) $\log\left|\frac{1}{2}\sec(x^2+1)\right|+C$ (b) $\log\left|\sec\left(\frac{x^2+1}{2}\right)\right|+C$

(c) $\frac{1}{2}\log|\sec(x^2+1)|+C$ (d) None of these

15. $\int \frac{dx}{\cos(2x)\cos(4x)}$ is equal to

(a) $\frac{1}{2\sqrt{2}} \log \left| \frac{1+\sqrt{2}\sin 2x}{1-\sqrt{2}\sin 2x} \right| - \frac{1}{2}(\log|\sec 2x - \tan 2x|) + C$

(b) $\frac{1}{2\sqrt{2}} \log \left| \frac{1+\sqrt{2}\sin 2x}{1+\sqrt{2}\sin x} \right| - \frac{1}{2}(\log|\sec 2x - \tan 2x|) + C$

(c) $\frac{1}{\sqrt{2}} \log \left| \frac{1+\sqrt{2}\sin 2x}{1-\sqrt{2}\sin 2x} \right| - \frac{1}{2}(\log|\sec 2x - \tan 2x|) + C$

(d) None of the above

16. $\int \frac{1-7\cos^2 x}{\sin^7 x \cos^2 x} dx = \frac{f(x)}{(\sin x)^7} + C$, then $f(x)$ is equal to

(a) $\sin x$ (b) $\cos x$ (c) $\tan x$ (d) $\cot x$

17. $\int \frac{\sin^3 x}{(\cos^4 x + 3\cos^2 x + 1)\tan^{-1}(\sec x + \cos x)} dx$ is equal to

(a) $\tan^{-1}(\sec x + \cos x) + C$ (b) $\log_e |\tan^{-1}(\sec x + \cos x)| + C$

(c) $\frac{1}{(\sec x + \cos x)^2} + C$ (d) None of these

18. The primitive of the function $f(x) = x|\cos x|$, when $\frac{\pi}{2} < x < \pi$ is given by

(a) $\cos x + x\sin x + C$ (b) $-\cos x - x\sin x + C$

(c) $x\sin x - \cos x + C$ (d) None of these

19. The primitive of the function $f(x) = (2x+1)|\sin x|$, when $\pi < x < 2\pi$ is

(a) $-(2x+1)\cos x + 2\sin x + C$ (b) $(2x+1)\cos x - 2\sin x + C$

(c) $(x^2 + x)\cos x + C$ (d) None of these

20. Given, $f(x) = \begin{vmatrix} 0 & x^2 - \sin x & \cos x - 2 \\ \sin x - x^2 & 0 & 1 - 2x \\ 2 - \cos x & 2x - 1 & 0 \end{vmatrix}$, then $\int f(x)\, dx$ is equal to

(a) $\frac{x^3}{3} - x^2 \sin x + \sin 2x + C$ (b) $\frac{x^3}{3} - x^2 \sin x - \cos 2x + C$

(c) $\frac{x^3}{3} - x^2 \cos x - \cos 2x + C$ (d) None of these

Type 2 : More Than One Correct Options

21. $\int \frac{dx}{(x+1)(x-2)} = A\log(x+1) + B\log(x-2) + C$, where

(a) $A + B = 0$ (b) $AB = 0$ (c) $A/B = -1$ (d) None of these

22. If $\int \frac{dx}{(x^2+1)(x^2+4)} = k\tan^{-1} x + l\tan^{-1}\frac{x}{2} + C$, then

(a) $k = \frac{1}{3}$ (b) $l = \frac{2}{3}$ (c) $k = -\frac{1}{3}$ (d) $l = -\frac{1}{6}$

23. If $\int x \log(1+x^2)\,dx = \phi(x)\log(1+x^2) + x(\psi) + C$, then

(a) $\phi(x) = \dfrac{1+x^2}{2}$ (b) $\psi(x) = \dfrac{1+x^2}{2}$ (c) $\psi(x) = -\dfrac{1+x^2}{2}$ (d) $\phi(x) = -\dfrac{1+x^2}{2}$

24. If $\int \dfrac{4e^x + 6e^{-x}}{9e^x - 4e^{-x}}\,dx = Ax + B\log_e(9e^{2x} - 4) + C$, then

(a) $A = \dfrac{3}{2}$ (b) $B = \dfrac{35}{36}$ (c) C is indefinite (d) $A + B = -\dfrac{19}{36}$

25. If $\int \tan^5 x\,dx = A\tan^4 x + B\tan^2 x + g(x) + C$, then

(a) $A = \dfrac{1}{4}, B = -\dfrac{1}{2}$ (b) $g(x) = \ln|\sec x|$ (c) $g(x) = \ln|\cos x|$ (d) $A = -\dfrac{1}{4}, B = \dfrac{1}{3}$

Type 3 : Assertion and Reason

Directions
(Q. Nos. 26 to 30)

For the following questions, choose the correct answers from the codes (a), (b), (c) and (d) defined as follows :

(a) Statement I is true, Statement II is also true; Statement II is the correct explanation of Statement I.
(b) Statement I is true, Statement II is also true; Statement II is not the correct explanation of Statement I.
(c) Statement I is true, Statement II is false.
(d) Statement I is false, Statement II is true.

26. Statement I If y is a function of x such that $y(x-y)^2 = x$, then

$$\int \frac{dx}{x-3y} = \frac{1}{2}[\log(x-y)^2 - 1]$$

Statement II $\int \dfrac{dx}{x-3y} = \log(x-3y) + C$

27. Statement I Integral of an even function is not always an odd function.

Statement II Integral of an odd function is an even function.

28. Statement I If $a > 0$ and $b^2 - 4ac < 0$, then the value of the integral $\int \dfrac{dx}{ax^2 + bx + c}$ will be of the type $\mu \tan^{-1}\dfrac{x+A}{B} + C$, where A, B, C, μ are constants.

Statement II If $a > 0$, $b^2 - 4ac < 0$, then $ax^2 + bx + C$ can be written as sum of two squares.

29. Statement I $\int \left(\dfrac{1}{1+x^4}\right) dx = \tan^{-1}(x^2) + C$

Statement II $\int \dfrac{1}{1+x^2}\,dx = \tan^{-1} x + C$

30. Statement I $\int 2^{\tan^{-1} x} d(\cot^{-1} x) = \dfrac{2^{\tan^{-1} x}}{\ln 2} + C$

Statement II $\dfrac{d}{dx}(a^x + C) = a^x \ln a$

Type 4 : Linked–Comprehension Based Questions

Passage I
(Q. Nos. 31 to 33)

Let us consider the integral of the following forms

$$f(x_1, \sqrt{mx^2 + nx + p})^{1/2}$$

Case I *If* $m > 0$, *then put* $\sqrt{mx^2 + nx + C} = u \pm x\sqrt{m}$

Case II *If* $p > 0$, *then put* $\sqrt{mx^2 + nx + C} = ux \pm \sqrt{p}$

Case III *If quadratic equation* $mx^2 + nx + p = 0$ *has real roots* α *and* β *there put*

$$\sqrt{mx^2 + nx + p} = (x - \alpha)u \text{ or } (x - \beta)u$$

31. If $I = \int \dfrac{dx}{x - \sqrt{9x^2 + 4x + 6}}$ to evaluate I, one of the most proper substitution could be

(a) $\sqrt{9x^2 + 4x + 6} = u \pm 3x$ (b) $\sqrt{9x^2 + 4x + 6} = 3u \pm x$

(c) $x = \dfrac{1}{t}$ (d) $9x^2 + 4x + 6 = \dfrac{1}{t}$

32. $\int \dfrac{(x + \sqrt{1 + x^2})^{15}}{\sqrt{1 + x^2}} dx$ is equal to

(a) $\dfrac{(x + \sqrt{1 + x^2})^{16}}{10} + C$ (b) $\dfrac{1}{15(\sqrt{1 + x^2} + x)} + C$

(c) $\dfrac{15}{(\sqrt{1 + x^2} - x)} + C$ (d) $\dfrac{(x + \sqrt{1 + x^2})^{15}}{15} + C$

33. To evaluate $\int \dfrac{dx}{(x-1)\sqrt{-x^2 + 3x - 2}}$ one of the most suitable substitution could be

(a) $\sqrt{-x^2 + 3x - 2} = u$ (b) $\sqrt{-x^2 + 3x - 2} = (ux\sqrt{2})$

(c) $\sqrt{-x^2 + 3x - 2} = u(1 - x)$ (d) $\sqrt{-x^2 + 3x - 2} = u(x - 2)$

Passage II
(Q. Nos. 34 to 36)

Let $I_{n, m} = \int \sin^n x \cos^m x \, dx$. *Then, we can relate* $I_{n, m}$ *with each of the following :*

(i) $I_{n-2, m}$ (ii) $I_{n+2, m}$ (iii) $I_{n, m-2}$ (iv) $I_{n, m+2}$

(v) $I_{n-2, m+2}$ (vi) $I_{n+2, m-2}$

Suppose we want to establish a relation between $I_{n, m}$ *and* $I_{n, m-2}$, *then we get*

$$P(x) = \sin^{n+1} x \cos^{m-1} x \quad \text{...(i)}$$

In $I_{n,m}$ and $I_{n,m-2}$ the exponent of $\cos x$ is m and $m-2$ respectively, the minimum of the two is $m-2$, adding 1 to the minimum we get $m-2+1=m-1$. Now, choose the exponent $m-1$ of $\cos x$ in $P(x)$. Similarly, choose the exponent of $\sin x$ for $P(x)$.

Now, differentiating both the sides of Eq. (i), *we get*

$$P'(x)=(n+1)\sin^n x\cos^m x-(m-1)\sin^{n+2}x\cos^{m-2}x$$
$$=(n+1)\sin^n x\cos^m x-(m-1)\sin^n x(1-\cos^2 x)\cos^{m-2}x$$
$$=(n+1)\sin^n x\cos^m x-(m-1)\sin^n x\cos^{m-2}x+(m-1)\sin^n x\cos^n x$$
$$=(n+m)\sin^n x\cos^m x-(m-1)\sin^n x\cos^{m-2}x$$

Now, integrating both the sides, we get

$$\sin^{n+1}x\cos^{m-1}x=(n+m)I_{n,m}-(m-1)I_{n,m-2}$$

Similarly, we can establish the other relations.

34. The relation between $I_{4,2}$ and $I_{2,2}$ is

(a) $I_{4,2}=\frac{1}{6}(-\sin^3 x\cos^3 x+3I_{2,2})$
(b) $I_{4,2}=\frac{1}{6}(\sin^3 x\cos^3 x+3I_{2,2})$
(c) $I_{4,2}=\frac{1}{6}(\sin^3 x\cos^3 x-3I_{2,2})$
(d) $I_{4,2}=\frac{1}{4}(-\sin^3 x\cos^3 x+2I_{2,2})$

35. The relation between $I_{4,2}$ and $I_{6,2}$ is

(a) $I_{4,2}=\frac{1}{5}(\sin^5 x\cos^3 x+8I_{6,2})$
(b) $I_{4,2}=\frac{1}{5}(-\sin^5 x\cos^3 x+8I_{6,2})$
(c) $I_{4,2}=\frac{1}{5}(\sin^5 x\cos^3 x-8I_{6,2})$
(d) $I_{4,2}=\frac{1}{6}(\sin^5 x\cos^3 x+8I_{6,2})$

36. The relation between $I_{4,2}$ and $I_{4,4}$ is

(a) $I_{4,2}=\frac{1}{3}(\sin^5 x\cos^3 x+8I_{4,4})$
(b) $I_{4,2}=\frac{1}{3}(-\sin^5 x\cos^3 x+8I_{4,4})$
(c) $I_{4,2}=\frac{1}{3}(\sin^5 x\cos^3 x-8I_{4,4})$
(d) $I_{4,2}=\frac{1}{3}(\sin^5 x\cos^3 x+6I_{4,4})$

Type 5 : **Match the Columns**

37. Match the following :

Column I	Column II
(A) If $I=\int\frac{\sin x-\cos x}{\lvert\sin x-\cos x\rvert}dx$, where $\frac{\pi}{4}<x<\frac{3\pi}{8}$, then I is equal to	(p) $\sin x$
(B) If $\int\frac{x^2}{(x^3+1)(x^3+2)}dx=\frac{1}{3}f\left(\frac{x^3+1}{x^3+2}\right)+C$, then $f(x)$ is equal to	(q) $x+C$
(C) If $\int\sin^{-1}x\cdot\cos^{-1}x\,dx=f^{-1}(x)\left[\frac{\pi}{2}x-xf^{-1}(x)-2\sqrt{1-x^2}\right]+2x+C$, then $f(x)$ is equal to	(r) $\ln\lvert x\rvert$
(D) If $\int\frac{dx}{xf(x)}=f(f(x))+C$, then $f(x)$ is equal to	(s) $\sin^{-1}x$

38. Match the following :

Column I	Column II
(A) If $\int \left(\dfrac{x^2+\cos^2 x}{1+x^2}\right)\operatorname{cosec}^2 x\,dx = A\cot^{-1} x + B\,\dfrac{\operatorname{cosec} x}{\sec x}$, then	(p) $A = 1$
(B) If $\int \sqrt{x+\sqrt{x^2+2}}\,dx = \dfrac{A}{3}(x+\sqrt{x^2+2})^{3/2} - \dfrac{B}{(x+\sqrt{x^2+2})}$, then	(q) $B = -1$
(C) If $\int \dfrac{\sqrt{2-x-x^2}}{x^2}\,dx = A\sqrt{\dfrac{2-x-x^2}{x}} + \dfrac{B}{4\sqrt{2}}\log\left(\dfrac{4-x+4\sqrt{2-x-x^2}}{x}\right) - \sin^{-1}\left(\dfrac{2x+1}{3}\right)$, then	(r) $B = 2$
(D) If $\int \dfrac{\sin 2x}{\sin^4 x + \cos^4 x}\,dx = B\cot^{-1}(\tan^2 x)$, then	(s) $A = -1$

Type 6 : Integer Answer Type Questions

39. If $\int \dfrac{(2x+3)\,dx}{x(x+1)(x+2)(x+3)+1} = C - \dfrac{1}{f(x)}$, where $f(x)$ is of the form of $ax^2 + bx + c$, then $(a+b+c)$ equals to ....

40. Let $F(x)$ be the primitive of $\dfrac{3x+2}{\sqrt{x-9}}$ w.r.t. x. If $F(10) = 60$, then the sum of digits of the value of $F(13)$, is ....

41. Let $u(x)$ and $v(x)$ are differentiable function such that $\dfrac{u(x)}{v(x)} = 7$. If $\dfrac{u'(x)}{v'(x)} = p$ and $\left(\dfrac{u(x)}{v(x)}\right)' = q$, then $\dfrac{p+q}{p-q}$ has the value equal to ....

42. If $\int \dfrac{1}{(x^2-1)}\ln\left(\dfrac{x-1}{x+1}\right)dx = 6A\left[\ln\left(\dfrac{x-1}{x+1}\right)\right]^2 + C$, then find $24A$.

43. If $\int \dfrac{e^x(2-x^2)}{(1-x)\sqrt{1-x^2}}\,dx = \mu e^x\left(\dfrac{1+x}{1-x}\right)^{\lambda} + C$, then $2(\lambda+\mu)$ is equal to ....

44. If $\int \dfrac{\cos x - \sin x + 1 - x}{e^x + \sin x + x}\,dx = \ln\{f(x)\} + g(x) + C$, where C is the constant of integrating and $f(x)$ is positive, then $\dfrac{f(x)+g(x)}{e^x + \sin x}$ is equal to ....

45. Suppose $A = \int \dfrac{dx}{x^2+6x+25}$ and $B = \int \dfrac{dx}{x^2-6x-27}$.

If $12(A+B) = \lambda\cdot\tan^{-1}\left(\dfrac{x+3}{4}\right) + \mu\cdot\ln\left|\dfrac{x-9}{x+3}\right| + C$, then the value of $(\lambda+\mu)$ is ... .

46. If $\int \frac{\cos 6x + \cos 9x}{1 - 2\cos 5x}\, dx = -\frac{\sin 4x}{k} - \sin x + C$, then the value of k is ....

47. The value of $\int \frac{\tan x}{1 + \tan x + \tan^2 x}\, dx = x - \frac{1}{\sqrt{A}} \tan^{-1}\left(\frac{2\tan x + 1}{\sqrt{A}}\right) + C$, then the value of A is......... .

48. $\int \sin^{5/2} x \cos^3 x\, dx = 2\sin^{A/2} x \left[\frac{1}{B} - \frac{1}{C}\sin^2 x\right] + C$, then the value of $(A + B) - C$ is equal to ....

Proficiency Exercise 2

Integrate the following :

1. $\frac{(x+a)^3}{x^3}+\frac{1-x^4}{1-x}-\frac{7}{x\sqrt{x^2-1}}+\frac{x+2}{(x+1)^2}+\frac{(1+x)^3}{\sqrt{x}}$

2. $\frac{5\cos^3 x+2\sin^3 x}{2\sin^2 x\cos^2 x}+\sqrt{1+\sin 2x}+\frac{1+2\sin x}{\cos^2 x}+\frac{1-\cos 2x}{1+\cos 2x}$

$+\sqrt{1+\sin\frac{x}{2}}+\frac{\sin^2 x}{(1+\cos x)^2}+(\tan x+\cot x)^2$

3. $\sec^2(2-3x)+\frac{1}{\sin^2(3-4x)}+\frac{\sin(2-3x)}{\cos^2(2-3x)}+\cos x\cdot\operatorname{cosec}^2 x$

$+\operatorname{cosec}(3x-2)\cot(3x-2)$

4. $\frac{\sin x}{1+\sin x}+\sin x\sin 2x\sin 3x+\sec^2 x\cos^2 2x+\sin^4 x\cos^4 x$

5. $\frac{\sin x\cos x}{a^2\cos^2 x+b^2\sin^2 x}$

6. $\frac{1}{e^x+1}$

7. $\frac{1}{\cos^2 x(1-\tan x)^2}+\frac{10x^9+10^x\log_e 10}{10^x+x^{10}}$

8. $\sec^p x\cdot\tan x$

9. $\frac{1}{x\log x\log(\log x)}$

10. $\frac{1}{x^2(x^4+1)^{3/4}}$

11. $x\cdot\sqrt{\frac{2\sin(x^2-1)-\sin 2(x^2-1)}{2\sin(x^2-1)+\sin 2(x^2-1)}}$, where $x^2-1\neq n\pi$

12. $x^{13/2}(1+x^{5/2})^{1/2}$

13. $[1+2\tan x(\tan x+\sec x)]^{1/2}$

14. $\frac{1}{\sqrt{\sin^3 x\sin(x+a)}}$

15. $\frac{\sec x}{\sqrt{\sin(2x+\alpha)+\sin\alpha}}$

16. $\frac{\cos 2x-\cos 2\alpha}{\cos x-\cos\alpha}$

17. $\frac{1+\cos 4x}{\cot x-\tan x}$

18. $\sqrt{\frac{x}{a^3-x^3}}$

19. $\frac{\log x\sin\{1+(\log x)^2\}}{x}$

20. $\dfrac{a^{x+\tan^{-1} a^x}}{1+a^{2x}}$

21. $\dfrac{1}{(2ax+x^2)^{3/2}}$

22. $\dfrac{1}{(1+x^2)\sqrt{p^2+q^2(\tan^{-1}x)^2}}$

23. $x^3\sqrt{q^2x^8-p^2}$

24. $\dfrac{x}{\sqrt{9+8x-x^2}}$

25. $x\sqrt{\dfrac{a^2-x^2}{a^2+x^2}}$

26. $\dfrac{1}{\sin^2 x+\sin 2x}$

27. $\cos 2\theta \log\left(\dfrac{\cos\theta+\sin\theta}{\cos\theta-\sin\theta}\right)$

28. $\log\{x+\sqrt{x^2+a^2}\}$

29. $\dfrac{xe^x}{(x+1)^2}$

30. $\dfrac{\log x}{(1+\log x)^2}$

31. $\log(1+\cos x)-x\tan(x/2)$

32. $\dfrac{\log x}{x^3}$

33. $\dfrac{2x+\sin 2x}{1+\cos 2x}$

34. $\dfrac{x^4}{(x-1)(x^2+1)}$

35. $\dfrac{x}{\sqrt{(x+a)}+\sqrt{(x+b)}}$

36. $\dfrac{x^2+1}{(x^2+2)(2x^2+1)}$.

37. $\sqrt{\dfrac{1-\sqrt{x}}{1+\sqrt{x}}}$

38. $\dfrac{x^4-x^3+5x^2+x+3}{(x+1)(x^2-x+1)^2}$

39. $\dfrac{1}{(x+1)\sqrt{2+x-x^2}}$

40. $\dfrac{1}{(x^2-6x+9)\sqrt{x^2-6x+4}}$

41. $\left(\dfrac{1-\sqrt{x}}{1+\sqrt{x}}\right)^{1/2}\cdot\dfrac{1}{x}$

42. $\dfrac{1}{x^2-2x\cos\theta+1}$

43. $\dfrac{(\sin x-\cos x)}{(\sin x+\cos x)\sqrt{\sin x\cos x+\sin^2 x\cos^2 x}}$

44. $\dfrac{(\cos 2x)^{1/2}}{\sin x}$

45. $\dfrac{1}{\sqrt{\sin^3 x\cos^5 x}}$

46. $\dfrac{\sin\theta+\cos\theta}{\sqrt{\sin 2\theta}}$

47. $\dfrac{1}{(x+\alpha)^{8/7}(x-\beta)^{6/7}}$

48. $a^x\left[\log x+\log a\log\left(\dfrac{x^x}{e^x}\right)\right]$

49. $e^{(x\sin x+\cos x)}\left[\dfrac{x^4\cos^3 x-x\sin x+\cos x}{x^2\cos^2 x}\right]$

50. $\sin^{-1}\sqrt{\dfrac{x}{a+x}}$

51. Let $f:[0, \pi/2] \to R$, such that

$f(0)=3$ and $f'(x)=\dfrac{1}{1+\cos x}$, then prove that $3+\dfrac{\pi}{4} \le f(\pi/2) \le 3+\dfrac{\pi}{2}$

52. If $\int \dfrac{4e^x+6e^{-x}}{9e^x-4e^{-x}}\,dx = Ax + B\log(9e^{2x}-4)+C$, then find A, B, C.

53. If $\int f(x)\sin x\cos x\,dx = \dfrac{1}{2(b^2-a^2)}\log f(x)+C$, then find $f(x)$.

54. If $l^r(x)$ means log log log ... x, the log being repeated r times. Then, $\int \{x\,l(x)\,l^2(x)\,l^3(x)\ldots l^r(x)\}^{-1}\,dx$

55. Evaluate $\int \dfrac{dx}{(x-b)^3(x-a)^2}$ (without using partial fractions)

56. Evaluate $\int \sqrt{x+\sqrt{x^2+2}}\,dx$.

57. If $I=\int(\sqrt{\tan x}+\sqrt{\cot x})\,dx = f(x)+C$, then find $f(x)$.

58. Evaluate $\int \dfrac{dx}{(\sqrt{(x-\alpha)^2-\beta^2})(ax+b)}$.

59. Evaluate $\int \dfrac{\sqrt{1+\sqrt[3]{x}}}{\sqrt[3]{x^2}}\,dx$.

60. Evaluate $\int \tan(x-\alpha)\cdot\tan(x+\alpha)\cdot\tan 2x\,dx$.

61. The estimated value of $\dfrac{1}{1000}+\dfrac{1}{1001}+\dfrac{1}{1002}+\dfrac{1}{1003}+\ldots+\dfrac{1}{2000}$.

62. Evaluate $\int \dfrac{\sin^3(\theta/2)\,d\theta}{\cos\theta/2\sqrt{\cos^3\theta+\cos^2\theta+\cos\theta}}$.

63. Evaluate $\int x^{x+1}\log x\,(1+\log x)\,dx$.

64. Evaluate $\int \dfrac{(2\sin\theta+\sin 2\theta)\,d\theta}{(\cos\theta-1)\sqrt{\cos\theta+\cos^2\theta+\cos^3\theta}}$.

65. Connect $\int x^{m-1}(a+bx^n)^p\,dx$ with $\int x^{m-n-1}(a+bx^n)^p\,dx$ and evaluate $\int \dfrac{x^8\,dx}{(1-x^3)^{1/3}}$.

66. Evaluate $\int \dfrac{(\sin^{3/2}\theta+\cos^{3/2}\theta)\,d\theta}{\sqrt{\sin^3\theta\cdot\cos^3\theta\cdot\sin(\theta+\alpha)}}$.

67. Evaluate $\int \operatorname{cosec}^2 x\,\ln(\cos x+\sqrt{\cos 2x})\,dx$.

68. Evaluate $\int \dfrac{dx}{(\sin x + a \sec x)^2}$, $a \in N$.

69. Evaluate $\int \dfrac{dx}{x - \sqrt{x^2 + 2x + 4}}$.

70. Evaluate $\int \dfrac{dx}{1 + \sqrt{x^2 + 2x + 2}}$.

71. Evaluate $\int \dfrac{x^4 + 1}{x^6 + 1}\, dx$.

72. Evaluate $\int \dfrac{dx}{(1 - x^2)^{1/3}}\, dx$.

73. Evaluate $\int \dfrac{(x + \sqrt{1 + x^2})^{15}}{\sqrt{1 + x^2}}\, dx$.

74. If $y^2 = ax^2 + 2bx + c$, and $u_n = \int \dfrac{x^n}{y}\, dx$, prove that $(n + 1)\, a\, u_{n+1} + (2n + 1)\, b\, u_n$ $+ {}^n c\, y_{n-1} = x^n y$ and deduce that
$a u_1 = y - b\, u_0$; $\quad 2a^2 u_2 = y\,(ax - 3b) - (ac - 3b^2)\, u_0$.

75. If $I(n) = \int \dfrac{x^n}{(ax^2 + bx + c)^{1/2}}\, dx$, $n \in N$, express $I(n + 1)$ in terms of $I(n)$ and $I(n - 1)$.

Answers

Target Exercise 1.1

1. $-\dfrac{2}{x} + \tan^{-1} x + C$

2. $\dfrac{2}{3}(x + 1)^{3/2} + \dfrac{2}{3} x^{3/2} + C$

3. $C - \dfrac{2}{x} + \dfrac{2}{3}\dfrac{1}{x^3} - \dfrac{3}{5}\dfrac{1}{x^5} - 2\tan^{-1} x$

4. $-\dfrac{1}{x} + \tan^{-1} x + C$

5. $\ln x + 2\tan^{-1} x + C$

6. $\dfrac{x^3}{3} - x + \tan^{-1} x + C$

7. $\dfrac{x^5}{5} - \dfrac{x^3}{3} + x - 2\tan^{-1} x + C$

8. $\dfrac{1}{2}\left[\dfrac{x^3}{3} + \tan^{-1} x\right] + C$

9. $\dfrac{x^2}{2} - x + C$

10. $\dfrac{4}{\sqrt{x}} - \dfrac{2}{3} x\sqrt{x} + C$

11. $x^2 + x + C$

12. $\tan x - \tan^{-1} x + C$

13. $\dfrac{2^x \cdot e^x}{1 + \ln 2} + C$

14. $\dfrac{e^{4x}}{4} + C$

15. $\dfrac{x^{a+1}}{a + 1} + \dfrac{a^x}{\ln a} + C$

16. $\tan x - \sec x + C$

17. $-\dfrac{1}{64}\cos 8x + C$

18. $\dfrac{1}{2}(\tan x + x) + C$

19. $\frac{1}{2}\sin 2x + C$

20. $-(\cot x + \tan x) + C$

21. $-\left[\frac{1}{9}\cos 9x + \frac{1}{10}\cos 10x + \frac{1}{11}\cos 11x + \frac{1}{12}\cos 12x\right] + C$

22. $\sin 2x + C$

23. $-\frac{\cos 3x}{3} + C$

24. $\frac{180}{\pi}\sin x^\circ + C$

25. $\tan x - x + C$

26. $x + 2\sin x + C$

27. $x + C$

28. $2(\sin x + x\cos\alpha) + C$

29. $\sec x - \operatorname{cosec} x + C$

30. $\tan x - \cot x + C$

31. $(\sin x + \cos x)\operatorname{sgn}(\cos x - \sin x) + C$

32. $\tan x - \cot x - 3x + C$

33. $-\sqrt{2}\cos\frac{x}{2} + C$

34. $-\frac{\cos 4x}{8} + C$

35. $\frac{1}{2}(x - \sin x) + C$

36. $-2\cos x + C$

37. $-\frac{\cot 8x}{8} + C$

38. $\frac{x}{\sqrt{2}} + C$

Target Exercise 1.2

1. (a)	**2.** (c)	**3.** (b)	**4.** (a)	**5.** (c)

Target Exercise 1.3

1. (c)	**2.** (b)	**3.** (b)	**4.** (c)	**5.** (b)

Proficiency Exercise 1

1. (b)	**2.** (a)	**3.** (b)	**4.** (c)	**5.** (b)
6. (b)	**7.** (b)	**8.** (d)	**9.** (b)	**10.** (d)
11. (a)	**12.** (a)	**13.** (c)	**14.** (b)	**15.** (a)
16. (c)	**17.** (b)	**18.** (b)	**19.** (b)	**20.** (d)
21. (a, c)	**22.** (a, d)	**23.** (a, c)	**24.** (b, c, d)	**25.** (a, b)
26. (c)	**27.** (c)	**28.** (a)	**29.** (d)	**30.** (d)
31. (a)	**32.** (d)	**33.** (c)	**34.** (a)	**35.** (a)

36. (b)

37. (A) → (q); (B) → (r); (C) → (p); (D) → (r)

38. (A) → (p, q); (B) → (p, r); (C) → (r); (D) → (q)

39. (5)	**40.** (6)	**41.** (1)	**42.** (1)	**43.** (3)
44. (1)	**45.** (4)	**46.** (4)	**47.** (3)	**48.** (3)

Proficiency Exercise 2

5. $\dfrac{1}{2(b^2-a^2)}\log|a^2\cos^2 x+b^2\sin^2 x|+C$

6. $x-\log|e^x+1|+C$

7. $\dfrac{1}{(1-\tan x)}+\log|10^x+x^{10}|+C$

8. $\dfrac{(\sec^p x)}{p}+C$

9. $\log\{\log(\log x)\}+C$

10. $-\dfrac{1}{x}(x^4+1)^{1/4}+C$

11. $\log\left|\sec\left(\dfrac{x^2-1}{2}\right)\right|+C$

13. $\log|\sec x(\sec x+\tan x)|+C$

14. $-2(\operatorname{cosec}a)\sqrt{\cos a+\sin a\cot x}+C$

15. $\dfrac{\sqrt{2}}{\cos\alpha}\sqrt{\cos\alpha\cdot\tan x+\sin\alpha}+C$

17. $-\dfrac{1}{8}\cos 4x+C$

19. $-\dfrac{1}{2}\cos[1+(\log x)^2]+C$

20. $\dfrac{a^{\tan^{-1}(a^x)}}{(\log a)^2}+C$

21. $-\dfrac{1}{a^2}\cdot\dfrac{x+a}{\sqrt{x^2+2ax}}+C$

24. $-\sqrt{9+8x-x^2}+4\sin^{-1}\left(\dfrac{x-4}{5}\right)+C$

26. $\dfrac{1}{2}\log\left|\dfrac{\tan x}{\tan x+2}\right|+C$

29. $e^x\cdot\dfrac{1}{x+1}+C$

30. $\dfrac{x}{1+\log x}+C$

31. $x\log(1+\cos x)+C$

32. $-\dfrac{1}{4x^2}(2\log_e x+1)+C$

33. $x\tan x+C$

39. $-\dfrac{2}{3}\sqrt{3\left(\dfrac{1}{x+1}\right)-1}+C$

40. $\dfrac{1}{5}\cdot\dfrac{\sqrt{x^2-6x+4}}{(x-3)}+C$

45. $\dfrac{2}{3}\left\{\dfrac{\tan^2 x+3}{\tan x}\right\}+C$

46. $\sin^{-1}(\sin\theta-\cos\theta)+C$

47. $\dfrac{7}{\alpha+\beta}\left(\dfrac{x-\beta}{x+\alpha}\right)^{1/7}+C$

48. $a^x\cdot x(\log x-1)+C$

50. $(a+x)\tan^{-1}\sqrt{\dfrac{x}{a}}-\sqrt{ax}+C$

52. $A=-\dfrac{3}{2}, B=\dfrac{35}{36}, C=-\dfrac{3}{2}\log 3$

53. $f(x)=\dfrac{2\sec(2x)}{(b^2-a^2)}$

54. $l^{r+1}(x)+C$

57. $\sqrt{2}\sin^{-1}(\sin x-\cos x)+C$

61. $\log 2$

62. $\tan^{-1}(\cos\theta+\sec\theta+1)^{1/2}+C$

63. $x^x(x\log x-1)+C$

64. $-\frac{2}{3}\log\left|\frac{\sqrt{\cos\theta+\sec\theta+1}-\sqrt{3}}{\sqrt{\cos\theta+\sec\theta+1}+\sqrt{3}}\right|+C$

65. $-\frac{1}{8}x^6(1-x^3)^{2/3}-\frac{3}{20}x^3(1-x^3)^{2/3}-\frac{9}{40}(1-x^3)^{2/3}+C$

66. $\frac{2}{\cos\alpha}\sqrt{\cos\alpha\cdot\tan\theta+\sin\alpha}-\frac{2}{\sin\alpha}\sqrt{\cos\alpha+\cot\theta\cdot\sin\alpha}+C$

67. $-\cot x\log(\cos x+\sqrt{\cos 2x})-\cot x-x+\sqrt{\cot^2 x-1}+C$

68. $\frac{1}{(4a^2-1)^{3/2}}\cdot\left\{2a\sin^{-1}\left(\frac{2a\sin 2x+1}{2a+\sin 2x}\right)+\sqrt{1-\left(\frac{2a\sin 2x+1}{2a+\sin 2x}\right)}\right\}$

$-\frac{1}{2a+\sin 2x}+C$

69. $2\log\{|\sqrt{x^2+2x+4}-x|\}-\frac{3}{2(\sqrt{x^2+2x+4}-(x+1))}$

$-\frac{3}{2}\log|\sqrt{x^2+2x+4}-(x+1)|+C$

70. $\log(x+\sqrt{x^2+2x+2})+\frac{2}{(x+2)+\sqrt{x^2+2x+2}}+C$

71. $\tan^{-1}(x-1/x)-\frac{2}{3}\tan^{-1}(x^3)+C$

72. $\frac{1}{3}\log\left|\frac{(1-x^3)^{1/3}+x}{x}\right|-\frac{1}{6}\log\left|\frac{(1-x^3)^{2/3}-x(1-x^3)^{1/3}+x^2}{x^2}\right|$

$-\frac{1}{\sqrt{3}}\tan^{-1}\left\{\frac{2(1-x^3)^{1/3}-x}{\sqrt{3}x}\right\}+C$

73. $\frac{(x+\sqrt{1+x^2})^{15}}{15}+C$

75. $2a(n+1)I_{(n+1)}+b(1+2n)I_{(n)}+2nc\,I_{(n-1)}=2x^n\sqrt{ax^2+bx}+C$

Solutions

(*Proficiency Exercise* 1)

Type 1 : Only One Correct Option

1. Putting, $x=\tan\theta$, $I=\int\frac{\tan^2\theta}{1+\sec\theta}d\theta=\int(\sec\theta-1)\,d\theta$

$=\log(\sec\theta+\tan\theta)-\theta+C$

$f(x)=\log_e(x+\sqrt{x^2+1})-\tan^{-1}x+C$

$f(0) = 0 \Rightarrow C = 0$

$\Rightarrow \quad f(1) = \log_e(1+\sqrt{2}) - \tan^{-1} 1 = \log_e(1+\sqrt{2}) - \frac{\pi}{4}$

2. We have, $\int f(x)\,dx = f(x)$

$\Rightarrow \quad \frac{d}{dx}\{f(x)\} = f(x)$

$\Rightarrow \quad \frac{1}{f(x)} d\{f(x)\} = dx$

$\Rightarrow \quad \log\{f(x)\} = x + \log C$

$\Rightarrow \quad f(x) = Ce^x \Rightarrow \{f(x)\}^2 = C^2 e^{2x}$

$\Rightarrow \quad \int \{f(x)\}^2\,dx = \int C^2 e^{2x}\,dx = \frac{C^2 e^{2x}}{2} = \frac{1}{2}\{f(x)\}^2$

3. We have, $\int f(x)\,dx = F(x)$

$\therefore \quad \int x^3 f(x^2)\,dx = \frac{1}{2}\int \underset{\text{I}}{x^2}\, \underset{\text{II}}{\underbrace{f(x^2)\,d(x^2)}}$

$= \frac{1}{2}[x^2 F(x^2) - \int F(x^2)\,d(x^2)]$

4. We have the following cases :

Case I When $x \geq 0$

In this case, we have

$\int |x^n|\,dx = \int |x|^n\,dx = \int x^n\,dx$ $\quad [\because |x| = x]$

$= \frac{x^{n+1}}{n+1} + C = \frac{|x|^n x}{n+1} + C$ $\quad [\because x \geq 0 \therefore |x| = x]$

Case II When $x \leq 0$

In this case, we have $|x| = -x$

$\therefore \quad \int |x^n|\,dx = \int |x|^n\,dx$

$= \int (-x)^n\,dx = -\int x^n\,dx$ $\quad [\because n \text{ in odd}]$

$= -\frac{x^{n+1}}{n+1} + C = \frac{(-x)^n x}{n+1} + C = \frac{|x|^n x}{n+1} + C$

Hence, $\int |x^n|\,dx = \frac{|x|^n x}{n+1} + C$

5. $F(x) = \int \frac{3x+2}{\sqrt{x-9}}\,dx$

Let $x - 9 = t^2 \Rightarrow dx = 2t\,dt$

$\therefore \quad F(x) = \int \left(\frac{3(t^2+9)+2}{t} \cdot 2t\right) dt$

$= 2\int (29 + 3t^2)\,dt = 2[29t + t^3]$

$F(x) = 2[29\sqrt{x-9} + (x-9)^{3/2}] + C$

Given, $F(10) = 60 = 2[29 + 1] + C \Rightarrow C = 0$

$\therefore$ $F(x) = 2[29\sqrt{x-9} + (x-9)^{3/2}]$

$F(13) = 2[29 \times 2 + 4 \times 2] = 132$

6. We have, $\int (x^x)^x (2x \log_e x + x)\, dx$

$$= \int x^{x^2} (2x \log_e x + x)\, dx$$

$$= \int 1 \cdot d(x^{x^2}) = x^{x^2} + C = (x^x)^x + C$$

7. We have, $\int x \log x (\log x - 1)\, dx$

$$= \int \log x (x \log x - x)\, dx$$

$$= \int (x \log x - x)\, d(x \log x - x)$$

$$= \frac{(x \log x - x)^2}{2} + C$$

8. $\int \frac{\left(\frac{1}{x^3} - \frac{1}{x^5}\right) dx}{\sqrt{2 - \frac{2}{x^2} + \frac{1}{x^4}}}$

Let $2 - \frac{2}{x^2} + \frac{1}{x^4} = z \Rightarrow \frac{1}{4}\int \frac{dz}{\sqrt{z}}$

$\Rightarrow$ $\frac{1}{2}\sqrt{z} + C \Rightarrow \frac{1}{2}\sqrt{2 - \frac{2}{x^2} + \frac{1}{x^4}} + C$

9. We have, $f''(x) = f(x)$

$\Rightarrow$ $2f'(x) f''(x) = 2f(x) f'(x)$

$\Rightarrow$ $\frac{d}{dx}\{f'(x)\}^2 = \frac{d}{dx}\{f(x)\}^2$

$\Rightarrow$ $\{f'(x)\}^2 = \{f(x)\}^2 + C$ ...(i)

Now, $f(0) = 2$ and $f'(0) = 3$. Therefore, from Eq. (i), we get

$$\{f'(0)\}^2 = \{f(0)\}^2 + C$$

$\Rightarrow$ $9 = 4 + C \Rightarrow C = 5$

$\therefore$ $\{f'(x)\}^2 = \{f(x)\}^2 + 5$

$\Rightarrow$ $f'(x) = \sqrt{5 + \{f(x)\}^2}$

$\Rightarrow$ $\int \frac{1}{\sqrt{(\sqrt{5})^2 + \{f(x)\}^2}}\, d\{f(x)\} = \int dx$

$\Rightarrow$ $\log\left| f(x) + \sqrt{5 + \{f(x)\}^2} \right| = x + C_1$

$\therefore$ $f(0) = 2 \Rightarrow \log|2 + 3| = C_1 \Rightarrow C_1 = \log 5$

$\therefore$ $\log|f(x) + \sqrt{5 + \{f(x)\}^2}| = x + \log 5$

$\Rightarrow$ $\log\left\{\frac{f(x) + \sqrt{5 + \{f(x)\}^2}}{5}\right\} = x$

$\Rightarrow$ $f(x) + \sqrt{5 + \{f(x)\}^2} = 5e^x$

$\Rightarrow \qquad \sqrt{5+\{f(x)\}^2}+f(x)=5e^x$ and $\sqrt{5+\{f(x)\}^2}-f(x)=5e^{-x}$

$\Rightarrow \qquad 2f(x)=5(e^x-e^{-x}) \qquad \Rightarrow \quad f(x)=\dfrac{5}{2}(e^x-e^{-x})$

$\Rightarrow \qquad f(4)=\dfrac{5}{2}(e^4-e^{-4})$

$\Rightarrow \qquad f(4)=\dfrac{5(e^8-1)}{2e^4}$

10. We have, $\displaystyle\int \frac{e^{x^2+4\ln x}-x^3e^{x^2}}{x-1}dx=\int\frac{e^{x^2}\cdot x^4-x^3e^{x^2}}{x-1}dx$

$$=\int x^3e^{x^2}dx=\frac{1}{2}\int te^t dt, \text{ where } t=x^2$$

$$=\frac{1}{2}(t-1)e^t+C=\frac{1}{2}(x^2-1)e^{x^2}+C$$

11. $\displaystyle\int \tan^4 x\,dx=\int(\tan^2 x\sec^2 x-\sec^2 x+1)\,dx=\frac{\tan^3 x}{3}-\tan x+x+C$

$\Rightarrow \qquad A=\dfrac{1}{3}, B=-1$ and $f(x)=x+C$

12. $\displaystyle\int\frac{\sin^4 x}{x}dx=f(x)$

$$\int\frac{\sin^4(p+q)x}{(p+q)x}dx=\frac{f\{(p+q)x\}}{p+q}$$

$$\int\frac{\sin^4(p+q)x}{x}dx=f\{(p+q)x\}$$

13. On solving, $\dfrac{\sin\theta}{\cos 3\theta}=\dfrac{1}{2}[\tan 3\theta-\tan\theta]$

$\dfrac{\sin 3\theta}{\cos 9\theta}=\dfrac{1}{2}[\tan 9\theta-\tan 3\theta]$

$\dfrac{\sin 9\theta}{\cos 27\theta}=\dfrac{1}{2}[\tan 27\theta-\tan 9\theta]$

$\therefore \displaystyle\int\left(\frac{\sin\theta}{\cos 3\theta}+\frac{\sin 3\theta}{\cos 9\theta}+\frac{\sin 9\theta}{\cos 27\theta}\right)d\theta$

$$=\frac{1}{2}\int(\tan 27\theta-\tan\theta)\,d\theta$$

$$=\frac{1}{2}\left\{\frac{1}{27}\log(\sec 127\theta)-\log(\sec\theta)\right\}+C$$

$$=\frac{1}{2}\log\frac{\sqrt[27]{\sec 27\theta}}{\sec\theta}+C$$

14. We have, $\displaystyle\int x\sqrt{\frac{2\sin(x^2+1)-\sin 2(x^2+1)}{2\sin(x^2+1)+\sin 2(x^2+1)}}\,dx$

$$=\int x\sqrt{\frac{2\sin(x^2+1)-2\sin(x^2+1)\cos(x^2+1)}{2\sin(x^2+1)+2\sin(x^2+1)\cos(x^2+1)}}\,dx$$

$$= \int x \sqrt{\frac{1-\cos(x^2+1)}{1+\cos(x^2+1)}}\, dx = \int x \tan\left(\frac{x^2+1}{2}\right) dx$$

$$= \int \tan\left(\frac{x^2+1}{2}\right) d\left(\frac{x^2+1}{2}\right) = \log\left|\sec\left(\frac{x^2+1}{2}\right)\right| + C$$

15. $\displaystyle\int \frac{\sin(4x-2x)\,dx}{\sin(2x)\cos(2x)\cos(4x)} = \int \frac{\sin(4x)\,dx}{\sin(2x)\cos(4x)} - \int \sec 2x\, dx$

$$= 2\int \frac{\cos 2x\, dx}{\cos 4x} - \frac{1}{2}(\log|\sec 2x - \tan 2x|)$$

On solving, we get solution.

16. $\displaystyle\int \frac{1-7\cos^2 x}{\sin^7 x\cos^2 x}\, dx = \int\left(\frac{\sec^2 x}{\sin^7 x} - \frac{7}{\sin^7 x}\right) dx = \int \frac{\sec^2 x}{\sin^7 x}\, dx - \int \frac{7}{\sin^7 x}\, dx = I_1 + I_2$

Now, $\displaystyle I_1 = \int \frac{\sec^2 x}{\sin^7 x}\, dx = \frac{\tan x}{\sin^7 x} + 7\int \frac{\tan x \cdot \cos x}{\sin^8 x}\, dx = \frac{\tan x}{\sin^7 x} + I_2$

$\therefore \quad \displaystyle I_1 + I_2 = \frac{\tan x}{\sin^2 x} + C$

$\Rightarrow \quad f(x) = \tan x$

17. We have, $\displaystyle\int \frac{\sin^3 x}{(\cos^4 x + 3\cos^2 x + 1)\tan^{-1}(\sec x + \cos x)}\, dx$

$$= \int \frac{\dfrac{\sin^3 x}{\cos^2 x}}{(\cos^2 x + 3 + \sec^2 x)\tan^{-1}(\sec x + \tan x)}\, dx$$

$$= \int \frac{1}{1+(\sec x + \cos x)^2} \times \frac{\sin x(1-\cos^2 x)}{\cos^2 x} \times \frac{1}{\tan^{-1}(\sec x + \tan x)}\, dx$$

$$= \int \frac{1}{\tan^{-1}(\sec x + \cos x)} \times \frac{1}{1+(\sec x + \cos x)^2}(\tan x \sec x - \sin x)\, dx$$

$$= \int \frac{1}{\tan^{-1}(\sec x + \cos x)}\, d\,|\tan^{-1}(\sec x + \cos x)|$$

$$= \log|\tan^{-1}(\sec x + \cos x)| + C$$

18. We have, $f(x) = x|\cos x|, \ \dfrac{\pi}{2} < x < \pi$

$\Rightarrow \quad f(x) = -x\cos x \qquad [\because \cos x < 0 \text{ for } x \in (\pi/2, \pi)]$

Hence, required primitive is given by

$$\int f(x)\, dx = -\int x\cos x\, dx + C = -x\sin x - \cos x + C$$

19. We have, $f(x) = (2x+1)|\sin x|, \ \pi < x < 2\pi$

$\Rightarrow \quad f(x) = -(2x+1)\sin x$

Hence, required primitive is given by

$$-\int \underset{\text{I}}{(2x+1)}\, \underset{\text{II}}{\sin x}\, dx = -[-(2x+1)\cos x + 2\sin x] + C$$

$$= (2x+1)\cos x - 2\sin x + C$$

20. We have, $f(x)=\begin{vmatrix} 0 & x^2-\sin x & \cos x-2 \\ \sin x - x^2 & 0 & 1-2x \\ 2-\cos x & 2x-1 & 0 \end{vmatrix}$

$$\Rightarrow \quad f(x)=\begin{vmatrix} 0 & \sin x - x^2 & 2-\cos x \\ x^2-\sin x & 0 & 2x-1 \\ \cos x-2 & 1-2x & 0 \end{vmatrix}$$

[Interchanging rows and columns]

$$\Rightarrow \quad f(x)=(-1)^3\begin{vmatrix} 0 & x^2-\sin x & \cos x-2 \\ \sin x - x^2 & 0 & 1-2x \\ 2-\cos x & 2x-1 & 0 \end{vmatrix}$$

[Taking (– 1) common from each column]

$$\Rightarrow \quad f(x)=-f(x) \quad \Rightarrow \quad f(x)=0$$

$$\Rightarrow \quad \int f(x)\,dx=0$$

Type 2 : More Than One Correct Options

21. $\int \frac{dx}{(x+1)(x-2)}=\int\left(-\frac{1}{3(x+1)}+\frac{1}{3(x-2)}\right)dx$

$$=-\frac{1}{3}\log(x+1)+\frac{1}{3}\log(x-2)+C$$

$$\therefore \quad A=-\frac{1}{3}, B=\frac{1}{3} \quad \therefore \ A+B=0$$

$$\Rightarrow \quad \frac{A}{B}=\frac{-\frac{1}{3}}{\frac{1}{3}}=-1$$

22. $\frac{1}{3}\int\left(\frac{1}{x^2+1}-\frac{1}{x^2+4}\right)dx$

$$\Rightarrow \quad \frac{1}{3}\tan^{-1}x-\frac{1}{6}\tan^{-1}\frac{x}{2}=k\tan^{-1}x+l\tan^{-1}\frac{x}{2}$$

$$\therefore \quad k=\frac{1}{3} \quad \text{and} \quad l=-\frac{1}{6}$$

23. $I=\int \underset{\text{I}}{\underbrace{x}}\ \underset{\text{II}}{\underbrace{\log(1+x^2)}}\,dx=\log(1+x^2)\cdot\frac{x^2}{2}-\int\frac{1}{1+x^2}\cdot 2x\cdot\frac{x^2}{2}\,dx$

$$=\frac{x^2}{2}\log(1+x^2)-\int\frac{x^3}{x^2+1}\,dx$$

$$I=\left(\frac{x^2+1}{2}\right)\log(1+x^2)-\left(\frac{x^2+1}{2}\right)+C$$

$$\therefore \quad \phi(x)=\frac{x^2+1}{2}, \ \psi(x)=-\left(\frac{1+x^2}{2}\right)$$

24. $I = \int \frac{4e^{2x}+6}{9e^{2x}-4}\,dx$, put $9e^{2x} - 4 = z$

Then, $I = \int \frac{1}{z}\left\{4\cdot\frac{z+4}{9}+6\right\}\frac{1}{18}\frac{dz}{\frac{z+4}{9}} = \int \frac{1}{z(z+4)}\left\{\frac{2z+8}{9}+3\right\}dz$

$$= \frac{1}{9}\int \frac{2z+35}{z(z+4)}\,dz = \frac{1}{9}\int \frac{2(z+4)+27}{z(z+4)}\,dz$$

$$= \frac{2}{9}\int \frac{dz}{z} + \frac{3}{4}\int\left(\frac{1}{z}-\frac{1}{z+4}\right)dz = \left(\frac{2}{9}+\frac{3}{4}\right)\log z - \frac{3}{4}\log(z+4) + C$$

$$= \frac{35}{36}\log(9e^{2x}-4) = \frac{3}{4}\log(e^{2x}) + C$$

$$= -\frac{3}{2}x + \frac{35}{36}\log(9e^{2x}-4) - \frac{3}{2}\log 3 + C$$

25. $\int \tan^3 x(\sec^2 x - 1)\,dx = \int \tan^3 x \sec^2 x\,dx - \int \tan^3 x\,dx$

$$I_1 = \int \tan^3 x \sec^2 x\,dx = \frac{\tan^4 x}{4} + C_1$$

$$I_2 = \int \tan x(\sec^2 x - 1)\,dx = \frac{\tan^2 x}{2} - \ln|\sec x| + C_2$$

Type 3 : **Assertion and Reason**

26. The Statement II is false since while writing

$$\int \frac{dx}{x-3y} = \log(x-3y) + C,$$

we are assuming that y is a constant. We will know prove the Statement I.
From the given relation $(x-y)^2 = \frac{x}{y}$ and $2\log(x-y) = \log x - \log y$.
Also, $\frac{dy}{dx} = \left(-\frac{y}{x}\right)\cdot\frac{x+y}{x-3y}$. To prove the integral relation it is sufficient to show that $\frac{d}{dx}\text{RHS} = \frac{1}{x-3y}$.

Now, $\text{RHS} = \frac{1}{2}\log\left[\frac{x}{y}-1\right]$ $\left[\because (x-y)^2 = \frac{x}{y}\right]$

$$= \frac{1}{2}[\log(x-y) - \log y]$$

$$= \frac{1}{2}\left[\frac{\log x - \log y}{2} - \log y\right]$$

$$= \frac{1}{4}[\log x - 3\log y]$$

$$\Rightarrow \frac{d}{dx}\text{RHS} = \frac{1}{4}\left[\frac{1}{x} - \frac{3}{y}\frac{dy}{dx}\right] = \frac{1}{4}\left[\frac{1}{x} - \frac{3}{y}\left(-\frac{y}{x}\right)\frac{x+y}{x-3y}\right] = \frac{1}{x-3y}$$

Thus, Statement I is true.
Hence, choice (c) is correct.

27. Let $g(x) = f(x) + f(-x)$

Assuming, $\int f(x)\,dx = F(x) + C$

$$\int g(x)\,dx = \int \{f(x) + f(-x)\}\,dx = \int f(x)\,dx + \int f(-x)\,dx$$

$$= F(x) + C + \{-F(-x) + C'\}$$

$$= F(x) - F(-x) + C + C'$$

which may be an odd function, if $C + C' = 0$.
Similarly, integral of an odd function is not always an even function.
Hence, Statement I is true and Statement II is false.

28. If $a > 0$ and $b^2 - 4ac < 0$, then

$$ax^2 + bx + c = a\left(x + \frac{b}{2a}\right)^2 + \frac{4ac - b^2}{4a}$$

$$\Rightarrow \quad \int \frac{dx}{ax^2 + bx + c} = \int \frac{dx}{a\left(x + \frac{b}{2a}\right)^2 + k^2}$$

where $k^2 = \dfrac{4ac - b^2}{4a} > 0$

which will have an answer of the type

$$\frac{1}{a} \cdot \frac{1}{k/\sqrt{a}} \tan^{-1} \frac{x + \frac{b}{2a}}{k/\sqrt{a}} + C$$

or $\quad \mu \tan^{-1} \dfrac{x + A}{B} + C$

Thus, choice (a) is correct.

29. $$I = \int \frac{dx}{x^4 + 1}$$

$$I = \frac{1}{2} \int \frac{\frac{2}{x^2}}{x^2 + \frac{1}{x^2}}\,dx = \frac{1}{2} \int \frac{\left(1 + \frac{1}{x^2}\right)}{\left(x - \frac{1}{x}\right)^2 + 2}\,dx - \frac{1}{2} \int \frac{\left(1 - \frac{1}{x^2}\right)}{\left(x + \frac{1}{x}\right)^2 - 2}\,dx$$

$$I = \frac{1}{2} \cdot \frac{1}{\sqrt{2}} \tan^{-1} \left(\frac{x - \frac{1}{x}}{2}\right) + \frac{1}{2} \cdot \frac{1}{2} \log \left(\frac{x + \frac{1}{x} - 1}{x + \frac{1}{x} + 1}\right) + C$$

∴ Statement I is false.

30. Since, $\quad \cot^{-1} x = \dfrac{\pi}{2} - \tan^{-1} x,$

∴ $\quad d(\cot^{-1} x) = -\,d\,(\tan^{-1} x)$

Thus, $\quad \displaystyle\int 2^{\tan^{-1} x}\, d(\cot^{-1} x) = -\int 2^{\tan^{-1} x}\, d(\tan^{-1} x) = -\frac{2^{\tan^{-1} x}}{\ln 2} + C$

∴ Statement I is false.

Statement II is true.

Type 4 : Linked–Comprehension Based Questions

31. As $m = 9 > 0$, hence, we can substitute

$$\sqrt{9x^2 + 4x + 6} = u \pm 3x$$

32. Here, as per notations given, we can substitute

$$\sqrt{1 + x^2} = (u - x)$$

As $m = 1 > 0$ and $p = 1 > 0$

$$\Rightarrow \quad I = \int \frac{u^{15}}{u}\, du = \int u^{14} du = \frac{1}{15} u^{15} + C$$

$$= \frac{1}{15}(x + \sqrt{1 + x^2})^{15} + C$$

33. Here, $m = -1 < 0$

$p = -2 < 0$

Also, $-x^2 + 3x - 2 = -(x - 1)(x - 2)$

We can use case III

$\Rightarrow$ Putting, $\sqrt{-x^2 + 3x - 2} = u\,(x - 2)$

or $(x - 1)u$ or $u\,(1 - x)$

34. Let $P = \sin^3 x \cos^3 x$

$$\frac{dP}{dx} = 3 \sin^2 x \cos^4 x - 3 \sin^4 x \cos^2 x$$

$$= 3 \sin^2 x (1 - \sin^2 x) \cos^2 x - 3 \sin^4 x \cos^2 x$$

$$= 3 \sin^2 x \cos^2 x - 6 \sin^4 x \cos^2 x$$

$$\therefore \quad P = 3I_{2,\,2} - 6I_{4,\,2}$$

$$\therefore \quad I_{4,\,2} = \frac{1}{6}(-P + 3I_{2,\,2})$$

35. Let $P = \sin^5 x \cos^3 x$

$$\therefore \quad \frac{dP}{dx} = 5 \sin^4 x \cos^4 x - 3 \sin^6 x \cos^2 x$$

$$= 5 \sin^4 x (1 - \sin^2 x) \cos^2 x - 3 \sin^6 x \cos^2 x$$

$$= 5 \sin^4 x \cos^2 x - 8 \sin^6 x \cos^2 x$$

$$\therefore \quad P = 5I_{4,\,2} - 8I_{6,\,2}$$

$$\therefore \quad I_{4,\,2} = \frac{1}{5}(P + 8I_{6,\,2})$$

36. Let $P = \sin^5 x \cos^3 x$

$$\therefore \quad \frac{dP}{dx} = 5 \sin^4 x \cos^4 x - 3 \sin^6 x \cos^2 x$$

$$= 5 \sin^4 x \cos^4 x - 3 \sin^4 x (1 - \cos^2 x) \cos^2 x$$

$$= 8 \sin^4 x \cos^4 x - 3 \sin^4 x \cos^2 x$$

$$\therefore \quad P = 8I_{4,\,4} - 3I_{4,\,2}$$

$$\therefore \quad I_{4,\,2} = \frac{1}{3}(-P + 8I_{4,\,4})$$

Type 5 : Match the Columns

37. (A) If $\frac{\pi}{4} < x < \frac{3\pi}{8}$, then $\sin x > \cos x$

$$\therefore \quad \int \frac{\sin x - \cos x}{|\sin x - \cos x|} dx = \int 1\, dx = x + C$$

(B) $$\int \frac{x^2 dx}{(x^3+1)(x^3+2)} = \frac{1}{3}\int 3x^2 \left(\frac{1}{x^3+1} - \frac{1}{x^3+2}\right) dx$$

$$= \frac{1}{3} \ln \left|\frac{x^3+1}{x^3+2}\right| + C$$

(C) $$\int \sin^{-1} x \cos^{-1} x\, dx = \int \left[\frac{\pi}{2}\sin^{-1} x - (\sin^{-1} x)^2\right] dx$$

$$\Rightarrow \frac{\pi}{2}(x \sin^{-1} x + \sqrt{1-x^2}) - \{x(\sin^{-1} x)^2 + \sin^{-1} x\sqrt{1-x^2} - x\} + C \qquad \text{(by parts)}$$

$$\Rightarrow \quad \sin^{-1} x \left[\frac{\pi}{2} x - x \sin^{-1} x - 2\sqrt{1-x^2}\right] + \frac{\pi}{2}\sqrt{1-x^2} + 2x + C$$

$$\therefore \quad f^{-1}(x) = \sin^{-1} x, f(x) = \sin x$$

(D) $$\int \frac{dx}{x \ln|x|} = \ln|\ln|x|| + C$$

$$\therefore \quad f(x) = \ln|x|$$

38. (A) $$I = \int \left(\frac{x^2 + \cos^2 x}{1+x^2}\right) \cdot \text{cosec}^2 x\, dx = \int \left(\frac{x^2 + 1 - \sin^2 x}{1+x^2}\right) \cdot \text{cosec}^2 x\, dx$$

$$= \int \left(\text{cosec}^2 x - \frac{1}{1+x^2}\right) dx = -\cot x + \cot^{-1} x + k$$

$$\Rightarrow \quad A = 1, B = -1$$

(B) $$I = \int \sqrt{x + \sqrt{x^2+2}}\, dx$$

Put $\sqrt{x^2+2} + x = t$

$$\Rightarrow \quad \sqrt{x^2+2} - x = \frac{2}{t}$$

$$\therefore \quad 2x = t - \frac{2}{t} \Rightarrow 2\,dx = \left(1 + \frac{2}{t^2}\right) dt$$

$$= \frac{1}{2}\int \left(\sqrt{t} + \frac{2}{t^{3/2}}\right) dt$$

$$\Rightarrow \quad I = \frac{1}{2} \cdot \frac{t^{3/2}}{\frac{3}{2}} + \frac{1}{-\frac{1}{2} t^{1/2}} + k$$

$$= \frac{1}{3}(x + \sqrt{x^2+2})^{3/2} - \frac{2}{\sqrt{x + \sqrt{x^2+2}}} + k$$

$$\Rightarrow \quad A = 1 \text{ and } B = 2$$

(C) $I = \int \frac{2 - x - x^2}{x^2 \sqrt{2 - x - x^2}} dx$

$$= -\int \frac{dx}{\sqrt{2 - x - x^2}} + 2\int \frac{dx}{x^2 \sqrt{2 - x - x^2}} - \int \frac{dx}{x\sqrt{2 - x - x^2}}$$

$$= -\int \frac{dx}{\sqrt{\frac{9}{4} - \left(x + \frac{1}{2}\right)^2}} + 2\int - \frac{dt}{\sqrt{2 - \frac{1}{t} - \frac{1}{t^2}}} + \int \frac{\frac{1}{t}dt}{\sqrt{2 - \frac{1}{t} - \frac{1}{t^2}}} \qquad \left(\text{put } x = \frac{1}{t}\right)$$

$$= -\sin^{-1}\left(\frac{2x + 1}{3}\right) - \frac{1}{2}\int \frac{(4t - 1)dt}{\sqrt{2t^2 - t - 1}} + \frac{1}{2}\int \frac{dt}{\sqrt{2t^2 - t - 1}}$$

$$= -\sin^{-1}\left(\frac{2x + 1}{3}\right) - \frac{1}{2}\frac{\sqrt{2t^2 - t - 1}}{\frac{1}{2}} + \frac{1}{2\sqrt{2}}\int \frac{dt}{\sqrt{t^2 - \frac{t}{2} - \frac{1}{2} + \frac{1}{16} - \frac{1}{16}}}$$

$$= -\sin^{-1}\left(\frac{2x + 1}{3}\right) - \frac{\sqrt{2 - x - x^2}}{x} + \frac{1}{2\sqrt{2}} \log\left|\left(t - \frac{1}{4}\right) + \sqrt{2t^2 - t - 1}\right| + K$$

$$= -\frac{\sqrt{2 - x - x^2}}{x} + \frac{1}{2\sqrt{2}} \log\left|\frac{(4 - x) + \sqrt{2 - x - x^2}}{4x}\right| - \sin^{-1}\left(\frac{2x + 1}{3}\right) + K$$

(D) $\int \frac{\sin 2x}{\sin^4 x + \cos^4 x} dx$, dividing N^r and D^r by $\cos^4 x$

$$I = \int \frac{2\tan x \cdot \sec^2 x}{\tan^4 x + 1} dx, \text{ put } \tan^2 x = t \Rightarrow 2\tan x \cdot \sec^2 x\, dx = dt$$

$$= \int \frac{dt}{t^2 + 1} = -\cot^{-1}(t) + C$$

$\Rightarrow \quad -\cot^{-1}(\tan^2 x) + C \quad \therefore \quad B = -1$

Type 6 : **Integer Answer Type Questions**

39. $\int \frac{2x + 3}{(x^2 + 3x)(x^2 + 3x + 2) + 1} dx$

Put $x^2 + 3x = t \Rightarrow (2x + 3)\, dx = dt$

$$\int \frac{dt}{t(t + 2) + 1} = \int \frac{dt}{(t + 1)^2} = C - \frac{1}{t + 1} = C - \frac{1}{x^2 + 3x + 1}$$

$\Rightarrow \quad a = 1, b = 3, c = 1$

$\Rightarrow \quad a + b + c = 5$

40. $F(x) = \int \frac{3x + 2}{\sqrt{x - 9}} dx$. Let $x - 9 = t^2 \Rightarrow dx = 2t\, dt$

$$\therefore \quad F(x) = \int \left(\frac{3(t^2 + 9) + 2}{t} \cdot 2t\right) dt$$

$$= 2\int (29 + 3t^2)\, dt = 2[29t + t^3]$$

$$F(x) = 2[29\sqrt{x-9} + (x-9)^{3/2}] + C$$

Given, $F(10) = 60 = 2[29 + 1] + C \Rightarrow C = 0$

$\therefore$ $F(x) = 2[29\sqrt{x-9} + (x-9)^{3/2}]$

$$F(13) = 2[29 \times 2 + 4 \times 2] = 4 \times 33 = 132$$

Hence, sum of digits $= 1 + 3 + 2 = 6$

41. $u(x) = 7v(x)$

$\Rightarrow$ $u'(x) = 7v'(x) \Rightarrow p = 7$ **(given)**

Again, $\dfrac{u(x)}{v(x)} = 7 \Rightarrow \left(\dfrac{u(x)}{v(x)}\right)' = 0$

$\Rightarrow$ $q = 0$

Now, $\dfrac{p+q}{p-q} = \dfrac{7+0}{7-0} = 1$

42. Let $t = \ln\left(\dfrac{x-1}{x+1}\right)$

$$\frac{dt}{dx} = \frac{2}{x^2-1}$$

$$I = \int \frac{1}{2} t\, dt = \frac{1}{4} t^2 + C$$

$$I = \frac{1}{4}\left[\ln\left(\frac{x-1}{x+1}\right)\right]^2 + C$$

$\Rightarrow$ $6A = \dfrac{1}{4} \Rightarrow 24A = 1$

43. $I = \int e^x \left(\dfrac{1}{(1-x)\sqrt{1-x^2}} + \sqrt{\dfrac{1+x}{1-x}}\right) dx$

As $\dfrac{d}{dx}\left(\sqrt{\dfrac{1+x}{1-x}}\right) = \dfrac{1}{(1-x)\sqrt{1-x^2}}$

$\therefore$ $I = e^x \sqrt{\dfrac{1+x}{1-x}} + C \Rightarrow I = e^x \left(\dfrac{1+x}{1-x}\right)^{1/2} + C$

$\Rightarrow$ $\mu = 1, \lambda = \dfrac{1}{2}$

$\Rightarrow$ $2(\mu + \lambda) = 2\left(1 + \dfrac{1}{2}\right) = 2 \times \dfrac{3}{2} = 3$

44. $I = \int \dfrac{(e^x + \cos x + 1) - (e^x + \sin x + x)}{e^x + \sin x + x} dx$

$= \ln(e^x + \sin x + x) - x + C$

$\therefore$ $f(x) = e^x + \sin x + x$ and $g(x) = -x$

$\Rightarrow$ $f(x) + g(x) = e^x + \sin x$

$\Rightarrow$ $\dfrac{f(x) + g(x)}{e^x + \sin x} = 1$

45. $12\left[\frac{1}{4}\tan^{-1}\frac{x+3}{4}+\frac{1}{2\cdot 6}\ln\left|\frac{x-9}{x+3}\right|\right]=3\tan^{-1}\left(\frac{x+3}{4}\right)+\ln\left|\frac{x-9}{x+3}\right|$

$\Rightarrow \quad \lambda=3, \mu=1 \;\Rightarrow\; \lambda+\mu=4$

46. $\dfrac{2\cos\frac{15}{2}x\cos\frac{3}{2}x}{1-2\left(2\cos^2\frac{5x}{2}-1\right)}=\dfrac{2\left(4\cos^3\frac{5x}{2}-3\cos\frac{5x}{2}\right)\cos\frac{3x}{2}}{3-4\cos^2\frac{5x}{2}}$

$$=-2\cos\frac{5x}{2}\cos\frac{3x}{2}=-(\cos 4x+\cos x)$$

$$I=-\frac{\sin 4x}{4}-\sin x+C$$

47. $I=\int\dfrac{\tan x}{1+\tan x+\tan^2 x}dx=\int\dfrac{\frac{\sin x}{\cos x}}{\frac{1}{\cos^2 x}+\frac{\sin x}{\cos x}}dx$

$=\int\dfrac{\sin 2x}{2+\sin 2x}dx=\int dx-2\int\dfrac{dx}{2+\sin 2x}$

$=x-2\int\dfrac{\sec^2 x}{2\sec^2 x+2\tan x}dx$

Let $t=\tan x$, $dt=\sec^2 x\,dx$

$$=x-\frac{2}{2}\int\frac{dt}{t^2+t+1}=x-\int\frac{dt}{\left(t+\frac{1}{2}\right)^2+\left(\frac{\sqrt{3}}{2}\right)^2}$$

$$I=x-\frac{1}{\sqrt{3}}\tan^{-1}\left(\frac{2\tan x+1}{\sqrt{3}}\right)+C$$

48. $\int\sin^{5/2}x\cos^3 x\,dx=\int\sin^2 x\sin^{1/2}x\cos^2 x\cos x\,dx$

$$=\int\sin^2 x\sin^{1/2}x(1-\sin^2 x)\cos x\,dx$$
$$=\int t^2 t^{1/2}(1-t^2)\,dt=\int[t^{5/2}-t^{9/2}]dt$$
$$=\int t^{5/2}dt-\int t^{9/2}dt=\frac{t^{7/2}}{7/2}-\frac{t^{11/2}}{11/2}+C$$
$$=\frac{2}{7}t^{7/2}-\frac{2}{11}t^{11/2}+C$$
$$=\frac{2}{7}\sin^{7/2}x-\frac{2}{11}\sin^{11/2}x+C$$
$$=2\sin^{7/2}x\left[\frac{1}{7}-\frac{1}{11}\sin^2 x\right]+C$$

$\Rightarrow \quad A=7, B=7, C=11$

$\Rightarrow \quad (A+B)-C=(7+7)-11=3$

(*Proficiency Exercise 2*)

1. $\int \left\{ \frac{(x+a)^3}{x^3} + \frac{1-x^4}{1-x} - \frac{7}{x\sqrt{x^2-1}} + \frac{x+2}{(x+1)^2} + \frac{(1+x)^3}{\sqrt{x}} \right\} dx$

$$\Rightarrow \int \left\{ \frac{x^3+3ax^2+3a^2x+a^3}{x^3} + \frac{(1-x)(1+x+x^2+x^3)}{(1-x)} - \frac{7}{x\sqrt{x^2-1}} + \frac{(x+1)}{(x+1)^2} + \frac{1}{(x+1)^2} \frac{1+3x+3x^2+x^3}{x^{1/2}} \right\} dx$$

$$\Rightarrow \quad x + 3a \log x - \frac{3a^2}{x} - \frac{a^3}{2x^2} + x + \frac{x^2}{2} + \frac{x^3}{3} + \frac{x^4}{4} - 7\sec^{-1} x + \log(x+1) - \frac{1}{x+1} + 2\sqrt{x} + 2x^{3/2} + \frac{6}{5}x^{5/2} + \frac{2}{7}x^{7/2} + C$$

2. Let $f(x) = \left\{ \frac{5\cos^3 x + 2\sin^3 x}{2\sin^2 x \cos^2 x} + \sqrt{1+\sin 2x} + \frac{1+2\sin x}{\cos^2 x} + \frac{1-\cos 2x}{1+\cos 2x} + \sqrt{1+\sin x/2} + \frac{\sin^2 x}{(1+\cos x)^2} + (\tan x + \cot x)^2 \right\}$

$$\Rightarrow f(x) = \left\{ \frac{5\cos x}{2\sin^2 x} + \frac{\sin x}{\cos^2 x} + \sqrt{\cos^2 x + \sin^2 x + 2\sin x \cos x} + \frac{1}{\cos^2 x} + \frac{2\sin x}{\cos^2 x} + \frac{2\sin^2 x}{2\cos^2 x} + \sqrt{\cos^2 x/4 + \sin^2 x/4 + 2\sin x/2 \cos x/2} + \frac{(1-\cos^2 x)}{(1+\cos x)^2} + \tan^2 x + \cot^2 x + 2 \right\}$$

$$f(x) = \left\{ \frac{5}{2}\operatorname{cosec} x \cdot \cot x + 3\tan x \cdot \sec x + (\cos x + \sin x) + 3\sec^2 x + \left(\cos\frac{x}{4} + \sin\frac{x}{4}\right) + \operatorname{cosec}^2 x + \sec^2 x/2 - 2 \right\}$$

$$\therefore \quad \int f(x)\,dx = -\frac{5}{2}\operatorname{cosec} x + 3\sec x + (\sin x - \cos x) + 3\tan x + 4\{\sin x/4 - \cos x/4\} - \cot x + 2\tan(x/2) - 2x + C$$

3. $\int \{\sec^2(2-3x) + \operatorname{cosec}^2(3-4x) + \tan(2-3x)\sec(2-3x) + \cot x \operatorname{cosec} x + \operatorname{cosec}(3x-2)\cot(3x-2)\}\,dx$

$$= -\frac{1}{3}\tan(2-3x) - \frac{1}{4}[-\cot(3-4x)] - \frac{1}{3}\sec(2-3x) - \operatorname{cosec} x - \frac{1}{3}\operatorname{cosec}(3x-2) + C$$

$$= -\frac{1}{3}[\tan(2-3x) + \sec(2-3x) + \operatorname{cosec}(3x-2)] + \frac{1}{4}\cot(3-4x) - \operatorname{cosec} x + C$$

4. Let $I = \int \left(\frac{\sin x}{1+\sin x} + \sin x \sin 2x \sin 3x + \sec^2 x \cos^2 2x + \sin^4 x \cos^4 x \right) dx$

$$= I_1 + I_2 + I_3 + I_4$$

where $I_1 = \int \frac{\sin x}{1+\sin x}\,dx = \int \frac{\sin x\,(1-\sin x)}{\cos^2 x}\,dx$

$$= \int (\tan x \sec x - \tan^2 x)\,dx = \sec x - \tan x + x \qquad \text{...(i)}$$

$$I_2 = \int (\sin x \sin 2x \sin 3x)\,dx = \int \frac{1}{2}(2\sin x \sin 3x)\sin 2x\,dx$$

$$= \frac{1}{2}\int (\cos 2x - \cos 4x)\sin 2x\,dx$$

$$= \frac{1}{4}\int (2\cos 2x \sin 2x - 2\cos 4x \sin 2x)\,dx$$

$$= \frac{1}{4}\int (\sin 2x + \sin 4x - \sin 6x)\,dx$$

$$= -\frac{1}{4}\left[\frac{\cos 2x}{2} + \frac{\cos 4x}{4} - \frac{\cos 6x}{6}\right] \qquad \text{...(ii)}$$

$$I_3 = \int (\sec^2 x \cdot \cos^2 2x)\,dx = \int \frac{1}{\cos^2 x}[2\cos^2 x - 1]^2\,dx$$

$$= \int \left(4\cos^2 x + \frac{1}{\cos^2 x} - 4\right)dx$$

$$= \int \{2(1+\cos 2x) + \sec^2 x - 4\}\,dx = \int (2\cos 2x + \sec^2 x - 2)\,dx$$

$$= (\sin 2x + \tan x - 2x) \qquad \text{...(iii)}$$

$$I_4 = \int \sin^4 x \cdot \cos^4 x\,dx = \frac{1}{16}\int \sin^4 2x\,dx = \frac{1}{16\cdot 4}\int (2\sin^2 2x)^2\,dx$$

$$= \frac{1}{64}\int (1-\cos 2x)^2\,dx = \frac{1}{128}\int (3 + \cos 4x - 4\cos 2x)\,dx$$

$$= \frac{1}{128}\left[3x + \frac{\sin 4x}{4} - \frac{4\sin 2x}{2}\right] \qquad \text{...(iv)}$$

Adding Eqs. (i), (ii), (iii) and (iv), we get

$$I = (x - \tan x + \sec x) + \frac{1}{4}\left(-\frac{\cos 2x}{2} - \frac{\cos 4x}{4} + \frac{\cos 6x}{6}\right)$$

$$+ (\sin 2x + \tan x - 2x) + \frac{1}{128}\left(3x + \frac{\sin 4x}{4} - 2\sin 2x\right)$$

5. $I = \int \frac{\sin x \cos x}{a^2\cos^2 x + b^2 \sin^2 x}\,dx$

Put $a^2\cos^2 x + b^2\sin^2 x = t$

$$[a^2(-2\cos x \sin x) + b^2(2\sin x \cos x)]\,dx = dt$$

$$\sin x \cos x\,dx = \frac{dt}{2(b^2 - a^2)}$$

$$\therefore \quad I = \int \frac{dt}{2(b^2-a^2)\,t}$$

$$= \frac{1}{2(b^2-a^2)}\log|t| + C$$

$$= \frac{1}{2(b^2-a^2)}\log|a^2\cos^2 x + b^2\sin^2 x| + C$$

6. $I = \int \left(\frac{1}{e^x + 1} \right) dx$

Put $e^x = t \quad \Rightarrow \quad e^x \, dx = dt \qquad$ or $\qquad dx = \frac{dt}{e^x} = \frac{dt}{t}$

$$\therefore \qquad I = \int \frac{1}{(t+1)} \cdot \frac{dt}{t} = \int \left(\frac{1}{t} - \frac{1}{t+1} \right) dt \qquad \text{(using partial fractions)}$$

$$= \log |t| - \log |t+1| + C$$

$$= \log e^x - \log |e^x + 1| + C$$

$$= x - \log |e^x + 1| + C$$

7. $I = \int \frac{1}{\cos^2 x \, (1 - \tan x)^2} dx + \int \frac{10x^9 + 10^x \log_e 10}{10^x + x^{10}} dx$

Put $10^x + x^{10} = t \quad \Rightarrow \quad (10^x \log 10 + 10 \cdot x^9) \, dx = dt$

$$= \int \frac{\sec^2 x}{(1 - \tan x)^2} dx + \int \frac{dt}{t}$$

Also, let $\quad 1 - \tan x = u \Rightarrow - \sec^2 x \, dx = du$

$$= \int - \frac{du}{u^2} + \log |t| + C$$

$$= \frac{1}{u} + \log |t| + C$$

$$= \frac{1}{(1 - \tan x)} + \log |10^x + x^{10}| + C$$

8. $I = \int (\sec^p x \tan x) \, dx = \int (\sec^{p-1} x \cdot \sec x \tan x) \, dx$

Put $\sec x = t \quad \Rightarrow \quad (\sec x \tan x) \, dx = dt$

$$\therefore \qquad I = \int t^{p-1} \, dt = \frac{t^p}{p} + C = \frac{(\sec^p x)}{p} + C$$

9. $I = \int \frac{dx}{x \cdot \log x \log (\log x)}$, put $\log (\log x) = t \Rightarrow \frac{1}{\log x} \cdot \frac{1}{x} dx = dt$

$$= \int \frac{dt}{t} = \log |t| + C$$

$$\therefore \qquad I = \log \{\log (\log x)\} + C$$

10. $I = \int \frac{dx}{x^2 (x^4 + 1)^{3/4}} = \int \frac{dx}{x^5 \left(1 + \frac{1}{x^4} \right)^{3/4}}$

Put $1 + \frac{1}{x^4} = t \quad \Rightarrow \quad - \frac{4}{x^5} dx = dt$

$$\therefore \qquad I = \int - \frac{dt}{4t^{3/4}} = - \frac{1}{4} \int t^{-3/4} \, dt = - \frac{1}{4} \cdot \frac{t^{1/4}}{1/4} + C$$

$$= - \left(1 + \frac{1}{x^4} \right)^{1/4} + C$$

$$= - \frac{1}{x} (x^4 + 1)^{1/4} + C$$

11. $I = \int x \cdot \sqrt{\dfrac{2\sin(x^2-1) - \sin 2(x^2-1)}{2\sin(x^2-1) + \sin 2(x^2-1)}}\, dx = \dfrac{1}{2}\sqrt{\dfrac{2\sin t - \sin 2t}{2\sin t + \sin 2t}}\, dt$

Put $x^2 - 1 = t \Rightarrow 2x\,dx = dt$

$\therefore$
$$I = \frac{1}{2}\int \sqrt{\frac{1-\cos t}{1+\cos t}}\, dt = \frac{1}{2}\int \tan\frac{t}{2}\, dt$$
$$= \frac{1}{2}\log\frac{|\sec t/2|}{1/2} + C = \log\left|\sec\left(\frac{x^2-1}{2}\right)\right| + C$$

12. $I = \int x^{13/2}(1+x^{5/2})^{1/2}\, dx = \int x^5 \cdot x^{3/2} \cdot (1+x^{5/2})^{1/2}\, dx$

Put $1 + x^{5/2} = t \Rightarrow \dfrac{5}{2}\cdot x^{3/2}\, dx = dt$

$\therefore$
$$I = \frac{2}{5}\int t^{1/2}(t-1)^2\, dt$$
$$= \frac{2}{5}\int t^{1/2}(t^2 - 2t + 1)\, dt$$
$$= \frac{2}{5}\int (t^{5/2} - 2t^{3/2} + t^{1/2})\, dt$$
$$= \frac{2}{5}\left[\frac{2}{7}t^{7/2} - \frac{4}{5}t^{5/2} + \frac{2}{3}t^{3/2}\right] + C$$
$$= \frac{4}{5}t^{3/2}\left[\frac{1}{7}t^2 - \frac{2}{5}t + \frac{1}{3}\right] + C$$
$$= \frac{4}{5}(1+x^{5/2})^{3/2}\left[\frac{1}{7}(1+x^{5/2})^2 - \frac{2}{5}(1+x^{5/2}) + \frac{1}{3}\right] + C$$

13. $I = \int (1 + 2\tan^2 x + 2\tan x \sec x)^{1/2} dx$
$$= \int (\sec^2 x + \tan^2 x + 2\tan x \sec x)^{1/2} dx$$
$$= \int (\sec x + \tan x)\, dx$$
$$= \log|\sec x + \tan x| + \log|\sec x| + C$$
$$= \log|\sec x(\sec x + \tan x)| + C$$

14. $I = \int \dfrac{dx}{\sqrt{\sin^3 x \sin(x+a)}} = \int \dfrac{dx}{\sqrt{\sin^3 x\{\sin x\cos a + \cos x \sin a\}}}$
$$= \int \frac{dx}{\sqrt{\sin^4 x\{\cos a + \cot x \sin a\}}}$$
$$= \int \frac{\operatorname{cosec}^2 x\, dx}{\sqrt{\cos a + \sin a \cot x}}$$

Put $\cos a + \sin a \cot x = t^2 \Rightarrow -\sin a \operatorname{cosec}^2 x\, dx = 2t\, dt$

$\therefore$
$$I = -\frac{2}{\sin a}\int \frac{t\, dt}{t}$$
$$= -\frac{2}{\sin a}\cdot t + C$$
$$= -2\operatorname{cosec} a\sqrt{\cos a + \sin a \cot x} + C$$

15. $I = \int \frac{\sec x}{\sqrt{\sin(2x+\alpha)+\sin\alpha}}\,dx$

$$= \int \frac{\sec x\,dx}{\sqrt{2\sin(x+\alpha)\cos(x)}} = \frac{1}{\sqrt{2}}\int \frac{\sec^2 x\,dx}{\sqrt{\tan x\cdot\cos\alpha+\sin\alpha}}$$

Put $\tan x\cdot\cos\alpha+\sin\alpha = t^2$

$\Rightarrow$ $(\sec^2 x\cdot\cos\alpha)\,dx = 2t\,dt$

$\therefore$ $$I = \frac{2}{\sqrt{2}\cos\alpha}\int\frac{t\,dt}{t} = \frac{\sqrt{2}}{\cos\alpha}\cdot t + C$$

$$= \frac{\sqrt{2}}{\cos\alpha}\sqrt{\cos\alpha\cdot\tan x+\sin\alpha} + C$$

16. $I = \int \frac{\cos 2x-\cos 2\alpha}{\cos x-\cos\alpha}\,dx = \int\frac{(2\cos^2 x-1)-(2\cos^2\alpha-1)}{\cos x-\cos\alpha}\,dx$

$$= 2\int(\cos x+\cos\alpha)\,dx = 2(\sin x + x\cdot\cos\alpha)+C$$

17. $I = \int\frac{1+\cos 4x}{\cot x-\tan x}\,dx = \int\frac{2\cos^2 2x}{\cos^2 x-\sin^2 x}\cdot\cos x\sin x\,dx$

$$= 2\int\cos 2x\cdot\frac{1}{2}\sin 2x\,dx$$

$$= \frac{1}{2}\int\sin 4x\,dx = -\frac{1}{8}\cos 4x + C$$

18. $I = \int\sqrt{\frac{x}{a^3-x^3}}\,dx = \int\frac{x^{1/2}}{\sqrt{a^3-(x^{3/2})^2}}\,dx$

Put $x^{3/2} = t \Rightarrow \frac{3}{2}x^{1/2}\,dx = dt$

$\therefore$ $$I = \frac{2}{3}\int\frac{dt}{\sqrt{(a^{3/2})^2-(t)^2}}$$

$$= \frac{2}{3}\sin^{-1}\left(\frac{t}{a^{3/2}}\right) = \frac{2}{3}\sin^{-1}\left(\frac{x^{3/2}}{a^{3/2}}\right)+C$$

19. $I = \int\log x\cdot\frac{\sin[1+(\log x)^2]}{x}\,dx$

Put $1+(\log x)^2 = t \quad\Rightarrow\quad 2\log x\cdot\frac{1}{x}\,dx = dt$

$\therefore$ $$I = \frac{1}{2}\int\sin(t)\,dt = \frac{1}{2}(-\cos t)+C$$

$$= -\frac{1}{2}\cos[1+(\log x)^2]+C$$

20. $I = \int\frac{a^{x+\tan^{-1}a^x}}{1+a^{2x}}\,dx = \int\frac{a^x\cdot a^{\tan^{-1}a^x}}{1+(a^x)^2}\,dx$

Put $a^x = t \Rightarrow a^x\cdot\log a\,dx = dt$

$\therefore$ $I = \int\frac{a^{\tan^{-1}t}}{1+t^2}\cdot\frac{dt}{\log a}$, now put $\tan^{-1}t = u \Rightarrow \frac{1}{1+t^2}dt = du$

$$= \frac{1}{\log a}\int a^u du = \frac{a^u}{(\log a)^2} + C = \frac{a^{\tan^{-1} t}}{(\log a)^2} + C$$

$$= \frac{a^{\tan^{-1}(a^x)}}{(\log a)^2} + C$$

21. $I = \int \frac{dx}{(2ax + x^2)^{3/2}} = \int \frac{dx}{\{(x+a)^2 - a^2\}^{3/2}}$

Put $(x + a) = a \sec\theta \Rightarrow dx = a \sec\theta \tan\theta \, d\theta$

$$\therefore \quad I = \int \frac{a \sec\theta \cdot \tan\theta \, d\theta}{\{a^2(\sec^2\theta - 1)\}^{3/2}} = \int \frac{a \sec\theta \cdot \tan\theta \, d\theta}{a^3 \tan^3\theta}$$

$$= \frac{1}{a^2}\int \frac{1}{\cos\theta} \cdot \frac{\cos^2\theta}{\sin^2\theta} d\theta = \frac{1}{a^2}\left(-\frac{1}{\sin\theta}\right) + C$$

$$= -\frac{1}{a^2} \cdot \frac{x + a}{\sqrt{x^2 + 2ax}} + C$$

22. $I = \int \frac{dx}{(x^2 + 1)\sqrt{p^2 + q^2(\tan^{-1} x)^2}}$

Put $(q \tan^{-1} x) = t \quad \Rightarrow \quad \frac{q}{1 + x^2} dx = dt$

$$\therefore \quad I = \frac{1}{q}\int \frac{dt}{\sqrt{p^2 + t^2}} = \frac{1}{q}\log|t + \sqrt{p^2 + t^2}| + C$$

$$= \frac{1}{q}\log|q(\tan^{-1} x) + \sqrt{p^2 + (q \tan^{-1} x)^2}| + C$$

23. $I = \int x^3 \sqrt{q^2 x^8 - p^2} \, dx$

Put $qx^4 = t \Rightarrow 4qx^3 \, dx = dt$

$$\therefore \quad I = \frac{1}{4q}\int \sqrt{t^2 - p^2} \, dt$$

$$= \frac{1}{4q}\left[\frac{t}{2}\sqrt{t^2 - p^2} - \frac{p^2}{2}\log|t + \sqrt{t^2 - p^2}|\right] + C \quad \text{where } t = qx^4$$

24. $I = \int \frac{x}{\sqrt{9 + 8x - x^2}} dx$

We write $x = A(8 - 2x) + B$, on comparing $A = -1/2, B = 4$

$$\therefore \quad I = \int -\frac{1}{2}\frac{(8 - 2x)}{\sqrt{9 + 8x - x^2}} dx + 4\int \frac{dx}{\sqrt{9 + 8x - x^2}}$$

$$= -\frac{1}{2}\frac{[\sqrt{9 + 8x - x^2}]}{1/2} + 4\int \frac{dx}{\sqrt{25 - (x^2 - 8x + 16)}}$$

$$= -\sqrt{9 + 8x - x^2} + 4\int \frac{dx}{\sqrt{5^2 - (x - 4)^2}}$$

$$= -\sqrt{9 + 8x - x^2} + 4\sin^{-1}\left(\frac{x - 4}{5}\right) + C$$

25. $I=\int x\sqrt{\dfrac{a^2-x^2}{a^2+x^2}}\,dx$

Put $x^2=t \Rightarrow 2x\,dx=dt$

$$\therefore \quad I=\frac{1}{2}\int\sqrt{\frac{a^2-t}{a^2+t}}\,dt=\frac{1}{2}\int\frac{a^2-t}{\sqrt{a^4-t^2}}\,dt$$

$$=\frac{1}{2}\int\frac{a^2}{\sqrt{a^4-t^2}}\,dt-\frac{1}{2}\int\frac{t}{\sqrt{a^4-t^2}}\,dt$$

$$=\frac{1}{2}a^2\cdot\sin^{-1}\left(\frac{t}{a^2}\right)+\frac{1}{2}\sqrt{a^4-t^2}+C$$

$$=\frac{a^2}{2}\sin^{-1}\left(\frac{x^2}{a^2}\right)+\frac{1}{2}\sqrt{a^4-x^4}+C$$

26. $I=\int\dfrac{1}{(\sin^2 x+\sin 2x)}\,dx=\int\dfrac{dx}{\sin^2 x+2\sin x\cos x}=\int\dfrac{\sec^2 x\,dx}{\tan^2 x+2\tan x}$

(dividing N^r and D^r by $\cos^2 x$)

Put $\tan x=t \Rightarrow \sec^2 x\,dx=dt$

$$\therefore \quad I=\int\frac{dt}{t^2+2t}=\frac{1}{2\cdot 1}\log\left|\frac{t+1-1}{t+1+1}\right|+C$$

$$=\frac{1}{2}\log\left|\frac{\tan x}{\tan x+2}\right|+C$$

27. $I=\int\cos 2\theta\cdot\log\left(\dfrac{\cos\theta+\sin\theta}{\cos\theta-\sin\theta}\right)d\theta$

$$=\int\cos 2\theta\cdot\log\left(\frac{1+\tan\theta}{1-\tan\theta}\right)d\theta$$

$=\int \underset{\text{I}}{\cos 2\theta}\cdot \underset{\text{II}}{\log|\tan(\pi/4+\theta)|}\,d\theta$, applying integration by parts, we get

$$=\log\{\tan(\pi/4+\theta)\}\cdot\frac{\sin 2\theta}{2}-\int\frac{1}{\tan(\pi/4+\theta)}\cdot\sec^2(\pi/4+\theta)\cdot\frac{\sin 2\theta}{2}\,d\theta$$

$$=\log\{\tan(\pi/4+\theta)\}\cdot\frac{\sin 2\theta}{2}-\int\frac{2}{\sin(\pi/2+2\theta)}\cdot\frac{\sin 2\theta}{2}\,d\theta$$

$$=\log\{\tan(\pi/4+\theta)\}\cdot\frac{\sin 2\theta}{2}-\int\frac{\sin 2\theta}{\cos 2\theta}\,d\theta$$

$$=\log\{\tan(\pi/4+\theta)\}\cdot\frac{\sin 2\theta}{2}-\frac{1}{2}\log|\sec 2\theta|+C$$

28. $I=\int\log\{x+\sqrt{x^2+a^2}\}\,dx$

We know, $\dfrac{d}{dx}\{\log(x+\sqrt{x^2+a^2})\}=\dfrac{1}{\sqrt{x^2+a^2}}$

$$\therefore \quad I=\int \underset{\text{II}}{1}\cdot\underset{\text{I}}{\log\{x+\sqrt{x^2+a^2}\}}\,dx$$

$$\log\{x+\sqrt{x^2+a^2}\}\cdot x-\frac{1}{2}\int\frac{2}{\sqrt{x^2+a^2}}\cdot x\,dx$$

$$=x\cdot\log\{x+\sqrt{x^2+a^2}\}-\sqrt{x^2+a^2}+C$$

As $\int\frac{f'(x)}{\sqrt{f(x)}}\,dx=2\sqrt{f(x)}+C$

29. $I=\int\frac{xe^x}{(1+x)^2}\,dx=\int e^x\left\{\frac{x+1-1}{(x+1)^2}\right\}dx$

$$=\int e^x\left\{\frac{1}{x+1}-\frac{1}{(x+1)^2}\right\}dx$$

[as $\int e^x\{f(x)+f'(x)\}\,dx=e^x f(x)+C$]

$\therefore\qquad I=e^x\cdot\frac{1}{x+1}+C$

30. $I=\int\frac{\log x}{(1+\log x)^2}\,dx,$ put $\log x=t,\ x=e^t\Rightarrow dx=e^t\,dt$

$$=\int\frac{t}{(1+t)^2}\cdot e^t\,dt=\int e^t\left\{\frac{t+1-1}{(t+1)^2}\right\}dt$$

$$=\frac{e^t}{(t+1)}+C=\frac{x}{1+\log x}+C$$

31. $I=\int\{\log(1+\cos x)-x\tan x/2\}\,dx$

$$=\int 1\cdot\log(1+\cos x)\,dx-\int x\tan\frac{x}{2}\,dx$$

$$=\{\log(1+\cos x)\}\cdot x-\int x\cdot\frac{1\cdot(-\sin x)}{1+\cos x}\,dx-\int x\tan\frac{x}{2}\,dx$$

$$=x\log(1+\cos x)+\int x\tan x/2\,dx-\int x\tan x/2\,dx+C$$
$$=x\log(1+\cos x)+C$$

32. $I=\int\frac{\log_e x}{x^3}\,dx=\int\log x\cdot x^{-3}\,dx$

I II

$$=(\log x)\left(-\frac{x^{-2}}{2}\right)-\int\frac{1}{x}\cdot\left(-\frac{x^{-2}}{2}\right)dx+C$$

$$=-(\log x)\cdot\frac{1}{2x^2}+\frac{1}{2}\int\frac{1}{x^3}\,dx+C$$

$$=-\frac{1}{2x^2}(\log x)-\frac{1}{4x^2}+C=-\frac{1}{4x^2}(2\log_e x+1)+C$$

33. $I=\int\left(\frac{2x+\sin 2x}{1+\cos 2x}\right)dx$

$$=\int x\cdot\sec^2 x\,dx+\int\frac{2\sin x\cos x}{2\cos^2 x}\,dx$$

$$=x\cdot\tan x-\int 1\cdot\tan x\,dx+\int\tan x\,dx=x\tan x+C$$

34.
$$I=\int\frac{x^4}{(x-1)(x^2+1)}\,dx$$
$$=\int\frac{x^4-1+1}{(x-1)(x^2+1)}\,dx$$
$$=\int\left[(x+1)+\frac{1}{(x-1)(x^2+1)}\right]dx$$
$$=\int\left[(x+1)+\frac{1}{2(x-1)}-\frac{(x+1)}{2(x^2+1)}\right]dx \quad \text{(using partial fractions)}$$
$$=\frac{1}{2}x^2+x+\frac{1}{2}\log|x-1|-\frac{1}{4}\log|x^2+1|-\frac{1}{2}\tan^{-1}x+C$$

35.
$$I=\int\frac{x}{(\sqrt{x+a}+\sqrt{x+b})}\,dx=\int\frac{x(\sqrt{x+a}-\sqrt{x+b})}{(x+a)-(x+b)}\,dx$$
$$=\frac{1}{a-b}\int\{(x+a-a)\sqrt{x+a}-(x+b-b)\sqrt{x+b}\}\,dx$$
$$=\frac{1}{a-b}\int\{(x+a)^{3/2}-(x+a)^{1/2}-b(x+b)^{3/2}+b(x+b)^{1/2}\}\,dx$$
$$=\frac{1}{a-b}\left[\frac{2}{5}(x+a)^{5/2}-\frac{2a}{3}(x+a)^{3/2}-\frac{2}{5}(x+b)^{5/2}+\frac{2}{3}b(x+b)^{3/2}\right]+C$$

36. $I=\int\frac{(x^2+1)\,dx}{(x^2+2)(2x^2+1)}$, write $x^2=t$ and split $\frac{t+1}{(t+2)(2t+1)}$ into partial fractions.

37. $I=\int\sqrt{\frac{1-\sqrt{x}}{1+\sqrt{x}}}\,dx=\int\frac{1-\sqrt{x}}{\sqrt{1-x}}\,dx$

Put $\sqrt{x}=\cos\theta$, $x=\cos^2\theta\Rightarrow dx=(2\cos\theta)(-\sin\theta)\,d\theta$

$\therefore$
$$I=\int\frac{1-\cos\theta}{\sin\theta}\cdot(-2\cos\theta\sin\theta)\,d\theta$$
$$=\int(-2\cos\theta+1+\cos2\theta)\,d\theta$$
$$=-2\sin\theta+\theta+\frac{1}{2}\sin2\theta+C$$
$$=-2\sqrt{1-x}+\cos^{-1}\sqrt{x}+\sqrt{x}\sqrt{1-x}+C$$

38. $I=\int\frac{x^4-x^3+5x^2+x+3}{(x+1)(x^2-x+1)^2}\,dx$

Let the given fraction be written as
$$=\frac{A}{x+1}+\frac{Bx+C}{x^2-x+1}+\frac{Dx+E}{(x^2-x+1)^2}$$

$\therefore$ $A=1,\ B=0,\ C=1,\ D=2,\ E=1$

$\therefore$
$$I=\int\left\{\frac{1}{x+1}+\frac{1}{(x-1/2)^2+(\sqrt{3}/2)^2}+\frac{2x+1}{(x^2-x+1)^2}\right\}dx$$
$$=\log|x+1|+\frac{2}{\sqrt{3}}\tan^{-1}\left(\frac{2x-1}{\sqrt{3}}\right)+\left(\int\frac{2x-1+2}{(x^2-x+1)^2}\,dx=I_1\right)$$

$$\Rightarrow \quad I_1 = -\frac{1}{(x^2 - x + 1)} + 2\int \frac{dx}{(x^2 - x + 1)^2}$$

$$= -\frac{1}{(x^2 - x + 1)} + 2\int \frac{dx}{[(x - 1/2)^2 + (\sqrt{3}/2)^2]^2},$$

(proceeding as in above, we get)

$$= -\frac{1}{(x^2 - x + 1)} + \frac{2}{2(\sqrt{3}/2)^3}\left[\tan^{-1}\left(\frac{2x-1}{\sqrt{3}}\right) + \frac{\frac{\sqrt{3}}{2}\left(x - \frac{1}{2}\right)}{x^2 - x + 1}\right]$$

$$\therefore \quad I = \log|x+1| + \frac{2}{\sqrt{3}}\tan^{-1}\left(\frac{2x-1}{\sqrt{3}}\right) - \frac{1}{(x^2 - x + 1)} + \frac{8}{3\sqrt{3}}\tan^{-1}\left(\frac{2x-1}{\sqrt{3}}\right) + \frac{2}{3}\cdot\frac{(2x-1)}{(x^2 - x + 1)}$$

$$= \log|x+1| + \frac{14}{3\sqrt{3}}\tan^{-1}\left(\frac{2x-1}{\sqrt{3}}\right) + \frac{(4x-5)}{3(x^2 - x + 1)} + C$$

39. $I = \int \frac{dx}{(1+x)\sqrt{2+x-x^2}}$, put $x + 1 = \frac{1}{t} \Rightarrow dx = -\frac{1}{t^2}\,dt$

$$\therefore \quad I = -\int \frac{dt}{\sqrt{3t-1}} = -2\left[\frac{1}{3}\sqrt{3t-1}\right] + C = -\frac{2}{3}\sqrt{3\left(\frac{1}{x+1}\right) - 1} + C$$

40. $I = \int \frac{dx}{(x^2 - 6x + 9)\sqrt{x^2 - 6x + 4}}$, put $x - 3 = \frac{1}{t} \Rightarrow dx = -\frac{1}{t^2}\,dt$

$$= \int \frac{dx}{(x-3)^2\sqrt{(x-3)^2 - 5}} = \int \frac{-1/t^2\,dt}{\frac{1}{t^2}\sqrt{1/t^2 - 5}}$$

$$= -\int \frac{t\,dt}{\sqrt{1-5t^2}} = \frac{2}{10}\sqrt{1-5t^2} + C$$

$$= \frac{1}{5}\cdot\frac{\sqrt{x^2 - 6x + 4}}{(x-3)} + C$$

41. $I = \int \left(\frac{1-\sqrt{x}}{1+\sqrt{x}}\right)^{1/2}\frac{dx}{x}$, put $x = t^2 \Rightarrow dx = 2t\,dt$

$$= \int \sqrt{\frac{1-t}{1+t}}\cdot\frac{2t}{t^2}\,dt = 2\int \frac{1-t}{t\sqrt{1-t^2}}\,dt$$

$$= -2\log\left\{\frac{1+\sqrt{1-x}}{\sqrt{x}}\right\} - 2\sin^{-1}\sqrt{x} + C$$

42. $I = \int \frac{dx}{x^2 - 2x\cos\theta + 1} = \int \frac{dx}{(x - \cos\theta)^2 + (1 - \cos^2\theta)}$

$$= \frac{1}{\sin\theta}\tan^{-1}\left(\frac{x - \cos\theta}{\sin\theta}\right) + C$$

43. $$I = \int \frac{(\sin x - \cos x)\, dx}{(\sin x + \cos x)\sqrt{\sin x \cos x + \sin^2 x \cos^2 x}}$$

$$= \int \frac{(\sin x - \cos x)\, dx}{(\sin x + \cos x)\sqrt{\sin x \cos x\,(1 + \sin x \cos x)}}$$

Put $\sin x + \cos x = t \Rightarrow (\cos x - \sin x)\, dx = dt$

$$\therefore \quad I = \int \frac{-dt}{t\sqrt{\frac{1}{2}(t^2-1)^2 + \frac{1}{4}(t^2-1)^2}}$$

$$I = -\int \frac{2\, dt}{t\sqrt{t^4 - 1}} = -\operatorname{cosec}^{-1}(t^2) + C$$

$$I = -\operatorname{cosec}^{-1}(1 + \sin 2x) + C$$

44. $$I = \int \left[\frac{(\cos 2x)^{1/2}}{\sin x}\right] dx$$

Put $\cos 2x = \dfrac{1 - \tan^2 x}{1 + \tan^2 x}$

$$\therefore \quad I = \int \frac{\sqrt{1 - \tan^2 x}}{\sin x \cdot \sec x}\, dx, \quad \text{put } \sqrt{1 - \tan^2 x} = t \Rightarrow -2\tan x \cdot \sec x\, dx = 2t\, dt$$

$$= -\int \frac{t^2}{(1-t^2)(2-t^2)}\, dt = -\int \left(\frac{1}{1-t^2} - \frac{2}{2-t^2}\right) dt \text{ (using partial fractions)}$$

$$= -\frac{1}{2}\log\left|\frac{1+t}{1-t}\right| - \frac{2\cdot 1}{2\sqrt{2}}\log\left|\frac{\sqrt{2}+1}{\sqrt{2}-t}\right| + C$$

where, $t = \sqrt{(1 - \tan^2 x)}$

45. $I = \int \dfrac{dx}{\sqrt{\sin^3 x \cos^5 x}}$, dividing N^r and D^r by $\cos^4 x$,

$$= \int \frac{\sec^2 x \cdot \sec^2 x}{\sqrt{\tan^3 x}}\, dx = \int \frac{(1 + t^2)}{t\sqrt{t}}\, dt$$

$$= \frac{2}{3}\left(\frac{t^2 + 3}{\sqrt{t}}\right) + C = \frac{2}{3}\left\{\frac{\tan^2 x + 3}{\tan x}\right\} + C$$

46. $$I = \int \frac{\sin\theta + \cos\theta}{\sqrt{\sin 2\theta}}\, d\theta$$

We know $(\sin\theta + \cos\theta)$ is differential coefficient of $(\sin\theta - \cos\theta)$

We put $(\sin\theta - \cos\theta) = t \Rightarrow (\cos\theta + \sin\theta)\, d\theta = dt$

On squaring $(\sin\theta - \cos\theta)^2 = t^2$

$$1 - \sin 2\theta = t^2$$

$$\therefore \quad I = \int \frac{dt}{\sqrt{1 - t^2}} = \sin^{-1}(t) + C = \sin^{-1}(\sin\theta - \cos\theta) + C$$

47. $I=\int \frac{dx}{(x+\alpha)^{8/7}(x-\beta)^{6/7}}=\int \frac{dx}{(x+\alpha)^2\left(\frac{x-\beta}{x+\alpha}\right)^{6/7}}$ put, $\frac{x-\beta}{x+\alpha}=t, \frac{(\alpha+\beta)\,dx}{(x+\alpha)^2}=dt$

$$=\int \frac{dt}{(\alpha+\beta)\,t^{6/7}}=\frac{1}{(\alpha+\beta)}\left[\frac{t^{1/7}}{1/7}\right]+C=\frac{7}{\alpha+\beta}\left(\frac{x-\beta}{x+\alpha}\right)^{1/7}+C$$

48. $I=\int a^x\left\{\log x+\log a\cdot\log\frac{x^x}{e^x}\right\}dx$; applying by parts, we get

$$=a^x\cdot x\,(\log x-1)+C$$

49. $I=\int e^{(x\sin x+\cos x)}\left\{\frac{x^4\cos^3 x-x\sin x+\cos x}{x^2\cos^2 x}\right\}dx$

$$=\int e^{(x\sin x+\cos x)}\cdot(x^2\cos x)\,dx-\int e^{(x\sin x+\cos x)}\cdot\frac{d}{dx}\left(\frac{1}{x\cos x}\right)dx$$

$$=e^{(x\sin x+\cos x)}\cdot\left(x-\frac{1}{x\cos x}\right)+C$$

50. $I=\int \sin^{-1}\sqrt{\frac{x}{a+x}}\,dx$

Put $x=a\tan^2\theta$, $dx=2a\tan\theta\cdot\sec^2\theta\,d\theta$

$$=a\int \theta\,(2\tan\theta\sec^2\theta)\,d\theta$$

$$=a\,[\theta\,(1+\tan^2\theta)-\tan\theta] \qquad \text{(using integration by parts)}$$

$$=(a+x)\tan^{-1}\sqrt{\frac{x}{a}}-\sqrt{ax}+C$$

51. $f'(x)=\frac{1}{1+\cos x}=\frac{1}{2}\sec^2\left(\frac{x}{2}\right)$

$\therefore$ $f(x)=\tan\frac{x}{2}+C \Rightarrow f(0)=\tan 0+C=3 \quad \therefore \quad C=3$

$f(x)=\tan\left(\frac{x}{2}\right)+3$

$\therefore$ $f(\pi/2)=1+3=4$ ...(i)

Now, $\pi=\frac{22}{7} \quad \therefore \quad 3+\frac{\pi}{4}=3+\frac{22}{8}=3.78$ ...(ii)

$3+\frac{\pi}{2}=3+\frac{11}{7}=4.57$ ...(iii)

The result is obvious from Eqs. (i), (ii) and (iii)

ie, $3.78\le 4\le 4.57$

$$3+\frac{\pi}{4}\le f\left(\frac{\pi}{2}\right)\le 3+\frac{\pi}{2}$$

53. Since, $\int f(x)\sin x\cos x\,dx=\frac{1}{2(b^2-a^2)}\log(f(x))+C$

Differentiating both the sides w.r.t. x, we get

$$f(x)\sin x\cos x = \frac{1}{2(b^2-a^2)}\cdot\frac{1}{f(x)}\cdot f'(x)$$

$$\Rightarrow \quad 2(b^2-a^2)\sin x\cos x = \frac{f'(x)}{[f(x)]^2}$$

$$\Rightarrow \quad 2b^2\sin x\cos x - 2a^2\sin x\cos x = \frac{f'(x)}{[f(x)]^2}$$

Integrating both the sides w.r.t. x, we get

$$\Rightarrow \quad -b^2\cos^2 x - a^2\sin^2 x = -\frac{1}{f(x)}$$

$$\Rightarrow \quad f(x) = \frac{1}{(a^2\sin^2 x + b^2\cos^2 x)}$$

54. Putting, $l^{r+1}(x) = t$

Then, $$\frac{1}{x\,(l(x))\,l^2(x)\,l^3(x)\ldots l^r(x)}dx = dt$$

$$I = \int \frac{1}{x\,l(x)\,l^2(x)\ldots l^r(x)} = t + C$$

$$I = l^{r+1}(x) + C$$

55. $I = \int \frac{dx}{(x-b)^3(x-a)^2}$

Put $(x-b) = t(x-a) \Rightarrow x = \frac{at-b}{t-1} \Rightarrow dx = \frac{b-a}{(t-1)^2}dt$

Also, $(x-a) = \frac{a-b}{t-1}$ and $(x-b) = \frac{(a-b)t}{(t-1)}$

Hence, $$I = \int \frac{(t-1)^3}{(a-b)^3 t^3}\cdot\frac{(t-1)^2}{(a-b)^2}\cdot\frac{b-a}{(t-1)^2}dt$$

$$= -\frac{1}{(a-b)^4}\int \frac{t^3-3t^2+3t-1}{t^3}dt$$

$$= -\frac{1}{(a-b)^4}\cdot\int\left(1-\frac{3}{t}+\frac{3}{t^2}-\frac{1}{t^3}\right)dt$$

$$= -\frac{1}{(a-b)^4}\cdot\left[t - 3\log|t| - \frac{3}{t} + \frac{1}{2t^2}\right], \text{ where } t = \frac{x-b}{x-a}.$$

56. $I = \int \sqrt{x+\sqrt{x^2+2}}\,dx$

Let $x + \sqrt{x^2+2} = p$ or $x^2+2 = p^2+x^2-2px$

$\Rightarrow$ $x = \frac{p^2-2}{2p}$ or $dx = \frac{(p^2+2)\,dp}{2p^2}$

$$I = \int \frac{p^{1/2}(p^2+2)}{2p^2}dp = \frac{1}{2}\int p^{1/2}\,dp + \int p^{-3/2}\,dp$$

$$= \frac{1}{3}(x+\sqrt{x^2+2})^{3/2} - 2\frac{1}{\sqrt{x+\sqrt{x^2+2}}} + C$$

57. $I = \int (\sqrt{\tan x} + \sqrt{\cot x})\, dx$

$$= \int \sqrt{2} \cdot \frac{\sin x + \cos x}{\sqrt{2}\sqrt{\sin x}\sqrt{\cos x}}\, dx = \sqrt{2} \int \frac{(\sin x + \cos x)}{\sqrt{1 - (\sin x - \cos x)^2}}\, dx$$

Put $(\sin x - \cos x) = t \Rightarrow (\cos x + \sin x)\, dx = dt$

$\therefore$ $$I = \sqrt{2} \int \frac{dt}{\sqrt{1 - t^2}} = \sqrt{2} \sin^{-1}(t) + C$$

$$= \sqrt{2} \sin^{-1}(\sin x - \cos x) + C$$

58. $I = \int \dfrac{dx}{(ax + b)\sqrt{(x - \alpha)^2 - \beta^2}}$

Put $(x - \alpha) = \beta \sec \theta \Rightarrow dx = \beta \sec \theta \tan \theta\, d\theta$

$\therefore$ $$I = \int \frac{d\theta}{a(\alpha \cos \theta + \beta) + b \cos \theta} = \int \frac{d\theta}{(a\alpha + b)\cos \theta + a\beta}$$

$$= \frac{1}{(a\alpha + b)} \int \frac{d\theta}{\cos \theta + \left(\dfrac{a\beta}{a\alpha + b}\right)} \quad \text{if} \left|\frac{a\beta}{a\alpha + b}\right| < 1 \quad \ldots\text{(i)}$$

Then, $$I = \frac{1}{(a\alpha + b)} \cdot \operatorname{cosec} \alpha^1 \log \left|\frac{\cot \dfrac{\alpha^1}{2} + 1}{\cot \dfrac{\alpha^1}{2} - 1}\right| + C$$

where $\dfrac{a\beta}{a\alpha + b} = \cos \alpha^1, \ \tan \dfrac{\theta}{2} = t$

Again, if $\left|\dfrac{a\beta}{a\alpha + b}\right| > 1$

$$I = \frac{1}{a\alpha + b} \cot \alpha^1 \tan^{-1}\left(t \tan \frac{\alpha^1}{2}\right)$$

where $\sec \alpha^1 = \dfrac{a\beta}{a\alpha + b}$

$\tan \dfrac{\theta}{2} = t$

59. $I = \int \dfrac{\sqrt{1 + \sqrt[3]{x}}}{\sqrt[3]{x^2}}\, dx = \int x^{-2/3}(1 + x^{1/3})^{1/2}\, dx, \ m = -\dfrac{2}{3}, n = \dfrac{1}{3}, p = \dfrac{1}{2}$

$\therefore$ $\dfrac{m + 1}{n} = 1$ *ie*, integer

$\therefore$ Let us make the substitution,

$$1 + x^{1/3} = t^2 \quad \therefore \quad \frac{1}{3} x^{-2/3}\, dx = 2t\, dt$$

Hence, $$I = 6 \int t^2\, dt = 2t^3 + C = 2(1 + x^{1/3})^{3/2} + C$$

60. $I = \int \tan(x - \alpha) \cdot \tan(x + \alpha) \cdot \tan 2x\, dx$

$$\tan 2x = \tan(\overline{x + \alpha} + \overline{x - \alpha}) = \frac{\tan(x + \alpha) + \tan(x - \alpha)}{1 - \tan(x + \alpha)\tan(x - \alpha)}$$

or $\tan 2x - \tan 2x \cdot \tan(x+\alpha)\tan(x-\alpha) = \tan(x+\alpha) + \tan(x-\alpha)$

$\therefore$ $\tan(x-\alpha)\tan(x+\alpha)\tan(2x) = \tan 2x - \tan(x+\alpha) - \tan(x-\alpha)$

$$\Rightarrow \quad I = \int \{\tan(x-\alpha)\tan(x+\alpha)\tan(2x)\}\,dx$$

$$= \int \{\tan 2x - \tan(x+\alpha) - \tan(x-\alpha)\}\,dx$$

$$= \frac{1}{2}\log|\sec 2x| - \log|\sec(x+\alpha)| - \log|\sec(x-\alpha)| + C$$

$$= \log|\sqrt{\sec 2x}\cdot\cos(x+\alpha)\cdot\cos(x-\alpha)| + C$$

61. $\frac{1}{1000} + \frac{1}{1001} + \frac{1}{1002} + \frac{1}{1003} + \ldots + \frac{1}{2000} = \log 2$

62. $I = \int \frac{\sin^3(\theta/2)}{\cos\theta/2\sqrt{\cos^3\theta + \cos^2\theta + \cos\theta}}\,d\theta$

$$= \frac{1}{2}\int \frac{2\sin\theta/2\cdot\cos\theta/2\cdot 2\sin^2\theta/2}{2\cos^2\theta/2\sqrt{\cos^3\theta + \cos^2\theta + \cos\theta}}\,d\theta$$

$$= \frac{1}{2}\int \frac{\sin\theta(1-\cos\theta)}{(1+\cos\theta)\sqrt{\cos^3\theta + \cos^2\theta + \cos\theta}}\,d\theta$$

Put $\cos\theta = t \Rightarrow -\sin\theta\,d\theta = dt$

$$\therefore \quad I = \frac{1}{2}\int \frac{(t-1)}{(t+1)\sqrt{t^3+t^2+t}}\,dt$$

$$= \frac{1}{2}\int \frac{t^2-1}{(t+1)^2\sqrt{t^3+t^2+t}}\,dt = \frac{1}{2}\int \frac{(1-1/t^2)}{(t+\frac{1}{t}+2)\sqrt{t+1+\frac{1}{t}}}\,dt$$

Put $t + 1 + \frac{1}{t} = u^2 \quad \Rightarrow (1 - \frac{1}{t^2})\,dt = 2u\,du$

$$I = \frac{1}{2}\int \frac{2u\,du}{(u^2+1)\cdot u} = \int \frac{du}{u^2+1} = \tan^{-1}(u) + C$$

$$= \tan^{-1}\left(t + 1 + \frac{1}{t}\right)^{1/2} + C = \tan^{-1}(\cos\theta + \sec\theta + 1)^{1/2} + C$$

63. $I = \int x^{x+1}\log x(1+\log x)\,dx$

We know, $\frac{d}{dx}(x^x) = x^x(1+\log x) \quad \Rightarrow \quad \int x^x(1+\log x)\,dx = x^x$

$$\therefore \quad I = \int (x\log x)\{x^x(1+\log x)\}\,dx$$

$$= (x\log x)\cdot x^x - \int\left(x\cdot\frac{1}{x} + \log x\cdot 1\right)\cdot x^x\,dx \qquad \text{(applying by parts)}$$

$$= x^{x+1}\log x - \int x^x(1+\log x)\,dx$$

$$= x^{x+1}\log x - x^x + C$$

$$= x^x(x\log x - 1) + C$$

64. $I = \int \frac{(2\sin\theta + \sin 2\theta)\,d\theta}{(\cos\theta - 1)\sqrt{\cos\theta + \cos^2\theta + \cos^3\theta}}$

Put $\cos\theta = x^2 \Rightarrow -\sin\theta\,d\theta = 2x\,dx$

$$= 2\int \frac{(1+x^2)}{(1-x^2)} \cdot \frac{2x\,dx}{\sqrt{x^2+x^4+x^6}}$$

$$= 4\int \frac{(1+1/x^2)\,dx}{(1/x-x)\sqrt{(1/x-x)^2+3}}, \text{ put } \frac{1}{x}-x=t \Rightarrow \left(-\frac{1}{x^2}-1\right)dx = dt$$

$$= -4\int \frac{dt}{t\sqrt{t^2+3}}$$

Again, put $t^2+3=u^2 \;\Rightarrow\; 2t\,dt = 2u\,du$

$$\therefore \quad I = 4\int \frac{-u\,du}{u(u^2-3)} = -4\int \frac{du}{u^2-3}$$

$$= -\frac{2}{\sqrt{3}}\log\left|\frac{u-\sqrt{3}}{u+\sqrt{3}}\right| + C$$

$$= -\frac{2}{\sqrt{3}}\log\left|\frac{\sqrt{t^2+3}-\sqrt{3}}{\sqrt{t^2+3}+\sqrt{3}}\right| + C$$

$$= -\frac{2}{\sqrt{3}}\log\left|\frac{\sqrt{x^2+1/x^2+1}-\sqrt{3}}{\sqrt{x^2+1/x^2+1}+\sqrt{3}}\right| + C$$

$$= -\frac{2}{3}\log\left|\frac{\sqrt{\cos\theta+\sec\theta+1}-\sqrt{3}}{\sqrt{\cos\theta+\sec\theta+1}+\sqrt{3}}\right| + C$$

65. Let $\quad I_{m-1} = \int x^{m-1}(a+bx^n)^p\,dx$

and $\quad I_{m-n-1} = \int x^{m-n-1}(a+bx^n)^p\,dx$

Let $\quad p = x^{\lambda+1}(a+bx^n)^{\mu+1}$

where λ and μ are the smaller indices of x and $(a+bx^n)$.

Here, $\lambda = m-n-1$, $\mu = p$

$$\therefore \quad p = x^{m-n}(a+bx^n)^{p+1}$$

Differentiating w.r.t. x, we have

$$\frac{dp}{dx} = x^{m-n}(p+1)(a+bx^n)^p(n\,bx^{n-1}) + (a+bx^n)^{p+1}(m-n)x^{m-n-1}$$

$$= nb(p+1)x^{m-1}(a+bx^n)^p + (m-n)x^{m-n-1}(a+bx^n)^p\{a+bx^n\}$$

$$= nb(p+1)x^{m-1}(a+bx^n)^p + a(m-n)x^{m-n-1}(a+bx^n)^p + b(m-n)x^{m-1}(a+bx^n)^p$$

$$= b(np+m)x^{m-1}(a+bx^n)^p + a(m-n)x^{m-n-1}(a+bx^n)^p$$

Integrating both the sides w.r.t. x, we get

$$p = b(np+m)I_{m-1} + a(m-n)I_{m-n-1}$$

$$\therefore \quad x^{m-n}(a+bx^n)^{p+1} = b(np+m)I_{m-1} + a(m-n)I_{m-n-1}$$

or $\quad I_{m-1} = \dfrac{x^{m-n}(a+bx^n)^{p+1}}{b(np+m)} - \dfrac{a(m-n)}{b(np+m)} \cdot I_{m-n-1}$ **Hence Proved**

Again, $\quad \int \dfrac{x^8}{(1-x^3)^{1/3}}\,dx = \int x^{9-1}(1-x^3)^{-1/3}\,dx$

Here, $m = 9, \quad b = -1, \quad n = 3, \quad p = -1/3, \quad a = 1$

$$\therefore \quad I_8 = \frac{(x^6 (1 - x^3)^{2/3})}{-8} + \frac{6}{8} I_5 \qquad \text{...(i)}$$

$$\Rightarrow \quad I_5 = \frac{x^3 (1 - x^3)^{2/3}}{-5} + \frac{3}{5} I_2 \qquad \text{(here } m = 6)$$

$$\Rightarrow \quad I_2 = \frac{(1 - x^3)^{2/3}}{2}$$

Hence, $$I_8 = -\frac{1}{8} x^6 (1 - x^3)^{2/3} - \frac{3}{20} x^3 (1 - x^3)^{2/3} - \frac{9}{40} (1 - x^3)^{2/3} + C$$

66. $$I = \int \frac{(\sin^{3/2} \theta + \cos^{3/2} \theta)\, d\theta}{\sqrt{\sin^3 \theta \cos^3 \theta \sin (\theta + \alpha)}}$$

$$= \int \frac{d\theta}{\sqrt{\cos^3 \theta (\sin \theta \cos \alpha + \cos \theta \sin \alpha)}} + \int \frac{d\theta}{\sqrt{\sin^3 \theta (\sin \theta \cos \alpha + \cos \theta \sin \alpha)}}$$

$$= \int \frac{(\sec^2 \theta)\, d\theta}{\sqrt{(\cos \alpha \cdot \tan \theta + \sin \alpha)}} + \int \frac{\operatorname{cosec}^2 \theta\, d\theta}{\sqrt{\cos \alpha + \cot \theta \cdot \sin \alpha}}$$

Put, $\cos \alpha \tan \theta + \sin \alpha = t^2$, in first integral and $\cos \alpha + \cot \theta \sin \alpha = u^2$ in second integral.

$$\Rightarrow \quad I = \int \frac{2t\, dt}{(\cos \alpha)\, t} - \int \frac{2u\, du}{(\sin \alpha)\, u} = \frac{2}{\cos \alpha} \cdot t - \frac{2}{\sin \alpha} \cdot u + C$$

$$= \frac{2}{\cos \alpha} \cdot \sqrt{\cos \alpha \tan \theta + \sin \alpha} - \frac{2}{\sin \alpha} \sqrt{\cos \alpha + \cot \theta \cdot \sin \alpha} + C$$

67. Let $I = \int \operatorname{cosec}^2 x \ln (\cos x + \sqrt{\cos 2x})\, dx$

$$= \int \operatorname{cosec}^2 x \cdot \ln \{\sin x (\cot x + \sqrt{\cot^2 x - 1})\}\, dx$$

$$= \int \operatorname{cosec}^2 x \ln (\sin x)\, dx + \int \operatorname{cosec}^2 x \cdot \ln (\cot x + \sqrt{\cot^2 x - 1})\, dx$$

In second integral put $\cot x = 1$

$$\therefore \quad \operatorname{cosec}^2 x\, dx = dt$$

$$\therefore \quad I = \int \operatorname{cosec}^2 x \cdot \ln (\sin x)\, dx - \int \ln (t + \sqrt{t^2 - 1})\, dt$$

In first integral (integrating by parts taking $\operatorname{cosec}^2 x$ as second integral) and in second integral (integration by parts taking unity as second function).

We have, $(\ln \sin x)(-\cot x) - \int \cot x (-\cot x)\, dx - \ln (t + \sqrt{t^2 - 1})\, t + \int \frac{t\, dt}{\sqrt{t^2 - 1}}$

$$= -\cot x (\ln \sin x) - \cot x - x - t \ln (t + \sqrt{t^2 - 1}) + \sqrt{t^2 - 1} + C$$

$$= -\cot x (\ln \sin x) - \cot x - x - \cot x \{\ln (\cot x + \sqrt{\cot^2 x - 1})\} + \sqrt{\cot^2 x - 1} + C$$

$$= -\cot x \cdot \ln (\cos x + \sqrt{\cos 2x}) - \cot x - x + \sqrt{\cot^2 x - 1} + C$$

68. $$I = \int \frac{dx}{(\sin x + a \sec x)^2}\; a \in N$$

$$= \int \frac{\cos^2 x\, dx}{(a + \sin x \cos x)^2} = \int \frac{\cos^2 x\, dx}{(a^2 + \sin^2 x \cdot \cos^2 x + 2a \sin x \cdot \cos x)}$$

$$= \int \frac{4\cos^2 x\, dx}{4a^2 + \sin^2 2x + 4a \sin 2x} = 2\int \frac{1 + \cos 2x}{(2a + \sin 2x)^2} dx$$

$$= 2\int \frac{1}{(2a + \sin 2x)^2} dx + 2\int \frac{\cos 2x}{(2a + \sin 2x)^2} dx$$

$$= 2I_1 - \frac{1}{(2a + \sin 2x)} \quad \ldots(i)$$

$$I_1 = \int \frac{dx}{(2a + \sin 2x)^2}, \qquad \text{we know}$$

$$\therefore \quad I_1 = \int \frac{du}{(4a^2 - 1)\dfrac{\sqrt{4a^2 - 1}\sqrt{1 - u^2}}{(2a - u)}} \qquad u = 2a\,\frac{\sin 2x + 1}{2a + \sin 2x}$$

$$= \frac{1}{(4a^2 - 1)^{3/2}} \int \frac{(2a - u)}{\sqrt{1 - u^2}} du \qquad \Rightarrow \frac{du}{dx} = \frac{(4a^2 - 1)\cos 2x}{(2a + \sin 2x)^2}$$

$$\Rightarrow \quad \frac{1}{(4a^2 - 1)^{3/2}} [2a \sin^{-1} u + \sqrt{1 - u^2}] = I_1 \qquad \text{and} \quad \sin 2x = \frac{2au - 1}{2a - u}$$

$$\therefore \quad I = \frac{1}{(4a^2 - 1)^{3/2}} \left[2a \sin^{-1}\left(\frac{2a \sin 2x + 1}{2a + \sin 2x}\right) + \sqrt{1 - \left(\frac{2a \sin 2x + 1}{2a + \sin 2x}\right)} \right] - \frac{1}{(2a + \sin 2x)} + C$$

69. $I = 2 \ln |\sqrt{x^2 + 2x + 4} - x| - \dfrac{3}{2(\sqrt{x^2 + 2x + 4} - x - 1)} - \dfrac{3}{2} \ln |\sqrt{x^2 + 2x + 4} - x - 1| + C$

70. $I = \displaystyle\int \frac{dx}{1 + \sqrt{x^2 + 2x + 2}}$

$\sqrt{x^2 + 2x + 2} = t - x$, **squaring both the sides, we get**

$$x^2 + 2x + 2 = t^2 + x^2 - 2tx$$

$$2x + 2tx = t^2 - 2$$

$$x = \frac{t^2 - 2}{2(1 + t)} \Rightarrow dx = \frac{t^2 + 2t + 2}{2(1 + t)^2} dt$$

$$\therefore \quad 1 + \sqrt{x^2 + 2x + 2} = 1 + t - \frac{t^2 - 2}{2(1 + t)} = \frac{t^2 + 4t + 4}{2(1 + t)}$$

$$\Rightarrow \quad I = \int \frac{2(1 + t)(t^2 + 2t + 2)}{(t^2 + 4t + 4) \cdot 2 \cdot (1 + t)^2} dt = \int \frac{(t^2 + 2t + 2)}{(1 + t)(t + 2)^2} dt$$

Using partial fractions, we get

$$I = \int \frac{dt}{t + 1} - 2\int \frac{dt}{(t + 2)^2} = \ln|t + 1| + \frac{2}{(t + 2)} + C$$

$$I = \ln(x + 1 + \sqrt{x^2 + 2x + 2}) + \frac{2}{(x + 2) + \sqrt{x^2 + 2x + 2}} + C$$

71. $I=\int \frac{x^4+1}{x^6+1}\,dx=\int \frac{(x^2+1)^2-2x^2}{(x^2+1)(x^4-x^2+1)}\,dx$

$$=\int \frac{(1+1/x^2)\,dx}{(x^2+1/x^2-1)}-2\int \frac{x^2\,dx}{(x^3)^2+1}=\int \frac{(1+1/x^2)\,dx}{(x-1/x)^2+1}-2\int \frac{x^2\,dx}{(x^3)^2+1}$$

In first integral put $x-1/x=t$ and in second integral put $x^3=u$

$$=\int \frac{dt}{t^2+1}-\frac{2}{3}\int \frac{du}{u^2+1}=\tan^{-1}(t)-\frac{2}{3}\tan^{-1}(u)+C$$

$$=\tan^{-1}(x-1/x)-\frac{2}{3}\tan^{-1}(x^3)+C$$

72. $I=\int \frac{dx}{(1-x^3)^{1/3}}$

Put $x=1/t$, $dx=-1/t^2\,dt$

$\therefore$ $$I=-\int \frac{dt}{t^2(1-1/t^3)^{1/3}}=-\int \frac{dt}{t(t^3-1)^{1/3}}$$

Again, put $t^3-1=u^3 \Rightarrow 3t^2\,dt=3u^2\,du$

$$=-\int \frac{u^2\,du}{(1+u^3)\cdot u}=-\int \frac{u\,du}{(1+u)(1-u+u^2)}$$

$$=\frac{1}{3}\int \frac{du}{1+u}-\frac{1}{3}\int \frac{u+1}{u^2-u+1}\,du \qquad \text{(using partial fractions)}$$

$$=\frac{1}{3}\int \frac{du}{u+1}-\frac{1}{3}\int \frac{1/2\,(2u-1)+3/2}{(u^2-u+1)}\,du$$

$$=\frac{1}{3}\int \frac{du}{u+1}-\frac{1}{6}\int \frac{2u-1}{u^2-u+1}\,du-\frac{1}{2}\int \frac{du}{(u-1/2)^2+3/4}$$

$$=\frac{1}{3}\log|u+1|-\frac{1}{6}\log|u^2-u+1|-\frac{1}{2}\cdot\frac{1}{\sqrt{3}/2}\tan^{-1}\left\{\frac{2u-1}{\sqrt{3}}\right\}+C$$

$$=\frac{1}{3}\log|(t^3-1)^{1/3}+1|-\frac{1}{6}\log|(t^3-1)^{2/3}-(t^3-1)^{1/3+1}|$$
$$-\frac{1}{\sqrt{3}}\tan^{-1}\left\{\frac{2(t^3-1)^{1/3}-1}{\sqrt{3}}\right\}+C$$

$$=\frac{1}{3}\log\left|\frac{(1-x^3)^{1/3}+x}{x}\right|-\frac{1}{6}\ln\left|\frac{(1-x^3)^{2/3}-x(1-x^3)^{1/3}+x^2}{x^2}\right|$$
$$-\frac{1}{\sqrt{3}}\tan^{-1}\left\{\frac{2(1-x^3)^{1/3}-x}{\sqrt{3}x}\right\}+C$$

73. $I=\int \frac{(x+\sqrt{1+x^2})^{15}}{\sqrt{1+x^2}}\,dx$

Put $(x+\sqrt{1+x^2})=t$ $\therefore$ $\left(1+\frac{x}{\sqrt{1+x^2}}\right)dx=dt$ or $\frac{t\,dx}{\sqrt{1+x^2}}=dt$

$\therefore$ $$I=\int t^{15}\cdot\frac{dt}{t}=\int t^{14}\,dt=\frac{t^{15}}{15}+C=\frac{(x+\sqrt{1+x^2})^{15}}{15}+C$$

74. $\because u_{n+1} = \int \frac{x^{n+1}}{y}\,dx = \int \frac{x^{n+1}}{\sqrt{ax^2+2bx+c}}\,dx = \frac{1}{2a}\int \frac{x^n(2ax+2b)-2bx^n}{\sqrt{ax^2+2bx+c}}\,dx$

$$= \frac{1}{2a}\int x^n \frac{(2ax+2b)\,dx}{\sqrt{ax^2+2bx+c}} - \frac{b}{a}\int \frac{x^n}{\sqrt{ax^2+2bx+c}}\,dx$$

$$u_{n+1} = \frac{1}{2a}\int \underset{\text{I}}{x^n}\ \underset{\text{II}}{\frac{(2ax+2b)}{\sqrt{ax^2+2bx+c}}} - \frac{b}{a}u_n$$

$$u_{n+1} = \frac{1}{2a}[x^n \cdot 2\sqrt{ax^2+2bx+c} - \int nx^{n-1}\cdot 2\sqrt{(ax^2+bx+c)}\,dx] - \frac{b}{a}u_n$$

$$= \frac{1}{a}x^n\,y - \frac{n}{a}\int x^{n-1}\cdot\sqrt{ax^2+2bx+c}\,dx - \frac{b}{a}u_n$$

$$au_{n+1} = x^n\,y - n\int \frac{x^{n-1}(ax^2+2bx+c)}{\sqrt{ax^2+2bx+c}}\,dx - \frac{b}{a}u_n$$

$$au_{n+1} = x^n\,y - n\,[au_{n+1} + 2bu_n + cu_{n-1}] - bu_n$$

$$\Rightarrow \qquad (n+1)\,a\,u_{n+1} + (2n+1)\,bu_n + nc\cdot u_{n-1} = x^n\,y \qquad \text{...(i)}$$

Now, putting, $n = 0$ in both the sides, we get

$$au_1 + bu_0 = x^0 y$$

$$au_1 = y - bu_0 \qquad \text{...(ii)}$$

Putting $n = 1$ in Eq. (i), we get

$$2au_2 + 3bu_1 + cu_0 = xy$$

$$2au_2 + 3b\left(\frac{y - bu_0}{a}\right) + cu_0 = xy \qquad \text{[from Eq. (ii)]}$$

$$\Rightarrow \qquad 2a^2u_2 + 3by - 3b^2u_0 + acu_0 = axy$$

$$\Rightarrow \qquad 2a^2u_2 = y\,(ax - 3b) + (3b^2 - ac)\,u_0$$

75. $$I(n) = \int \frac{x^n}{\sqrt{ax^2+bx+c}}\,dx$$

$$\therefore \qquad I(n+1) = \int \frac{x^{n+1}}{\sqrt{ax^2+bx+c}}\,dx = \int \frac{x^n\cdot x}{\sqrt{ax^2+bx+c}}\,dx$$

$$= \frac{1}{2a}\int \frac{x^n\,(2ax+b-b)\,dx}{ax^2+bx+c}$$

$$\Rightarrow\ 2aI(n+1) = \int \frac{x^n\,(2ax+b)}{\sqrt{ax^2+bx+c}}\,dx - b\int \frac{x^n\,dx}{\sqrt{ax^2+bx+c}}$$

$$= x^n\cdot 2\sqrt{ax^2+bx+c} - 2n\int x^{n-1}\sqrt{ax^2+bx+c}\,dx - b\,I(n)$$

$$= 2x^n\sqrt{ax^2+bx+c} - 2n\int \frac{x^{n-1}(ax^2+bx+c)}{\sqrt{ax^2+bx+c}} - bI(n)$$

$$= 2x^n\sqrt{ax^2+bx+c} - 2n\,(aI(n-1) + bI(n) + c\cdot I(n-1)) - bI(n)$$

$$\Rightarrow\quad 2a\,(n+1)\,I(n+1) = 2x^n\sqrt{ax^2+bx+c} - 2nbI(n) - 2ncI(n-1) - bI(n)$$

$$\Rightarrow\quad 2a\,(n+1)\,I(n+1) + b\,(1+2n)\,I(n) + 2ncI(n-1) = 2x^n\sqrt{ax^2+bx+c}$$

2

Definite Integrals

Chapter in a Snapshot

What is Definite Integral ?

Let f be a function of x defined in the closed interval $[a, b]$ and ϕ be another function, such that $\phi'(x) = f(x)$ for all x in the domain of f, then

$$\int_a^b f(x)\,dx = \left[\phi(x) + c\right]_a^b = \phi(b) - \phi(a)$$

is called the definite integral of the function $f(x)$ over the interval $[a, b]$, a and b are called the limits of integration, a being the lower limit and b be the upper limit.

Points to Consider

In definite integrals constant of integration is never present.

Working Rule

To evaluate definite integral $\int_a^b f(x)\,dx$.

1. First evaluate the indefinite integral $\int f(x)\,dx$ and suppose the result is $g(x)$.
2. Next find $g(b)$ and $g(a)$.
3. Finally, the value of the definite integral is obtained by subtracting $g(a)$ from $g(b)$.

Thus, $$\int_a^b f(x)\,dx = [\,g(x)]_a^b = g(b) - g(a)$$

Illustration 1 Evaluate

(i) $\int_0^1 \frac{1}{3+4x}\,dx$ (ii) $\int_0^{\pi/2} \sin^4 x\,dx$

Solution. (i) Here, $I = \int_0^1 \frac{1}{3+4x}\,dx = \left[\frac{\ln(3+4x)}{4}\right]_0^1$

$$= \frac{1}{4}[\ln 7 - \ln 3] = \frac{1}{4}\ln\left(\frac{7}{3}\right)$$

(ii) Let $$I = \int_0^{\pi/2} \sin^4 x\,dx$$

$$= \frac{1}{4}\int_0^{\pi/2}(2\sin^2 x)^2\,dx = \frac{1}{4}\int_0^{\pi/2}(1-\cos 2x)^2\,dx$$

$$= \frac{1}{4}\int_0^{\pi/2}(1 - 2\cos 2x + \cos^2 2x)\,dx$$

$$= \frac{1}{4}\int_0^{\pi/2}\left(1 - 2\cos 2x + \frac{1+\cos 4x}{2}\right)dx$$

$$= \frac{1}{4}\int_0^{\pi/2}\left(\frac{3 - 4\cos 2x + \cos 4x}{2}\right)dx$$

$$= \frac{1}{8}\left[3x - \frac{4}{2}\sin 2x + \frac{\sin 4x}{4}\right]_0^{\pi/2}$$

$$= \frac{1}{8}\left[\left(\frac{3\pi}{2} - 2\sin\pi + \frac{1}{4}\sin 2\pi\right) - 0\right]$$

$$= \frac{1}{8}\left(\frac{3\pi}{2} - 0 + 0\right) = \frac{3\pi}{16}$$

Illustration 2 Let $I_1 = 2\int_a^x f(t)\,dt$, $I_2 = \int_b^c f(t)\,dt$, $I_3 = 2\int_b^a f(t)\,dt$. If I_1, I_2, I_3 are in AP, $\forall\ x \in R$ and a, b, c are positive real numbers, $f(x)$ is a continuous function, then $f(x)$ is

(a) a constant function (b) a power function

(c) a linear function (d) cannot discuss

Solution. Here, $2I_2 = I_1 + I_3$

Let $\int f(t)\,dt = F(t) + K \quad \therefore \quad 2I_2 = I_1 + I_3$

$\Rightarrow \quad 2[F(c) - F(b)] = 2[F(x) - F(a)] + 2[F(a) - F(b)]$

$\Rightarrow \quad F(x) = F(c), \forall\ x \in R$

Hence, $f(x) = F'(x) = 0, \forall\ x \in R$

$\therefore f(x)$ is a constant function.

Hence, (a) is the correct answer.

Illustration 3 If $f : R \to R; f(x) = \sin x + x$, then the value of $\int_0^\pi [f^{-1}(x)]\,dx$, is equal to

(a) $\frac{\pi^2}{2} - 2$ (b) π^2 (c) $\frac{\pi^2}{2} + 2$ (d) None of these

Solution. Let $I = \int_0^x f^{-1}(x)\,dx$, put $x = f(t)$

$$= \int_{f^{-1}(0)}^{f^{-1}(x)} t \cdot f'(t)\,dt = [t\,f(t)]_{f^{-1}(0)}^{f^{-1}(x)} - \int_{f^{-1}(0)}^{f^{-1}(x)} f(t)\,dt$$

$$= [t \cdot (t + \sin t)]_{f^{-1}(0)}^{f^{-1}(x)} - \int_{f^{-1}(0)}^{f^{-1}(x)} f(t)\,dt$$

$$\therefore \quad \int_0^\pi f^{-1}(x)\,dx = [t^2 + t\sin t\,]_0^\pi - \int_0^\pi (t + \sin t)\,dt$$

$$= \pi^2 - \left(\frac{t^2}{2} - \cos t\right)_0^\pi = \pi^2 - \left(\frac{\pi^2}{2} + 1 - (0 - 1)\right)$$

$$= \pi^2 - \frac{\pi^2}{2} - 2 = \frac{\pi^2}{2} - 2$$

Hence, (a) is the correct answer.

Geometrical Interpretation of Definite Integral

If $f(x) > 0$ for all $x \in [a, b]$, then $\int_a^b f(x)\,dx$ is numerically equal to the area bounded by the curve $y = f(x)$, the x-axis and the straight lines $x = a$ and $x = b$

ie, $\int_a^b f(x)\,dx$

In general, $\int_a^b f(x)\,dx$ represents the algebraic sum of the areas of the figures bounded by the curve $y = f(x)$, the x-axis and the straight lines $x = a$ and $x = b$. The areas above x-axis are taken with plus sign and the areas below x-axis are taken with minus sign, *ie*,

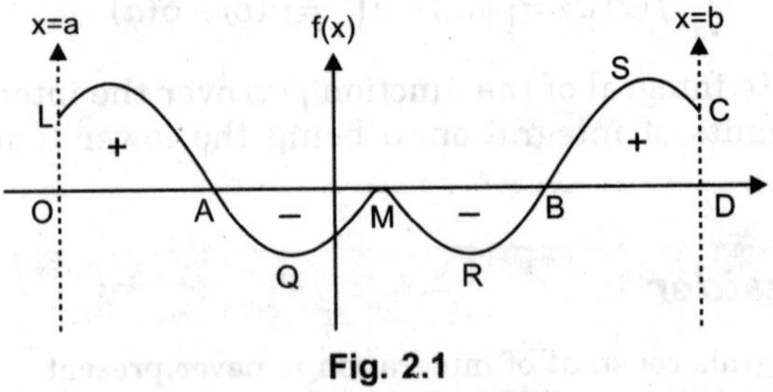

Fig. 2.1

ie, $\int_a^b f(x)\,dx = \text{area } OLA - \text{area } AQM - \text{area } MRB + \text{area } BSCD$

Points to Consider

$\int_a^b f(x)\,dx$, represents algebraic sum of areas—means that area of function $y = f(x)$ is asked between a to b.

$\Rightarrow$ Area bounded $= \int_a^b |f(x)|\,dx$ and not been represented by $\int_a^b f(x)\,dx$.

eg, if someone asks for the area of $y = x^3$ between -1 to 1, then, $y = x^3$ could be plotted as;

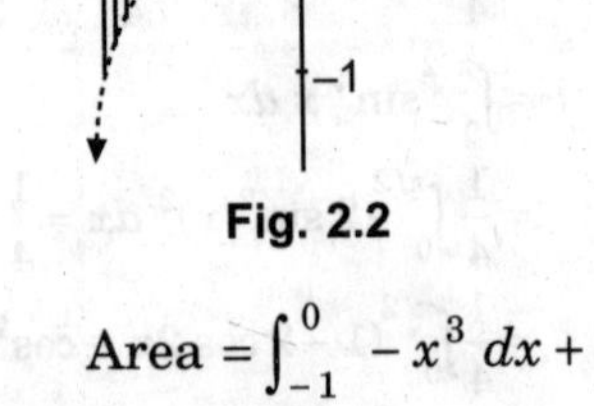

Fig. 2.2

$$\therefore \qquad \text{Area} = \int_{-1}^{0} -x^3\,dx + \int_0^1 x^3\,dx = \frac{1}{2}$$

or using above definition, Area $= \int_{-1}^{1} |x^3|\,dx = 2\int_0^1 x^3\,dx$

$$= 2\left[\frac{x^4}{4}\right]_0^1 = \frac{1}{2}$$

But, if we integrate x^3 between -1 to 1.

$\Rightarrow \qquad \int_{-1}^{1} x^3 \, dx = 0$ which does not represent the area.

Thus, students are adviced to make difference between area and definite integral.

Illustration 4 Evaluate $\int_0^3 |(x-1)(x-2)|\, dx$.

Solution. $I = \int_0^3 |(x-1)(x-2)|\, dx$

We know, $|(x-1)(x-2)| = \begin{cases} (x-1)(x-2), & x < 1 \text{ or } x > 2 \\ -(x-1)(x-2), & 1 < x < 2 \end{cases}$

Using number line rule,

$$\therefore \quad I = \int_0^3 |(x-1)(x-2)|\, dx$$

$$= \int_0^1 (x-1)(x-2)\, dx - \int_1^2 (x-1)(x-2)\, dx + \int_2^3 (x-1)(x-2)\, dx$$

$$= \int_0^1 (x^2 - 3x + 2)\, dx - \int_1^2 (x^2 - 3x + 2)\, dx + \int_2^3 (x^2 - 3x + 2)\, dx$$

$$= \left[\frac{x^3}{3} - \frac{3x^2}{2} + 2x\right]_0^1 - \left[\frac{x^3}{3} - \frac{3x^2}{2} + 2x\right]_1^2 + \left[\frac{x^3}{3} - \frac{3x^2}{2} + 2x\right]_2^3$$

$$= \left(\frac{1}{3} - \frac{3}{2} + 2\right) - \left(\frac{8}{3} - \frac{12}{2} + 4 - \frac{1}{3} + \frac{3}{2} - 2\right) + \left(\frac{27}{3} - \frac{27}{2} + 6 - \frac{8}{3} + \frac{12}{2} - 4\right)$$

$$= \frac{11}{6}$$

$$\therefore \qquad \int_0^3 |(x-1)(x-2)|\, dx = \frac{11}{6}$$

Evaluation of Definite Integrals By Substitution

Sometimes, the indefinite integral may need substitution say $x = \phi(t)$. Then, in that case don't forget to change the limits of integration a and b corresponding to the new variable t. The substitution $x = \phi(t)$ is not valid, if it is not continuous in the interval $[a, b]$.

Illustration 5 Show that

$$\int_0^{\pi/2} \frac{dx}{a^2 \cos^2 x + b^2 \sin^2 x} = \frac{\pi}{2ab}, \quad (a, b > 0)$$

Solution. Let $I = \int_{x=0}^{x=\pi/2} \frac{dx}{a^2 \cos^2 x + b^2 \sin^2 x}$

(divide numerator and denominator by $\cos^2 x$)

$$= \int_{x=0}^{x=\pi/2} \frac{\sec^2 x \, dx}{a^2 + b^2 \tan^2 x} \qquad \text{put } \tan x = t \Rightarrow \sec^2 x\, dx = dt$$

$$\therefore \qquad I = \int_{t=0}^{t=\infty} \frac{dt}{a^2 + b^2 t^2}$$

We find the new limits of integration $t = \tan x \Rightarrow t = 0$ when $x = 0$ and $t = \infty$ when $x = \pi/2$.

$$\Rightarrow \qquad I = \frac{1}{b^2}\int_0^{\infty} \frac{dt}{\left(\frac{a}{b}\right)^2 + t^2} = \frac{1}{b^2}\cdot\frac{1}{a/b}\cdot\left[\tan^{-1}\frac{bt}{a}\right]_0^{\infty}$$

$$= \frac{1}{ab}\left[\frac{\pi}{2} - 0\right] = \frac{\pi}{2ab}$$

Points to Consider

We have, $\displaystyle\int_0^{\pi/2} \frac{dx}{a^2\cos^2 x + b^2\sin^2 x} = \frac{\pi}{2ab} = f(a, b)(a, b > 0)$ ...(i)

Differentiating w.r.t. a on both the sides, we get

$$\Rightarrow \qquad -\int_0^{\pi/2} \frac{2a\cos^2 x}{(a^2\cos^2 x + b^2\sin^2 x)^2}\,dx = -\frac{\pi}{2b}\cdot\left(\frac{1}{a^2}\right)$$

$$\Rightarrow \qquad \int_0^{\pi/2} \frac{\cos^2 x}{(a^2\cos^2 x + b^2\sin^2 x)^2}\,dx = \frac{\pi}{4a^3 b} \qquad \text{...(ii)}$$

Similarly, differentiating Eq. (i) w.r.t. b both the sides, we get

$$\int_0^{\pi/2} \frac{\sin^2 x}{(a^2\cos^2 x + b^2\sin^2 x)^2}\,dx = \frac{\pi}{4ab^3} \qquad \text{...(iii)}$$

Adding Eqs. (ii) and (iii), we get

$$\int_0^{\pi/2} \frac{1\,dx}{(a^2\cos^2 x + b^2\sin^2 x)^2} = \frac{\pi}{4ab}\left(\frac{1}{a^2} + \frac{1}{b^2}\right)$$

Thus, it should be noted that if there is a parameter involved in the integrand, then sometimes we get useful results by differentiating w.r.t. that parameter and keeping the variable constant.

Illustration 6 Evaluate $\displaystyle\int_{-2}^{2} \frac{dx}{4 + x^2}$ directly as well as by the substitution $x = 1/t$. Examine as to why the answer don't tally?

Solution. $\displaystyle I = \int_{-2}^{2} \frac{dx}{4 + x^2}$

$$= \left[\frac{1}{2}\tan^{-1}\left(\frac{x}{2}\right)\right]_{-2}^{2} = \frac{1}{2}\left[\tan^{-1}(1) - \tan^{-1}(-1)\right]$$

$$= \frac{1}{2}\left[\frac{\pi}{4} - \left(-\frac{\pi}{4}\right)\right] = \frac{\pi}{4}$$

$$\Rightarrow \qquad I = \frac{\pi}{4}$$

On the other hand; if $x = 1/t$, then

$$I = \int_{-2}^{2} \frac{dx}{4 + x^2} = -\int_{-1/2}^{1/2} \frac{dt}{t^2(4 + 1/t^2)} = -\int_{-1/2}^{1/2} \frac{dt}{4t^2 + 1}$$

$$= -\left[\frac{1}{2}\tan^{-1}(2t)\right]_{-1/2}^{1/2}$$

$$= -\frac{1}{2}\tan^{-1}(1) - \left(-\frac{1}{2}\tan^{-1}(-1)\right)$$

$$= -\frac{\pi}{8} - \frac{\pi}{8} = -\frac{\pi}{4}$$

$$\therefore \quad I = -\frac{\pi}{4}, \text{ when } x = \frac{1}{t}$$

In above two results, $I = -\pi/4$, is wrong. Since, the integrand $\frac{1}{4+x^2} > 0$ and therefore the definite integral of this function cannot be negative. Since, $x = 1/t$ is discontinuous at $t = 0$, the substitution is not valid ($\therefore I = \pi/4$).

Points to Consider

It is important that the substitution must be continuous in the interval of integration.

Illustration 7 The value of $\int_0^{e-1} \frac{e^{\frac{x^2+2x-1}{2}}}{(x+1)}\,dx + \int_1^e x\log x \cdot e^{\frac{x^2-2}{2}}\,dx$, is equal to

(a) $(\sqrt{e})^{(e^2+1)}$ (b) $(\sqrt{e})^{e^2-1}$ (c) 0 (d) $(\sqrt{e})^{e^2-2}$

Solution. Let $I = \int_0^{e-1} \frac{e^{\frac{x^2+2x-1}{2}}}{(x+1)}\,dx + \int_1^e x\log x \cdot e^{\frac{x^2-2}{2}}\,dx$

Put $x + 1 = t$ in first integral

$$\therefore \quad I = \int_1^e \frac{e^{\frac{t^2-2}{2}}}{t}\,dt + \int_1^e x\log x \cdot e^{\frac{x^2-2}{2}}\,dx$$

$$= \int_1^e e^{\frac{t^2-2}{2}}\left\{\frac{1}{t} + t\cdot\log t\right\}dt$$

$$= \left(\log t \cdot e^{\frac{t^2-2}{2}}\right)_1^e = (\sqrt{e})^{e^2-2}$$

Hence, (d) is the correct answer.

Illustration 8 Let $f(x) = \int_2^x \frac{dt}{\sqrt{1+t^4}}$ and g be the inverse of f. Then, the value of $g'(0)$ is

(a) 1 (b) 17

(c) $\sqrt{17}$ (d) None of these

Solution. $f'(x) = \frac{1}{\sqrt{1+x^4}} = \frac{dy}{dx}$

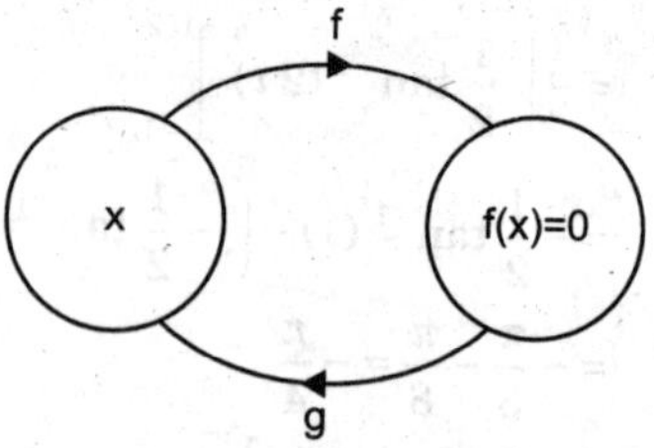

Now, $g'(x) = \dfrac{dx}{dy} = \sqrt{1+x^4}$

When $y = 0$ *ie*, $\displaystyle\int_2^x \frac{dt}{\sqrt{1+t^4}} = 0$, then $x = 2$

Therefore, $g'(0) = \sqrt{1+16} = \sqrt{17}$

Hence, (c) is the correct answer.

Illustration 9 Let $a_n = \displaystyle\int_0^{\pi/2} (1-\sin t)^n \sin 2t$, then $\displaystyle\lim_{n\to\infty} \sum_{n=1}^{n} \frac{a_n}{n}$ is equal to

(a) 1/2 (b) 1 (c) 4/3 (d) 3/2

Solution. $a_n = \displaystyle\int_0^{\pi/2} (1-\sin t)^n \sin 2t\, dt$

Let $1-\sin t = u \Rightarrow -\cos t\, dt = du$

$$= 2\int_0^1 u^n(1-u)du = 2\left(\int_0^1 u^n du - \int_0^1 u^{n+1} du\right)$$

$$= 2\left(\frac{1}{n+1} - \frac{1}{n+2}\right)$$

Therefore, $$\frac{a_n}{n} = 2\left(\frac{1}{n(n+1)} - \frac{1}{n(n+2)}\right)$$

$$\lim_{n\to\infty} \sum_{n=1}^{n} \frac{a_n}{n} = 2\left[\sum\left(\frac{1}{n} - \frac{1}{n+1}\right) - \frac{1}{2}\sum\left(\frac{1}{n} - \frac{1}{n+2}\right)\right]$$

$$= 2\sum_{n=1}^{n}\left(\frac{1}{n} - \frac{1}{n+1}\right) - \sum_{n=1}^{n}\left(\frac{1}{n} - \frac{1}{n+2}\right)$$

$$= 2(1) - \left[\left(1-\frac{1}{3}\right) \div \left(\frac{1}{2}-\frac{1}{4}\right) + \left(\frac{1}{3}-\frac{1}{5}\right) + \ldots\right]$$

$$= 2 - \frac{3}{2} = \frac{1}{2}$$

Hence, (a) is the correct answer.

Illustration 10 The value of $x > 1$ satisfying the equation $\displaystyle\int_1^x t \ln t\, dt = \frac{1}{4}$, is

(a) $\sqrt{e}$ (b) e (c) e^2 (d) $e-1$

Solution. $I = \displaystyle\int_1^x t \ln t\, dt = \left[\ln t \cdot \frac{t^2}{2}\right]_1^x - \frac{1}{2}\int_1^x \frac{1}{t} \cdot t^2\, dt = \frac{x^2}{2}\ln x - \frac{1}{2}\left[\frac{t^2}{2}\right]_1^x$

$$= \frac{x^2 \ln x}{2} - \frac{1}{4}[x^2 - 1] = \frac{1}{4}$$

$$\therefore \quad \frac{x^2 \ln x}{2} - \frac{1}{4}x^2 = 0 \Rightarrow [2\ln x - 1] = 0 \quad (\text{as } x > 1)$$

$$\ln x = \frac{1}{2} \Rightarrow x = \sqrt{e}$$

Hence, (a) is the correct answer.

Illustration 11 If $\lim_{a \to \infty} \frac{1}{a} \int_0^\infty \frac{x^2 + ax + 1}{1 + x^4} \cdot \tan^{-1}\left(\frac{1}{x}\right) dx$ is equal to $\frac{\pi^2}{k}$, where $k \in N$ equals to

(a) 4 (b) 8 (c) 16 (d) 32

Solution. Let $I = \int_0^\infty \frac{x^2 + ax + 1}{1 + x^4} \cdot \tan^{-1}\left(\frac{1}{x}\right) dx$

Put $x = 1/t$ and adding, we get [using $\tan^{-1}(1/x) + \cot^{-1} x = \pi/2$]

$$I = \frac{\pi}{4} \int_0^\infty \frac{(x^2 + 1) + ax}{1 + x^4} dx$$

$$= \frac{\pi}{4}\left[\int_0^\infty \frac{(x^2 + 1)}{1 + x^4} dx + a \int_0^\infty \frac{x\,dx}{1 + x^4}\right]$$

$$= \frac{\pi}{4}\left[\frac{\pi}{2\sqrt{2}} + \frac{a\pi}{4}\right] = \left[\frac{\pi^2}{8\sqrt{2}} + \frac{\pi^2 a}{16}\right]$$

$$\therefore \quad l = \lim_{a \to \infty} \frac{1}{a}\left[\frac{\pi^2}{8\sqrt{2}} + \frac{\pi^2 a}{16}\right] = \lim_{a \to \infty}\left[\frac{\pi^2}{(8\sqrt{2})a} + \frac{\pi^2}{16}\right] = \frac{\pi^2}{16} \Rightarrow k = 16$$

Hence, (c) is the correct answer.

Illustration 12 If the value of definite integral $\int_1^a x \cdot a^{-[\log_a x]} dx$, where $a > 1$ and $[x]$ denotes the greatest integer, is $\frac{e-1}{2}$, then the value of 'a' equals to

(a) $\sqrt{e}$ (b) e (c) $\sqrt{e+1}$ (d) $e - 1$

Solution. $\int_1^a x \cdot a^{-[\log_a x]} dx$

Put $\log_a x = t \Rightarrow a^t = x$

$$I = \ln a \cdot \int_0^1 (a^t \cdot a^{-[t]} \cdot a^t)\, dt = \ln a \cdot \int_0^1 (a^{t-[t]} \cdot a^t) dt$$

$$= \ln a \cdot \int_0^1 (a^{\{t\}} \cdot a^t) dt = \ln a \cdot \int_0^1 a^{2t}\, dt$$

$$= \left[\frac{\ln a}{2} \cdot \frac{a^{2t}}{\ln a}\right]_0^1 = \frac{1}{2}(a^2 - 1) \quad [\text{as } \{t\} = t, \text{ if } t \in (0,1)]$$

$$\therefore \quad \frac{1}{2}(a^2 - 1) = \frac{e-1}{2} \Rightarrow a = \sqrt{e}$$

Aliter $x \in (1, a) \Rightarrow \log_a x \in (0,1) \Rightarrow [\log_a x] = 0$

$$I = \int_1^a x\,dx = \frac{1}{2}(a^2 - 1) = \frac{e-1}{2} \Rightarrow a = \sqrt{e}$$

Hence, (a) is the correct answer.

Illustration 13 The value of $\int_0^1 \left(\prod_{r=1}^{n}(x+r)\right)\left(\sum_{k=1}^{n}\frac{1}{x+k}\right) dx$ equals to

(a) n (b) $n!$ (c) $(n+1)!$ (d) $n \cdot n!$

Solution. The given integrand is perfect coefficient of $\prod_{r=1}^{n}(x+r)$

$$\Rightarrow \quad I = \left[\prod_{r=1}^{n}(x+r)\right]_0^1 = (n+1)! - n! = n \cdot n!$$

Aliter $(x+1)(x+2)(x+3)...(x+n) = e^t$

So that, when $x = 0$, $t = \ln n!$; when $x = 1$, $t = \ln(n+1)!$

$$[\ln(x+1) + \ln(x+2) + ... + \ln(x+n)] = t$$

$$\therefore \quad \left(\frac{1}{(x+1)} + \frac{1}{(x+2)} + ... + \frac{1}{(x+n)}\right) dx = dt$$

Therefore, $I = \int_{\ln n!}^{\ln(n+1)!} e^t dt = [e^t]_{\ln n!}^{\ln(n+1)!}$

$$= e^{\ln(n+1)!} - e^{\ln n!} = (n+1)! - n! = n \cdot n!$$

Hence, (d) is the correct answer.

Illustration 14 The true set of values of 'a' for which the inequality $\int_a^0 (3^{-2x} - 2 \cdot 3^{-x})\, dx \geq 0$ is true, is

(a) $[0, 1]$ (b) $(-\infty, -1]$

(c) $[0, \infty)$ (d) $(-\infty, -1] \cup [0, \infty)$

Solution. $\int_a^0 3^{-x}(3^{-x} - 2)\, dx \geq 0$, put $3^{-x} = t \Rightarrow 3^{-x} \ln 3\, dx = -dt$

$$\ln 3 \int_1^{3^{-a}} (t-2)dt \geq 0 = \left[\frac{t^2}{2} - 2t\right]_1^{3^{-a}} \geq 0$$

$$\left(\frac{3^{-2a}}{2} - 2 \cdot 3^{-a}\right) - \left(\frac{1}{2} - 2\right) \geq 0$$

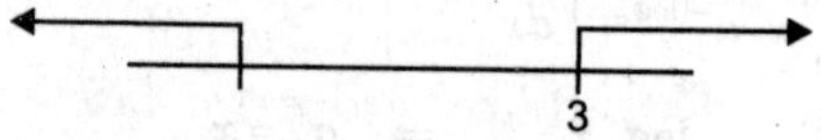

$$3^{-2a} - 4 \times 3^{-a} + 3 > 0$$

$$(3^{-a} - 3)(3^{-a} - 1) > 0$$

$$3^{-a} > 3^1 \Rightarrow a < 1$$

or $$3^{-a} < 3^0 \Rightarrow a > 0$$

Thus, $a \in (-\infty, -1] \cup [0, \infty)$

Hence, (d) is the correct answer.

Properties of Definite Integrals

Property I. $\int_a^b f(x)\, dx = \int_a^b f(t)\, dt$

ie, The integration is independent of the change of variable.

Proof Let $\phi(x)$ be a primitive of $f(x)$, then

$$\frac{d}{dx}[\phi(x)] = f(x) \Rightarrow \frac{d}{dt}[\phi(t)] = f(t)$$

Therefore, $$\int_a^b f(x)\,dx = [\phi(x)]_a^b = \phi(b) - \phi(a) \quad \ldots\text{(i)}$$

and $$\int_a^b f(t)\,dt = [\phi(t)]_a^b = \phi(b) - \phi(a) \quad \ldots\text{(ii)}$$

From Eqs. (i) and (ii), we have

$$\int_a^b f(x)\,dx = \int_a^b f(t)\,dt$$

Property II. $$\int_a^b f(x)\,dx = -\int_b^a f(x)\,dx$$

ie, if the limits of definite integral are interchanged, then its value changes by minus sign only.

Proof Let $\phi(x)$ be a primitive of $f(x)$, then

$$\int_a^b f(x)\,dx = \phi(b) - \phi(a)$$

and $$-\int_b^a f(x)\,dx = -[\phi(a) - \phi(b)] = \phi(b) - \phi(a)$$

$\therefore$ $$\int_a^b f(x)\,dx = -\int_b^a f(x)\,dx$$

Property III. $$\int_a^b f(x)\,dx = \int_a^c f(x)\,dx + \int_c^b f(x)\,dx, \text{ where } c \leftrightarrow R.$$

Proof Let $\phi(x)$ be primitive of $f(x)$, then

$$\int_a^b f(x)\,dx = \phi(b) - \phi(a) \quad \ldots\text{(i)}$$

and $$\int_a^c f(x)\,dx + \int_c^b f(x)\,dx = [\phi(c) - \phi(a)] + [\phi(b) - \phi(c)] = \phi(b) - \phi(a) \quad \ldots\text{(ii)}$$

From Eqs. (i) and (ii), we get $\int_a^b f(x)\,dx = \int_a^c f(x)\,dx + \int_c^b f(x)\,dx$

Generalization The above property can be generalized into the following form,

$$\int_a^b f(x)\,dx = \int_a^{c_1} f(x)\,dx + \int_{c_1}^{c_2} f(x)\,dx + \ldots + \int_{c_n}^b f(x)\,dx$$

Where $a < c_1 < c_2 < \ldots < c_{n-1} < c_n < b$

Illustration 15 The value of the definite integral $\int_0^{\pi/2} \sin|2x - \alpha|\,dx$, where $\alpha \in [0, \pi]$, is

(a) 1 (b) $\cos\alpha$ (c) $\frac{1+\cos\alpha}{2}$ (d) $\frac{1-\cos\alpha}{2}$

Solution. $I = \int_{-\alpha}^{(\pi-\alpha)} \sin|t|\,dt$, where $2x - \alpha = t \Rightarrow dx = \frac{dt}{2}$

$$= \frac{1}{2}\int_{-\alpha}^{0} -\sin t\,dt + \frac{1}{2}\int_0^{\pi-\alpha} \sin t\,dt = \left[\frac{1}{2}\cos t\right]_{-\alpha}^{0} - \left[\frac{1}{2}\cos t\right]_0^{\pi-\alpha}$$

$$= \frac{1}{2}[1-\cos\alpha] - \frac{1}{2}[-\cos\alpha - 1]$$
$$= \frac{1}{2}(1-\cos\alpha) + \frac{1}{2}(1+\cos\alpha) = 1$$
$$= \frac{1}{2}(1-\cos\alpha) + \frac{1}{2}(1+\cos\alpha) = 1$$

Hence, (a) is the correct answer.

Illustration 16 Let f be a continuous functions satisfying

$$f'(\ln x) = \begin{cases} 1 & \text{for } 0 < x \le 1 \\ e^x - 1 & \text{for } x > 1 \end{cases}$$

and $f(0) = 0$, then $f(x)$ can be defined as

(a) $f(x) = \begin{cases} 1, & \text{if } x \le 0 \\ 1-e^x, & \text{if } x > 0 \end{cases}$ (b) $f(x) = \begin{cases} 1, & \text{if } x \le 0 \\ e^x - 1, & \text{if } x > 0 \end{cases}$

(c) $f(x) = \begin{cases} x, & \text{if } x < 0 \\ e^x, & \text{if } x > 0 \end{cases}$ (d) $f(x) = \begin{cases} x, & \text{if } x \le 0 \\ e^x - 1, & \text{if } x > 0 \end{cases}$

Solution. $f'(\ln x) = \begin{cases} 1, & \text{for } 0 < x \le 1 \\ x, & \text{for } x > 1 \end{cases}$

Put $\ln x = t \Rightarrow x = e^t$

For $x > 1$; $f'(t) = e^t$ for $t > 0$

Integrating $f(t) = e^t + C$; $f(0) = e^0 + C \Rightarrow C = -1$ [given, $f(0) = 0$]

$\therefore$ $f(t) = e^t - 1$, for $t > 0$ (corresponding to $x > 1$)

Therefore, $f(x) = e^x - 1$, for $x > 0$ ...(i)

Again, for $0 < x \le 1$

$f'(\ln x) = 1$ $(x = e^t)$

$f'(t) = 1$, for $t \le 0$

$f(t) = t + C$

$f(0) = 0 + C \Rightarrow C = 0 \Rightarrow f(t) = t$, for $t \le 0$

$\Rightarrow$ $f(x) = x$, for $x \le 0$

Hence, (d) is the correct answer.

Illustration 17 The integral, $\int_{\pi/4}^{5\pi/4} (|\cos t|\sin t + |\sin t|\cos t)\, dt$ has the value equal to

(a) 0 (b) 1/2 (c) $1/\sqrt{2}$ (d) 1

Solution. $I = \int_{\pi/4}^{\pi/2} 2\sin t\cos t\, dt + \underbrace{\int_{\pi/2}^{\pi} [(-\sin t\cos t) + (\sin t\cos t)]\, dt}_{\text{zero}}$

$$+ \int_{\pi}^{5\pi/4} -2\sin t\cos t\, dt$$

$$= \int_{\pi/4}^{\pi/2} \sin 2t\, dt - \int_{\pi}^{5\pi/4} \sin 2t\, dt$$

These two integrals cancels $\Rightarrow$ Zero.

Hence, (a) is the correct answer.

Property IV. $\int_0^a f(x)\,dx = \int_0^{a/2} f(x)\,dx + \int_0^{a/2} f(a-x)\,dx$

Proof As we know, $\int_0^a f(x)\,dx = \int_0^{a/2} f(x)\,dx + \int_{a/2}^a f(x)\,dx$

Put $x = a - t$ and $dx = -\,dt$ in the second integral also, when $x = a/2$, then $t = a/2$ and when $x = a$, then $t = 0$.

$$\therefore \quad \int_0^a f(x)\,dx = \int_0^{a/2} f(x)\,dx + \int_{a/2}^0 f(a-t)(-\,dt)$$

$$= \int_0^{a/2} f(x)\,dx + \int_0^{a/2} f(a-t)\,dt$$

$$\int_0^a f(x)\,dx = \int_0^{a/2} f(x)\,dx + \int_0^{a/2} f(a-x)\,dx$$

Property V. $\int_a^b f(x)\,dx = \begin{cases} 0, & \text{if } f(a+x) = -f(b-x) \\ 2\int_a^{\frac{a+b}{2}} f(x)\,dx, & \text{if } f(a+x) = f(b-x) \end{cases}$

Proof Let us consider the function $f(x)$ on $[a, b]$ when $f(a+x) = f(b-x)$, then f is even symmetric about the mid-point $x = \dfrac{a+b}{2}$ on the interval $[a, b]$ when $f(a+x) = -f(b-x)$, then f is odd symmetric about the mid-point $x = \dfrac{a+b}{2}$ of the interval $[a, b]$.

$$\therefore \text{ Using } \quad \int_{-a}^a f(x)\,dx = \begin{cases} 0, & \text{if } f \text{ is an odd function} \\ 2f(x)\,dx, & \text{if } f \text{ is an even function} \end{cases}$$

$$\Rightarrow \quad \int_a^b f(x)\,dx = \begin{cases} 0, & \text{if } f(a+x) = -f(b-x) \\ 2\int_a^{\frac{a+b}{2}} f(x)\,dx, & \text{if } f(a+x) = f(b-x) \end{cases}$$

Illustration 18 Given function, $f(x) = \begin{cases} x^2, & \text{for } 0 \le x < 1 \\ \sqrt{x}, & \text{for } 1 \le x \le 2 \end{cases}$.

Evaluate $\int_0^2 f(x)\,dx$.

Solution. Here, $\int_0^2 f(x)\,dx = \int_0^1 f(x)\,dx + \int_1^2 f(x)\,dx$

$$\therefore \quad \int_0^2 f(x)\,dx = \int_0^1 x^2\,dx + \int_1^2 \sqrt{x}\,dx = \left[\frac{x^3}{3}\right]_0^1 + \left[\frac{x^{3/2}}{3/2}\right]_1^2$$

$$\therefore \quad \int_0^2 f(x)\,dx = \left[\frac{x^3}{3}\right]_0^1 + \left[\frac{2}{3}x\sqrt{x}\right]_1^2 = \left[\frac{1}{3} - 0\right] + \frac{2}{3}[2\sqrt{2} - 1]$$

$$= \frac{1}{3} + \frac{4\sqrt{2}}{3} - \frac{2}{3} = \frac{4\sqrt{2}}{3} - \frac{1}{3}$$

$$\therefore \quad \int_0^2 f(x)\,dx = \frac{1}{3}(4\sqrt{2} - 1)$$

Illustration 19 Evaluate the integral $I = \int_0^2 |1 - x|\, dx.$

Solution. By definition $|a - b| = \begin{cases} b - a, & \text{if } a < b \\ a - b, & \text{if } a | b \end{cases}$

Then, $|1 - x| = \begin{cases} (1 - x), & 0 \le x \le 1 \\ (x - 1), & 1 \le x \le 2 \end{cases}$

$$\Rightarrow \int_0^2 |1 - x|\, dx = \int_0^1 (1 - x)\, dx + \int_1^2 (x - 1)\, dx = \left[x - \frac{x^2}{2} \right]_0^1 + \left[\frac{x^2}{2} - x \right]_1^2$$

$$= \left\{ \left(1 - \frac{1}{2}\right) - (0 - 0) \right\} + \left\{ \left(\frac{4}{2} - 2\right) - \left(\frac{1}{2} - 1\right) \right\} = \frac{1}{2} + \left\{ 0 + \frac{1}{2} \right\} = 1$$

Illustration 20 Evaluate

(i) $\int_0^{\pi} |\cos x|\, dx$ (ii) $\int_0^2 |x^2 + 2x - 3|\, dx$

Solution. (i) $\int_0^{\pi} |\cos x|\, dx = \int_0^{\pi/2} |\cos x|\, dx + \int_{\pi/2}^{\pi} |\cos x|\, dx$

$$= \int_0^{\pi/2} (\cos x)\, dx - \int_{\pi/2}^{\pi} (\cos x)\, dx$$

$$= [\sin x]_0^{\pi/2} - [\sin x]_{\pi/2}^{\pi}$$

$$= (1 - 0) - (0 - 1) = 2$$

(ii) $$\int_0^2 |x^2 + 2x - 3|\, dx = \int_0^1 |x^2 + 2x - 3|\, dx + \int_1^2 |x^2 + 2x - 3|\, dx \quad \text{...(i)}$$

We have, $x^2 + 2x - 3 = (x + 3)(x - 1)$

$\therefore$ $x^2 + 2x - 3 > 0$ for $x < -3$ or $x > 1$

and $x^2 + 2x - 3 < 0$ for $-3 < x < 1$

So, $|x^2 + 2x - 3| = \begin{cases} (x^2 + 2x - 3), & \text{for } x < -3 \text{ or } x > 1 \\ -(x^2 + 2x - 3), & \text{for } -3 < x < 1 \end{cases}$

$\therefore$ Eq. (i) becomes,

$$I = \int_0^1 -(x^2 + 2x - 3)\, dx + \int_1^2 (x^2 + 2x - 3)\, dx$$

$$= -\left[\frac{x^3}{3} + x^2 - 3x \right]_0^1 + \left[\frac{x^3}{3} + x^2 - 3x \right]_1^2$$

$$I = 4$$

Illustration 21 Evaluate $\int_{-1}^1 (x - [x])\, dx$; (where [.] denotes the greatest integral part of x.)

Solution.

$$I = \int_{-1}^1 (x - [x])\, dx = \int_{-1}^1 x \cdot dx - \int_{-1}^1 [x]\, dx$$

$$= \left(\frac{x^2}{2} \right)_{-1}^1 - \left(\int_{-1}^0 [x]\, dx + \int_0^1 [x]\, dx \right)$$

$$= \frac{1}{2}(1 - 1) - \left(\int_{-1}^0 -1\, dx + \int_0^1 0\, dx \right)$$

$$= 0 + (x)_{-1}^1 - 0 = 1$$

Illustration 22 Evaluate $\int_0^2 \{x\}\, dx$, (where $\{x\}$ denotes the fractional part of x.)

Solution. We have,

$$\int_0^2 \{x\}\, dx = \int_0^2 (x - [x])\, dx = \int_0^2 x\, dx - \int_0^2 [x]\, dx$$

$$= \left(\frac{x^2}{2}\right)_0^2 - \left(\int_0^1 [x]\, dx + \int_1^2 [x]\, dx\right)$$

$$= \frac{1}{2}(4-0) - \left(\int_0^1 0\, dx + \int_1^2 1\, dx\right)$$

$$= 2 - (x)_1^2 = 2 - (1) = 1$$

Points to Consider

In above illustration, for greatest integer less than or equal to x, it is compulsory to break it at integral limits.

Illustration 23 Evaluate $\int_0^9 \{\sqrt{x}\}\, dx$, where $\{x\}$ denotes the fractional part of x.

Solution. $\int_0^9 \{\sqrt{x}\}\, dx = \int_0^9 \left(\sqrt{x} - [\sqrt{x}]\right) dx$

$$= \int_0^9 (x^{1/2})\, dx - \int_0^9 [\sqrt{x}]\, dx$$

$$= \frac{2}{3}(x^{3/2})_0^9 - \int_0^9 [\sqrt{x}]\, dx$$

$$= \frac{2}{3}[27] - \int_0^9 [\sqrt{x}]\, dx$$

As $\int_0^9 [\sqrt{x}]\, dx \Rightarrow 0 \le x \le 9$

and $0 \le \sqrt{x} \le 3$

Thus, it should be divided into three parts $0 \le \sqrt{x} \le 1, 1 \le \sqrt{x} \le 2, 2 \le \sqrt{x} \le 3$

ie, $I = 2(9) - \int_0^9 [\sqrt{x}]\, dx$

$$= 18 - \left[\int_0^1 [\sqrt{x}]\, dx + \int_1^4 [\sqrt{x}]\, dx + \int_4^9 [\sqrt{x}]\, dx\right]$$

$$= 18 - \left(\int_0^1 0\, dx + \int_1^4 1\, dx + \int_4^9 2\, dx\right)$$

$$= 18 - \left(0 + (x)_1^4 + (2x)_4^9\right) = 18 - (3 + 10)$$

$$I = 5$$

Illustration 24 If for a real number y, $[y]$ is the greatest integer less than or equal to y, then find the value of the integral $\int_{\pi/2}^{3\pi/2} [2 \sin x]\, dx$.

Solution. We know, $-1 \le \sin x \le 1$ as $x \in [\pi/2, 3\pi/2]$

$\Rightarrow$ $$-2 \le 2 \sin x \le 2$$

∴ 2 sin x must be divided (or broken) at $x = 5\pi/6, \pi, 7\pi/6$.

As $2\sin x = +1, 0, -1$ at these points.

$$\therefore \quad \int_{\pi/2}^{3\pi/2} [2\sin x]\,dx$$

$$= \int_{\pi/2}^{5\pi/6} [2\sin x]\,dx + \int_{5\pi/6}^{\pi} [2\sin x]\,dx + \int_{\pi}^{7\pi/6} [2\sin x]\,dx + \int_{7\pi/6}^{3\pi/2} [2\sin x]\,dx$$

$$= \int_{\pi/2}^{5\pi/6} 1\,dx + \int_{5\pi/6}^{\pi} 0\,dx + \int_{\pi}^{7\pi/6} -1\,dx + \int_{7\pi/6}^{3\pi/2} -2\,dx$$

$$= (5\pi/6 - \pi/2) + 0 - (7\pi/6 - \pi) - 2(3\pi/2 - 7\pi/6) = -\frac{\pi}{2}$$

Illustration 25 The value of $\int_0^{100} [\tan^{-1} x]\,dx$ is equal to (where [.] denotes the greatest integer function.)

(a) $\tan 1 - 100$ (b) $\pi/2 - \tan 1$ (c) $100 - \tan 1$ (d) None of these

Solution. Let $I = \int_0^{100} [\tan^{-1} x]\,dx$

where $[\tan^{-1} x]$ is shown as;

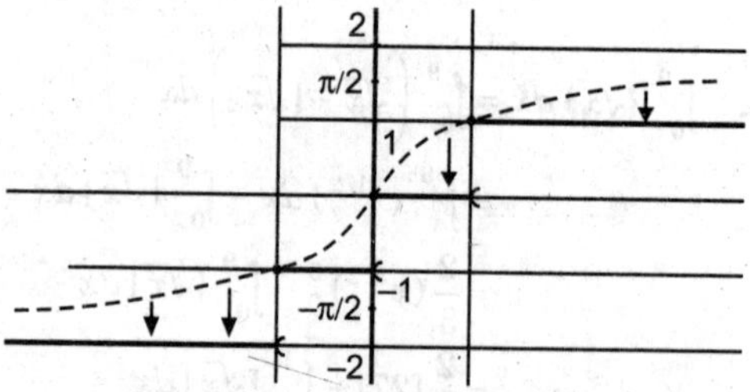

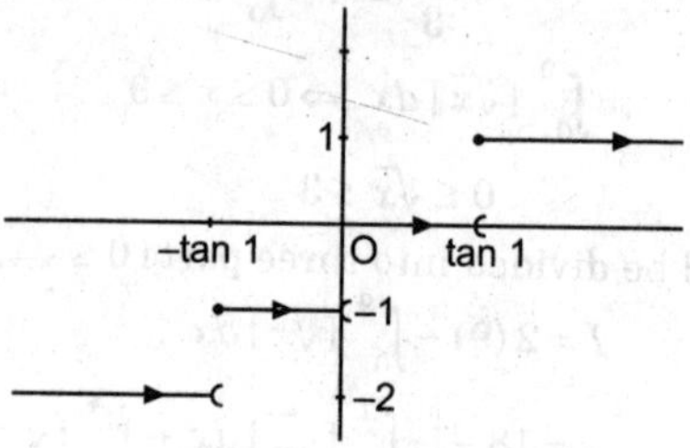

$$\therefore \quad \int_0^{100} [\tan^{-1} x]\,dx, \text{ is shown as;}$$

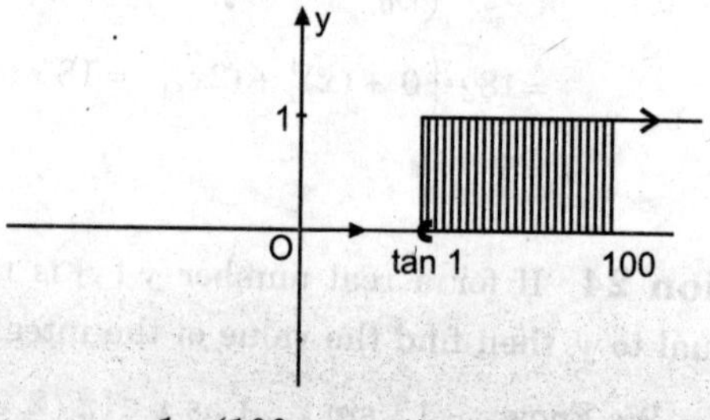

$$= 1 \times (100 - \tan 1)$$

Hence, (c) is the correct answer.

Illustration 26 The value of $\int_{-2}^{2} \min\{x-[x], -x-[-x]\}\, dx$, is equal to (where [.] denotes the greatest integer function.)

(a) 1/2 (b) 1 (c) 3/2 (d) 2

Solution. Let $f(x) = \min(x-[x], -x-[-x])$

$$= \min(\{x\}, \{-x\})$$

Graphically, $\{x\}$ and $\{-x\}$ could be plotted as;

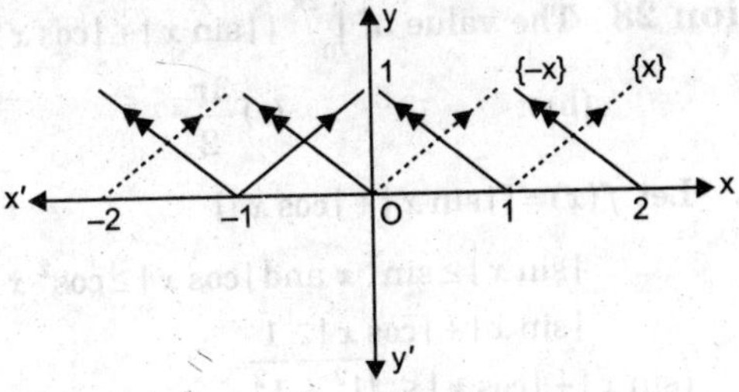

Since, from above graph, we need $\int_{-2}^{2} \min(\{x\}, \{-x\})$ shown as;

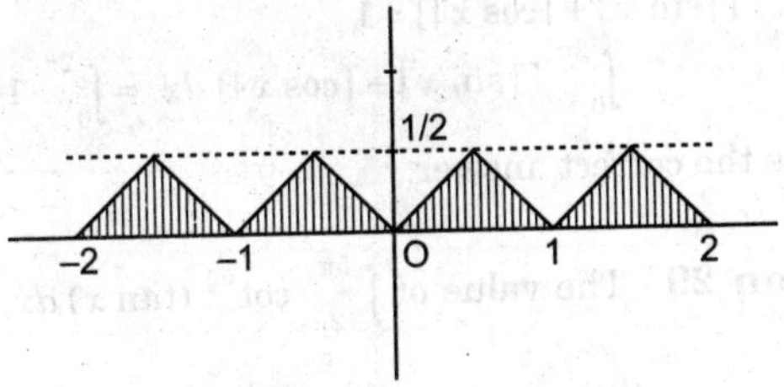

$$\therefore \quad \int_{-2}^{2} f(x)\, dx = 4\int_{0}^{1} f(x)\, dx = 4 \times \frac{1}{2} \times 1 \times \frac{1}{2} = 1$$

Hence, (b) is the correct answer.

Illustration 27 The value of $\int_{1}^{2} (x^{[x^2]} + [x^2]^x)\, dx$, where [.] denotes the greatest integer function, is equal to

(a) $\frac{5}{4} + \sqrt{3} + (2^{\sqrt{3}} - 2^{\sqrt{2}}) + \frac{1}{\log 3}(9 - 3^{\sqrt{3}})$

(b) $\frac{5}{4} + \sqrt{3} + \frac{\sqrt{2}}{3} + \frac{1}{\log 2}(2^{\sqrt{3}} - 2^{\sqrt{2}}) + \frac{1}{\log 3}(9 - 3^{\sqrt{3}})$

(c) $\frac{5}{4} + \frac{\sqrt{2}}{3} + \frac{1}{\log 2}(2^{\sqrt{3}} - 2^{\sqrt{2}}) + \frac{1}{\log 3}(9 - 3^{\sqrt{3}})$

(d) None of the above

Solution. We have,

$$I = \int_{1}^{2} (x^{[x^2]} + [x^2]^x)\, dx$$

$$= \int_{1}^{\sqrt{2}} (x+1)\, dx + \int_{\sqrt{2}}^{\sqrt{3}} (x^2 + 2^x)\, dx + \int_{\sqrt{3}}^{2} (x^3 + 3^x)\, dx$$

$$=\left[\frac{x^2}{2}+x\right]_1^{\sqrt{2}}+\left[\frac{x^3}{3}+\frac{2^x}{\log 2}\right]_{\sqrt{2}}^{\sqrt{3}}+\left[\frac{x^4}{4}+\frac{3^x}{\log 3}\right]_{\sqrt{3}}^{2}$$

$$=\frac{5}{4}+\sqrt{3}+\frac{\sqrt{2}}{3}+\frac{1}{\log 2}(2^{\sqrt{3}}-2^{\sqrt{2}})+\frac{1}{\log 3}(3^2-3^{\sqrt{3}})$$

Hence, (b) is the correct answer.

Illustration 28 The value of $\int_0^{2\pi}[|\sin x|+|\cos x|]\,dx$, is equal to

(a) $\frac{\pi}{2}$ (b) π (c) $\frac{3\pi}{2}$ (d) 2π

Solution. Let $f(x)=[|\sin x|+|\cos x|]$

As, $|\sin x|\geq \sin^2 x$ and $|\cos x|\geq \cos^2 x$

$\therefore$ $|\sin x|+|\cos x|\geq 1$

and $|\sin x|+|\cos x|\leq \sqrt{1^2+1^2}$

$\Rightarrow$ $1\leq |\sin x|+|\cos x|\leq \sqrt{2}$

Thus, $[|\sin x|+|\cos x|]=1$

$\therefore$ $\int_0^{2\pi}[|\sin x|+|\cos x|]\,dx=\int_0^{2\pi}1\,dx=2\pi$

Hence, (d) is the correct answer.

Illustration 29 The value of $\int_{-2\pi}^{5\pi}\cot^{-1}(\tan x)\,dx$, is equal to

(a) $\frac{7\pi}{2}$ (b) $\frac{7\pi^2}{2}$ (c) $\frac{3\pi}{2}$ (d) π^2

Solution. Let $I=\int_{-2\pi}^{5\pi}\cot^{-1}(\tan x)\,dx$

$$I=7\int_0^{\pi}\cot^{-1}\left(\cot\left(\frac{\pi}{2}-x\right)\right)dx \quad \ldots(i)$$

$$\left[\because \int_{nT}^{mT}f(x)\,dx=(m-n)\int_0^{T}f(x)\,dx\right]$$

As we know, $\cot^{-1}(\cot x)=\begin{cases} x, & 0<x<\pi/2 \\ \pi+x, & \pi/2<x<\pi \end{cases}$

$$\therefore \quad I=7\left\{\int_0^{\pi/2}\left(\frac{\pi}{2}-x\right)dx+\int_{\pi/2}^{\pi}\left(\pi+\frac{\pi}{2}-x\right)dx\right\}$$

$$=7\left\{\left(\frac{\pi}{2}x-\frac{x^2}{2}\right)_0^{\pi/2}+\left(\frac{3\pi}{2}x-\frac{x^2}{2}\right)_{\pi/2}^{\pi}\right\}$$

$$=7\left\{\left(\frac{\pi^2}{4}-\frac{\pi^2}{8}\right)+\left(\frac{3\pi^2}{2}-\frac{\pi^2}{2}-\frac{3\pi^2}{4}+\frac{\pi^2}{8}\right)\right\}$$

$$=\frac{7\pi^2}{2}$$

Hence, (b) is the correct answer.

Illustration 30 The value of $\int_0^2 f(x)\,dx$, where

$$f(x)=\begin{cases}0, & \text{when } x=\dfrac{n}{n+1},\ n=1,2,3,\ldots\\ 1, & \text{else where}\end{cases}$$

is equal to

(a) 1 (b) 2 (c) 3 (d) None of these

Solution. Here,

$$\int_0^2 f(x)\,dx=\int_0^{1/2}1\,dx+\int_{1/2}^{2/3}1\,dx+\int_{2/3}^{3/4}1\,dx+\ldots+\int_{\frac{n-1}{n}}^{\frac{n}{n+1}}1\,dx+\ldots+\int_1^2 1\,dx$$

$$=\left(\frac{1}{2}\right)+\left(\frac{2}{3}-\frac{1}{2}\right)+\left(\frac{3}{4}-\frac{2}{3}\right)+\ldots+\left(\frac{n}{n+1}-\frac{n-1}{n}\right)+\ldots+1$$

$$=\frac{n}{n+1}+\ldots..+1,\ \text{as } n\to\infty$$

We take, limit $n\to\infty$

$\therefore$ We have, $\int_0^2 f(x)\,dx=1+1=2$

Hence, (b) is the correct answer.

Target Exercise 2.1

1. The value of $\int_{-1}^{3}\{|x-2|+[x]\}\,dx$, where [.] denotes the greatest integer function, is equal to

(a) 5 (b) 6 (c) 7 (d) None of these

2. The value of $\int_{-1}^{3}(|x|+|x-1|)\,dx$, is equal to

(a) 9 (b) 6 (c) 3 (d) None of these

3. Let $f(x)=x-[x]$, for every real number x (where $[x]$ is integral part of x). Then the value of $\int_{-1}^{1} f(x)\,dx$, is equal to

(a) 0 (b) 1 (c) 2 (d) None of these

4. The value of $\int_0^2 [x+[x+[x]]]\,dx$, (where [.] denotes the greatest integer function), is equal to

(a) 2 (b) 3 (c) –3 (d) None of these

5. The value of $\int_0^{[x]}\frac{2^x}{2^{[x]}}\,dx$, is equal to (where [.] denotes the greatest integer function)

(a) $\frac{[x]}{\log 2}$ (b) $\frac{[x]}{2\log 2}$ (c) $\frac{[x]}{4\log 2}$ (d) None of these

6. The value of $\int_0^4 \{x\}\,dx$, (where {.} denotes fractional part of x) is equal to

(a) $\frac{4}{3}$ (b) $\frac{5}{3}$ (c) $\frac{7}{3}$ (d) None of these

7. The value of $\int_1^4 \{x\}^{[x]}\, dx$, (where [.] and {.} denote the greatest integer and fractional part of x) is equal to

(a) $\frac{11}{12}$ (b) $\frac{13}{12}$ (c) $\frac{7}{12}$ (d) $\frac{19}{12}$

8. The value of $\int_0^x [t+1]^3\, dt$, (where [.] denotes the greatest integer function of x) is equal to

(a) $\left(\frac{[x]([x]+1)}{2}\right)^2 + ([x]+1)^3 \{x\}$ (b) $\left(\frac{[x]([x]+1)}{2}\right)^3 + ([x]+1)^3 \{x\}$

(c) $\left(\frac{[x]([x]+1)}{2}\right)^3 + ([x]+1)^2 \{x\}$ (d) None of these

Property VI. $\int_0^a f(x)\, dx = \int_0^a f(a-x)\, dx$

Proof On RHS put $(a-x) = t$, so that $dx = -dt$

Also, when $x = 0, t = a$ and when $x = a, t = 0$

$$\therefore \quad \int_0^a f(a-x)\, dx = -\int_a^0 f(t)\, dt = \int_0^a f(t)\, dt = \int_0^a f(x)\, dx$$

$$\therefore \quad \int_0^a f(a-x)\, dx = \int_0^a f(x)\, dx$$

Points to Consider

This property is useful to evaluate a definite integral without first finding the corresponding indefinite integrals which may be difficult or sometimes impossible to find.

Illustration 31 Show that

(i) $\int_0^{\pi/2} f(\sin x)\, dx = \int_0^{\pi/2} f(\cos x)\, dx$

(ii) $\int_0^{\pi/2} f(\tan x)\, dx = \int_0^{\pi/2} f(\cot x)\, dx$

(iii) $\int_0^{\pi/2} f(\sin 2x) \sin x\, dx = \int_0^{\pi/2} \sqrt{2}\, f(\cos 2x) \cdot \cos x\, dx = \int_0^{\pi/2} f(\sin 2x) \cdot \cos x\, dx$

(iv) $\int_0^{\pi} x f(\sin x)\, dx = \frac{\pi}{2} \int_0^{\pi} f(\sin x)\, dx$ **[IIT JEE 1982]**

Solution. (i) We know,

$$\int_0^{\pi/2} f(\sin x)\, dx = \int_0^{\pi/2} f\left(\sin\left(\frac{\pi}{2} - x\right)\right) dx$$

$$\left[\text{using } \int_0^a f(x)\, dx = \int_0^a f(a-x)\, dx\right]$$

$$= \int_0^{\pi/2} f(\cos x)\, dx$$

$$\therefore \quad \int_0^{\pi/2} f(\sin x)\, dx = \int_0^{\pi/2} f(\cos x)\, dx$$

(ii) $\int_0^{\pi/2} f(\tan x)\,dx = \int_0^{\pi/2} f\left(\tan\left(\frac{\pi}{2} - x\right)\right) dx$

$$\left[\text{using} \int_0^a f(x)\,dx = \int_0^a f(a-x)\,dx\right]$$

$$= \int_0^{\pi/2} f(\cot x)\,dx$$

$\therefore \quad \int_0^{\pi/2} f(\tan x)\,dx = \int_0^{\pi/2} f(\cot x)\,dx$

(iii) We know, $I = \int_0^{\pi/2} f(\sin 2x)\sin x\,dx$ ...(i)

$$= \int_0^{\pi/2} f\left(\sin 2\left(\frac{\pi}{2} - x\right)\right)\cdot \sin\left(\frac{\pi}{2} - x\right) dx$$

$$\left[\text{using} \int_0^a f(x)\,dx = \int_0^a f(a-x)\,dx\right]$$

$$= \int_0^{\pi/2} f(\sin(\pi - 2x))\cos x\,dx$$

$I = \int_0^{\pi/2} f(\sin 2x)\cos x\,dx$ ...(ii)

Adding Eqs. (i) and (ii), we get

$$2I = \int_0^{\pi/2} f(\sin 2x)(\sin x + \cos x)\,dx$$

$$= \sqrt{2}\int_0^{\pi/2} f(\sin 2x)\sin\left(x + \frac{\pi}{4}\right) dx$$

Put $\quad x + \frac{\pi}{4} = \left(\frac{\pi}{2} - \theta\right) \quad \left(ie,\ x = \frac{\pi}{4} - \theta\right)$

$\Rightarrow \quad 2I = -\sqrt{2}\int_{\pi/4}^{-\pi/4} f(\cos 2\theta)\cos\theta\,d\theta$

$$= \sqrt{2}\int_{-\pi/4}^{\pi/4} f(\cos 2\theta)\cos\theta\,d\theta$$

$$= 2\sqrt{2}\int_0^{\pi/4} f(\cos 2\theta)\cos\theta\,d\theta \quad \text{(since it is even)}$$

$\therefore \quad I = \sqrt{2}\int_0^{\pi/4} f(\cos 2\theta)\cos\theta\,d\theta$

(iv) Let $\quad I = \int_0^{\pi} x f(\sin x)\,dx$ ...(i)

Replacing x by $(\pi - x)$

$$I = \int_0^{\pi} (\pi - x) f(\sin(\pi - x))\,dx$$

$I = \int_0^{\pi} (\pi - x) f(\sin x)\,dx$ ...(ii)

Adding Eqs. (i) and (ii), we get

$$2I = \int_0^{\pi} \pi f(\sin x)\,dx$$

$$I = \frac{\pi}{2}\int_0^{\pi} f(\sin x)\,dx$$

$\therefore \quad \int_0^{\pi} x f(\sin x)\,dx = \frac{\pi}{2}\int_0^{\pi} f(\sin x)\,dx$

Illustration 32 If f and g are continuous functions satisfying

$$f(x) = f(a-x) \quad \text{and} \quad g(x) + g(a-x) = 2,$$

then show that $\int_0^a f(x)\,g(x)\,dx = \int_0^a f(x)\,dx.$

Solution. Let $I = \int_0^a f(x)\,g(x)\,dx = \int_0^a f(a-x)\,g(a-x)\,dx$

$$= \int_0^a f(x)\cdot[2-g(x)]\,dx \quad \left\{ \begin{array}{r} \text{using} \\ \int_0^a f(x)\,dx = \int_0^a f(a-x)\,dx \end{array} \right.$$

$\because$ $\quad f(x) = f(a-x)$ and $g(a-x) + g(x) = 2$ (given)

$\therefore$ $\quad \int_0^a f(x)\cdot g(x)\,dx = 2\int_0^a f(x)\,dx - \int_0^a f(x)\cdot g(x)\,dx$

or $\quad 2\int_0^a f(x)\cdot g(x)\,dx = 2\int_0^a f(x)\,dx$

$\Rightarrow$ $\quad \int_0^a f(x)\,g(x)\,dx = \int_0^a f(x)\,dx$

Illustration 33 Evaluate

(i) $\int_0^{\pi/2} \frac{dx}{1+\sqrt{\tan x}}$ (ii) $\int_0^{\pi/2} \log(\tan x)\,dx$

(iii) $\int_0^{\pi/4} \log(1+\tan x)\,dx$ (iv) $\int_0^{\pi/2} \frac{\sin x - \cos x}{1+\sin x\cos x}\,dx$

Solution. (i) $I = \int_0^{\pi/2} \frac{dx}{1+\sqrt{\tan x}} = \int_0^{\pi/2} \frac{\sqrt{\cos x}}{\sqrt{\cos x}+\sqrt{\sin x}}\,dx$ ...(i)

Then, $\quad I = \int_0^{\pi/2} \frac{\sqrt{\cos(\pi/2 - x)}}{\sqrt{\cos(\pi/2-x)}+\sqrt{\sin(\pi/2-x)}}\,dx$

$$= \int_0^{\pi/2} \frac{\sqrt{\sin x}}{\sqrt{\sin x}+\sqrt{\cos x}}\,dx \qquad \text{...(ii)}$$

Adding Eqs. (i) and (ii), we get

$$2I = \int_0^{\pi/2} \frac{\sqrt{\sin x}}{\sqrt{\sin x}+\sqrt{\cos x}}\,dx + \int_0^{\pi/2} \frac{\sqrt{\cos x}}{\sqrt{\sin x}+\sqrt{\cos x}}\,dx$$

$$= \int_0^{\pi/2} \frac{\sqrt{\sin x}+\sqrt{\cos x}}{\sqrt{\sin x}+\sqrt{\cos x}}\,dx = \int_0^{\pi/2} 1\,dx$$

$$= [x]_0^{\pi/2} = \frac{\pi}{2} - 0$$

$\therefore$ $\quad 2I = \frac{\pi}{2},\ I = \frac{\pi}{4}$

(ii) $\quad I = \int_0^{\pi/2} \log(\tan x)\,dx$ ...(i)

Then, $\quad I = \int_0^{\pi/2} \log\left\{\tan\left(\frac{\pi}{2}-x\right)\right\}dx$

$$I = \int_0^{\pi/2} \log(\cot x)\,dx \qquad \text{...(ii)}$$

$$2I = \int_0^{\pi/2} \log(\tan x)\,dx + \int_0^{\pi/2} \log(\cot x)\,dx$$

$$= \int_0^{\pi/2} \{ \log \tan x + \log \cot x \} dx$$

$$= \int_0^{\pi/2} \log (\tan x \cdot \cot x) dx = \int_0^{\pi/2} \log (1) dx$$

$$2I = 0 \Rightarrow I = 0$$

(iii) $$I = \int_0^{\pi/4} \log (1 + \tan x) dx \quad \text{...(i)}$$

$$I = \int_0^{\pi/4} \log (1 + \tan (\pi / 4 - x)) dx$$

$$= \int_0^{\pi/4} \log \left(1 + \frac{\tan \pi / 4 - \tan x}{1 + \tan (\pi / 4) \cdot \tan x}\right) dx$$

$$= \int_0^{\pi/4} \log \left(\frac{1 + \tan x + 1 - \tan x}{1 + \tan x}\right) dx$$

$$= \int_0^{\pi/4} \log \left(\frac{2}{1 + \tan x}\right) dx$$

$$= \int_0^{\pi/4} \log (2) dx - \int_0^{\pi/4} \log (1 + \tan x) dx$$

$$I = (\log 2) (x)_0^{\pi/4} - I \quad \text{[using Eq. (i)]}$$

$$2I = \frac{\pi}{4} \log 2 \Rightarrow I = \frac{\pi}{8} \log 2$$

(iv) $$I = \int_0^{\pi/2} \frac{\sin x - \cos x}{1 + \sin x \cos x} dx \quad \text{...(i)}$$

Then, $$I = \int_0^{\pi/2} \frac{\sin\left(\frac{\pi}{2} - x\right) - \cos\left(\frac{\pi}{2} - x\right)}{1 + \sin\left(\frac{\pi}{2} - x\right) \cos\left(\frac{\pi}{2} - x\right)} dx$$

$$I = \int_0^{\pi/2} \frac{\cos x - \sin x}{1 + \cos x \cdot \sin x} dx \quad \text{...(ii)}$$

Adding Eqs. (i) and (ii), we get

$$2I = \int_0^{\pi/2} \frac{\sin x - \cos x}{1 + \sin x \cos x} dx + \int_0^{\pi/2} \frac{\cos x - \sin x}{1 + \sin x \cos x} dx$$

$\therefore$ $$2I = 0 \Rightarrow I = 0$$

Illustration 34 The value of $\int_{-a}^{a} (\cos^{-1} x - \sin^{-1} \sqrt{1 - x^2}) dx$ is $(a > 0)$ (where $\int_0^a \cos^{-1} x \, dx = A$) is

(a) $\pi a - A$ (b) $\pi a + 2A$ (c) $\pi a - 2A$ (d) $\pi a + A$

Solution. Let $$I = \int_{-a}^{a} (\cos^{-1} x - \sin^{-1} \sqrt{1 - x^2}) dx \quad (\text{as } a > 0)$$

$$= \int_{-a}^{0} (\cos^{-1} x - \sin^{-1} \sqrt{1 - x^2}) dx + \int_0^a (\cos^{-1} x - \sin^{-1} \sqrt{1 - x^2}) dx$$

$$= \int_{-a}^{0} \cos^{-1} x \, dx + A - 2 \int_0^a \sin^{-1} \sqrt{1 - x^2} \, dx$$

$$= - \int_a^0 (\pi - \cos^{-1} x) dx + A - 2A = \pi a - \int_0^a \cos^{-1} x \, dx - A$$

$$= \pi a - A - A = \pi a - 2A$$

Hence, (c) is the correct answer.

Illustration 35 The value of $\int_0^a \log(\cot a + \tan x)\,dx$, where $a \in (0, \pi/2)$ is equal to

(a) $a \log(\sin a)$ (b) $-a \log(\sin a)$

(c) $-a \log(\cos a)$ (d) None of these

Solution. Let $I = \int_0^a \log(\cot a + \tan x)\,dx$

$$= \int_0^a \log\left(\frac{\cos a}{\sin a} + \frac{\sin x}{\cos x}\right) dx$$

$$= \int_0^a \log\left(\frac{\cos(a-x)}{\sin a \cos x}\right) dx$$

$$= \int_0^a \log(\cos(a-x))\,dx - \int_0^a \log(\sin a)\,dx - \int_0^a \log(\cos x)\,dx$$

$$= \int_0^a \log\cos(x)\,dx - \int_0^a \log(\sin a)\,dx - \int_0^a \log(\cos x)\,dx$$

[using $\int_0^a f(x)\,dx = \int_0^a f(a-x)\,dx$ to first integral]

$$= -\log(\sin a)\int_0^a dx$$

$$= -a\log(\sin a)$$

Hence, (b) is the correct answer.

Property VII. $\int_{-a}^{a} f(x)\,dx = \begin{cases} 2\int_0^a f(x)\,dx, & \text{if } f(x) \text{ is an even function} \\ 0, & \text{if } f(x) \text{ is an odd function} \end{cases}$

Proof We know,

$$\int_a^b f(x)\,dx = \int_a^c f(x)\,dx + \int_c^b f(x)\,dx, \text{ if } a < c < b$$

$$\therefore \quad \int_{-a}^{a} f(x)\,dx = \int_{-a}^{0} f(x)\,dx + \int_0^a f(x)\,dx \quad \text{...(i)}$$

Now, $\int_{-a}^{0} f(x)\,dx = \int_a^0 f(-t)(-dt)$, where $t = -x$

$$= -\int_a^0 f(-t)\,dt = \int_0^a f(-t)\,dt$$

$$= \int_0^a f(-x)\,dx \quad \text{(using property II and I)}$$

$$= \begin{cases} \int_0^a f(x)\,dx, & \text{if } f(x) \text{ is even} \\ -\int_0^a f(x)\,dx, & \text{if } f(x) \text{ is odd} \end{cases} \quad \text{...(ii)}$$

$\therefore$ From Eqs. (i) and (ii), we get

$$\int_{-a}^{a} f(x)\,dx = \begin{cases} 2\int_0^a f(x)\,dx, & f(x) \text{ is even} \\ 0, & f(x) \text{ is odd} \end{cases}$$

Illustration 36 Evaluate $\int_{-\pi/4}^{\pi/4} x^3 \sin^4 x \, dx$.

Solution. Let $f(x) = x^3 \sin^4 x$, then

$$f(-x) = (-x)^3 \sin^4(-x) = -x^3 (\sin(-x))^4$$
$$= -x^3(-\sin x)^4 = -x^3 \sin^4 x = -f(x)$$

So, $f(x)$ is an odd function,

Hence, $\int_{-\pi/4}^{\pi/4} f(x)\,dx = 0 \quad ie, \int_{-\pi/4}^{\pi/4} x^3 \sin^4 x \, dx = 0$

Illustration 37 Evaluate $\int_{-\pi/2}^{\pi/2} \sin^2 x \, dx$.

Solution. Let $f(x) = \sin^2 x$, then

$$f(-x) = \sin^2(-x) = (\sin(-x))^2 = (-\sin x)^2$$
$$= \sin^2 x = f(x)$$

So, $f(x)$ is an even function, hence

$$\int_{-\pi/2}^{\pi/2} (\sin^2 x)\,dx = 2\int_0^{\pi/2} (\sin^2 x)\,dx = 2\int_0^{\pi/2} \frac{1-\cos 2x}{2}\,dx$$
$$= \left(x - \frac{\sin 2x}{2} \right)_0^{\pi/2} = \frac{\pi}{2}$$

$$\therefore \quad \int_{-\pi/2}^{\pi/2} \sin^2 x \, dx = \frac{\pi}{2}$$

Illustration 38 The value of $\int_{-1}^{1} \frac{d}{dx}\left(\tan^{-1}\frac{1}{x} \right) dx$, is equal to

(a) $\pi/2$ (b) $-\pi/2$ (c) $\pi/4$ (d) $3\pi/4$

Solution. We have, $\frac{d}{dx}\left(\tan^{-1}\frac{1}{x} \right) = \frac{d}{dx}(\cot^{-1} x) = -\frac{1}{1+x^2}$

$$\therefore \quad \int_{-1}^{1} \frac{d}{dx}\left(\tan^{-1}\frac{1}{x} \right) dx = \int_{-1}^{1} -\frac{1}{(1+x^2)}\,dx = -2\int_0^1 \frac{1}{1+x^2}\,dx$$
$$= -2\left(\tan^{-1} x \right)_0^1 = -2\left(\frac{\pi}{4} - 0 \right) = -\frac{\pi}{2}$$

$\Rightarrow \int_{-1}^{1} \tan^{-1}\left(\frac{1}{x} \right) dx = \frac{\pi}{2}$, but it is wrong as $\frac{1}{x}$ is not defined at $x = 0 \in (-1, 1)$.

$\therefore$ Above method is wrong.

Hence, (b) is the correct answer.

Points to Consider

In above illustration, if

$$\int_{-1}^{1} \frac{d}{dx}\left(\tan^{-1}\frac{1}{x} \right) dx = \left(\tan^{-1}\frac{1}{x} \right)_{-1}^{1}$$
$$= \tan^{-1}(1) - \tan^{-1}(-1)$$
$$= \pi/4 - (-\pi/4) = \pi/2$$

Illustration 39 The value of $\int_{-1}^{1} \log\left(\frac{2-x}{2+x}\right) dx$, is equal to

(a) $\frac{1}{2}$ (b) 1 (c) –1 (d) 0

Solution. Let $f(x) = \log\left(\frac{2-x}{2+x}\right)$

Now, $$f(-x) = \log\left(\frac{2+x}{2-x}\right) = \log\left(\frac{2-x}{2+x}\right)^{-1} = -\log\left(\frac{2-x}{2+x}\right)$$

$\therefore$ $$f(-x) = -\log\left(\frac{2-x}{2+x}\right) = -f(x)$$

ie, $f(x)$ is odd function.

So, $$\int_{-1}^{1} f(x)\,dx = \int_{-1}^{1} \log\left(\frac{2-x}{2+x}\right) dx = 0$$

$$\left[\because \int_{-a}^{a} f(x)\,dx = \begin{cases} 0, & \text{if } f(x) \text{ is odd} \\ 2\int_0^a f(x)\,dx, & \text{if } f(x) \text{ is even} \end{cases}\right].$$

Hence, (d) is the correct answer.

Illustration 40 The value of $\int_0^{\pi} \frac{x \sin(2x) \cdot \sin\left(\frac{\pi}{2}\cos x\right)}{(2x-\pi)} dx$, is equal to

(a) $\frac{8}{\pi}$ (b) $\frac{\pi}{8}$ (c) $\frac{8}{\pi^2}$ (d) $\frac{\pi^2}{8}$

Solution. Let $I = \int_0^{\pi} \frac{x \sin(2x) \cdot \sin\left(\frac{\pi}{2}\cos x\right)}{(2x-\pi)} dx$ ...(i)

$$I = \int_0^{\pi} \frac{(\pi - x) \cdot \sin 2(\pi - x) \cdot \sin\left(\frac{\pi}{2}\cos(\pi - x)\right)}{2(\pi - x) - \pi} dx$$

$$I = \int_0^{\pi} \frac{(\pi - x) \cdot \sin(2x) \cdot \sin\left(\frac{\pi}{2}\cos x\right)}{-(2x-\pi)} dx \quad \text{...(ii)}$$

Adding Eqs. (i) and (ii), we get

$$2I = \int_0^{\pi} \frac{(2x-\pi)\sin 2x \cdot \sin\left(\frac{\pi}{2}\cos x\right)}{(2x-\pi)} dx$$

$\therefore$ $$I = \frac{1}{2}\int_0^{\pi} 2\sin x \cos x \cdot \sin\left(\frac{\pi}{2}\cos x\right) dx$$

(put $\cos x = t \therefore -\sin x\,dx = dt$)

$\therefore$ $$I = -\int_1^{-1} t \sin\left(\frac{\pi}{2}t\right) dt$$

$$= 2\int_0^1 t \cdot \sin\left(\frac{\pi}{2}t\right) dt \quad \text{(using by parts)}$$

$$= 2\left\{\left(t \cdot \frac{\cos\left(\frac{\pi}{2}t\right)}{-\frac{\pi}{2}}\right)_0^1 - \int_0^1 1 \cdot \frac{\cos\left(\frac{\pi}{2}t\right)}{-\frac{\pi}{2}}\, dt\right\}$$

$$= 2\left\{0 + \frac{2}{\pi} \cdot \left(\frac{\sin\left(\frac{\pi}{2}t\right)}{\frac{\pi}{2}}\right)_0^1\right\} = \frac{8}{\pi^2}$$

Hence, (c) is the correct answer.

Illustration 41 If $f(x) = \begin{vmatrix} \cos x & e^{x^2} & 2x\cos^2 x/2 \\ x^2 & \sec x & \sin x + x^3 \\ 1 & 2 & x + \tan x \end{vmatrix}$, then the value of $\int_{-\pi/2}^{\pi/2} (x^2 + 1)(f(x) + f''(x))\, dx$, is equal to

(a) 1 (b) –1 (c) 2 (d) None of these

Solution. As, $f(x) = \begin{vmatrix} \cos x & e^{x^2} & 2x\cos^2 x/2 \\ x^2 & \sec x & \sin x + x^3 \\ 1 & 2 & x + \tan x \end{vmatrix}$

$\Rightarrow$ $f(-x) = -f(x)$

$\Rightarrow$ $f(x)$ is odd.

$\Rightarrow$ $f'(x)$ is even.

$\Rightarrow$ $f''(x)$ is odd.

Thus, $f(x) + f''(x)$ is odd function, let

$$\phi(x) = (x^2 + 1) \cdot \{f(x) + f''(x)\}$$

$\Rightarrow$ $\phi(-x) = -\phi(x)$,

ie, $\phi(x)$ is odd.

$\therefore$ $\int_{-\pi/2}^{\pi/2} \phi(x)\, dx = 0$

Hence, (d) is the correct answer.

Illustration 42 The value of $\int_{-1}^{1} \frac{x}{\sqrt{1-x^2}} \cdot \sin^{-1}(2x\sqrt{1-x^2})\, dx$, is equal to

(a) $4\sqrt{2}$ (b) $4(\sqrt{2} - 1)$ (c) $4(\sqrt{2} + 1)$ (d) None of these

Solution. Let $I = \int_{-1}^{1} \frac{x}{\sqrt{1-x^2}} \cdot \sin^{-1}(2x\sqrt{1-x^2})\, dx$

$$= 2\int_0^1 \frac{x}{\sqrt{1-x^2}} \cdot \sin^{-1}(2x\sqrt{1-x^2})\, dx$$

[using $\int_{-a}^{a} f(x)\, dx = 2\int_0^a f(x)\, dx$, if $f(-x) = f(x)$]

Put $x = \sin\theta \Rightarrow dx = \cos\theta\, d\theta$

$$\therefore \quad I = 2\int_0^{\pi/2} \frac{\sin\theta}{\sqrt{1-\sin^2\theta}} \cdot \sin^{-1}(2\sin\theta\cos\theta)\cdot\cos\theta\, d\theta$$

$$= 2\int_0^{\pi/4} \sin\theta \cdot 2\theta\, d\theta + 2\int_{\pi/4}^{\pi/2} (\pi - 2\theta)\sin\theta\, d\theta$$

$$\left[\text{using } \sin^{-1}(\sin 2\theta) = \begin{cases} 2\theta, & 0<\theta<\pi/4 \\ \pi - 2\theta, & \pi/4<\theta<\pi/2 \end{cases}\right]$$

$$= 4\int_0^{\pi/4} \theta\cdot\sin\theta\, d\theta + 2\pi\int_{\pi/4}^{\pi/2} \sin\theta\, d\theta - 4\int_{\pi/4}^{\pi/2} \theta\cdot\sin\theta\, d\theta$$

$$= 4\{\theta(-\cos\theta)\}_0^{\pi/4} - 4\int_0^{\pi/4} 1\cdot(-\cos\theta)\, d\theta + 2\pi(-\cos\theta)_{\pi/4}^{\pi/2}$$

$$- 4\{\theta(-\cos\theta)\}_{\pi/4}^{\pi/2} + 4\int_{\pi/4}^{\pi/2} (-\cos\theta)\, d\theta$$

$$= -\frac{\pi}{\sqrt{2}} + 2\sqrt{2} + \sqrt{2}\pi - \frac{\pi}{\sqrt{2}} - 4 + 2\sqrt{2} = 4\sqrt{2} - 4 = 4(\sqrt{2}-1)$$

Hence, (b) is the correct answer.

Illustration 43 Suppose the function $g_n(x) = x^{2n+1} + a_n x + b_n (n \in N)$ satisfies the equation $\int_{-1}^{1} (px+q) g_n(x)dx = 0$ for all linear functions $(px+q)$, then

(a) $a_n = b_n = 0$ (b) $b_n = 0;\ a_n = -\dfrac{3}{2n+3}$

(c) $a_n = 0;\ b_n = -\dfrac{3}{2n+3}$ (d) $a_n = \dfrac{3}{2n+3};\ b_n = -\dfrac{3}{2n+3}$

Solution. We have, $\int_{-1}^{1} (px+q)(x^{2n+1} + a_n x + b_n)dx = 0$

Equating the odd component to be zero and integrating, we get

$$\frac{2p}{2n+3} + \frac{2a_n p}{3} + 2b_n q = 0 \text{ for all } p, q$$

Therefore, $b_n = 0$ and $a_n = -\dfrac{3}{2n+3}$

Hence, (b) is the correct answer.

Illustration 44 Let $u = \int_0^{\pi/2} \cos\left(\frac{2\pi}{3}\sin^2 x\right) dx$ and $v = \int_0^{\pi/2} \cos\left(\frac{\pi}{3}\sin x\right) dx$, then the relation between u and v is

(a) $2u = v$ (b) $2u = 3v$ (c) $u = v$ (d) $u = 2v$

Solution.

$$u = \int_0^{\pi/2} \cos\left(\frac{2\pi}{3}\sin^2 x\right) dx$$

$$u = \int_0^{\pi/2} \cos\left(\frac{2\pi}{3}\cos^2 x\right) dx$$

$$\left[\text{using } \int_0^a f(x)dx = \int_0^a f(a-x)dx\right]$$

$$\therefore \quad 2u = \int_0^{\pi/2} \cos\left(\frac{2\pi}{3}\sin^2 x\right) + \cos\left(\frac{2\pi}{3}\cdot\cos^2 x\right) dx$$

On adding, $2u = \int_0^{\pi/2} 2\cos\frac{\pi}{3}\cdot\cos\left(\frac{\pi}{3}\cos 2x\right) dx$ (Using $\cos C + \cos D$)

$$= \frac{1}{2}\int_0^{\pi} \cos\left(\frac{\pi}{3}\cos t\right) dt \qquad \text{(Put } 2x = t\text{)}$$

$$= \int_0^{\pi/2} \cos\left(\frac{\pi}{3}\cos t\right) dt$$

$$= \int_0^{\pi/2} \cos\left(\frac{\pi}{3}\sin t\right) dt = v$$

Hence, (a) is the correct answer.

Illustration 45 Let $f(x) = \dfrac{\sin x}{x}$, then $\int_0^{\pi/2} f(x) f\left(\dfrac{\pi}{2} - x\right) dx$ is equal to

(a) $\dfrac{2}{\pi}\int_0^{\pi} f(x)\,dx$ (b) $\int_0^{\pi} f(x)\,dx$ (c) $\pi\int_0^{\pi} f(x)\,dx$ (d) $\dfrac{1}{\pi}\int_0^{\pi} f(x)\,dx$

Solution. $I = \int_0^{\pi/2} \dfrac{\sin x \cos x}{x\left(\dfrac{\pi}{2} - x\right)} dx = \int_0^{\pi/2} \dfrac{\sin 2x}{x(\pi - 2x)} dx$; put $2x = t$

$$I = \int_0^{\pi} \frac{\sin t}{t(\pi - t)} dt = \frac{1}{\pi}\int_0^{\pi}\left(\frac{\sin t}{t} + \frac{\sin t}{(\pi - t)}\right) dt$$

$$= \frac{1}{\pi}\int_0^{\pi} \frac{\sin t}{t} dt + \frac{1}{\pi}\int_0^{\pi} \frac{\sin t}{\pi - t} dt$$

$$= \frac{1}{\pi}\int_0^{\pi} \frac{\sin t}{t} dt + \frac{1}{\pi}\int_0^{\pi} \frac{\sin t}{t} dt$$

$$= \frac{2}{\pi}\int_0^{\pi} \frac{\sin t}{t} dt = \frac{2}{\pi}\int_0^{\pi} f(x) dx$$

Hence, (a) is the correct answer.

Property VIII. $\int_0^{2a} f(x)\,dx = \begin{cases} 2\int_0^{a} f(x)\,dx, & \text{if } f(2a - x) = f(x) \\ 0, & \text{if } f(2a - x) = -f(x) \end{cases}$

Proof We know, $\int_0^{2a} f(x)\,dx = \int_0^{a} f(x)\,dx + \int_a^{2a} f(x)\,dx$ ...(i)

Consider the integral $\int_0^{2a} f(x)\,dx$; putting $x = 2a - t$, so that $dx = -dt$

Also, when $x = a$, $t = a$ and when $x = 2a$, $t = 0$.

$$\therefore \quad \int_a^{2a} f(x)\,dx = \int_a^{0} f(2a - t)(-dt) = -\int_a^{0} f(2a - t)\,dt$$

$$= \int_0^{a} f(2a - t)\,dt = \int_0^{a} f(2a - x)\,dx$$

$$= \begin{cases} \int_0^{a} f(x)\,dx, & \text{if } f(2a - x) = f(x) \\ -\int_0^{a} f(x)\,dx, & \text{if } f(2a - x) = -f(x) \end{cases} \qquad \text{...(ii)}$$

Thus, from Eqs. (i) and (ii), we have

$$\int_0^{2a} f(x)\,dx = \begin{cases} 2\int_0^{a} f(x)\,dx, & \text{if } f(2a - x) = f(x) \\ 0, & \text{if } f(2a - x) = -f(x) \end{cases}$$

Illustration 46 Evaluate $\int_0^{\pi} \frac{x}{1+\cos^2 x}\,dx$.

Solution. Let $I = \int_0^{\pi} \frac{x}{1+\cos^2 x}\,dx$

$$= \int_0^{\pi} \frac{(\pi - x)\,dx}{1+\cos^2(\pi - x)} = \int_0^{\pi} \frac{\pi\,dx}{1+\cos^2 x} - \int_0^{\pi} \frac{x\,dx}{1+\cos^2 x}$$

$\therefore$ $$I = \pi \int_0^{\pi} \frac{dx}{1+\cos^2 x} - I$$

$\Rightarrow$ $$2I = \pi \int_0^{\pi} \frac{dx}{1+\cos^2 x} = 2\pi \int_0^{\pi/2} \frac{dx}{1+\cos^2 x}$$

$$\left[\text{using } \int_0^{2a} f(x)\,dx = \begin{cases} 0, & f(2a-x) = -f(x) \\ 2\int_0^a f(x)\,dx, & f(2a-x) = +f(x) \end{cases}\right]$$

$\Rightarrow$ $$I = \pi \int_0^{\pi/2} \frac{\sec^2 x}{\sec^2 x + 1}\,dx \qquad (\text{dividing } N^r \text{ and } D^r \text{ by } \cos^2 x)$$

$$I = \pi \int_0^{\pi/2} \frac{\sec^2 x}{2+\tan^2 x}\,dx$$

Let $\tan x = t, \sec^2 x\,dx = dt$

So that, for $x \to 0, t \to 0$ and for $x \to \pi/2, t \to \infty$

Hence, $$I = \pi \int_0^{\infty} \frac{dt}{2+t^2} = \frac{\pi}{\sqrt{2}}\left(\tan^{-1}\frac{t}{\sqrt{2}}\right)_0^{\infty}$$

$$= \frac{\pi}{\sqrt{2}}\left(\frac{\pi}{2} - 0\right)$$

$$I = \frac{\pi^2}{2\sqrt{2}}$$

Illustration 47 Prove that

$$\int_0^{\pi/2} \log(\sin x)\,dx = \int_0^{\pi/2} \log(\cos x)\,dx = -\frac{\pi}{2}\log 2$$

Solution. Let $I = \int_0^{\pi/2} \log(\sin x)\,dx$ ...(i)

Then, $I = \int_0^{\pi/2} \log\sin(\pi/2 - x)\,dx = \int_0^{\pi/2} \log(\cos x)\,dx$ ...(ii)

Adding, (i) and (ii), we get

$$2I = \int_0^{\pi/2} \log\sin x\,dx + \int_0^{\pi/2} \log\cos x\,dx$$

$$= \int_0^{\pi/2} (\log\sin x + \log\cos x)\,dx$$

$\Rightarrow$ $$2I = \int_0^{\pi/2} \log(\sin x\cos x)\,dx = \int_0^{\pi/2} \log\left(\frac{2\sin x\cos x}{2}\right)dx$$

$$= \int_0^{\pi/2} \log\left(\frac{\sin 2x}{2}\right)dx$$

$$= \int_0^{\pi/2} \log(\sin 2x)\,dx - \int_0^{\pi/2} (\log 2)\,dx$$

$$= \int_0^{\pi/2} \log \sin 2x \, dx - (\log 2)(x)_0^{\pi/2}$$

$$\Rightarrow \quad 2I = \int_0^{\pi/2} \log (\sin 2x) \, dx - \frac{\pi}{2} \log 2 \qquad \ldots\text{(iii)}$$

Let $\quad I_1 = \int_0^{\pi/2} \log (\sin 2x) \, dx$

Putting $2x = t$, we get

$$I_1 = \int_0^{\pi} \log \sin t \, \frac{dt}{2} = \frac{1}{2} \int_0^{\pi} \log \sin t \, dt$$

$$= \frac{1}{2} \cdot 2 \int_0^{\pi/2} \log (\sin t) \, dt$$

$$I_1 = \int_0^{\pi/2} \log (\sin x) \, dx$$

$\therefore$ Eq. (iii) becomes, $\quad 2I = I - \frac{\pi}{2} \log 2$

Hence, $\quad \int_0^{\pi/2} \log \sin x \, dx = -\frac{\pi}{2} \log 2.$

Illustration 48 If $f(x) = -\int_0^{x} \log (\cos t) \, dt$, then the value of

$$f(x) - 2f\left(\frac{\pi}{4} + \frac{x}{2}\right) + 2f\left(\frac{\pi}{4} - \frac{x}{2}\right), \text{ is equal to}$$

(a) $-x \log 2$ (b) $\frac{x}{2} \log 2$ (c) $\frac{x}{3} \log 2$ (d) None of these

Solution. Here, $f\left(\frac{\pi}{4} + \frac{x}{2}\right) = -\int_0^{\pi/4 + x/2} \log (\cos t) \, dt$

$$= -\int_0^{\pi/4} \log (\cos t) \, dt - \int_{\pi/4}^{\pi/4 + x/2} \log (\cos t) \, dt \qquad \ldots\text{(i)}$$

$$f\left(\frac{\pi}{4} - \frac{x}{2}\right) = -\int_0^{\pi/4 - x/2} \log (\cos t) \, dt$$

$$= -\int_0^{\pi/4} \log (\cos t) \, dt - \int_{\pi/4}^{\pi/4 - x/2} \log (\cos t) \, dt \qquad \ldots\text{(ii)}$$

$$\therefore \quad 2f\left(\frac{\pi}{4} + \frac{x}{2}\right) - 2f\left(\frac{\pi}{4} - \frac{x}{2}\right)$$

$$= 2 \int_{\pi/4}^{\pi/4 - x/2} \log (\cos t) \, dt - 2 \int_{\pi/4}^{\pi/4 + x/2} \log (\cos t) \, dt$$

Put $t = \frac{\pi}{4} - z$ in first integral and $t = \frac{\pi}{4} + z$ in second integral

$$= -2 \int_0^{x/2} \log \cos\left(\frac{\pi}{4} - z\right) dz - 2 \int_0^{x/2} \log \cos\left(z + \frac{\pi}{4}\right) dz$$

$$= -2 \int_0^{x/2} \log \left(\frac{1}{2}(\cos^2 z - \sin^2 z)\right) dz$$

$$= 2 \int_0^{x/2} (\log 2) \, dx - 2 \int_0^{x/2} \log (\cos 2z) \, dz$$

$$= x \log 2 - 2 \int_0^{x/2} \log (\cos 2z) \, dz$$

$$\therefore \quad 2f\left(\frac{\pi}{4} + \frac{x}{2}\right) - 2f\left(\frac{\pi}{4} - \frac{x}{2}\right) = x (\log 2) - 2 \int_0^{x/2} \log (\cos 2z) \, dz$$

$$\Rightarrow \quad 2f\left(\frac{\pi}{4} + \frac{x}{2}\right) - 2f\left(\frac{\pi}{4} - \frac{x}{2}\right) = x (\log 2) + f(x)$$

or $$f(x) - 2f\left(\frac{\pi}{4} + \frac{x}{2}\right) + 2f\left(\frac{\pi}{4} - \frac{x}{2}\right) = -x \log 2$$

Hence, (a) is the correct answer.

Illustration 49 If $\int_0^{\pi}\left(\frac{x}{1+\sin x}\right)^2 dx = A$, then the value for $\int_0^{\pi}\frac{2x^2 \cdot \cos^2 x/2}{(1+\sin x)^2} dx$, is equal to

(a) $A + 2\pi - \pi^2$ (b) $A - 2\pi + \pi^2$ (c) $2\pi - A - \pi^2$ (d) None of these

Solution. Let $$B = \int_0^{\pi}\frac{2x^2 \cos^2 x/2}{(1+\sin x)^2} dx$$

$\therefore$ $$B - A = \int_0^{\pi}\frac{x^2(2\cos^2 x/2 - 1)}{(1+\sin x)^2} dx$$

$$= \int_0^{\pi}\frac{x^2 \cdot \cos x}{(1+\sin x)^2} dx$$

Using by parts

$$B - A = \left\{-\frac{x^2}{1+\sin x}\right\}_0^{\pi} + 2\int_0^{\pi}\frac{x}{(1+\sin x)} dx$$

$$B - A = -\pi^2 + 2K \qquad \text{...(i)}$$

Where, $$K = \int_0^{\pi}\frac{x\,dx}{1+\sin x} = \int_0^{\pi}\frac{(\pi - x)\,dx}{1+\sin x}$$

$\therefore$ $$K = \pi\int_0^{\pi}\frac{dx}{1+\sin x} - K$$

$\Rightarrow$ $$2K = 2\pi\int_0^{\pi/2}\frac{dx}{1+\sin x} = 2\pi\int_0^{\pi/2}\frac{dx}{1+\cos x}$$

$$= 2\pi\int_0^{\pi/2}\frac{1}{2}\sec^2\frac{x}{2}\,dx = 2\pi\left\{\tan\frac{x}{2}\right\}_0^{\pi/2} = 2\pi$$

$\therefore$ $$K = \pi$$

Thus, $$B - A = -\pi^2 + 2K$$

$\Rightarrow$ $$B = A - \pi^2 + 2\pi$$

Hence, (a) is the correct answer.

Property IX. $$\int_a^b f(x)\,dx = \int_a^b f(a+b-x)\,dx$$

Proof Putting $x = a + b - t$, $dx = -dt$, when $x = a$, $t = b$, $x = b$, $t = a$, we get

$$\int_a^b f(x)\,dx = \int_b^a f(a+b-t)(-dt) = -\int_b^a f(a+b-t)\,dt$$

$$= \int_a^b f(a+b-t)\,dt = \int_a^b f(a+b-x)\,dx$$

$\Rightarrow$ $$\int_a^b f(x)\,dx = \int_a^b f(a+b-x)\,dx$$

Illustration 50 Evaluate $\int_{\pi/6}^{\pi/3} \frac{dx}{1+\sqrt{\tan x}}$.

Solution. $I = \int_{\pi/6}^{\pi/3} \frac{dx}{1+\sqrt{\tan x}} = \int_{\pi/6}^{\pi/3} \frac{\sqrt{\cos x}\cdot dx}{\sqrt{\cos x}+\sqrt{\sin x}}$ ... (i)

$$I = \int_{\pi/6}^{\pi/3} \frac{\sqrt{\cos(\pi/2 - x)}}{\sqrt{\cos(\pi/2-x)}+\sqrt{\sin(\pi/2-x)}}\,dx \text{ (since } a+b=\pi/2)$$

$$I = \int_{\pi/6}^{\pi/3} \frac{\sqrt{\sin x}}{\sqrt{\sin x}+\sqrt{\cos x}}\,dx \quad \text{...(ii)}$$

Adding Eqs. (i) and (ii), we get

$$2I = \int_{\pi/6}^{\pi/3} 1\,dx = \frac{\pi}{3} - \frac{\pi}{6} = \frac{\pi}{6}$$

$$\Rightarrow \quad I = \frac{\pi}{12}$$

Illustration 51 Prove that $\int_a^b \frac{f(x)}{f(x)+f(a+b-x)}\,dx = \frac{b-a}{2}$.

Solution. Let $I = \int_a^b \frac{f(x)}{f(x)+f(a+b-x)}\,dx$ ...(i)

$$I = \int_a^b \frac{f(a+b-x)}{f(a+b-x)+f(a+b-(a+b-x))}\,dx$$

$$\Rightarrow \quad I = \int_a^b \frac{f(a+b-x)}{f(a+b-x)+f(x)}\,dx \quad \text{...(ii)}$$

Adding Eqs. (i) and (ii), we get

$$2I = \int_a^b \frac{f(a+b-x)+f(x)}{f(a+b-x)+f(x)}\,dx$$

$$2I = \int_a^b 1\,dx = (b-a)$$

$$I = \frac{b-a}{2}$$

Illustration 52 Solve $I = \int_{\cos^4 t}^{-\sin^4 t} \frac{\sqrt{f(z)}\,dz}{\sqrt{f(\cos 2t - z)}+\sqrt{f(z)}}$.

Solution. Here, $I = \int_{\cos^4 t}^{-\sin^4 t} \frac{\sqrt{f(z)}\,dz}{\sqrt{f(\cos 2t - z)}+\sqrt{f(z)}}$ ...(i)

$$I = \int_{\cos^4 t}^{-\sin^4 t} \frac{\sqrt{f(\cos 2t - z)}\,dz}{\sqrt{f(\cos 2t - z)}+\sqrt{f(z)}} \quad \text{...(ii)}$$

[using $\int_a^b f(x)\,dx = \int_a^b f(a+b-x)\,dx$]

Adding Eqs. (i) and (ii), we get

$$2I = \int_{\cos^4 t}^{-\sin^4 t} dz$$

$$2I = (z)_{\cos^4 t}^{-\sin^4 t}$$

$$I = -\frac{1}{2}(\sin^4 t + \cos^4 t) = -\frac{1}{2}(1 - 2\sin^2 t \cos^2 t)$$

$$I = -\frac{1}{2}\left(1 - \frac{1}{2}\sin^2 2t\right)$$

$$I = -\frac{1}{2} + \frac{1}{4}\sin^2 2t$$

Property X. If $f(x)$ is a periodic function with period T, then

$$\int_0^{nT} f(x)\, dx = n \int_0^T f(x)\, dx$$

Proof We have, $\int_0^{nT} f(x)\, dx = \sum_{r=1}^{n} \int_{(r-1)T}^{rT} f(x)\, dx$

$$= \sum_{r=1}^{n} \int_0^T f\{(r-1)T + y\}\, dy$$

where $x = (r-1)T + y$

$$= \sum_{r=1}^{n} \int_0^T f(y)\, dy$$

$\left\{ \begin{array}{l} \because f(x) \text{ is periodic with period } T \\ \because f((r-1)T + y) = f(y) \end{array} \right.$

$$= n \int_0^T f(y)\, dy = n \int_0^T f(x)\, dx$$

Property XI. If $f(x)$ is periodic with period T, then

$$\int_a^{a+nT} f(x)\, dx = n \int_0^T f(x)\, dx, \text{ where } n \in I$$

(i) In particular, if $a = 0$

$$\int_0^{nT} f(x)\, dx = n \int_0^T f(x)\, dx, \text{ where } n \in I$$

(ii) If $n = 1$

$$\int_a^{a+T} f(x)\, dx = \int_0^T f(x)\, dx$$

Property XII. If $f(x)$ is periodic with period T, then

$$\int_{mT}^{nT} f(x)\, dx = (n-m) \int_0^T f(x)\, dx, \quad \text{where } n, m \in I$$

Property XIII. If $f(x)$ is periodic with period T, then

$$\int_{a+nT}^{b+nT} f(x)\, dx = \int_a^b f(x)\, dx, \quad \text{where } n \in I$$

Illustration 53 Show $\int_0^{n\pi + V} |\sin x|\, dx = (2n+1) - \cos V$, where n is positive integer and $0 \le V < \pi$. **[IIT JEE 1994, 2004]**

Solution. $\int_0^{n\pi + V} |\sin x|\, dx = \int_0^V |\sin x|\, dx + \int_V^{n\pi + V} |\sin x|\, dx$

$$= \int_0^V \sin x\, dx + n \int_0^{\pi} |\sin x|\, dx$$

(using property XI)

$$= \left(-\cos x\right)_0^V + n\int_0^{\pi} \sin x\, dx$$

$$= (-\cos V + 1) + n\left(-\cos x\right)_0^{\pi}$$

$$= -(\cos V) + 1 + n(1+1)$$

$$= (2n+1) - \cos V$$

$$\Rightarrow \quad \int_0^{n\pi + V} |\sin x|\, dx = (2n+1) - \cos V,$$

where n is positive integer and $0 \le V < \pi$.

Illustration 54 Evaluate $\int_0^{4\pi} |\cos x|\, dx$.

Solution. Note that $|\cos x|$ is a periodic with period π.

Hence, the given integral, $I = 4\int_0^{\pi} |\cos x|\, dx$ (using property XI)

$$= 4\left\{\int_0^{\pi/2} \cos x\, dx - \int_{\pi/2}^{\pi} \cos x\, dx\right\}$$

$$= 4\left\{\left(\sin x\right)_0^{\pi/2} - \left(\sin x\right)_{\pi/2}^{\pi}\right\}$$

$$= 4\{1+1\} = 8$$

Illustration 55 Prove that $\int_0^{25} e^{x-[x]}\, dx = 25(e-1)$.

Solution. Since, $x - [x]$ is periodic function with period one.

$\therefore$ $e^{x-[x]}$ has period one

$\therefore$ $I = \int_0^{25\times 1} e^{x-[x]}\, dx = 25\int_0^1 e^{x-[x]}\, dx$

$$= 25\int_0^1 e^{x-0}\, dx = 25(e^x)_0^1 = 25(e^1 - e^0)$$

$$I = 25(e-1)$$

Illustration 56 The value of $\int_{-5}^{5} f(x)\, dx$; where $f(x) =$ minimum $(\{x+1\}, \{x-1\})$, $\forall\, x \in R$ and $\{.\}$ denotes fractional part of x, is equal to

(a) 3 (b) 4 (c) 5 (d) 6

Solution. We know, $\{x+1\} = \{x-1\} = \{x\}$

Thus, $f(x) = \text{minimum}(\{x+1\}, \{x-1\}) = \{x\}$

$\Rightarrow$ $\int_{-5}^{5} f(x)\, dx = \int_{-5}^{5} \{x\}\, dx = (5-(-5))\int_0^1 \{x\}\, dx$

{as $\{x\}$ is periodic with period '1'}

$$= 10\int_0^1 (x - [x])\, dx = 10\int_0^1 x\, dx$$

$$= 10\left(\frac{x^2}{2}\right)_0^1 = 5$$

Hence, (c) is the correct answer.

Illustration 57 The value of $\int_0^{2n\pi} [\sin x + \cos x]\, dx$, is equal to

(a) $-n\pi$ (b) $n\pi$ (c) $-2n\pi$ (d) None of these

(where [.] denotes the greatest integer function)

Solution. Let $I = \int_0^{2n\pi} [\sin x + \cos x]\, dx$

We know, $\sin x + \cos x = \sqrt{2} \sin\left(x + \frac{\pi}{4}\right)$ ...(i)

$$\therefore \quad [\sin x + \cos x] = \begin{cases} 1, & 0 \le x \le \pi/2 \\ 0, & \pi/2 < x \le 3\pi/4 \\ -1, & 3\pi/4 < x \le \pi \\ -2, & \pi < x < 3\pi/2 \\ -1, & 3\pi/2 \le x < 7\pi/4 \\ 0, & 7\pi/4 \le x \le 2\pi \end{cases}$$

$$\therefore \int_0^{2\pi} [\sin x + \cos x] dx = \int_0^{\pi/2} (1)\, dx + \int_{\pi/2}^{3\pi/4} (0)\, dx + \int_{3\pi/4}^{\pi} (-1)\, dx$$

$$+ \int_{\pi}^{3\pi/2} (-2)\, dx + \int_{3\pi/2}^{7\pi/4} (-1)\, dx + \int_{7\pi/4}^{2\pi} (0)\, dx$$

$$= \left(\frac{\pi}{2}\right) + (0) - \left(\pi - \frac{3\pi}{4}\right) - 2\left(\frac{3\pi}{2} - \pi\right) - 1\left(\frac{7\pi}{4} - \frac{3\pi}{2}\right) + 0$$

$$= -\pi$$

Since, $\sin x + \cos x$ has the period 2π.

So, $I = \int_0^{2n\pi} [\sin x + \cos x]\, dx = n \int_0^{2\pi} [\sin x + \cos x]\, dx = -n\pi$

Hence, (a) is the correct answer.

Illustration 58 Let $g(x)$ be a continuous and differentiable function such that $\int_0^2 \left\{ \int_{\sqrt{2}}^{\sqrt{5/2}} [2x^2 - 3]\, dx \right\} \cdot g(x)\, dx = 0$, then $g(x) = 0$ when $x \in (0, 2)$ has (where [.] denotes the greatest integer function)

(a) exactly one real root (b) at least one real root

(c) no real root (d) None of these

Solution. As, $1 < 2x^2 - 3 < 2,\ \forall\, x \in (\sqrt{2}, \sqrt{5/2})$

$\Rightarrow \quad \int_{\sqrt{2}}^{\sqrt{5/2}} [2x^2 - 3]\, dx > 0,\ \forall\, x \in (0, 2)$

$\Rightarrow \quad g(x) = 0$ should have at least one root in (0, 2). $[\because\ g'(x) \ne 0]$

Hence, (b) is the correct answer.

Illustration 59 The value of x satisfying $\int_0^{2[x+14]} \left\{\frac{x}{2}\right\} dx = \int_0^{\{x\}} [x + 14]\, dx$, is equal to (where [.] and {.} denotes the greatest integer and fractional part of x)

(a) $[-14, -13)$ (b) $(0, 1)$

(c) $(-15, -14]$ (d) None of these

Solution. $\int_0^{2[x+14]} \left\{\frac{x}{2}\right\} dx = \int_0^{\{x\}} [x+14]\, dx$

$\Rightarrow \int_0^{28+2[x]} \left\{\frac{x}{2}\right\} dx = \int_0^{\{x\}} (14+[x])\, dx$

$\Rightarrow \int_0^{28} \left\{\frac{x}{2}\right\} dx + \int_{28}^{28+2[x]} \left\{\frac{x}{2}\right\} dx = (14+[x])\{x\}$

$\Rightarrow 14\int_0^{2} \left\{\frac{x}{2}\right\} dx + \int_0^{2[x]} \left\{\frac{x}{2}\right\} dx = (14+[x])\{x\}$

$$\left[\begin{array}{l}\text{using } \int_0^{nT} f(x)\,dx = n\int_0^{T} f(x)\,dx \\ \text{and } \int_a^{a+nT} f(x)\,dx = \int_0^{nT} f(x)\,dx; \text{ where } T \text{ is period of } f(x)\end{array}\right]$$

$\Rightarrow \quad 14 + [x] = (14+[x])\{x\}$

$\Rightarrow \quad (14+[x])(1-\{x\}) = 0$

$\Rightarrow \quad [x] = -14$

$\Rightarrow \quad x \in [-14, -13)$

Hence, (a) is the correct answer.

Target Exercise 2.2

1. The value of $\int_0^{\pi/4} \log(1+\tan\theta)\, d\theta$, is equal to

(a) $\frac{\pi}{2}\log 2$ (b) $-\frac{\pi}{4}\log 2$ (c) $\frac{\pi}{8}\log 2$ (d) None of these

2. Let $f: R \to R$ and $g: R \to R$ be continuous functions, then the value of $\int_{-\pi/2}^{\pi/2} (f(x)+f(-x))(g(x)-g(-x))\, dx$, is equal to

(a) –1 (b) 0 (c) 1 (d) None of these

3. For any integer n, the value of $\int_0^{\pi} e^{\cos^2 x} \cdot \cos^3 (2n+1)x \cdot dx$ is equal to

(a) 0 (b) 1 (c) –1 (d) None of these

4. The value of $\int_2^3 \frac{\sqrt{x}}{\sqrt{5-x}+\sqrt{x}}\, dx$, is equal to

(a) 1/2 (b) 1/3 (c) 1/4 (d) None of these

5. The value of $\int_0^2 \frac{dx}{(17+8x-4x^2)(e^{6(1-x)}+1)}$, is equal to

(a) $-\frac{1}{8\sqrt{21}}\log\left|\frac{2-\sqrt{21}}{2+\sqrt{21}}\right|$

(b) $-\frac{1}{8\sqrt{21}}\log\left|\frac{2+\sqrt{21}}{\sqrt{21}-2}\right|$

(c) $-\frac{1}{8\sqrt{21}}\left\{\log\left|\frac{2-\sqrt{21}}{2+\sqrt{21}}\right| - \log\left|\frac{2+\sqrt{21}}{\sqrt{21}-2}\right|\right\}$

(d) None of the above

6. The value of $\int_{-1}^{1} x\,|x|\, dx$, is equal to

(a) 1 (b) 1/2 (c) 0 (d) None of these

7. The value of $\int_{-1}^{1}\left(\frac{x^2+\sin x}{1+x^2}\right)dx$, is equal to

(a) $2-\pi$ (b) $\pi - 2$ (c) $2-\pi/2$ (d) None of these

8. If f is an odd function, then the value of $\int_{-a}^{a}\frac{f(\sin x)}{f(\cos x)+f(\sin^2 x)}dx$, is equal to

(a) 0 (b) $f(\cos x)+f(\sin x)$
(c) 1 (d) None of these

9. The value of $\int_{-1}^{10}\operatorname{sgn}(x-[x])\,dx$, is equal to (where [.] denotes the greatest integer function)

(a) 9 (b) 10 (c) 11 (d) None of these

10. If $n \in N$, then the value of $\int_{-n}^{n}(-1)^{[x]}\,dx$ (where [.] denotes the greatest integer function), is equal to

(a) 0 (b) 1 (c) $2n$ (d) n

11. The value of $\int_{0}^{[x]}(x-[x])\,dx$, (where [.] denotes the greatest integer function) is equal to

(a) $[x]$ (b) $\frac{[x]}{2}$ (c) $x\,[x]$ (d) None of these

12. The value of $\int_{-\pi/4}^{n\pi-\pi/4}|\sin x+\cos x|\,dx$, is equal to

(a) $2\sqrt{2}n$ (b) $\sqrt{2}n$
(c) $\frac{1}{2\sqrt{2}}n$ (d) None of these

Property XIV. Leibnitz's Rule for the Differentiation Under the Integral Sign

If the function $\phi(x)$ and $\psi(x)$ are defined on $[a, b]$ and are differentiable at a point $x \in (a, b)$ and $f(x, t)$ is continuous, then

(i) $$\frac{d}{dx}\left[\int_{\phi(x)}^{\psi(x)} f(x,t)\,dt\right]$$

$$=\int_{\phi(x)}^{\psi(x)}\frac{\partial}{\partial x}f(x,t)\,dt+\left\{\frac{d\psi(x)}{dx}\right\}f^{(x,\,\psi(x))}-\left\{\frac{d\phi(x)}{dx}\right\}f^{(x,\,f(x))}$$

If the function $\phi(x)$ and $\psi(x)$ are defined on $[a, b]$ and differentiable at a point $x \in (a, b)$ and $f(t)$ is continuous on $[\phi(a), \phi(b)]$, then

(ii) $$\frac{d}{dx}\left(\int_{\phi(x)}^{\psi(x)} f(t)\,dt\right)=\frac{d}{dx}\{\psi(x)\}\,f(\psi(x))-\frac{d}{dx}\{\phi(x)\}\,f(\phi(x))$$

Illustration 60 Find the derivative of the following with respect to 'x'

(i) $\int_0^x \cos t\,dt$ (ii) $\int_0^{x^2}\cos t^2\,dt$

Solution. (i) Let $f(x) = \int_0^x \cos t\, dt$

$$\therefore \quad \frac{d}{dx}(f(x)) = \cos(x)\left\{\frac{d}{dx}(x)\right\} - \cos 0 \cdot \left\{\frac{d}{dx}(0)\right\} = \cos x$$

$$\Rightarrow \quad \frac{d}{dx}\{\int_0^x \cos t\, dt\} = \cos x$$

(ii) Let $\quad f(x) = \int_0^{x^2} \cos t^2\, dt$

$$\frac{d}{dx}(f(x)) = \cos(x^2)^2 \cdot \left\{\frac{d}{dx}(x^2)\right\} - \cos(0)^2\left\{\frac{d}{dx}(0)\right\} = 2x \cdot \cos x^4$$

$$\therefore \quad \frac{d}{dx}\left(\int_0^{x^2} (\cos t^2)\, dt\right) = 2x \cos x^4$$

Illustration 61 Evaluate $\frac{d}{dx}\left(\int_{1/x}^{\sqrt{x}} \cos t^2\, dt\right)$

Solution. Let $f(x) = \int_{1/x}^{\sqrt{x}} \cos t^2\, dt$

$$\therefore \quad \frac{d}{dx}(f(x)) = \cos(\sqrt{x})^2 \cdot \left\{\frac{d}{dx}(\sqrt{x})\right\} - \cos\left(\frac{1}{x}\right)^2 \left\{\frac{d}{dx}\left(\frac{1}{x}\right)\right\}$$

$$= \frac{1}{2\sqrt{x}} \cos x + \frac{1}{x^2} \cdot \cos\left(\frac{1}{x^2}\right)$$

$$\Rightarrow \frac{d}{dx}\left(\int_{1/x}^{\sqrt{x}} \cos t^2\, dt\right) = \frac{1}{2\sqrt{x}} \cos x + \frac{1}{x^2} \cos\left(\frac{1}{x^2}\right)$$

Illustration 62 If $\frac{d}{dx}\left(\int_0^y e^{-t^2}\, dt + \int_0^{x^2} \sin^2 t\, dt\right) = 0$, find $\frac{dy}{dx}$

Solution. We know, $\frac{d}{dx}\left(\int_0^y e^{-t^2}\, dt + \int_0^{x^2} \sin^2 t\, dt\right) = 0$

$$\Rightarrow \quad e^{-y^2} \cdot \left\{\frac{d}{dx}(y)\right\} - e^{-0}\left\{\frac{d}{dx}(0)\right\} + \sin^2(x^2)\left\{\frac{d}{dx}(x^2)\right\} - \sin^2 0\left\{\frac{d}{dx}(0)\right\} = 0$$

$$\Rightarrow \quad e^{-y^2}\frac{dy}{dx} + 2x \sin^2 x^2 = 0 \;\Rightarrow\; \frac{dy}{dx} = -2x\, e^{y^2} \sin^2 x^2$$

Illustration 63 Find the points of maxima/minima of $\int_0^{x^2} \frac{t^2 - 5t + 4}{2 + e^t}\, dt$.

Solution. Let $f(x) = \int_0^{x^2} \frac{t^2 - 5t + 4}{2 + e^t} \cdot dt$

$$f'(x) = \frac{x^4 - 5x^2 + 4}{2 + e^{x^2}} \cdot 2x - 0 = \frac{(x-1)(x+1)(x-2)(x+2) \cdot 2x}{2 + e^{x^2}}$$

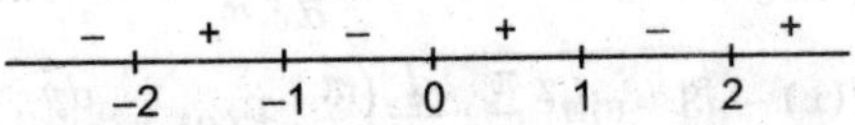

From the wavy curve, it is clear that $f'(x)$ changes its sign at $x = \pm 2, \pm 1, 0$ and hence the points of maxima are – 1, 1 (as sign changes from + ve to – ve) and of the minima are –2, 0, 2 (as sign changes from – ve to + ve).

Illustration 64 Find $\frac{d}{dx}\left(\int_{x^2}^{x^3} \frac{1}{\log t}\,dt\right)$

Solution. $$\frac{d}{dx}\left(\int_{x^2}^{x^3} \frac{1}{\log t}\,dt\right) = \frac{1}{\log x^3}\cdot\frac{d}{dx}(x^3) - \frac{1}{\log x^2}\cdot\frac{d}{dx}(x^2)$$

$$= \frac{3x^2}{3\log x} - \frac{2x}{2\log x}$$

$$\therefore \quad \frac{d}{dx}\left(\int_{x^2}^{x^3} \frac{1}{\log t}\,dt\right) = \frac{1}{\log x}(x^2 - x)$$

Illustration 65 If $y = \int_0^x f(t)\sin\{K(x-t)\}\,dt$, then prove that

$$\frac{d^2y}{dx^2} + K^2 y = K f(x).$$

Solution. $$y = \int_0^x f(t)\sin\{K(x-t)\}\,dt$$

Differentiating w.r.t. 'x', we get

$$\frac{dy}{dx} = \int_0^x \frac{\partial}{\partial x}\{f(t)\sin K(x-t)\}\,dt + \frac{d}{dx}(x)\cdot\{f(x)\sin K(x-x)\}$$
$$- \frac{d}{dx}(0)\{f(0)\sin K(x-0)\}$$

$$= K\int_0^x f(t)\cos K(x-t)\,dt + 0 - 0$$

$$\Rightarrow \quad \frac{dy}{dx} = K\int_0^x f(t)\cos K(x-t)\,dt$$

Again, differentiating both the sides w.r.t. x, we get

$$\frac{d^2y}{dx^2} = K\left\{\int_0^x \frac{\partial}{\partial x}\{f(t)\cos K(x-t)\}\,dt + \left\{f(x)\cdot\cos K(x-x)\cdot\frac{d}{dx}(x)\right\}\right.$$
$$\left. - \{f(0)\cos K(x-0)\cdot\frac{d}{dx}(0)\}\right\}$$

$$= K\left[-K\int_0^x f(t)\sin K(x-t)\,dt + f(x) - 0\right]$$

$$= -K^2\int_0^x f(t)\sin K(x-t)\,dt + Kf(x)$$

$$\frac{d^2y}{dx^2} = -K^2y + Kf(x) \Rightarrow \frac{d^2y}{dx^2} + K^2y = Kf(x)$$

Illustration 66 If $\int_{\pi/3}^{x}\sqrt{3-\sin^2 t}\,dt + \int_0^y \cos t\,dt = 0$, then evaluate $\frac{dy}{dx}$.

Solution. Differentiating w.r.t. 'x', we have

$$\frac{d}{dx}\left(\int_{\pi/3}^{x}\sqrt{3-\sin^2 t}\,dt\right) + \frac{d}{dx}\int_0^y (\cos t)\,dt = 0$$

$$\Rightarrow \sqrt{3-\sin^2 x}\cdot\frac{d}{dx}(x) - \sqrt{3-\sin^2\frac{\pi}{3}}\cdot\frac{d}{dx}\left(\frac{\pi}{3}\right) + \cos y\cdot\frac{dy}{dx} - \cos(0)\cdot\frac{d}{dx}(0) = 0$$

$$\Rightarrow \quad \sqrt{3-\sin^2 x} + \cos y\,\frac{dy}{dx} = 0 \Rightarrow \frac{dy}{dx} = -\frac{\sqrt{3-\sin^2 x}}{\cos y}$$

Illustration 67 Let $\frac{d}{dx}(F(x)) = \frac{e^{\sin x}}{x}, x > 0.$

If $\int_1^4 \frac{2e^{\sin x^2}}{x} dx = F(K) - F(1)$, then the possible value of K is

(a) 10 (b) 14 (c) 16 (d) 18

Solution. We have, $\frac{d}{dx}(F(x)) = \frac{e^{\sin x}}{x}$

$$\Rightarrow \quad \int \frac{e^{\sin x}}{x} dx = F(x) \qquad \text{...(i)}$$

Now, $\int_1^4 \frac{2e^{\sin x^2}}{x} dx = \int_1^4 \frac{e^{\sin (x^2)}}{x^2} d(x^2) = \int_1^{16} \frac{e^{\sin t}}{t} dt \qquad (\because \text{ where } x^2 = t)$

$$= (F(t))_1^{16} = F(16) - F(1)$$

$$\Rightarrow \quad K = 16$$

Hence, (c) is the correct answer.

Illustration 68 The function $f(x) = \int_0^{x \in (0, 2\pi)} \log_{|\sin t|}\left(\sin t + \frac{1}{2}\right) dt,$ strictly increases in the interval

(a) $\left(\frac{\pi}{6}, \frac{5\pi}{6}\right)$ (b) $\left(\frac{5\pi}{6}, 2\pi\right)$ (c) $\left(\frac{\pi}{6}, \frac{7\pi}{6}\right)$ (d) $\left(\frac{5\pi}{6}, \frac{7\pi}{6}\right)$

Solution. Here, $f'(x) = \log_{|\sin x|}\left(\sin x + \frac{1}{2}\right) > 0$

$$\Rightarrow \quad \sin x + \frac{1}{2} < 1 \text{ and } \left(\sin x + \frac{1}{2}\right) > 0$$

$$\Rightarrow \quad 0 < \sin x + \frac{1}{2} < 1$$

$$\Rightarrow \quad -1/2 < \sin x < 1/2$$

$$\Rightarrow \quad x \in \left(\frac{5\pi}{6}, \frac{7\pi}{6}\right) \text{ as } x \in (0, 2\pi)$$

Hence, (d) is the correct answer.

Illustration 69 Let $f : (0, \infty) \to R$ and $F(x) = \int_0^x t\, f(t)\, dt.$

If $F(x^2) = x^4 + x^5$, then $\sum_{r=1}^{12} f(r^2)$ is equal to

(a) 216 (b) 219 (c) 221 (d) 223

Solution. Here, $F(x^2) = x^4 + x^5 = \int_0^{x^2} t\, f(t)\, dt$. Differentiating w.r.t. 'x'

$$x^2 f(x^2) \cdot 2x = 4x^3 + 5x^4$$

$$\Rightarrow \quad f(x^2) = 2 + \frac{5}{2} x$$

$$f(r^2) = 2 + \frac{5}{2} r$$

$$\therefore \quad \sum_{r=1}^{n} f(r^2) = 2n + \frac{5}{2} n(n+1) = \frac{n(5n+13)}{4}$$

$$\Rightarrow \qquad \sum_{r=1}^{12} f(r^2) = \frac{12\,(60+13)}{4} = 219$$

Hence, (b) is the correct answer.

Illustration 70 A function $f(x)$ satisfies

$$f(x) = \sin x + \int_0^x f'(t)(2\sin t - \sin^2 t)dt, \text{ then } f(x) \text{ is}$$

(a) $\dfrac{x}{1-\sin x}$ (b) $\dfrac{\sin x}{1-\sin x}$ (c) $\dfrac{1-\cos x}{\cos x}$ (d) $\dfrac{\tan x}{1-\sin x}$

Solution. Differentiating both the sides w.r.t. x, we get

$$f'(x) = \cos x + f'(x)(2\sin x - \sin^2 x)$$

$$\therefore \qquad (1+\sin^2 x - 2\sin x) f'(x) = \cos x$$

$$f'(x) = \frac{\cos x}{1+\sin^2 x - 2\sin x} = \frac{\cos x}{(1-\sin x)^2}$$

Integrating, $\quad f(x) = \int \dfrac{\cos x}{(1-\sin x)^2} dx$

Put $1-\sin x = t$, $\quad f(x) = -\int \dfrac{dt}{t^2} = \dfrac{1}{t} = \dfrac{1}{1-\sin x} + C$

Also, $f(0) = 0$, hence $C = -1$

$$f(x) = \frac{1}{1-\sin x} - 1 = \frac{1-1+\sin x}{1-\sin x} = \frac{\sin x}{1-\sin x}$$

Hence, (b) is the correct answer.

Illustration 71 If $F(x) = \int_1^x f(t)dt$, where $f(t) = \int_1^{t^2} \dfrac{\sqrt{1+u^4}}{u} du$, then the value of $F''(2)$ equals to

(a) $\dfrac{7}{4\sqrt{17}}$ (b) $\dfrac{15}{\sqrt{17}}$ (c) $\sqrt{257}$ (d) $\dfrac{15\sqrt{17}}{68}$

Solution. $\quad f'(t) = \dfrac{\sqrt{1+t^8}\cdot 2t}{t^2} = \dfrac{2\sqrt{1+t^8}}{t}$ ...(i)

Now, $\quad F(x) = \int_1^x f(t)dt \Rightarrow F'(x) = f(x)$

$$F''(x) = f'(x) \Rightarrow F''(2) = f'(2)$$

From Eq. (i), $\quad f'(2) = \sqrt{256+1} = \sqrt{257}$

Hence, (c) is the correct answer.

Property XV. Let a function $f(x, \alpha)$ be continuous for $a \le x \le b$ and $c \le \alpha \le d$, then for any $\alpha \in [c, d]$, if $I(\alpha) = \int_a^b f(x, \alpha)\, dx$, then

$$\frac{dI(\alpha)}{d\alpha} = \int_a^b \frac{\partial f(x, \alpha)}{\partial \alpha} dx$$

Illustration 72 Evaluate $I(b) = \int_0^1 \dfrac{x^b - 1}{\ln x} dx, (b \ge 0)$.

Solution. $\quad I(b) = \int_0^1 \dfrac{x^b - 1}{\ln x} dx$

$$\Rightarrow \quad \frac{d}{db}(I(b)) = \int_0^1 \frac{\partial}{\partial b}\left(\frac{x^b - 1}{\ln x}\right) dx + 0 - 0 = \int_0^1 \frac{x^b \ln x}{\ln x} dx$$

$$= \int_0^1 (x^b)\, dx = \left(\frac{x^{b+1}}{b+1}\right)_0^1 = \frac{1}{b+1}[1-0] = \frac{1}{b+1}$$

$$\therefore \quad \frac{d}{dx}(I(b)) = \frac{1}{b+1}$$

Integrating both the sides w.r.t. 'b', we get

$$I(b) = \log(b+1) + C \qquad \text{...(i)}$$

Given, $I(b) = \int_0^1 \frac{x^b - 1}{\ln x} dx$

$\therefore$ $I(0) = 0$ (when $b = 0$) ...(ii)

Also, from Eq. (i), $I(0) = \log(1) + C$

$\therefore$ $I(0) = C$ ...(iii)

From Eqs. (ii) and (iii), $C = 0$

$\Rightarrow$ $I(b) = \log(b+1)$

Illustration 73 Prove that

$$\int_0^{\pi/2} \frac{dx}{(a^2 \sin^2 x + b^2 \cos^2 x)^2} = \frac{\pi(a^2 + b^2)}{4a^3 b^3}.$$

Solution. Let $I = \int_0^{\pi/2} \frac{1\, dx}{a^2 \sin^2 x + b^2 \cos^2 x}$.

Then, $I = \int_0^{\pi/2} \frac{\sec^2 x}{a^2 \tan^2 x + b^2} dx = \int_0^{\infty} \frac{dt}{a^2 t^2 + b^2}$ (where $t = \tan x$)

$$\Rightarrow \quad I = \frac{1}{ab}\left[\tan^{-1}\left(\frac{at}{b}\right)\right]_0^{\infty} = \frac{1}{ab}\left(\frac{\pi}{2} - 0\right) = \frac{\pi}{2ab}$$

Thus, $\int_0^{\pi/2} \frac{1}{a^2 \sin^2 x + b^2 \cos^2 x} dx = \frac{\pi}{2ab} = f(a, b)$ ...(i)

Differentiating both the sides w.r.t 'a', we get

$$\int_0^{\pi/2} \frac{-2a \sin^2 x}{(a^2 \sin^2 x + b^2 \cos^2 x)^2} dx = -\frac{\pi}{2a^2 b}$$

$$\Rightarrow \quad \int_0^{\pi/2} \frac{\sin^2 x}{(a^2 \sin^2 x + b^2 \cos^2 x)^2} dx = \frac{\pi}{4a^3 b} \qquad \text{...(ii)}$$

Similarly, by differentiating Eq. (i) w.r.t a, we get

$$\int_0^{\pi/2} \frac{\cos^2 x}{(a^2 \sin^2 x + b^2 \cos^2 x)^2} dx = \frac{\pi}{4ab^3} \qquad \text{...(iii)}$$

Adding, Eqs. (ii) and (iii), we get

$$\int_0^{\pi/2} \frac{(\sin^2 x + \cos^2 x)\, dx}{(a^2 \sin^2 x + b^2 \cos^2 x)^2} = \frac{\pi}{4a^3 b} + \frac{\pi}{4ab^3}$$

$$\Rightarrow \quad \int_0^{\pi/2} \frac{dx}{(a^2 \sin^2 x + b^2 \cos^2 x)^2} = \frac{\pi}{4a^3 b^3}(a^2 + b^2)$$

Illustration 74 The value of $\int_0^{\pi/2} \frac{\log(1 + x\sin^2\theta)}{\sin^2\theta} d\theta; x \geq 0$ is equal to

(a) $\pi(\sqrt{1+x} - 1)$ (b) $\pi(\sqrt{1+x} - 2)$ (c) $\sqrt{\pi}(\sqrt{1+x} - 1)$ (d) None of these

Solution. Given, $f(x) = \int_0^{\pi/2} \frac{\log(1 + x\sin^2\theta)}{\sin^2\theta} d\theta, x \geq 0$

As above integral is function of x. Thus, differentiating both the sides w.r.t. 'x', we get

$$f'(x) = \int_0^{\pi/2} \frac{1}{1 + x\sin^2\theta} \cdot \frac{\sin^2\theta}{\sin^2\theta} d\theta + 0 - 0$$

(using Newton Leibnitz formula)

$$= \int_0^{\pi/2} \frac{d\theta}{1 + x\sin^2\theta} = \int_0^{\pi/2} \frac{\operatorname{cosec}^2\theta\, d\theta}{\cot^2\theta + (1 + x)}$$

$$= \left(-\frac{1}{\sqrt{1+x}} \cdot \tan^{-1}\frac{\cot\theta}{\sqrt{1+x}}\right)_0^{\pi/2} = \frac{\pi}{2\sqrt{1+x}}$$

$\therefore$ $$f'(x) = \frac{\pi}{2\sqrt{1+x}} \Rightarrow f(x) = \pi\sqrt{1+x} + C$$

Where $f(0) = 0 \quad \therefore \quad C = -\pi$

$\Rightarrow$ $$f(x) = \pi\sqrt{1+x} - \pi = \pi(\sqrt{1+x} - 1)$$

Hence, (a) is the correct answer.

Illustration 75 Let $f(x)$ be a continuous functions for all x, such that

$$(f(x))^2 = \int_0^x f(t) \cdot \frac{2\sec^2 t}{4 + \tan t} dt \text{ and } f(0) = 0, \text{ then}$$

(a) $f\left(\frac{\pi}{4}\right) = \log\frac{5}{4}$ (b) $f\left(\frac{\pi}{4}\right) = \frac{3}{4}$ (c) $f\left(\frac{\pi}{2}\right) = 2$ (d) None of these

Solution. Here, $(f(x))^2 = \int_0^x f(t) \cdot \frac{2\sec^2 t}{4 + \tan t} dt,$

On differentiating both the sides w.r.t. x, we get

$\Rightarrow$ $$2f(x) \cdot f'(x) = f(x) \cdot \frac{2\sec^2 x}{4 + \tan x}$$

$\Rightarrow$ $$f'(x) = \frac{\sec^2 x}{4 + \tan x}$$

Integrating both the sides, we get

$$f(x) = \int \frac{\sec^2 x}{4 + \tan x} dx = \log(4 + \tan x) + C$$

Since, $f(0) = 0 \Rightarrow 0 = \log(4) + C$

$\Rightarrow$ $C = -\log 4$

$\therefore$ $f(x) = \log(4 + \tan x) - \log(4)$

$\Rightarrow$ $f(\pi/4) = \log(4 + 1) - \log(4) = \log\frac{5}{4}$

Hence, (a) is the correct answer.

Property XVI. If $f(x)$ is periodic function with period T, then $\int_a^{a+T} f(x)\,dx$ is independent of a.

Proof Let $$\phi(a) = \int_a^{a+T} f(x)\,dx$$

Then, $$\frac{d}{da}(\phi(a)) = \frac{d(a+T)}{da} f(a+T) - \frac{da}{da} f(a)$$

$$= f(a+T) - f(a) = 0$$

$[\because f$ is periodic with period $T \therefore f(a+T) = f(a)]$

Thus, $\frac{d(\phi(a))}{da} = 0 \Rightarrow \phi(a)$ is independent of a.

Illustration 76 If $f(x)$ is a function satisfying $f(x+a)+f(x)=0$ for all $x \in R$ and constant a such that $\int_b^{c+b} f(x)\,dx$ is independent of b, then find the least positive value of c.

Solution. We have, $f(x+a)+f(x)=0$, for all $x \in R$. ...(i)

$\Rightarrow$ $f(x+a+a)+f(x+a)=0$ (replacing x by $x+a$)

$\Rightarrow$ $f(x+2a)+f(x+a)=0$ ...(ii)

Subtracting Eq. (i) from Eq. (ii), we get

$f(x+2a)-f(x)=0$, for all $x \in R$

$\Rightarrow$ $f(x+2a)=f(x)$, for all $x \in R$

So, $f(x)$ is periodic with period $2a$.

It is given that $\int_b^{c+b} f(x)\,dx$ is independent of b.

$\therefore$ The minimum value of 'c' is equal to the period of $f(x)$, *ie*, $2a$.

Property XVII. If $\int_a^b f(x)\,dx = (b-a)\int_0^1 f((b-a)x+a)\,dx$

Illustration 77 Evaluate $\int_{-4}^{-5} e^{(x+5)^2}\,dx + 3\int_{1/3}^{2/3} e^{9\left(x-\frac{2}{3}\right)^2}\,dx$.

Solution. Here, we know $\int e^{x^2}\,dx$ cannot be evaluated by indefinite integral. Thus,

Let $$I_1 = \int_{-4}^{-5} e^{(x+5)^2}\,dx = (-5+4)\int_0^1 e^{((-5+4)x-4+5)^2}\,dx$$

$\therefore$ $$I_1 = -\int_0^1 e^{(x-1)^2}\,dx \quad \text{...(i)}$$

Again, let $$I_2 = \int_{1/3}^{2/3} e^{9(x-2/3)^2}\,.dx$$

$$= \left(\frac{2}{3}-\frac{1}{3}\right)\int_0^1 e^{9\left[\left(\frac{2}{3}-\frac{1}{2}\right)x+\frac{1}{3}-\frac{2}{3}\right]^2}\,dx$$

$$= \frac{1}{3}\int_0^1 e^{(x-1)^2}\,dx = \frac{1}{3}(-I_1) \quad \text{...(ii)}$$

Where, $I = I_1 + 3I_2 = I_1 + 3\left(-\frac{I_1}{3}\right) = I_1 - I_1$

$I = 0$

$\therefore \quad \int_{-4}^{-5} e^{(x+5)^2}\, dx + 3\int_{1/3}^{2/3} e^{9(x-2/3)^2}\, dx = 0$

Points to Consider

Sometimes you are asked to prove inequalities involving definite integrals or to estimate the upper and lower boundary values of definite integral, where the exact value of the definite integral is difficult to find. Under these circumstances, we use the following properties (*ie*, from property XVIII to property XXII).

Property XVIII. If $f(x)$ is defined on $[a, b]$, then $\left|\int_a^b f(x)\, dx\right| \le \int_a^b |f(x)|\, dx$

Equality sign holds where $f(x)$ is entirely of the same sign on $[a, b]$.

Illustration 78 Estimate the absolute value of the integral $\int_{10}^{19} \frac{\sin x}{1+x^8}\, dx$.

Solution. To find, $I = \left|\int_{10}^{19} \frac{\sin x}{1+x^8}\, dx\right| \le \int_{10}^{19} \left|\frac{\sin x}{1+x^8}\right| dx$ ...(i)

(using property XVIII)

Since, $|\sin x| \le 1$ for $x \ge 10$

The inequality $\left|\frac{\sin x}{1+x^8}\right| \le \frac{1}{|1+x^8|}$ ...(ii)

Also, $10 \le x \le 19 \Rightarrow 1 + x^8 > 10^8$

$\Rightarrow \quad \frac{1}{1+x^8} < \frac{1}{10^8}$ or $\frac{1}{|1+x^8|} < 10^{-8}$ ...(iii)

From Eqs. (ii) and (iii), $\left|\frac{\sin x}{1+x^8}\right| < 10^{-8}$ is fulfilled.

$$\left|\int_{10}^{19} \frac{\sin x}{1+x^8}\, dx\right| < \int_{10}^{19} 10^{-8}\, dx$$

$$\therefore \quad \left|\int_{10}^{19} \frac{\sin x}{1+x^8}\, dx\right| < (19-10) \cdot 10^{-8} < 10^{-7}$$

($\because$ The true value of integral $\approx 10^{-8}$)

Illustration 79 The minimum odd value of 'a' $(a > 1)$ for which $\int_{10}^{19} \frac{\sin x}{1+x^a}\, dx < \frac{1}{9}$, is equal to

(a) 1 (b) 3 (c) 5 (d) 9

Solution. Let $I = \int_{10}^{19} \frac{\sin x\, dx}{1+x^a} < \int_{10}^{19} \frac{dx}{1+x^a}$

$\left[\sin x < 1 \Rightarrow \frac{\sin x}{1+x^a} < \frac{1}{1+x^a}\right]$

$$\therefore \qquad I < \int_{10}^{19} \frac{dx}{1+x^a} < \int_{10}^{19} \frac{dx}{1+10^a}$$

$$(\because\ 10 < x < 19 \Rightarrow 10^a + 1 < 1 + x^a < 19^a + 1)$$

$$\Rightarrow \qquad I < \frac{9}{1+10^a}$$

$$\therefore \qquad \frac{9}{1+10^a} < \frac{1}{9} \Rightarrow 1 + 10^a > 81$$

$$\text{or} \qquad 10^a > 80$$

$$ie, \qquad a = 2, 3, 4, 5, \ldots$$

$\therefore$ Minimum odd value of 'a' is 3.

Hence, (b) is the correct answer.

Property XIX. $\left| \int_a^b f(x)\, g(x)\, dx \right| \le \sqrt{\left(\int_a^b f^2(x)\, dx \right)\left(\int_a^b g^2(x)\, dx \right)}$,

where $f^2(x)$ and $g^2(x)$ are integrable on $[a, b]$.

Illustration 80 Prove that $\int_0^1 \sqrt{(1+x)(1+x^3)}\, dx$ cannot exceed $\sqrt{15/8}$.

Solution. $\int_0^1 \sqrt{(1+x)(1+x^3)}\, dx \le \sqrt{\left(\int_0^1 (1+x)\, dx \right)\left(\int_0^1 (1+x^3)\, dx \right)}$

$$\le \sqrt{\left(x + \frac{x^2}{2} \right)_0^1 \left(x + \frac{x^4}{4} \right)_0^1} \le \sqrt{\frac{3}{2} \cdot \frac{5}{4}} \le \sqrt{\frac{15}{8}}$$

Property XX. If $f(x) | g(x)$ on $[a, b]$, then $\int_a^b f(x)\, dx \ge \int_a^b g(x)\, dx$. In particular, if $f(x) \ge 0$, then $\int_a^b f(x)\, dx | 0$.

Illustration 81 If $f(x)$ is a continuous function such that

$f(x) | 0, \forall x \in [2, 10]$ and $\int_4^8 f(x)\, dx = 0$, then find $f(6)$.

Solution. $f(x)$ is above the x-axis or on the x-axis for all $x \in [2, 10]$.

If $f(x)$ is greater than zero at any sub-interval of [4, 8],then $\int_4^8 f(x)\, dx$ must be greater than zero.

But $\int_4^8 f(x)\, dx = 0$, which shows $f(x)$ can't have any value greater than zero in the sub-interval [4, 8].

$\Rightarrow f(x)$ is constant in the sub-interval [4, 8] and has to be zero at all points, $x \in [4, 8]$.

$$\Rightarrow \qquad f(x) = 0, \quad \forall x \in [4, 8] \Rightarrow f(6) = 0$$

Property XXI. For a given function $f(x)$ continuous on $[a, b]$, if you are able to find two continuous functions $f_1(x)$ and $f_2(x)$ on $[a, b]$ such that $f_1(x) \le f(x) \le f_2(x), \forall x \in [a, b]$, then

$$\int_a^b f_1(x)\, dx \le \int_a^b f(x)\, dx \le \int_a^b f_2(x)\, dx$$

Illustration 82 Prove that $\dfrac{\pi}{6} \le \displaystyle\int_0^1 \dfrac{dx}{\sqrt{4 - x^2 - x^3}} \le \dfrac{\pi}{4\sqrt{2}}$.

Solution. Since, $4 - x^2 \mid 4 - x^2 - x^3 \mid 4 - 2x^2 > 1, \forall\, x \in [0, 1]$

$$\sqrt{4 - x^2} \mid \sqrt{4 - x^2 - x^3} \mid \sqrt{4 - 2x^2} > 1, \forall\, x \in [0, 1]$$

$$\Rightarrow \quad \frac{1}{\sqrt{4 - x^2}} \le \frac{1}{\sqrt{4 - x^2 - x^3}} \le \frac{1}{\sqrt{4 - 2x^2}}, \forall\, x \in [0, 1]$$

$$\Rightarrow \quad \int_0^1 \frac{dx}{\sqrt{4 - x^2}} \le \int_0^1 \frac{dx}{\sqrt{4 - x^2 - x^3}} \le \int_0^1 \frac{1}{\sqrt{4 - 2x^2}}\, dx, \forall\, x \in [0, 1]$$

$$\Rightarrow \quad \left(\sin^{-1} \frac{x}{2} \right)_0^1 \le \int_0^1 \frac{dx}{\sqrt{4 - x^2 - x^3}} \le \frac{1}{\sqrt{2}} \left(\sin^{-1} \frac{x}{\sqrt{2}} \right)_0^1$$

$$\Rightarrow \quad \frac{\pi}{6} \le \int_0^1 \frac{dx}{\sqrt{4 - x^2 - x^3}} \le \frac{\pi}{4\sqrt{2}}$$

Property XXII. If m and M be global minimum and global maximum of $f(x)$ respectively in $[a, b]$, then $m\,(b - a) \le \displaystyle\int_a^b f(x)\, dx \le M(b - a)$.

Proof We have, $m \le f(x) \le M$ for all $x \in [a, b]$

$$\Rightarrow \quad \int_a^b m\, dx \le \int_a^b f(x)\, dx \le \int_a^b M\, dx$$

$$\Rightarrow \quad m\,(b - a) \le \int_a^b f(x)\, dx \le M\,(b - a)$$

Illustration 83 Prove that $4 \le \displaystyle\int_1^3 \sqrt{3 + x^3}\, dx \le 2\sqrt{30}$.

Solution. Since, the function $f(x) = \sqrt{3 + x^3}$ increases monotonically on the interval [1, 3].

$\therefore \quad M = \text{maximum value of } \sqrt{3 + x^3} = \sqrt{3 + 3^3} = \sqrt{30}$

$\quad\;\; m = \text{minimum value of } \sqrt{3 + x^3} = \sqrt{3 + 1^3} = 2$

$\therefore \quad m = 2, M = \sqrt{30}, b - a = 2$

Hence, $\quad 2 \cdot 2 \le \displaystyle\int_1^3 \sqrt{3 + x^3}\, dx \le 2\sqrt{30}$

$$\Rightarrow \quad 4 \le \int_1^3 \sqrt{3 + x^3}\, dx \le 2\sqrt{30}$$

Illustration 84 Prove that $1 \le \displaystyle\int_0^1 e^{x^2}\, dx \le e$.

Solution. For $0 \le x \le 1$, we have $e^0 \le e^{x^2} \le e^1$

$$\therefore \quad e^0\,(1 - 0) \le \int_0^1 e^{x^2}\, dx \le e^1\,(1 - 0)$$

$$\Rightarrow \quad 1 \le \int_0^1 e^{x^2} dx \le e$$

Property XXIII. If $f(t)$ is an odd function, then $\phi(x) = \int_a^x f(t)\, dt$ is an even function.

Proof We have, $\phi(-x) = \int_a^{-x} f(t)\, dt$

$\Rightarrow \quad \phi(-x) = \int_a^{-a} f(t)\, dt + \int_{-a}^{-x} f(t)\, dt$

$\Rightarrow \quad \phi(-x) = 0 + \int_{-a}^{-x} f(t)\, dt$

$[\because f(t) \text{ is odd function} \Rightarrow \int_{-a}^{a} f(t)\, dt = 0]$

$\Rightarrow \quad \phi(-x) = -\int_a^x f(-y)\, dy$, where $t = -y$

$\Rightarrow \quad \phi(-x) = +\int_a^x f(y)\, dy$

$[\because f \text{ is an odd function} \therefore f(-y) = -f(y)]$

$\Rightarrow \quad \phi(-x) = \int_a^x f(t)\, dt \Rightarrow \phi(-x) = \phi(x)$

Hence, $\phi(x) = \int_a^x f(t)\, dt$ is even function, if $f(t)$ is odd.

Illustration 85 If $f(x) = \int_0^x \log\left(\frac{1-t}{1+t}\right) dt$, then discuss whether even or odd?

Solution. Let $f(t) = \log\left(\frac{1-t}{1+t}\right)$

$\therefore \quad f(-t) = \log\left(\frac{1+t}{1-t}\right) = -\log\left(\frac{1-t}{1+t}\right) = -f(t)$

$\Rightarrow \quad f(-t) = -f(t)$, *ie*, $f(t)$ is odd function

$\therefore \quad \phi(x) = \int_0^x \log\left(\frac{1-t}{1+t}\right) dt$ is an even function.

Property XXIV. If $f(t)$ is an even function, then $\phi(x) = \int_0^x f(t)\, dt$ is an odd function.

Proof We have, $\phi(-x) = \int_0^{-x} f(t)\, dt = -\int_0^x f(-y)\, dy$, where $t = -y$

$\Rightarrow \quad \phi(-x) = -\int_0^x f(y)\, dy$

$[\because f \text{ is even function} \Rightarrow f(-y) = f(y)]$

$\Rightarrow \quad \phi(-x) = -\int_0^x f(t)\, dt \Rightarrow \phi(-x) = -\phi(x)$

Hence, $\phi(x)$ is an odd function.

Points to Consider

If $f(t)$ is an even function, then for non-zero 'a', $\int_a^x f(t)\, dt$ is not necessarily an odd function. It will be an odd function, if $\int_0^a f(t)\, dt = 0$, because, if $\phi(x) = \int_a^x f(t)\, dt$, it is an odd function.

$$\Rightarrow \quad \phi(-x) = -\phi(x)$$

$$\Rightarrow \quad \int_a^{-x} f(t)\,dt = -\int_a^x f(t)\,dt$$

$$\Rightarrow \quad \int_a^0 f(t)\,dt + \int_0^{-x} f(t)\,dt = -\int_a^0 f(t)\,dt - \int_0^x f(t)\,dt$$

$$\Rightarrow \quad \int_a^0 f(t)\,dt - \int_0^x f(-y)\,dy = -\int_0^a f(t)\,dt - \int_0^x f(t)\,dt$$

$$\left(\begin{array}{r}\text{where } y = -t \text{ in second integral of LHS}\\ \because f(-y) = f(y)\end{array}\right)$$

$$\Rightarrow \quad 2\int_a^0 f(t)\,dt = \int_0^x f(y)\,dy - \int_0^x f(t)\,dt$$

$$\Rightarrow \quad 2\int_a^0 f(t)\,dt = 0$$

$$\Rightarrow \quad -\int_0^a f(t)\,dt = 0 \Rightarrow \int_0^a f(t)\,dt = 0$$

or $\phi(x) = \int_a^x f(t)\,dt$ is odd function (when $f(t)$ is even)

Only, if $\int_0^a f(t)\,dt = 0$

Property XXV. If $f(x)$ is continuous on $[a, \infty]$, then $\int_a^\infty f(x)\,dx$ is called an improper integral and is defined as;

$$\int_a^\infty f(x)\,dx = \lim_{b\to\infty} \int_a^b f(x)\,dx \qquad \text{...(i)}$$

If there exists a finite limit on the right hand side of Eq. (i), then we say that the improper integral is convergent; otherwise it is divergent.

Points to Consider

1. Geometrically, for $f(x) > 0$ the improper integral $\int_a^\infty f(x)\,dx$ gives area of the figure bounded by the curve $y = f(x)$, the x-axis and the straight line $x = a$.
2. $\int_{-\infty}^b f(x)\,dx = \lim_{a\to-\infty} + \int_a^b f(x)\,dx$
3. $\int_{-\infty}^{+\infty} f(x)\,dx = \int_{-\infty}^0 f(x)\,dx + \int_0^{+\infty} f(x)\,dx$

Illustration 86 Prove that $\int_0^\infty \dfrac{dx}{(x+\sqrt{x^2+1})^n} = \dfrac{n}{n-1}$, $(n > 1)$.

Solution. Let $I = \int_0^\infty \dfrac{dx}{(x+\sqrt{x^2+1})^n}$, put $x + \sqrt{x^2+1} = t$

$$\Rightarrow \quad \left(1 + \frac{x}{\sqrt{x^2+1}}\right) dx = dt$$

$$\Rightarrow \quad \frac{t}{\sqrt{x^2+1}}\,dx = dt \quad \text{or} \quad dx = \sqrt{1+x^2}\cdot\frac{dt}{t}$$

Also, $\sqrt{1+x^2} - x = 1/t$ and $\sqrt{x^2+1} + x = t$

$$\Rightarrow \quad 2\sqrt{1+x^2} = t + \frac{1}{t}$$

$$\therefore \quad dx = \frac{t^2+1}{2t^2}\, dt$$

when $x \to 0, t \to 1$ and when $x \to \infty, t \to \infty$.

$$\Rightarrow \quad I = \int_1^{\infty} \frac{t^2+1}{2t^2 \cdot t^n}\, dt = \frac{1}{2}\int_1^{\infty}\left(\frac{1}{t^n} + \frac{1}{t^{n+2}}\right) dt$$

$$= \frac{1}{2} \lim_{a \to +\infty} \left[\frac{-1}{(n-1)t^{n-1}} - \frac{1}{(n+1)t^{n+1}}\right]_1^a$$

$$= \frac{1}{2}\left[-0-0-\left(\frac{-1}{n-1} - \frac{1}{n+1}\right)\right] = \frac{n}{n^2-1}$$

Points to Consider

Another type of improper integral is that in which the integrand is not defined for one or both the limits of integration. Here, we take limiting values as shown in the following Illustration.

Illustration 87 Evaluate $\int_0^1 \ln(x)\, dx$.

Solution. Note that $\ln(x)$ is not defined for $x = 0$. Hence, we can write;

$$I = \lim_{h \to 0^+} \int_h^1 \ln x\, dx$$

$$= \lim_{h \to 0^+} [x \ln x - x]_h^1 \qquad \text{(integration by parts)}$$

ie,
$$I = \lim_{h \to 0^+} [x \ln x - x]_h^1$$

$$= \{0-1\} - \lim_{h \to 0^+} \{h \ln h - h\} = -1 - \lim_{h \to 0} \{h \ln h\}$$

Now,
$$\lim_{h \to 0^+} h \ln h \qquad [0 \times (-\infty) \text{ form}]$$

$$\Rightarrow \quad \lim_{h \to 0^+} \frac{\ln h}{(1/h)} \qquad \left[-\frac{\infty}{\infty} \text{ form}\right]$$

$$\Rightarrow \quad \lim_{h \to 0^+} \frac{(1/h)}{(-1/h^2)} \qquad \text{(using L'Hospital's rule)}$$

$$\Rightarrow \quad \lim_{h \to 0^+} (-h) = 0$$

Hence, $I = -1$

Mean Value of a Function Over an Interval

Let $f(x)$ be a continuous function defined on $[a, b]$, then there exists a point $c \in (a, b)$ such that $\int_a^b f(x)\, dx = f(c)(b-a)$

$\therefore \quad f(c) = \frac{1}{b-a}\int_a^b f(x)\, dx$, is called the mean value of $f(x)$ over $[a, b]$.

Illustration 88 Find the mean value of $\left(\dfrac{1}{x+1}\right)$ over the interval [0, 2].

Solution. Here, $f(x) = \dfrac{1}{x+1}$

$$\therefore \text{ Mean value of } f(x) \text{ over } [0, 2] = \frac{1}{(2-0)} \int_0^2 \frac{1}{x+1}\, dx$$

$$= \frac{1}{2} [\log (x+1)]_0^2 = \frac{1}{2} \log 3$$

Illustration 89 Find the mean value of $f(x) = x\,[2x]\, \text{sgn}\,(x-2)$ on [1, 3], where $[x]$ denotes the greatest integer function.

Solution. Here, mean value of $f(x)$ over [1, 3]

$$= \frac{1}{(3-1)} \int_1^3 f(x)\, dx$$

$$= \frac{1}{2} \int_1^3 x\,[2x]\, \text{sgn}\,(x-2)\, dx$$

$$= \frac{1}{2} \left\{ \int_1^{1.5} x\,[2x]\, \text{sgn}\,(x-2)\, dx + \int_{1.5}^2 x\,[2x]\, \text{sgn}\,(x-2)\, dx \right.$$

$$\left. + \int_2^{2.5} x\,[2x]\, \text{sgn}\,(x-2)\, dx + \int_{2.5}^3 x\,[2x]\, \text{sgn}\,(x-2)\, dx \right\}$$

$$= \frac{1}{2} \left\{ \int_1^{1.5} -2x\, dx + \int_{1.5}^2 -3x\, dx + \int_2^{2.5} 4x\, dx + \int_{2.5}^3 5x\, dx \right\}$$

$$= \frac{15}{4}$$

Definite Integral as Limit of a Sum

An alternative way of describing $\int_a^b f(x)\, dx$ is that the definite integral $\int_a^b f(x)\, dx$ is a limiting case of summation of an infinite series, provided $f(x)$ is continuous on $[a, b]$, *ie*, $\int_a^b f(x)\, dx = \lim\limits_{n \to \infty} h \sum\limits_{r=1}^{n-1} f(a + rh)$, where $h = \dfrac{b-a}{n}$. The converse is also true, *ie*, if we have an infinite series of the above form, it can be expressed as definite integral.

Method to express the infinite series as definite integral

(i) Express the given series in the form $\Sigma\, \dfrac{1}{n} f\left(\dfrac{r}{n}\right)$.

(ii) Then the limit is its sum when $n \to \infty$, *ie*, $\lim\limits_{n \to \infty} \Sigma\, \dfrac{1}{n} f\left(\dfrac{r}{n}\right)$.

(iii) Replace $\dfrac{r}{n}$ by x and $\dfrac{1}{n}$ by (dx) and $\lim\limits_{n \to \infty} \Sigma$ by the sign of $\int$.

(iv) The lower and the upper limit of integration are limiting values of $\dfrac{r}{n}$ for the first and the last term of r respectively.

Some particular cases of the above are

(a) $\lim_{n\to\infty} \sum_{r=1}^{n} \frac{1}{n} f\left(\frac{r}{n}\right)$ or $\lim_{n\to\infty} \sum_{r=1}^{n-1} \frac{1}{n} f\left(\frac{r}{n}\right) = \int_0^1 f(x)\,dx$

(b) $\lim_{n\to\infty} \sum_{r=1}^{Pn} \frac{1}{n} f\left(\frac{r}{n}\right) = \int_\alpha^\beta f(x)\,dx$

where $\alpha = \lim_{n\to\infty} \frac{r}{n} = 0$ (as $r = 1$)

$\beta = \lim_{n\to\infty} \frac{r}{n} = p$ (as $r = pn$)

Points to Consider

The method of evaluate the integral, as limit of the sum of an infinite series is known as integration by first principles.

Illustration 90 Evaluate the following :

(i) $\lim_{n\to\infty} \left(\frac{1}{n^2} + \frac{2}{n^2} + \frac{3}{n^2} + \ldots + \frac{n-1}{n^2}\right)$

(ii) $\lim_{n\to\infty} \left(\frac{1}{n+1} + \frac{1}{n+2} + \ldots + \frac{1}{2n}\right)$

(iii) $\lim_{n\to\infty} \left(\frac{n}{n^2+1^2} + \frac{n}{n^2+2^2} + \ldots + \frac{n}{2n^2}\right)$

(iv) $\lim_{n\to\infty} \frac{(1^P + 2^P + \ldots + n^P)}{n^{P+1}}, P > 0$

Solution. Let

(i) $S_n = \frac{1}{n^2} + \frac{2}{n^2} + \frac{3}{n^2} + \ldots + \frac{n-1}{n^2}$

$= \sum_{r=2}^{n} \frac{r-1}{n^2} = \sum_{r=2}^{n} \frac{1}{n}\left(\frac{r-1}{n}\right)$

$\Rightarrow \quad S = \lim_{n\to\infty} \sum_{r=2}^{n} \frac{1}{n}\left(\frac{r}{n}\right)$ $\left[\text{as } n \to \infty \Rightarrow \frac{1}{n} \to 0\right]$

$= \int_0^1 x\,dx = \frac{1}{2}$

(ii) Let $S_n = \frac{1}{n+1} + \frac{1}{n+2} + \ldots + \frac{1}{2n} = \sum_{r=1}^{n} \frac{1}{n+r}$

$= \sum_{r=1}^{n} \frac{1}{n\left(1+\frac{r}{n}\right)}$

Hence, $S = \lim_{n\to\infty} S_n = \lim_{n\to\infty} \frac{1}{n} \sum_{r=1}^{n} \frac{1}{1+\frac{r}{n}}$

$= \int_0^1 \frac{dx}{1+x} = [\log(1+x)]_0^1 = \log 2$

(iii) Let $S_n = \frac{n}{n^2+1^2} + \frac{n}{n^2+2^2} + \ldots + \frac{n}{2n^2}$

$$= \sum_{r=1}^{n} \frac{n}{n^2+r^2} = \sum_{r=1}^{n} \frac{1}{n(1+r^2/n^2)}$$

Hence, $S = \lim_{n\to\infty} S_n = \lim_{n\to\infty} \frac{1}{n}\sum_{r=1}^{n} \frac{1}{1+\frac{r^2}{n^2}}$

$$= \int_0^1 \frac{dx}{1+x^2} = \left[\tan^{-1} x\right]_0^1 = \frac{\pi}{4}$$

(iv) Let $S_n = \frac{1^P + 2^P + \ldots + n^P}{n^{P+1}} = \sum_{r=1}^{n} \frac{r^P}{n^{P+1}} = \sum_{r=1}^{n} \frac{1}{n}\left(\frac{r}{n}\right)^P$

$\therefore$ $S = \lim_{n\to\infty} S_n = \lim_{n\to\infty} \frac{1}{n}\sum_{r=1}^{n}\left(\frac{r}{n}\right)^P$

$$= \int_0^1 x^P\,dx = \left[\frac{x^{P+1}}{P+1}\right]_0^1 = \frac{1}{P+1}$$

Illustration 91 Evaluate $S = \sum_{r=0}^{n-1} \frac{1}{\sqrt{4n^2-r^2}}$ as $n \to \infty$.

Solution. Let $S_n = \sum_{r=0}^{n-1} \frac{1}{\sqrt{4n^2-r^2}} = \sum_{r=0}^{n-1} \frac{1}{n\sqrt{4-(r/n)^2}}$

$$= \frac{1}{n}\sum_{r=0}^{n-1} \frac{1}{\sqrt{4-(r/n)^2}}$$

Hence, $S = \lim_{n\to\infty} S_n = \lim_{n\to\infty} \frac{1}{n}\sum_{r=0}^{n-1} \frac{1}{\sqrt{4-(r/n)^2}}$

$$= \int_0^1 \frac{dx}{\sqrt{4-x^2}} = \left[\sin^{-1}\frac{x}{2}\right]_0^1$$

$\Rightarrow$ $S = \sin^{-1}\left(\frac{1}{2}\right) = \frac{\pi}{6}$

$\therefore$ $S = \frac{\pi}{6}$

Illustration 92 Evaluate $\lim_{n\to\infty}\left(\frac{1}{2n+1} + \frac{1}{2n+2} + \ldots + \frac{1}{6n}\right)$.

Solution. Let $S_n = \left(\frac{1}{2n+1} + \frac{1}{2n+2} + \ldots + \frac{1}{6n}\right)$

$$= \sum_{r=1}^{4n} \frac{1}{2n+r} = \sum_{r=1}^{4n} \frac{1}{n}\cdot\frac{1}{2+(r/n)}$$

$\Rightarrow$ $S = \lim_{n\to\infty} S_n = \int_0^4 \frac{dx}{2+x}$

$$= [\ln|2+x|]_0^4 = \log 6 - \log 2 = \log 3$$

$\therefore$ $S = \ln 3$

Illustration 93 Evaluate $\int_1^4 (ax^2 + bx + c)\,dx$ from the first principle.

Solution. Let $x = 1 + rh$, where $h = \frac{4-1}{n} = \frac{3}{n}$

As $x \to 1, r \to 0$ and $x \to 4, r \to n$

$$\therefore \int_1^4 (ax^2 + bx + c)\,dx = \lim_{n\to\infty} \sum_{r=0}^{n} h\, f(1 + rh)$$

$$= \lim_{n\to\infty} \sum_{r=0}^{n} h\,[a(1+rh)^2 + b(1+rh) + c]$$

$$= \lim_{n\to\infty} \sum_{r=0}^{n} \frac{3}{n}\left[a\left(1+\frac{3r}{n}\right)^2 + b\left(1+\frac{3r}{n}\right) + c\right]$$

$$= 3\lim_{n\to\infty} \sum_{r=0}^{n} \frac{1}{n}\left[a\left(1+\frac{9r^2}{n^2}+\frac{6r}{n}\right) + b\left(1+\frac{3r}{n}\right) + c\right]$$

$$= 3\lim_{n\to\infty} \frac{1}{n}\left[a\left(n+\frac{9n(n+1)(2n+1)}{6n^2}+\frac{6n(n+1)}{2n}\right) + b\left(n+\frac{3n(n+1)}{2n}\right) + cn\right]$$

$$= 3\,[a(1+3+3) + b(1+3/2) + c] = 21a + \frac{15}{2}b + 3c$$

Illustration 94 The value of

$$\lim_{n\to\infty}\left(\sin\frac{\pi}{2n}\cdot\sin\frac{2\pi}{2n}\cdot\sin\frac{3\pi}{2n}\ldots\sin\frac{(n-1)\pi}{n}\right)^{1/n}, \text{ is equal to}$$

(a) $\frac{1}{2}$ (b) $\frac{1}{3}$ (c) $\frac{1}{4}$ (d) None of these

Solution. Let $A = \left\{\lim_{n\to\infty} \sin\frac{\pi}{2n}\cdot\sin\frac{2\pi}{2n}\cdot\sin\frac{3\pi}{2n}\ldots\sin\frac{2(n-1)\pi}{2n}\right\}^{1/n}$

$$\therefore \quad \log A = \lim_{n\to\infty} \frac{1}{n}\log\left(\sin\frac{\pi}{2n}\cdot\sin\frac{2\pi}{2n}\ldots\sin\frac{2(n-1)\pi}{2n}\right)$$

$$= \lim_{n\to\infty} \frac{1}{n}\sum_{r=1}^{2(n-1)} \log\sin\frac{r\pi}{2n}$$

$$= \int_0^2 \log\sin\left(\frac{\pi x}{2}\right)dx$$

$$\left[\text{using, } \lim_{n\to\infty}\sum\frac{1}{n}f\left(\frac{r}{n}\right) = \int_a^b f(x)\,dx\right]$$

$$= \int_0^{\pi} \log(\sin t)\cdot\frac{2}{\pi}\,dt \qquad \text{putting, } \frac{\pi x}{2} = t$$

$$= \frac{2\cdot 2}{\pi}\int_0^{\pi/2} \log(\sin t)\,dt$$

$$\left[\text{using } \int_0^{2a} f(x)\,dx = 2\int_0^a f(x)\,dx, \text{ if } f(2a-x) = f(x)\right]$$

$$= \frac{4}{\pi}\left\{-\frac{\pi}{2}\log 2\right\} \quad \left[\text{using } \int_0^{\pi/2}\log(\sin x)\,dx = -\frac{\pi}{2}\log 2\right]$$

$$= -2\log 2$$

$$\therefore \quad \log A = \log(1/4) \Rightarrow A = 1/4$$

Hence, (c) is the correct answer.

Illustration 95 The interval [0,4] is divided into n equal sub-intervals by the points $x_0, x_1, x_2, ..., x_{n-1}, x_n$ where $0 = x_0 < x_1 < x_2 < x_3 < ... < x_n = 4$. If $\delta x = x_i - x_{i-1}$ for $i = 1,2,3, ..., n$, then $\lim_{\delta x \to 0} \sum_{i=1}^{n} x_i \delta x$ is equal to

(a) 4 (b) 8 (c) $\frac{32}{3}$ (d) 16

Solution. $\lim_{\delta x \to 0} \delta x(x_1 + x_2 + x_3 + ... + x_n)$

$$= \lim_{n \to \infty} \frac{4}{n}\left[\frac{4}{n} + \frac{8}{n} + \frac{12}{n} + ... + 4 \cdot \frac{n}{n}\right] \qquad \left(\delta x = \frac{4}{n}\right)$$

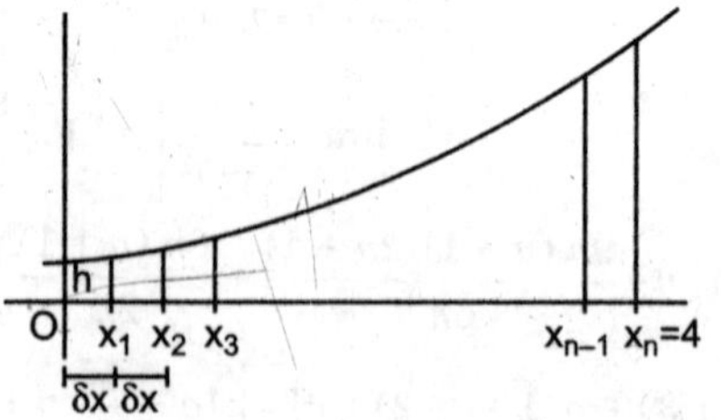

$$= \lim_{n \to \infty} \frac{16}{n^2}(1 + 2 + 3 + ... + n) = \lim_{n \to \infty} \frac{16}{n^2} \cdot \frac{n(n+1)}{2} = 8$$

Hence, (b) is the correct answer.

Gamma Function

If n is a positive number, then the improper integral $\int_0^\infty e^{-x} x^{n-1} dx$ is defined as Gamma function and is denoted by Γn.

ie, $\Gamma n = \int_0^\infty e^{-x} x^{n-1} dx$, where $x \in Q^+$.

Illustration 96 Evaluate

(i) $\Gamma 1$ (ii) $\Gamma 2$

Solution. (i) $\Gamma 1 = \int_0^\infty e^{-x} . x^{1-1} dx = \lim_{b \to \infty} \int_0^b e^{-x} dx$

$$= \lim_{b \to \infty} \left[- e^{-x}\right]_0^b = \lim_{b \to \infty} (- e^{-b} + e^0) = 0 + 1$$

$\therefore$ $\Gamma 1 = 1$

(ii) $\Gamma 2 = \int_0^\infty e^{-\infty} x^{2-1} dx = \lim_{b \to \infty} \int_0^b e^{-x} \cdot x \, dx$

$$= \lim_{b \to \infty} \left[- x e^{-x} - e^{-x}\right]_0^b = \lim_{b \to \infty} [(- b e^{-b} - e^{-b}) - (0 - 1)]$$

$$= \lim_{b \to \infty} \left[\frac{-b}{e^b} - \frac{1}{e^b} + 1\right] = \lim_{b \to \infty} \frac{-(b+1)}{e^b} + 1$$

$$= \left(\lim_{b \to \infty} - \frac{1}{e^b}\right) + 1 \qquad \text{(using L'Hospital's rule)}$$

$$= 1$$

Properties of Gamma Function

Gamma function has following properties :

(i) $\Gamma 1 = 1, \Gamma 0 = \infty$ and $\Gamma(n+1) = n\,\Gamma n$

eg, $\Gamma 5 = 4\,\Gamma 4 = 4\times 3\,\Gamma 3 = 4\times 3\times 2\,\Gamma 2 = 4\times 3\times 2\times 1\,\Gamma 1 = 4\times 3\times 2\times 1$

$\therefore \qquad \Gamma 5 = 24$

(ii) If $n \in N$, then $\Gamma(n+1) = (n)!$

(iii) $\Gamma(1/2) = \sqrt{\pi}$

(iv) $\displaystyle\int_0^{\pi/2} \sin^m x \cdot \cos^n x\,dx = \frac{\Gamma\left(\frac{m+1}{2}\right)\Gamma\left(\frac{n+1}{2}\right)}{2\Gamma\left(\frac{m+n+2}{2}\right)}$

(v) $\Gamma n\,\Gamma(1-n) = \dfrac{\pi}{\sin n\pi}, 0 < n < 1$

(vi) $\Gamma m\,\Gamma\left(m+\frac{1}{2}\right) = \dfrac{\sqrt{\pi}}{2^{2m-1}}\,\Gamma 2m$

(vii) $\Gamma\frac{1}{n}\,\Gamma\left(\frac{2}{n}\right)\ldots\Gamma\left(\frac{n-1}{n}\right) = \dfrac{(2\pi)^{\frac{n-1}{2}}}{n^{1/2}}$

Illustration 97 Evaluate $\int_0^\infty e^{-x} x^3\,dx$.

Solution. By definition of Gamma function

$$\int_0^\infty e^{-x} x^3\,dx = \int_0^\infty e^{-x} x^{4-1}\,dx = \Gamma 4$$

$$\therefore \qquad \int_0^\infty e^{-x} x^3\,dx = 6$$

Illustration 98 Evaluate $\displaystyle\int_0^1 \left(\log\frac{1}{x}\right)^{n-1} dx$.

Solution. Let $\displaystyle I = \int_0^1 \left(\log\frac{1}{x}\right)^{n-1} dx$, put $\log\frac{1}{x} = t \Rightarrow dx = -e^{-t}\,dt$

$$\Rightarrow \qquad I = \int_\infty^0 t^{n-1}(-e^{-t})\,dt$$

$$= \int_0^\infty e^{-t}\,t^{n-1}\,dt$$

$$\therefore \qquad \int_0^1 \left(\log\frac{1}{x}\right)^{n-1} dx = \Gamma n$$

Beta Function

The beta function is;

$$B(m,n) = \int_0^1 x^{m-1}(1-x)^{n-1}\,dx, \text{ where } m, n > 0$$

Properties of Beta Function

(i) $B(m, n) = B(n, m)$, where $m, n > 0$

(ii) $B(m, n) = \dfrac{\Gamma m \Gamma n}{\Gamma(m+n)}$, where $m, n > 0$

(iii) $B(m, n) = \int_0^{\infty} \dfrac{x^{m-1}}{(1+x)^{m+n}}\, dx$

Illustration 99 Evaluate $\int_0^1 x^6 \sqrt{(1-x^2)}\, dx$.

Solution. Here, $I = \int_0^1 x^6 \sqrt{(1-x^2)}\, dx$, let $x^2 = t \Rightarrow 2x\, dx = dt$

$$\therefore \quad I = \frac{1}{2}\int_0^1 t^{5/2} \sqrt{(1-t)}\, dt$$

$$I = \frac{1}{2} B\left(\frac{7}{2}, \frac{3}{2}\right) = \frac{1}{2} \cdot \frac{\Gamma 7/2 \Gamma 3/2}{\Gamma\left(\frac{3}{2}+\frac{7}{2}\right)}$$

$$= \frac{1}{2} \frac{\frac{5}{2}\cdot\frac{3}{2}\cdot\frac{1}{2}\cdot\sqrt{\pi}\cdot\frac{1}{2}\cdot\sqrt{\pi}}{4\cdot 3\cdot 2\cdot 1} = \frac{5\pi}{256}$$

$$\therefore \quad \int_0^1 x^6 \sqrt{1-x^2}\, dx = \frac{5\pi}{256}$$

Walli's Formula

An easy way to evaluate $\int_0^{\pi/2} \sin^m x \cdot \cos^n x\, dx$, where $m, n \in I_+$.

We have, $$\int_0^{\pi/2} \sin^m x \cdot \cos^n x\, dx = \int_0^{\pi/2} \sin^n x \cdot \cos^m x\, dx$$

$$= \frac{(m-1)(m-3)\ldots(1 \text{ or } 2).(n-1)(n-3)\ldots(1 \text{ or } 2)}{(m+n)(m+n-2)\ldots(1 \text{ or } 2)} \cdot \frac{\pi}{2},$$

when both m and $n \in$ even integer.

$$= \frac{(m-1)(m-3)\ldots(1 \text{ or } 2)\cdot(n-1)(n-3)\ldots(1 \text{ or } 2)}{(m+n)(m+n-2)\ldots(1 \text{ or } 2)},$$

when either of m or $n \in$ odd integer.

Points to Consider

If n be a positive integer, then

$$\int_0^{\pi/2} \sin^n x\, dx = \int_0^{\pi/2} \cos^n x\, dx$$

$$= \frac{n-1}{n}\cdot\frac{n-3}{n-2}\cdot\frac{n-5}{n-4}\ldots\frac{3}{4}\cdot\frac{1}{2}\cdot\frac{\pi}{2}, \ n \text{ is even.}$$

$$= \frac{n-1}{n}\cdot\frac{n-3}{n-2}\cdot\frac{n-5}{n-4}\ldots\frac{4}{5}\cdot\frac{2}{3}\cdot 1, \ n \text{ is odd.}$$

Illustration 100 Evaluate $\int_0^{\pi/2} \sin^4 x \cdot \cos^6 x\, dx$.

Solution. Using Gamma function, we have

$$\int_0^{\pi/4} \sin^4 x \cdot \cos^6 x\, dx = \frac{\Gamma(5/2)\,\Gamma(7/2)}{2\,\Gamma\left(\dfrac{4+6+2}{2}\right)} = \frac{\Gamma(5/2)\,\Gamma(7/2)}{2\,\Gamma 6}$$

$$= \frac{\left(\dfrac{3}{2}\times\dfrac{1}{2}\times\Gamma(1/2)\right)\left(\dfrac{5}{2}\times\dfrac{3}{2}\times\dfrac{1}{2}\times\Gamma(1/2)\right)}{2\times 5\,!} = \frac{3\pi}{512}$$

Illustration 101 The value of $\int_0^{\infty} e^{-a^2x^2}\, dx$, is equal to

(a) $\dfrac{\sqrt{\pi}}{2a}$ (b) $\dfrac{\pi}{2a}$ (c) $\dfrac{\pi}{\sqrt{2a}}$ (d) None of these

Solution. Let $I = \int_0^{\infty} e^{-a^2x^2}\, dx$ (put, $a^2x^2 = t \Rightarrow 2a^2x\, dx = dt$)

$$\therefore \quad I = \int_0^{\infty} e^{-t}\cdot\frac{1}{2a\sqrt{t}}\, dt = \frac{1}{2a}\int_0^{\infty} e^{-t}\cdot t^{-1/2}\, dt$$

$$= \frac{1}{2a}\int_0^{\infty} t^{1/2-1}\cdot e^{-t}\cdot dt \quad \left(\text{using } \int_0^{\infty} e^{-x}x^{n-1}\, dx = \Gamma n\right)$$

$$= \frac{1}{2a}\,\Gamma\frac{1}{2} = \frac{1}{2a}\sqrt{\pi} \quad \left(\text{using } \Gamma\frac{1}{2} = \sqrt{\pi}\right)$$

$$\Rightarrow \quad \int_0^{\infty} e^{-a^2x^2}\, dx = \frac{\sqrt{\pi}}{2a}$$

Hence, (a) is the correct answer.

Illustration 102 If $f : [0, \pi] \to R$ is continuous and

$$\int_0^{\pi} f(x)\sin x\, dx = \int_0^{\pi} f(x)\cos x\, dx = 0,$$

then the number of roots of $f(x)$ in $(0, \pi)$ is

(a) Zero (b) exactly one (c) exactly two (d) at least two

Solution. Here, $\int_0^{\pi} f(x)\sin x\, dx = \int_0^{\pi} f(x)\cos x\, dx = 0$

Clearly, $f(x)$ must have at least one sign change in $(0, \pi)$.

Let $f(x)$ has only one sign change in $(0, \pi)$ at $x = K$.

$\Rightarrow$ $\sin(x - K)$, where $K \in (0, \pi)$ also has only one sign change in $(0, \pi)$ at $x = K$.

$$\Rightarrow \int_0^{\pi} f(x)\sin(x - K)\, dx = \cos K\int_0^{\pi} f(x)\sin x\, dx - \sin K\int_0^{\pi} f(x)\cos x\, dx \neq 0$$

which is contradiction

$\Rightarrow$ $f(x)$ has at least two sign change in $(0, \pi)$.

Hence, (d) is the correct answer.

Illustration 103 Let $f(x)$ be a quadratic expression with positive integral coefficients such that for every $\alpha, \beta \in R, \beta > \alpha$, $\int_{\alpha}^{\beta} f(x)\, dx > 0$.
Let $g(t) = f''(t)\cdot f(t)$ and $g(0) = 12$, then

(a) 16 such quadratics are possible
(b) $f(x) = 0$ has either no real or distinct roots
(c) minimum value of $f(1)$ is 6
(d) maximum value of $f(1)$ is 11

Solution. Let $f(x) = ax^2 + bx + c$ given $\int_\alpha^\beta f(x)\,dx > 0$ for all $\beta > \alpha$

Thus, $f(x)$ is shown as;

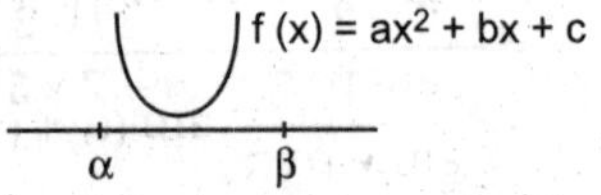

So, $f(x)$ has no real roots.

Now, $g(0) = f''(0) \cdot f(0) = 2ac$ [where $g(0) = 12$]

$\therefore$ $ac = 6$

$\because$ a and c are positive integers. So, the possible values are $(a = 6, c = 1); (a = 1, c = 6); (a = 2, c = 3); (a = 3, c = 2)$.

Again, $b^2 < 4ac$ or $b^2 < 24$. So, b can be 1, 2, 3, 4.

$\therefore$ In all $4 \times 4 = 16$ such quadratic are possible.

Now, $f(1) = a + b + c$, then maximum value of $f(1) = 11$ and minimum value of $f(1) = 6$.

Hence, (a), (b), (c) and (d) are the correct answers.

Target Exercise 2.3

1. The value of $f(x) = \int_0^{\pi/2} \frac{\log(1 + x \sin^2 \theta)}{\sin^2 \theta} d\theta$; $x \ge 0$ is equal to

(a) $\frac{1}{\pi}(\sqrt{1+x} - 1)$ (b) $\sqrt{\pi}(\sqrt{1+x} - 1)$ (c) $\pi(\sqrt{1+x} - 1)$ (d) None of these

2. The value of $\lim_{n \to \infty} \frac{1}{n} \sum_{r=1}^{n} \left(\frac{r}{n+r}\right)$, is equal to

(a) $1 - \log 2$ (b) $\log 4 - 1$ (c) $\log 2$ (d) None of these

3. The value of $\lim_{n \to \infty} \frac{1}{n}\{(n+1)(n+2)(n+3)\ldots(n+n)\}^{1/n}$, is equal to

(a) $4e$ (b) $\frac{e}{4}$ (c) $\frac{4}{e}$ (d) None of these

4. If $m, n \in N$, the value of $\int_a^b (x-a)^m (b-x)^n\,dx$, is equal to

(a) $\frac{(b-a)^{m+n} \cdot m!\,n!}{(m+n)!}$ (b) $\frac{(b-a)^{m+n+1} \cdot m!\,n!}{(m+n+1)!}$

(c) $\frac{(b-a)^m \cdot m!}{m!}$ (d) None of these

5. The value of $\lim_{n \to \infty} \left(\frac{n!}{n^n}\right)^{\frac{2n^4+1}{5n^5+1}}$, is equal to

(a) e (b) $\frac{2}{e}$ (c) $\left(\frac{1}{e}\right)^{\frac{2}{5}}$ (d) None of these

6. The value of $\lim_{n \to \infty} n \left\{ \frac{1}{3n^2 + 8n + 4} + \frac{1}{3n^2 + 16n + 16} + \ldots + \text{to } n \text{ terms} \right\}$ is equal to

(a) $\frac{1}{2} \log \left(\frac{9}{5} \right)$ (b) $\frac{1}{3} \log \left(\frac{9}{5} \right)$

(c) $\frac{1}{4} \log \left(\frac{9}{5} \right)$ (d) None of these

7. The value of $\lim_{n \to \infty} \frac{1}{n^2} \left\{ \sin^3 \frac{\pi}{4n} + 2 \sin^3 \frac{2\pi}{4n} + \ldots + n \sin^3 \frac{n\pi}{4n} \right\}$, is equal to

(a) $\frac{\sqrt{2}}{9\pi^2}(52 - 15n)$ (b) $\frac{2}{9\pi^2}(52 - 15n)$

(c) $\frac{1}{9\pi^2}(15n - 15)$ (d) None of these

8. The value of $\int_0^{\pi/2} \log (\sin^2 \theta + k^2 \cos^2 \theta) \, d\theta$, is equal to

(a) $\pi \log (1 + K) - \pi \log 2$ (b) $\pi \log 2 - \log (1 + K)$

(c) $\log (1 + K) - \pi \log 2$ (d) None of these

9. If $m, n \in N$, then $I_{m,n} = \int_0^1 x^m (1 - x)^n \, dx$, is equal to

(a) $\frac{m!n!}{(m + n + 2)!}$ (b) $\frac{2m!n!}{(m + n + 1)!}$

(c) $\frac{m!n!}{(m + n + 1)!}$ (d) None of these

10. The value of $\int_0^{\pi} \frac{\sin^2 n\theta}{\sin^2 \theta} \, d\theta$, is $(\forall \ n \in N)$

(a) $n\pi$ (b) $\frac{n\pi}{2}$ (c) $\frac{n\pi}{4}$ (d) None of these

Some Important Results to Remember

(i) $\sum_{r=1}^{n} r = \frac{n(n+1)}{2}$

(ii) $\sum_{r=1}^{n} r^2 = \frac{n(n+1)(2n+1)}{6}$

(iii) $\sum_{r=1}^{n} r^3 = \frac{n^2 (n+1)^2}{4}$

(iv) In GP, sum of n terms, $S_n = \begin{cases} \frac{a(r^n - 1)}{(r-1)}, & |r| > 1 \\ an, & r = 1 \\ \frac{a(1 - r^n)}{(1 - r)}, & |r| < 1 \end{cases}$

(v) $\sin \alpha + \sin (\alpha + \beta) + \sin (\alpha + 2\beta) + \ldots + \sin (\alpha + \overline{(n-1)}\beta)$

$$= \frac{\sin n\beta/2}{\sin \beta/2} \cdot \sin (\alpha + (n-1)\beta/2)$$

(vi) $\cos\alpha + \cos(\alpha+\beta) + \cos(\alpha+2\beta) + \ldots + \cos(\alpha + (\overline{n-1})\beta)$

$$= \frac{\sin n\beta/2}{\sin\beta/2}\cdot\cos(\alpha + (n-1)\beta/2)$$

(vii) $1 - \frac{1}{2^2} + \frac{1}{3^2} - \frac{1}{4^2} + \frac{1}{5^2} - \frac{1}{6^2} + \ldots \infty = \frac{\pi^2}{12}$

(viii) $1 + \frac{1}{2^2} + \frac{1}{3^2} + \frac{1}{4^2} + \frac{1}{5^2} + \frac{1}{6^2} + \ldots \infty = \frac{\pi^2}{6}$

(ix) $1 + \frac{1}{3^2} + \frac{1}{5^2} + \ldots \infty = \frac{\pi^2}{8}$

(x) $\frac{1}{2^2} + \frac{1}{4^2} + \frac{1}{6^2} + \ldots \infty = \frac{\pi^2}{24}$

(xi) $\cos\theta = \frac{e^{i\theta} + e^{-i\theta}}{2}$, and $\sin\theta = \frac{e^{i\theta} - e^{-i\theta}}{2i}$

(xii) $\cos h\,\theta = \frac{e^{\theta} + e^{-\theta}}{2}$, and $\sin h\,\theta = \frac{e^{\theta} - e^{-\theta}}{2}$

Worked Examples

Type 1 : Subjective Type Questions

Example 1 Find the error in steps to evaluate the following integral;

$$\int_0^{\pi} \frac{dx}{1+2\sin^2 x} = \int_0^{\pi} \frac{\sec^2 x\, dx}{\sec^2 x + 2\tan^2 x} = \int_0^{\pi} \frac{\sec^2 x\, dx}{1+3\tan^2 x}$$

$$= \frac{1}{\sqrt{3}} (\tan^{-1}(\sqrt{3}\tan x))_0^{\pi} = 0.$$

Solution. Here, the Newton-Leibnitz formula for evaluating the definite integrals is not applicable here because the anti-derivative, $f(x) = \frac{1}{\sqrt{3}} (\tan^{-1}(\sqrt{3}\tan x))$ has a discontinuity at the point $x = \frac{\pi}{2}$ which lies in the interval $[0, \pi]$.

$$\text{LHL} = \lim_{h \to 0} \frac{1}{\sqrt{3}} \tan^{-1} \sqrt{3} \left(\tan\left(\frac{\pi}{2} - h\right)\right) \quad \text{at } x = \frac{\pi}{2}$$

$$= \lim_{h \to 0} \frac{1}{\sqrt{3}} \tan^{-1}(\sqrt{3}\cot h)$$

$$= \lim_{h \to 0} \frac{1}{\sqrt{3}} \tan^{-1}(\to \infty) = \frac{\pi}{2\sqrt{3}} \quad \text{...(i)}$$

Also,

$$\text{RHL} = \lim_{h \to 0} \frac{1}{\sqrt{3}} \tan^{-1} \sqrt{3} \left(\tan\left(\frac{\pi}{2} + h\right)\right) \quad \text{at } x = \frac{\pi}{2}$$

$$= \lim_{h \to 0} \frac{1}{\sqrt{3}} \tan^{-1}(-\sqrt{3}\cot h)$$

$$= \lim_{h \to 0} \frac{1}{\sqrt{3}} \tan^{-1}(-\infty) = -\frac{\pi}{2\sqrt{3}} \quad \text{...(ii)}$$

From Eqs. (i) and (ii), LHL $\neq$ RHL at $x = \pi/2$.

$\Rightarrow$ Anti-derivative, $f(x)$ is discontinuous at $x = \pi/2$.

So, the correct solution for above integral;

$$I = \int_0^{\pi} \frac{dx}{1+2\sin^2 x} = \int_0^{\pi} \frac{\sec^2 x\, dx}{1+3\tan^2 x} \quad \text{...(iii)}$$

Using $\int_0^{2a} f(x)\,dx = \begin{cases} 0, f(2a-x) = -f(x) \\ 2\int_0^{a} f(x)\,dx,\ f(2a-x) = f(x) \end{cases}$

We know, if $f(x) = \frac{\sec^2 x}{1+3\tan^2 x}$ $f(\pi - x) = f(x)$

$\therefore$ Eq. (iii) reduces to; $I = 2\int_0^{\pi/2} \frac{\sec^2 x\, dx}{1+3\tan^2 x}$

$$= 2 \cdot \frac{1}{\sqrt{3}} \int_0^{\infty} \frac{dt}{1+t^2}, \quad \text{Let } \sqrt{3}\tan x = t \Rightarrow \sqrt{3}\sec^2 x\, dx = dt$$

$$= \frac{2}{\sqrt{3}} [\tan^{-1}(t)]_0^{\infty}$$

$$= \frac{2}{\sqrt{3}}(\tan^{-1}\infty - \tan^{-1}0) = \frac{2}{\sqrt{3}}\left(\frac{\pi}{2} - 0\right) = \frac{\pi}{\sqrt{3}}$$

$$\Rightarrow \quad \int_0^{\pi} \frac{dx}{1 + 2\sin^2 x} = \frac{\pi}{\sqrt{3}}$$

Points to Consider

Students are advised to check continuity of anti-derivatives before substitution of integral limits.

Example 2 If $\int_a^b |\sin x|\, dx = 8$ and $\int_0^{a+b} |\cos x|\, dx = 9$, then find the value of $\int_a^b x \sin x\, dx$.

Solution. We know, $\int_a^b |\sin x|\, dx$ represents the area under the curve from $x = a$ to $x = b$.

We also know, area from $x = a$ to $x = a + \pi$ is 2.

$$\therefore \quad \int_a^b |\sin x|\, dx = 8$$

$$\Rightarrow \quad b - a = \frac{8\pi}{2} \quad \text{...(i)}$$

Similarly, $$\int_0^{a+b} |\cos x|\, dx = 9$$

$$\Rightarrow \quad a + b - 0 = \frac{9\pi}{2} \quad \text{...(ii)}$$

From Eqs. (i) and (ii), $a = \pi/4,\ b = 17\pi/4$

Hence, $$\int_a^b x\sin x\, dx = \int_{\pi/4}^{17\pi/4} x \sin x\, dx = [-x\cos x]_{\pi/4}^{17\pi/4} + \int_{\pi/4}^{17\pi/4} \cos x\, dx$$

$$= -\frac{17\pi}{4}\cos\frac{17\pi}{4} + \frac{\pi}{4}\cos\frac{\pi}{4} + [\sin x]_{\pi/4}^{17\pi/4} = -\frac{4\pi}{\sqrt{2}}$$

$$\therefore \quad \int_a^b x \sin x\, dx = -2\sqrt{2}\pi$$

Example 3 Evaluate $\int_{\cos(\cos^{-1}\alpha)}^{\sin(\sin^{-1}\beta)} \left|\frac{\cos(\cos^{-1}x)}{\sin(\sin^{-1}x)}\right| dx$.

Solution. We know, $\cos(\cos^{-1}x)$ and $\sin(\sin^{-1}x)$ could be plotted as;

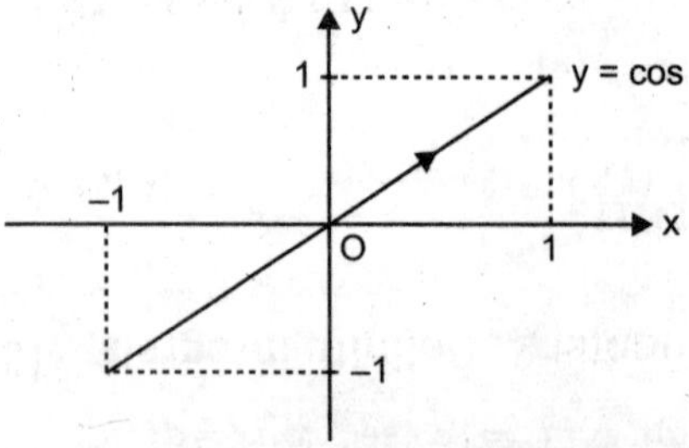

and

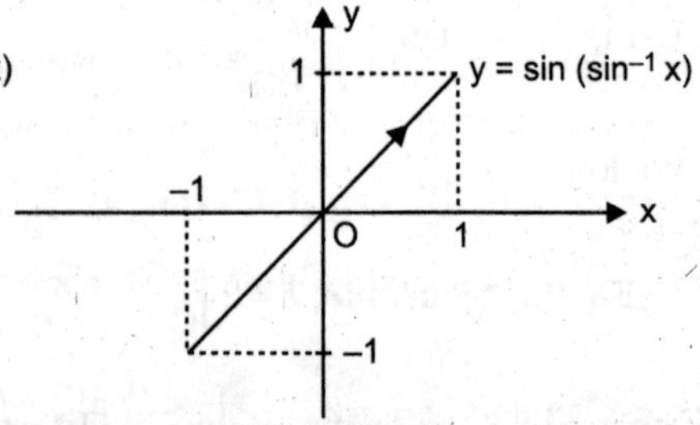

$\therefore$ $\sin(\sin^{-1}x)$ and $\cos(\cos^{-1}x)$ are identical functions.

$$\therefore \quad \int_{\cos(\cos^{-1}\alpha)}^{\sin(\sin^{-1}\beta)} \frac{\cos(\cos^{-1}x)}{\sin(\sin^{-1}x)} dx = \int_{\cos(\cos^{-1}\alpha)}^{\sin(\sin^{-1}\beta)} 1\, dx = (\beta - \alpha)$$

Example 4 Evaluate $\int_0^{10\pi} [\tan^{-1} x]\, dx$.

Solution. As we know

$$-\frac{\pi}{2} < \tan^{-1} x < \frac{\pi}{2}, \forall\, x \in R$$

$$\therefore \quad 0 < \tan^{-1} x < \frac{\pi}{2}, \text{ when } x \in [0, 10\pi]$$

$$ie, \quad \begin{cases} 0 < \tan^{-1} x < 1; & \text{when } 0 < x < \tan 1 \\ 1 < \tan^{-1} x < \dfrac{\pi}{2}; & \text{when } \tan 1 < x < 10\pi \end{cases}$$

$$\therefore \quad I = \int_0^{10\pi} [\tan^{-1} x]\, dx = \int_0^{\tan 1} 0\, dx + \int_{\tan 1}^{10\pi} 1\, dx = 10\pi - \tan 1$$

Example 5 Evaluate $\int_0^{\pi} (x \sin [x^2 - \pi])\, dx$, where [.] denotes the greatest integer function.

Solution. Here, $y = x^2 - \pi$ could be plotted as;

$$\Rightarrow [x^2 - \pi] = \begin{cases} -4, & 0 < x < \sqrt{(\pi - 3)} \\ -3, & \sqrt{(\pi - 3)} < x < \sqrt{(\pi - 2)} \\ -2, & \sqrt{(\pi - 2)} < x < \sqrt{(\pi - 1)} \\ -1, & \sqrt{(\pi - 1)} < x < \sqrt{\pi} \\ 0, & \sqrt{\pi} < x < \sqrt{(\pi + 1)} \\ 1, & \sqrt{(\pi + 1)} < x < \sqrt{(\pi + 2)} \\ 2, & \sqrt{(\pi + 2)} < \sqrt{(\pi + 3)} \\ 3, & \sqrt{(\pi + 3)} < x < \sqrt{(\pi + 4)} \\ 4, & \sqrt{(\pi + 4)} < x < \sqrt{(\pi + 5)} \\ 5, & \sqrt{(\pi + 5)} < x < \sqrt{(\pi + 6)} \\ 6, & \sqrt{(\pi + 6)} < x < \pi \end{cases}$$

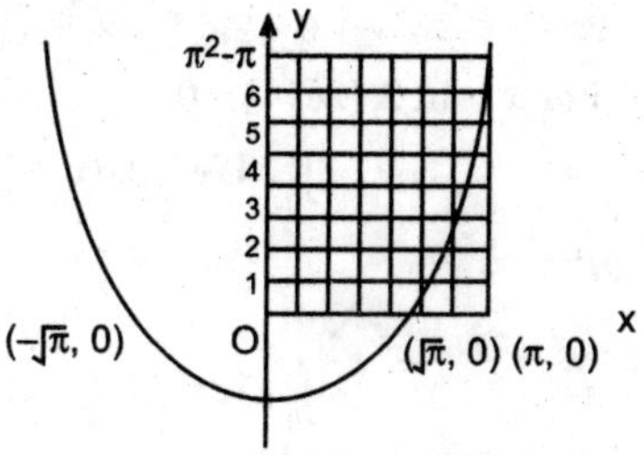

Hence, $\int_0^{\pi} x (\sin [x^2 - \pi])\, dx$

$$= \int_0^{\sqrt{(\pi-3)}} x \sin(-4)\, dx + \int_{\sqrt{(\pi-3)}}^{\sqrt{(\pi-2)}} x \sin(-3)\, dx + \int_{\sqrt{(\pi-2)}}^{\sqrt{(\pi-1)}} x \sin(-2)\, dx$$

$$+ \int_{\sqrt{(\pi-1)}}^{\sqrt{\pi}} x \sin(-1)\, dx + \int_{\sqrt{\pi}}^{\sqrt{(\pi+1)}} x \sin 0\, dx + \int_{\sqrt{(\pi+1)}}^{\sqrt{(\pi+2)}} x \sin(1)\, dx$$

$$+ \int_{\sqrt{(\pi+2)}}^{\sqrt{(\pi+3)}} x \sin(2)\, dx + \int_{\sqrt{(\pi+3)}}^{\sqrt{(\pi+4)}} x \sin(3)\, dx + \int_{\sqrt{(\pi+4)}}^{\sqrt{(\pi+5)}} x \sin(4)\, dx$$

$$+ \int_{\sqrt{(\pi+5)}}^{\sqrt{(\pi+6)}} x \sin(5)\, dx + \int_{\sqrt{(\pi+6)}}^{\pi} x \sin(6)\, dx$$

$$\Rightarrow \quad \frac{1}{2}\{(4 - \pi)\sin 4 + \sin 5 - (6 + \pi - \pi^2)\sin 6\}$$

Example 6 If $f(x) = \min\{|x - 1|, |x|, |x + 1|\}$, then evaluate $\int_{-1}^{1} f(x)\, dx$.

Solution. Here, $f(x) = \min\{|x - 1|, |x|, |x + 1|\}$

Graphically it could be shown as;

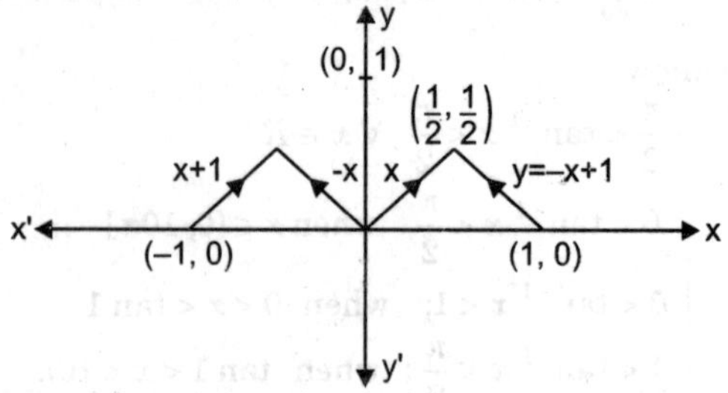

$$\therefore \qquad \int_{-1}^{1} f(x)\,dx = 2\int_0^1 f(x)\,dx = 2\left(\frac{1}{2}\times 1\times\frac{1}{2}\right) = \frac{1}{2}$$

Example 7 Evaluate $\int_0^{\infty} [2e^{-x}]\,dx$, where $[x]$ represents the greatest integer function.

Solution. Since, $2e^{-x}$ is decreasing for $x \in [0, \infty)$.

$$\therefore \qquad 0 < 2e^{-x} \le 2, \ \forall\, x \in [0, \infty)$$

For $x > \ln 2$, $[2e^{-x}] = 0$

$$\Rightarrow \qquad \int_0^{\infty} [2e^{-x}]\,dx = \int_0^{\ln 2} [2e^{-x}]\,dx + \int_{\ln 2}^{\infty} [2e^{-x}]\,dx$$

$$= \int_0^{\ln 2} 1\,dx + \int_{\ln 2}^{\infty} 0\,dx = (x)_0^{\ln 2} = \ln 2$$

$$\Rightarrow \qquad \int_0^{\infty} [2e^{-x}]\,dx = \ln 2$$

Example 8 If $[x]$ stands for the greatest integer function, then evaluate

$$\int_4^{10} \frac{[x^2]\,dx}{[x^2 - 28x + 196] + [x^2]}.$$

Solution. Let $$I = \int_4^{10} \frac{[x^2]\,dx}{[(14-x)^2] + [x^2]} \qquad \ldots\text{(i)}$$

Then, $$I = \int_4^{10} \frac{[(14-x)^2]\,dx}{[x^2] + [(14-x)^2]} \qquad \ldots\text{(ii)}$$

$$\left[\because \int_a^b f(x)\,dx = \int_a^b f(a+b-x)\,dx\right]$$

Adding Eqs. (i) and (ii), $$2I = \int_4^{10} 1\cdot dx = 6$$

$$\therefore \qquad I = 3$$

Example 9 Evaluate $\dfrac{\int_0^n [x]\,dx}{\int_0^n \{x\}\,dx}$, (where $[x]$ and $\{x\}$ are integral and fractional parts of x and $n \in N$).

Solution. We have, $\int_0^n [x]\,dx$

$$= \int_0^1 0\,dx + \int_1^2 1\,dx + \int_2^3 2\,dx + \ldots + \int_{n-1}^{n} (n-1)\,dx$$

$$= 0 + 1(2-1) + 2(3-2) + \ldots + (n-1)(n-(n-1))$$

$$= 1 + 2 + 3 + \ldots + (n-1) = \frac{n(n-1)}{2} \quad \ldots(i)$$

and

$$\int_0^n \{x\}\, dx = \int_0^n (x - [x])\, dx = \int_0^n x\, dx - \int_0^n [x]\, dx$$

$$= \frac{n^2}{2} - \frac{n(n-1)}{2} \quad \text{[using Eq. (i)]}$$

$$= \frac{n}{2}(n - n + 1) = \frac{n}{2} \quad \ldots(ii)$$

$$\therefore \quad \frac{\int_0^n [x]\, dx}{\int_0^n \{x\}\, dx} = \frac{\frac{n(n-1)}{2}}{\frac{n}{2}} = (n-1)$$

$$\Rightarrow \quad \frac{\int_0^n [x]\, dx}{\int_0^n \{x\}\, dx} = (n-1)$$

Example 10 Evaluate $\int_1^{10\pi} ([\sec^{-1} x] + [\cot^{-1} x])\, dx$, where $[\cdot]$ denotes the greatest integer function.

Solution. Here, $I = \int_0^{10\pi} ([\sec^{-1} x] + [\cot^{-1} x])\, dx$

$$= \int_0^{\sec 1} ([\sec^{-1} x] + [\cot^{-1} x])\, dx + \int_{\sec 1}^{10\pi} ([\sec^{-1} x] + [\cot^{-1} x])\, dx$$

$$= \int_0^{\sec 1} (0 + 0) dx + \int_{\sec 1}^{10\pi} (1 + 0)\, dx$$

$$= 0 + (x)_{\sec 1}^{10\pi} = 10\pi - \sec 1$$

Example 11 Evaluate $\int_{-\pi/2}^{2\pi} [\cot^{-1} x]\, dx$, where $[\cdot]$ denotes the greatest integer function.

Solution. Here, $I = \int_{-\pi/2}^{2\pi} [\cot^{-1} x]\, dx$

$$= \int_{-\pi/2}^{\cot 2} [\cot^{-1} x]\, dx + \int_{\cot 2}^{\cot 1} [\cot^{-1} x]\, dx + \int_{\cot 1}^{2\pi} [\cot^{-1} x]\, dx$$

$$= \int_{-\pi/2}^{\cot 2} 2 dx + \int_{\cot 2}^{\cot 1} 1 dx + \int_{\cot 1}^{2\pi} 0 dx$$

$$= \pi + \cot 1 + \cot 2$$

Example 12 Evaluate $\int_0^x [\cos t]\, dt$, where $n \in \left(2n\pi, (4n+1)\frac{\pi}{2}\right)$; $n \in N$ and $[\cdot]$ denotes the greatest integer function.

Solution. Let $I = \int_0^x [\cos t]\, dt$

$$= \int_0^{2n\pi} [\cos t]\, dt + \int_{2n\pi}^x [\cos t]\, dt$$

$$= n \int_0^{2\pi} [\cos t]\, dt + \int_{2n\pi}^x [\cos t]\, dt$$

$$= n\left(\int_0^{\pi/2}[\cos t]\,dt + \int_{\pi/2}^{3\pi/2}[\cos t]\,dt + \int_{3\pi/2}^{2\pi}[\cos t]\,dt\right) + \int_{2n\pi}^{x}[\cos t]\,dt$$

$$= n\left(\int_0^{\pi/2} 0\,dt + \int_{\pi/2}^{3\pi/2}(-1)\,dt + \int_{3\pi/2}^{2\pi} 0\,dt\right) + \int_{2n\pi}^{x} 0\,dt$$

$$= n\left(0 - \left(\frac{3\pi}{2} - \frac{\pi}{2}\right) + 0\right) + 0 = -n\pi$$

$$\therefore \quad \int_0^x [\cos t]\,dt = -n\pi$$

Example 13 Evaluate $\int_0^{\pi/4} \{\tan^n (x - [x]) + \tan^{n-2}(x - [x])\}\,dx$; (where $[\cdot]$ denotes the greatest integer function.)

Solution. Let $I = \int_0^{\pi/4} \{\tan^n (x - [x]) + \tan^{n-2}(x - [x])\}\,dx$

$$\left[\because 0 < x \le \frac{\pi}{4} \Rightarrow [x] = 0\right]$$

$$\therefore \quad I = \int_0^{\pi/4} \{\tan^n x + \tan^{n-2} x\}\,dx$$

$$= \int_0^{\pi/4} \tan^{n-2} x\,(\tan^2 x + 1)\,dx$$

$$= \int_0^{\pi/4} \tan^{n-2} x \cdot \sec^2 x\,dx$$

$$= \left(\frac{\tan^{n-1}(x)}{n-1}\right)_0^{\pi/4} = \frac{1}{(n-1)}\left\{\tan^{n-1}\left(\frac{\pi}{4}\right)\right\} - 0$$

$$= \frac{1}{n-1}$$

Example 14 Evaluate $\int_0^a [x^n]\,dx$.

Solution. Let K be a natural number such that $K \le a^n < K + 1$.

Then, $$\int_0^a [x^n]\,dx = \int_0^1 [x^n]\,dx + \int_1^{2^{1/n}} [x^n]\,dx + \int_{2^{1/n}}^{3^{1/n}} [x^n]\,dx + \ldots$$

$$\ldots + \int_{(K-1)^{1/n}}^{K^{1/n}} [x^n]\,dx + \int_{K^{1/n}}^{(a^n)^{1/n}} [x^n]\,dx$$

$$= \int_0^1 0\,dx + \int_1^{2^{1/n}} 1\,dx + \int_{2^{1/n}}^{3^{1/n}} 2\,dx + \ldots + \int_{(K-1)^{1/n}}^{K^{1/n}} (K-1)\,dx$$

$$+ \int_{K^{1/n}}^{(a^n)^{1/n}} K\,dx$$

$$= 0 + 1\cdot(2^{1/n} - 1) + 2\cdot(3^{1/n} - 2^{1/n}) + \ldots +$$

$$\ldots + (K-1)(K^{1/n} - (K-1)^{1/n}) + K(a - K^{1/n})$$

$$\Rightarrow \quad \left(\sum_{r=1}^{K} (r-1)[r^{1/n} - (r-1)^{1/n}]\right) + K(a - K^{1/n})$$

Example 15 Prove that $\int_0^x [x]\,dx = x[x] - \frac{1}{2}[x]([x]+1)$.

Solution. Let K be an integer such that; $K \le x < K + 1$.

Then, $$\int_0^x [x]\,dx = \int_0^K [x]\,dx + \int_K^x [x]\,dx$$

$$= \int_0^1 [x]\,dx + \int_1^2 [x]\,dx + \int_2^3 [x]\,dx + \ldots + \int_{K-1}^{K} [x]\,dx + \int_K^x [x]\,dx$$

$$= 0 + 1\,(2-1) + 2\,(3-2) + \ldots + (K-1)\,[K-(K-1)] + K\,(x-K)$$

$$= [1 + 2 + 3 + \ldots + (K-1)] + Kx - K^2$$

$$= Kx + \frac{K\,(K-1)}{2} - K^2 = Kx - \frac{K\,(K+1)}{2}$$

$$= [x]\,x - \frac{[x]([x]+1)}{2}$$

Example 16 Evaluate $\int_0^2 [x^2 - x + 1]\,dx$; where $[\cdot]$ denotes the greatest integer function.

Solution. We know $[x^2 - x + 1]$, $\forall\ x \in [0, 2]$ could be plotted as;

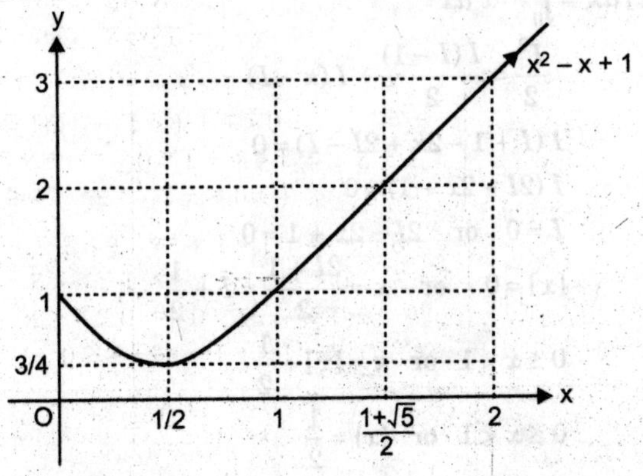

$\therefore$ $[x^2 - x + 1]$ is shown as;

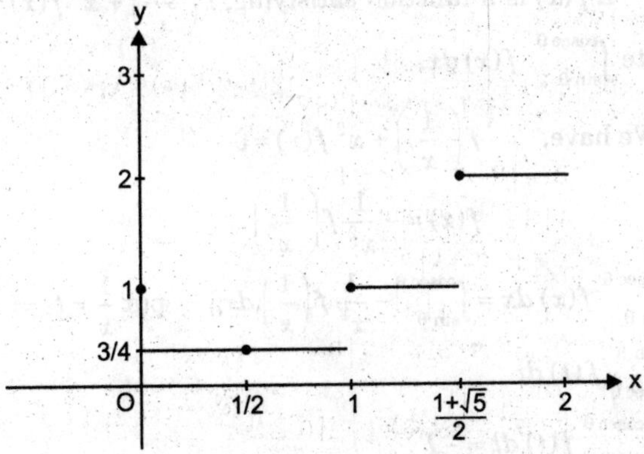

$$\therefore \quad I = \int_0^2 [x^2 - x + 1]\,dx$$

$$= \int_0^1 0\,dx + \int_1^{(1+\sqrt{5})/2} 1\,dx + \int_{(1+\sqrt{5})/2}^{2} 2\,dx$$

$$= \left(\frac{1+\sqrt{5}}{2} - 1\right) + 2\left(2 - \frac{1+\sqrt{5}}{2}\right) = \frac{5-\sqrt{5}}{2}$$

Example 17 If $\int_0^x [x]\,dx = \int_0^{[x]} x\,dx$, then prove that either x is purely fractional or x is such that $\{x\} = \frac{1}{2}$, (where $[\cdot]$ and $\{\cdot\}$ denote the greatest integer and fractional parts respectively.)

Solution. Let $x = I + f \Rightarrow [x] = I$ ...(i)

Now, $\int_0^{[x]} x\,dx = \int_0^I x\,dx = \frac{I^2}{2}$

and $\int_0^x [x]\,dx = \int_0^{I+f} [x]\,dx = \int_0^1 0\,dx + \int_1^2 1\,dx + \ldots + \int_{I-1}^I (I-1)\,dx + \int_I^{I+f} I\,dx$

$$= \{1 + 2 + 3 + \ldots + (I-1)\} + I(I + f - I)$$

$$= \frac{I(I-1)}{2} + I(f) = \frac{I(I-1)}{2} + I(x - I) \quad \text{[using Eq. (i)]}$$

Given, $\int_0^x [x]\,dx = \int_0^{[x]} x\,dx$

$$\Rightarrow \quad \frac{I^2}{2} = \frac{I(I-1)}{2} + I(x - I)$$

$$\Rightarrow \quad I(I + 1 - 2x + 2I - I) = 0$$

$$\Rightarrow \quad I(2I - 2x + 1) = 0$$

$$\Rightarrow \quad I = 0 \quad \text{or} \quad 2I - 2x + 1 = 0$$

ie, $$[x] = 0 \quad \text{or} \quad x = \frac{2I + 1}{2} = I + \frac{1}{2}$$

$$\Rightarrow \quad 0 \le x < 1 \quad \text{or} \quad x = [x] + \frac{1}{2}$$

$$\Rightarrow \quad 0 \le x < 1 \quad \text{or} \quad \{x\} = \frac{1}{2}$$

Example 18 If $f(x)$ is a function satisfying, $f\left(\frac{1}{x}\right) + x^2 f(x) = 0$ for all non-zero x, then evaluate $\int_{\sin\theta}^{\operatorname{cosec}\theta} f(x)\,dx$.

Solution. We have, $f\left(\frac{1}{x}\right) + x^2 f(x) = 0$

$$\Rightarrow \quad f(x) = -\frac{1}{x^2} f\left(\frac{1}{x}\right)$$

$$\therefore \quad I = \int_{\sin\theta}^{\operatorname{cosec}\theta} f(x)\,dx = \int_{\sin\theta}^{\operatorname{cosec}\theta} -\frac{1}{x^2} f\left(\frac{1}{x}\right) dx, \quad \text{put } \frac{1}{x} = t \Rightarrow -\frac{1}{x^2}\,dx = dt$$

$$= \int_{\operatorname{cosec}\theta}^{\sin\theta} f(t)\,dt$$

$$= -\int_{\sin\theta}^{\operatorname{cosec}\theta} f(t)\,dt = -I$$

$$\therefore \quad 2I = 0 \Rightarrow I = 0$$

Example 19 Evaluate $\int_0^{\pi/4} \frac{\sin x + \cos x}{\cos^2 x + \sin^4 x}\,dx$.

Solution. $$I = \int_0^{\pi/4} \frac{\sin x + \cos x}{\cos^2 x + \sin^4 x}\,dx$$

$$= \int_0^{\pi/4} \frac{\sin x + \cos x}{1 - \sin^2 x\,(1 - \sin^2 x)}\, dx = \int_0^{\pi/2} \frac{\sin x + \cos x}{1 - \sin^2 x \cdot \cos^2 x}\, dx$$

$$= 4\int_0^{\pi/4} \frac{\sin x + \cos x}{4 - 4\sin^2 x \cos^2 x}\, dx$$

$$I = 4\int_0^{\pi/4} \frac{\sin x + \cos x}{4 - \{1 - (\sin x - \cos x)^2\}^2}\, dx \quad \text{put, } \sin x - \cos x = t$$

$$\Rightarrow (\cos x + \sin x)\, dx = dt$$

$$I = 4\int_{-1}^{0} \frac{dt}{4 - (1 - t^2)^2} = 4\int_{-1}^{0} \frac{dt}{(2 - 1 + t^2)(2 + 1 - t^2)}$$

$$= 4\int_{-1}^{0} \frac{dt}{(1 + t^2)(3 - t^2)} = 4\int_{-1}^{0} \left\{ \frac{1}{4(3 - t^2)} + \frac{1}{4(1 + t^2)} \right\} dt$$

(using partial fraction)

$$= \int_{-1}^{0} \frac{dt}{3 - t^2} + \int_{-1}^{0} \frac{dt}{1 + t^2}$$

$$= \left\{ \frac{1}{2\sqrt{3}} \log\left(\frac{\sqrt{3} + t}{\sqrt{3} - t} \right) \right\}_{-1}^{0} + \{\tan^{-1} t\}_{-1}^{0}$$

$$= \frac{1}{2\sqrt{3}} \left[0 - \log\left(\frac{\sqrt{3} - 1}{\sqrt{3} + 1} \right) \right] + [0 - \tan^{-1}(-1)]$$

$$I = -\frac{1}{2\sqrt{3}} \log\left(\frac{\sqrt{3} - 1}{\sqrt{3} + 1} \right) + \frac{\pi}{4}$$

Example 20 Evaluate $\int_0^{\pi/4} \frac{\sin x + \cos x}{9 + 16\sin 2x}\, dx.$

Solution. Let $I = \int_0^{\pi/4} \frac{\sin x + \cos x}{9 + 16\{1 - (\sin x - \cos x)^2\}}\, dx$

$\therefore$ $$I = \int_0^{\pi/4} \frac{\sin x + \cos x}{25 - 16(\sin x - \cos x)^2}\, dx, \quad \text{put } 4(\sin x - \cos x) = t$$

$$\Rightarrow 4(\cos x + \sin x)\, dx = dt$$

$$= \frac{1}{4}\int_{-4}^{0} \frac{dt}{25 - t^2}$$

$$I = \frac{1}{4} \cdot \frac{1}{2(5)} \left(\log\left| \frac{5 + t}{5 - t} \right| \right)_{-4}^{0}$$

(Note : Don't forget to write modulus for log function)

$$= \frac{1}{40}\left[\log\left| \frac{5 + 0}{5 - 0} \right| - \log\left| \frac{5 - 4}{5 + 4} \right| \right]$$

$$= \frac{1}{40}\left[\log 1 - \log\left(\frac{1}{9} \right) \right], \qquad \text{(where } \log 1 = 0\text{)}$$

$$= \frac{1}{40}[\log 9] = \frac{2}{40}\log 3 = \frac{1}{20}(\log 3)$$

Example 21 Evaluate $\int_{\alpha}^{\beta} \sqrt{\frac{x-\alpha}{\beta-x}}\, dx.$

Solution. Let $I = \int_{\alpha}^{\beta} \sqrt{\frac{x-\alpha}{\beta-x}}\, dx$ ...(i)

Put $x = \alpha \cos^2 t + \beta \sin^2 t$

$\therefore$ $x - \alpha = \alpha \cos^2 t + \beta \sin^2 t - \alpha = \beta (\sin^2 (t) + \alpha (\cos^2 t - 1))$

$= \beta \sin^2 t - \alpha \sin^2 t \Rightarrow (x - \alpha) = (\beta - \alpha) \sin^2 t$

Similarly, $\beta - x = (\beta - \alpha) \cos^2 t$

$\therefore$ $dx = 2(\beta - \alpha) \sin t \cos t\, dt$

where, $x = \alpha \Rightarrow \sin^2 t = 0 \therefore t = 0$ $(\because \alpha < \beta)$

and $x = \beta \Rightarrow \cos^2 t = 0 \therefore t = \pi/2$

$\therefore$ Eq. (i) reduces to;

$$I = \int_0^{\pi/2} \sqrt{\frac{(\beta-\alpha)\sin^2 t}{(\beta-\alpha)\cos^2 t}} \cdot 2(\beta-\alpha)\sin t \cos t\, dt$$

$$= \int_0^{\pi/2} \frac{\sin t}{\cos t} \cdot 2(\beta-\alpha)\sin t \cos t\, dt$$

$$\left[\because \sqrt{\frac{\sin^2 t}{\cos^2 t}} = + \frac{\sin t}{\cos t} \text{ as } t \in [0, \pi/2]\right]$$

$$= (\beta-\alpha) \int_0^{\pi/2} 2\sin^2 t\, dt$$

$$= (\beta-\alpha) \int_0^{\pi/2} (1 - \cos 2t)\, dt$$

$$= (\beta-\alpha) \left(t - \frac{\sin 2t}{2} \right)_0^{\pi/2}$$

$$= (\beta-\alpha) \left[\left(\frac{\pi}{2} - \frac{1}{2}\sin \pi \right) - \left(0 - \frac{1}{2}\sin 0 \right) \right]$$

$$= (\beta-\alpha) \left[\frac{\pi}{2} \right]$$

$$\therefore \quad I = \frac{\pi}{2}(\beta - \alpha)$$

Aliter $I = \int_{\alpha}^{\beta} \sqrt{\frac{x-\alpha}{\beta-x}}\, dx$

$$\left[\begin{array}{l} \text{put } \sqrt{\beta - x} = t \\ \therefore \beta - x = t^2 \\ -dx = 2t\, dt \end{array}\right]$$

when $x = \alpha$ $\quad x = \beta$

$\beta - \alpha = t^2$ $\quad 0 = t^2$

$\sqrt{\beta - \alpha} = t$ $\quad 0 = t$

$$\therefore \quad I = \int_{\sqrt{\beta-\alpha}}^{0} \frac{\sqrt{(\beta - t^2) - \alpha}}{t} (-2t)\, dt$$

$$= -2 \int_{\sqrt{\beta-\alpha}}^{0} \sqrt{(\beta-\alpha) - t^2}\, dt$$

$$I = 2 \int_0^{\sqrt{\beta-\alpha}} \sqrt{(\sqrt{\beta-\alpha})^2 - t^2}\, dt$$

$$= 2\left\{\frac{t}{2}\sqrt{(\beta-\alpha)-t^2}+\frac{(\beta-\alpha)}{2}\sin^{-1}\left(\frac{t}{\sqrt{\beta-\alpha}}\right)\right\}_0^{\sqrt{\beta-\alpha}}$$

$$I = 2\left[\left\{0+\frac{(\beta-\alpha)}{2}\sin^{-1}(1)\right\}-\left\{0+\frac{(\beta-\alpha)}{2}\sin^{-1}(0)\right\}\right]$$

$$= \frac{2(\beta-\alpha)}{2}\cdot\frac{\pi}{2}$$

$\therefore$ $$I = \frac{\pi}{2}(\beta-\alpha)$$

Example 22 Evaluate $\displaystyle\int_{\sqrt{(3a^2+b^2)/4}}^{\sqrt{(a^2+b^2)/2}} \frac{x}{\sqrt{(x^2-a^2)(b^2-x^2)}}\,dx.$

Solution. Let $\displaystyle\int_{\sqrt{(3a^2+b^2)/4}}^{\sqrt{(a^2+b^2)/2}} \frac{x}{\sqrt{(x^2-a^2)(b^2-x^2)}}\,dx$ ...(i)

Put $$x^2 = a^2\cos^2 t + b^2\sin^2 t$$

$\Rightarrow$ $$2x\,dx = [2a^2\cos t(-\sin t)+2b^2\sin t(\cos t)]\,dt$$

$\Rightarrow$ $$x\,dx = \frac{1}{2}(b^2-a^2)\sin 2t\,dt$$

For $$x^2 = \frac{a^2+b^2}{2} = a^2\cos^2 t + b^2\sin^2 t$$

$\Rightarrow$ $$a^2+b^2 = 2(1-\sin^2 t)a^2 + 2b^2\sin^2 t$$

or $$(a^2+b^2) = 2a^2 + 2(b^2-a^2)\sin^2 t$$

$\Rightarrow$ $$\sin^2 t = \frac{1}{2} \Rightarrow \cos 2t = 0$$

$\Rightarrow$ $$t = \frac{\pi}{4}$$

For $$x^2 = \frac{3a^2+b^2}{4} = a^2\cos^2 t + b^2\sin^2 t$$

$$3a^2+b^2 = 4a^2 + 4(b^2-a^2)\sin^2 t$$

$$\sin^2 t = \frac{1}{4} \Rightarrow \cos 2t = \frac{1}{2} \Rightarrow t = \frac{\pi}{6}$$

$\therefore$ Eq. (i) reduces to;

$$I = \int_{\pi/6}^{\pi/4}\frac{1}{2}\cdot\frac{(b^2-a^2)\sin 2t}{\sqrt{(b^2-a^2)\sin^2 t\,(b^2-a^2)\cos^2 t}}\,dt$$

$$= \int_{\pi/6}^{\pi/4} dt = \left(t\right)_{\pi/6}^{\pi/4} = \left(\frac{\pi}{4}-\frac{\pi}{6}\right)$$

$\therefore$ $$I = \frac{\pi}{12}$$

Example 23 Evaluate $\displaystyle\int_0^{\pi}\frac{x\,dx}{1+\cos\alpha\cdot\sin x}\,(0<\alpha<\pi)$

Solution. Let $\displaystyle I = \int_0^{\pi}\frac{x\,dx}{1+\cos\alpha\cdot\sin x}$ ...(i)

$$I = \int_0^{\pi} \frac{(\pi - x)\,dx}{1 + \cos\alpha \cdot \sin(\pi - x)}$$

$$I = \int_0^{\pi} \frac{(\pi - x)\,dx}{1 + \cos\alpha \cdot \sin x} \quad \ldots\text{(ii)}$$

Adding Eqs. (i) and (ii), $2I = \int_0^{\pi} \left(\frac{x}{1 + \cos\alpha \cdot \sin x} + \frac{\pi - x}{1 + \cos\alpha \cdot \sin x} \right) dx$

$$= \pi \int_0^{\pi} \frac{1}{1 + \cos\alpha \cdot \sin x}\,dx$$

$$= \pi \int_0^{\pi} \frac{\sec^2 x/2}{1 + \tan^2 x/2 + 2\cos\alpha \tan x/2}\,dx \qquad \text{put } \tan\frac{x}{2} = t$$

$$\Rightarrow \quad 2I = \pi \int_0^{\infty} \frac{2\,dt}{(1 + t^2) + 2t\cos\alpha}$$

$$2I = 2\pi \int_0^{\infty} \frac{dt}{(t + \cos\alpha)^2 + \sin^2\alpha}$$

$$I = \frac{\pi}{\sin\alpha} \left[\tan^{-1}\left(\frac{t + \cos\alpha}{\sin\alpha} \right) \right]_0^{\infty}$$

$$= \frac{\pi}{\sin\alpha} [\tan^{-1}\infty - \tan^{-1}(\cot\alpha)]$$

$$= \frac{\pi}{\sin\alpha} \left[\frac{\pi}{2} - \left(\frac{\pi}{2} - \alpha \right) \right]$$

$$I = \frac{\alpha\pi}{\sin\alpha}$$

Example 24 Evaluate $\int_0^{\pi/4} \frac{e^{\sec x}\left(\sin\left(x + \frac{\pi}{4}\right)\right)}{\cos x\,(1 - \sin x)}\,dx.$

Solution. Let $I = \int_0^{\pi/4} \frac{e^{\sec x}\left(\sin\left(x + \frac{\pi}{4}\right)\right)}{\cos x\,(1 - \sin x)}\,dx$

$$= \frac{1}{\sqrt{2}} \int_0^{\pi/4} \frac{e^{\sec x}(\sin x + \cos x)(1 + \sin x)}{\cos x\,(1 - \sin^2 x)}\,dx$$

$$= \frac{1}{\sqrt{2}} \int_0^{\pi/4} \frac{e^{\sec x}(\sin x + \cos x)(1 + \sin x)}{\cos^2 x \cdot \cos x}\,dx$$

$$= \frac{1}{\sqrt{2}} \int_0^{\pi/4} e^{\sec x} \cdot (\sec x \tan x + \sec x)(\sec x + \tan x)\,dx$$

$$= \frac{1}{\sqrt{2}} \int_0^{\pi/4} e^{\sec x} \cdot (\sec x \tan x\,(\sec x + \tan x) + \sec^2 x + \sec x \tan x)\,dx$$

$$= \frac{1}{\sqrt{2}} \int_0^{\pi/4} e^{\sec x} \cdot (\sec x \tan x\,(\sec x + \tan x + 1) + \sec^2 x)\,dx$$

$$= \frac{1}{\sqrt{2}} \int_0^{\pi/4} e^{\sec x} \cdot \sec x \tan x\,(\sec x + \tan x + 1) + \frac{1}{\sqrt{2}} \int_0^{\pi/4} e^{\sec x} \cdot \sec^2 x\,dx$$

Applying integration by parts

$$= \frac{1}{\sqrt{2}}\{(\sec x + \tan x + 1)\, e^{\sec x}\}_0^{\pi/4} - \frac{1}{\sqrt{2}}\int_0^{\pi/4} (\sec x \tan x + \sec^2 x) \cdot e^{\sec x} dx + \frac{1}{\sqrt{2}}\int_0^{\pi/4} e^{\sec x} \cdot \sec^2 x\, dx$$

$$= \frac{1}{\sqrt{2}}\{(\sqrt{2}+1+1)\, e^{\sqrt{2}} - (1+1)\, e\} - \frac{1}{\sqrt{2}}[e^{\sec x}]_0^{\pi/4}$$

$$= \frac{1}{\sqrt{2}}(\sqrt{2}+2)\, e^{\sqrt{2}} - \sqrt{2}e - \frac{1}{\sqrt{2}}(e^{\sqrt{2}} - e) = \frac{1}{\sqrt{2}}(1+\sqrt{2})\, e^{\sqrt{2}} - \frac{e}{\sqrt{2}}$$

Example 25 Evaluate $\displaystyle\int_0^{\pi/4} \frac{x^2 (\sin 2x - \cos 2x)\, dx}{(1 + \sin 2x)\cos^2 x}$.

Solution. Let $\displaystyle I = \int_0^{\pi/4} \frac{x^2 (\sin 2x - \cos 2x)\, dx}{(1+\sin 2x)\cos^2 x}$, put $2x = t$

$$= \frac{1}{8}\int_0^{\pi/2} \frac{t^2 (\sin t - \cos t)}{(1+\sin t)(\cos^2 t/2)}\, dt$$

$$I = \frac{1}{4}\int_0^{\pi/2} \frac{t^2 (\sin t - \cos t)}{(1+\sin t)(1+\cos t)}\, dt \qquad \text{...(i)}$$

$$\left[\text{applying, } \int_0^a f(x)\, dx = \int_0^a f(a-x)\, dx\right]$$

$$I = \frac{1}{4}\int_0^{\pi/2} \frac{\left(\frac{\pi}{2} - t\right)^2 \cdot (\cos t - \sin t)}{(1+\cos t)(1+\sin t)}\, dt \qquad \text{...(ii)}$$

Adding Eqs. (i) and (ii), we get

$$2I = \frac{1}{4}\int_0^{\pi/2} \frac{\left(\pi t - \frac{\pi^2}{4}\right)(\sin t - \cos t)}{(1+\cos t)(1+\sin t)}\, dt \qquad \text{...(iii)}$$

where, $\displaystyle\int_0^{\pi/2} \frac{\sin t - \cos t}{(1+\cos t)(1+\sin t)}\, dt = 0 \quad \left[\text{as } \int_0^{2a} f(x)\, dx = 0, \text{ if } f(2a-x) = -f(x)\right]$

$\therefore$ Eq. (iii) becomes,

$$I = \frac{1}{8}\int_0^{\pi/2} \frac{\pi t (\sin t - \cos t)\, dt}{(1+\cos t)(1+\sin t)} + 0$$

$$= \frac{\pi}{8}\int_0^{\pi/2} \frac{t\{(1+\sin t) - (1+\cos t)\}\, dt}{(1+\cos t)(1+\sin t)}$$

$$= \frac{\pi}{8}\int_0^{\pi/2} \frac{t\, dt}{1+\cos t} - \frac{\pi}{8}\int_0^{\pi/2} \frac{t\, dt}{1+\sin t}$$

$$= \frac{\pi}{8}\int_0^{\pi/2} \frac{t\, dt}{1+\cos t} - \frac{\pi}{8}\int_0^{\pi/2} \frac{(\pi/2 - t)\, dt}{1+\cos t}$$

$$= \frac{2\pi}{8}\int_0^{\pi/2} \frac{t\, dt}{1+\cos t} - \frac{\pi^2}{16}\int_0^{\pi/2} \frac{dt}{1+\cos t}$$

$$= \frac{\pi}{8}\int_0^{\pi/2} t(\sec^2 t/2)\, dt - \frac{\pi^2}{16}\int_0^{\pi/2} (\sec^2 t/2)\, dt$$

$$= \frac{\pi}{8}\left\{\left(2t \cdot \tan\frac{t}{2}\right)_0^{\pi/2} - \int_0^{\pi/2} 1 \cdot 2\tan\frac{t}{2}\,dt\right\} - \frac{\pi^2}{16}\left(2\tan\frac{t}{2}\right)_0^{\pi/2}$$

$$= \frac{\pi}{8}\left\{(\pi - 0) + 4\left(\log\cos\frac{t}{2}\right)_0^{\pi/2}\right\} - \frac{\pi^2}{8}\{1 - 0\}$$

$$= \frac{\pi}{8}\left\{\pi + 4\log\frac{1}{\sqrt{2}}\right\} - \frac{\pi^2}{8} = \frac{\pi^2}{8} + \pi\log\frac{1}{\sqrt{2}} - \frac{\pi^2}{8}$$

$$I = \frac{\pi^2}{8} - \frac{1}{2}\pi\log 2$$

$$\Rightarrow \quad I = \frac{\pi^2}{16} - \frac{\pi}{4}\log 2$$

Example 26 Evaluate $\int_0^{\pi} \frac{x^2\cos x}{(1+\sin x)^2}\,dx.$

Solution. Let $I = \int_0^{\pi} \frac{x^2\cos x}{(1+\sin x)^2}\,dx$

$$= \int_0^{\pi} x^2\{(1+\sin x)^{-2}\cos x\}\,dx$$

Applying integration by parts,

$$I = \left(x^2\frac{(1+\sin x)^{-1}}{-1}\right)_0^{\pi} - \int_0^{\pi} 2x \cdot \frac{(1+\sin x)^{-1}}{-1}\,dx$$

$$= (-\pi^2 + 0) + 2\int_0^{\pi}\frac{x}{1+\sin x}\,dx \qquad \text{...(i)}$$

$$\left[\text{using } \int_0^a f(x)\,dx = \int_0^a f(a-x)\,dx\right]$$

$$I = -\pi^2 + 2\int_0^{\pi}\frac{(\pi - x)}{1+\sin x}\,dx \qquad \text{...(ii)}$$

Adding Eqs. (i) and (ii), we get

$$2I = -2\pi^2 + 2\int_0^{\pi}\frac{\pi}{1+\sin x}\,dx$$

$$2I = -2\pi^2 + 2\pi\int_0^{\pi}\frac{1-\sin x}{\cos^2 x}\,dx$$

$$= -2\pi^2 + 2\pi\,[\tan x - \sec x]_0^{\pi}$$

$$= -2\pi^2 + 2\pi\,[0 - (-1 - 1)]$$

$$2I = -2\pi^2 + 4\pi$$

$$I = -\pi^2 + 2\pi$$

Example 27 Compute the following integrals :

(i) $\int_0^{\infty} f(x^n + x^{-n})\ln x\,\frac{dx}{x} = 0$ (ii) $\int_0^{\infty} f(x^n + x^{-n})\ln x\,\frac{dx}{1+x^2} = 0$

Solution. (i) Let $t = \ln x \Rightarrow x = e^t$

$$dx = e^t\,dt \quad \text{or} \quad \frac{dx}{x} = dt$$

Also, $\quad x = 0 \;\Rightarrow\; t = -\infty$

and $\quad x = \infty \;\Rightarrow\; t = \infty$

Thus, Eq. (i) $\displaystyle\int_0^{\infty} f(x^n + x^{-n}) \ln x \frac{dx}{x} = \int_{-\infty}^{\infty} f(e^{nt} + e^{-nt}) \cdot t\, dt = 0$

(∵ integrand is an odd function of t)

(ii) $\displaystyle\int_0^{\infty} f(x^n + x^{-n}) \ln x \cdot \frac{dx}{1 + x^2} = \int_{-\infty}^{\infty} f(e^{nt} + e^{-nt}) \cdot t \cdot \frac{e^t}{1 + e^{2t}} \cdot dt$

Now, if $\quad h(t) = f(e^{nt} + e^{-nt}) \cdot \dfrac{t\, e^t}{1 + e^{2t}}$

Then, $\quad h(-t) = f(e^{-nt} + e^{nt}) \cdot (-t) \cdot \dfrac{1}{e^t + e^{-t}} \qquad \left[\because \dfrac{e^t}{1 + e^{2t}} = \dfrac{1}{e^{-t} + e^t}\right]$

$h(-t) = -h(t)$

Thus, integrand is an odd function and hence

$$\int_0^{\infty} f(x^n + x^{-n}) \ln x \cdot \frac{dx}{1 + x^2} = 0$$

Example 28 Show that

(a) $\displaystyle\int_0^{\infty} \sin x\, dx = 1$ $\qquad$ (b) $\displaystyle\int_0^{\infty} \cos x\, dx = 0$

Solution. Let us first evaluate;

$$I = \int e^{-sx} \sin x\, dx \quad \text{and} \quad J = \int e^{-sx} \cos x\, dx$$

Using integer by parts, we get

$$I = -e^{-sx} \cos x - s\, J \qquad \text{...(i)}$$

$$J = e^{-sx} \sin x + s\, I \qquad \text{...(ii)}$$

Subtracting Eqs. (i) and (ii), we get

$$I = -e^{-sx} \left| \frac{\cos x + s \sin x}{1 + s^2} \right|$$

$$\Rightarrow \quad J = e^{-sx} \left| \sin x - \frac{s^2}{s^2 + 1} \sin x - \frac{s}{1 + s^2} \cos x \right|.$$

$$= e^{-sx} \left| \frac{\sin x - s \cos x}{1 + s^2} \right|$$

Thus, $\quad \displaystyle\int_0^{\infty} e^{-sx} \sin x\, dx = \frac{1}{s^2 + 1}$

$$\int_0^{\infty} e^{-sx} \cos x\, dx = \frac{s}{s^2 + 1}$$

Now, $\quad \displaystyle\int_0^{\infty} \sin x\, dx = \lim_{s \to 0} \int_0^{\infty} e^{-sx} \sin x\, dx = \lim_{s \to 0} \frac{1}{s^2 + 1} = 1$

and $\quad \displaystyle\int_0^{\infty} \cos x\, dx = \lim_{s \to 0} \int_0^{\infty} e^{-sx} \cos x\, dx$

$$= \lim_{s \to 0} \frac{s}{s^2 + 1} = 0$$

Example 29 Evaluate $\int_0^{(2n-1)\pi} [\sin x]\, dx,\ n \in N.$

Solution. The graph of $[\sin x]$ for $0 \le x \le 2\pi$ is shown below :

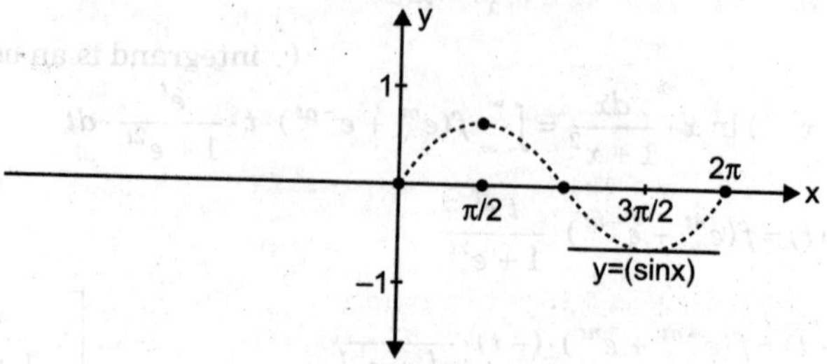

It is obvious that

$$f(x) = [\sin x] = \begin{cases} 1, & \text{for } x = (4n+1)\pi/2 \\ 0, & \text{for } x \in [2n\pi, (4n+1)\pi/2) \cup ((4n+1)\pi/2), (2n+1)\pi] \\ -1, & \text{for } x \in ((2n+1)\pi, (2n+2)\pi) \end{cases}$$

Hence, $$\int_0^{(2n-1)\pi} [\sin x]\, dx = \sum_{r=1}^{n-1} \int_{(2r-1)\pi}^{2r\pi} (-1)\, dx$$

$$= (-1) \sum_{r=1}^{n-1} (2r\pi - (2r-1)\pi) = (-1) \sum_{r=1}^{n-1} \pi$$

$$= (1-n)\pi$$

$\Rightarrow$ $$\int_0^{(2n-1)\pi} [\sin x]\, dx = (1-n)\pi$$

Example 30 Find a function $g : R \to R$, continuous in $[0, \infty)$ and positive in $(0, \infty)$ satisfying $g(0) = 1$ and $\frac{1}{2}\int_0^x g^2(t)\, dt = \frac{1}{x}\left(\int_0^x g(t)\, dt\right)^2.$

Solution. Let $$f(x) = \int_0^x g(t)\, dt$$

$\Rightarrow$ $$f'(x) = g(x)$$

$\Rightarrow$ $$\frac{1}{2}\int_0^x (f'(t))^2\, dt = \frac{1}{2}\int_0^x g^2(t)\, dt \qquad \text{...(i)}$$

$\Rightarrow$ $$\frac{1}{2}\left[\int_0^x g(t)\right]^2 = \frac{1}{2}(f(x))^2$$

$\Rightarrow$ $$\frac{1}{2}\frac{\left[\int_0^x g(t)\right]^2}{x} = \frac{1}{2}\frac{(f(x))^2}{x}$$

$\Rightarrow$ $$\frac{1}{2}\cdot\frac{1}{2}\left[\int_0^x g^2(t)\right] dt = \frac{1}{2}\frac{(f(x))^2}{x}$$

$\Rightarrow$ $$\frac{1}{2}\int_0^x (f'(t))^2\, dt = \frac{1}{2}\frac{(f(x))^2}{x} \qquad \text{[using Eq. (i)]}$$

Differentiating both the sides w.r.t. x, we get

$$\frac{1}{2}[f'(x)]^2 = \frac{x\, 2f(x)f'(x) - (f(x))^2}{x^2}$$

$\Rightarrow$ $$\frac{1}{2}[g(x)]^2 x^2 = 2x\, f(x)(g(x)) - (f(x))^2$$

$$\Rightarrow \quad \left(\frac{x\,g(x)}{f(x)}\right)^2 = 4\left(\frac{x\,g(x)}{f(x)}\right) - 2$$

$$\Rightarrow \quad t^2 - 4t + 2 = 0, \text{ where } t = \frac{x\,g(x)}{f(x)}$$

$$\Rightarrow \quad t = \frac{4 \pm \sqrt{8}}{2} = 2 \pm \sqrt{2}$$

or
$$\frac{x\,f'(x)}{f(x)} = 2 \pm \sqrt{2}$$

$$\Rightarrow \quad \ln f(x) = 2 \pm \sqrt{2}\ln x + \text{constant}$$

$$\Rightarrow \quad f(x) = c\,x^{2+\sqrt{2}} \quad \text{or} \quad c\,x^{2-\sqrt{2}}$$

$$\Rightarrow \quad g(x) = f'(x) = c_1\,x^{1+\sqrt{2}} \quad \text{or} \quad c_2\,x^{1-\sqrt{2}}$$

Where c_1 and c_2 are constants of integration.

But g is continuous on $[0, \infty)$, then $c_2\,x^{1-\sqrt{2}}$ is ruled out.

Hence, $\quad g(x) = c_1\,x^{1+\sqrt{2}}$

Also, $\quad g(0) = c^1 = 1$

$$\Rightarrow \quad g(x) = x^{1+\sqrt{2}}$$

Example 31 Let $I_n = \int_0^{\pi/4} \tan^n x\,dx$ ($n > 1$ and is an integer). Show that,

(i) $I_n + I_{n-2} = \dfrac{1}{n-1}$ (ii) $\dfrac{1}{2(n+1)} < I_n < \dfrac{1}{2(n-1)}$

Solution. (i) $I_n = \int_0^{\pi/4} \tan^n x\,dx$

$$= \int_0^{\pi/4} \tan^{n-2} x \cdot \tan^2 x\,dx$$

$$= \int_0^{\pi/4} \sec^2 x \cdot \tan^{n-2} x\,dx - \int_0^{\pi/4} \tan^{n-2} x\,dx$$

$$= \int_0^1 t^{n-2}\,dt - I_{n-2}, \qquad \text{where } t = \tan x$$

$$I_n + I_{n-2} = \left(\frac{t^{n-1}}{n-1}\right)_0^1$$

$$\therefore \quad I_n + I_{n-2} = \frac{1}{n-1}$$

(ii) For $0 < x < \pi/4$, we have $0 < \tan^n x < \tan^{n-2} x$

So that; $\quad 0 < I_n < I_{n-2}$

$$\therefore \quad I_n + I_{n+2} < 2I_n < I_n + I_{n-2}$$

$$\Rightarrow \quad \frac{1}{n+1} < 2I_n < \frac{1}{n-1}$$

$$\Rightarrow \quad \frac{1}{2(n+1)} < I_n < \frac{1}{2(n-1)}$$

Example 32 If $U_n = \int_0^\pi \frac{1-\cos nx}{1-\cos x}\,dx$, where n is positive integer or zero, then show that $U_{n+2} + U_n = 2U_{n+1}$. Hence, deduce that $\int_0^{\pi/2} \frac{\sin^2 n\theta}{\sin^2\theta} = \frac{1}{2}n\pi$.

Solution.

$$U_{n+2} - U_{n+1} = \int_0^\pi \frac{(1-\cos(n+2)x) - (1-\cos(n+1)x)}{1-\cos x}\,dx$$

$$= \int_0^\pi \frac{\cos(n+1)x - \cos(n+2)x}{1-\cos x}\,dx$$

$$= \int_0^\pi \frac{2\sin\left(n+\frac{3}{2}\right)x \cdot \sin\frac{x}{2}}{2\sin^2\frac{x}{2}}\,dx$$

$$U_{n+2} - U_{n+1} = \int_0^\pi \frac{\sin\left(n+\frac{3}{2}\right)x}{\sin\frac{x}{2}}\,dx \quad \text{...(i)}$$

$$\Rightarrow \quad U_{n+1} - U_n = \int_0^\pi \frac{\sin\left(n+\frac{1}{2}\right)x}{\sin\frac{x}{2}}\,dx \quad \text{...(ii)}$$

From Eqs. (i) and (ii), we get

$$(U_{n+2} - U_{n+1}) - (U_{n+1} - U_n) = \int_0^\pi \frac{\sin\left(n+\frac{3}{2}\right)x - \sin\left(n+\frac{1}{2}\right)x}{\sin\frac{x}{2}}\,dx$$

$$U_{n+2} + U_n - 2U_{n+1} = \int_0^\pi \frac{2\cos(n+1)x \cdot \sin x/2}{\sin x/2}\,dx$$

$$= 2\int_0^\pi \cos(n+1)x\,dx = 2\left(\frac{\sin(n+1)x}{n+1}\right)_0^\pi = 0$$

$$\Rightarrow \quad U_{n+2} + U_n = 2U_{n+1}$$

$\Rightarrow$ U_n, U_{n+1}, U_{n+2} are in AP.

Now,

$$U_0 = \int_0^\pi \frac{1-1}{1-\cos x}\,dx = 0$$

$$U_1 = \int_0^\pi \frac{1-\cos x}{1-\cos x}\,dx = \pi$$

$$U_1 - U_0 = \pi \quad \text{(common difference)}$$

$$U_n = U_0 + n\pi = n\pi$$

$$U_n = n\pi$$

Now,

$$I_n = \int_0^{\pi/2} \frac{\sin^2 n\theta}{\sin^2\theta}\,d\theta = \int_0^{\pi/2} \frac{1-\cos 2n\theta}{1-\cos 2\theta}\,d\theta$$

$$= \frac{1}{2}\int_0^\pi \frac{1-\cos nx}{1-\cos x}\,dx$$

$$I_n = \frac{1}{2}n\pi$$

Example 33 Evaluate $\int_{-1/\sqrt{3}}^{1/\sqrt{3}} \frac{\cos^{-1}\left[\frac{2x}{x^2+1}\right] + \tan^{-1}\left[\frac{2x}{1-x^2}\right]}{e^x+1}\, dx.$

Solution. Let $\tan^{-1} x = \theta$ as $-\frac{\pi}{6} \le x \le \frac{\pi}{6}$ and $-\frac{1}{\sqrt{3}} \le x \le \frac{1}{\sqrt{3}}$

Now, $\cos^{-1}\left(\frac{2x}{x^2+1}\right) + \tan^{-1}\left(\frac{2x}{1-x^2}\right)$

$$= \cos^{-1}\left(\frac{2\tan\theta}{1+\tan^2\theta}\right) + \tan^{-1}\left(\frac{2\tan\theta}{1-\tan^2\theta}\right)$$

$$= \cos^{-1}(\sin 2\theta) + \tan^{-1}(\tan 2\theta)$$

$$= \left(\frac{\pi}{2} - 2\theta\right) + (2\theta); \quad \left(\text{as } -\frac{\pi}{3} \le 2\theta \le \frac{\pi}{3} \text{ and } \frac{\pi}{6} \le \frac{\pi}{2} - 2\theta \le \frac{5\pi}{6}\right)$$

$$= \frac{\pi}{2}$$

So, $$I = \int_{-1/\sqrt{3}}^{1/\sqrt{3}} \frac{\cos^{-1}\left(\frac{2x}{x^2+1}\right) + \tan^{-1}\left(\frac{2x}{1-x^2}\right)}{e^x+1}\, dx$$

$$= \frac{\pi}{2}\int_{-1/\sqrt{3}}^{1/\sqrt{3}} \frac{dx}{e^x+1} = \frac{\pi}{2}\left[\int_0^{1/\sqrt{3}} \frac{dx}{e^x+1} + \int_0^{1/\sqrt{3}} \frac{dx}{e^{-x}+1}\right]$$

$$= \frac{\pi}{2}\left[\int_0^{1/\sqrt{3}} dx\right] = \frac{\pi}{2} \times \frac{1}{\sqrt{3}}$$

Example 34 Prove that for any positive integer K,

$$\frac{\sin 2Kx}{\sin x} = 2[\cos x + \cos 3x + \ldots + \cos(2K-1)x]$$

Hence, prove that $\int_0^{\pi/2} \sin 2Kx \cdot \cot x\, dx = \frac{\pi}{2}$.

Solution. To prove; $\sin 2Kx = 2\sin x[\cos x + \cos 3x + \cos 5x + \ldots + \cos(2K-1)x]$

Taking RHS

$$[2\sin x\cos x + 2\sin x\cos 3x + 2\sin x\cos 5x + \ldots + 2\sin x\cos(2K-1)x]$$

$\Rightarrow$ $(\sin 2x) + (\sin 4x - \sin 2x) + (\sin 6x - \sin 4x) + \ldots + (\sin 2Kx - \sin(2K-2)x)$

$\Rightarrow$ $\sin 2Kx$

Now, $$\int_0^{\pi/2} \sin 2Kx \cdot \cot x\, dx = \int_0^{\pi/2}\left(\frac{\sin 2Kx}{\sin x}\right)\cdot \cos x\, dx$$

$$= \int_0^{\pi/2} 2\cos x[\cos x + \cos 3x + \ldots + \cos(2K-1)x]\, dx$$

$$= \int_0^{\pi/2}(1+\cos 2x)\, dx + \int_0^{\pi/2}(\cos 4x + \cos 2x)\, dx$$

$$+ \ldots + \int_0^{\pi/2}(\cos 2Kx + \cos(2K-2)x)\, dx$$

But we know that, $\int_0^{\pi/2} (\cos 2n\, x)\, dx = 0, \forall\, n \in I$ and $n \neq 0$

$$\Rightarrow \quad \int_0^{\pi/2} \sin 2K\, x \cdot \cot x\, dx = \int_0^{\pi/2} 1\, dx + 0 = \frac{\pi}{2}$$

Example 35 Evaluate $\int_0^{\sqrt{3}} \frac{1}{1+x^2} \cdot \sin^{-1}\left(\frac{2x}{1+x^2}\right) dx.$

Solution. Let $I = \int_0^{\sqrt{3}} \frac{1}{1+x^2} \sin^{-1}\left(\frac{2x}{1+x^2}\right) dx$

Now, $\sin^{-1}\left(\frac{2x}{1+x^2}\right) = \begin{cases} 2\tan^{-1} x, & \text{if } -1 \le x \le 1 \\ \pi - 2\tan^{-1} x, & \text{if } x > 1 \end{cases}$

$$I = \int_0^1 \frac{1}{1+x^2} \sin^{-1}\left(\frac{2x}{1+x^2}\right) dx + \int_1^{\sqrt{3}} \frac{1}{1+x^2} \cdot \sin^{-1}\left(\frac{2x}{1+x^2}\right) dx$$

$$= \int_0^1 \frac{2\tan^{-1} x}{1+x^2} dx + \int_1^{\sqrt{3}} \frac{\pi - 2\tan^{-1} x}{1+x^2} dx$$

$$= 2\int_0^1 \frac{\tan^{-1} x}{1+x^2} dx + \pi \int_1^{\sqrt{3}} \frac{1}{1+x^2} dx - 2\int_1^{\sqrt{3}} \frac{\tan^{-1} x}{1+x^2} dx$$

$$= 2\int_0^{\pi/4} t\, dt + \pi\, (\tan^{-1} x)_1^{\sqrt{3}} - 2\int_{\pi/4}^{\pi/3} t\, dt, \qquad (\text{put } \tan^{-1} x = t)$$

$$= 2\left(\frac{t^2}{2}\right)_0^{\pi/4} + \pi\, \{\tan^{-1}\sqrt{3} - \tan^{-1} 1\} - 2\left(\frac{t^2}{2}\right)_{\pi/4}^{\pi/3}$$

$$= \frac{\pi^2}{16} + \pi\left\{\frac{\pi}{3} - \frac{\pi}{4}\right\} - \left\{\frac{\pi^2}{9} - \frac{\pi^2}{16}\right\} = \frac{7}{72}\pi^2$$

Example 36 Prove that $\int_0^x e^{xt} \cdot e^{-t^2}\, dt = e^{x^2/4} \int_0^x e^{-t^2/4}\, dt.$

Solution. Let $I = \int_0^x e^{xt} \cdot e^{-t^2}\, dt$

Put $t = \frac{x+z}{2},\ dt = \frac{1}{2} dz$

$$\therefore \quad I = \int_0^x e^{tx} \cdot e^{-t^2}\, dt = \frac{1}{2}\int_{-x}^x e^{\left(\frac{x+z}{2}\right)x} \cdot e^{-\left(\frac{x+z}{2}\right)^2} dz$$

$$= \frac{1}{2}\int_{-x}^x e^{\left(\frac{x+z}{2}\right)\left(x - \frac{x+z}{2}\right)} dz = \frac{1}{2}\int_{-x}^x e^{\left(\frac{x+z}{2}\right)\left(\frac{x-z}{2}\right)} dz$$

$$= \frac{1}{2}\int_{-x}^x e^{\frac{x^2}{4} - \frac{z^2}{4}} dz = \frac{1}{2} e^{x^2/4} \cdot \int_{-x}^x e^{-z^2/4}\, dz$$

$$= \frac{1}{2} e^{x^2/4} \int_{-x}^x e^{-t^2/4}\, dt = \frac{1}{2} e^{x^2/4} \cdot 2\int_0^x e^{-t^2/4}\, dt$$

$$\left[\text{using } \int_{-a}^a f(x)\, dx = 2\int_0^a f(x)\, dx, \text{ when } f(-x) = f(x)\right]$$

$$= e^{x^2/4} \int_0^x e^{-t^2/4}\, dt$$

$$\Rightarrow \quad \int_0^x e^{xt} \cdot x^{-t^2}\, dt = e^{x^2/4} \int_0^x e^{-t^2/4}\, dt$$

Example 37 If $f(x) = e^x + \int_0^1 (e^x + te^{-x}) f(t)\, dt$, find $f(x)$.

Solution. We can write $f(x) = Ae^x + Be^{-x}$, where

$$A = 1 + \int_0^1 f(t)\, dt \quad \text{and} \quad B = \int_0^1 t\, f(t)\, dt$$

$$\therefore \quad A = 1 + \int_0^1 (Ae^t + Be^{-t})\, dt$$

$$= 1 + (Ae^t - Be^{-t})_0^t$$

$$A = 1 + A(e^1 - 1) - B(e^{-1} - 1)$$

$$\Rightarrow \quad (2 - e)A + (e^{-1} - 1)B = 1 \qquad \text{...(i)}$$

$$B = \int_0^1 t(Ae^t + Be^{-t})\, dt$$

$$= A(te^t - e^t)_0^t + B(-te^{-t} - e^{-t})_0^1$$

$$B = A + B(1 - 2e^{-1})$$

$$\Rightarrow \quad A - 2e^{-1}B = 0 \qquad \text{...(ii)}$$

From Eqs. (i) and (ii), we get

$$A = \frac{2(e-1)}{4e - 2e^2},\ B = \frac{e-1}{4-2e}$$

Hence,
$$f(x) = \frac{2(e-1)}{4e - 2e^2} \cdot e^x + \frac{e-1}{4-2e} \cdot e^{-x}$$

Example 38 If f, g, h be continuous functions on $[0, a]$ such that

$$f(a - x) = f(x),\ g(a - x) = -g(x) \text{ and } 3h(x) - 4h(a - x) = 5,$$

then prove $\int_0^a f(x) \cdot g(x) \cdot h(x) = 0.$

Solution.
$$I = \int_0^a f(x) \cdot g(x) \cdot h(x)\, dx$$

$$= \int_0^a f(a - x) \cdot g(a - x) \cdot h(a - x)\, dx$$

$$= -\int_0^a f(x) \cdot g(x)\, h(a - x)\, dx$$

$$7I = 3I + 4I$$

$$= \int_0^a f(x)\, g(x)\, \{3h(x) - 4h(a - x)\}\, dx$$

$$= 5\int_0^a f(x)\, g(x)\, dx = 0$$

Since, $f(a - x)\, g(a - x) = -f(x)\, g(x)$

$$\Rightarrow \quad I = 0$$

Example 39 Find all possible values of negative real number a such that

$$\int_a^0 (9^{-2t} - 2(9^{-t}))\, dt \,|\, 0.$$

Solution. Given inequality, $\int_a^0 (9^{-2t} - 2(9)^{-t})\, dt \,|\, 0$

$$\Rightarrow \quad \left(\frac{9^{-2t}}{-2 \ln 9} - \frac{2 \cdot 9^{-t}}{-\ln 9} \right)_a^0 \,|\, 0$$

$$\Rightarrow \qquad \left(-9^{-2t}+4(9)^{-t}\right)_a^0 \mid 0$$

$$\Rightarrow \qquad 9^{-2a}-4\cdot 9^{-a}+3 \mid 0$$

$$t^2-4t+3 \mid 0, \quad \text{where} \quad t=9^{-a}\in(1,\infty)$$

$$t\le 1 \quad \text{or} \quad t \mid 3$$

$\Rightarrow$ $\qquad t \mid 3$ is possible.

So, $\qquad 9^{-a} \mid 3 \Rightarrow a \le -1/2$

Example 40 If $f(x)=\begin{cases} 1-|x|, & |x|\le 1 \\ |x|-1, & |x|>1 \end{cases}$ and $g(x)=f(x-1)+f(x+1)$. Find the value of $\int_{-3}^{5} g(x)\,dx$.

Solution. Given, $f(x)=\begin{cases} -x-1, & x<-1 \\ 1+x, & -1\le x<0 \\ 1-x, & 0\le x\le 1 \\ x-1, & x>1 \end{cases}$

$$\Rightarrow \qquad f(x-1)=\begin{cases} -x, & x-1<-1 \Rightarrow x<0 \\ x, & -1\le x-1<0 \Rightarrow 0\le x<1 \\ 2-x, & 0\le x-1\le 1 \Rightarrow 1\le x\le 2 \\ x-2, & x-1>1 \Rightarrow x>2 \end{cases}$$

Similarly, $$f(x+1)=\begin{cases} -x-2, & x+1<-1 \Rightarrow x<-2 \\ x+2, & -1\le x+1<0 \Rightarrow -2\le x<-1 \\ -x, & 0\le x+1\le 1 \Rightarrow -1\le x\le 0 \\ x, & x+1>1 \Rightarrow x>0 \end{cases}$$

$$\Rightarrow \qquad g(x)=f(x-1)+f(x+1)=\begin{cases} -2x-2, & x<-2 \\ 2, & -2\le x<-1 \\ 2x, & 0<x<1 \\ 2, & 1<x\le 2 \\ 2x-2, & 2<x \end{cases}$$

Clearly, $g(x)$ is even.

Now, $$\int_{-3}^{5} g(x)\,dx = 2\int_0^3 g(x)\,dx + \int_3^5 g(x)\,dx$$

$$=2\left(\int_0^1 2x\,dx + 1\int_1^2 2dx + \int_2^3 (2x-2)\,dx\right) + \int_3^5 (2x-2)\,dx$$

$$\int_{-3}^{5} g(x)\,dx = 24$$

Example 41 If $|a|<1$, show that $\int_0^{\pi} \frac{\log(1+a\cos x)}{\cos x}\,dx = \pi\sin^{-1} a$.

Solution. Given, $\qquad |a|<1$

Let $$f(a)=\int_0^{\pi} \frac{\log(1+a\cos x)}{\cos x}\,dx$$

$$\therefore \qquad f'(a)=\int_0^{\pi} \frac{\cos x}{\cos x\cdot(1+a\cos x)}\,dx$$

(Differentiating w.r.t. 'a' using Leibnitz rule)

$$= \int_0^{\pi} \frac{dx}{1 + a\cos x}$$

$$\left[\text{put } \tan\frac{x}{2} = t,\ \frac{dx}{dt} = \frac{2}{1+t^2},\ \text{where } x = 0 \to t = 0 \text{ and } x = \pi \Rightarrow t \to \infty\right]$$

$$\Rightarrow \quad f'(a) = \int_0^{\infty} \frac{\dfrac{2dt}{1+t^2}}{1 + a\left(\dfrac{1-t^2}{1+t^2}\right)}$$

$$= \int_0^{\infty} \frac{2dt}{(1+t^2) + a(1-t^2)} = \int_0^{\infty} \frac{2dt}{(1-a)t^2 + (1+a)}$$

$$= \frac{2}{1-a}\int_0^{\infty} \frac{dt}{t^2 + \left(\sqrt{\dfrac{1+a}{1-a}}\right)^2}$$

$$= \frac{2}{1-a}\cdot\frac{1}{\sqrt{(1+a)/(1-a)}}\cdot\left(\tan^{-1} t\sqrt{\frac{1-a}{1+a}}\right)_0^{\infty} = \frac{2}{\sqrt{1-a^2}}\cdot\frac{\pi}{2} = \frac{\pi}{\sqrt{1-a^2}}$$

(Integrating both the sides w.r.t. 'a')

$$\Rightarrow \quad f(a) = \pi \sin^{-1} a + c$$

But $\quad f(0) = \int_0^{\pi} \frac{\log(1 + 0\cos x)\,dx}{\cos x} = 0$

$$\Rightarrow \quad c = 0 \quad \therefore \ f(a) = \pi\sin^{-1} a$$

Example 42 Evaluate $\int_0^{\pi/2} \operatorname{cosec}\theta \tan^{-1}(c\sin\theta)\,d\theta$.

Solution. Let $f(c) = \int_0^{\pi/2} \operatorname{cosec}\theta \cdot \tan^{-1}(c\sin\theta)\,d\theta$

$$\therefore \quad f'(c) = \int_0^{\pi/2} \operatorname{cosec}\theta \cdot \frac{(\sin\theta)}{1 + c^2\sin^2\theta}\,d\theta + 0 - 0$$

$$= \int_0^{\pi/2} \frac{d\theta}{1 + c^2\sin^2\theta}$$

$$= \int_0^{\pi/2} \frac{\operatorname{cosec}^2\theta\,d\theta}{c^2 + \operatorname{cosec}^2\theta} = \int_0^{\pi/2} \frac{\operatorname{cosec}^2\theta\,d\theta}{(c^2+1) + \cot^2\theta}$$

$$= -\frac{1}{\sqrt{c^2+1}}\left(\tan^{-1}\frac{\cot\theta}{\sqrt{c^2+1}}\right)_0^{\pi/2}$$

$$f'(c) = \frac{\pi}{2\sqrt{c^2+1}}$$

Integrating both the sides, we get

$$f(c) = \int \frac{\pi\,dc}{2\sqrt{c^2+1}} = \frac{\pi}{2}\log(c + \sqrt{c^2+1}) + A$$

where, $\quad f(0) = A$

But, $\quad f(0) = \int_0^{\pi/2} \operatorname{cosec}\theta \tan^{-1}(0)\,d\theta = 0$

$$\Rightarrow \quad f(c) = \frac{\pi}{2}\log(c + \sqrt{c^2+1})$$

Example 43 Evaluate $\int_0^{\pi/2} \sec\theta \cdot \tan^{-1}(a\cos\theta)\, d\theta$.

Solution. The given definite integral is a function of 'a'. Let its value be $I(a)$.

Then, $$I(a) = \int_0^{\pi/2} \sec\theta \cdot \tan^{-1}(a\cos\theta)\, d\theta$$

$$\Rightarrow \quad I'(a) = \int_0^{\pi/2} \sec\theta \cdot \frac{1}{1 + a^2\cos^2\theta} \cdot \cos\theta\, d\theta + 0 - 0 \quad \text{(using Leibnitz rule)}$$

$$= \int_0^{\pi/2} \frac{1}{1 + a^2\cos^2\theta}\, d\theta = \int_0^{\pi/2} \frac{\sec^2\theta}{\sec^2\theta + a^2}\, d\theta$$

$$= \int_0^{\pi/2} \frac{\sec^2\theta}{1 + \tan^2\theta + a^2}\, d\theta \qquad \text{put } \tan\theta = t$$

$$= \int_0^{\infty} \frac{dt}{t^2 + (a^2 + 1)} = \frac{1}{\sqrt{a^2+1}}\left(\tan^{-1}\frac{t}{\sqrt{a^2+1}}\right)_0^{\infty}$$

$$= \frac{1}{\sqrt{a^2+1}}\left(\frac{\pi}{2} - 0\right)$$

$$I'(a) = \frac{\pi}{2\sqrt{a^2+1}}$$

Integrating both the sides w.r.t. 'a', we get

$$\Rightarrow \quad I(a) = \frac{\pi}{2}\log|a + \sqrt{a^2+1}| + C$$

Since, $$I(0) = 0 + C$$

$$\Rightarrow \quad I(a) = \frac{\pi}{2}\log|a + \sqrt{1 + a^2}|$$

Example 44 Let f be a continuous function on $[a, b]$. Prove that there exists a number $x \in [a, b]$ such that $\int_a^x f(t)\, dt = \int_x^b f(t)\, dt$.

Solution. Let $g(x) = \int_a^x f(t)\, dt - \int_x^b f(t)\, dt,\ x \in [a, b]$

We have, $$g(a) = -\int_a^b f(t)\, dt \quad \text{and} \quad g(b) = \int_a^b f(t)\, dt$$

$$\Rightarrow \quad g(a)\cdot g(b) = -\left(\int_a^b f(t)\, dt\right)^2 \le 0$$

Clearly, $g(x)$ is continuous in $[a, b]$ and $g(a)\cdot g(b) \le 0$.

It implies that $g(x)$ will become zero at least once in $[a, b]$.

Hence, $\int_a^x f(t)\, dt = \int_x^b f(t)\, dt$ for at least one value of $x \in [a, b]$.

Example 45 If $f(x) = x + \int_0^1 (xy^2 + x^2y)(f(y))\, dy$, find $f(x)$.

Solution. Given, $$f(x) = x + x\int_0^1 y^2 f(y)\, dy + x^2 \int_0^1 y f(y)\, dy$$

$$= x\left(1 + \int_0^1 y^2 f(y)\, dy\right) + x^2\left(\int_0^1 y f(y)\, dy\right)$$

$\Rightarrow$ $f(x)$ is a quadratic expression;

$\Rightarrow \quad f(x) = ax + bx^2 \quad$ or $\quad f(y) = ay + by^2$ ...(i)

where $\quad a = 1 + \int_0^1 y^2 f(y)\, dy$

$$= 1 + \int_0^1 y^2 (ay + by^2)\, dy$$

$$= 1 + \left(\frac{ay^4}{4} + \frac{by^5}{5} \right)_0^1$$

$$a = 1 + \left(\frac{a}{4} + \frac{b}{5} \right)$$

$\Rightarrow \quad 20a = 20 + 5a + 4b$

or $\quad 15a - 4b = 20$ ...(ii)

and $\quad b = \int_0^1 y f(y)\, dy = \int_0^1 y(ay + by^2)\, dy$

$$= \left(\frac{a y^3}{3} + \frac{b y^4}{4} \right)_0^1$$

$$b = \frac{a}{3} + \frac{b}{4}$$

$\Rightarrow \quad 12b = 4a + 3b$

or $\quad 9b - 4a = 0$ ...(iii)

From Eqs. (ii) and (iii),

$$a = \frac{180}{119} \quad \text{and} \quad b = \frac{80}{119}$$

$\therefore$ Eq. (i) reduces to

$\therefore \quad f(x) = \dfrac{80x^2 + 180x}{119}$

Example 46 Prove that identity :

$$\int_0^x \{\int_0^u f(t)\, dt\}\, du = \int_0^x (x - u) f(u)\, du.$$

Solution. Here, applying integration by parts to,

$$\underbrace{\int_0^x 1}_{\text{II}} \cdot \underbrace{\{\int_0^u f(t)\, dt\}\, du}_{\text{I}},$$

ie, taking '1' as second function and $\int_0^u f(t)\, dt$ as first function we have,

$$\int_0^x 1 \cdot \{\int_0^u f(t)\, dt\}\, du = \{\int_0^u f(t)\, dt\}_0^x \cdot (u)_0^x - \int_0^x f(u) \cdot u\, du$$

$$= \left(u \int_0^u f(t)\, dt \right)_0^x - \int_0^x u f(u)\, du$$

$$= x \int_0^x f(t)\, dt - \int_0^x u f(u)\, du$$

$$= \int_0^x (x - u) f(u)\, du$$

Example 47 Evaluate $\int_0^{3\pi/2} (\log|\sin x|)(\cos(2nx))\,dx;\ n \in N$

Solution. Let $I(n) = \int_0^{3\pi/2} (\log|\sin x|)(\cos 2nx)\,dx$, using integration by parts,

$$I(n) = \left(\underset{\text{I}}{\log|\sin x|} \cdot \underset{\text{II}}{\frac{\sin 2nx}{2n}}\right)_0^{3\pi/2} - \int_0^{3\pi/2} \cot x \cdot \frac{\sin 2nx}{2n}\,dx$$

$$I(n) = 0 - \frac{1}{2n}\int_0^{3\pi/2} \frac{\sin 2nx \cdot \cos x}{\sin x}\,dx = 0 - \frac{1}{2}I_1(n) \qquad \text{...(i)}$$

Let $\quad I_1(n) = \int_0^{3\pi/2} \frac{\sin(2nx)\cos x}{\sin x}\,dx$

$$I_1(n+1) = \int_0^{3\pi/2} \frac{\sin(2n+2)x \cdot \cos x}{\sin x}\,dx$$

$\therefore$
$$I_1(n+1) - I_1(n) = \int_0^{3\pi/2} \frac{(\sin(2n+2)x - \sin 2nx)\cdot\cos x}{\sin x}\,dx$$

$$= \int_0^{3\pi/2} \frac{2\cos(2n+1)x \cdot \sin x \cdot \cos x}{\sin x}\,dx$$

$$= \int_0^{3\pi/2} (\cos(2n+2)x + \cos 2nx)\,dx$$

$$= \left(\frac{\sin(2n+2)x}{(2n+2)} + \frac{\sin(2n)x}{2n}\right)_0^{3\pi/2}$$

$$= 0 + 0$$

$\Rightarrow \quad I_1(n+1) = I_1(n) = \ldots = I_1(1)$

$\therefore$
$$I_1(1) = \int_0^{3\pi/2} \frac{\sin 2x \cdot \cos x}{\sin x}\,dx$$

$$= \int_0^{3\pi/2} \frac{2\sin x\cos x \cdot \cos x}{\sin x}\,dx$$

$$= \int_0^{3\pi/2} 2\cos^2 x\,dx = \int_0^{3\pi/2} (1 + \cos 2x)\,dx$$

$$= \left(x + \frac{\sin 2x}{2}\right)_0^{3\pi/2}$$

$\Rightarrow \quad I_1(1) = \frac{3\pi}{2} = I_1(n)$

$\Rightarrow \quad I(n) = -\frac{1}{2n}I_1(n)$ [using Eq. (i)]

$$I(n) = -\frac{3\pi}{4n}$$

Example 48 It is known that $f(y)$ is an odd function in the interval $\left[-\frac{T}{2}, \frac{T}{2}\right]$ and has a period equal to T. Prove that $\int_a^y f(t)\,dt$ is also periodic with period T.

Solution. It is given that, $f(-y) = -f(y)$

and $\quad f(y+T) = f(y)$

Let $g(y)=\int_a^y f(t)\,dt$

$$\Rightarrow \quad g(y+T)=\int_a^{y+T} f(t)\,dt=\int_a^y f(t)\,dt+\int_y^{y+T} f(t)\,dt$$

$$=\int_a^y f(t)\,dt+\int_y^{T/2} f(t)\,dt+\int_{T/2}^{y+T} f(t)\,dt$$

Put $t=x+T$ in III integral,

$$=\int_a^y f(t)\,dt+\int_y^{T/2} f(t)\,dt+\int_{-T/2}^{y} f(x+T)\,dx$$

$$=\int_a^y f(t)\,dt+\int_y^{T/2} f(t)\,dt+\int_{-T/2}^{y} f(x)\,dx$$

[since $f(x+T)=f(x)$]

$$=\int_a^y f(t)\,dt+\int_{-T/2}^{y} f(t)\,dt+\int_y^{T/2} f(t)\,dt$$

$$=\int_a^y f(t)\,dt+\int_{-T/2}^{T/2} f(t)\,dt$$

$$\Rightarrow \quad g(y+T)=\int_a^y f(t)\,dt+0 \qquad \left[\text{as } f(t) \text{ is odd } \therefore \int_{-a}^{a} f(t)\,dt=0\right]$$

$$\Rightarrow \quad g(y+T)=g(y)$$

$\Rightarrow$ $g(y)$ is a periodic function with period T.

Example 49 If $2f(x)+f(-x)=\frac{1}{x}\sin\left(x-\frac{1}{x}\right)$, then find the value of $\int_{1/e}^{e} f(x)\,dx$.

Solution. Since, $2f(x)+f(-x)=\frac{1}{x}\sin\left(x-\frac{1}{x}\right)$ ...(i)

$\therefore$ $2f(-x)+f(x)=\frac{1}{x}\sin\left(x-\frac{1}{x}\right)$ (replace x by $-x$) ...(ii)

$\Rightarrow$ $f(x)=f(-x)$ [subtracting Eqs. (i) and (ii)]

$\therefore$ $3f(x)=\frac{1}{x}\sin\left(x-\frac{1}{x}\right)$

Hence, $$I=\int_{1/e}^{e} f(x)\,dx=\frac{1}{3}\int_{1/e}^{e}\frac{1}{x}\sin\left(x-\frac{1}{x}\right)dx$$

Now, put $x=\frac{1}{t}$; $dx=-\frac{1}{t^2}\,dt$

$$\therefore \quad I=\frac{1}{3}\int_e^{1/e} t\sin\left(\frac{1}{t}-t\right)\cdot\left(-\frac{1}{t^2}\right)dt$$

$$=\frac{1}{3}\int_e^{1/e}\frac{1}{t}\sin\left(t-\frac{1}{t}\right)dt$$

$$=-\frac{1}{3}\int_{1/e}^{e}\frac{1}{t}\sin\left(t-\frac{1}{t}\right)dt$$

$$I=-I$$

$$\Rightarrow \quad 2I=0$$

$$\Rightarrow \quad I=0$$

$$\therefore \quad \int_{1/e}^{e} f(x)=0$$

Example 50 Evaluate $\int_0^\infty e^{-x} \sin^n x \, dx$, if n is an even integer.

Solution. Here, $I_n = \int_0^\infty e^{-x} \sin^n x \, dx$

$$= (-\sin^n x \, e^{-x})_0^\infty + n \int_0^\infty \sin^{n-1} x \cos x \, e^{-x} \, dx$$

$$= n \int_0^\infty \underset{\text{I}}{(\sin^{n-1} x \cdot \cos x)} \underset{\text{II}}{(e^{-x})} \, dx \quad [\text{where, } (-\sin^n x \, e^{-x})_0^\infty = 0]$$

$$\Rightarrow \quad I_n = n \left[(-\sin^{n-1} x \cdot \cos x \, e^{-x})_0^\infty + (n-1) \int_0^\infty \sin^{n-2} x \cos^2 x \, e^{-x} dx - \int_0^\infty \sin^n x \, e^{-x} \, dx \right]$$

$$I_n = n(n-1) \int_0^\infty e^{-x} \sin^{n-2} x \, dx - n^2 \int_0^\infty e^{-x} \sin^n x \, dx$$

$$= n(n-1) I_{n-2} - n^2 I_n$$

$$\Rightarrow \quad (n^2 + 1) I_n = n(n-1) I_{n-2}$$

$$\therefore \quad I_n = \frac{n(n-1)}{n^2+1} I_{n-2}$$

Replacing n by $(n-2), (n-4), \ldots, 2$

$$\Rightarrow \quad I_{n-2} = \frac{(n-2)(n-3)}{(n-2)^2+1} I_{n-4} = \frac{(n-1)(n-3)}{(n-2)^2+1} \cdot \frac{(n-4)(n-5)}{(n-4)^2+1} I_{n-6} \ldots \text{ and so on.}$$

$$I_2 = \frac{2(2-1)}{2^2+1} I_0$$

$$\therefore \quad I_n = \frac{n(n-1)}{n^2+1} \cdot \frac{(n-2)(n-3)}{(n-2)^2+1} \cdot \frac{(n-4)(n-5)}{(n-3)^2+1} \cdots \frac{2(2-1)}{2^2+1} I_0$$

$$= \frac{n!}{\prod_{r=1}^{n/2} \{1 + (2r)^2\}} \cdot I_0 = \frac{n!}{\prod_{r=1}^{n/2} 1 + 4r^2} \qquad \left(\because I_0 = \int_0^\infty e^{-x} \, dx = 1 \right)$$

Example 51 Evaluate $\int_0^1 (tx + 1 - x)^n \, dx$, $n \in N$ and t is independent of x. Hence, show $\int_0^1 x^k (1-x)^{n-k} \, dx = \frac{1}{{}^nC_k (n+1)}$.

Solution. Here, $I = \int_0^1 (tx + 1 - x)^n \, dx = \int_0^1 ((t-1)x + 1)^n \, dx$

$$= \left\{ \frac{((t-1)x + 1)^{n+1}}{(n+1)(t-1)} \right\}_0^1$$

$$= \frac{1}{n+1} \{t^n + t^{n-1} + t^{n-2} + \ldots + t + 1\} \qquad \ldots \text{(i)}$$

Again, $I = \int_0^1 (tx + 1 - x)^n \, dx = \int_0^1 \{(1-x) + tx\}^n \, dx$

$$= \int_0^1 \{{}^nC_0 (1-x)^n + {}^nC_1 (1-x)^{n-1} (tx) + {}^nC_2 (1-x)^{n-2} (tx)^2 + \ldots + {}^nC_r (1-x)^{n-r} (tx)^r + \ldots + {}^nC_n (tx)^n\} \, dx$$

$$= \int_0^1 \left\{ \sum_{r=0}^{n} {}^nC_r (1-x)^{n-r} (tx)^r \right\} dx$$

$$= \sum_{r=0}^{n} {}^nC_r\, t^r \int_0^1 \{(1-x)^{n-r}\, x^r\}\, dx \qquad \text{...(ii)}$$

From Eqs. (i) and (ii),

$$\sum_{n=0}^{n} {}^nC_r\, t^r \int_0^1 (1-x)^{n-r}\, x^r\, dx = \frac{1}{n+1}\{t^n + t^{n-1} + t^{n-2} + \dots + t + 1\}$$

Equating coefficient of t^k on both the sides, we get

$${}^nC_k \int_0^1 (1-x)^{n-k}\, x^k\, dx = \frac{1}{n+1}$$

$$\Rightarrow \qquad \int_0^1 (1-x)^{n-k}\, x^k\, dx = \frac{1}{{}^nC_k\,(n+1)}$$

Example 52 Given a real-valued function $f(x)$ which is monotonic and differentiable, prove that for any real numbers a and b;

$$\int_a^b \{f^2(x) - f^2(a)\}\, dx = \int_{f(a)}^{f(b)} 2x\,\{b - f^{-1}(x)\}\, dx$$

Solution. As, $f(x)$ is differentiable and monotonic.

$\Rightarrow$ $\qquad f^{-1}(x)$ exists.

Let $\qquad f^{-1}(x) = t \Rightarrow x = f(t)$ or $dx = f'(t)\, dt$

As, $\qquad x = f(a) \Rightarrow f^{-1}\{f(a)\} = t \Rightarrow t = a$

$\qquad x = f(b) \Rightarrow f^{-1}\{f(b)\} = t \Rightarrow t = b$

$$\therefore \quad \int_{f(a)}^{f(b)} 2x\,\{b - f^{-1}(x)\}\, dx = \int_a^b 2f(t)(b-t)\, f'(t)\, dt$$

$$= \int_a^b \underset{\text{I}}{(b-t)}\, \underset{\text{II}}{\{2f(t)\, f'(t)\}}\, dt$$

$$= ((b-t)\{f(t)\}^2)_a^b + \int_a^b \{f(t)\}^2\, dt$$

$$= -(b-a)\{f(a)\}^2 + \int_a^b \{f(t)\}^2\, dt$$

$$= \int_a^b (\{f(t)\}^2 - \{f(a)\}^2)\, dt$$

$$= \int_a^b \{f^2(x) - f^2(a)\}\, dx$$

Example 53 Evaluate $\displaystyle\int_0^1 \frac{\sin\theta\,(\cos^2\theta - \cos^2 \pi/5)(\cos^2\theta - \cos^2 2\pi/5)}{\sin 5\theta}\, d\theta.$

Solution. We know,

$$z^{10} - 1 = (z^2 - 1)\left(z^2 - 2\cos\left(\frac{\pi}{5}\right) z + 1\right)\left(z^2 - 2\left(\cos\frac{2\pi}{5}\right) z + 1\right) \times \left(z^2 - 2\cos\frac{4\pi}{5}\, z + 1\right)\left(z^2 - 2\cos\frac{6\pi}{5}\, z + 1\right)$$

or

$$z^5 - \frac{1}{z^5} = \left(z - \frac{1}{z}\right)\left(z - 2\cos\frac{\pi}{5} + \frac{1}{z}\right)\left(z - 2\cos\frac{2\pi}{5} + \frac{1}{z}\right) \times \left(z - 2\cos\frac{4\pi}{5} + \frac{1}{z}\right)\left(z - 2\cos\frac{6\pi}{5} + \frac{1}{z}\right)$$

Put $\quad z = \cos\theta + i\sin\theta$

$$\Rightarrow \quad 2i\sin 5\theta = (2i\sin\theta)(2\cos\theta - 2\cos\pi/5)(2\cos\theta - 2\cos 2\pi/5)$$
$$\times(2\cos\theta - 2\cos 4\pi/5)(2\cos\theta - 2\cos 6\pi/5)$$

$$\Rightarrow \quad \sin 5\theta = 16\sin\theta(\cos^2\theta - \cos^2\pi/5)(\cos^2\theta - \cos^2 2\pi/5)$$

$$\Rightarrow \quad \frac{\sin\theta(\cos^2\theta - \cos^2\pi/5)(\cos^2\theta - \cos^2 2\pi/5)}{\sin 5\theta} = \frac{1}{16}$$

$$\therefore \quad \int_0^1 \frac{\sin\theta(\cos^2\theta - \cos^2\pi/5)(\cos^2\theta - \cos^2 2\pi/5)}{\sin 5\theta}\,d\theta = \int_0^1 \frac{1}{16}\,d\theta = \frac{1}{16}$$

Example 54 Evaluate

$$\int_0^{2\pi} \{\cos(x)\cos(2x)\cos(2^2 x)\ldots\cos(2^{n-1}x)\cos(2^n - 1)x\}\,dx.$$

Solution. Let $I = \int_0^{2\pi} \{\cos x\cos 2x\cos 2^2 x\ldots\cos(2^{n-1})x\}\cos(2^n - 1)x\,dx$

$$\therefore \quad I = \int_0^{2\pi} \frac{\sin 2^n x}{2^n \sin x}\cos(2^n - 1)x\,dx$$

$$= \frac{1}{2^{n+1}}\int_0^{2\pi} \frac{(e^{2^n xi} - e^{-2^n xi})(e^{(2^n-1)xi} + e^{-(2^n-1)xi})}{(e^{ix} - e^{-ix})}\,dx$$

$$\left[\text{using } \cos x = \frac{e^{ix} + e^{-ix}}{2} \text{ and } \sin x = \frac{e^{ix} - e^{-ix}}{2i}\right]$$

$$= \frac{1}{2^{n+1}}\int_0^{2\pi} \frac{\{e^{(2^{n+1}-1)xi} + e^{xi} - e^{-ix} - e^{-(2^{n+1}-1)xi}\}}{e^{ix} - e^{-ix}}\,dx$$

$$= \frac{1}{2^{n+1}}\int_0^{2\pi} \left\{1 + \frac{(e^{(2^{n+1}-1)xi} - e^{-(2^{n+1}-1)xi})}{e^{ix} - e^{-ix}}\right\}dx$$

$$= \frac{1}{2^{n+1}}\int_0^{2\pi} \{1 + e^{(2^{n+1}-2)xi} + e^{(2^{n+1}-3)xi} + \ldots$$
$$+ e^{ix} + 1 + e^{-ix} + \ldots + e^{-(2^{n+1}-3)xi} + e^{-(2^{n+1}-2)xi}\}\,dx \quad \ldots\text{(i)}$$

$$\left[\text{using } \int_0^{2\pi} e^{imx}\,dx = 0, \text{ whenever } m \neq 0\right]$$

$$\therefore \quad I = \frac{(2\pi)\cdot 2}{2^{n+1}} = \frac{\pi}{2^{n-1}}$$

Example 55 Show that $\lim\limits_{n\to\infty} \sum\limits_{k=0}^{n} \frac{{}^nC_K}{n^K(K+3)} = e - 2.$

Solution. Here, $\lim\limits_{n\to\infty} \sum\limits_{K=0}^{n} \frac{{}^nC_k}{n^K(K+3)}$

$$\lim_{n\to\infty} \sum_{K=0}^{n} \frac{1}{K+3}\,{}^nC_K\cdot\frac{1}{n^K}$$

$$= \lim_{n\to\infty} \sum_{k=0}^{n} {}^nC_K\cdot\frac{1}{n^K}\int_0^1 x^{K+2}\,dx \quad \because \frac{1}{K+3} = \int_0^1 x^{K+2}\,dx$$

$$= \int_0^1 \left(x^2 \lim_{n \to \infty} \sum_{k=0}^{n} {}^nC_K \cdot \left(\frac{x}{n}\right)^K \right) dx$$

$$= \int_0^1 x^2 \left\{ \lim_{n \to \infty} \left(1 + \frac{x}{n}\right)^n \right\} dx$$

$$= \int_0^1 x^2 \cdot e^x dx \qquad \left[\because \lim_{n \to \infty} \left(1 + \frac{x}{n}\right)^n = e^x\right]$$

$$= (x^2 \cdot e^x)_0^1 - \int_0^1 2x \cdot e^x dx = e - 2\left\{ (x\, e^x)_0^1 - \int_0^1 e^x\, dx\right\}$$

$$= e - 2\{e - e + 1\} = e - 2$$

Example 56 Let $I = \int_0^{\pi/2} \frac{\cos x}{a\cos x + b\sin x} dx$ and $J = \int_0^{\pi/2} \frac{\sin x}{a\cos x + b\sin x} dx$, where $a > 0$ and $b > 0$. Compute the values of I and J.

Solution. $$aI + bJ = \frac{\pi}{2} \qquad \text{...(i)}$$

and $$bI - aJ = \int_0^{\pi/2} \frac{b\cos x - a\sin x}{a\cos x + b\sin x} dx$$

$\therefore$ $$bI - aJ = \ln[a\cos x + b\sin x]_0^{\pi/2}$$

$\Rightarrow$ $$bI - aJ = \ln\left(\frac{b}{a}\right) \qquad \text{...(ii)}$$

From Eqs. (i) and (ii),

$$a^2 I + abJ = \frac{a\pi}{2}$$

$$b^2 I - abJ = b\ln(b/a)$$

$$I = \frac{1}{a^2 + b^2}\left(\frac{a\pi}{2} + b\ln\left(\frac{b}{a}\right)\right)$$

Again, $$abI + b^2 I = \frac{b\pi}{2}$$

and $$abI - a^2 J = a\ln(b/a)$$

Subtracting $$J = \frac{1}{a^2 + b^2}\left(\frac{b\pi}{2} - a\ln\left(\frac{b}{a}\right)\right)$$

Aliter Convert $a\cos x + b\sin x$ into a single cosine say $\cos(x + \phi)$ and put $(x + \phi) = t$.

Example 57 Evaluate $\int_0^{\infty} \frac{\ln x\, dx}{x^2 + 2x + 4}$.

Solution. $I = \int_0^{\infty} \frac{\ln x\, dx}{x^2 + 2x + 4}$;

(Put $x = 2t$; $dx = 2dt$ to make coefficient of x^2 and constant term same

$$= 2\int_0^{\infty} \frac{\ln 2 + \ln t}{4(t^2 + t + 1)} dt = \frac{\ln 2}{2} \underbrace{\int_0^{\infty} \frac{dt}{t^2 + t + 1}}_{I_1} + \frac{1}{2} \underbrace{\int_0^{\infty} \frac{\ln t\, dt}{t^2 + t + 1}}_{I_2 = \text{zero}}$$

$$I_2 = \int_0^{\infty} \frac{\ln t\, dt}{t^2 + t + 1}$$ (Put $t = \frac{1}{y}$; $dt = -\frac{1}{y^2}$

$$I_2 = \int_{\infty}^{0} \frac{\ln y \cdot (+1)}{\left(\frac{1}{y^2} + \frac{1}{y} + 1\right) y^2} dy$$

$$I_2 = \int_{\infty}^{0} \frac{\ln y\, dy}{y^2 + y + 1} = -\int_{\infty}^{0} \frac{\ln y\, dy}{y^2 + y + 1} = -I_2$$

Now, $$I_1 = \int_0^{\infty} \frac{dt}{\left(t + \frac{1}{2}\right)^2 + \left(\frac{\sqrt{3}}{2}\right)^2} = \left[\frac{2}{\sqrt{3}} \tan^{-1} \frac{(t + (1/2))^2}{\sqrt{3}}\right]_0^{\infty}$$

$$= \frac{2}{\sqrt{3}}\left[\frac{\pi}{2} - \frac{\pi}{6}\right] = \frac{2\pi}{3\sqrt{3}}$$

$\therefore$ $$I = \frac{\ln 2}{2} \cdot \frac{2\pi}{3\sqrt{3}} = \frac{\pi \ln 2}{3\sqrt{3}}$$

Note : For $a > 0$, $$I = \int_0^{\infty} \frac{\ln x\, dx}{ax^2 + bx + a} = 0$$

[**Hint** : By putting $x = 1/t$, we get $I = -I \therefore I = 0$]

Example 58 Find a function f, continuous for all x (and not zero everywhere), such that $f^2(x) = \int_0^x \frac{f(t)\sin t}{2 + \cos t} dt$.

Solution. $$f^2(x) = \int_0^x \frac{f(t)\sin t}{2 + \cos t} dt$$

(Note that $f^2(x)$, being an integral function of a continuous function is continuous and differentiable)

$$2f(x)f'(x) = \frac{f(x)\cdot \sin x}{2 + \cos x}$$

Integrating, $$2f(x) = C - \ln(2 + \cos x)$$

$$x = 0;\quad f(0) = 0 \Rightarrow C = \ln 3$$

$$f(x) = \frac{1}{2} \ln \frac{3}{2 + \cos x}$$

Example 59 Evaluate $\int_0^{\infty} \frac{\tan^{-1} ax - \tan^{-1} x}{x} dx$, where a is a parameter.

Solution. Let $$I = \int_0^{\infty} \frac{\tan^{-1} ax - \tan^{-1} x}{x} dx$$

$$\frac{dI}{da} = \int_0^{\infty} \frac{1 \cdot x}{(1 + a^2x^2)x} dx = \int_0^{\infty} \frac{dx}{1 + a^2x^2}$$

$$= \frac{1}{a^2} \int_0^{\infty} \frac{dx}{x^2 + \frac{1}{a^2}} = \frac{1}{a^2} [a \tan^{-1} x]_0^{\infty} = \frac{\pi}{2a}$$

$$\int dI = \frac{\pi}{2} \int \frac{da}{a};\quad I = \frac{\pi}{2} \ln a + C$$

When $$a = 1, I = 0 \Rightarrow C = 0$$

Hence, $$I = \frac{\pi}{2} \ln a$$

Example 60 $\int_0^1 \frac{\ln(1-a^2x^2)}{x^2\sqrt{(1-x^2)}}dx$ $(a^2 < 1)$

Solution. $I = \int_0^1 \frac{\ln(1-a^2x^2)}{x^2\sqrt{(1-x^2)}}dx, |a| < 1$

$$\frac{dI}{da} = \int_0^1 \frac{-2ax^2}{(1-a^2x^2)}\cdot\frac{dx}{x^2\sqrt{1-x^2}} \quad \text{(Put } x = \sin\theta\text{)}$$

$$= \int_0^{\pi/2} \frac{2a\,d\theta}{a^2\sin^2\theta - 1} = \int_0^{\pi/2} \frac{2a\sec^2\theta\,d\theta}{a^2\tan^2\theta - (1+\tan^2\theta)}$$

$$= -\int_0^{\pi/2} \frac{2a\sec^2\theta\,d\theta}{(1-a^2)\tan^2\theta + 1} \quad \text{(Put } \tan\theta = 1\text{)}$$

$$= \int_0^{\infty} \frac{2a\,dt}{(1-a^2)t^2+1} \text{ or } -\frac{2a}{(1-a^2)}\int_0^{\infty}\frac{dt}{t^2+\left(\frac{1}{\sqrt{1-a^2}}\right)^2}$$

$$\frac{dI}{da} = -\frac{2a}{\sqrt{1-a^2}}[\tan^{-1}t\sqrt{1-a^2}]_0^{\infty} = -\frac{\pi\cdot 2a}{2\sqrt{1-a^2}} = -\frac{\pi a}{\sqrt{1-a^2}}$$

$$I = \pi\sqrt{1-a^2} + C$$

If $a = 0; I = 0 \Rightarrow C = -\pi \therefore I = \pi[\sqrt{1-a^2} - 1]$

Example 61 $\int_0^{\pi/2} \ln\left(\frac{1+a\sin x}{1-a\sin x}\right)\frac{dx}{\sin x}$, $(|a| < 1)$

Solution. $I = \int_0^{\pi/2} \ln\left(\frac{1+a\sin x}{1-a\sin x}\right)\frac{dx}{\sin x}$ $(|a| < 1)$

$$\frac{dI}{da} = \int_0^{\pi/2} \frac{2\sin x}{(1-a^2\sin^2 x)}\cdot\frac{dx}{\sin x}$$

$$= \int_0^{\pi/2} \frac{2\sec^2 x}{1+\tan^2 x - a^2\tan^2 x}dx$$

$$= \int_0^{\pi/2} \frac{2\sec^2 x}{(1-a^2)\tan^2 x + 1}dx \quad \text{(Put } \tan x = t\text{)}$$

$$= \int_0^{\infty} \frac{2\,dt}{(1-a^2)t^2+1} = \frac{2}{(1-a^2)}\int_0^{\infty}\frac{dt}{t^2+\left(\frac{1}{\sqrt{1-a^2}}\right)^2}$$

$$= \frac{2}{\sqrt{1-a^2}}[\tan^{-1}(t\sqrt{1-a^2})]_0^{\infty} = \frac{\pi}{\sqrt{1-a^2}}$$

$dI = \frac{\pi\,da}{\sqrt{1-a^2}}$ when $a = 0 \Rightarrow I = 0 \therefore C = 0 \Rightarrow I = \pi\sin^{-1}(a)$

$$I = \pi\sin^{-1}a + C$$

Type 2 : **Only One Correct Option**

Example 62 The value of the definite integral $\int_0^{2n\pi} \max(\sin x, \sin^{-1}(\sin x))dx$ equals to
(where $n \in I$)

(a) $\frac{n(\pi^2-4)}{2}$ (b) $\frac{n(\pi^2-4)}{4}$ (c) $\frac{n(\pi^2-8)}{4}$ (d) $\frac{n(\pi^2-2)}{4}$

Solution. $I = \int_0^{2n\pi} \max(\sin x, \sin^{-1}(\sin x))dx$

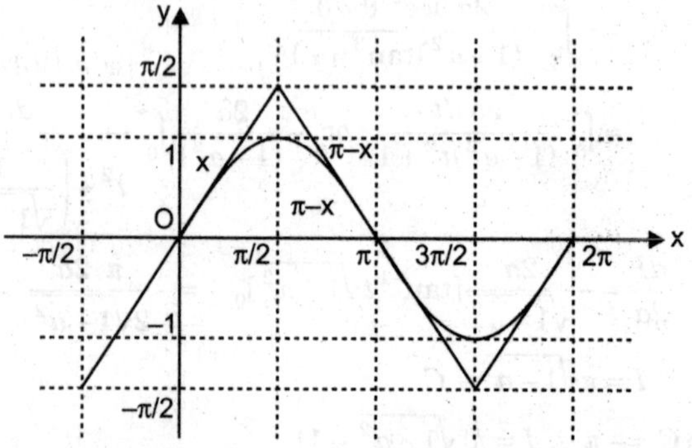

$$= n\left[\int_0^{\pi/2} x\,dx + \int_{\pi/2}^{\pi} (\pi - x)dx + \int_{\pi}^{2\pi} (\sin x)\,dx\right]$$

$$= n\left[\frac{\pi^2}{8} + \frac{\pi^2}{2} - \frac{1}{2}\left(\pi^2 - \frac{\pi^2}{4}\right) - 2\right]$$

$$= n\left[\frac{\pi^2}{8} + \frac{\pi^2}{2} - \frac{3\pi^2}{8} - 2\right] = \frac{n(\pi^2-8)}{4}$$

Hence, (c) is the correct answer.

Example 63 $\lim_{n\to\infty} n^{-\frac{1}{2}\left(1+\frac{1}{n}\right)}.(1^1 \cdot 2^2 \cdot 3^3 .....n^n)^{\frac{1}{n^2}}$ is equal to

(a) $\sqrt{e}$ (b) $\frac{1}{\sqrt{e}}$ (c) $\frac{1}{\sqrt[4]{e}}$ (d) $\sqrt[4]{e}$

Solution. Let $L = \lim_{n\to\infty} n^{-\frac{1}{2}\left(1+\frac{1}{n}\right)}.(1^1 \cdot 2^2 \cdot 3^3 ....n^n)^{\frac{1}{n^2}}$

$$\ln L = \lim_{n\to\infty} -\frac{1}{2}\left(\frac{n+1}{n}\right)\ln n + \frac{1}{n^2}\sum_{k=1}^{n} k \ln k$$

$$= \lim_{n\to\infty} -\frac{1}{2}\left(\frac{n+1}{n}\right)\ln n + \frac{1}{n^2}\sum_{k=1}^{n} (k\ln k - k\ln n + k\ln n)$$

$$= \lim_{n\to\infty} -\frac{1}{2}\left(\frac{n+1}{n}\right)\ln n + \frac{1}{n^2}\sum_{k=1}^{n} k\ln\frac{k}{n} + \frac{\ln n}{n^2}\sum_{k=1}^{n} k$$

$$= \lim_{n\to\infty} -\frac{1}{2}\left(\frac{n+1}{n}\right)\ln n + \frac{1}{n}\sum_{k=1}^{n}\frac{k}{n}\ln\frac{k}{n} + \frac{\ln n}{n^2}\cdot\frac{n(n+1)}{2}$$

$$= -\frac{1}{2}\left(\frac{n+1}{n}\right)\ln n + \int_0^1 x\ln x\,dx + \frac{1}{2}\left(\frac{n+1}{n}\right)\ln n$$

$$= \int_0^1 \underbrace{x}_{\text{II}}\,\underbrace{\ln x}_{\text{I}}\,dx = -\frac{1}{4};\quad \therefore\ L = e^{-\frac{1}{4}}$$

Hence, (c) is the correct answer.

Example 64 $\lim\limits_{x\to\frac{\pi}{4}} n\dfrac{\int_2^{\sec^2 x} f(t)dt}{x^2 - \frac{\pi^2}{16}}$ **is equal to** **[IIT JEE 2007]**

(a) $\frac{8}{\pi}f(2)$ (b) $\frac{2}{\pi}f(2)$ (c) $\frac{2}{\pi}f\left(\frac{1}{2}\right)$ (d) $4f(2)$

Solution. $\lim\limits_{x\to\frac{\pi}{4}} \dfrac{\int_2^{\sec^2 x} f(t)dt}{x^2 - \frac{\pi^2}{16}}$ **Applying L'Hospital rule**

$$= \lim_{x\to\frac{\pi}{4}} \frac{f(\sec^2 x)\cdot 2\sec^2 x\tan x - 0}{2x} = \frac{f(2)\cdot 4}{\pi/2} = \frac{8f(2)}{\pi}$$

Hence, (a) is the correct answer.

Example 65 **Let f be a non-negative function defined on the interval [0,1]. If $\int_0^x \sqrt{1-(f'(t))^2}dx = \int_0^x f(t)dt, 0 \le x \le 1$ and $f(0) = 0$, then** **[IIT JEE 2009]**

(a) $f\left(\frac{1}{2}\right) < \frac{1}{2}$ and $f\left(\frac{1}{3}\right) > \frac{1}{3}$ (b) $f\left(\frac{1}{2}\right) > \frac{1}{2}$ and $f\left(\frac{1}{3}\right) > \frac{1}{3}$

(c) $f\left(\frac{1}{2}\right) < \frac{1}{2}$ and $f\left(\frac{1}{3}\right) < \frac{1}{3}$ (d) $f\left(\frac{1}{2}\right) > \frac{1}{2}$ and $f\left(\frac{1}{3}\right) < \frac{1}{3}$

Solution. **Given,** $\int_0^x \sqrt{1-(f'(t))^2}dt = \int_0^x f(t)dt, 0 \le x \le 1$

Applying Leibnitz theorem, we get

$$\sqrt{1-(f'(x))^2} = f(x)$$

$$\Rightarrow \quad 1-(f'(x))^2 = f^2(x)$$

$$\Rightarrow \quad (f'(x))^2 = 1 - f^2(x) \Rightarrow f'(x) = \pm\sqrt{1-f^2(x)}$$

$$\Rightarrow \quad \frac{dy}{dx} = \pm\sqrt{1-y^2}, \text{ where } y = f(x)$$

$$\Rightarrow \quad \frac{dy}{\sqrt{1-y^2}} = \pm\, dx$$

Integrating both the sides, $\sin^{-1}(y) = \pm x + C$

$$\therefore \quad f(0) = 0 \Rightarrow C = 0$$

$$\therefore \quad y = \pm\sin x$$

$$y = \sin x = f(x) \text{ given } f(x) \ge 0 \text{ for } x \in [0,1].$$

Its known that $\sin x < x, \forall x \in R^+$

$$\sin\left(\frac{1}{2}\right) < \frac{1}{2} \Rightarrow f\left(\frac{1}{2}\right) < \frac{1}{2} \text{ and } \sin\left(\frac{1}{3}\right) < \frac{1}{3}$$

$$\Rightarrow \quad f\left(\frac{1}{3}\right) < \frac{1}{3}$$

Hence, (c) is the correct answer.

Example 66 The value of $\lim_{x\to 0} \frac{1}{x^3} \int_0^x \frac{t\ln(1+t)}{t^4+4} dt$ **[IIT JEE 2010]**

(a) 0 (b) $\frac{1}{12}$ (c) $\frac{1}{24}$ (d) $\frac{1}{64}$

Solution. $\lim_{x\to 0} \frac{1}{x^3} \int_0^x \frac{t\log(1+t)}{4+t^4} dt$

Using L'Hospital's rule,

$$\Rightarrow \quad \lim_{x\to 0} \frac{\frac{x\log(1+x)}{4+x^4}}{3x^2} = \lim_{x\to 0} \frac{\log(1+x)}{3x} \cdot \frac{1}{4+x^4}$$

$$= \frac{1}{12} \quad \left[\text{using } \lim_{x\to 0} \frac{\log(1+x)}{x} = 1\right]$$

Hence, (b) is the correct answer.

Example 67 The value(s) of $\int_0^1 \frac{x^4(1-x)^4}{1+x^2} dx$ is(are) **[IIT JEE 2010]**

(a) $\frac{22}{7} - \pi$ (b) $\frac{2}{105}$ (c) 0 (d) $\frac{71}{15} - \frac{3\pi}{2}$

Solution. $I = \int_0^1 \frac{x^4(1-x)^4}{1+x^2} dx$

$$= \int_0^1 \frac{(x^4-1)(1-x)^4 + (1-x)^4}{(1+x^2)} dx$$

$$= \int_0^1 (x^2-1)(1-x)^4 dx + \int_0^1 \frac{(1+x^2-2x)^2}{(1+x^2)} dx$$

$$= \int_0^1 \left\{(x^2-1)(1-x)^4 + (1+x^2) - 4x + \frac{4x^2}{(1+x^2)}\right\} dx$$

$$= \int_0^1 \left((x^2-1)(1-x)^4 + (1+x^2) - 4x + 4 - \frac{4}{1-x^2}\right) dx$$

$$= \int_0^1 \left(x^6 - 4x^5 + 5x^4 + 4 - \frac{4}{1+x^2}\right) dx$$

$$= \frac{1}{7} - \frac{4}{6} + \frac{5}{5} - \frac{4}{3} + 4 - 4\left(\frac{\pi}{4} - 0\right) = \frac{22}{7} - \pi$$

Hence, (a) is the correct answer.

Example 68 Let f be a real-valued function defined on the interval (–1, 1) such that $e^{-x}f(x) = 2 + \int_0^x \sqrt{t^4+1}\,dt$, for all $x \in (-1, 1)$ and let f^{-1} be the inverse function of f. Then, $(f^{-1})'(2)$ is equal to **[IIT JEE 2010]**

(a) 1 (b) 1/3 (c) 1/2 (d) $1/e$

Solution. We have, $e^{-x}f(x) = 2 + \int_0^x \sqrt{t^4+1}\,dt\ x \in (-1, 1)$

Differentiating w.r.t. x, we get

$$e^{-x}(f'(x) - f(x)) = \sqrt{x^4+1}$$

$$\Rightarrow \quad f'(x) = f(x) + \sqrt{x^4+1}\,e^x$$

$\therefore$ f^{-1} is the inverse of f

$$\therefore \quad f^{-1}(f(x)) = x \Rightarrow f^{-1\prime}(f(x))f'(x) = 1$$

$$\Rightarrow \quad f^{-1\prime}(f(x)) = \frac{1}{f'(x)}$$

$$\Rightarrow \quad f^{-1\prime}(f(x)) = \frac{1}{f(x) + \sqrt{x^4+1}\,e^x}$$

At $x = 0$, $\quad f(x) = 2$

$$f^{-1\prime}(2) = \frac{1}{2+1} = \frac{1}{3}$$

Hence, (b) is the correct answer.

Type 3 : More Than One Correct Options

Example 69 Let $S_n = \sum_{k=1}^{n} \frac{n}{n^2+kn+k^2}$ and $T_n = \sum_{k=1}^{n-1} \frac{n}{n^2+kn+k^2}$, for $n = 1, 2, 3, \ldots$. Then, **[IIT JEE 2008]**

(a) $S_n < \frac{\pi}{3\sqrt{3}}$ (b) $S_n > \frac{\pi}{3\sqrt{3}}$ (c) $T_n < \frac{\pi}{3\sqrt{3}}$ (d) $T_n > \frac{\pi}{3\sqrt{3}}$

Solution.

$$S_n = \sum_{k=1}^{n} \frac{n}{n^2+kn+k^2}.$$

$$= \sum_{k=1}^{n} \frac{1}{n} \cdot \frac{1}{1+\left(\frac{k}{n}\right)+\left(\frac{k}{n}\right)^2} = \frac{1}{n}\sum_{k=1}^{n} \frac{1}{1+\left(\frac{k}{n}\right)+\left(\frac{k}{n}\right)^2}$$

$$\Rightarrow \quad \lim_{n\to\infty} S_n = \lim_{n\to\infty} \frac{1}{n}\sum_{k=1}^{n} \frac{1}{1+\left(\frac{k}{n}\right)+\left(\frac{k}{n}\right)^2}$$

$$= \int_0^1 \frac{dx}{1+x+x^2} = \int_0^1 \frac{dx}{\left(x+\frac{1}{2}\right)^2+\frac{3}{4}} = \left[\frac{2}{\sqrt{3}}\tan^{-1}\left(\frac{2x+1}{\sqrt{3}}\right)\right]_0^1$$

$$= \frac{2}{\sqrt{3}}\left(\tan^{-1}\sqrt{3} - \tan^{-1}\frac{1}{\sqrt{3}}\right) = \frac{\pi}{3\sqrt{3}}$$

Thus, $\quad S_n < \frac{\pi}{3\sqrt{3}}$

Similarly, $\quad T_n < \frac{\pi}{3\sqrt{3}}$.

Hence, (a) and (c) are the correct answers.

Example 70 If $I_n = \int_{-\pi}^{\pi} \frac{\sin nx}{(1+\pi^x)\sin x} dx$ $n = 0, 1, 2, \ldots$, then

(a) $I_n = I_{n+2}$ (b) $\sum_{m=1}^{10} I_{2m+1} = 10\pi$ (c) $\sum_{m=1}^{10} I_{2m} = 0$ (d) $I_n = I_{n+1}$

Solution. (a) $I_n = \int_{-\pi}^{\pi} \frac{\sin x}{(1+\pi^x)\sin x} dx$

$$I_n = \int_{-\pi}^{\pi} \frac{\pi^x \sin nx}{(1+\pi^x)\sin x} dx \quad \left[\text{by property } \int_a^b f(x)dx = \int_a^b f(a+b-x)dx\right]$$

$$2I_n = \int_{-\pi}^{\pi} \frac{\sin nx}{x} dx$$

$$2I_n = 2\int_0^{\pi} \frac{\sin nx}{\sin x} dx$$

$$I_n = \int_0^{\pi} \frac{\sin nx}{\sin x} dx$$

$$I_{n+2} - I_n = \int_0^{\pi} \frac{\sin(n+2)x - \sin nx}{\sin x} dx$$

$$= \int_0^{\pi} \frac{2\cos(n+1)x \sin x}{\sin x} dx$$

$$= 2\left[\frac{\sin(n+1)x}{(n+1)}\right]_0^{\pi} = 0$$

$\Rightarrow$ $$I_{n+2} = I_n$$

(b) $$I_3 = I_5 = \ldots = I_{21}$$

$\therefore$ $$\sum_{m=1}^{10} I_{2m+1} = 10I_3 = 10\int_0^{\pi} \frac{\sin 3x}{\sin x} dx$$

$$= 10\int_0^{\pi} (3 - 4\sin^2 x)dx$$

$$= 10[3x - 2x + 2\sin 2x]_0^{\pi} = 10\pi$$

(c) $$I_2 = I_4 = \ldots = I_{20}$$

$$\sum_{m=1}^{10} I_{2m} = 10\int_0^{\pi} \frac{\sin 2x}{\sin x} dx = 20[\sin x]_0^{\pi} = 0$$

Hence, (a), (b) and (c) are the corect answers.

Type 4 : **Assertion and Reason**

Directions

(Q. Nos. 71 to 73)

For the following questions, choose the correct answers from the codes (a), (b), (c) and (d) defined as follows :

(a) Statement I is true, Statement II is also true; Statement II is the correct explanation of Statement I.

(b) Statement I is true, Statement II is also true; Statement II is not the correct explanation of Statement I.

(c) Statement I is true, Statement II is false.

(d) Statement I is false, Statement II is true.

71 Statement I If $f(x)=\int_0^1 (xf(t)+1)dt$, then $\int_0^3 f(x)dx=12$

Because

Statement II $f(x)=3x+1$

Solution. Let $\int_0^1 f(t)dt=k$, So, $f(x)=xk+1$

Now, $\int_0^1 (kt+1)dt=k \Rightarrow \frac{k}{2}+1=k$, so $k=2$

$\therefore$ $f(x)=2x+1$

Also $\int_0^3 f(x)dx=12$

Hence, (c) is the correct answer.

72 Statement I The function $f(x)=\int_0^x \sqrt{1+t^2}dt$ is an odd function and $g(x)=f'(x)$ is an even function.

Because

Statement II For a differentiable function $f(x)$ if $f'(x)$ is an even function, then $f(x)$ is an odd function.

Solution. If $f(x)$ is of odd $\Rightarrow f'(x)$ is even but converse is not true *eg*,

if $f'(x)=x\sin x$, then $f(x)=\sin x - x\cos x + C$

$f(-x)=-\sin x + x\cos x + C$

On adding, $f(x)+f(-x)=$ constant which need not to be zero.

For Statement I $f(x)=\int_0^x \sqrt{1+t^2}dt;\ g(x)=\sqrt{1+x^2}$

$f(-x)=\int_0^{-x}\sqrt{1+t^2}dt;\ t=-y$

$f(-x)=-\int_0^x \sqrt{1+y^2}dy$

$\therefore$ $f(x)+f(-x)=0 \Rightarrow f$ is odd and g is obviously even.

Hence, (c) is the correct answer.

73 Given, $f(x)=\sin^3 x$ and $P(x)$ is a quadratic polynomial with leading coefficient unity.

Statement I $\int_0^{2\pi} P(x)\cdot f''(x)dx$ vanishes.

Because

Statement II $\int_0^{2\pi} f(x)dx$ vanishes.

Solution. $P(x)=ax^2+bx+c;\ f(x)=\sin^3 x$

$$I=\int_0^{2\pi} \underbrace{P(x)}_{\text{I}}\cdot\underbrace{f''(x)}_{\text{II}}dx$$

Using I.B.P. $\underbrace{P(x)\cdot f'(x)\,|_0^{2\pi}}_{\text{zero}}-\int_0^{2\pi}\underbrace{P'(x)}_{\text{I}}\cdot\underbrace{f'(x)}_{\text{II}}dx$

$$=-\left[[P'(x)\cdot f(x)]_0^{2\pi}-\int_0^{2\pi}P''(x)\cdot f(x)dx\right]$$

$$=\int_0^{2\pi}P''(x)\cdot f(x)dx=2\int_0^{2\pi}\sin^3 x\,dx=0$$

Hence, (a) is the correct answer.

Type 5 : Linked–Comprehension Based Questions

Passage I
(Q. Nos. 74 to 76)

Let the function f satisfies

$$f(x)\cdot f'(-x)=f(-x)\cdot f'(x) \text{ for all } x \text{ and } f(0)=3.$$

74 The value of $f(x)\cdot f(-x)$ for all x, is

(a) 4 (b) 9 (c) 12 (d) 16

Solution. Given, $f(x)\cdot f'(-x)=f(-x)\cdot f'(x)$

$$\frac{f'(x)}{f(x)}=\frac{f'(-x)}{f(-x)}$$

Integrating $\ln f(x)=-f(-x)+C$

$$\ln(f(x)\cdot f(-x))=C$$

$$f(x)\cdot f(-x)=C$$

But $f(0)=3 \therefore f^2(0)=C \therefore C=9$

$\therefore$ $f(x)\cdot f(-x)=9$

Aliter $f(x)\cdot f'(x)-f(-x)\cdot f'(x)=0$

$\Rightarrow$ $\frac{d}{dx}(f(-x)\cdot f(x))=0$

Integrating $f(x)\cdot f(-x)=\text{constant}$

Hence, (b) is the correct answer.

75 $\int_{-51}^{51}\frac{dx}{3+f(x)}$ has the value equal to

(a) 17 (b) 34 (c) 102 (d) 0

Solution. Let $I=\int_{-51}^{51}\frac{dx}{3+f(x)}=\int_{-51}^{51}\frac{dx}{3+f(-x)}$

Using, $\int_a^b f(x)dx=\int_a^b f(a+b-x)dx$

$$2I=\int_{-51}^{51}\frac{6+f(x)+f(-x)}{(3+f(x))(3+f(-x))}dx$$

$$=\int_{-51}^{51}\frac{6+f(x)+f(-x)}{9+3(f(x)+f(-x))+f(x)\cdot f(-x)}dx$$

$$=\int_{-51}^{51}\frac{6+f(x)+f(-x)}{18+3(f(x)+f(-x))}dx$$

$$=\frac{1}{3}\int_{-51}^{51}dx=\frac{2\cdot 51}{3}$$

$\Rightarrow$ $I=\frac{51}{3}=17$

Hence, (a) is the correct answer.

76 Number of roots of $f(x)=0$ in $[-2, 2]$ is

(a) 0 (b) 1 (c) 2 (d) 4

Solution. Let $x=\alpha$ be the root of $f(x)=0$

$$\therefore \quad f(\alpha)=0$$

$$f(x)\cdot f(-x)=9$$

Put $x=\alpha \Rightarrow \quad 0=9$ impossible

Therefore, $f(x)$ has no root but $f(0)=3$

Therefore, $f(x)>0, \forall x \in R$ as f is continuous possible function $f(x)=3e^{-x}$

Hence, (a) is the correct answer.

Passage II
(Q. Nos. 77 to 78)

Consider the function defined on $[0, 1] \to R$

$$f(x)=\frac{\sin x - x\cos x}{x^2}, \text{ if } x \neq 0 \text{ and } f(0)=0$$

77 $\int_0^1 f(x)dx$ is equal to

(a) $1-\sin(1)$ (b) $\sin(1)-1$ (c) $\sin(1)$ (d) $-\sin(1)$

Solution. $\int_0^1 \frac{\sin x}{x^2}dx - \int_0^1 \frac{\cos x}{x}dx = \left[\sin x\left(-\frac{1}{x}\right)\right]_0^1 + \int_0^1 \cos x\left(\frac{1}{x}\right)dx - \int_0^1 \frac{\cos x}{x}dx$

$$= -\left[\frac{\sin x}{x}\right]_0^1 = (1)-\sin(1)$$

Hence, (a) is the correct answer.

78 $\lim_{t\to 0} \frac{1}{t^2}\int_0^t f(x)\,dx$ is equal to

(a) 1/3 (b) 1/6 (c) 1/12 (d) 1/24

Solution. $\lim_{t\to 0} \frac{\int_0^t f(x)dx}{t^2} = \lim_{t\to 0} \frac{\int_0^t \frac{\sin x - x\cos x}{x^2}dx}{t^2}$

Using L'Hospital's rule,

$$l = \lim_{t\to 0}\frac{\sin t - t\cos t}{t^2\cdot 2t} = \lim_{t\to 0}\frac{\cos t(\tan t - t)}{2t^3} = \frac{1}{2}\lim_{t\to 0}\frac{\sec^2 t - 1}{3t^2} = \frac{1}{6}$$

Hence, (b) is the correct answer.

Passage III
(Q. Nos. 79 to 81)

Suppose we define the definite integral using the following formula

$\int_a^b f(x)dx = \frac{b-a}{2}(f(a)+f(b))$, *for more accurate result for*

$$c \in (a, b),\ F(c) = \frac{c-a}{2}(f(a)+f(c)) + \frac{b-c}{2}(f(b)+f(c)).$$

When $c=\frac{a+b}{2}$, $\int_a^b f(x)dx = \frac{b-a}{4}(f(a)+f(b)+2f(c))$. **[IIT JEE 2007]**

79 $\int_0^{\pi/2} \sin x\, dx$ is equal to

(a) $\frac{\pi}{8}(1+\sqrt{2})$ (b) $\frac{\pi}{4}(1+\sqrt{2})$ (c) $\frac{\pi}{8\sqrt{2}}$ (d) $\frac{\pi}{4\sqrt{2}}$

Solution. $\int_0^{\pi/2} \sin x\, dx = \frac{\frac{\pi}{2}-0}{4}\left(\sin(0)+\sin\left(\frac{\pi}{2}\right)+2\sin\left(\frac{0+\frac{\pi}{2}}{2}\right)\right) = \frac{\pi}{8}(1+\sqrt{2})$

Hence, (a) is the correct answer.

80 If $f(x)$ is a polynomial and if $\lim_{t\to a}\frac{\int_a^t f(x)dx - \frac{(t-a)}{2}(f(t)+f(a))}{(t-a)^3} = 0$ for all a, then the degree of $f(x)$ can atmost be

(a) 1 (b) 2 (c) 3 (d) 4

Solution. Applying L' Hospital's rule

$$\lim_{t\to a}\frac{f(t)-\frac{1}{2}(f(t)+f(a))-\frac{(t-a)}{2}f'(t)}{3(t-a)^2} = 0$$

$$\Rightarrow \lim_{t\to a}\frac{f(t)-\frac{1}{2}(f(t)+f(a))-\frac{(t-a)}{2}f'(t)}{3(t-a)^2} = 0$$

$$\Rightarrow \lim_{t\to a}\frac{f'(t)-\frac{1}{2}(f(t)+f(a))-\frac{(t-a)}{2}f'(t)}{12(t-a)} = 0$$

$$\Rightarrow \lim_{t\to a}\frac{f''(t)}{12} = 0$$

$$\Rightarrow \lim_{t\to a}\frac{f'(t)-\frac{1}{2}(f(t)+f(a))-\frac{(t-a)}{2}f'(t)}{12(t-a)} = 0$$

$$\Rightarrow \lim_{t\to a}\frac{f''(t)}{12} = 0$$

$f''(a) = 0$ for any a.

$\Rightarrow$ $f(a)$ is atmost of degree 1.

Hence, (a) is the correct answer.

81 If $f''(x) < 0, \forall\, x \in (a,b)$ and c is a point such that $a < c < b$ and $(c, f(c))$ is the point on the curve for which $F(c)$ is maximum, then $f'(c)$ is equal to

(a) $\frac{f(b)-f(a)}{b-a}$ (b) $\frac{2(f(b)-f(a))}{b-a}$ (c) $\frac{2f(b)-f(a)}{2b-a}$ (d) 0

Solution. $F'(c) = (b-a)'(c) + f(a) - f(b)$

$F''(c) = f''(c)(b-a) < 0$

$\Rightarrow$ $F'(c) = 0 \rightarrow f'(c) = \frac{f(b)-f(a)}{b-a}$

Hence, (a) is the correct answer.

Type 6 : Match the Columns

Example 82 Match the statements of Column I with values of Column II.

Column I	Column II
(A) Suppose, $f(n)=\log_2(3)\cdot\log_3(4)\cdot\log_4(5)......\log_{n-1}(n)$, then the sum $\sum_{k=2}^{100} f(2^k)$ is equal to	(p) 5010
(B) Let $f(x)=\sqrt{1+x\sqrt{1+(x+1)\sqrt{1+(x+2)(x+4)}}}$, then $\int_0^{100} f(x)dx$ is	(q) 5050
(C) In an AP the series containing 99 terms, the sum of all the odd numbered terms is 2550. The sum of all the 99 terms of the AP is	(r) 5100
(D) $\lim_{x\to 0}\dfrac{\prod_{r=1}^{100}(1+rx)-1}{x}$ is equal to	(s) 5049

Solution. (A) → (s), (B) → (r), (C) → (s), (D) → (q)

(A)
$$f(n)=\frac{\log 3}{\log 2}\cdot\frac{\log 4}{\log 3}\cdots\frac{\log n}{n-1}$$

$$f(n)=\frac{\log n}{\log 2}=\log_2(n)$$

$$\therefore \quad f(2^k)=\log_2(2^k)=k$$

$$\therefore \quad \sum_{k=2}^{100} f(2^k)=\sum_{k=2}^{100} k=2+3+4+\ldots+100=5050-1=5049$$

(B)
$$f(x)=\sqrt{1+x\sqrt{1+(x+1)\sqrt{1+(x+2)(x+4)}}}$$

$$=\sqrt{1+x\sqrt{1+(x+1)\sqrt{x^2+6x+9}}}$$

$$=\sqrt{1+x\sqrt{1+(x+1)(x+3)}}=\sqrt{1+x\sqrt{x^2+4x+4}}$$

$$=\sqrt{1+x(x+2)}=\sqrt{x^2+2x+1}=(x+1)$$

$$\therefore \quad I=\int_0^{100}(x+1)\,dx\left[=\frac{(x+1)^2}{2}\right]_0^{100}$$

$$=\frac{(101)^2-1^2}{2}=\frac{100\times 102}{2}=5100$$

(C) AP is $a,(a+d),(a+2d)\ldots(a+98d)$ sum of odd terms $=2550$

$$\frac{a+(a+2d)+(a+4d)+\ldots+(a+98d)}{50 \text{ terms}}=2500$$

$$\frac{50}{2}[2a+98d]=2500 \text{ or } 50[a+49d]=2500 \text{ or } a+49d=51$$

This is the 50th terms of the AP.

Hence, $S_{99}=51\times 99=5049$

Example 83 Match the following :

Column I	Column II
(A) The function $f(x)=\dfrac{e^{x\cos x}-1-x}{\sin x^2}$ is not defined at $x=0$. The value of $f(0)$, so that f is continuous at $x=0$ is	(p) -1
(B) The value of the definite integral $\int_0^1 \dfrac{dx}{\sqrt{x}+\sqrt[3]{x}}$ is equal to $a+b\ln 2$ where a and b are integers, then $(a+b)$ is equal to	(q) 0
(C) Given, $e^n \int_0^n \dfrac{\sec^2\theta-\tan\theta}{e^\theta}\,d\theta=1$, then the value of $\tan(n)$ is equal to	(r) 1/2
(D) Let $a_n=\int_{\frac{1}{n+1}}^{\frac{1}{n}} \tan^{-1}(nx)\,dx$ and $b_n=\int_{\frac{1}{n+1}}^{\frac{1}{n}} \sin^{-1}(nx)\,dx$, then $\lim\limits_{n\to\infty}\dfrac{a_n}{b_n}$ has the value equal to	(s) 1

Solution. (A) → (r), (B) → (p), (C) → (s), (D) → (r)

(A)
$$\lim_{x\to 0}\frac{e^x\cos x-1-x}{x^2\cdot\dfrac{\sin x^2}{x^2}}=\lim_{x\to 0}\frac{e^{x\cos x}-1-x}{x^2}$$

(Applying L'Hospital's twice)

$$=\lim_{x\to 0}\frac{e^{x\cos x}\cdot(-x\sin x+\cos x)-1}{2x}$$

$$=\lim_{x\to 0}\frac{e^{x\cos x}(-x\cos x-\sin x-\sin x)+e^{x\cos x}(-x\sin x+\cos x)}{2}=\frac{1}{2}$$

(B) $x=u^6\Rightarrow dx=6u^5\,du$

$$I=\int_0^1\frac{6u^5\,du}{u^3+u^2}=6\int_0^1\frac{u^3+1-1}{u+1}\,du=5-6\ln 2\Rightarrow a+b=5-6=-1$$

(C) $e^n\int_0^n e^{-\theta}(\sec^2\theta-\tan\theta)\,d\theta=1$

Put $-\theta=t;\ d\theta=-dt$

$$-e^n\int_0^{-n}e^t(\sec^2 t+\tan t)\,dt=1\ [\text{Use}\int e^x(f(x)+f'(x))=e^xf(x)]$$

$$\Rightarrow\quad -e^n[e^t\tan t]_0^{-n}=1\Rightarrow -e^n[-e^{-n}\tan n]=1\Rightarrow\tan n=+1$$

(D)
$$\lim_{n\to\infty}\frac{\int_{1/n+1}^{1/n}\tan^{-1}(nx)\,dx}{\int_{1/n+1}^{1/n}\sin^{-1}(nx)\,dx}$$

(put $nx=t$)

$$\therefore\quad \lim_{n\to\infty}\frac{\frac{1}{n}\int_{n/n+1}^{1}\tan^{-1}(t)\,dt}{\frac{1}{n}\int_{n/n+1}^{1}\sin^{-1}(t)\,dt}\quad\left(\frac{0}{0}\text{ form}\right)$$

Use L'Hospital's rule $\lim\limits_{n\to\infty}\dfrac{\tan^{-1}\left(\dfrac{n}{n+1}\right)}{\sin^{-1}\left(\dfrac{n}{n+1}\right)}=\dfrac{\frac{\pi}{4}}{\frac{\pi}{2}}=\dfrac{1}{2}$

Example 84 Match the following : **[IIT JEE 2006]**

	Column I		Column II
(A)	$\int_{-1}^{1}\frac{dx}{1+x^2}$	(p)	$\frac{1}{2}\log\left(\frac{2}{3}\right)$
(B)	$\int_{0}^{1}\frac{dx}{\sqrt{1-x^2}}$	(q)	$\frac{1}{2}\log\left(\frac{3}{2}\right)$
(C)	$\int_{2}^{3}\frac{dx}{1-x^2}$	(r)	$\frac{\pi}{3}$
(D)	$\int_{1}^{2}\frac{dx}{x\sqrt{x^2-1}}$	(s)	$\frac{\pi}{2}$

Solution. (A) → (s), (B) → (s), (C) → (p), (D) → (r)

(A) $\int_{-1}^{1}\frac{dx}{1+x^2}$

$$f(x)=\frac{1}{1+x^2} \text{ is an even function}$$

$$\therefore \quad I=2\int_0^1\frac{dx}{1+x^2}=[2(\tan^{-1}x)]_0^1=\frac{\pi}{2}$$

(B) $$\int_0^1\frac{dx}{\sqrt{1-x^2}}=[\sin^{-1}x]_0^1=\frac{\pi}{2}$$

(C) $$\int_2^3\frac{dx}{1-x^2}=-\int_2^3\frac{dx}{x^2-1}=-\left[\frac{1}{2}\ln\left|\frac{x-1}{x+1}\right|\right]_2^3$$

$$=-\frac{1}{2}\ln\frac{2}{3}=\frac{1}{2}\ln\left(\frac{3}{2}\right)$$

(D) $$I=\int_1^2\frac{dx}{x\sqrt{x^2-1}} \qquad \text{Put } x=\sec\theta,\ dx=\sec\theta\tan\theta\,d\theta$$

$$=\int_0^{\pi/3}\frac{\sec\theta\tan\theta\,d\theta}{\sec\theta\tan\theta}=\int_0^{\pi/3}1\,d\theta=\frac{\pi}{3}$$

Type 7 : Integer Answer Type Questions

Example 85 Let $f:R\to R$ be a continuous function which satisfies $f(x)=\int_0^x f(t)dt$. Then, the value of $f(\ln 5)$ is____. **[IIT JEE 2009]**

Solution. (0) From given integral equation, $f(0)=0$

Also, differentiating the given integral equation w.r.t. x

$$f'(x)=f(x)$$

If $$f(x)\neq 0 \Rightarrow \frac{f'(x)}{f(x)}=1 \Rightarrow f(x)=e^c e^x$$

$$\because \quad f(0)=0 \Rightarrow e^c=0, \text{ a contradiction}$$

$$\therefore \quad f(x)=0,\forall x\in R \Rightarrow f(\ln 5)=0$$

Example 86 For any real number x, let $[x]$ denotes the largest integer less than or equal to x. Let f be a real valued function defined on the interval $[-10, 10]$ by $f(x)=\begin{cases} x-[x] & \text{if}[x] \text{ is odd} \\ 1+[x]-x & \text{if}[x] \text{ is even}\end{cases}$. Then, the value of $\frac{\pi^2}{10}\int_{-10}^{10} f(x)\cos \pi x\, dx$.

[IIT JEE 2010]

Solution. (4) $f(x)=\begin{cases} x-[x] & \text{if } [x] \text{ is odd} \\ 1+[x]-x & \text{if } [x] \text{ is even}\end{cases}$

$f(x)$ and $\cos \pi x$ are both periodic with period 2 and are both even.

$$\therefore \quad \int_{-10}^{10} f(x)\cos \pi x\, dx = 2\int_0^{10} f(x)\cos \pi x\, dx = 10\int_0^2 f(x)\cos \pi x\, dx$$

$$\int_0^1 f(x)\cos \pi x\, dx = \int_0^1 (1-x)\cos \pi x\, dx = -\int_0^1 u\cos \pi u\, du$$

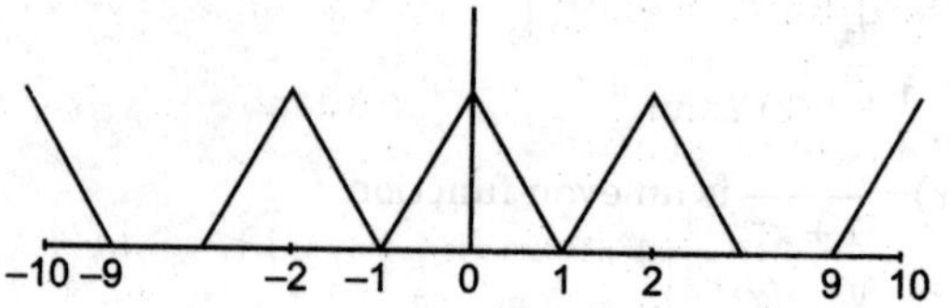

$$\int_1^2 f(x)\cos \pi x\, dx = \int_1^2 (x-1)\cos \pi x\, dx = -\int_0^1 u\cos \pi u\, du$$

$$\therefore \quad \int_{-10}^{10} f(x)\cos \pi x\, dx = -20\int_0^1 u\cos \pi u\, du = \frac{40}{\pi^2}$$

$$\Rightarrow \quad \frac{\pi^2}{10}\int_{-10}^{10} f(x)\cos \pi x\, dx = 4$$

Proficiency Exercise 1

Type 1 : Only One Correct Option

1. The value of the definite integral $\int_0^{\infty} \frac{dx}{(1+x^a)(1+x^2)} (a > 0)$ is

(a) $\frac{\pi}{4}$ (b) $\frac{\pi}{2}$ (c) π (d) Some function of a

2. The value of the definite integral $\int_0^{3\pi/4} ((1+x)\sin x + (1-x)\cos x)\,dx$, is

(a) $2\tan\frac{3\pi}{8}$ (b) $2\tan\frac{\pi}{4}$ (c) $2\tan\frac{\pi}{8}$ (d) 0

3. Let $C_n = \int_{1/n+1}^{1/n} \frac{\tan^{-1}(nx)}{\sin^{-1}(nx)}\,dx$, then $\lim_{n \to \infty} n^2 \cdot C_n$ is equal to

(a) 1 (b) 0 (c) −1 (d) $\frac{1}{2}$

4. If x satisfies the equation $\left(\int_0^1 \frac{dt}{t^2 + 2t\cos\alpha + 1}\right)x^2 - \left(\int_{-3}^3 \frac{t^2 \sin 2t}{t^2+1}\,dt\right)x - 2 = 0$ $(0 < \alpha < \pi)$, then the value of x is

(a) $\pm\sqrt{\frac{\alpha}{2\sin\alpha}}$ (b) $\pm\sqrt{\frac{2\sin\alpha}{\alpha}}$ (c) $\pm\sqrt{\frac{\alpha}{\sin\alpha}}$ (d) $\pm 2\sqrt{\frac{\sin\alpha}{\alpha}}$

5. If $f(x) = e^{g(x)}$ and $g(x) = \int_2^x \frac{t\,dt}{1+t^4}$, then $f'(2)$ is equal to

(a) 2/17 (b) 0
(c) 1 (d) Cannot be determined

6. If a, b and c are real numbers, then the value of $\lim_{t \to 0} \ln\left(\frac{1}{t}\int_0^t (1 + a\sin bx)^{c/x}\,dx\right)$ equals

(a) abc (b) $\frac{ab}{c}$ (c) $\frac{bc}{a}$ (d) $\frac{ca}{b}$

7. The value of $\lim_{n \to \infty} \sum_{r=1}^{r=4n} \frac{\sqrt{n}}{\sqrt{r}\,(3\sqrt{r} + 4\sqrt{n})^2}$ is equal to

(a) $\frac{1}{35}$ (b) $\frac{1}{14}$ (c) $\frac{1}{10}$ (d) $\frac{1}{5}$

8. Let $f(x) = \int_{-1}^x e^{t^2}\,dt$ and $h(x) = f(1 + g(x))$, where $g(x)$ is defined for all x, $g'(x)$ exists for all x, and $g(x) < 0$ for $x > 0$. If $h'(1) = e$ and $g'(1) = 1$, then the possible values which $g(1)$ can take

(a) 0 (b) −1 (c) −2 (d) −4

9. Let $f(x)$ be a function satisfying $f'(x) = f(x)$ with $f(0) = 1$ and g be the function satisfying $f(x) + g(x) = x^2$. The value of the integral $\int_0^1 f(x)\, g(x)\, dx$ is

(a) $e - \frac{1}{2}e^2 - \frac{5}{2}$ (b) $e - e^2 - 3$

(c) $\frac{1}{2}(e - 3)$ (d) $e - \frac{1}{2}e^2 - \frac{3}{2}$

10. Let $f(x) = \int_0^{g(x)} \frac{dt}{\sqrt{1+t^2}}$ where $g(x) = \int_0^{\cos x} (1 + \sin t^2)\, dt$. Also $h(x) = e^{-|x|}$ and $f(x) = x^2 \sin \frac{1}{x}$, if $x \neq 0$ and $f(0) = 0$, then $f'\left(\frac{\pi}{2}\right)$ is equal to

(a) $l'(0)$ (b) $h'(0^-)$ (c) $h'(0^+)$ (d) $\lim_{x \to 0} \frac{1 - \cos x}{x \sin x}$

11. For $f(x) = x^4 + |x|$, let $I_1 = \int_0^{\pi} f(\cos x)\, dx$ and $I_2 = \int_0^{\pi/2} f(\sin x)\, dx$, then $\frac{I_1}{I_2}$ has the value equal to

(a) 1 (b) 1/2 (c) 2 (d) 4

12. Let f be a positive function. Let $I_1 = \int_{1-k}^{k} (x) f(x(1-x))\, dx$; $I_2 = \int_{1-k}^{k} f(x(1-x))\, dx$, where $2k - 1 > 0$. Then, $\frac{I_2}{I_1}$ is

(a) k (b) 1/2 (c) 1 (d) 2

13. Suppose that the quadratic function $f(x) = ax^2 + bx + c$ is non-negative on the interval $[-1, 1]$. Then the area under the graph of f over the interval $[-1, 1]$ and the x-axis is given by the formula

(a) $A = f(-1) + f(1)$ (b) $A = f\left(-\frac{1}{2}\right) + f\left(\frac{1}{2}\right)$

(c) $A = \frac{1}{2}[f(-1) + 2f(0) + f(1)]$ (d) $A = \frac{1}{3}[f(-1) + 4f(0) + f(1)]$

14. Let $I(a) = \int_0^{\pi} \left(\frac{x}{a} + a \sin x\right)^2 dx$ where 'a' is positive real. The value of 'a' for which $I(a)$ attains its minimum value is

(a) $\sqrt{\pi\sqrt{\frac{2}{3}}}$ (b) $\sqrt{\pi\sqrt{\frac{3}{2}}}$ (c) $\sqrt{\frac{\pi}{16}}$ (d) $\sqrt{\frac{\pi}{13}}$

15. The set of values of 'a' which satisfy the equation $\int_0^2 (t - \log_2 a)\, dt = \log_2\left(\frac{4}{a^2}\right)$ is

(a) $a \in R$ (b) $a \in R^+$ (c) $a < 2$ (d) $a > 2$

16. $\lim_{x \to \infty} \left(x^3 \int_{-1/x}^{1/x} \frac{\ln(1+t^2)}{1+e^t}\, dt\right)$ is equal to

(a) $\frac{1}{3}$ (b) $\frac{2}{3}$ (c) 1 (d) 0

17. The value of $\sqrt{\pi\left(\int_0^{2008} x\,|\sin \pi x|\, dx\right)}$ is equal to

(a) $\sqrt{2008}$ (b) $\pi\sqrt{2008}$ (c) 1004 (d) 2008

18. $\lim_{n \to \infty} \sum_{k=1}^{n} \frac{n}{n^2 + k^2x^2}$, $x > 0$ is equal to

(a) $x \tan^{-1}(x)$ (b) $\tan^{-1}(x)$ (c) $\frac{\tan^{-1}(x)}{x}$ (d) $\frac{\tan^{-1}(x)}{x^2}$

19. Let $a > 0$ and $f(x)$ is monotonic increasing such that $f(0) = 0$ and $f(a) = b$, then $\int_0^a f(x)\,dx + \int_0^b f^{-1}(x)\,dx$ is equal to

(a) $a + b$ (b) $ab + b$ (c) $ab + a$ (d) ab

20. If $\int_{-1/\sqrt{3}}^{1/\sqrt{3}} \frac{x^4}{1 - x^4} \cos^{-1} \frac{2x}{1 + x^2}\,dx = k \int_0^{1/\sqrt{3}} \frac{x^4}{1 - x^4}$, then '$k$' is equal to

(a) π (b) 2π (c) 2 (d) 1

21. $\int_0^\infty f\left(x + \frac{1}{x}\right) \cdot \frac{\ln x}{x}\,dx$ is equal to

(a) 0 (b) 1

(c) $\frac{1}{2}$ (d) Cannot be evaluated

22. $\lim_{\lambda \to 0} \left(\int_0^1 (1 + x)^\lambda\,dx\right)^{1/\lambda}$ is equal to

(a) $2 \ln 2$ (b) $\frac{4}{e}$ (c) $\ln \frac{4}{e}$ (d) 4

23. If $g(x)$ is the inverse of $f(x)$ and $f(x)$ has domain $x \in [1, 5]$, where $f(1) = 2$ and $f(5) = 10$, then the values of $\int_1^5 f(x)\,dx + \int_2^{10} g(y)\,dy$ is equal to

(a) 48 (b) 64 (c) 71 (d) 52

24. The vlaue of the definite integral $\int_0^{\pi/2} \sin x \sin 2x \sin 3x\,dx$ is equal to

(a) $\frac{1}{3}$ (b) $-\frac{2}{3}$ (c) $-\frac{1}{3}$ (d) $\frac{1}{6}$

25. If $f(x) = \int_0^x (f(t))^2\,dt$, $f : R \to R$ be differentiable function and $f(g(x))$ is differentiable function at $x = a$, then

(a) $g(x)$ must be differentiable at $x = a$

(b) $g(x)$ may be non-differentiable at $x = a$

(c) $g(x)$ may be discontinuous at $x = a$

(d) None of the above

26. The number of integral solutions of equation $4 \int_0^\infty \frac{\ln t\,dt}{x^2 + t^2} - \pi \ln 2 = 0$; $x > 0$, is

(a) 0 (b) 1 (c) 2 (d) 3

27. $\int_0^{16n^2/\pi} \cos \frac{\pi}{2} \left[\frac{x\pi}{n}\right] dx$ is

(a) 0 (b) 1 (c) 2 (d) 3

28. If $\int_{-2}^{-1} (ax^2 - 5)\,dx = 0$ and $5 + \int_1^2 (bx + c)\,dx = 0$, then

(a) $ax^2 - bx + c = 0$ has at least one root in $(1, 2)$

(b) $ax^2 - bx + c = 0$ has at least one root in $(-2, -1)$

(c) $ax^2 + bx + c = 0$ has at least one root in $(-2, -1)$

(d) None of the above

29. The value of $\int_0^6 (\sqrt{x + \sqrt{12x - 36}} + \sqrt{x - \sqrt{12x - 36}})\, dx$ is equal to

(a) $6\sqrt{3}$ (b) $4\sqrt{3}$ (c) $12\sqrt{3}$ (d) None of these

30. If $I_n = \int_{-n}^{n} (\{x+1\}\{x^2+2\} + \{x^2+3\}\{x^3+4\})\, dx$, (where {} denotes the fractional part), then I_1 is equal to

(a) $-\frac{1}{3}$ (b) $-\frac{2}{3}$ (c) $\frac{1}{3}$ (d) None of these

31. Number of positive continuous functions $f(x)$ defined in [0, 1] for which $\int_0^1 f(x)\, dx = 1, \int_0^1 xf(x)\, dx = 2, \int_0^1 x^2 f(x)\, dx = 4$, is

(a) 1 (b) 4 (c) infinite (d) None of these

32. Let $I_n = \int_0^{\pi/2} \frac{\sin(2n-1)x}{\sin x} dx, J_n = \int_0^{\pi/2} \frac{\sin^2 nx}{\sin^2 x}, n \in N$, then

(a) $J_{(n+1)} - J_n = I_n$ (b) $J_{(n+1)} - J_n = I_{(n+1)}$

(c) $J_{n+1} + J_n = J_n$ (d) $J_{n+1} - J_{n+1} = J_n$

Type 2 : More Than One Correct Options

33. If $f(x) = [\sin^{-1}(\sin 2x)]$ ([] denotes the integer function), then

(a) $\int_0^{\pi/2} f(x)\, dx = \frac{\pi}{2} - \sin^{-1}(\sin 1)$ (b) $f(x)$ is periodic with period π

(c) $\lim_{x \to \frac{\pi}{2}} f(x) = -1$ (d) None of these

34. Which of the following definite integral(s) vanishes?

(a) $\int_0^{\pi/2} \ln(\cot x)\, dx$ (b) $\int_0^{2\pi} \sin^3 x\, dx$

(c) $\int_{1/e}^{e} \frac{dx}{x(\ln x)^{1/3}}$ (d) $\int_0^{\pi} \sqrt{\frac{1+\cos 2x}{2}}\, dx$

35. The equation $10x^4 - 3x^2 - 1 = 0$ has

(a) at least one root in $(-1, 0)$ (b) at least one root in $(0, 1)$

(c) at least who roots in $(-1, 1)$ (d) no root in $(-1, 1)$

36. Suppose $I_1 = \int_0^{\pi/2} \cos(\pi \sin^2 x)\, dx$;

$I_2 = \int_0^{\pi/2} \cos(2\pi \sin^2 x)\, dx$ and $I_3 = \int_0^{\pi/2} \cos(\pi \sin x)\, dx$, then

(a) $I_1 = 0$ (b) $I_2 + I_3 = 0$ (c) $I_1 + I_2 + I_3 = 0$ (d) $I_2 = I_3$

37. Let $f(x) = \int_{-1}^{1} (1 - |t|) \cos(xt)\, dt$, then which of the following hold true ?

(a) $f(0)$ is not defined

(b) $\lim_{x \to 0} f(x)$ exists and equals to 2

(c) $\lim_{x \to 0} f(x)$ exists and is equal to 1

(d) $f(x)$ is continuous at $x = 0$

38. The function f is continuous and has the property $f(f(x)) = 1 - x$ for all $x \in [0, 1]$ and $J = \int_0^1 f(x)\,dx$, then

(a) $f\left(\frac{1}{4}\right) + f\left(\frac{3}{4}\right) = 1$

(b) the value of J equals to 1/2

(c) $f\left(\frac{1}{3}\right) \cdot f\left(\frac{2}{3}\right) = 1$

(d) $\int_0^{\pi/2} \frac{\sin x\,dx}{(\sin x + \cos x)^3}$ has the same value as J

39. Let $f(x)$ is a real valued function defined by :

$$f(x) = x^2 + x^2 \int_{-1}^{1} t \cdot f(t)\,dt + x^3 \int_{-1}^{1} f(t)\,dt,$$

then which of the following hold(s) good ?

(a) $\int_{-1}^{1} t \cdot f(t)\,dt = \frac{10}{11}$ (b) $f(1) + f(-1) = \frac{30}{11}$

(c) $\int_{-1}^{1} t \cdot f(t)\,dt > \int_{-1}^{1} f(t)\,dt$ (d) $f(1) - f(-1) = \frac{20}{11}$

40. Let $f(x)$ and $g(x)$ are differentiable functions such that $f(x) + \int_0^x g(t)\,dt = \sin x\,(\cos x - \sin x)$ and $(f'(x))^2 + (g(x))^2 = 1$, then $f(x)$ and $g(x)$ respectively, can be

(a) $\frac{1}{2}\sin 2x, \sin 2x$ (b) $\frac{\cos 2x}{2}, \cos 2x$ (c) $\frac{1}{2}\sin 2x, -\sin 2x$ (d) $-\sin^2 x, \cos 2x$

41. Let $f(x) = \int_{-x}^{x} (t \sin at + bt + c)\,dt$ where a, b, c are non-zero real numbers, then $\lim_{x \to 0} \frac{f(x)}{x}$ is

(a) independent of a

(b) independent of a and b and has the value equals to c

(c) independent a, b and c

(d) dependent only on c

42. Let $L = \lim_{n \to \infty} \int_a^{\infty} \frac{n\,dx}{1 + n^2x^2}$ where $a \in R$, then L can be

(a) π (b) $\pi/2$ (c) 0 (d) 1

Type 3 : **Assertion and Reason**

Directions

(Q. Nos. 43 to 46)

For the following questions, choose the correct answers from the codes (a), (b), (c) and (d) defined as follows :

(a) Statement I is true, Statement II is true and Statement II is the correct explanation for Statement I.

(b) Statement I is true, Statement II is true and Statement II is not the correct explanation for Statement I.

(c) Statement I is true, Statement II is false.
(d) Statement I is false, Statement II is true.

43. **Statement I** If $n > 1$, then $\int_0^\infty \frac{dx}{1+x^n} = \int_0^1 \frac{dx}{(1-x^n)^{1/n}}$

Statement II $\int_a^b f(x)\,dx = \int_a^b f(a+b-x)\,dx$

44. **Statement I** If $n \in I^+$, then $\int_0^{n\pi} \left|\frac{\sin x}{x}\right| dx > \frac{2}{\pi}\left[1 + \frac{1}{2} + \frac{1}{3} + \dots + \frac{1}{n}\right]$

Statement II $\frac{\sin x}{x} > \frac{2}{\pi}$ in $\left(0, \frac{\pi}{2}\right)$.

45. Consider $I = \int_{-\pi/4}^{\pi/4} \frac{dx}{1 - \sin x}$

Statement I $I = 0$

Statement II $\int_{-a}^{a} f(x)\,dx = 0$, wherever $f(x)$ is an odd function.

46. **Statement I** If $2f(x) + f(-x) = \frac{1}{x}\sin\left(x - \frac{1}{x}\right)$, then value of $I = \int_{1/e}^{e} f(x)\,dx = 0$.

Statement II If $f(2a - x) = -f(x)$, then $\int_0^{2a} f(x)\,dx = 0$.

Type 4 : Linked–Comprehension Based Questions

Passage I
(Q. Nos. 47 to 49)

Suppose $\lim_{x \to 0} \frac{\int_0^x \frac{t^2\,dt}{(a + t^r)^{1/p}}}{bx - \sin x} = l$ *where* $p \in N$, $p \geq 2$, $a > 0$, $r > 0$ *and* $b \neq 0$.

47. If l exists and is non-zero, then

(a) $b > 1$ (b) $0 < b < 1$ (c) $b < 0$ (d) $b = 1$

48. If $p = 3$ and $l = 1$, then the value of 'a' is equal to

(a) 8 (b) 3 (c) 6 (d) 3/2

49. If $p = 2$ and $a = 9$ and l exists, then the value of l is equal to

(a) $\frac{3}{2}$ (b) $\frac{2}{3}$ (c) $\frac{1}{3}$ (d) $\frac{7}{9}$

Passage II
(Q. Nos. 50 to 52)

Suppose $f(x)$ *and* $g(x)$ *are two continuous functions defined for* $0 \leq x \leq 1$. *Given,* $f(x) = \int_0^1 e^{x+1} \cdot f(t)\,dt$ *and* $g(x) = \int_0^1 e^{x+t} \cdot g(t)\,dt + x$

50. The value of $f(1)$ equals

(a) 0 (b) 1 (c) e^{-1} (d) e

51. The value of $g(0)-f(0)$ equals

(a) $\frac{2}{3-e^2}$ (b) $\frac{3}{e^2-2}$ (c) $\frac{1}{e^2-1}$ (d) 0

52. The value of $\frac{g(0)}{g(2)}$ equals

(a) 0 (b) $\frac{1}{3}$ (c) $\frac{1}{e^2}$ (d) $\frac{2}{e^2}$

Passage III

(Q. Nos. 53 to 55)

Suppose a and b are positive real numbers such that $ab=1$. Let for any real parameter t, the distance from the origin to the line $(ae^t)x+(be^{-t})y=1$ be denoted by D(t), then

53. The value of the definite integral $I=\int_0^1 \frac{dt}{(D(t))^2}$ is equal to

(a) $\frac{e^2-1}{2}\left(b^2+\frac{a^2}{e^2}\right)$ (b) $\frac{e^2+1}{2}\left(a^2+\frac{b^2}{e^2}\right)$

(c) $\frac{e^2-1}{2}\left(a^2+\frac{b^2}{e^2}\right)$ (d) $\frac{e^2+1}{2}\left(b^2+\frac{a^2}{e^2}\right)$

54. The value of 'b' at which I is minimum, is

(a) e (b) $\frac{1}{e}$ (c) $\frac{1}{\sqrt{e}}$ (d) $\sqrt{e}$

55. Minimum value of I is

(a) $e-1$ (b) $e-\frac{1}{e}$ (c) e (d) $e+\frac{1}{e}$

Passage IV

(Q. Nos. 56 to 58)

We are given the curves $y=\int_{-\infty}^{x} f(t)\,dt$ through the point $\left(0,\frac{1}{2}\right)$ any $y=f(x)$, where $f(x)>0$ and $f(x)$ is differentiable $\forall\, x \in R$ through (0, 1). If tangents drawn to both the curves at the points with equal abscissae intersect on the same point on the x-axis, then

56. Number of solutions $f(x)=2ex$ is equal to

(a) 0 (b) 1 (c) 2 (d) None of these

57. $\lim_{x\to\infty} (f(x))^{f(-x)}$ is

(a) 3 (b) 6 (c) 1 (d) None of these

58. The function $f(x)$ is

(a) increasing for all x (b) non-monotonic

(c) decreasing for all x (d) None of these

Passage V
(Q. Nos. 59 to 61)

$f(x)=\int_0^x (4t^4-at^3)\,dt$ and $g(x)$ is quadratic satisfying $g(0)+6=g'(0)-c=g''(c)+2b=0$. If $y=h(x)$ and $y=g(x)$ intersect in 4 distinct points with abscissae x_i; $i=1,2,3,4$ such that $\sum \frac{i}{x_i}=8$, $a,b,c\in R^+$ and $h(x)=f'(x)$, then

59. Abscissae of point of intersection are in

(a) AP (b) GP (c) HP (d) None of these

60. 'a' is equal to

(a) 6 (b) 8 (c) 20 (d) 12

61. 'c' is equal to

(a) 25 (b) 25/2 (c) 25/4 (d) 25/8

Passage VI
(Q. Nos. 62 to 64)

Let $y=\int_{u(x)}^{v(x)} f(t)\,dt$, let us define $\frac{dy}{dx}$ as $\frac{dy}{dx}=v'(x)f^2(v(x))-u'(x)f^2(u(x))$ and the equation of the tangent at (a,b) and $y-b=\left(\frac{dy}{dx}\right)_{(a,\,b)}(x-a)$.

62. If $y=\int_x^{x^2} t^2\,dt$, then equation of tangent at $x=1$ is

(a) $x+y=1$ (b) $y=x-1$ (c) $y=x$ (d) $y=x+1$

63. If $y=\int_{x^2}^{x^4}(\ln t)dt$, then $\lim_{x\to 0^+}\frac{dy}{dx}$ is

(a) 0 (b) 1 (c) 2 (d) −1

64. If $f(x)=\int_1^x e^{t^2/2}(1-t^2)\,dt$, then $\frac{d}{dx}f(x)$ at $x=1$ is

(a) 0 (b) 1 (c) 2 (d) −1

Passage VII
(Q. Nos. 65 to 67)

Let n be non-negative integer, $I_n=\int x^n\sqrt{a^2-x^2}\,dx$, $a>0$. Relation between I_{n-2}, I_{n-1}, I_n can be obtained by integrating by parts.

Clearly, $I_1=-\frac{1}{3}(a^2-x^2)^{3/2}$.

65. If $I_n=\frac{-x^{n-1}(a^2-x^2)^{3/2}}{A}+a^2BI_{n-2}$, where A and B are constants, then A must be equal to

(a) $n+1$ (b) $n-1$ (c) $n+2$ (d) n

66. In the above question, B is equal to

(a) $\frac{n+1}{n+2}$ (b) $\frac{n}{n+2}$ (c) $\frac{n+2}{n+1}$ (d) $\frac{n-1}{n+2}$

67. The value of the integral $\int_0^a x^4 \sqrt{a^2 - x^2}\, dx$ is equal to

(a) $\frac{\pi a^6}{32}$ (b) $\frac{\pi a^4}{16}$ (c) $\frac{\pi a^4}{64}$ (d) $\frac{\pi a^2}{4}$

Passage VIII
(Q. Nos. 68 to 70)

Suppose in the definite integral $\int_a^b f(x)\,dx$ the upper limit b is ∞, then to obtain the value of $\int_a^b f(x)\,dx$, we say that $\int_a^b f(x)\,dx = \lim_{t \to \infty} \int_a^t f(x)\,dx$, where $t > a$. If the limit exists finitely, then the value of the limit is defined as the value of the symbol a $\int_a^\infty f(x)\,dx$ otherwise the integral is supposed to be non-existent. We can have another case of improper integrals. It may happen that $f(x) \to \infty$ as $x \to a$ or $x \to b$. In this case the principal value of the definite integral $\int_a^b f(x)\,dx$ is defined as $\int_a^b f(x)\,dx = \lim_{h \to \infty} \int_{a+h}^b f(x)\,dx$. In this limit exists finitely, the value of the limit is defined as the value of the integral.

68. $\int_0^\infty \frac{\sin^2 x}{x^2}\,dx$ must be same as

(a) $\int_0^\infty \frac{\sin x}{x}\,dx$ (b) $\left(\int_0^\infty \frac{\sin x}{x}\,dx\right)^2$

(c) $\int_0^\infty \frac{\cos^2 x}{x^2}\,dx$ (d) None of these

69. $\lim_{x \to 0} \int_0^{x^2} \frac{\sin \sqrt{x}}{x^3}\,dx$ must be

(a) $\frac{1}{2}$ (b) $\frac{1}{3}$ (c) $\frac{2}{3}$ (d) $\frac{1}{\sqrt{3}}$

70. If $I = \int_0^{\pi/2} \log \cos x \cos 2nx\,dx$, then I is same as

(a) $\int_0^{\pi/2} \cot x \cos(2nx)\,dx$ (b) $\int_0^{\pi/2} \cot x \sin(2nx)\,dx$

(c) $\int_0^{\pi/2} \tan x \cos(2nx)\,dx$ (d) $\frac{1}{2n}\int_0^{\pi/2} \tan x \sin(2nx)\,dx$

Type 5 : Match the Columns

71. Let $\lim_{T \to \infty} \frac{1}{T} \int_0^T (\sin x + \sin ax)^2\,dx = L$, then

Column I	Column II
(A) for $a = 0$, the value of L is	(p) 0
(B) for $a = 1$ the value of L is	(q) 1/2
(C) for $a = -1$ the value of L is	(r) 1
(D) $\forall a \in R - \{-1, 0, 1\}$ the value of L is	(s) 2

72. Let $f(\theta) = \int_0^1 (x + \sin\theta)^2\, dx$ and $g(\theta) = \int_0^1 (x + \cos\theta)^2\, dx$ where $\theta \in [0, 2\pi]$. The quantity $f(\theta) - g(\theta)\ \forall\ \theta$ in the interval given in column I, is

Column I	Column II
(A) $\left(\frac{\pi}{4}, \frac{3\pi}{4}\right)$	(p) negative
(B) $\left(\frac{3\pi}{4}, \pi\right]$	(q) positive
(C) $\left[\frac{3\pi}{2}, \frac{7\pi}{4}\right)$	(r) non-negative
(D) $\left(0, \frac{\pi}{4}\right) \cup \left(\frac{7\pi}{4}, 2\pi\right)$	(s) non-positive

73. Match the following :

Column I	Column II
(A) $\int_0^1 (1 + 2008\, x^{2008})\, e^{x^{2008}}\, dx$ equals	(p) e^{-1}
(B) The value of the definite integral $\int_0^1 e^{-x^2}\, dx + \int_1^{1/e} \sqrt{-\ln x}\, dx$ is equal to	(q) $e^{-1/4}$
(C) $\lim_{n \to \infty} \left(\frac{1^1 \cdot 2^2 \cdot 3^3 \ldots (n-1)^{n-1} \cdot n^n}{n^{1+2+3+\ldots+n}}\right)^{\frac{1}{n^2}}$ equals	(r) $e^{1/2}$
	(s) e

74. Match the following :

Column I	Column II
(A) Let $f(x) = \int x^{\sin x}(1 + x\cos x \cdot \ln x + \sin x)\, dx$ and $f\left(\frac{\pi}{2}\right) = \frac{\pi^2}{4}$, then the value of $f(\pi)$ is	(p) rational
(B) Let $g(x) = \int \frac{1 + 2\cos x}{(\cos x + 2)^2}\, dx$ and $g(0) = 0$, then the value of $g\left(\frac{\pi}{2}\right)$ is	(q) irrational
(C) If real numbers x and y satisfy $(x+5)^2 + (y-12)^2 = (14)^2$, then the minimum value of $\sqrt{(x^2 + y^2)}$ is	(r) integral
(D) Let $k(x) = \int \frac{(x^2+1)\, dx}{\sqrt[3]{x^3 + 3x + 6}}$ and $k(-1) = \frac{1}{\sqrt[3]{2}}$, then the value of $k(-2)$ is	(s) prime

75. Match the following :

Column I	Column II
(A) If $f(x)=\int_0^{g(x)}\frac{dt}{\sqrt{1+t^3}}$, where $g(x)=\int_0^{\cos x}(1+\sin t^2)\,dt$, then value of $f'\left(\frac{\pi}{2}\right)$ is	(p) −2
(B) If $f(x)$ is a non-zero differentiable function such that $\int_0^x f(t)\,dt=\{f(x)\}^2, \forall x\in R$, then $f(2)$ is equal to	(q) 2
(C) If $\int_a^b (2+x-x^2)\,dx$ is maximum, then $a+b$ is equal to	(r) 1
(D) If $\lim_{x\to 0}\left(\frac{\sin 2x}{x^3}+a+\frac{b}{x^2}\right)=0$, then $3a+b$ has the value	(s) −1

76. Match the following :

Column I	Column II
(A) The area of the figure bounded by $y=x^2$ and $y=\sqrt{x}$ is	(p) 4/3
(B) $\int_0^4 \{\sqrt{x}\}\,dx$ has the value ($\{x\}$ denotes fractional part of x)	(q) 5/3
(C) The area of the region for which $0<y<3-2x-x^2$ and $x>0$	(r) 7/3
(D) $\int_{-\pi/2}^{\pi/2}\sqrt{\cos x-\cos^2 x}\,dx$ is equal to	(s) 1/3

Proficiency Exercise 2

1. Evaluate the following :

(i) $\int_0^{\pi/4} \cos^2 x\, dx$

(ii) $\int_0^{\pi/2} \frac{dx}{1+\cos x}$

(iii) $\int_0^{\pi/2} \sqrt{1+\cos x}\; dx$

(iv) $\int_0^{\pi/6} \sin 2x \cdot \cos x\, dx$

(v) $\int_1^{2} \frac{dx}{\sqrt{x}-\sqrt{x-1}}$

(vi) $\int_0^{1/2} \frac{x \sin^{-1} x}{\sqrt{1-x^2}}\, dx$

(vii) $\int_0^{1} \log x\, dx$

(viii) $\int_{-2}^{2} |1-x^2|\, dx$

(ix) $\int_0^{2} (x\,[x])\, dx$

(x) $\int_0^{2} f(x)\,dx$ where $f(x) = \max(x, x^2)$

(xi) $\int_0^{\pi} \frac{e^{\cos x}}{e^{\cos x}+e^{-\cos x}}\, dx$

(xii) $\int_{\infty}^{0} \frac{z\, e^{-z}}{\sqrt{1-e^{-2z}}}\, dz$

(xiii) $\int_0^{1} \tan^{-1}\left(\frac{1}{1-x+x^2}\right) dx$

(xiv) $\int_0^{100\pi} \sqrt{1-\cos 2x}\; dx$

(xv) $\int_1^{1.5} [x^2]\, dx$

(xvi) $\int_a^{b} \frac{|x|}{x}\, dx$

(xvii) $\int_{1/2}^{4} [\log_2 x]\, dx$

(xviii) $\int_0^{2} \left[2\cos\left(\frac{\pi x}{2}\right)\right] dx$

(xix) $\int_0^{\pi/3} [\tan^2 x]\, dx$

(xx) $\int_0^{[x]} \left(\int_0^{[x]} [x] - \left[x - \frac{1}{2}\right] dx \right) dx$

2. Prove that $\int_0^{\pi} x\, f(\sin x)\, dx = \frac{\pi}{2}\int_0^{\pi} f(\sin x)\, dx$.

3. Evaluate $\int_0^{\pi} \frac{x^2 \sin 2x \sin\left(\frac{\pi}{2}\cos x\right) dx}{2x-\pi}$.

4. Let $a+b=4$, where $a<2$ and let $g(x)$ be a differentiable function. If $dg/dx > 0 \ \forall\, x$. Prove that $\int_0^{a} g(x)\,dx + \int_0^{b} g(x)\,dx$ increases as $(b-a)$ increases.

5. Evaluate $\lim_{n\to\infty} \left(\frac{n^{1/2}}{n^{3/2}} + \frac{n^{1/2}}{(n+3)^{3/2}} + \frac{n^{1/2}}{(n+6)^{3/2}} + \ldots + \frac{n^{1/2}}{\{n+3(n-1)\}^{3/2}} \right)$

6. Evaluate

$$\lim_{n\to\infty} \left[\frac{1}{n^2}\sec^2\left(\frac{\pi}{3n^2}\right) + \frac{2}{n^2}\sec^2\left(\frac{4\pi}{3n^2}\right) + \frac{3}{n^2}\sec^2\left(\frac{9\pi}{3n^2}\right) + \ldots + \frac{1}{n}\sec^2\left(\frac{\pi}{3}\right) \right]$$

7. If for non-zero x, $a\,f(x) + b\,f\left(\frac{1}{x}\right) = \frac{1}{x} - 5$, where $a \neq b$, then find $\int_1^2 f(x)\,dx$.

8. Evaluate $\int_0^{2\pi} \frac{6\sin^2\theta}{a - b\cos\theta}\,d\theta$, $a > b > 0$.

9. Evaluate $\int_{-1}^{3/2} |x \sin \pi x|\,dx$.

10. Find the value of $\int_0^{n\pi} \sin\left[\frac{2x}{\pi}\right] dx$, where [.] denotes the greatest integer function.

11. (a) If $\int_1^{\infty} \frac{(x^3+3)\,dx}{x^6(x^2+1)} = \frac{a + b\pi}{c}$, find the values of a, b and c.

(b) If $\int_0^1 \frac{(1-x^2)dx}{1+x^2+x^4} = a\ln(b)$, find the values of a and b.

12. Let $f(x) = \int_{-2}^{x} |t+1|\,dt$. Examine the continuity and differentiability of $f(x)$ in the interval $\left[-\frac{3}{2}, 1\right]$.

13. If $f(x)$ is a continuous function with $\int_0^x f(t)\,dt \to \infty$ as $|x| \to \infty$, then show that every line $y = mx$ intersects the curve $y^2 + \int_0^x f(t)\,dt = 2$.

14. Evaluate $\int_0^1 \frac{x^4(1-x)^4}{1+x^4}\,dx$.

15. Show that

(a) $\frac{P}{P+1} < \int_0^1 \frac{dx}{1+x^P} < 1$, $P \in R_+$ (b) $\left(\frac{1}{e}\right)^{1/e} < \int_0^1 x^x\,dx < 1$.

16. If $I(u) = \int_0^{\pi} \ln(1 - 2u\cos x + u^2)\,dx$, then prove that

(a) $I(u) = I(-u)$ (b) $I(u) + I(-u) = I(u^2)$

17. Evaluate $\int_0^{\infty} e^{-x}\sin^n x\,dx$; if n is even integer.

18. Let $I_n = \int_0^{\pi/4} \tan^n x\,dx$, show that $\frac{1}{I_2+I_4}, \frac{1}{I_3+I_5}, \frac{1}{I_4+I_6}, \frac{1}{I_5+I_7}$, form an AP.

Find common difference of this progression.

19. Evaluate $I = \int_0^1 2\sin(pt)\sin(qt)\,dt$ if,

(i) p and q are different roots of the equation $\tan x = x$.
(ii) p and q are equal and either is roots of equation $\tan x = x$.

20. Assuming $\int_0^{\pi} \log \sin x\,dx = -\pi \log 2$, show that

$\int_0^{\pi} \theta^3 \log \sin\theta\,d\theta = \frac{3\pi}{2}\int_0^{\pi} \theta^2 \log(\sqrt{2}\sin\theta)\,d\theta$.

21. Evaluate $\int_0^2 \max\{\log(1+x^2), 1\}\,dx$.

22. Consider the function $f(x) = \begin{cases} x - [x] - \dfrac{1}{2}, \text{ if } x \approx 1 \\ 0, \text{ if } x \in 1 \end{cases}$, where $[x]$ denotes the greatest integer function and I be the set of integers. If $g(x) = \max\{x^2, f(x) | x |\}, -10 \le x \le 10$, then find the value of $\int_{-2}^{2} g(x)\, dx$.

23. A periodic function with period "1" is integrable over any finite interval. Also for two real numbers a, b and for two unequal non-zero positive integers m and n. $\int_a^{a+n} f(x)\, dx = \int_b^{b+m} f(x)\, dx$. Calculate the value of $\int_m^n f(x)\, dx$.

24. Show that for a differentiable function $f(x)$

$$\int_0^n f'(x) \left\{[x] - x + \frac{1}{2}\right\} dx = \int_0^n f(x)\, dx + \frac{1}{2} f(0) + \frac{1}{2} f(n) - \sum_{r=0}^{n} f(r),$$

where [.] denotes the greatest integer function and $n \in N$.

25. Let $f : R \to R$ be defined by $f(x) = \int_1^3 \frac{1}{|t - x| + 1}\, dt$, then find the value of $\int_1^3 f(t)\, dt$.

26. Find the value of $\int_0^1 (\{2x\} - 1)(\{3x\} - 1)\, dx$, where $\{\cdot\}$ denotes the fractional part.

27. Let $g(t) = |t - 1| - |t| + |t + 1|, \forall\, t \in R$ and $f(x) = \max\{g(t) : -3/2 \le t \le x\}$ $\forall\, x \in [-3/2, \infty)$. Find the value of $\int_{-3/2}^{5} f(x)\, dx$.

28. Evaluate $\int_0^a \ln(\cot a + \tan x)\, dx,\ a \in \left(0, \frac{\pi}{2}\right)$.

29. Evaluate $\int_0^{\pi/4} [\sin x + [\cos x + [\tan x + [\sec x]]]]\, dx$, where [.] denotes the greatest integer function.

30. Let $f : R^+ \to R$ be a differentiable function with $f(1) = 3$ and satisfying :

$$\int_1^{xy} f(t)\, dt = y \int_1^x f(t)\, dt + x \int_1^y f(t)\, dt;\ \forall\, x, y \in R^+.$$ Find $f(x)$.

Answers

Target Exercise 2.1

1. (c)	2. (a)	3. (b)	4. (b)	5. (a)	6. (d)	7. (b)
8. (a)						

Target Exercise 2.2

1. (c)	2. (b)	3. (a)	4. (a)	5. (c)	6. (c)	7. (c)
8. (a)	9. (c)	10. (a)	11. (b)	12. (a)		

Target Exercise 2.3

1. (c)	2. (a)	3. (c)	4. (b)	5. (c)	6. (c)	7. (a)
8. (a)	9. (c)	10. (a)				

Proficiency Exercise 1

1. (a)	**2.** (a)	**3.** (d)	**4.** (d)	**5.** (a)	**6.** (a)	**7.** (c)
8. (c)	**9.** (d)	**10.** (c)	**11.** (c)	**12.** (d)	**13.** (d)	**14.** (a)
15. (b)	**16.** (a)	**17.** (d)	**18.** (c)	**19.** (d)	**20.** (a)	**21.** (a)
22. (b)	**23.** (a)	**24.** (d)	**25.** (a)	**26.** (c)	**27.** (a)	**28.** (b)
29. (a)	**30.** (b)	**31.** (d)	**32.** (b)	**33.** (a,b,c)	**34.** (a,b,c)	
35. (a,b,c)	**36.** (a,b,c)	**37.** (c,d)	**38.** (a,b,d)	**39.** (b,d)	**40.** (c,d)	
41. (a,d)	**42.** (a,b,c)	**43.** (b)	**44.** (b)	**45.** (d)	**46.** (b)	**47.** (d)
48. (a)	**49.** (b)	**50.** (a)	**51.** (a)	**52.** (b)	**53.** (c)	**54.** (d)
55. (b)	**56.** (b)	**57.** (c)	**58.** (a)	**59.** (a)	**60.** (c)	**61.** (a)
62. (b)	**63.** (a)	**64.** (a)	**65.** (c)	**66.** (d)	**67.** (a)	**68.** (a)
69. (c)	**70.** (d)	**71.** (A) → (q), (B) → (s), (C) → (p), (D) → (r)				

72. (A) → (q), (B) → (r), (C) → (s), (D) → (p)

73. (A) → (s), (B) → (p), (C) → (q)

74. (A) → (q), (B) → (p), (C) → (p,r), (D) → (p,r,s)

75. (A) → (s), (B) → (r), (C) → (r), (D) → (q)

76. (A) → (s), (B) → (r), (C) → (q), (D) → (p)

Proficiency Exercise 2

1. (i) $\left(\frac{\pi+2}{8}\right)$ (ii) 1 (iii) 2

(iv) $\frac{2}{3}-\frac{\sqrt{3}}{4}$ (v) $\frac{4\sqrt{2}}{3}$ (vi) $\frac{1}{2}-\frac{\pi\sqrt{3}}{12}$

(vii) -1 (viii) 4 (ix) $\frac{3}{2}$

(x) $\frac{17}{6}$ (xi) π (xii) $-\frac{\pi}{2}\log 2$

(xiii) $\frac{\pi}{2}-\frac{1}{2}\log 2$ (xiv) $200\sqrt{2}$ (xv) $2-\sqrt{2}$

(xvi) $(|b|-|a|)$ where $a<b$ (xvii) $\frac{3}{2}$ (xviii) -1

(xix) $\frac{5\pi}{12}-\tan^{-1}\sqrt{2}$ (xx) $\frac{[x]([x]-1)}{4}$

3. $\frac{16}{\pi}$ **5.** $\frac{1}{3}$ **6.** $\frac{3\sqrt{3}}{2\pi}$

7. $\frac{1}{a^2-b^2}\left[a(\ln 2-5)+\frac{7b}{2}\right]$ **8.** $\frac{2\pi}{b^2}[a-\sqrt{a^2-b^2}]$

9. $\frac{\pi+1}{\pi^2}$ **10.** $\frac{\pi}{2}\cdot\left[\frac{\sin(n)\cdot\sin(n-1/2)}{\sin(1/2)}\right]$

11. (a) $a=40-6\ln 2,\ b=-9,\ c=12$ (b) $a=1/2,\ b=3$

12. $f(x)$ is continuous and differentiable in $[(-3/2)\,1]$ except $x=-1$.

14. $\frac{365}{105}-\pi$ **17.** $\frac{n!}{\prod_{r=1}^{n/2}[(2r)^2+1]}$ **18.** Common difference = 1

19. (i) $I=0$ (ii) $I=\frac{p^2}{1+p^2}$

21. $I = \sqrt{e-1} + \log\left(\frac{25}{e^{\sqrt{e-1}}}\right) - 2\,[2 - \tan^{-1} 2 - \sqrt{e-1} + \tan^{-1}(\sqrt{e-1})]$

22. $\frac{275}{48}$ sq units **23.** 0 **25.** $6 \ln 3 - 4$

26. $19/72$ **27.** $135/8$ **28.** $-a \ln(\sin a)$

29. $\pi/4$ **30.** $3 \ln(e\,x)$

Solutions

(*Proficiency Exercise* 1)

Type 1 : Only One Correct Option

1. Put $x = \tan\theta$

$$I = \int_0^{\pi/2} \frac{d\theta}{1 + (\tan\theta)^a} = \int_0^{\pi/2} \frac{(\cos\theta)^a}{(\sin\theta)^a + (\cos\theta)^a}\, d\theta \Rightarrow I = \frac{\pi}{4}$$

2. $I = \int_0^{3\pi/4} (\sin x + \cos x)\, dx + \int_0^{3\pi/4} \underset{\text{I}}{x}\, \underbrace{(\sin x - \cos x)}_{\text{II}}\, dx$

$$= \int_0^{3\pi/4} (\sin x + \cos x)\, dx + \underbrace{[x(-\cos x - \sin x)]_0^{3\pi/4}}_{\text{zero}} + \int_0^{3\pi/4} (\sin x + \cos x)\, dx$$

$$= 2\int_0^{3\pi/4} (\sin x + \cos x)\, dx = 2(\sqrt{2} + 1)$$

3. $C_n = \int_{\frac{1}{n+1}}^{\frac{1}{n}} \frac{\tan^{-1}(nx)}{\sin^{-1}(nx)}\, dx$ (put $nx = t$) $\Rightarrow C_n = \frac{1}{n}\int_{\frac{n}{n+1}}^{1} \frac{\tan^{-1}(t)}{\sin^{-1}(t)}\, dt$

$$L = \lim_{n\to\infty} n^2 \cdot C_n = \lim_{n\to\infty} n \int_{\frac{n}{n+1}}^{1} \frac{\tan^{-1} t}{\sin^{-1} t}\, dt\ (\infty \times 0);\ L = \frac{\int_{\frac{n}{n+1}}^{1} \frac{\tan^{-1} t}{\sin^{-1} t}\, dt}{\frac{1}{n}}\ \left(\frac{0}{0}\text{ form}\right)$$

Applying, Leibnitz rule
$$L = \lim_{n\to\infty} \frac{0 - \frac{\tan^{-1}\frac{n}{n+1}}{\sin^{-1}\frac{n}{n+1}}\left(\frac{1}{(n+1)^2}\right)}{-\frac{1}{n^2}} = \frac{\pi}{4}\cdot\frac{2}{\pi} = \frac{1}{2}$$

4. $\int_{-3}^{3} \frac{t^2 \sin 2t}{t^2 + 1}\, dt = 0$ as the integrand is an odd function.

Also,
$$\int_0^1 \frac{dt}{t^2 + 2t\cos\alpha + 1} = \frac{1}{\sin\alpha}\left|\tan^{-1}\frac{t + \cos\alpha}{\sin\alpha}\right|_0^1 = \frac{\alpha}{2\sin\alpha}$$

Thus, the given equation reduces to

$$x^2\,\frac{\alpha}{2\sin\alpha} - 2 = 0 \quad\Rightarrow\quad x = \pm 2\sqrt{\frac{\sin\alpha}{\alpha}}$$

5. $f'(x) = e^{g(x)} \cdot g'(x);\ g'(x) = \dfrac{x}{1+x^4};\ f'(x) = e^{g(x)} \cdot \dfrac{x}{1+x^4};\ e^{g(2)} = e^0 = 1$

Hence, $\quad f'(2) = e^{g(2)} \cdot g'(2) = e^0 \cdot \dfrac{2}{17} = \dfrac{2}{17}$

7. $T_r = \dfrac{1}{\sqrt{\dfrac{r}{n}} \cdot n\left(3\sqrt{\dfrac{r}{n}}+4\right)^2}$

$$S = \frac{1}{n}\sum_{1}^{4n} \frac{1}{\left(3\sqrt{\dfrac{r}{n}}+4\right)^2 \cdot \sqrt{\dfrac{r}{n}}} = \int_0^4 \frac{dx}{\sqrt{x}\,(3\sqrt{x}+4)^2}$$

Put, $3\sqrt{x}+4 = t \Rightarrow \dfrac{3}{2}\dfrac{1}{\sqrt{x}}\,dx = dt$

$$= \frac{2}{3}\int_4^{10} \frac{dt}{t^2} = \frac{2}{3}\left[\frac{1}{t}\right]_{10}^{4} = \frac{2}{3}\left[\frac{1}{4} - \frac{1}{10}\right] = \frac{2}{3} \cdot \frac{6}{40} = \frac{1}{10}$$

8. Given, $f(x) = \int_{-1}^{x} e^{t^2}\,dt;\ h(x) = f(1+g(x));\ g(x) < 0$ for $x > 0$

Now, $\quad h(x) = \int_{-1}^{1+g(x)} e^{t^2}\,dt = f(1+g(x))$ (given)

Differentiating $\quad h'(x) = e^{(1+g(x))^2} \cdot g'(x)$

$h'(1) = e$ (given)

$e^{(1+g(1))^2} \cdot g'(1) = e$

$\therefore \quad (1+g(1))^2 = 1$

$1 + g(1) = \pm 1$

$\Rightarrow g(1) = 0$ (not possible) or $g(1) = -2$

9. $f'(x) = f(x) \Rightarrow f(x) = C\,e^x$ and since $f(0) = 1$

$\therefore \quad 1 = f(0) = C$

$\therefore\ f(x) = e^x$ and hence $g(x) = x^2 - e^x$

Thus, $\quad \int_0^1 f(x)\,g(x)\,dx = \int_0^1 (x^2e^x - e^{2x})\,dx$

$$= [x^2e^x]_0^1 - 2\int_0^1 xe^x\,dx - \left[\frac{e^{2x}}{2}\right]_0^1$$

$$= (e-0) - 2\,[(xe^x)_0^1 - (e^x)_0^1\,] - \frac{1}{2}(e^2-1)$$

$$= (e-0) - 2\,[(e-0) - (e-1)] - \frac{1}{2}(e^2-1)$$

$$= e - \frac{1}{2}e^2 - \frac{3}{2}$$

10. $\quad f'(x) = \dfrac{1}{\sqrt{1+g^2(x)}}\,g'(x);$

$f'\left(\dfrac{\pi}{2}\right) = \dfrac{g'(\pi/2)}{\sqrt{1+g^2(\pi/2)}}; \qquad g\left(\dfrac{\pi}{2}\right) = 0$

$= g'\left(\dfrac{\pi}{2}\right)$

But $g'(x) = [1 + \sin(\cos^2 x)](-\sin x)$

$$g'\left(\frac{\pi}{2}\right) = 1(-1) = -1$$

Hence, $f'\left(\frac{\pi}{2}\right) = -1$ as $h'(0^+) = -1$

11. Clearly f is an even function, hence

$$I_1 = \int_0^\pi f(\cos(\pi - x))\,dx = \int_0^\pi f(-\cos x)\,dx = \int_0^\pi f(\cos x)\,dx$$

$$\therefore \quad I_1 = 2\int_0^{\pi/2} f(\cos x)\,dx \cdot 2\int_0^{\pi/2} f(\sin x)\,dx = 2I_2 \Rightarrow \frac{I_1}{I_2} = 2$$

Aliter Let $u = \cos x \Rightarrow du = -\sin x\,dx$

$$\therefore \quad I_1 = \int_{-1}^{1} \frac{f(u)}{\sqrt{1-u^2}}\,du \Rightarrow 2\int_0^1 \frac{f(u)}{\sqrt{1-u^2}}\,du \qquad \text{...(i)}$$

Similarly, with $\sin t = t$, $\quad I_2 = \int_0^1 \frac{f(t)}{\sqrt{1-t^2}}\,dt \qquad$ ...(ii)

From Eqs. (i) and (ii), $\frac{I_1}{I_2} = 2$

12. $I_1 = \int_{1-k}^{k} x f(x(1-x))\,dx$; $I_2 = \int_{1-k}^{k} f(x(1-x))\,dx$

Using king, $\quad I_1 = \int_{1-k}^{k} (1-x) f(x(1-x))\,dx$

$$2I_1 = \int_{1-k}^{k} f(x(1-x))\,dx = I_2$$

$$\therefore \quad \frac{I_2}{I_1} = 2$$

13.

$$A = \int_{-1}^{1} (ax^2 + bx + c)\,dx$$

$$= 2\int_0^1 (ax^2 + c)\,dx$$

$$= 2\left[\frac{a}{3} + c\right] = \frac{1}{3}[2a + 6c]$$

$$\therefore \quad A = \frac{1}{3}[f(-1) + 4f(0) + f(1)]$$

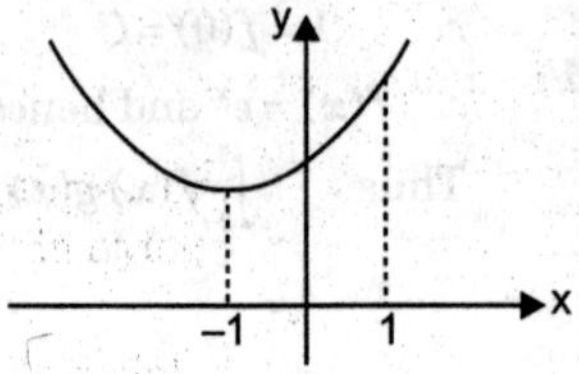

14.

$$I(a) = \int_0^\pi \left(\frac{x^2}{a^2} + a^2 \sin^2 x + 2x \sin x\right) dx \qquad \left(\int_0^\pi x \sin x\,dx = \pi\right)$$

$$\therefore \quad I(a) = \frac{\pi^3}{3a^2} + \frac{\pi a^2}{2} + 2\pi = \pi\left[\frac{\pi^2}{3a^2} + \frac{a^2}{2}\right] + 2\pi$$

$$= \pi\left[\left(\frac{\pi}{\sqrt{3}a} - \frac{a}{\sqrt{2}}\right)^2 + \frac{2\pi}{\sqrt{6}}\right] + 2\pi$$

$I(a)$ is minimum when $\frac{\pi}{\sqrt{3}a} = \frac{a}{\sqrt{2}} \Rightarrow a^2 = \pi\sqrt{\frac{2}{3}} \Rightarrow a = \sqrt{\pi\sqrt{\frac{2}{3}}}$

Also, $\quad I(a)|_{\min} = 2\pi + \pi^2\sqrt{\frac{2}{3}}$

15. $$\left[\frac{t^2}{2} - \log_2 a \cdot t\right]_0^2 = 2 - \log_2 (a^2) \qquad \text{(note } a > 0)$$

$$(2 - 2\log_2 a) = 2 - 2\log_2 a$$

$$2\log_2 a = 2\log_2 a \quad \Rightarrow \quad a \in R^+$$

16. Consider $$I = \int_{-1/x}^{1/x} \frac{\ln(1+t^2)}{1+e^t}\, dt \qquad \text{...(i)}$$

$$= \int_{-1/x}^{1/x} \frac{\ln(1+t^2)}{1+e^{-t}}\, dt \qquad \text{(Using King)}$$

$$I = \int_{-1/x}^{1/x} \frac{\ln(1+t^2)\, e^t}{1+e^t}\, dt \qquad \text{...(ii)}$$

On adding Eqs. (i) and (ii) $$2I = \int_{-1/x}^{1/x} \ln(1+t^2)\, dt = 2\int_0^{1/x} \ln(1+t^2)\, dt$$

$\Rightarrow$ $$I = \int_0^{1/x} \ln(1+t^2)\, dt$$

Hence, $$l = \lim_{x\to\infty} x^3 \int_0^{1/x} \ln(1+t^2)\, dt$$

$$= \lim_{x\to\infty} \frac{\int_0^{1/x} \ln(1+t^2)\, dt}{x^{-3}} \qquad \left(\frac{0}{0} \text{ form}\right)$$

Using L' Hospital's Rule

$$l = \lim_{x\to 0} \frac{x^4 \ln\left(1 + \frac{1}{x^2}\right)\cdot\left(-\frac{1}{x^2}\right)}{-3}$$

$$= \frac{1}{3}\lim_{x\to\infty} x^2 \ln\left(1 + \frac{1}{x^2}\right) = \frac{1}{3}\lim_{x\to\infty} \ln\left(1 + \frac{1}{x^2}\right)^{x^2} \qquad (1^\infty \text{ form})$$

$$= \lim_{x\to\infty} \frac{1}{3} x^2 \left(1 + \frac{1}{x^2} - 1\right) = \frac{1}{3}$$

17. Put $\pi x = t \Rightarrow dx = \frac{dt}{\pi}$

$$I = \frac{1}{\pi}\frac{\pi}{\pi}\int_0^{2008\pi} t\,|\sin t|\, dt = \frac{1}{\pi}\int_0^{2008\pi} t\,|\sin t|\, dt \qquad \text{...(i)}$$

$$I = \frac{1}{\pi}\int_0^{2008\pi} (2008\pi - t)\,|\sin t|\, dt \qquad \text{...(ii)}$$

On adding Eqs. (i) and (ii),

$$2I = \frac{2008\pi}{\pi}\int_0^{2008\pi} |\sin t|\, dt = (2008)^2 \cdot \int_0^{\pi} |\sin t|\, dt$$

$I = (2008)^2$; Hence, $\sqrt{I} = 2008$

18. $$T_n = \frac{n}{n^2 + k^2x^2} = \frac{1}{n(1 + (k/n)^2 x^2)};$$

$$S = \frac{1}{n}\sum_{k=1}^{n} \frac{1}{1 + \underbrace{(k/n)^2}_{t} x^2} = \int_0^1 \frac{dt}{1 + t^2x^2}$$

$$= \frac{1}{x^2}\int_0^1 \frac{dt}{t^2 + (1/x^2)} = \left[\frac{1}{x}\tan^{-1} tx\right]_0^1 = \frac{\tan^{-1} x}{x}$$

19. $y=f(x) \Rightarrow x=g(y)$ and $dy=f'(x)\,dx$

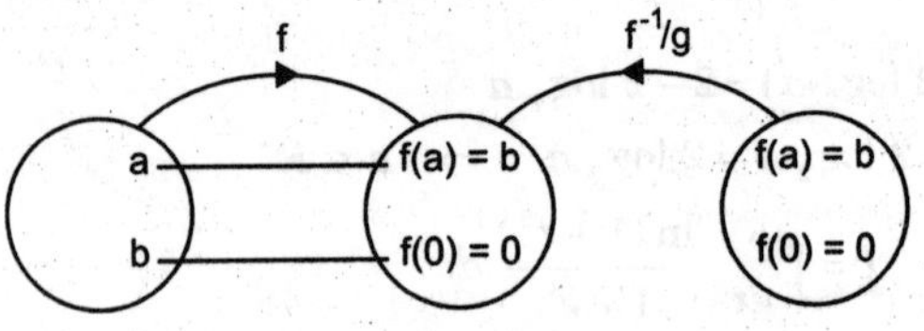

$$I=\int_0^a f(x)\,dx+\int_0^b g(y)\,dy; \quad y=f(x) \Rightarrow x=f^{-1}(y)=g(y)$$

$$=\int_0^a f(x)dx+\int_0^a xf'(x)\,dx$$

$$=\int_0^a (f(x)+xf'(x))\,dx=[xf(x)]_0^a=a\,f(a)=ab$$

20.
$$I=\int_{-1/\sqrt{3}}^{1/\sqrt{3}} \frac{x^4}{1-x^4}\cos^{-1}\frac{2x}{1+x^2}\,dx. \qquad \text{...(i)}$$

$$I=\int_{-1/\sqrt{3}}^{1/\sqrt{3}} \frac{x^4}{1-x^4}\cos^{-1}\left(\frac{-2x}{1-x^4}\right)dx \qquad \text{(using King)}$$

$$I=\int_{-1/\sqrt{3}}^{1/\sqrt{3}} \frac{x^4}{1-x^4}\left(\pi-\cos^{-1}\frac{2x}{1-x^4}\right)dx \qquad \text{...(ii)}$$

On adding Eqs. (i) and (ii),

$$\therefore \qquad 2I=\pi\int_{-1/\sqrt{3}}^{1/\sqrt{3}} \frac{x^4}{1-x^4}\,dx$$

$$2I=2\pi\int_0^{1/\sqrt{3}} \frac{x^4}{1-x^4}\,dx$$

$$\therefore \qquad k=\pi$$

21. Put $\ln x=t \Rightarrow x=e^t \Rightarrow dx=e^t\,dt$

$$I=\int_{-\infty}^{\infty} f(e^t+e^{-t})\frac{t}{e^t}e^t\,dt=\int_{-\infty}^{\infty} f(e^t+e^{-t})\,t\,dt=0 \qquad \text{(as the function is odd)}$$

***Aliter* 1** Put $x=\tan\theta$; $\displaystyle\int_0^{\pi/2} f\left(\tan\theta+\frac{1}{\tan\theta}\right)\frac{\ln\tan\theta}{\tan\theta}\cdot\sec^2\theta\,d\theta$

$$=\int_0^{\pi/2} f\left(\tan\theta+\frac{1}{\tan\theta}\right)\frac{\ln\tan\theta}{\sin\theta\cos\theta}\,d\theta$$

***Aliter* 2** Put $x=1/t \Rightarrow I=-1 \Rightarrow 2I=0 \Rightarrow I=0$

22.
$$\lim_{\lambda\to 0}\left(\int_0^1 (1+x)^\lambda\,dx\right)^{1/\lambda}=\lim_{\lambda\to 0}\left(\left|\frac{(1+x)^{\lambda+1}}{\lambda+1}\right|_0^1\right)^{1/\lambda}=\lim_{\lambda\to 0}\left(\frac{2^{\lambda+1}-1}{\lambda+1}\right)^{1/\lambda} \qquad (1^\infty \text{ form})$$

$$=e^{\lim_{\lambda\to 0}\frac{1}{\lambda}\left(\frac{2^{\lambda+1}-1-\lambda-1}{\lambda+1}\right)}=e^{\lim_{\lambda\to 0}\left(\frac{2^{\lambda+1}-2-\lambda}{\lambda(\lambda+1)}\right)}$$

$$=e^{\lim_{\lambda\to 0}\left(\frac{2(2^\lambda-1)}{\lambda}-1\right)}=e^{2\ln 2-1}=e^{\ln\left(\frac{4}{e}\right)}=\frac{4}{e}$$

23. $y=f(x) \Rightarrow x=f^{-1}(y)=g(y)$

$$dy=f'(x)\,dx$$

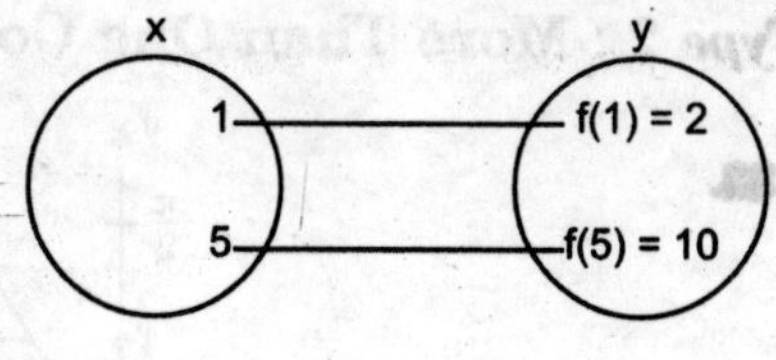

$$\therefore \qquad I = \int_1^5 f(x)\,dx + \int_1^5 xf'(x)\,dx$$

where y is 2, then $x = 1$

y is 1, then $x = 5$

$$\therefore \qquad I = \int_1^5 (f(x) + x\,f'(x))dx$$

$$= |x\,f(x)|_1^5 = 5\,f(5) - f(1) = 5 \cdot 10 - 2 = 48$$

24. Use $\int_0^a f(x) = \int_0^a (a - x)\,dx$ and add two integrals.

25. Here, $\quad f'(x) = (f(x))^2 > 0;\ \frac{d}{dx} f(g(x))|_{x=a}$

$$= f'(g(x)) \lim_{x \to a} \frac{g(x) - g(a)}{x - a}$$

26. We have on putting $t = \frac{x^2}{z}$ and solving

$$\int_0^\infty \frac{\ln x}{x^2 + t^2}\,dt = \frac{2\pi \ln x}{x}$$

$$\Rightarrow \qquad \frac{\ln x}{x} = \frac{\ln 2}{x}$$

$\Rightarrow \qquad x = 2$ and 4; two solutions.

27. Let $\frac{x\pi}{n} = t \Rightarrow \int_0^{\frac{16n^2}{\pi}} \cos\frac{\pi}{2}\left[\frac{\pi x}{n}\right] dx = \frac{n}{\pi}\int_0^{16n} \cos\frac{\pi}{2}[t]\,dt$

$$= \frac{4n^2}{\pi}\int_0^4 \cos\frac{\pi}{2}[t]\,dt = 0$$

28. We have $\int_{-2}^{-1} (ax^2 - 5)\,dx + \int_1^2 (bx + c)\,dx + 5$

$$= \int_{-2}^{-1} (ax^2 - 5 - bx + c + 5)\,dx = 0$$

$\Rightarrow ax^2 - bx + c = 0$ has at least one root in $(-2, -1)$.

29. $I = \int_3^6 ((\sqrt{x-3} + \sqrt{3}) + (\sqrt{3} - \sqrt{x-3}))\,dx = 6\sqrt{3}$

30. $I_1 = \int_{-1}^1 (\{x\} + \{x^3\})\{x^2\}\,dx = -2\int_0^1 \{x^2\}\,dx = \left[-2 \times \frac{x^3}{3}\right]_0^1 = -\frac{2}{3}$

31. Multiplying these three integral by 4, – 4, 1 and adding, we get

$$\int_0^1 f(x)(x - 2)^2\,dx = 0.$$

Hence, there does not exist any function satisfying these conditions.

32. $J_n - J_{n-1} = \int_0^{\pi/2} \frac{\sin^2 nx - \sin^2(n-1)x}{\sin^2 x}\,dx$

$$= \int_0^{\pi/2} \frac{\sin(2n-1)x \cdot \sin x}{\sin^2 x}\,dx = I_n$$

ie, $\qquad J_n - J_{n-1} = I_n$

$\Rightarrow \qquad J_{n+1} - J_n = I_{n+1}$

Type 2 : More Than One Correct Options

33.

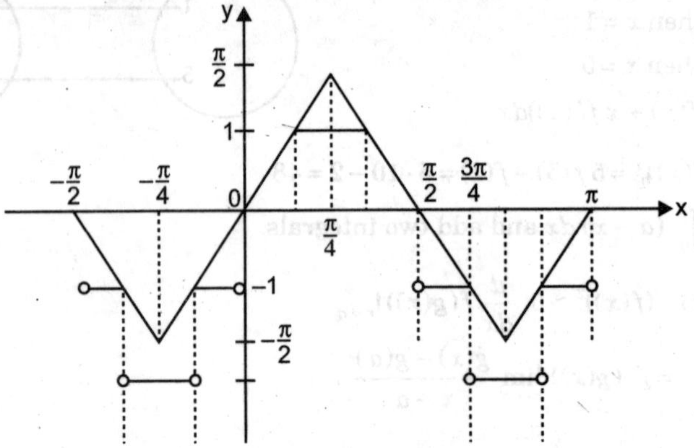

34. (a) $I=\int_0^{\pi/2}\ln(\cot x)\,dx \Rightarrow I=\int_0^{\pi/2}\ln(\tan x)\,dx$

$I=-\int_0^{\pi/2}\ln(\cot x)\,dx \Rightarrow I=-I \Rightarrow I=0$

(b) $I=\int_0^{2\pi}\sin^3 x\,dx=-\int_0^{2\pi}\sin^3 x\,dx \Rightarrow I=0$

(c) at $x=1/t$, $I=\int_e^{1/e}\dfrac{-(1/t^2)\,dt}{-1/t\,(\ln t)^{1/3}}=-\int_{1/e}^{e}\dfrac{dt}{t\,(\ln t)^{1/3}}$

$I=-1 \Rightarrow I=0$

(d) $\sqrt{\dfrac{1+\cos 2x}{2}}>0 \Rightarrow \int_0^{\pi}\sqrt{\dfrac{1+\cos 2x}{2}}\,dx>0$

35. $I=\int_0^1 (10x^4-3x^2-1)\,dx=[2x^5-x^3-x]_0^1=0$

Since, $f(x)$ is even, hence must have a root in $(-1, 0)$.

36. $I_1=\int_0^{\pi/2}\cos(\pi\sin^2 x)\,dx$

$I_1=\int_0^{\pi/2}\cos(\pi\cos^2 x)\,dx$

On adding, $2I_1=\int_0^{\pi/2}\cos(\pi\sin^2 x)+\cos(\pi\cos^2 x)\,dx$

$=\int_0^{\pi/2}2\cos\left(\dfrac{\pi}{2}\right)\cdot\cos\left(\dfrac{\pi}{2}\cos 2x\right)dx=0$

$\Rightarrow \quad I_1=0$ ...(i)

$I_2=\int_0^{\pi/2}\cos\{\pi(1-\cos 2x)\}\,dx$

$=-\int_0^{\pi/2}\cos(\pi\cos 2x)\,dx=-\dfrac{1}{2}\int_0^{\pi}\cos(\pi\cos t)\,dt$ [Put $2x=t$]

$=-\dfrac{2}{2}\int_0^{\pi/2}\cos(\pi\cos t)\,dt=I_3 \Rightarrow I_2+I_3=0$

$I_3=-\int_0^{\pi/2}\cos(\pi\sin t)\,dt$

$\therefore \quad I_2+I_3=0$ ...(ii)

Hence, $I_1+I_2+I_3=0$

37. $f(x)=2\int_0^1 \underbrace{(1-t)}_{\text{I}}\underbrace{\cos(xt)}_{\text{II}}\,dt=2\left[(1-t)\frac{\sin xt}{x}\Big|_0^t+\frac{1}{x}\int_0^1 \sin xt\,dt\right]$

$$=2\left[\left|0-\frac{1}{x^2}\cos xt\right|_0^1\right]$$

$$f(x)=2\left[\frac{1-\cos x}{x^2}\right] \qquad (x\neq 0)$$

If $x=0$, then $f(x)=\int_{-1}^1 (1-|t|)\,dt=2\int_0^1 (1-t)\,dt=1\Rightarrow$ (c) is correct.

Hence, $$f(x)=\begin{cases}2\left(\dfrac{1-\cos x}{x^2}\right), & \text{if } x\neq 0\\ 1, & \text{if } x=0\end{cases}$$

$\therefore f$ is continuous at $x=0\Rightarrow$ (d) is correct.

38. Given, $f(f(x))=-x+1$

Replacing, $x\to f(x)$

$$f(f(f(x)))=-f(x)+1$$
$$f(1-x)=-f(x)+1$$
$$f(x)+f(1-x)=1 \qquad \text{...(i)}$$

Now, $J=\int_0^1 f(x)\,dx=\int_0^1 f(1-x)\,dx$ (Using King)

$$2J=\int_0^1 (f(x)+f(1-x))\,dx;\;\; 2J=\int_0^1 dx=1\Rightarrow J=\frac{1}{2}$$

39. We have, $f(x)=x^2+ax^2+bx^3$

Where $a=\int_{-1}^1 t\cdot f(t)\,dt$ and $b=\int_{-1}^1 f(t)\,dt$

Now, $a=\int_{-1}^1 t\,[(a+1)t^2+bt^3]\,dt$

$$a=2b\int_0^1 t^4dt=\frac{2b}{5} \qquad \text{...(i)}$$

Again, $b=\int_{-1}^1 f(t)\,dt=\int_{-1}^1 ((a+1)t^2+bt^3)\,dt$

$$=2\int_0^1 (a+1)t^2\,dt\Rightarrow b=\frac{2(a+1)}{3} \qquad \text{...(ii)}$$

From Eqs. (i) and (ii), $\frac{5a}{2}=\frac{2(a+1)}{3}$

$$\left(\frac{5}{2}-\frac{2}{3}\right)a=\frac{2}{3}\Rightarrow \frac{11}{6}a=\frac{2}{3}$$

$$a=\frac{4}{11} \text{ and } b=\frac{10}{11}$$

Hence, $\int_{-1}^1 t\cdot f(t)\,dt=\frac{4}{11}$ and $\int_{-1}^1 f(t)\,dt=\frac{10}{11}$

$\therefore$
$$f(x)=(a+1)x^2+bx^3$$
$$f(1)=(a+1)+b$$
$$f(-1)=(a+1)-b$$

$\Rightarrow \quad f(1)+f(-1)=2(a+1)=\dfrac{30}{11}$

and $\quad f(1)-f(-1)=2b=\dfrac{20}{11}$

40. Given, $\quad (f'(x))^2+(g(x))^2=1$

$$f(x)+\int_0^x g(t)\,dt=\sin x(\cos x-\sin x)$$

Differentiating both sides,

$$f'(x)+g(x)=\cos 2x-\sin 2x \quad \text{...(i)}$$

Squaring Eq. (i)

$$(f'(x))^2+(g(x))^2=2f'(x)\cdot g(x)=1-\sin 4x$$

$$1+2f'(x)\cdot g(x)=1-\sin 4x$$

$\therefore \quad 2f'(x)\,g(x)=-\sin 4x$

Now, substituting, $g(x)=-\dfrac{\sin 4x}{2f'(x)}$ in Eq. (i)

$$f'(x)-\frac{\sin 4x}{2f'(x)}=\cos 2x-\sin 2x$$

Put $\quad f'(x)=t$

$$2t^2-2(\cos 2x-\sin 2x)t-\sin 4x=0$$

$\Rightarrow \quad t=\dfrac{2(\cos 2x-\sin 2x)\pm\sqrt{4(1-\sin 4x)+8\sin 4x}}{4}$

$\therefore \quad 4t=2(\cos 2x-\sin 2x)\pm\sqrt{4(1-\sin 4x)+8\sin 4x}$

$\Rightarrow \quad 2t=(\cos 2x-\sin 2x)\pm\sqrt{1+\sin 4x}$

Taking (+) ve sign, $2t=\cos 2x-\sin 2x+\cos 2x+\sin 2x \Rightarrow t=\cos 2x$

Taking (–) ve sign, $t=-\sin 2x$

Since, $\quad f'(x)=\cos 2x$ or $f'(x)=-\sin 2x$

$$f(x)=\frac{1}{2}\sin 2x+C_1 \quad \text{or} \quad f(x)=\frac{\cos 2x}{2}+C_2$$

$$f(0)=0 \Rightarrow C_1=0 \text{ and } C_2=-1/2$$

$\therefore \quad f(x)=\dfrac{1}{2}\sin 2x$ or $f(x)=\dfrac{\cos 2x-1}{2}$

If $f'(x)=\cos 2x$, then $g(x)=-\sin 2x$

If $f'(x)=-\sin 2x$, then $g(x)=\cos 2x$

ie, $f(x)=\dfrac{1}{2}\sin 2x$ and $g(x)=-\sin 2x$

$f(x)=\dfrac{\cos 2x-1}{2}$ and $g(x)=\cos 2x$

41. Consider $f(x)=\int_{-x}^{x}(\underbrace{t\sin at}_{\text{even}}+\underbrace{bt}_{\text{odd}}+\underbrace{c}_{\text{even}})dt=2\int_0^x(t\sin at+c)\,dt$

$$=2\left[\left|-t\frac{\cos at}{a}\right|_0^x+\int_0^x\frac{\cos at}{a}\,dt+|ct|_0^x\right] \quad \text{(Using I.B.P.)}$$

$$= 2\left[\frac{-x\cos ax}{a} + \frac{1}{a^2}\sin ax + cx\right]$$

$$\therefore \quad \lim_{x\to 0}\frac{f(x)}{x} = \lim_{x\to 0} 2\left[-\frac{\cos ax}{a} + \frac{\sin ax}{a\cdot ax} + c\right] = 2\left[-\frac{1}{a} + \frac{1}{a} + c\right] = 2c$$

42. Consider $I = \int_a^\infty \frac{n\,dx}{n^2\left(x^2 + \frac{1}{n^2}\right)}$

$$= \frac{1}{n}\cdot n\,(\tan^{-1} nx)_0^\infty = \left(\frac{\pi}{2} - \tan^{-1} an\right)$$

$$\therefore \quad L = \lim_{n\to\infty}\left(\frac{\pi}{2} - \tan^{-1} an\right) = \begin{bmatrix} \pi, & \text{if } a < 0 \\ \pi/2, & \text{if } a = 0 \\ 0, & \text{if } a > 0 \end{bmatrix}$$

Type 3 : **Assertion and Reason**

43. The Statement I can be proved by showing that both integrals are equal to a third integral. If we put $x^n = \tan^2\theta$ in the integral on LHS and $x^2 = \sin^2\theta$ in the integral on RHS, then both integrals will be equal to $\frac{2}{n}\int_0^{\pi/2}\tan^{(2/n)-1}\phi\, d\phi$ and $\frac{2}{n}\int_0^{\pi/2}\tan^{(2/n)-1}\theta\, d\theta$ respectively.Since, the last two integrals are equal Statement I is proved but correct Statement II has no role to play here.

44. $\int_0^{n\pi}\left|\frac{\sin x}{x}\right| dx = \int_0^{\pi}\left|\frac{\sin x}{x}\right| dx + \int_\pi^{2\pi}\left|\frac{\sin x}{x}\right| dx + \ldots + \int_{(n-1)\pi}^{n\pi}\left|\frac{\sin x}{x}\right| dx$

$$= \int_0^\pi \frac{\sin x}{x}\,dx + \int_0^\pi \left|\frac{\sin(t+\pi)}{t+\pi}\right| dt + \int_0^\pi \left|\frac{\sin(u+2\pi)}{u+2\pi}\right| du + \ldots$$

$$= \sum_{t=1}^{n}\int_0^\pi \frac{\sin x}{x+(r-1)\pi}\,dx > \sum_{t=1}^{n}\int_0^\pi \frac{\sin x}{\pi+(r-1)\pi}\,dx$$

$$= \sum_{r=1}^{n}\int_0^\pi \frac{\sin x}{\pi r}\,dx = \sum_{r=1}^{n}\frac{2}{\pi r} = \frac{2}{\pi}\left(1 + \frac{1}{2} + \frac{1}{3} + \ldots + \frac{1}{n}\right)$$

45. $I = \int_{-\pi/4}^{\pi/4}\frac{dx}{1+\sin x}$

$$2I = \int_{-\pi/4}^{\pi/4}\frac{2\,dx}{1-\sin^2 x} \quad \Rightarrow \quad I = \int_{-\pi/4}^{\pi/4}\frac{dx}{\cos^2 x}$$

$$I = 2\int_0^{\pi/4}\sec^2 x\,dx \neq 0$$

$\Rightarrow$ Statement I is false.

46. (b) $2f(x) + f(-x) = 2f(-x) + f(x) \Rightarrow f(x) = f(-x)$

$$f(x) = \frac{1}{3x}\sin\left(x - \frac{1}{x}\right)$$

$$I = \int_{1/e}^{e}\frac{1}{3x}\sin\left(x - \frac{1}{x}\right)dx = -\int_{1/e}^{e}\frac{1}{3t}\sin\left(t - \frac{1}{t}\right) = -1$$

$$\Rightarrow \quad I = 0$$

Type 4 : Linked–Comprehension Based Questions

47. $\lim_{x \to 0} \dfrac{\int_0^x \frac{t^2}{(a+t^r)^{1/p}}}{bx - \sin x}$

Using L'Hospital Rule

$$\lim_{x \to 0} \frac{\frac{x^2}{(a+x^r)^{1/p}}}{b - \cos x}$$

For existence of limit $\lim_{x \to 0} D^r \to 0$

$\therefore$ $b - 1 = 0 \Rightarrow b = 1$

48. $l = \lim_{x \to 0} \dfrac{x^2}{(a+x^r)^{1/p}} \cdot \dfrac{x^2}{(1-\cos x)} \cdot \dfrac{1}{x^2} = 2 \lim_{x \to 0} \dfrac{1}{(a+x^r)^{1/p}} = \dfrac{2}{a^{1/p}}$

If $p = 3$ and $l = 1$

$\therefore$ $1 = \dfrac{2}{a^{1/3}} \Rightarrow a = 8$

49. Again $p = 2$ and $a = 9$, then $l = \dfrac{2}{9^{1/2}} = \dfrac{2}{3}$

50. $f(x) = e^x \underbrace{\int_0^1 e^t \cdot f(t)\,dt}_{A \text{ say}}$

$f(x) = Ae^x$ ...(i)

$\Rightarrow$ $f(t) = Ae^t$

Where, $A = \int_0^1 e^t \cdot f(t)\,dt$

$\Rightarrow$ $A = \int_0^1 e^t \cdot Ae^t\,dt;\ A = A\int_0^1 e^{2t}\,dt$

Now, $A\left[\int_0^1 e^{2t}dt - 1\right] = 0 \Rightarrow A = 0$ as $\int_0^1 e^{2t}\,dt \neq 0$

Hence, $f(x) = 0 \Rightarrow f(1) = 0$

51. Again, $g(x) = e^x \int_0^1 e^t\,g(t)\,dt + x$

$g(x) = Be^x + x$ ...(ii)

$\Rightarrow$ $g(t) = Be^t + t$

Where $B = \int_0^1 e^t\,g(t)\,dt; B = \int_0^t e^t\,(Be^t + t)\,dt; B = B\int_0^1 e^{2t}\,dt + \int_0^1 e^t \cdot t\,dt$

But $\int_0^1 e^{2t}\,dt = \dfrac{1}{2}(e^2 - 1)$ and $\int_0^1 te^t\,dt = 1$

$\therefore$ $B = \dfrac{B}{2}(e^2 - 1) + 1 \Rightarrow 2B = B(e^2 - 1) + 2$

$\Rightarrow$ $3B = Be^2 + 2 \Rightarrow B = \dfrac{2}{3 - e^2}$

$\therefore$ From Eq. (ii)

$$g(x) = \left(\frac{2}{3 - e^2}\right)e^x + x;\ g(0) = \frac{2}{3 - e^2}$$

52. $g(2)=\frac{2e^2}{3-e^2}+2=\frac{6}{3-e^2}$; $\frac{g(0)}{g(2)}=\frac{2}{3-e^2}\cdot\frac{3-e^2}{6}=\frac{1}{3}$

53. $D(t)=\left|\frac{1}{\sqrt{(ae^t)^2+(Be^{-t})^2}}\right|=\frac{1}{\sqrt{a^2e^{2t}+b^2e^{-2t}}}$

$$\frac{1}{(D(t))^2}=(a^2e^{2t}+b^2e^{-2t})$$

$\therefore$ $$I=\int_0^1(a^2e^{2t}+b^2e^{-2t})\,dt=\left[\frac{a^2e^{2t}}{2}-\frac{b^2e^{-2t}}{2}\right]_0^1$$

$$=\left(\frac{a^2e^2-b^2e^{-2}}{2}\right)-\left(\frac{a^2-b^2}{2}\right)$$

$$=\frac{a^2(e^2-1)-b^2(e^{-2}-1)}{2}=\frac{a^2(e^2-1)+\frac{b^2}{e^2}(e^2-1)}{2}$$

$$=\frac{e^2-1}{2}\left(a^2+\frac{b^2}{e^2}\right)$$

54. Now put $b=\frac{1}{a}$

$$I=\frac{e^2-1}{2}\left(a^2+\frac{1}{a^2e^2}\right)=\frac{e^2-1}{2}\left(\left(a-\frac{1}{ae}\right)^2+\frac{2}{e}\right)$$

I is minimum if $a=\frac{1}{ae}\Rightarrow a^2=\frac{1}{e}\Rightarrow a=\frac{1}{\sqrt{e}}\Rightarrow b=\sqrt{e}$

55. $I_{\min}=\frac{e^2-1}{2}\cdot\frac{2}{e}=e-\frac{1}{e}$

Solution (Q. Nos. 56 to 58)

We have the equations of the tangents to the curve $y=\int_{-\infty}^x f(t)dt$ and $y=f(x)$ at arbitrary points on them are

$$Y-\int_{-x}^x f(t)\,dt=f(x)(X-x) \qquad \text{...(i)}$$

and $$Y-f(x)=f'(x)(X-x) \qquad \text{...(ii)}$$

As Eqs. (i) and (ii) intersect at the same point on the x-axis

$\therefore$ Putting $Y=0$ and equating X-coordinates, we have

$$x-\frac{f(x)}{f'(x}=x-\frac{\int_{-\infty}^x f(t)\,dt}{f(x)}$$

$\Rightarrow$ $$\frac{f(x)}{\int_{-\infty}^x f(t)\,dt}=\frac{f'(x)}{f(x)}$$

$\Rightarrow$ $$\int_{-\infty}^x f(t)\,dt=cf(x) \qquad \text{...(iii)}$$

As, $$f(0)=1\Rightarrow\int_{-\infty}^0 f(t)\,dt=c\times1\Rightarrow c=\frac{1}{2}$$

$\Rightarrow \quad \int_{-\infty}^{x} f(t)\, dt = \frac{1}{2} f(x)$; differentiating both sides and on integrating and using boundary conditions, we get $f(x) = e^{2x}$; $y = 2ex$ is tangent to $y = e^{2x} \Rightarrow$ Number of solutions = 1.

Celarly, $f(x)$ is increasing for all x.

$$\lim_{x \to \infty} (e^{2x})^{e^{-2x}} = 1 \qquad (\infty^0 \text{ form})$$

Solution (Q. Nos. 59 to 61)

We have, $\quad g(x) = g(0) + xg'(0) + \frac{x^2}{2} g''(0) = -bx^2 + cx - 6$

$h(x) = g(x) = 4x^4 - ax^3 + bx^2 - cx + 6 = 0$ has 4 distinct real roots.

Using Descarte's rule of signs.

$\Rightarrow$ Given biquadratic equation has 4 distinct positive roots.

Let x_1, x_2, x_3 and x_4.

Now,
$$\frac{\frac{1}{x_1} + \frac{2}{x_2} + \frac{3}{x_3} + \frac{4}{x_4}}{4} \geq \sqrt[4]{\frac{24}{x_1 x_2 x_3 x_4}}$$

$$\Rightarrow \quad 2 \geq 2 \Rightarrow \frac{1}{x_1} = \frac{2}{x_2} = \frac{3}{x_3} = \frac{4}{x_4} = k$$

$$\Rightarrow \quad \frac{1}{x_1} \cdot \frac{2}{x_2} \cdot \frac{3}{x_3} \cdot \frac{4}{x_4} = k^4$$

$$\Rightarrow \quad \frac{24}{3/2} = k^4 \Rightarrow k = 2$$

$\Rightarrow$ Roots are $\frac{1}{2}, 1, \frac{3}{2}, 2$.

$$a = 20; c = 25$$

62. At $x = 1, y = 0$

$$\frac{dy}{dx} = 2x \cdot (x^4)^2 - (x^2)^2 = 1$$

Equation of tangent is $y = x - 1$

63. $\frac{dy}{dx} = 4x^3 (\ln x^4)^2 - 3x^2 (\ln x^3)^2 = 64x^3 (\ln x)^2 - 27x^2 (\ln x)^2$

$$\lim_{x \to 0^+} \frac{dy}{dx} = 64 \lim_{x \to 0^+} x^3 (\ln x)^2 - 27 \lim_{x \to 0^+} x^2 (\ln x)^2 = 0$$

64.
$$f(x) = \int_1^x e^{t^2/2} (1 - t^2)\, dt$$

$$f'(x) = [e^{x^2/2} (1 - x^2)]^2$$

$$f'(1) = e^{1/2} \cdot 0 = 0$$

Solution (Q. Nos. 65 to 67)

$$I_n = -\frac{1}{2} \int x^{n-1} (-2x) \sqrt{a^2 - x^2}\, dx$$

$$= -\frac{1}{2} \left[x^{n-1} \frac{2}{3} (a^2 - x^2)^{3/2} - \int (n-1) x^{n-2} \frac{2}{3} (a^2 - x^2)^{3/2}\, dx \right]$$

$$= \frac{-x^{n-1}(a^2-x^2)^{3/2}}{3} + \frac{n-1}{3}\int (a^2x^{n-2}\sqrt{a^2-x^2} - x^4\sqrt{a^2-x^2})\,dx$$

$$= \frac{-x^{n-1}(a^2-x^2)^{3/2}}{3} + \frac{n-1}{3}a^2 I_{n-2} - \frac{n-1}{3} I_n$$

$$\therefore I_n = \frac{-x^{n-1}(a^2-x^2)^{3/2}}{n+2} + \frac{n-1}{n+2}a^2 I_{n-2}$$

$$I_4 = \frac{-x^3(a^2-x^2)^{3/2}}{6} + \frac{1}{2}a^2 I_2$$

$$= \frac{-x^3(a^2-x^2)^{3/2}}{6} + \frac{a^2}{2}\left[\frac{-x(a^2-x^2)^{3/2}}{4} + \frac{1}{4}a^2\int\sqrt{a^2-x^2}\,dx\right]$$

$$= \frac{-x^3(a^2-x^2)^{3/2}}{6} - \frac{a^2x(a^2-x^2)^{3/2}}{8} + \frac{a^4}{8}\left[\frac{x\sqrt{a^2-x^2}}{2} + \frac{a^2}{2}\sin^{-1}\frac{x}{a}\right]$$

$$\therefore \quad \int_0^a x^4\sqrt{a^2-x^2}\,dx = \frac{\pi a^6}{32}$$

68. On integration by parts,

$$\int_0^\infty \frac{\sin^2 x}{x^2}dx = \left(\sin^2 x\left(-\frac{1}{x}\right)\right)_0^\infty - \int_0^\infty 2\sin x\cos x\left(-\frac{1}{x}\right)dx = \int_0^\infty \frac{\sin 2x}{x}dx$$

Put $2x = t \Rightarrow dx = \frac{dt}{2}$

$$\int_0^\infty \frac{\sin 2x}{x}dx = \frac{1}{2}\int_0^\infty \frac{\sin t}{t/2}dt = \int_0^\infty \frac{\sin x}{x}dx$$

69. $\lim\limits_{x\to 0} \dfrac{\int_0^{x^2} \frac{\sin\sqrt{x}}{x^3}dx}{x^3} = \lim\limits_{x\to 0} \dfrac{\sin\sqrt{x^2}\cdot 2x}{3x^2} = \dfrac{2}{3}$

70. $I = \int_0^{\pi/2} \log\cos x\cos 2nx\,dx = \left(\dfrac{\log\cos x\sin 2nx}{2n}\right)_0^{\pi/2} - \int_0^{\pi/2}\dfrac{(-\tan x)\sin 2nx}{2n}dx$

$$= \frac{1}{2n}\int_0^{\pi/2}\tan x\sin(2nx)\,dx$$

Type 5 : **Match the Columns**

71. (A) For $a = 0$, $I = \int_0^T \sin^2 x\,dx = \int_0^T \frac{1-\cos 2x}{2}dx = \left[\frac{x}{2} - \frac{1}{4}\sin 2x\right]_0^T = \frac{T}{2} - \frac{1}{4}\sin 2T$

$$\therefore \quad L = \frac{1}{2} - \lim_{T\to\infty}\frac{1}{4}\frac{\sin 2T}{T} = \frac{1}{2}$$

(B) For $a = 1$, $\int_0^T 4\sin^2 x\,dx \Rightarrow L = 2$

(C) $a = -1$, $\int_0^T 0\,dx = 0 \Rightarrow L = 0$

(D) For $a \neq 0, -1, 1$

$$I = \int_0^T (\sin^2 x + \sin^2 ax + 2\sin x\cdot\sin ax)\,dx$$

$$= \int_0^T \left(\frac{1-\cos 2x}{2} + \frac{1-\cos 2ax}{2} + \cos(a-1)x - \cos(a+1)x \right) dx$$

$$= \left[x - \frac{1}{4}\sin 2x - \frac{1}{4a}\sin 2ax \frac{\sin(a-1)x}{a-1} - \frac{\sin(a+1)x}{a+1} \right]_0^T$$

$$\therefore \quad L = \lim_{T\to\infty} \frac{T}{T} - \lim_{T\to\infty} \frac{1}{T}\left[\frac{1}{4}\sin 2x - \frac{1}{4a}\sin 2ax + \frac{\sin(a-1)x}{a-1} - \frac{\sin(a+1)x}{a+1} \right]_0^T$$

$$L = 1$$

72. $f(\theta) = \left[\frac{(x+\sin\theta)^3}{3} \right]_0^1 = \frac{(1+\sin\theta)^3 - \sin^3\theta}{3}$

and $\quad g(\theta) = \left[\frac{(x+\cos\theta)^3}{3} \right]_0^1 = \frac{(1+\cos\theta)^3 - \cos^3\theta}{3}$

$$f(\theta) = \frac{1 + 3\sin\theta + 3\sin^2\theta}{3};\ g(\theta) = \frac{1+3\cos\theta + 3\cos^2\theta}{3}$$

$$f(\theta) - g(\theta) = (\sin\theta - \cos\theta) + (\sin^2\theta - \cos^2\theta)$$

$$f(\theta) - g(\theta) = (\sin\theta - \cos\theta)(1 + \sin\theta + \cos\theta)$$

Now verify all matchings.

73. (A) Let $a = 2008$

$$I = \int_0^1 (1 + ax^a)e^{x^a}\, dx;\ I = \int_0^1 (e^{x^a} + ax^a e^{x^a})\, dx \text{ (Note : } ax^a = ax \cdot x^{a-1})$$

$$\therefore \quad I = \int_0^1 (e^{x^a} + e^{x^a} \cdot x \cdot ax^{a-1})\, dx$$

$$= \int_0^1 (f(x) + x f'(x))\, dx \text{ where } f(x) = e^{x^a}$$

Hence, $\quad I = [xe^{x^a}]_0^1 = e$

(B) $I = I_1 + I_2$

Consider $I_2 = \int_1^{1/e} \sqrt{-\ln x}\, dx$

Put $\sqrt{-\ln x} = t \Rightarrow -\ln x = t^2 \Rightarrow x = e^{-t^2} \Rightarrow dx = -2t\, e^{-t^2}\, dt$

$$\therefore \quad I_2 = \int_0^1 \underbrace{t}_{\text{I}} \cdot \underbrace{(-2te^{-t^2})}_{\text{II}}\, dt = [te^{-t^2}]_0^1 - \int_0^1 e^{-t^2} dt = \frac{1}{e} - \int_0^1 e^{-t^2} dt$$

Hence, $I_2 = \int_0^1 e^{-x}\, dx + \frac{1}{e} - \int_0^1 e^{-t^2}\, dt = \frac{1}{e} = e^{-1}$

Note that if $f(x) = e^{-x^2}$, then $f^{-1}(x) = \sqrt{-\ln x}$

(C) $L = \lim_{n\to\infty} \left(\left(\frac{1}{n}\right)^1 \cdot \left(\frac{2}{n}\right)^2 \cdot \left(\frac{3}{n}\right)^3 \cdots \left(\frac{n}{n}\right)^n \right)^{\frac{1}{n^2}}$

$$\ln L = \lim_{n\to\infty} \frac{1}{n^2}\left(1\cdot \ln\left(\frac{1}{n}\right) + 2\cdot\ln\left(\frac{2}{n}\right) + 3\cdot\ln\left(\frac{3}{n}\right) + \ldots + n\cdot\ln\left(\frac{n}{n}\right) \right)$$

General term of $\ln L = \frac{r}{n^2} \ln \frac{r}{n}$

$$\text{Sum} = \frac{1}{n}\sum_{r=1}^{n} \frac{r}{n} \ln\left(\frac{r}{n}\right)$$

$$\ln L = \int_0^1 x \ln x \, dx = \left[\frac{x^2 \cdot \ln x}{2}\right]_0^1 - \frac{1}{2}\int_0^1 x^2 \frac{1}{x} \, dx = \left[0 - \frac{1}{2}\frac{x^2}{2}\right]_0^1 = -\frac{1}{4}$$

$$L = e^{-1/4}$$

74. (A) $f(x) = \int x^{\sin x}(1 + x\cos x \cdot \ln x + \sin x)\, dx$

If $\quad F(x) = x^{\sin x} = e^{\sin x \ln x}$

$\therefore \quad x\, F'(x) = x^{\sin x}(x \cos x \ln x + \sin x)$

$\therefore \quad f(x) = \int (F(x) + x\, F'(x)) = x\, F(x) + C$

$f(x) = x \cdot x^{\sin x} + C$

$f\left(\frac{\pi}{2}\right) = \frac{\pi}{2} \cdot \frac{\pi}{2} + C \Rightarrow C = 0$

$\therefore \quad f(x) = x\,(x)^{\sin x};\ f(\pi) = \pi(\pi)^0 = \pi$ (irrational)

(B) $g(x) = \int \dfrac{\cos x(\cos x + 2) + \sin^2 x}{(\cos x + 2)^2}\, dx$

$$= \int \underbrace{\cos x}_{\text{II}} \cdot \underbrace{\frac{1}{(\cos x + 2)}}_{\text{I}}\, dx = \int \frac{\sin^2 x}{\cos x + 2}\, dx$$

$$= \frac{1}{\cos x + 2} \cdot \sin x - \int \frac{\sin^2 x}{(\cos x + 2)^2}\, dx + \int \frac{\sin^2 x}{(\cos x + 2)^2}\, dx$$

$$g(x) = \frac{\sin x}{\cos x + 2} + C$$

Aliter Consider $\dfrac{d}{dx}\left(\dfrac{\sin x}{\cos x + 2}\right) = \dfrac{(\cos x + 2)\cos x + \sin^2 x}{(\cos x + 2)^2} = \dfrac{2\cos x + 1}{(\cos x + 2)^2}$

$\therefore \quad \int \dfrac{2\cos x + 1}{(\cos x + 2)^2}\, dx = \dfrac{\sin x}{\cos x + 2} + C$

$\therefore \quad g(x) = \dfrac{\sin x}{\cos x + 2} + C$

$g(0) = 0 \Rightarrow C = 0$

$\therefore \quad g(x) = \dfrac{\sin x}{\cos x + 2} \Rightarrow g\left(\dfrac{\pi}{2}\right) = \dfrac{1}{2}$ (rational)

(C) Let $\quad x + 5 = 14 \cos\theta$

$y - 12 = 14 \sin\theta$

$\therefore \quad x^2 + y^2 = (14\cos\theta - 5)^2 + (14\sin\theta + 12)^2$

$= 196 + 25 + 144 + 28(12\sin\theta - 5\cos\theta)$

$= 365 + 28(12\sin\theta - 5\cos\theta)$

$\therefore \quad \left|\sqrt{(x^2 + y^2)}\right|_{\text{min}} = \sqrt{365 - 28 \times 13} = \sqrt{365 - 364} = 1$

(D) $k(x) = \int \dfrac{(x^2 + 1)\, dx}{(x^3 + 3x + 6)^{1/3}}$

Put $x^3 + 3x + 6 = t^3 \quad \Rightarrow \quad 3(x^2 + 1)\, dx = 3t^2\, dt$

$$k(x) = \int \frac{t^2\, dt}{t} = \frac{t^2}{2} + C$$

$$k(x) = \frac{1}{2}(x^3 + 3x + 6)^{2/3} + C$$

$$k(-1) = \frac{1}{2}(2)^{2/3} + C \Rightarrow C = 0$$

$$\therefore \quad k(x) = \frac{1}{2}(x^3 + 3x + 6)^{2/3}$$

$$f(-2) = \frac{1}{2}(-8)^{2/3} = \frac{1}{2}[(-2)^3]^{2/3} = 2$$

75. (A) We have, $f'(x) = \dfrac{g'(x)}{\sqrt{1 + g^3(x)}}$ and $g'(x) = (1 + \sin(\cos^2 x))(-\sin x)$

Hence, $f'(x) = \dfrac{(1 + \sin(\cos^2 x))(-\sin x)}{\sqrt{1 + g^3(x)}}$

$$f'\left(\frac{\pi}{2}\right) = \frac{1 + 0}{\sqrt{1 + g^3\left(\frac{\pi}{2}\right)}} = -1,\ g\left(\frac{\pi}{2}\right) = 0$$

$$\therefore \quad f'\left(\frac{\pi}{2}\right) = -1$$

(B) We have, $\int_0^x f(x)\, dx = \{f(x)\}^2$

Differentiating both sides, we get

$$f(x) = 2f(x) \cdot f'(x) \Rightarrow f'(x) = \frac{1}{2}$$

Integrating both sides, $f(x) = \frac{1}{2}x + C$

where, $f(0) = 0 \quad \therefore \quad C = 0$

$\Rightarrow \quad f(x) = \frac{x}{2} \Rightarrow f(2) = 1$

(C) Maximum when $a = -1, b = 2 \Rightarrow a + b = 1$

(D) If $\lim_{x \to 0} \frac{\sin 2x}{x^3} + a + \frac{b}{x^2} = 0 \Rightarrow \lim_{x \to 0} \frac{\sin 2x + ax^3 + bx}{x^3} = 0$

For limit to exist $2 + b = 0 \Rightarrow b = -2$

$$\therefore \quad \lim_{x \to 0} \frac{\sin 2x + ax^3 - 2x}{x^3} = 0$$

Using Left Hand Rule and solving, we get $a = \frac{4}{3}$

$$\therefore \quad 3a + b = 2$$

76. (A) Required area $= \int_0^1 (\sqrt{x} - x^2)\, dx = \frac{1}{3}$

(B) $\int_0^4 \{\sqrt{x}\}\, dx = \int_0^4 (\sqrt{x} - [\sqrt{x}])\, dx$

$$= \int_0^4 \sqrt{x}\, dx - \int_0^1 [\sqrt{x}]\, dx - \int_1^4 [\sqrt{x}]\, dx = \frac{7}{3}$$

(C) Area $= \int_0^1 (3 - 2x - x^2)\, dx = \left[3x - x^2 - \dfrac{x^3}{3}\right]_0^1 = \dfrac{5}{3}$

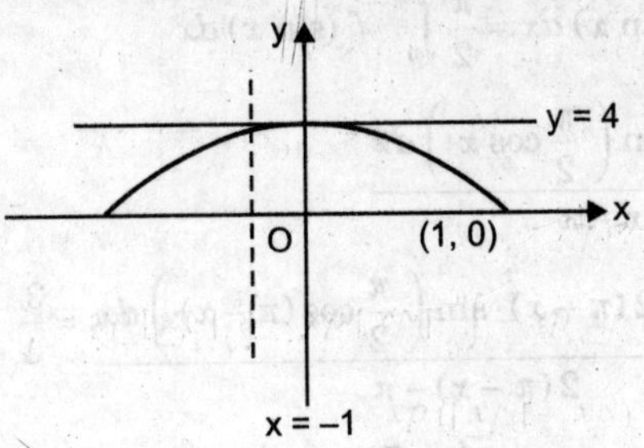

(D) $2\int_0^{\pi/2} \sqrt{\cos x}\, \sin x\, dx = \dfrac{4}{3}$

(Proficiency Exercise 2)

1. (i) $\left(\dfrac{\pi + 2}{8}\right)$ (ii) (1)

(iii) (2) (iv) $\left(\dfrac{2}{3} - \dfrac{\sqrt{3}}{4}\right)$

(v) $\left(\dfrac{4\sqrt{2}}{3}\right)$ (vi) $\left(\dfrac{1}{2} - \dfrac{\pi\sqrt{3}}{12}\right)$

(vii) (– 1) (viii) (4)

(ix) $\left(\dfrac{3}{2}\right)$ (x) $\left(\dfrac{17}{6}\right)$

(xi) (π) (xii) $\left(-\dfrac{\pi}{2}\log 2\right)$

(xiii) $\left(\dfrac{\pi}{2} - \dfrac{1}{2}\log 2\right)$ (xiv) $(200\sqrt{2})$

(xv) $(2 - \sqrt{2})$ (xvi) $(|b| - |a|)$ where $a < b$

(xvii) $\left(\dfrac{3}{2}\right)$ (xviii) (– 1)

(xix) $\left(\dfrac{5\pi}{12} - \tan^{-1}\sqrt{2}\right)$ (xx) $\left(\dfrac{[x]([x]-1)}{4}\right)$

2. Let $I = \int_0^{\pi} x\, f(\sin x)\, dx$

Replacing x by $(\pi - x)$ we get;

$$I = \int_0^{\pi} (\pi - x) f(\sin(\pi - x))\, dx$$

$$I = \int_0^{\pi} f(\sin x)\, dx \qquad \text{...(ii)}$$

Adding Eqs. (i) and (ii), we get

$$I + I = \int_0^{\pi} \pi f(\sin x)\, dx$$

$$\therefore \quad I = \frac{\pi}{2}\int_0^{\pi} f(\sin x)\,dx$$

$$\Rightarrow \quad \int_0^{\pi} xf(\sin x)\,dx = \frac{\pi}{2}\int_0^{\pi} f(\sin x)\,dx$$

3. $$I = \int_0^{\pi} \frac{x^2 \sin 2x \cdot \sin\left(\frac{\pi}{2}\cos x\right) dx}{2x - \pi}$$

$$= \int_0^{\pi} \frac{(\pi - x)^2 \sin 2(\pi - x) \cdot \sin\left(\frac{\pi}{2}\cos(\pi - x)\right) dx}{2(\pi - x) - \pi}$$

$$= \int_0^{\pi} - \frac{(\pi - x)^2 \sin(2x) \cdot \sin\left(-\frac{\pi}{2}\cos x\right) dx}{(\pi - 2x)}$$

$$I = -\int_0^{\pi} \frac{(\pi - x)^2 \cdot \sin(2x)\sin\left(\frac{\pi}{2}\cos x\right) dx}{(2x - \pi)}$$

Adding the above expansion

$$2I = \int_0^{\pi} \frac{\{x^2 - (\pi - x)^2\} \cdot \sin 2x \sin\left(\frac{\pi}{2}\cos x\right) dx}{(2x - \pi)}$$

$$2I = \int_0^{\pi} \pi \cdot 2\sin x \cos x \sin\left(\frac{\pi}{2}\cos x\right) dx$$

Put $\frac{\pi}{2}\cos x = t \Rightarrow -\frac{\pi}{2}\sin x\,dx = dt$

Also when $x = 0$; $t = \frac{\pi}{2}$ and when $x = \pi$; $t = -\frac{\pi}{2}$

$$\therefore \quad 2I = \int_{\pi/2}^{-\pi/2} 2 \cdot \frac{2t}{\pi} \cdot \sin t\,(-2dt) \Rightarrow I = \frac{16}{\pi}$$

4. *Graphical method* :

$g'(x) > 0 \Rightarrow g(x)$ is an increasing function and $a < 2$, hence $b > 2 \Rightarrow b > a$. Thus the graph of $g(x)$ is ahead.

Let $$A = \int_0^{a} g(x)\,dx + \int_0^{b} g(x)\,dx$$

From the graph it is clear that $(b - a)$ indicates the gap between the right ordinates of two areas. If in order to increase the gap 'a' is decreased to '$a - h$' (*ie*, ordinate AP is shifted leftward by h), then b is increased to '$b + h$' (*ie*, ordinate BQ is shifted rightwards by h), then the final area.

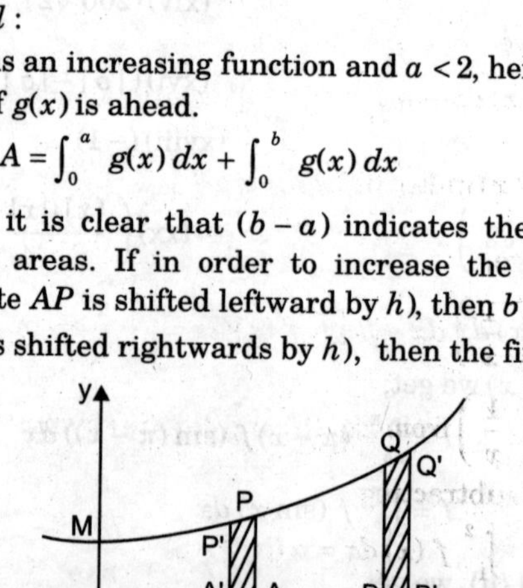

$$A' = \int_0^{a-h} g(x)\,dx + \int_0^{b+h} g(x)\,dx$$

$$= \int_0^{a} g(x)\,dx - \int_{a-h}^{a} g(x)\,dx + \int_0^{b} g(x)\,dx + \int_b^{b+h} g(x)\,dx$$

$$= A + \int_b^{b+h} g(x)\,dx - \int_{a-h}^{a} g(x)\,dx$$

From the figure it is clear that

$$\int_b^{b+h} g(x)\,dx > \int_{a-h}^{a} g(x)\,dx$$

$\Rightarrow$ **$A' > A \Rightarrow A$ increases if the gap $(b-a)$ increases.**

5. **Here, rth term** $= \dfrac{\sqrt{n}}{\{n+3(r-1)\}^{3/2}} = \dfrac{1}{n}\cdot\dfrac{1}{\left\{1+3\left(\dfrac{r-1}{n}\right)\right\}^{3/2}}$

Since, the rth term contains $(r-1)$, so let us consider the $(r+1)$ th term :

The $(r+1)$ th term $= \dfrac{\sqrt{n}}{\{n+3r\}^{3/2}} = \dfrac{1/n}{\{1+3r/n\}^{3/2}}$

Thus, the given limit $= \lim\limits_{n\to\infty} \sum\limits_{r=0}^{n-1} \dfrac{1/n}{\left\{1+3\left(\dfrac{r}{n}\right)\right\}^{3/2}} = \int_0^1 \dfrac{dx}{(1+3x)^{3/2}} = \dfrac{1}{3}$

6. **Here,**

$$\lim_{n\to\infty}\left[\frac{1}{n^2}\sec^2\left(\frac{\pi}{3n^2}\right) + \frac{2}{n^2}\sec^2\left(\frac{4\pi}{3n^2}\right) + \frac{3}{n^2}\sec^2\left(\frac{9\pi}{3n^2}\right) + \ldots + \ldots + \frac{1}{n}\sec^2\frac{\pi}{3}\right]$$

$$\Rightarrow \quad \lim_{n\to\infty}\frac{1}{n}\left[\sum_{r=0}^{n-1}\frac{r}{n}\sec^2\frac{\pi}{3}\cdot\left(\frac{r^2}{n^2}\right)\right]$$

$$\Rightarrow \quad \int_0^1 x\sec^2\frac{\pi x^2}{3}\,dx \quad \Rightarrow \quad \frac{3\sqrt{3}}{2\pi}$$

7. $af(x) + bf\left(\dfrac{1}{x}\right) = \dfrac{1}{x} - 5$ ...(i)

Integrating, we have

$$a\int_1^2 f(x)\,dx + b\int_1^2 f\left(\frac{1}{x}\right)dx = (\ln x - 5x)_1^2 = \ln 2 - 5 \quad \ldots\text{(ii)}$$

Replacing (x) by $(1/x)$ in Eq. (i), we get; $a\,f\left(\dfrac{1}{x}\right) + b\,f(x) = x - 5$

Integrating, we have,

$$a\int_1^2 f\left(\frac{1}{x}\right)dx + b\int_1^2 f(x)\,dx = \left[\left(\frac{x^2}{2} - 5x\right)\right]_1^2 = -\frac{7}{2} \quad \ldots\text{(iii)}$$

Eliminating $\int_1^2 f\left(\dfrac{1}{x}\right)$ **from Eq. (ii) and Eq. (iii) by multiplying Eq. (ii) by 'a' and (iii) by 'b' and subtracting,**

$$(a^2 - b^2)\int_1^2 f(x)\,dx = a(\ln 2 - 5) + \frac{7b}{2}$$

$$\Rightarrow \quad \int_1^2 f(x)\,dx = \frac{1}{a^2 - b^2}\left[a(\ln 2 - 5) + \frac{7b}{2}\right]$$

8. $$I=\int_0^{2\pi}\frac{\sin^2\theta}{a-b\cos\theta}\,d\theta$$

$$I=2\int_0^{\pi}\frac{\sin^2\theta}{a-b\cos\theta}\,d\theta \qquad \text{...(i)}$$

Also, $$I=2\int_0^{\pi}\frac{\sin^2\theta}{a+b\cos\theta}\,d\theta \qquad \text{...(ii)}$$

Adding Eqs. (i) and (ii), $2I=2\int_0^{\pi}\frac{2a\sin^2\theta}{a^2-b^2\cos^2\theta}\,d\theta$

$$\Rightarrow \quad 2I=8a\int_0^{\pi/2}\frac{\sin^2\theta}{a^2-b^2\cos^2\theta}\,d\theta=8a\int_0^{\pi/2}\frac{\tan^2 d\theta}{a^2\sec^2\theta-b^2}$$

$$2I=8a\int_0^{\pi/2}\frac{\tan^2\theta\,d\theta}{(a^2-b^2)+a^2\tan^2\theta}$$

Put $a\tan\theta=t$

$$\therefore \quad I=4\int_0^{\infty}\frac{(t^2/a^2)\,a^2\,dt}{[(a^2-b^2)+t^2][a^2+t^2]}$$

$$=4\int_0^{\infty}\left[\frac{a^2}{b^2(a^2+t^2)}-\frac{a^2-b^2}{b^2(a^2-b^2+t^2)}\right]dt$$

$$=\frac{4}{b^2}\left[a^2\cdot\frac{1}{a}\cdot\tan^{-1}\frac{t}{a}-\frac{(a^2-b^2)}{\sqrt{a^2-b^2}}\cdot\tan^{-1}\left(\frac{t}{\sqrt{a^2-b^2}}\right)\right]_0^{\infty}$$

$$=\frac{4}{b^2}\left[a\,\frac{\pi}{2}-\sqrt{a^2-b^2}\,\frac{\pi}{2}\right]_0^{\infty}$$

$$I=\frac{2\pi}{b^2}[a-\sqrt{a^2-b^2}]$$

9. Here, $\sin\pi x$ is negative, if $1<x<3/2$.

$$\Rightarrow \quad \int_1^{3/2}|x\sin\pi x|\,dx=\int_1^{3/2}x(-\sin\pi x)\,dx=\left(\frac{\pi+1}{\pi^2}\right)$$

10. $$\int_0^{n\pi}\sin\left[\frac{2x}{\pi}\right]dx$$

$$=\int_0^{\pi/2}\sin\left[\frac{2x}{\pi}\right]dx+\int_{\pi/2}^{\pi}\sin\left[\frac{2x}{\pi}\right]dx+\ldots+\int_{(2n-1)\pi/2}^{n\pi}\sin\left[\frac{2x}{\pi}\right]dx$$

$$=\int_0^{\pi/2}\sin(0)\,dx+\int_{\pi/2}^{\pi}\sin(1)\,dx+\int_{\pi}^{3\pi/2}\sin(2)\,dx+\ldots$$

$$+\int_{(2n-1)\pi/2}^{n\pi}\sin(2n-1)\,dx$$

$$=\frac{\pi}{2}[\sin(1)+\sin(2)+\ldots+\sin(2n-1)]$$

$$=\frac{\pi}{2}\left(\frac{\sin(n)\cdot\sin(n-1/2)}{\sin(1/2)}\right)$$

11. (a) $$\int_1^{\infty}\frac{(x^3+3)\,dx}{x^6(x^2+1)}=\frac{a+b\pi}{c}$$

$$\int_1^{\infty}\frac{(x^3+3)\,dx}{x^6(x^2+1)}=\int_1^{\infty}\frac{dx}{x^3(x^2+1)}+\int_1^{\infty}\frac{3\,dx}{x^6(x^2+1)}$$

Put, $x = \tan\theta \Rightarrow dx = \sec^2\theta\, d\theta$

$$\therefore \quad = \int_{\pi/4}^{\pi/2} \cot^3\theta\, d\theta + 3\int_{\pi/4}^{\pi/2} \cot^6\theta\, d\theta$$

Using reduction formula $\int \cot^n x\, dx = -\dfrac{\cot^{n-1} x}{n-1} - \int \cot^{n-2} x\, dx$

We have $a = 40 - 6\ln 2,\ b = -9,\ c = 12$

(b) $\displaystyle\int_0^1 \frac{(1-x^2)\,dx}{1+x^2+x^4} = a\ln(b)$

$$\therefore \quad \int_0^1 \frac{1-x^2}{1+x^2+x^4}\,dx = \int_0^1 \frac{1/x^2-1}{1/x^2+1+x^2}\,dx$$

Put $x + 1/x = t$

$$\Rightarrow \quad = \int_2^\infty \frac{dt}{t^2-1} = \frac{1}{2}\left(\ln\left|\frac{t-1}{t+1}\right|\right)_2^\infty = \frac{1}{2}\ln 3$$

$$\Rightarrow \quad a = 1/2,\ b = 3$$

12. If $-2 \le x \le -1$, then ; $f(x) = \displaystyle\int_{-2}^{x} -(t+1)\,dt = -\left[\frac{t^2}{2}+t\right]_{-2}^{x}$

$$= -\frac{x^2}{2} - x + \left(\frac{4}{2}-2\right) = -\frac{x^2}{2} - x$$

If $-1 \le x \le 1$ then; $f(x) = \displaystyle\int_{-2}^{x} |t+1|\,dt = \int_{-2}^{-1} |t+1|\,dt + \int_{-1}^{x} |t+1|\,dt$

$$= \int_{-2}^{-1} -(t+1)\,dt + \int_{-1}^{x} (t+1)\,dt$$

$$= -\left[\frac{t^2}{2}+t\right]_{-2}^{-1} + \left[\frac{t^2}{2}+t\right]_{-1}^{x} = \frac{x^2}{2}+x+1$$

$$\Rightarrow \quad f(x) = \begin{cases} -\dfrac{x^2}{2} - x, & -2 \le x \le -1 \\ \dfrac{x^2}{2} + x + 1, & -1 \le x \le 1 \end{cases}$$

As both $\left(-\dfrac{x^2}{2} - x\right)$ and $\left(\dfrac{x^2}{2} + x + 1\right)$ are polynomial functions which are continuous and differentiable everywhere, $f(x)$ is continuous and differentiable in $[-3/2, 1]$ except perhaps at the turning point of definition *ie*, $x = -1$.

13. Putting $y = mx$ in $y^2 + \int_0^x f(t)\,dt$ we get $m^2x^2 + \int_0^x f(t)\,dt$.

Consider the function $\phi(x) = m^2x^2 + \int_0^x f(t)\,dt$

As $f(x)$ is continuous and $\dfrac{d}{dx}\displaystyle\int_0^x f(t)\,dt = f(x) \frac{d}{dx}\int_0^x f(t)\,dt$ is also continuous.

The derivative of a discontinuous function cannot be continuous so $\int_0^x f(t)\,dt$ is also continuous, again m^2x^2 is also continuous, being a polynomial function. As the sum of continuous function is also continuous, $\phi(x)$ is also continuous.

Now, $\phi(0) = 0 + \int_0^0 f(t)\,dt = 0$

$\phi(x) \to \infty$ as $|x| \to \infty$, because $m^2x^2 \to \infty$ and $\int_0^x f(t)\,dt \to \infty$ when $|x| \to \infty$

For all real values of x between $-\infty$ and ∞, $\phi(x)$ attains values between 0 to ∞.

$\therefore$ $\phi(x)$ will also attain the value 2 for some real x and real m.

$\therefore$ $\phi(x)=2$, ie. $m^2x^2+\int_0^x f(t)\,dt=2$

ie, $$y^2+\int_0^x f(t)\,dt=2$$

will have a real solution.

Points to Consider

If $\phi(a)=p$ and $\phi(b)=q$ and $\phi(x)$ is continuous, then every value between p and q will be attained by $\phi(x)$ for x between 'a' and 'b' this is clear from the graph of a continuous function.

14. $$I=\int_0^1 \frac{x^4[(1+x^2)-(2x)]^2}{1+x^2}\,dx$$

$$=\int_0^1 x^4(1+x^2)\,dx-4\int_0^1 x^5\,dx+4\int_0^1\frac{x^6}{1+x^2}\,dx$$

{applying $(a-b)^2$ formula and simplifying}

$$=\left[\frac{x^5}{5}+\frac{x^7}{7}\right]_0^1-4\left[\frac{x^6}{6}\right]_0^1+4\int_0^1-\frac{dx}{1+x^2}+4\int_0^1\frac{(x^2+1)(x^4+1-x^2)}{1+x^2}\,dx$$

$$=\left(\frac{1}{5}+\frac{1}{7}\right)-4\left(\frac{1}{6}\right)-4(\tan^{-1}x)_0^1+4\left(\frac{1}{5}+1-\frac{1}{3}\right)$$

$$=\frac{1}{5}+\frac{1}{7}-\frac{2}{3}-\pi+\frac{52}{15}=\frac{365}{105}-\pi$$

15. (a) $f(x)=\dfrac{1}{1+x^P}\cdot f'(x)=-\dfrac{Px^{P-1}}{(1+x^P)^2}<0,$

$\Rightarrow$ Decreasing in (0, 1)

$\Rightarrow$ $m=f(1)$ and $M=f(0)$

or $m=1/2$ and $M=1$

$\Rightarrow$ $$\frac{1}{2}<\int_0^1\frac{1}{1+x^P}\,dx<1 \quad \text{...(i)}$$

But LHS must be $\dfrac{P}{P+1}$

$\therefore$ $$1-x^{2P}<1\Rightarrow(1-x^P)(1+x^P)<1$$

$\Rightarrow$ $$(1-x^P)<\frac{1}{(1+x^P)}$$

Integrate both sides from (0, 1),

$$\int_0^1(1-x^P)\,dx<\int_0^1\frac{dx}{1+x^P}$$

$\Rightarrow$ $$1-\frac{1}{P+1}<\int_0^1\frac{dx}{1+x^P}$$

$\Rightarrow$ $$\frac{P}{P+1}<\int_0^1\frac{dx}{1+x^P} \quad \text{...(ii)}$$

Comparing (i) and (ii), we get

$$\frac{P}{P+1} < \int_0^1 \frac{dx}{1+x^8} < 1$$

(b) Let $f(x) = x^x = e^{x \log x}$

$$f'(x) = e^{x \log x}(1 + \log x) = 0 \Rightarrow x = \frac{1}{e} \text{ (L. Min.)}$$

$M = f(1)$, and $m = f(1/e)$, ... $f(x)$ is decreasing in $(0, 1/e)$.

$$\Rightarrow \qquad m(b-a) < \int_a^b f(x)\,dx < M(b-a)$$

$$\Rightarrow \qquad \left(\frac{1}{e}\right)^{1/e} < \int_0^1 x^x\,dx < 1$$

16. $I(-u) = \int_0^\pi \ln(1 + 2u\cos x + u^2)\,dx$

$$= \int_0^\pi \ln[1 + 2u\cos(\pi - x) + u^2]\,dx$$

$$= \int_0^\pi \ln[1 - 2u\cos x + u^2]\,du = I(u)$$

$$\Rightarrow \qquad I(u) = I(-u)$$

$$\therefore \qquad I(u) + I(-u) = \int_0^\pi \ln[(1+u^2)^2 - 4u^2\cos^2 x]\,dx$$

$$= \int_0^\pi \ln[1 + u^4 + 2u^2 - 4u^2\cos^2 x]\,dx$$

$$= \int_0^\pi \ln[1 + u^4 - 2u^2\cos 2x]\,dx$$

Let $\qquad 2x = t \Rightarrow 2dx = dt$

$$I(u) + I(-u) = \frac{1}{2}\int_0^{2\pi} \ln[1 + u^4 - 2u^2\cos t]\,dt$$

$$= \int_0^\pi \ln[1 + u^4 - 2u^2\cos t]\,dt = I(u^2).$$

Hence, $\quad I(u) + I(-u) = I(u^2)$.

17. $I_n = \int_0^\infty e^{-x}\sin^n x\,dx = \sin^n x \cdot \int_0^\infty e^{-x}\,dx + \int_0^\infty e^{-x} \cdot n\sin^{n-1} x \cdot \cos x\,dx$

$$= (-e^{-x}\sin^n x)_0^\infty + n\int_0^\infty e^{-x}\sin^{n-1} x\cos x\,dx$$

$$I_n = n\int_0^\infty e^{-x}\sin^{n-1} x \cdot \cos x\,dx$$

Applying by parts again;

$$I_n = n\left[\sin^{n-1} x\cos x \int_0^\infty e^x\,dx + \int_0^\infty ((n-1)\sin^{n-2} x\cos x - \sin^n x)e^{-x}\,dx\right]$$

$$= (-n\sin^{n-1} x\cos x \cdot e^{-x})_0^\infty + n(n-1)\int_0^\infty \sin^{n-2} x \cdot e^{-x}\,dx - n^2 I_n$$

$$\therefore \qquad I_n = n(n-1)I_{n-2} - n^2 I_n$$

$$I_n = \frac{n(n-1)}{(n^2+1)} I_{n-2}$$

Replace n by $n-2$, n by $n-4$, ... multiply all relationships;

$$I_n = \frac{n(n-1)(n-2)(n-3)}{(n^2+1)[(n-2)^2+1]} \cdots \frac{2 \cdot 1}{(2^2+1)} I_0$$

$$I_0 = \int_0^\infty e^{-x}\, dx = 1$$

$$\therefore \quad \frac{n(n-1)\ldots 1}{(n^2+1)[(n-2)^2+1]\ldots(2^2+1)} = \frac{n!}{\prod_{r=1}^{n/2}[(2r)^2+1]}$$

18. $I_n = \int_0^{\pi/4} \tan^n x\, dx = \int_0^{\pi/4} \tan^{n-2} x\,(\sec^2 x - 1)\, dx$

$$\Rightarrow \quad I_n = \frac{1}{n-1} - I_{n-2}$$

$$\Rightarrow \quad I_n + I_{n-2} = \frac{1}{n-1}$$

$$\Rightarrow \quad n - 1 = \frac{1}{I_n + I_{n-2}} \qquad \text{putting } n = 4, 5, 6, 7, \ldots$$

$$\frac{1}{I_4 + I_2} = 3, \frac{1}{I_5 + I_3} = 4, \frac{1}{I_6 + I_4} = 5, \ldots \text{ and So on.}$$

$$\Rightarrow \quad \frac{1}{I_4 + I_2}, \frac{1}{I_5 + I_3}, \frac{1}{I_6 + I_4}, \ldots \text{ forms an AP}$$

With common difference = 1

19. (i) $I = \left[-2\sin qt\, \frac{\cos pt}{p}\right]_0^1 + \int_0^1 2q\cos qt \cdot \frac{\cos pt}{p}\, dt$

$$= -\frac{2}{p}\sin q\cos p + \frac{2q}{p}\int_0^1 \cos qt \cdot \cos pt \cdot dt$$

$$I = -\frac{2}{p}\sin q\cos p + \frac{2q}{p}\left[\frac{\sin p\cos q}{p} + \frac{q}{p}\frac{I}{2}\right]$$

or $\quad I\left(1 - \frac{q^2}{p^2}\right) = -\frac{2}{p}\sin q\cos p + \frac{2q}{p^2}\sin p\cos q$

$$= \cos p\cos q\left[\frac{2q}{p^2}\tan p - \frac{2}{p}\tan q\right]$$

As p and q are different roots of $\tan x = x$

$$\therefore \quad I = 0$$

(ii) $\quad I = \int_0^1 (2\sin^2 pt)\, dt \quad \text{as } p = q$

$$= \int_0^1 (1 - \cos 2pt)\, dt = \left(t - \frac{\sin 2pt}{2p}\right)_0^1 = 1 - \frac{\sin 2p}{2p}$$

$$= 1 - \frac{2\tan p}{2p(1+\tan^2 p)} \qquad \{\text{as } \tan p = p\}$$

$$= 1 - \frac{2p}{2p(1+p^2)} = \frac{p^2}{1+p^2}$$

20. $I = \int_0^\pi \theta^3 \log\sin\theta\, d\theta = \int_0^\pi (\pi - \theta)^3 \log\sin(\pi - \theta)\, d\theta$

$$= \int_0^\pi (\pi^3 - \theta^3 - 3\pi^2\theta + 3\pi\theta^2)\log\sin\theta\, d\theta$$

$$= \pi^3(-\pi\log 2) - I - 3\pi^2\int_0^\pi \theta\log\sin\theta\, d\theta + 3\pi\int_0^\pi \theta^2\log\sin\theta\, d\theta$$

$$2I = -\pi \log 2 - 3\pi^2 (I_1) + 3\pi \int_0^{\pi} \theta^2 \log (\sin \theta)\, d\theta \qquad \text{...(i)}$$

Now, $\quad I_1 = \int_0^{\pi} \theta \log (\sin \theta)\, d\theta$

$$= \int_0^{\pi} (\pi - \theta) \log (\sin (\pi - \theta))\, d\theta$$

$$= \pi (-\pi \log 2) - I_1$$

$$I_1 = -\left(\frac{\pi^2}{2} \log 2 \right)$$

$\therefore \quad$ Eq. (i) becomes,

$$2I = -\pi^4 \log 2 + \frac{3}{2}\pi^4 \log 2 + 3\pi \int_0^{\pi} \theta^2 \log (\sin \theta)\, d\theta$$

$$I = \frac{3\pi}{2}\left[\frac{\pi^3 \log 2}{6} + \int_0^{\pi} \theta^2 \log \sin \theta\, d\theta \right]$$

$$= \frac{3\pi}{2}\left[\int_0^{\pi} \theta^2 \log \sqrt{2}\, d\theta + \int_0^{\pi} \theta^2 \log \sin \theta\, d\theta \right]$$

$$= \frac{3\pi}{2}\left[\int_0^{\pi} \theta^2 \log (\sqrt{2} \sin \theta)\, d\theta \right]$$

21. $I = \int_0^2 \max.\{\log (1 + x^2), 1\}\, dx = \int_0^{\sqrt{e-1}} dx + \int_{\sqrt{e-1}}^{2} \ln (1 + x^2)\, dx = \sqrt{e-1} + I_2$

$$I_2 = \int_{\sqrt{e-1}}^{2} \ln (1 + x^2)\, dx = \left| x \ln (1 + x^2) \right|_{\sqrt{e-1}}^{2} - \int_{\sqrt{e-1}}^{2} \frac{x \cdot 2x\, dx}{x^2 + 1}$$

$$= 2 \log 5 = (\sqrt{e-1}) \log e - \int_{\sqrt{e-1}}^{2} \frac{2x^2\, dx}{x^2 + 1}$$

$$= \log \frac{25}{e^{\sqrt{e-1}}} - 2\left[x - \tan^{-1} x\right]_{\sqrt{e-1}}^{2}$$

$$I_2 = \log \frac{25}{e^{\sqrt{e-1}}} - 2\left[2 - \tan^{-1} 2 - \sqrt{e-1} + \tan^{-1} (\sqrt{e-1})\right]$$

$\therefore \quad I = \sqrt{e-1} + I_2$

22. $g(x) = \begin{cases} x^2, & -2 \le x \le -1 \\ -x, & -1 \le x \le -1/4 \\ x + \frac{1}{2}, & -\frac{1}{4} \le x < 0 \\ x, & 0 \le x \le 1 \\ x^2, & 1 \le x \le 2 \end{cases}$

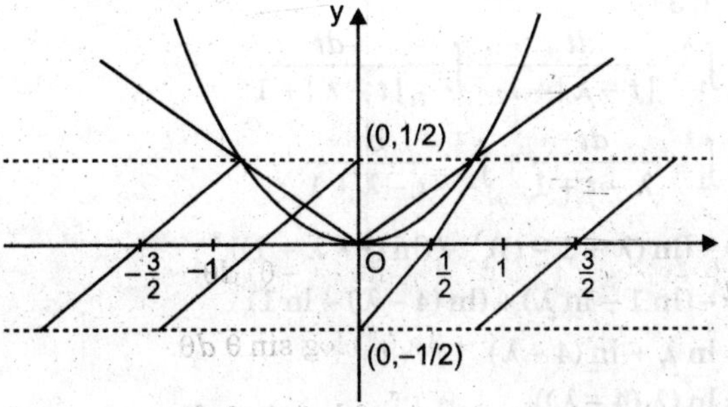

$$\therefore \quad \int_{-2}^{2} g(x)\,dx = \int_{-2}^{-1} x^2\,dx + \int_{-1}^{-1/4} (-x)\,dx + \int_{-1/4}^{0} \left(x + \frac{1}{2}\right) dx + \int_0^1 (x)\,dx + \int_1^2 x^2\,dx$$

$$= \left[\frac{x^3}{3}\right]_{-2}^{-1} - \left[\frac{x^2}{2}\right]_{-1}^{-1/4} + \left[\frac{x^2}{2} + \frac{1}{2}x\right]_{-1/4}^{0} + \left[\frac{x^2}{2}\right]_0^1 + \left[\frac{x^3}{3}\right]_1^2$$

$$= \frac{275}{48} \text{ sq units}$$

23. Given, $f(x+1) = f(x)$

$$\Rightarrow \quad \int_c^{c+n} f(x)\,dx = n\int_0^1 f(x)\,dx \qquad \text{for any real } C \qquad \text{...(i)}$$

Now, $\int_a^{a+n} f(x)\,dx = \int_b^{b+m} f(x)\,dx \Rightarrow n\int_0^1 f(x)\,dx = m\int_0^1 f(x)\,dx$

$$\Rightarrow \quad \int_0^1 f(x)\,dx = 0 \quad \text{as} \quad (m \neq n) \qquad \text{...(ii)}$$

$$\Rightarrow \quad \int_m^n f(x)\,dx = \int_m^{m+1} f(x)\,dx + \int_{m+1}^{m+2} f(x)\,dx + \ldots + \int_{n-1}^{n} f(x)\,dx$$

Assuming $n > m$.

$$\Rightarrow \quad \int_m^n f(x)\,dx = \int_0^1 f(x)\,dx + \int_0^1 f(x)\,dx + \ldots + \int_0^1 f(x)\,dx = 0$$

[using Eq. (ii)]

24. $I = \int_0^n f'(x)\cdot[x]\,dx - \int_0^1 xf'(x)\,dx + \frac{1}{2}\int_0^n f'(x)\,dx$

$$= \sum_{r=1}^{n} \int_{r-1}^{r} f'(x)[x]\,dx - \{x\cdot f(x)\}_0^n - \int_0^n f(x)\,dx\} + \frac{1}{2}(f(x))_0^n$$

$$= \sum_{r=1}^{n} (r-1)\int_{r-1}^{r} f'(x)\,dx - nf(n) + \frac{1}{2}f(n) - \frac{1}{2}f(0) + \int_0^n f(x)\,dx$$

$$= \sum_{r=1}^{n} (r-1)\{f(r) - f(r-1)\} - nf(n) + \frac{1}{2}f(n) + \int_0^n f(x)\,dx - \frac{1}{2}f(0)$$

$$= -f(1) - f(2) \ldots - f(n-1) - f(n) + \frac{1}{2}f(n) + \frac{1}{2}f(0) + \int_0^n f(x)\,dx$$

$$\Rightarrow \quad I = -\sum_{r=1}^{n} f(r) + \frac{1}{2}f(n) + \frac{1}{2}f(0) + \int_0^n f(x)\,dx$$

25. $f(x) = \int_1^3 \frac{dt}{|t-x|+1}, \; x \in [1, 3]$

Let $x = \lambda \in (1, 3)$

$$f(\lambda) = \int_1^{\lambda} \frac{dt}{|t-\lambda|+1} + \int_{\lambda}^{3} \frac{dt}{|t-\lambda|+1}$$

$$= \int_1^{\lambda} \frac{dt}{\lambda - t + 1} + \int_{\lambda}^{3} \frac{dt}{t - \lambda + 1}$$

$$= -(\ln(\lambda + 1 - t))_1^{\lambda} + (\ln(t - \lambda + 1))_{\lambda}^{3}$$

$$= -(\ln 1 - \ln \lambda) + (\ln(4-\lambda) - \ln 1)$$

$$= \ln \lambda + \ln(4 - \lambda)$$

$$= \ln(\lambda(4 - \lambda))$$

$$= \int_1^3 f(x)\,dx = \int_1^3 \ln(x(4-x))\,dx$$

$$= (\ln(x(4-x))\,x)_1^3 - \int_1^3 \frac{(4-2x)\,x}{x(4-x)}\,dx$$

$$= 3\ln 3 - \ln 3 - 2\int_1^3 \frac{x(x-2)}{x(x-4)}\,dx$$

$$\Rightarrow \quad 2\ln 3 - 2(2 - 2\ln 3) = 6\ln 3 - 4$$

26. $\int_0^1 (\{2x\}-1)(\{3x\}-1)\,dx$

$$= \int_0^{1/3} (\{2x\}-1)(\{3x\}-1)\,dx + \int_{1/3}^{1/2} (\{2x\}-1)(\{3x\}-1)\,dx$$

$$+ \int_{1/3}^{2/3} (\{2x\}-1)(\{3x\}-1)\,dx + \int_{2/3}^{1} (\{2x\}-1)(\{3x\}-1)\,dx$$

$$= \int_0^{1/3} (2x-1)(3x-1)\,dx + \int_{1/3}^{1/2} (2x-1)(3x-2)\,dx$$

$$+ \int_{1/2}^{2/3} (2x-2)(3x-2)\,dx + \int_{2/3}^{1} (2x-2)(3x-2)\,dx$$

$$= \int_0^{1/3} (6x^2-5x+1)\,dx + \int_{1/3}^{1/2} (6x^2-7x+2)\,dx$$

$$+ \int_{1/2}^{2/3} (6x^2-10x+4)\,dx + \int_{2/3}^{1} (6x^2-12x+6)\,dx$$

$$= \frac{19}{72}$$

27. $g(t)\,|t-1| - |t| + |t+1|$

$$g(x) = \begin{cases} -t, & t \le -1 \\ 2+t, & -1 < t \le 0 \\ 2-t, & 0 < t \le 1 \\ t, & 1 < t \end{cases}$$

Hence,
$$f(x) = \begin{cases} 3/2, & -3/2 \le x - 1/2 \\ 2+x, & -1/2 < x \le 0 \\ 2, & 0 < x \le 2 \\ x, & 2 < x \end{cases}$$

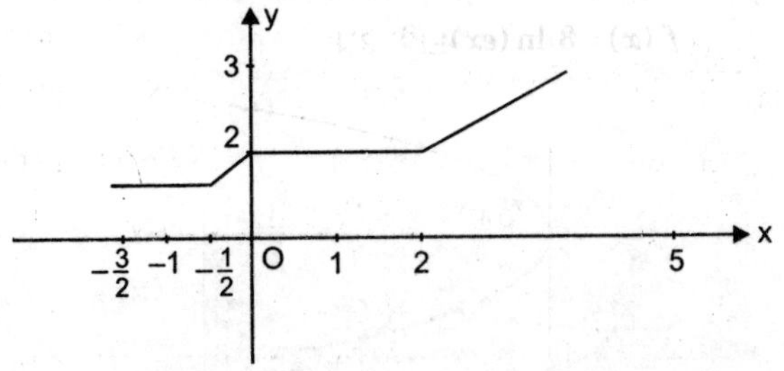

$\Rightarrow \int_{-3/2}^{5} f(x)\,dx$ = Area bounded by $y = f(x)$; x-axis and between the ordinates $x = -3/2$ and $x = 5$.

$$= \frac{3}{2}(1) + \left(2 + \frac{3}{2}\right)\cdot\frac{1}{4} + 2\cdot 2 + (5+2)\cdot\frac{3}{2} = \frac{135}{8} \text{ units}$$

28. $$I=\int_0^a \ln\left(\frac{\cos a}{\sin a}+\frac{\sin a}{\cos a}\right)dx$$

$$=\int_0^a \ln\left(\frac{\cos a\cos x+\sin a\sin x}{\sin a\cos x}\right)dx$$

$$=\int_0^a \ln\left(\frac{\cos(a-x)}{\sin a\cos x}\right)dx$$

$$=\int_0^a \ln\{\cos(a-x)\}\,dx-\int_0^a \ln(\sin a)\,dx-\int_0^a \ln(\cos x)\,dx$$

$$=\int_0^a \ln\cos[a-(a-x)]\,dx-\int_0^a \ln(\sin a)\,dx-\int_0^a \ln(\cos x)\,dx$$

$$=\int_0^a \ln\cos x\,dx-\int_0^a \ln\sin a\,dx-\int_0^a \ln\cos x\,dx=-a\ln(\sin a)$$

29. $$\int_0^{1/4}[\sin x+[\cos x+[\tan x+[\sec x]]]]\,dx$$

$$=\int_0^{\pi/4}[\sin x]\,dx+\int_0^{\pi/4}[\cos x]\,dx+\int_0^{\pi/4}[\tan x]\,dx+\int_0^{\pi/4}[\sec x]\,dx$$

$$=0+0+0+\int_0^{\pi/4}1\,dx=\pi/4$$

30. We have $\int_1^{xy} f(t)\,dt=y\int_1^x f(t)\,dt+x\int_1^y f(t)\,dt$

By differentiating both sides w.r.t. x; we get

$$f(xy)\cdot y=yf(x)+\int_1^y f(t)\,dt \quad \text{...(i)}$$

Putting $x=1$, we get

$$yf(y)=yf(1)+\int_1^y f(t)\,dt=3y+\int_1^y f(t)\,dt$$

By differentiating both sides w.r.t. 'y', we get

$$yf'(y)+f(y)=3+f(y)$$

$\Rightarrow$ $$\frac{df}{dy}=\frac{3}{y}\Rightarrow f(y)=3\ln y+c$$

Since, $$f(1)=3\Rightarrow c=3\Rightarrow f(y)=3\ln y+3$$

$\Rightarrow$ $$f(x)=3\ln(ex)$$

3

Area of Bounded Regions

Chapter in a Snapshot

- Curve Sketching
- Sketching of Some Common Curves
- Some More Curves Which Occur Frequently in Mathematics in Standard Forms
- Asymptotes
- Areas of Curves Given by Cartesian Equations
- Sign Convention For Finding the Areas Using Integration
- Area Between the Given Curves
- Standard Areas to Be Remembered For the Purpose of Solving Objective Questions
- Average Value of a Function
- Variable Area, Greatest and Least Values

Curve Sketching

For the evaluation of area of bounded regions it is very essential to know the rough sketch of the curves. The following points are very useful to draw a rough sketch of a curve.

(i) Symmetry

(a) **Symmetry about x-axis** If all powers of y in the equation of the given curve are even, then it is symmetric about x-axis, *ie*, the shape of the curve above x-axis is exactly identical to its shape below x-axis.

For example $y^2 = 4ax$ is symmetric about x-axis.

(b) **Symmetry about y-axis** If all powers of x in the equation of the given curve are even, then it is symmetric about y-axis.

For example $x^2 = 4ay$ is symmetric about y-axis.

(c) **Symmetry in opposite quadrants** If by putting $-x$ for x and $-y$ for y, the equation of curve remains same, then it is symmetric in opposite quadrants.

For example $xy = c^2, x^2 + y^2 = a^2$ are symmetric in opposite quadrants.

(d) **Symmetric about the line $y = x$** If the equation of a given curve remains unaltered by interchanging x and y, then it is symmetric about the line $y = x$ which passes through the origin and makes an angle of 45° with positive direction of x-axis.

(ii) Origin and Tangents at the Origin

See whether the curve passes through origin or not. If the point (0, 0) satisfies the equation of the curve, then it passes through the origin and in such a case to find the equations of the tangents at the origin, equate the lowest degree term to zero.

For example $y^2 = 4ax$ passes through the origin. The lowest degree term in this equation is $4ax$. Equating $4ax$ to zero, we get $x = 0$.

So, $x = 0$ *ie*, y-axis is tangent at the origin to $y^2 = 4ax$.

(iii) Points of Intersection of the Curve with the Coordinate Axes

By putting $y = 0$ in the equation of the given curve, find points where the curve crosses the x-axis. Similarly, by putting $x = 0$ in the equation of the given curve we can find points where the curve crosses the y-axis.

For example To find the points where the curve $xy^2 = 4a^2 (2a - x)$ meets x-axis, we put $y = 0$ in the equation which gives $4a^2 (2a - x) = 0$ or $x = 2a$. So the curve $xy^2 = 4a^2 (2a - x)$, meets x-axis at $(2a, 0)$. This curve does not intersect y-axis, because by putting $x = 0$ in the equation of the given curve get an absurd result.

(iv) Regions Where the Curve Does Not Exist

Determine the regions in which the curve does not exist. For this, find the value of y in terms of x from the equation of the curve and find the value of x for which y is imaginary. Similarly, find the value of x in terms of y and determine the values of y for which x is imaginary. The curve does not exist for these values of x and y.

For example The values of y obtained from $y^2 = 4ax$ are imaginary for negative values of x. So, the curve does not exist on the left side of y-axis. Similarly, the curve $a^2 y^2 = x^2 (a - x)$ does not exist for $x > a$ as the values of y are imaginary for $x > a$.

(v) Special Points

Find the points at which $\frac{dy}{dx} = 0$. At these points the tangent to the curve is parallel to x-axis.

Find the points at which $\frac{dx}{dy} = 0$. At these points the tangents to the curve is parallel to y-axis.

(vi) Sign of $\frac{dy}{dx}$ and Points of Maxima and Minima

Find the interval in which $\frac{dy}{dx} > 0$. In this interval, the function is monotonically increasing, find the interval in which $\frac{dy}{dx} < 0$. In this interval, the function is monotonically decreasing. Put $\frac{dy}{dx} = 0$ and check the sign of $\frac{d^2 y}{dx^2}$ at the points so obtained to find the points of maxima and minima.

Keeping the above facts in mind and plotting some points on the curve one can easily have a rough sketch of the curve. Following examples will clear the procedure.

Illustration 1 Sketch the curve $y = x^3$.

Solution. We observe the following points about the given curve :

(i) The equation of the curve remains unchanged, if x is replaced by $-x$ and y by $-y$. So, it is symmetric in opposite quadrants. Consequently, the shape of the curve is similar in the first and the third quadrants.

(ii) The curve passes through origin. Equating lowest degree term y to zero, we get $y = 0$ *ie*, x-axis is the tangent at the origin.

(iii) Putting $y = 0$ in the equation of the curve, we get $x = 0$. Similarly, when $x = 0$, we get $y = 0$. So, the curve meets the coordinate axes at (0, 0) only.

(iv) $y = x^3 \Rightarrow \frac{dy}{dx} = 3x^2, \frac{d^2y}{dx^2} = 6x$ and $\frac{d^3y}{dx^3} = 6$

Clearly, $\frac{dy}{dx} = 0 = \frac{d^2y}{dx^2}$ at the origin but $\frac{d^3y}{dx^3} \neq 0$. So, origin is a point of inflexion.

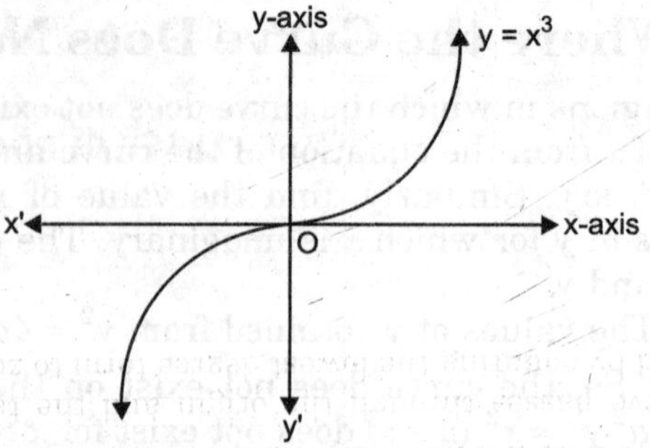

(v) As x increases from 0 to ∞, y also increases from 0 to ∞.

Keeping all the above points in mind, we obtain a sketch of the curve as shown in figure.

Illustration 2 Sketch the curve $y = x^3 - 4x$.

Solution. We note the following points about the curve :

(i) The equation of the curve remains same if (x) is replaced by $(-x)$ and (y) by $(-y)$, so it is symmetric in opposite quadrants.

Consequently, the curve in the first quadrant is identical to the curve in third quadrant and the curve in second quadrant is similar to the curve in fourth quadrant.

(ii) The curve passes through the origin. Equating the lowest degree term $y + 4x$ to zero, we get $y + 4x = 0$ or $y = -4x$. So, $y = -4x$ is tangent to the curve at the origin.

(iii) Putting $y = 0$ in the equation of the curve, we obtain $x^3 - 4x = 0 \Rightarrow x = 0, \pm 2$. So, the curve meets x-axis at (0, 0), (2, 0), (–2, 0).

Putting $x = 0$ in the equation of the curve, we get $y = 0$. So, the curve meets y-axis at (0, 0) only.

(iv) $y = x^3 - 4x \Rightarrow \dfrac{dy}{dx} = 3x^2 - 4$

Now, $\dfrac{dy}{dx} > 0 \Rightarrow 3x^2 - 4 > 0 \Rightarrow \left(x - \dfrac{2}{\sqrt{3}}\right)\left(x + \dfrac{2}{\sqrt{3}}\right) > 0$

$\Rightarrow \quad x < -\dfrac{2}{\sqrt{3}}$ or $x > \dfrac{2}{\sqrt{3}}$ (using number line rule)

and $\dfrac{dy}{dx} < 0 \Rightarrow -\dfrac{2}{\sqrt{3}} < x < \dfrac{2}{\sqrt{3}}$

So, the curve is decreasing in the interval $(-2/\sqrt{3}, 2/\sqrt{3})$ and increasing for $x > \dfrac{2}{\sqrt{3}}$ or $x < -\dfrac{2}{\sqrt{3}}$.

$x = -\dfrac{2}{\sqrt{3}}$ is a point of local maximum and $x = \dfrac{2}{\sqrt{3}}$ is point of local minimum.

When $x = \dfrac{2}{\sqrt{3}} \Rightarrow y = -\dfrac{16}{3\sqrt{3}}$

When $x = -\dfrac{2}{\sqrt{3}} \Rightarrow y = \dfrac{16}{3\sqrt{3}}$

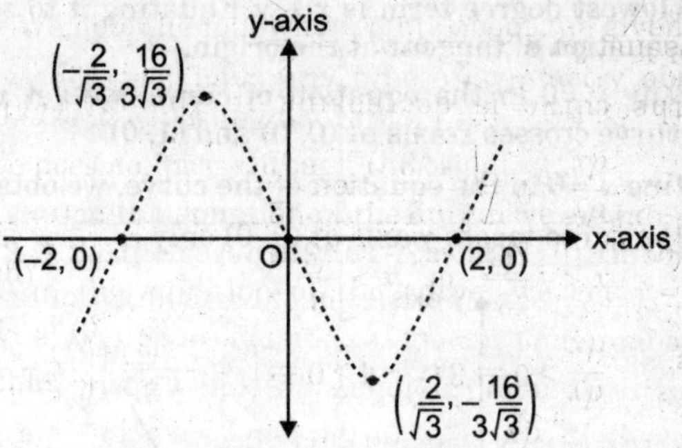

Keeping above points in mind, we sketch the curve as shown in figure.

Illustration 3 Sketch the curve $y = (x-1)(x-2)(x-3)$.

Solution. We note the following points about the given curve :

(i) The curve does not have any type of symmetry about the coordinate axes and also in opposite quadrants.

(ii) The curve does not pass through the origin.

(iii) Putting $y = 0$ in the equation of the curve, we get $(x-1)(x-2)(x-3) = 0$ $\Rightarrow x = 1, 2, 3$. So, the curve meets x-axis at (1, 0), (2, 0) and (3, 0).

Putting $x = 0$ in the equation of the curve, we get $y = -6$. So, the curve crosses y-axis at (0, – 6).

We observe that $x < 1 \Rightarrow y < 0$

$1 < x < 2 \Rightarrow y > 0$

$2 < x < 3 \Rightarrow y < 0$

and $x > 3 \Rightarrow y > 0$

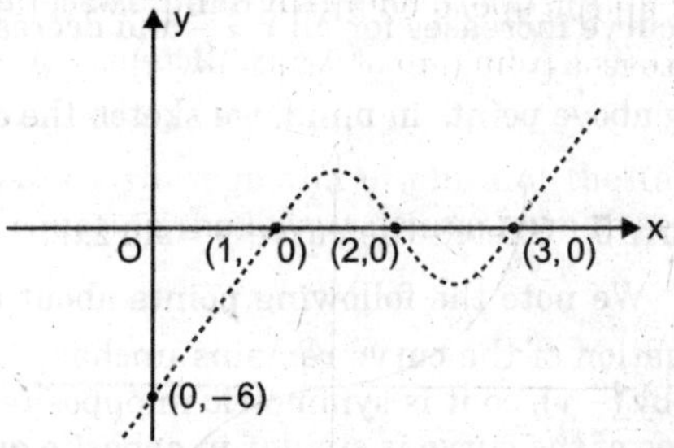

Clearly, y decreases as x decreases for all $x < 1$ and y increases as x increases for $x > 3$.

Keeping all the above points in mind, we sketch the curve as shown in figure.

Illustration 4 Sketch the graph for $y = x^2 - x$.

Solution. We note the following points about the curve :

(i) The curve does not have any kind of symmetry.

(ii) The curve passes through the origin and the tangent at the origin is obtained by equating the lowest degree term to zero.

The lowest degree term is $x + y$. Equating it to zero, we get $x + y = 0$ as the equation of tangent at the origin.

(iii) Putting $y = 0$ in the equation of curve, we get $x^2 - x = 0 \Rightarrow x = 0, 1$. So, the curve crosses x-axis at (0, 0) and (1, 0).

Putting $x = 0$ in the equation of the curve, we obtain $y = 0$.

So, the curve meets y-axis at (0, 0) only.

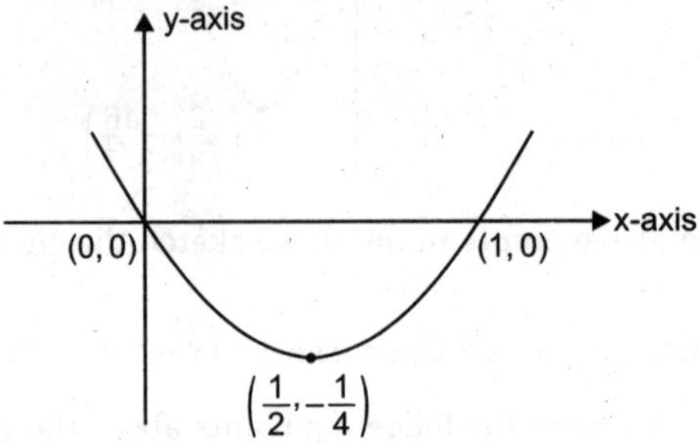

(iv) $y = x^2 - x \Rightarrow \dfrac{dy}{dx} = 2x - 1$ and $\dfrac{d^2y}{dx^2} = 2$

Now, $\quad \dfrac{dy}{dx} = 0 \Rightarrow x = \dfrac{1}{2}$

At $x = \dfrac{1}{2}$, $\quad \dfrac{d^2y}{dx^2} > 0$

So, $x = \dfrac{1}{2}$ is point of local minima.

(v) $\dfrac{dy}{dx} > 0 \Rightarrow 2x - 1 > 0 \Rightarrow x > \dfrac{1}{2}$

So, the curve increases for all $x > \dfrac{1}{2}$ and decreases for all $x < \dfrac{1}{2}$.

Keeping above points in mind, we sketch the curve as shown in figure.

Illustration 5 Sketch the curve $y = \sin 2x$.

Solution. We note the following points about the curve :

(i) The equation of the curve remains unchanged, if (x) is replaced by $(-x)$ and (y) by $(-y)$, so it is symmetric in opposite quadrants. Consequently; the shape of the curve is similar in opposite quadrants.

(ii) The curve passes through origin.

(iii) Putting $x = 0$ in the equation of the curve, we get $y = 0$. So, the curve crosses the y-axis at (0, 0) only.

Putting $y = 0$ in the equation of the curve, we get

$$\sin 2x = 0 \Rightarrow 2x = n\pi, \ n \in Z$$

$$\Rightarrow \quad x = \frac{n\pi}{2}, n \in Z$$

So, the curve cuts the x-axis at the points

$$\ldots, (-\pi, 0), (-\pi/2, 0), (0, 0), (\pi/2, 0), (\pi, 0), \ldots$$

(iv) $y = \sin 2x \Rightarrow \frac{dy}{dx} = 2\cos 2x$ and $\frac{d^2y}{dx^2} = -4\sin 2x$

Now, $\frac{dy}{dx} = 0 \Rightarrow \cos 2x = 0$

$\Rightarrow \quad 2x = \pm\frac{\pi}{2}, \pm\frac{3\pi}{2}, \ldots$

$\Rightarrow \quad x = \pm\frac{\pi}{4}, \pm\frac{3\pi}{4}, \ldots$

Clearly, $\frac{d^2y}{dx^2} < 0$ at $x = \frac{\pi}{4}, \frac{5\pi}{4}, \ldots$ and at $x = -\frac{3\pi}{4}, -\frac{7\pi}{4}, \ldots$

and $\frac{d^2y}{dx^2} > 0$ at $x = \frac{3\pi}{4}, \frac{7\pi}{4}, \ldots$ and at $x = -\frac{\pi}{4}, -\frac{5\pi}{4}, \ldots$

So, the points $x = \frac{\pi}{4}, \frac{5\pi}{4}, \frac{7\pi}{4}, \ldots$ and $x = -\frac{3\pi}{4}, -\frac{7\pi}{4}, \ldots$ are points of local maximum and local maximum values at these points are 1.

Similarly, $x = -\frac{\pi}{4}, -\frac{5\pi}{4}, \ldots$ and $x = \frac{3\pi}{4}, \frac{7\pi}{4}, \ldots$ are points of local minimum and local minimum value at these points is (-1).

(v) $\sin 2(x + \pi) = \sin 2x$ for all x. So, the periodicity of the function is π. This means that the pattern of the curve repeats at intervals of length π.

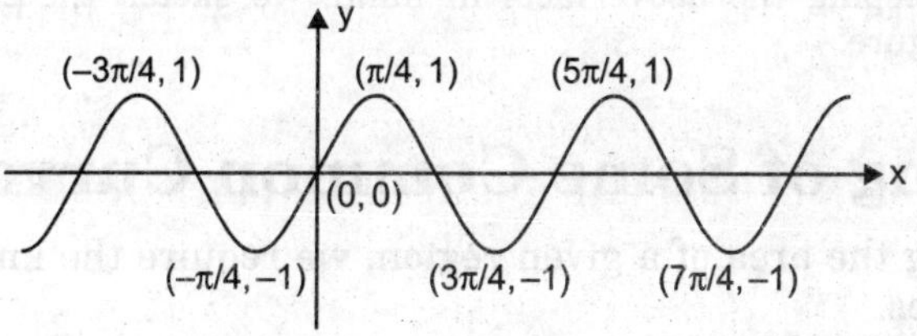

Thus, keeping in mind, we sketch the curve as shown in figure.

Illustration 6 Sketch the curve $y = \sin^2 x$.

Solution. We note the following points about the curve :

(i) The equation of the curve remains same, if x is replaced by $-x$. So, the curve is symmetric about y-axis, *ie*, the curve on the left side of y-axis is identical to the curve on its right side.

(ii) The curve meets the coordinate axes at the same points where $y = \sin x$ meets them.

(iii) $y = \sin^2 x \Rightarrow \frac{dy}{dx} = \sin 2x$ and $\frac{d^2y}{dx^2} = 2\cos 2x$

Now, $\frac{dy}{dx} = 0 \Rightarrow \sin 2x = 0 \Rightarrow 2x = n\pi, n \in Z$

$\Rightarrow \quad x = \frac{n\pi}{2}, \; n \in Z \Rightarrow x = \pm\pi/2, \pm\pi, \pm 3\pi/2, \pm 2\pi, \ldots$

Clearly, $\frac{d^2y}{dx^2} < 0$ at $x = \pm\pi/2, \pm 3\pi/2, \pm 5\pi/2, \ldots$

and $\frac{d^2y}{dx^2} > 0$ at $x = \pm\pi, \pm 2\pi, \pm 3\pi \ldots$

So, $x = \pm \pi/2, \pm 3\pi/2, \pm 5\pi/2, \ldots$ are the points of local maximum and local maximum value at these points is 1. Points $x = \pm \pi, \pm 2\pi, \pm 3\pi, \ldots$ are points of local minimum and the local minimum value at these points is 0.

(iv) $y = \sin^2 x \Rightarrow \frac{dy}{dx} = \sin 2x$

Clearly, $\frac{dy}{dx} > 0$ when $0 < x < \frac{\pi}{2}$ and $\frac{dy}{dx} < 0$ when $\frac{\pi}{2} < x < \pi$.

So, the given curve is increasing in the interval $[0, \pi/2]$ and decreasing in $[\pi/2, \pi]$.

(v) $\sin^2(\pi + x) = \sin^2 x$ for all x. So, the periodicity of the function is π. This means that the shape of the curve repeats at the interval of length π.

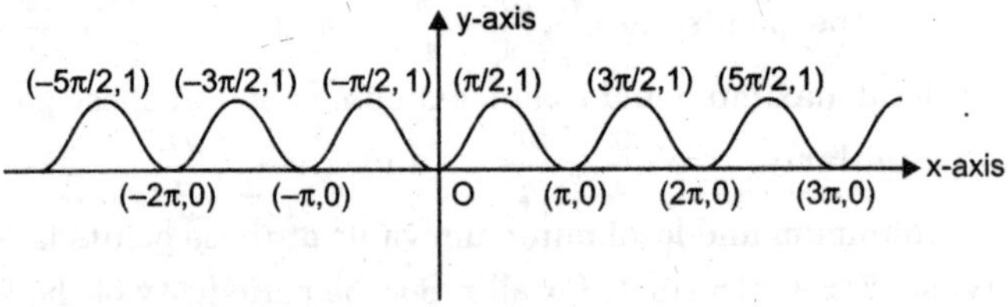

Keeping the above facts in mind, we sketch the curve as shown in figure.

Sketching of Some Common Curves

For finding the area of a given region, we require the knowledge of some standard curves.

(i) Straight Line

Every first degree equation in x, y represents a straight line. So, the general equation of a line is $ax + by + c = 0$. To draw a straight line find the points where it meets with the coordinate axes by putting $y = 0$ and $x = 0$ respectively in its equation. By joining these two points we get the sketch of the line. Sometimes the equation of a line is given in the form $y = mx$. This equation represents a line passing through the origin and inclined at an angle $\tan^{-1} m$ with the positive direction of x-axis. The equation of the form $x = a$ and $y = b$ represents straight lines parallel to y-axis and x-axis respectively.

Region Represented by a Linear Inequality

To find the region represented by linear inequality $ax + by \le c$ and $ax + by \ge c$, we proceed as follows :

(i) Convert the inequality into equality to obtain a linear equation in x, y.

(ii) Draw the straight line represented by it.

(iii) The straight line obtained in (ii) divides the xy-plane in two parts.

To determine the region represented by the inequality choose some convenient points; *eg*, origin or some points on the coordinate axes.

If the coordinates of a point satisfy the inequality, then region containing the points is the required region, otherwise the region not containing the point is required region.

Illustration 7 **Mark the region represented by $3x + 4y \le 12$.**

Solution. **Converting the inequality into equation, we get $3x + 4y = 12$. This line meets the coordinate axes at (4, 0) and (0, 3) respectively. Join these points to obtain straight line represented by $3x + 4y = 12$.**

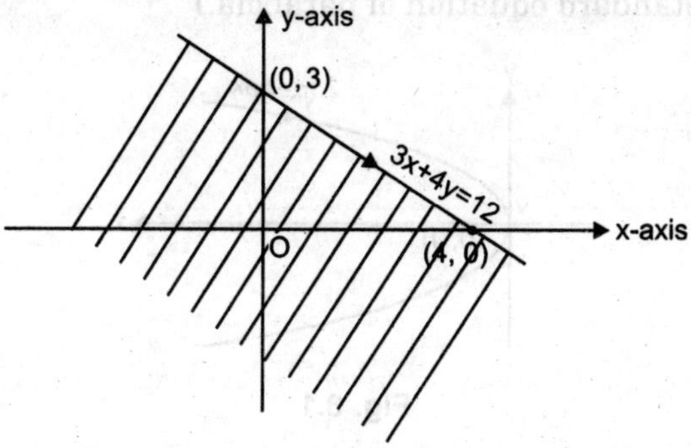

This straight line divides the plane in two parts. One part contains the origin and the other does not contains the origin. Clearly, (0, 0) satisfy the inequality $3x + 4y < 12$. So, the region represented by $3x + 4y < 12$ is region containing the origin as shown in figure.

(ii) Circle

The general equation of a circle is

$$x^2 + y^2 + 2gx + 2fy + c = 0$$

$\therefore$ The second degree equation in x, y such that coeff. of x^2 = coeff. of y^2 and there is no term containing xy; it always represents a circle. To draw a sketch of a circle, we write the equation in the standard form $(x-h)^2 + (y-k)^2 = r^2$, whose centre is (h, k) and radius is r.

Points to Consider

1. **The inequality $(x-a)^2 + (y-b)^2 < r^2$ represents the interior of a circle.**
2. **The inequality $(x-a)^2 + (y-b)^2 > r^2$ represents the exterior of a circle. (*ie*, region lying outside the circle).**

(iii) Parabola

It is the locus of points such that its distance from a fixed point is equal to its distance from a fixed straight line.

Taking the fixed straight line $x = -a, a > 0$ and fixed point $(a, 0)$, we get the equation of parabola $y^2 = 4ax$.

Nature of the Curve

(i) It passes through (0, 0).

(ii) It is symmetrical about axis of x.

(iii) No part of the curve lies on the negative side of axis of x.

(iv) Curve turns at (0, 0) which is called the vertex of the curve.

(v) The curve extends to infinity. It is not a closed curve.

(1) $y^2 = 4ax$ (Standard equation of parabola.)

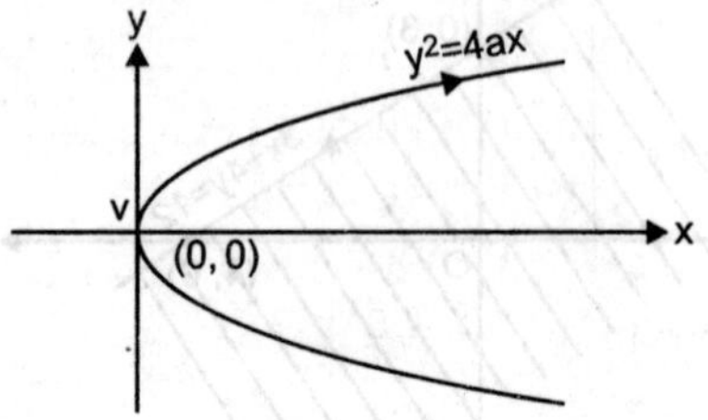

Fig. 3.1

(2) $y^2 = 4a(x-h)$; where a and h are positive.

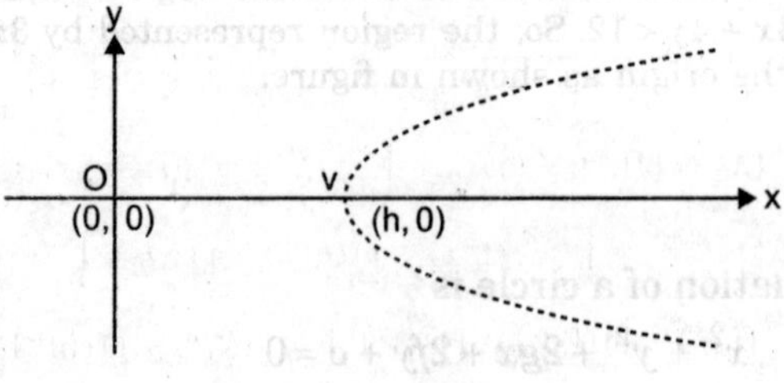

Fig. 3.2

(3) $y^2 = 4a(x+h)$; where a and h are positive.

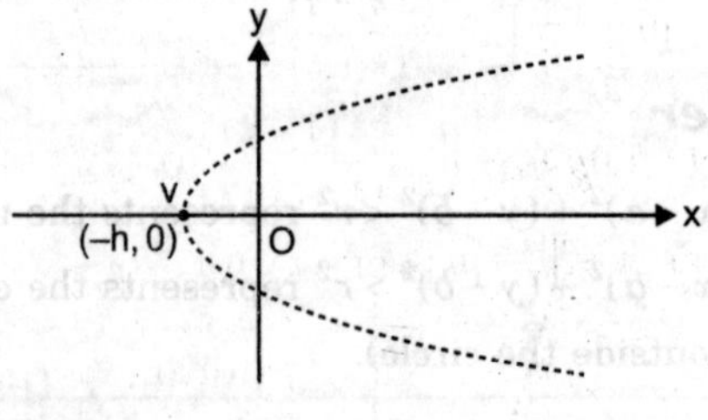

Fig. 3.3

(4) $y^2 = -4ax; a > 0$

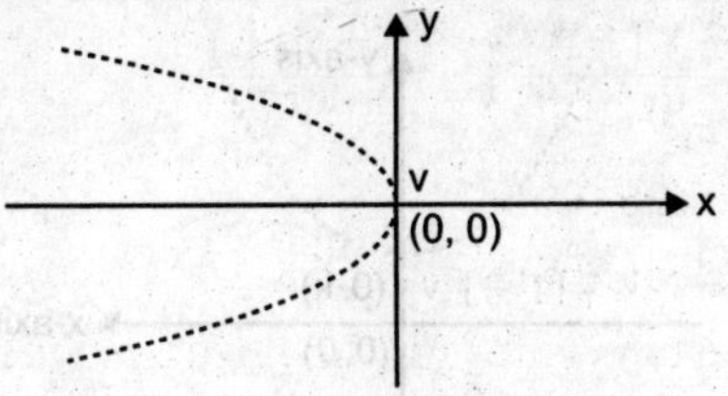

Fig. 3.4

(5) $y^2 = -4a(x - h); a, h > 0$

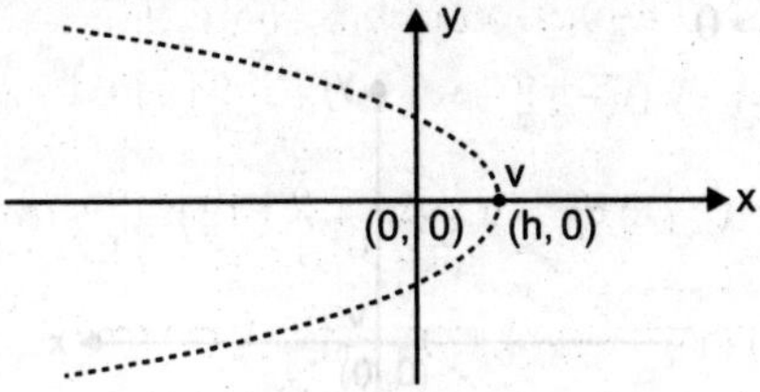

Fig. 3.5

(6) $y^2 = -4a(x + h); a, h > 0$

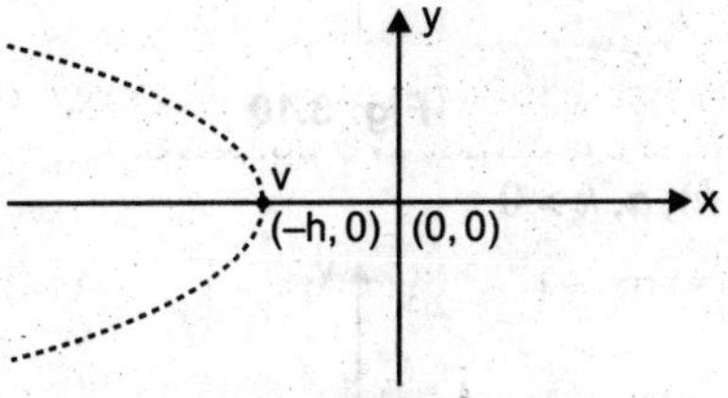

Fig. 3.6

(7) $x^2 = 4ay; a > 0$

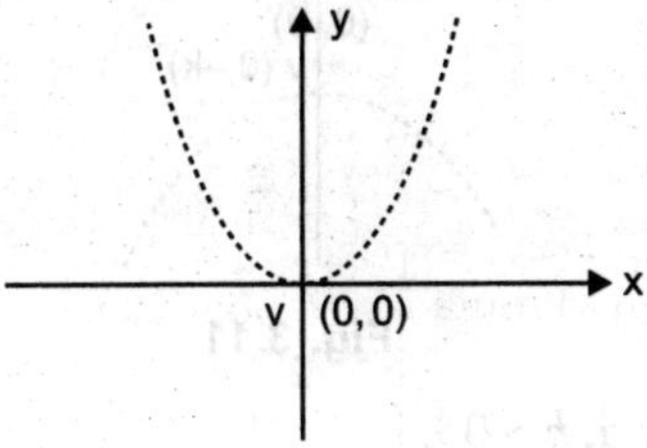

Fig. 3.7

(8) $x^2 = 4a(y + k); a > 0, k > 0$

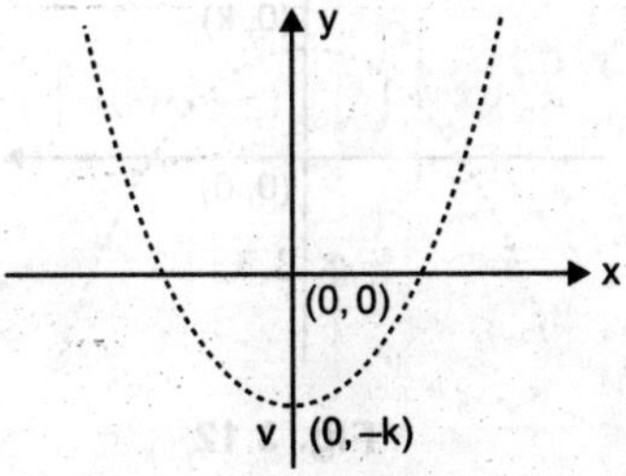

Fig. 3.8

(9) $x^2 = 4a(y - k); a, k > 0$

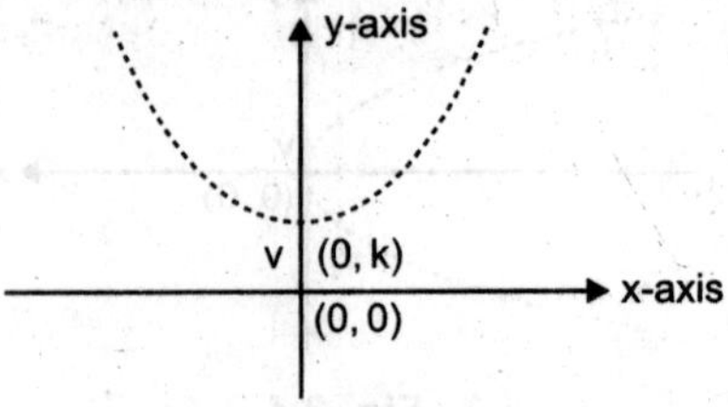

Fig. 3.9

(10) $x^2 = -4ay; a > 0$

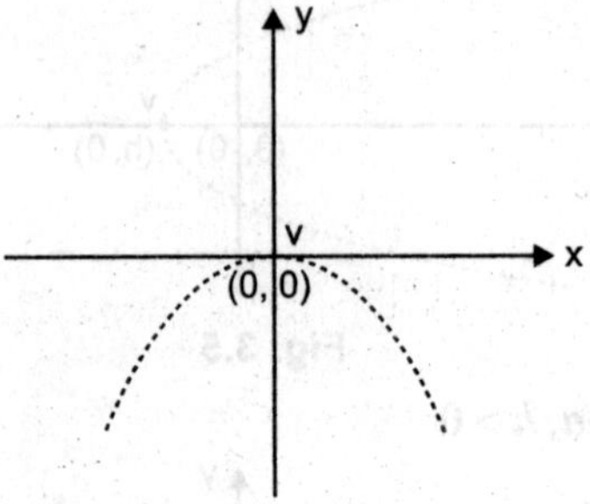

Fig. 3.10

(11) $x^2 = -4a(y + k); a, k > 0$

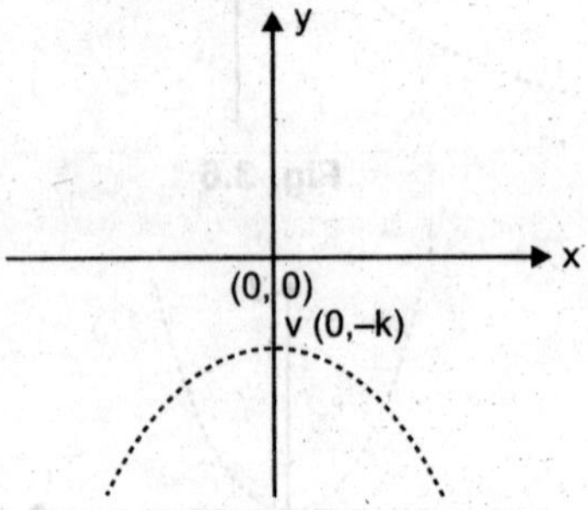

Fig. 3.11

(12) $x^2 = -4a(y - k); a, k > 0$

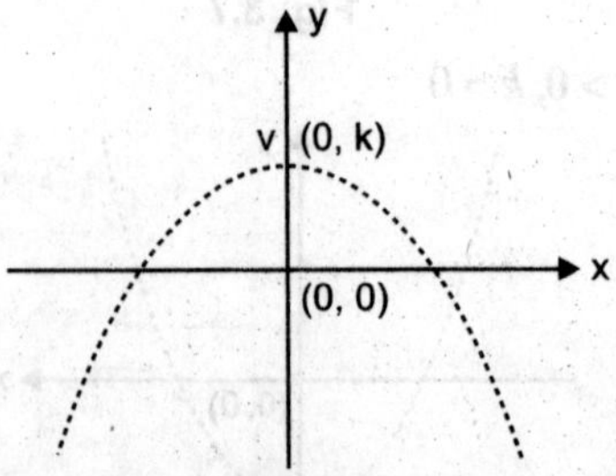

Fig. 3.12

(iv) Ellipse

(1) $\dfrac{x^2}{a^2} + \dfrac{y^2}{b^2} = 1; a > 0, b > 0$ (Standard equation of the ellipse)

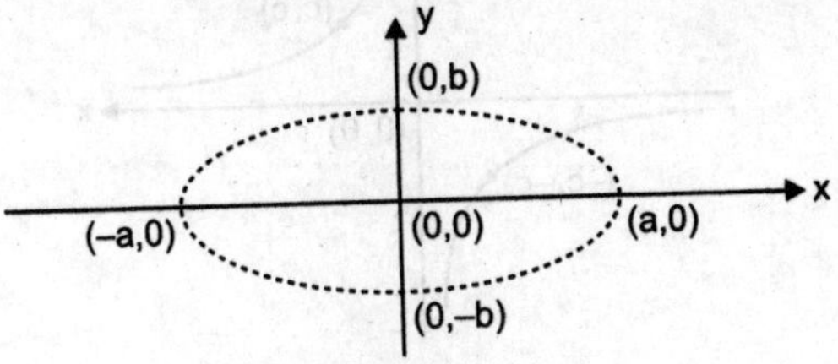

Fig. 3.13

(2) $\dfrac{x^2}{a^2} + \dfrac{y^2}{b^2} = 1; b > a > 0$ (Conjugate ellipse)

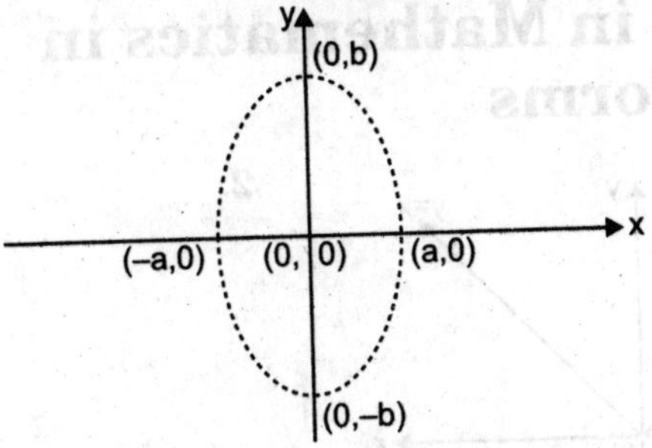

Fig. 3.14

(v) Hyperbola

(1) $\dfrac{x^2}{a^2} - \dfrac{y^2}{b^2} = 1$ (Standard equation of hyperbola)

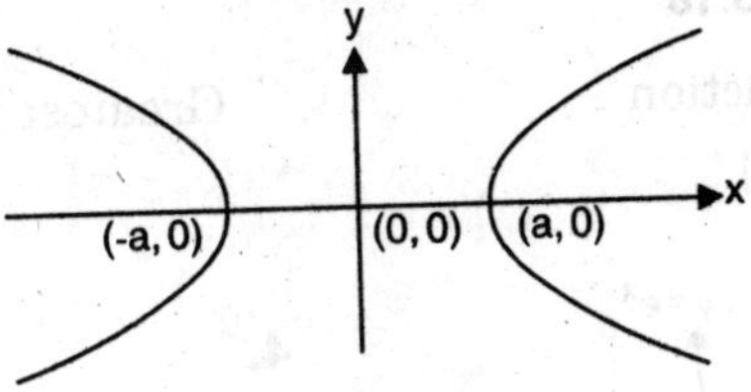

Fig. 3.15

(2) $\dfrac{y^2}{b^2} - \dfrac{x^2}{a^2} = 1$ (Conjugate hyperbola)

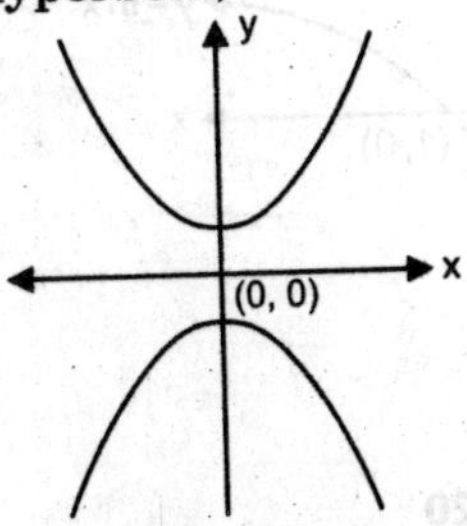

Fig. 3.16

(3) $xy = c^2$ **(Rectangular hyperbola)**

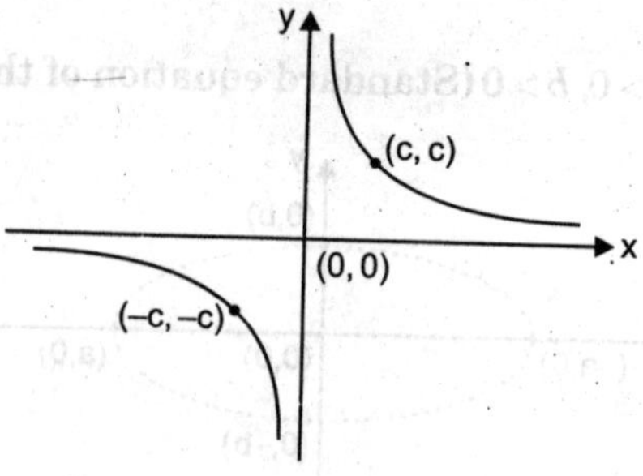

Fig. 3.17

Some More Curves Which Occur Frequently in Mathematics in Standard Forms

1.

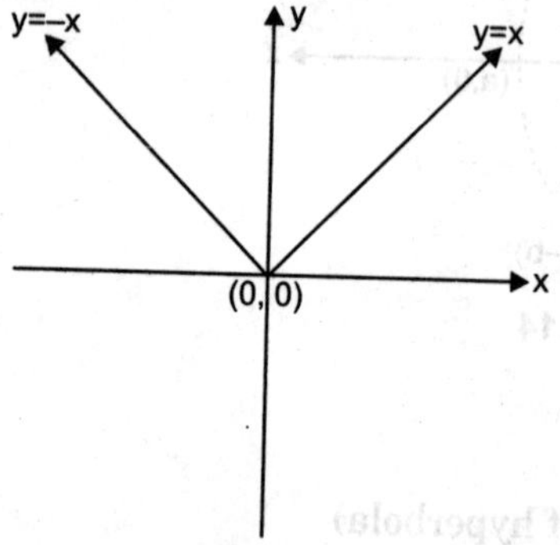

Fig. 3.18

Modulus function

$y = |x|$

2.

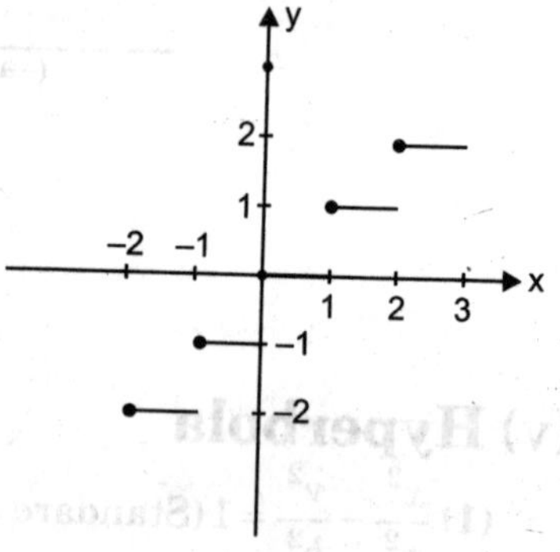

Fig. 3.19

Greatest integer function

$y = [x]$

3.

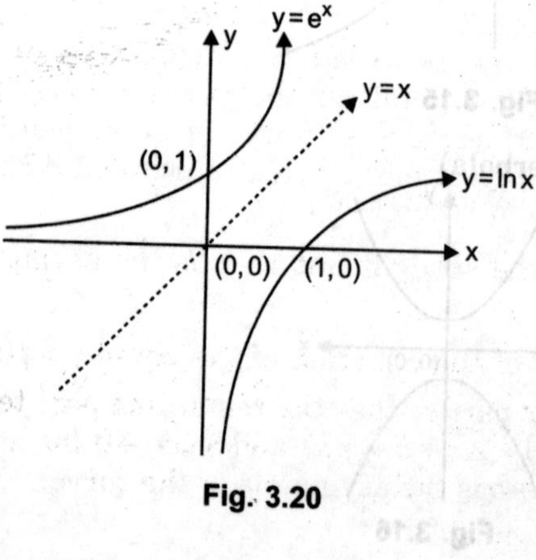

Fig. 3.20

4.

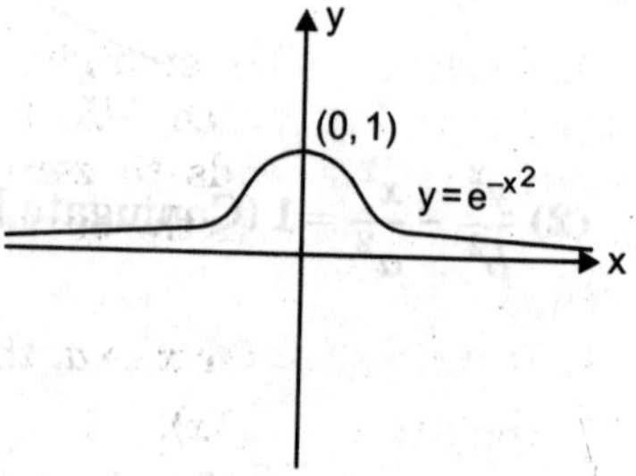

Fig. 3.21

5.

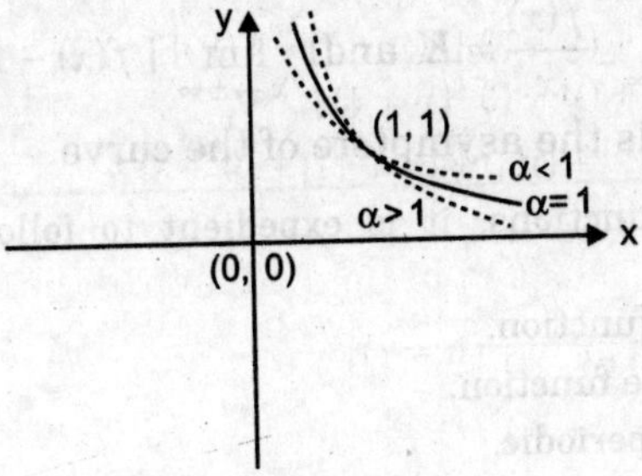

Fig. 3.22

$$y = \frac{1}{x^{\alpha}}, \alpha > 0$$

6.

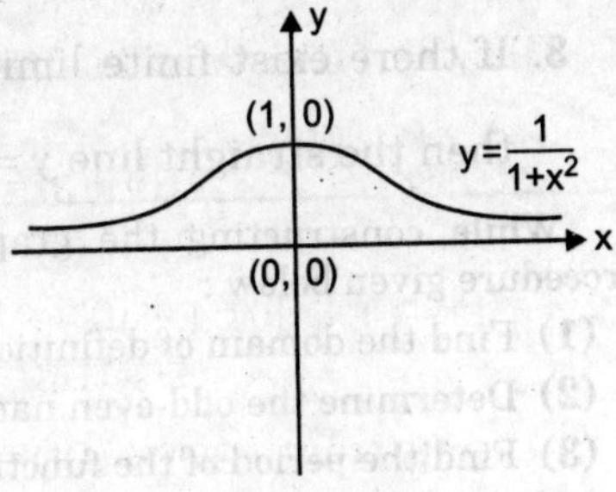

Fig. 3.23

7. The Astroid

Its cartesian equation is

$$x^{2/3} + y^{2/3} = a^{2/3}$$

Its parametric equation is

$$x = a\cos^3 t, y = a\sin^3 t$$

and it could be plotted as :

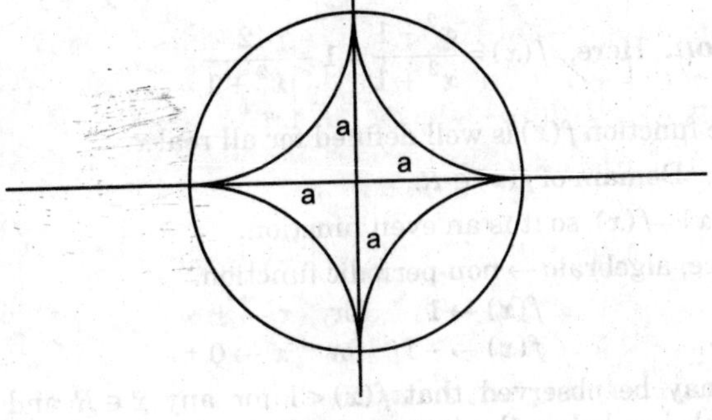

Fig. 3.24

Asymptotes

Definition The straight line *AB* is called the asymptote of curve $y = f(x)$, if the distance *MK* from *M* a point on the curve $y = f(x)$ to the straight line *AB* tends to zero as *M* recedes infinity. In other words, the straight line *AB* meets the curve $y = f(x)$ at infinity (*K* is a point on *AB*).

Thus,

1. If $f(x) \to \pm\infty$ for $x \to a$, then the straight line $x = a$ is the asymptote of the curve $y = f(x)$.
2. If in the right hand member of the equation of the curve $y = f(x)$ it is possible to single out a linear part so that the remaining part tends to zero as $x \to \pm\infty$, *ie*, if $y = f(x) = Kx + b + g(x)$ and $g(x) \to 0$ for $x \to \pm\infty$, then the straight line $y = Kx + b$ is the asymptote of the curve.

3. If there exist finite limits $\lim_{x \to \pm\infty} \frac{f(x)}{x} = K$ and $\lim_{x \to \pm\infty} [f(x) - Kx] = b$, then the straight line $y = Kx + b$ is the asymptote of the curve.

While constructing the graphs of functions, it is expedient to follow the procedure given below :

(1) Find the domain of definition of the function.

(2) Determine the odd-even nature of the function.

(3) Find the period of the function if its periodic.

(4) Find the asymptotes of the function.

(5) Check the behaviour of the function for $x \to 0\pm$

(6) Find the values of x, if possible for which $f(x) \to 0$.

(7) The interval of increase and decrease of the function in its range. Hence, determine the greatest and the least values of the function if any.

Points to Consider

(5), (6) and (7) gives the points where the function cuts the coordinate axes.

Illustration 8 Construct the graph for $f(x) = \frac{x^2 - 1}{x^2 + 1}$.

Solution. Here, $f(x) = \frac{x^2 - 1}{x^2 + 1} = 1 - \frac{2}{x^2 + 1}$

(1) The function $f(x)$ is well defined for all real x.

$\Rightarrow$ Domain of $f(x) \in R$.

(2) $f(-x) = f(x)$, so it is an even function.

(3) Since, algebraic $\to$ non-periodic function.

$$f(x) \to 1 \quad \text{for} \quad x \to \pm\infty$$

and

$$f(x) \to -1 \quad \text{for} \quad x \to 0\pm$$

It may be observed that $f(x) < 1$ for any $x \in R$ and consequently its graph lies below the line $y = 1$ which is asymptote to the graph of the given function.

Again, $\frac{2}{x^2 + 1}$ decreases for $(0, \infty)$ and increases for $(-\infty, 0)$, thus $f(x)$ increases for $(0, \infty)$ and decreases for $(-\infty, 0)$ in its range.

(4) The greatest value $\to 1$ for $x \to \pm\infty$ and the least value is -1 for $x = 0$.

Thus, its graph is as shown in figure.

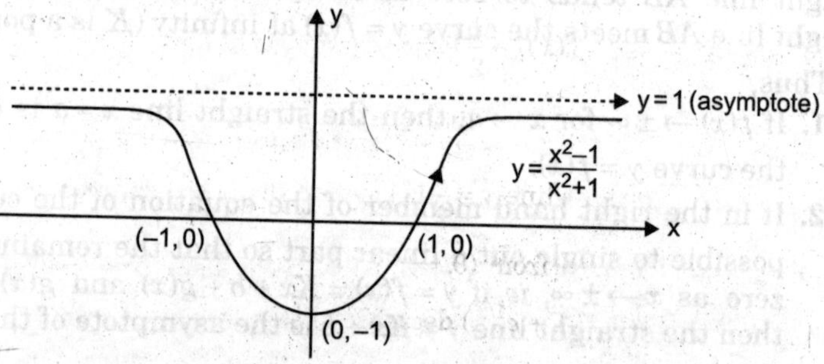

Illustration 9 Construct the graph for $f(x) = x + \frac{1}{x}$.

Solution. The function is defined for all x except for $x = 0$.

It is an odd function for $x \neq 0$.

It is not a periodic function.

For $x \to 0+$ $f(x) \to +\infty$, for $x \to 0-$ $f(x) \to -\infty$

For $x \to -\infty$ $f(x) \to -\infty$, for $x \to \infty$ $f(x) \to \infty$

$$\therefore \quad \lim_{x \to \pm\infty} (f(x) - x) = 0$$

$\therefore$ The straight lines $x = 0$ and $y = x$ are the asymptotes of the graph of the given function.

Now, consider $f(x_2) - f(x_1)$ (for $x_2 > x_1$)

$$= (x_2 - x_1) + \frac{1}{x_2} - \frac{1}{x_1}$$

$$= (x_2 - x_1)\left[1 - \frac{1}{x_1 x_2}\right] < 0 \text{ for } x_1\, x_2 \in (0,1]$$

and it is > 0 for $x_1\, x_2 \in [1, \infty)$.

Thus, $f(x)$ increases for $x \in [1, \infty)$ and decreases for $x \in (0, 1]$.

Thus, the least value of the function is at $x = 1$ which is $f(1) = 2$. Thus, its graph can be drawn as :

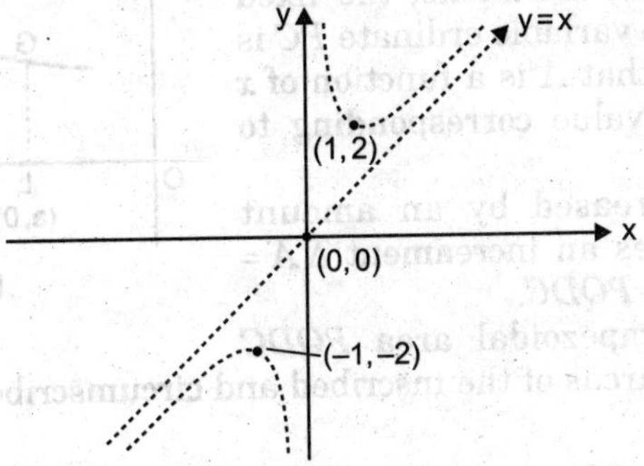

Illustration 10 Construct the graph for $f(x) = \frac{1}{1 + e^{1/x}}$.

Solution. The function is defined for all x except for $x = 0$. It is neither even nor an odd function. It is not a periodic function.

For $x \to 0+$ $f(x) \to 0$ for $x \to 0-$ $f(x) \to 1$

For $x \to \infty$ $f(x) \to \frac{1}{2}$ for $x \to -\infty$ $f(x) \to \frac{1}{2}$

$$\therefore \quad \lim_{x \to \pm\infty} f(x) = \frac{1}{2}$$

$\therefore$ The straight line $y = \frac{1}{2}$ is asymptote of the graph of the given function.

As x increases from $(0, \infty)$, $\frac{1}{x}$ decreases from $(0, \infty)$ and $e^{1/x}$ decreases from $(0, \infty)$. Thus, $(1 + e^{1/x})$ decreases from $(2, \infty)$.

$\therefore$ $f(x)$ increases from $\left(0, \frac{1}{2}\right)$ for $x \in (0, \infty)$.

Similarly, $f(x)$ increases from $(1/2, 1)$ for $x \to (-\infty, 0)$.

ie, $f(x)$ is an increasing function except for $x = 0$. Thus, its graph can be drawn as shown in figure.

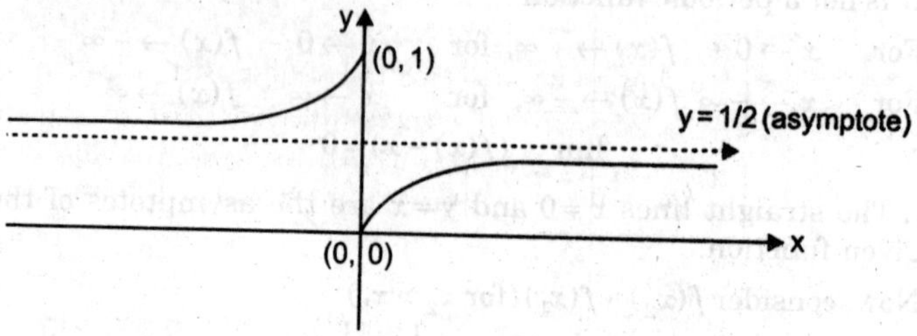

Areas of Curves Given by Cartesian Equations

Let $y = f(x)$ be a finite and continuous function in the interval $[a, b]$.

Let the variable area $GLCP$, bounded by the curve $y = f(x)$, the x-axis, the fixed ordinate GL and the variable ordinate PC is A. Thus, it is clear that A is a function of x and has a unique value corresponding to each value of x.

When x is increased by an amount $\Delta x (= CD)$, A assumes an increament $\Delta A =$ the trapezoidal area $PQDC$.

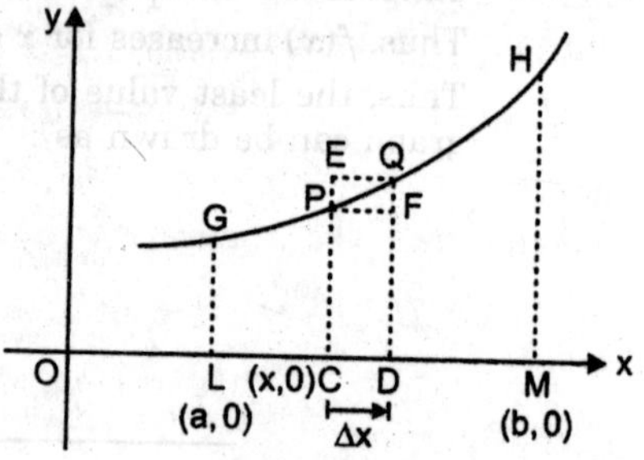

Fig. 3.25

Clearly, this trapezoidal area $PQDC$ lies in between the areas of the inscribed and circumscribed rectangles $PFDC$ and $EQDC$.

$$OL = a, OM = b \text{ and } OC = x$$

ie, $$\text{Ar}(PFDC) < \text{Ar}(PQDC) < \text{Ar}(EQDC)$$

ie, $$f(x) \cdot \Delta x < \Delta A < f(x + \Delta x) \cdot (\Delta x)$$

$\therefore$ $$f(x) < \frac{\Delta A}{\Delta x} < f(x + \Delta x)$$

Taking the limit $\Delta x \to 0$, we have

$$\lim_{\Delta x \to 0} f(x) < \lim_{\Delta x \to 0} \frac{\Delta A}{\Delta x} < \lim_{\Delta x \to 0} f(x + \Delta x)$$

ie, $$f(x) < \frac{dA}{dx} < f(x)$$

[because of the continuity of $f(x)$ at x, we have $\lim_{\Delta x \to 0} f(x + \Delta x) = f(x)$]

ie, $$\frac{dA}{dx} = f(x)$$

$\therefore$ By definition $dA = f(x)\, dx$

ie, $A = \int f(x)\,dx = \phi(x) + c$, where c is an arbitrary constant of integration and $\phi(x)$ is an indefinite integral of $f(x)$.

Now, when $x = a$, PC coincides with GL and the area becomes zero.

Also, when $x = OM = b$, the area A becomes the required area A_1

$$\therefore \quad 0 = \phi(a) + c \quad \text{and} \quad A_1 = \phi(b) + c$$

$$\therefore \quad A_1 = \phi(b) - \phi(a) = \int_a^b f(x)\,dx$$

Corollary In the same way it can be shown that the area bounded by any curve $x = g(y)$, y-axis and two abscissae $y = c$ and $y = d$ is

$$\int_c^d x\,dy, \quad ie, \int_c^d g(y)\,dy$$

Sign Convention For Finding the Areas Using Integration

We can tactfully apply the concept of definite integration to find the area enclosed between the curves. But then we must be very careful, while applying the discussed sign convention we will discuss the three cases :

Case I If in the expression $\int_a^b f(x)\,dx$, if $b > a$ and $f(x) > 0$ for all $a \le x \le b$, then this integration will give the area enclosed between the curve $f(x)$, x-axis and the line $x = a$ and $x = b$ which is positive. No need of any modification.

Case II If in the expression $\int_a^b f(x)\,dx$, if $b > a$ and $f(x) < 0$ for all $a \le x \le b$, then this integration will calculate to be negative. But the numerical or the absolute value is to be taken to mean the area enclosed between the curve $y = f(x)$, x-axis and the lines $x = a$ and $x = b$.

Case III If in the expression $\int_a^b f(x)\,dx$, where $b > a$ but $f(x)$ changes its sign a numbers of times in the interval $a \le x \le b$, then we must divide the region $[a, b]$ in such a way that we clearly get the points lying between $[a, b]$ where $f(x)$ changes its sign. For the region where $f(x) > 0$ we just integrate to get the area in that region and then add the absolute value of the integration calculated in the region where $f(x) < 0$ to get the desired area between the curve $y = f(x)$, x-axis and the line $x = a$ and $x = b$.

Hence, if $f(x)$ is as shown in Fig 3.26, the area enclosed by $y = f(x)$; x-axis and the lines $x = a$ and $x = b$ is given by :

$$A = \int_a^c f(x)\,dx + \left|\int_c^d f(x)\,dx\right| + \int_d^e f(x)\,dx + \left|\int_e^f f(x)\,dx\right| + \int_f^b f(x)\,dx$$

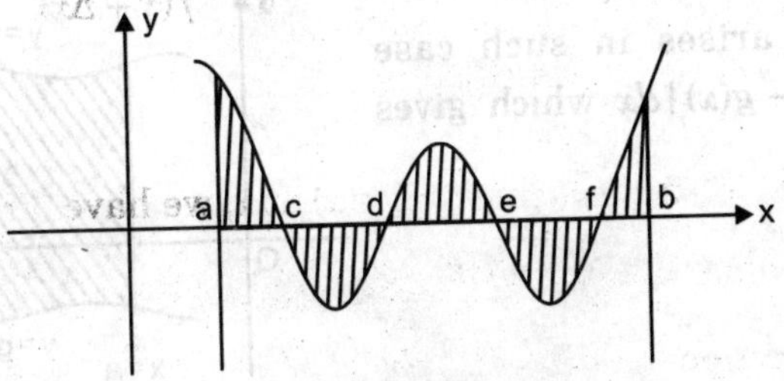

Fig. 3.26

Area Between the Given Curves

Area bounded by the curves $y = f(x)$, $y = g(x)$ and the lines $x = a$ and $y = b$.

Let the curves $y = f(x)$ and $y = g(x)$ be represented by AB and CD respectively. We assume that the two curves do not intersect each other in the interval $[a, b]$.

Thus, the shaded area = Area of curvilinear trapezoid $APQB$ − Area of curvilinear trapezoid $CPQD$.

$$= \int_a^b f(x)\,dx - \int_a^b g(x)\,dx = \int_a^b \{f(x) - g(x)\}\,dx$$

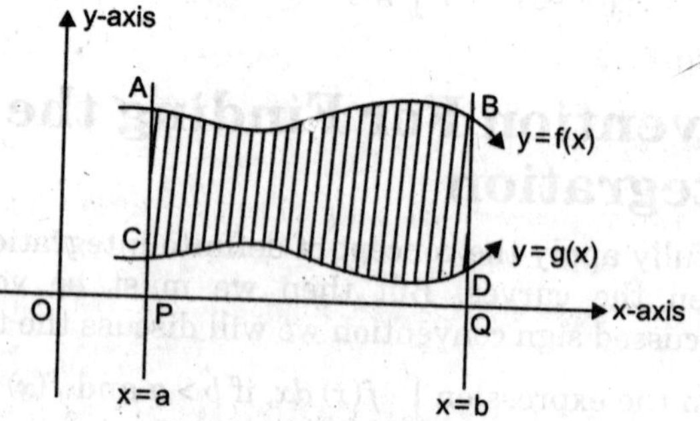

Fig. 3.27

Now, consider the case when $f(x)$ and $g(x)$ intersect each other in the interval $[a, b]$.

First of all we should find the intersection point of $y = f(x)$ and $y = g(x)$. For that we solve $f(x) = g(x)$. Let the root is $x = c$.

(We consider only one intersection point to illustrate the phenomenon).

Thus, the required (shaded) area $= \int_a^c \{f(x) - g(x)\}\,dx + \int_c^b \{g(x) - f(x)\}\,dx$

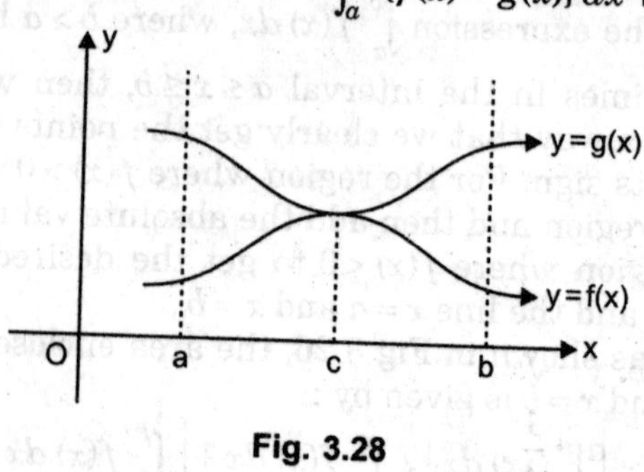

Fig. 3.28

If confusion arises in such case evaluate $\int_a^b |f(x) - g(x)|\,dx$ which gives the required area.

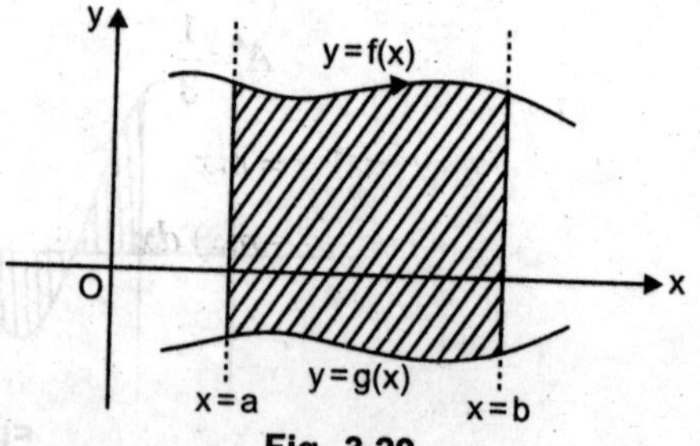

Fig. 3.29

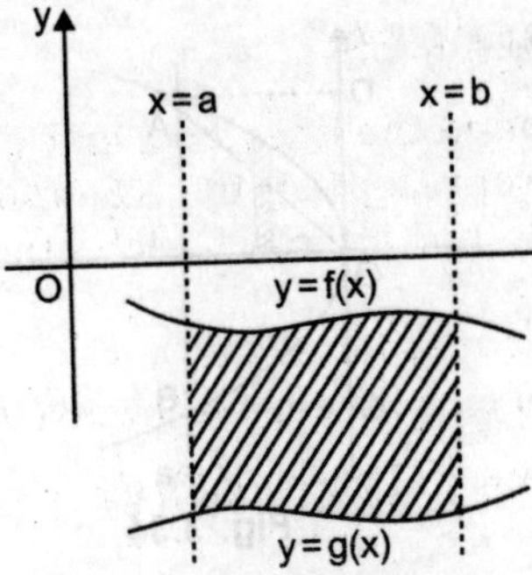

Fig. 3.30

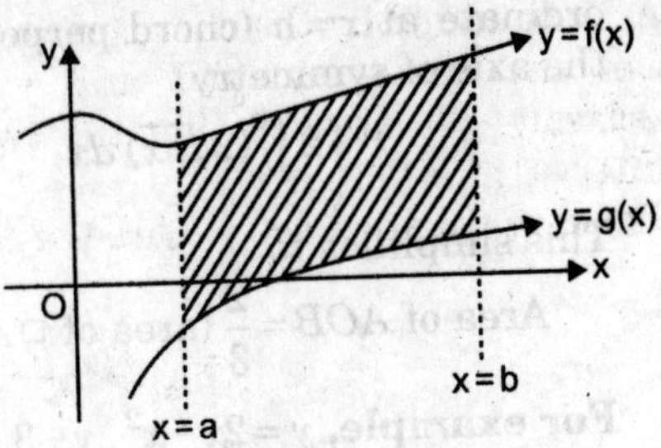

Fig. 3.31

Points to Consider

Area between two curves $y = f(x)$, $y = g(x)$ and the lines $x = a$ and $x = b$ is always given by $\int_a^b \{f(x) - g(x)\}\, dx$ provided $f(x) > g(x)$ in $[a, b]$; the position of the graph is immaterial. As shown in Fig. 3.29, Fig. 3.30, Fig. 3.31.

Standard Areas to Be Remembered For the Purpose of Solving Objective Questions

1. $y^2 = 4ax$, $x^2 = 4by$; $a > 0, b > 0$

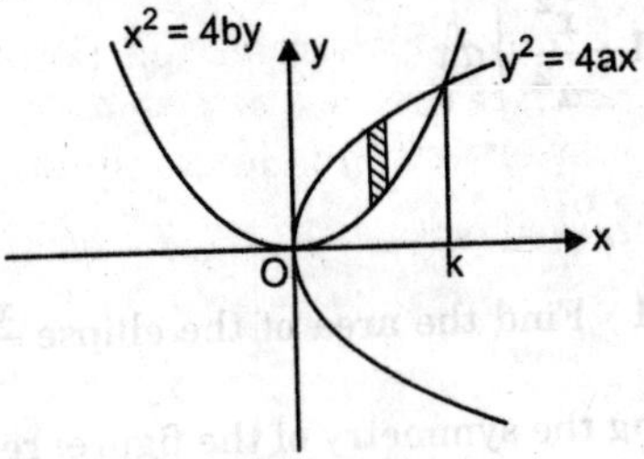

Fig. 3.32

$$A = \int_0^k \left(2\sqrt{a}\sqrt{x} - \frac{x^2}{4b}\right) dx \quad \Rightarrow \quad A = \frac{16ab}{3}$$

For example, $y = \sqrt{x}$, $x = \sqrt{y}$

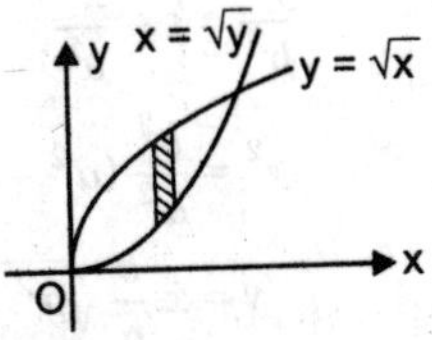

$$a = \frac{1}{4},\ b = \frac{1}{4}$$

$$A = \frac{1}{3}$$

2. $y^2 = 4ax$ and $y = mx$

$$A = \int_0^c (2\sqrt{a}\sqrt{x} - mx)\, dx$$

$$A = \frac{8a^2}{3m^3}$$

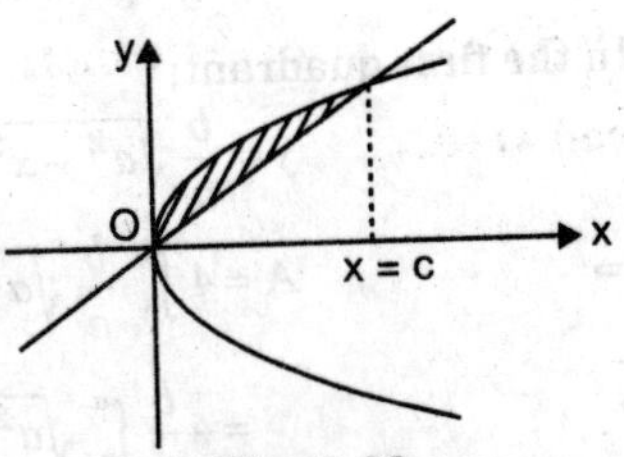

Fig. 3.33

3. Area enclosed by $y^2 = 4ax$ and its double ordinate at $x = a$ (chord perpendicular to the axis of symmetry)

$$2\int_0^a (2\sqrt{ax})\,dx$$

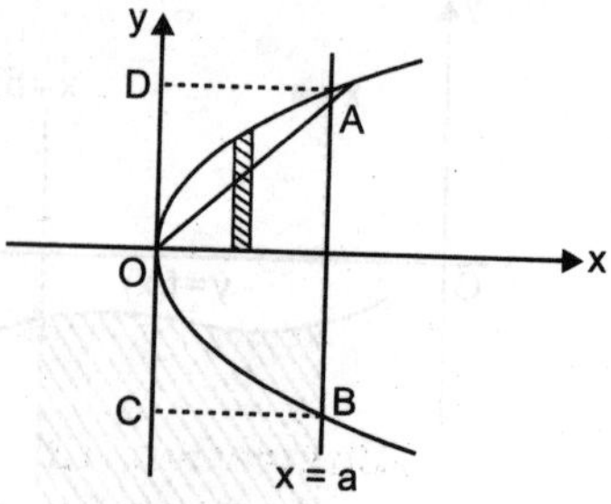

Fig. 3.34

This simplifies to

$$\text{Area of } AOB = \frac{2}{3}(\text{area of } \square ABCD)$$

For example, $y = 2x - x^2$, $y + 3 = 0$

Solving then $x = -1$ or 3

$$\text{Area } (\square ABCD) = 4 \times 4 = 16$$

$$\text{Required area} = \frac{2}{3} \times 16 = \frac{32}{3}$$

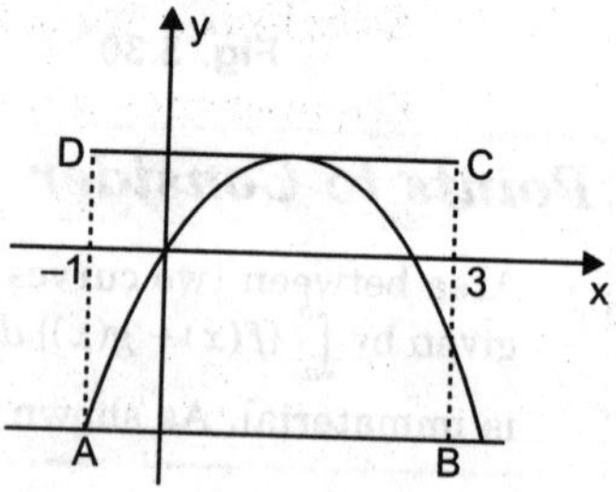

Aliter By integration

$$A = \int_{-1}^{3} [(2x - x^2) - (-3)]\,dx$$

4. Whole area of ellipse

$$\frac{x^2}{a^2} + \frac{y^2}{b^2} = 1$$

$$A = 4\int_0^a \left(b\sqrt{1 - \frac{x^2}{a^2}}\right) dx$$

$$= \pi a b$$

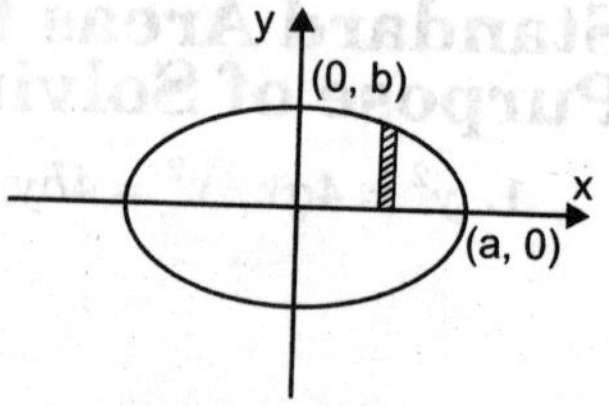

Fig. 3.35

Illustration 11 Find the area of the ellipse $\dfrac{x^2}{a^2} + \dfrac{y^2}{b^2} = 1$.

Solution. Using the symmetry of the figure; required area is given by;

$$A = 4\,(\text{area } OABO)$$

$$= 4\int_0^a y\,dx, \quad \text{where } \frac{x^2}{a^2} + \frac{y^2}{b^2} = 1$$

$$\therefore \quad \frac{y^2}{b^2} = 1 - \frac{x^2}{a^2}$$

$$y^2 = \frac{b^2}{a^2}(a^2 - x^2)$$

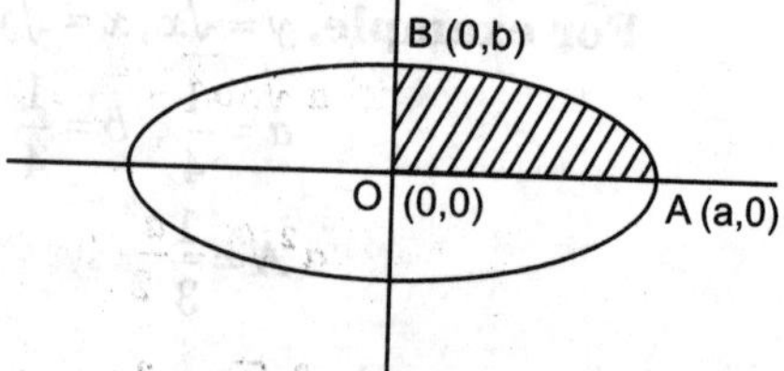

$$\therefore \quad y = \pm \frac{b}{a}\sqrt{a^2 - x^2}$$

In the first quadrant;

$$y = \frac{b}{a}\sqrt{a^2 - x^2}$$

$$\Rightarrow \quad A = 4\int_0^a \frac{b}{a}\sqrt{a^2 - x^2}\,dx$$

$$= 4\frac{b}{a}\int_0^a \sqrt{a^2 - x^2}\,dx = 4\frac{b}{a}\left\{\frac{x}{2}\sqrt{a^2 - x^2} + \frac{a^2}{2}\sin^{-1}\frac{x}{a}\right\}_0^a$$

$$= 4\frac{b}{a}\left[\left\{0 + \frac{a^2}{2}\sin^{-1}\frac{a}{a}\right\} - \{0 + 0\}\right] = \frac{4b}{a}\cdot\frac{a^2}{2}\sin^{-1}(1)$$

$$= 2ab\left(\frac{\pi}{2}\right)$$

$$A = \pi\, ab \text{ sq units}$$

Illustration 12 Find the area bounded by the hyperbola $x^2 - y^2 = a^2$ between the straight lines $x = a$ and $x = 2a$.

Solution. We use the symmetry of figure.

Required area $A = 2\int_a^{2a} y\,dx$, where $x^2 - y^2 = a^2$

ie, $x^2 - a^2 = y^2$

$\therefore$ $y = \pm\sqrt{x^2 - a^2}$

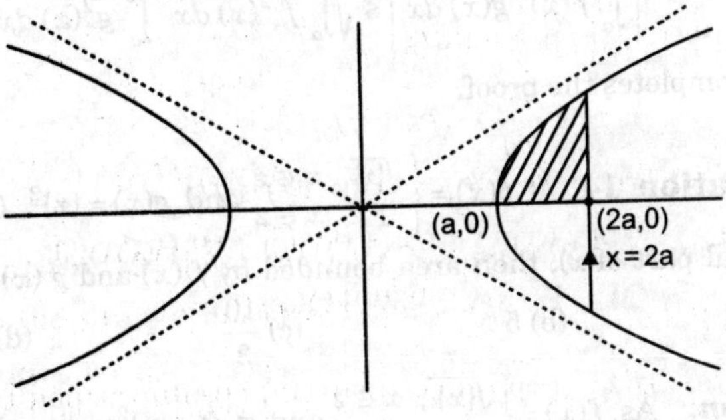

In the first quadrant; $y = +\sqrt{x^2 - a^2}$

$$A = 2\int_a^{2a}\sqrt{x^2 - a^2}\,dx$$

$$= 2\left[\frac{x}{2}\sqrt{x^2 - a^2} - \frac{a^2}{2}\log\left(x + \sqrt{x^2 - a^2}\right)\right]_a^{2a}$$

$$= 2\left[\left\{a\sqrt{4a^2 - a^2} - \frac{a^2}{2}\log\left(2a + \sqrt{4a^2 - a^2}\right)\right\} - \left\{0 - \frac{a^2}{2}\log a\right\}\right]$$

$$= 2\left[a\sqrt{3a^2} - \frac{a^2}{2}\log(2a + a\sqrt{3}) + \frac{a^2}{2}\log a\right]$$

$$= 2\left[a^2\sqrt{3} - \frac{a^2}{2}\log\left(\frac{2a + a\sqrt{3}}{a}\right)\right]$$

$\therefore$ $A = 2a^2\sqrt{3} - a^2\log(2 + \sqrt{3})$ sq units

Illustration 13 Prove that for any functions $f(x)$ and $g(x)$, integrable on the interval (a, b), the inequality takes place

$$\left|\int_a^b f(x)\,g(x)\,dx\right| \le \sqrt{\int_a^b f^2(x)\,dx \int_a^b g^2(x)\,dx}.$$

Solution. Consider the function; $F(x) = \{f(x) - \lambda g(x)\}^2$ where λ is any real number. Since, $F(x) \geq 0$, then

$$\int_a^b \{f(x) - \lambda g(x)\}^2 \geq 0$$

or
$$\lambda^2 \int_a^b g^2(x)\,dx - 2\lambda \int_a^b f(x)\cdot g(x)\,dx + \int_a^b f^2(x)\,dx \geq 0$$

The expression in the left side of the latter inequality is a quadratic trinomial with respect to λ. It follows from the inequality that at any λ this trinomial is non-negative.

Hence, its discriminant is non-positive, *ie*,

$$\left\{-2\int_a^b f(x)\cdot g(x)\,dx\right\}^2 - 4\int_a^b g^2(x)\,dx \cdot \int_a^b f^2(x)\,dx \leq 0$$

or
$$\left\{\int_a^b f(x)\cdot g(x)\,dx\right\}^2 \leq \int_a^b f^2(x)\,dx \cdot \int_a^b g^2(x)\,dx$$

Hence,
$$\left|\int_a^b f(x)\cdot g(x)\,dx\right| \leq \sqrt{\int_a^b f^2(x)\,dx \cdot \int_a^b g^2(x)\,dx}$$

which completes the proof.

Illustration 14 If $f(x) = \begin{cases} \sqrt{\{x\}}, & x \notin z \\ 1, & x \in z \end{cases}$ and $g(x) = \{x\}^2$, (where $\{.\}$ denotes fractional part of x), then area bounded by $f(x)$ and $g(x)$ for $x \in [0, 10]$ is

(a) $\frac{5}{3}$ (b) 5 (c) $\frac{10}{3}$ (d) None of these

Solution. As, $f(x) = \begin{cases} \sqrt{\{x\}}, & x \notin z \\ 1, & x \in z \end{cases}$ and $g(x) = \{x\}^2$, where both $f(x)$ and $g(x)$ are periodic with period '1' shown as;

Thus, required area $= 10\int_0^1 [\sqrt{\{x\}} - \{x\}^2]\,dx$

$$= 10\int_0^1 [(x)^{1/2} - x^2]\,dx$$

$$= 10\left[\frac{x^{3/2}}{3/2} - \frac{x^3}{3}\right]_0^1$$

$$= 10\left(\frac{2}{3} - \frac{1}{3}\right) = \frac{10}{3} \text{ sq units}$$

Hence, (c) is the correct answer.

Illustration 15 The area enclosed by the curve $|y| = \sin 2x$, when $x \in [0, 2\pi]$ is

(a) 1 (b) 2 (c) 3 (d) 4

Solution. As, we know $y = \sin 2x$ could be plotted as;

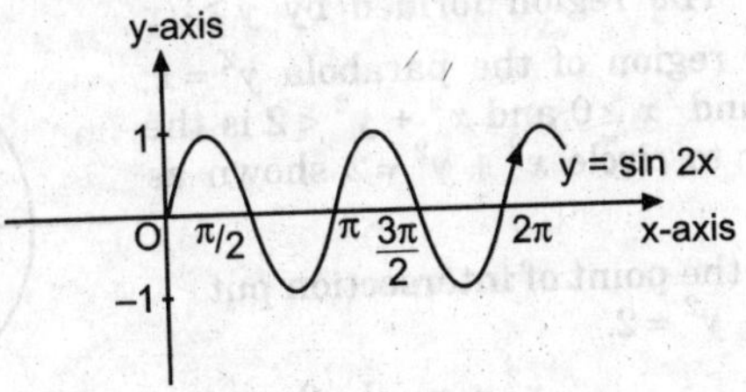

Thus, $|y| = \sin 2x$ is whenever positive, y can have both positive and negative values, *ie*, the curve is symmetric about the axes.
$\sin 2x$ is positive only in $0 \le x \le \frac{\pi}{2}$ and $\pi \le x \le \frac{3\pi}{2}$. Thus, the curve consists of two loops one in $\left[0, \frac{\pi}{2}\right]$ and another in $\left[\pi, \frac{3\pi}{2}\right]$.

Thus, $\quad \text{Area} = 4\int_0^{\pi/2} (\sin 2x)\, dx$

$$= 4\left(-\frac{\cos 2x}{2}\right)_0^{\pi/2}$$

$$= -2(\cos \pi - \cos 0)$$

$$= -2(-1-1) = 4$$

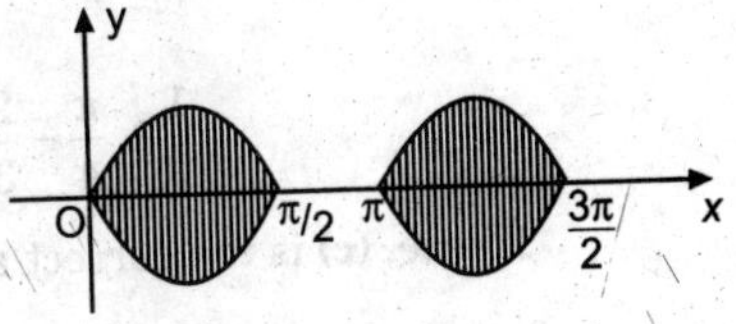

Hence, (d) is the correct answer.

Illustration 16 Let $f(x) = x^2$, $g(x) = \cos x$ and $\alpha, \beta\ (\alpha < \beta)$ be the roots of the equation $18x^2 - 9\pi x + \pi^2 = 0$. Then, the area bounded by the curves $y = fog(x)$, the ordinates $x = \alpha$, $x = \beta$ and the x-axis is

(a) $\frac{1}{2}(\pi - 3)$ (b) $\frac{\pi}{3}$ (c) $\frac{\pi}{4}$ (d) $\frac{\pi}{12}$

Solution. Here, $y = fog(x) = f\{g(x)\} = (\cos x)^2 = \cos^2 x$

Also, $\quad 18x^2 - 9\pi x + \pi^2 = 0$

$\Rightarrow \quad (3x - \pi)(6x - \pi) = 0$

$\Rightarrow \quad x = \frac{\pi}{6}, \frac{\pi}{3}$ (as α, β)

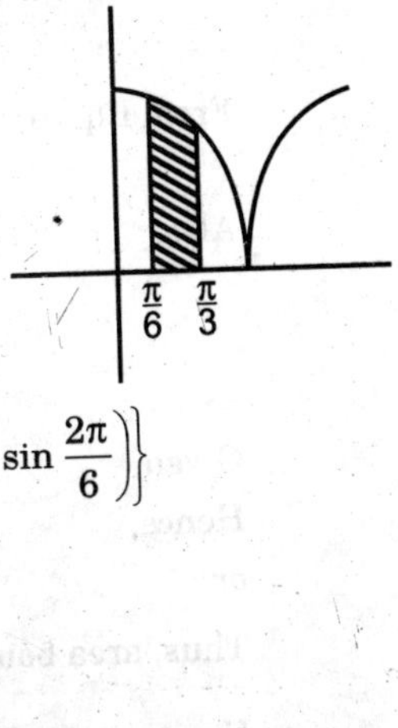

$\therefore$ Required area of curve is

$$= \int_{\pi/6}^{\pi/3} \cos^2 x\, dx = \frac{1}{2}\int_{\pi/6}^{\pi/3} (1 + \cos 2x)\, dx$$

$$= \frac{1}{2}\left\{x + \frac{\sin 2x}{2}\right\}_{\pi/6}^{\pi/3} = \frac{1}{2}\left\{\left(\frac{\pi}{3} - \frac{\pi}{6}\right) + \frac{1}{2}\left(\sin\frac{2\pi}{3} - \sin\frac{2\pi}{6}\right)\right\}$$

$$= \frac{1}{2}\left\{\frac{\pi}{6} + \frac{1}{2}\left(\frac{\sqrt{3}}{2} - \frac{\sqrt{3}}{2}\right)\right\} = \frac{\pi}{12}$$

Hence, (d) is the correct answer.

Illustration 17 The area common to the region determined by $y \geq \sqrt{x}$ and $x^2 + y^2 < 2$ has the value

(a) π (b) $(2\pi - 1)$ (c) $\frac{\pi}{4} - \frac{1}{6}$ (d) None of these

Solution. The region formed by $y \geq \sqrt{x}$ is the outer region of the parabola $y^2 = x$, when $y \geq 0$ and $x \geq 0$ and $x^2 + y^2 < 2$ is the region inner to circle $x^2 + y^2 = 2$ shown as in figure.

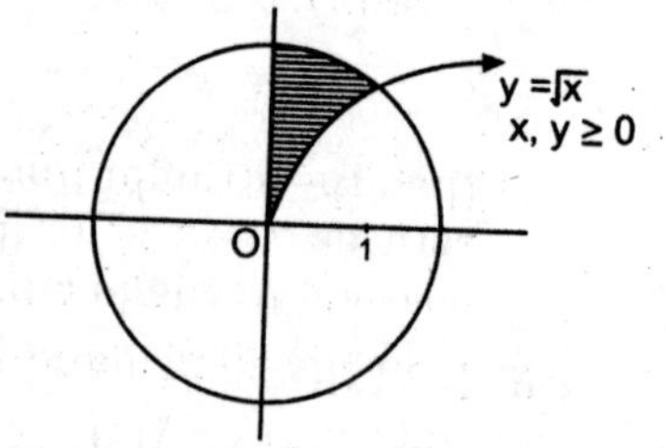

Now, to find the point of intersection put $y^2 = x$ in $x^2 + y^2 = 2$.

$\Rightarrow$ $$x^2 + x - 2 = 0$$

$\Rightarrow$ $$(x + 2)(x - 1) = 0$$

$\Rightarrow$ $$x = 1, \text{ as } x \geq 0$$

$\therefore$ The required area is

$$= \int_0^1 (\sqrt{2 - x^2} - \sqrt{x})\, dx = \left[\frac{x\sqrt{2 - x^2}}{2} + \sin^{-1}\frac{x}{\sqrt{2}} - \frac{2}{\sqrt{3}}x^{3/2}\right]_0^1$$

$$= \frac{1}{2} + \frac{\pi}{4} - \frac{2}{3} = \left(\frac{\pi}{4} - \frac{1}{6}\right)$$

Hence, (c) is the correct answer.

Illustration 18 Let 'n' be a positive integer and $f(n) = 1! + 2! + \ldots + n!$ also $p(x)$ and $q(x)$ are polynomials such that $f(n + 2) = p(n) f(n + 1) + q(n) f(n)$ for all $n \geq 1$, then

(a) area bounded by $p(x)$ and $q(x)$ and y-axis is $\frac{25}{4}$ sq units

(b) polynomial $p(x)$ is $(x - 3)$

(c) polynomial $p(x)$ is $-(x + 2)$

(d) polynomial $p(x)$ is $-(x + 3)$

Solution. Since, $f(n) = 1! + 2! + \ldots + n!$ ...(i)

$\therefore$ $$f(n + 1) = 1! + 2! + \ldots + (n + 1)! \quad \ldots\text{(ii)}$$

From Eq. (ii) – Eq. (i), we get

$$f(n + 1) - f(n) = (n + 1)! \quad \ldots\text{(iii)}$$

Also, $$f(n + 2) = 1! + 2! + \ldots + (n!) + (n + 1)! + (n + 2)!$$

$$= f(n + 1) + (n + 2) \cdot (n + 1)!$$

$$= f(n + 1) + (n + 2)\{f(n + 1) - f(n)\}$$

$\Rightarrow$ $$f(n + 2) = (n + 3) f(n + 1) - (n + 2) f(n) \quad \ldots\text{(iv)}$$

Given, $$f(n + 2) = p(n) f(n + 1) + q(n) f(n)$$

Hence, $$p(n) = n + 3 \text{ and } q(n) = -n - 2$$

or $$p(x) = x + 3 \text{ and } q(x) = -2 - x$$

Thus, area bounded by $p(x)$ and $q(x)$ and y-axis $= \frac{1}{2} \cdot 5 \cdot \frac{5}{2} = \frac{25}{4}$ sq units

Hence, (a) is the correct answer.

Illustration 19 The graph of $y^2 + 2xy + 40|x| = 400$ divides the plane into regions. The area of the bounded region is

(a) 400 sq units (b) 800 sq units

(c) 600 sq units (d) None of these

Solution. Here,

$$y^2 + 2xy + 40|x| = 400$$

can be factorised as,

$$\begin{cases}(y+20)(y+2x-20)=0, & \text{for } x \geq 0\\(y-20)(y+2x+20)=0, & \text{for } x < 0\end{cases}$$

The bounded region is parallelogram shown as in figure.

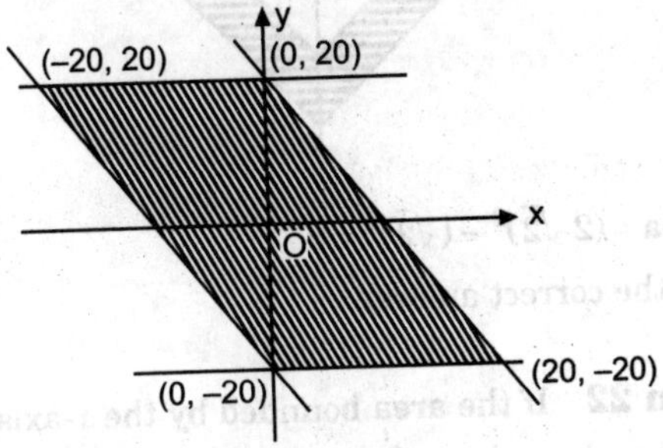

Thus, required area = $20 \times 40 = 800$ sq units

Hence, (b) is the correct answer.

Illustration 20 The area of the region defined by $||x| - |y|| \leq 1$ and $x^2 + y^2 \leq 1$ in the x-y coordinate plane is

(a) π (b) 2π

(c) 3π (d) 1

Solution. Here, $||x| - |y|| \leq 1$

$\Rightarrow$ $-1 \leq |x| - |y| \leq 1$ and $x^2 + y^2 \leq 1$ shown as

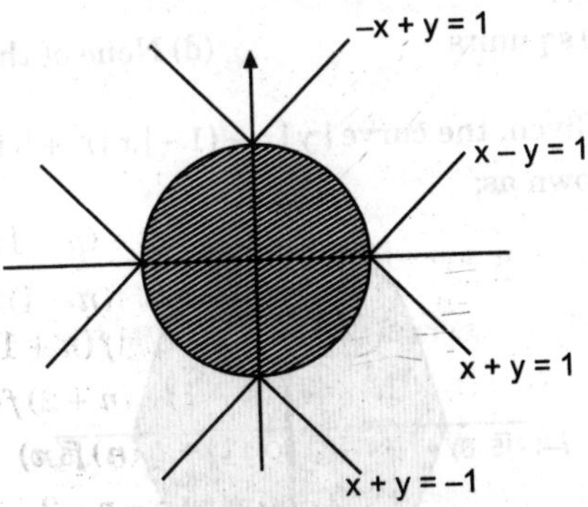

$\therefore$ Required area = area of circle = $\pi(1)^2 = \pi$

Hence, (a) is the correct answer.

Illustration 21 The area defined by $1 \leq |x-2| + |y+1| \leq 2$ is

(a) 2 (b) 4 (c) 6 (d) None of these

Solution. Put $x - 2 = x,\ y + 1 = y$

Then, the required region is defined by $1 \leq |x| + |y| \leq 2$.

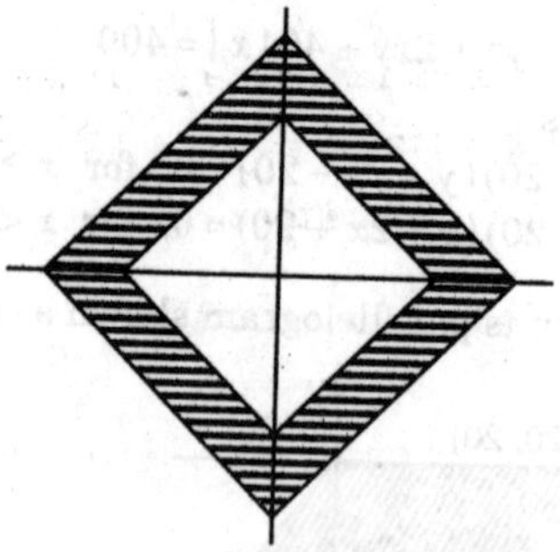

Required area $= (2\sqrt{2})^2 - (\sqrt{2})^2 = 6$

Hence, (c) is the correct answer.

Illustration 22 If the area bounded by the x-axis, the curve $y = f(x)$ and the lines $x = c$ and $x = d$ is independent of d, $\forall\ d > c$ (c is constant), then f is

(a) a non-zero constant function (b) the identity function

(c) the zero function (d) None of these

Solution. Here, required area $= \int_c^d y\,dx = \int_c^d f(x)\,dx$

which independent of d, it is possible only when $f(x) = 0$.

Hence, (c) is the correct answer.

Illustration 23 The area enclosed by the curve $|y| = -(1-|x|)^2 + 5$ is

(a) $\frac{8}{3}(7 + 5\sqrt{5})$ sq units (b) $\frac{2}{3}(7 + 5\sqrt{5})$ sq units

(c) $\frac{2}{3}(5\sqrt{5} - 7)$ sq units (d) None of these

Solution. Given, the curve $|y| = -(1-|x|)^2 + 5$ is symmetric about both and y-axes, shown as;

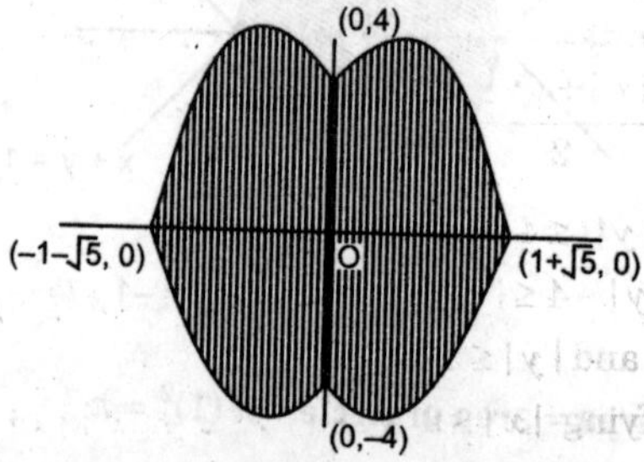

Thus, required area $= 4\int_0^{(\sqrt{5}+1)} y\,dx = 4\int_0^{\sqrt{5}+1} \{5-(x-1)^2\}\,dx$

$= \frac{8}{3}(7+5\sqrt{5})$ sq units

Hence, (a) is the correct answer.

Illustration 24 The area bounded by the curve

$$f(x) = ||\tan x + \cot x| - |\tan x - \cot x||$$

between the lines $x = 0$, $x = \frac{\pi}{2}$ and the x-axis, is

(a) $\log 4$ (b) $\log\sqrt{2}$ (c) $2\log 2$ (d) $\sqrt{2}\log 2$

Solution. Here,

$$f(x) = ||\tan x + \cot x| - |\tan x - \cot x|| = 2\cdot\min\{|\tan x|, |\cot x|\},$$

shown as;

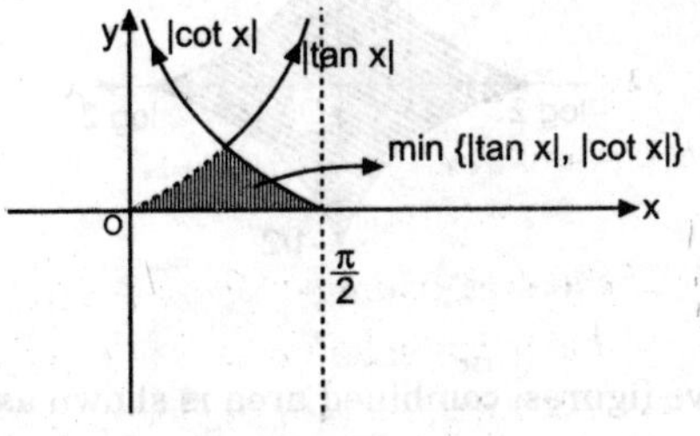

Thus, required area $= 2\int_0^{\pi/4}(\tan x)\,dx + 2\int_{\pi/4}^{\pi/2}(\cot x)\,dx$

$= 2(\log|\sec x|)_0^{\pi/4} + 2(\log|\sin x|)_{\pi/4}^{\pi/2}$

$= 2(\log\sqrt{2} - \log 1) + 2\left(\log 1 - \log\frac{1}{\sqrt{2}}\right)$

$= 2\left(\log\sqrt{2} - \log\frac{1}{\sqrt{2}}\right) = 2\log 2 = \log 4$

Hence, (a) is the correct answer.

Illustration 25 The area bounded by the curves,

$$|y| = e^{-|x|} - \frac{1}{2} \quad \text{and} \quad \frac{|x|+|y|}{2} + \left|\frac{|x|-|y|}{2}\right| \le 2 \text{ is}$$

(a) $(7+\log 4)$ sq units (b) $(7-\log 2)$ sq units
(c) $(14+2\log 2)$ sq units (d) $(14-2\log 2)$ sq units

Solution. Here, $\frac{|x|+|y|}{2} + \left|\frac{|x|-|y|}{2}\right| \le 2$

$\Rightarrow \quad ||x|-|y|| \le 4 - (|x|+|y|)$ (here, $|x|+|y| \le 4$)

$\Rightarrow \quad |x|+|y|-4 \le |x|-|y| \le 4-(|x|+|y|)$

$\Rightarrow \quad |x| \le 2$ and $|y| \le 2$

So, the curve satisfying $|x| \le 2$ and $|y| \le 2$ and $|x|+|y| \le 4$ is a square, shown as;

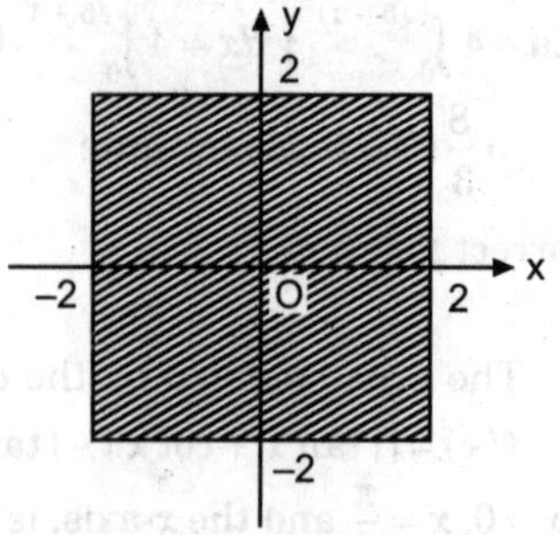

Also, the curve $|y| = e^{-|x|} - \frac{1}{2}$ is symmetric about x and y-axes. So, the curve is

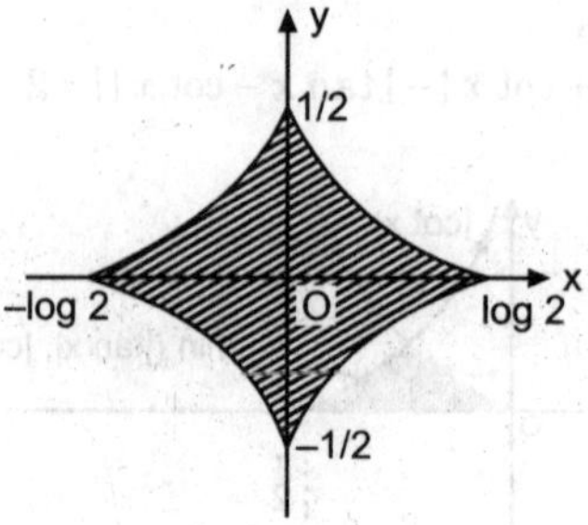

From above two figures, combined area is shown as;

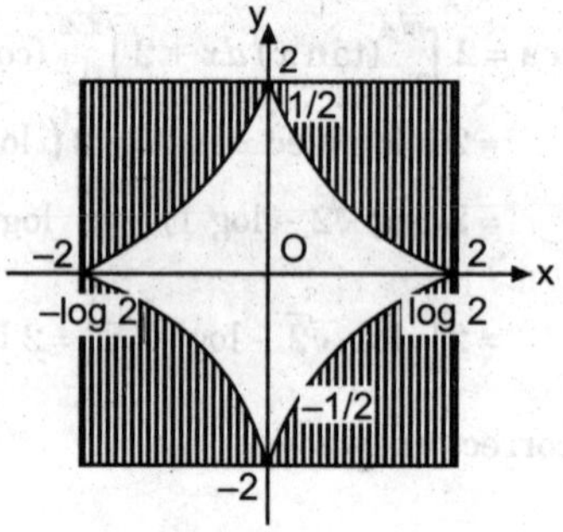

So, required area = Area of square $-4\int_0^{\log 2}\left(e^{-x} - \frac{1}{2}\right)dx$

$$= 16 - 4\left(-e^{-x} - \frac{1}{2}x\right)_0^{\log 2}$$

$$= (14 + \log 4) \text{ sq units.}$$

Hence, (c) is the correct answer.

Average Value of a Function

Average value of the function in $y = f(x)$ w.r.t. x over an interval $a \le x \le b$ is defined as

$$y_{av} = \frac{1}{b-a}\int_a^b f(x)\,dx$$

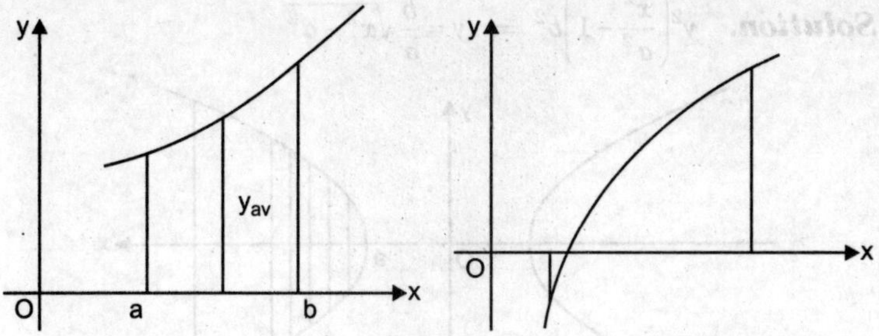

Fig. 3.36

AND

Root Mean Square Value (RMS) is defined as $\left[\rho = \frac{1}{b-a}\int_a^b f^2(x)\,dx\right]^{\frac{1}{2}}$

Points to Consider

1. Average value can be +ve, –ve or zero.
2. If the function is defined in $(0, \infty)$, then $y_{av} = \lim_{b \to \infty} \frac{1}{b}\int_0^b f(x)\,dx$ provided the limit exists.

Illustration 26 Compute the average value of

$$f(x) = \frac{\cos^2 x}{\sin^2 x + 4\cos^2 x} \text{ in } \left[0, \frac{\pi}{2}\right].$$

Solution. $y_{av} = \frac{1}{\frac{\pi}{2}-0}\int_0^{\pi/2} \frac{\cos^2 x}{\sin^2 x + 4\cos^2 x}dx$

$$= \frac{2}{\pi}\int_0^{\pi/2} \frac{\sec^2 x}{(4+\tan^2 x)(1+\tan^2 x)}dx$$

$$= \frac{2}{\pi}\int_0^{\infty} \frac{dt}{(4+t^2)(1+t^2)}$$

$$= \frac{2}{3\pi}\int_0^{\infty} \frac{(4+t^2)-(1+t^2)}{(4+t^2)(1+t^2)}dt$$

$$= \frac{2}{3\pi}\left\{\int_0^{\infty} \frac{1}{1+t^2}dt - \int_0^{\infty} \frac{1}{4+t^2}dt\right\}$$

$$= \frac{2}{3\pi}\left\{(\tan^{-1} t)_0^{\infty} - \frac{1}{2}\left(\tan^{-1}\frac{1}{2}\right)_0^{\infty}\right\}$$

$$= \frac{2}{3\pi}\left[\frac{\pi}{2} - \frac{1}{2}\cdot\frac{\pi}{2}\right] = \frac{1}{6}$$

Illustration 27 Find the average length of all vertical chords of the hyperbola $\frac{x^2}{a^2} - \frac{y^2}{b^2} = 1$ over the interval $a \le x \le 2a$.

Solution. $y^2\left(\frac{x^2}{a^2}-1\right)b^2 \Rightarrow y=\frac{b}{a}\sqrt{x^2-a^2}$

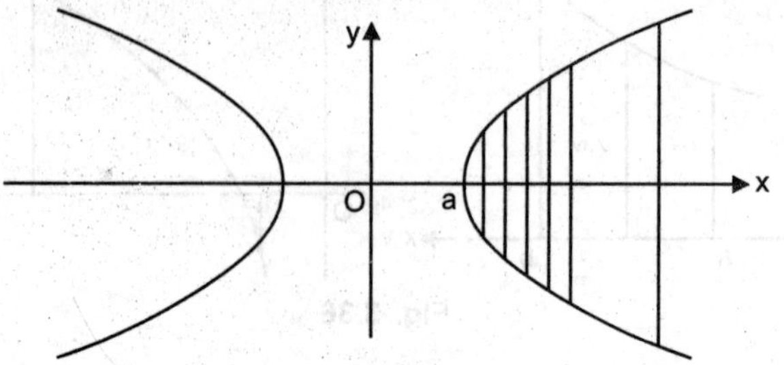

$$\frac{1}{2}y_{av}=\frac{1}{2a-a}\int_a^{2a}\frac{b}{a}\sqrt{x^2-a^2}\,dx$$

$$y_{av}=\frac{2b}{a^2}\int_a^{2a}\sqrt{x^2-a^2}\,dx$$

$$=\frac{2b}{a^2}\left[\frac{x}{2}\sqrt{x^2-a^2}-\frac{a^2}{2}\log\left(x+\sqrt{x^2+a^2}\right)\right]_a^{2a}$$

$$=\frac{2b}{a}\left\{\left(a\cdot\sqrt{3}a-\frac{a^2}{2}\log\left(2a+\sqrt{5}a\right)\right)-\left(0-\frac{a^2}{2}\log\left(a+\sqrt{2}a\right)\right)\right\}$$

$$=2ab\left\{\sqrt{3}+\frac{1}{2}\log\left(\frac{\sqrt{2}+1}{\sqrt{5}+2}\right)\right\}$$

Variable Area, Greatest and Least Values

An Important Concept

If $y=f(x)$ is a monotonic function in (a,b), then the area bounded by the ordinates at $x=a, x=b, y=f(x)$ and $y=f(c)$, [where $c\in(a,b)$] is minimum when $c=\frac{a+b}{2}$.

Proof $A=\int_a^c\{f(c)-f(x)\}\,dx+\int_c^b\{f(x)-f(c)\}\,dx$

$$=f(c)(c-a)-\int_a^c\{f(x)\}\,dx+\int_c^b\{f(x)\}\,dx-f(c)(b-c)$$

$$=\{(c-a)-(b-c)\}f(c)+\int_c^b\{f(x)\}\,dx-\int_a^c\{f(x)\}\,dx$$

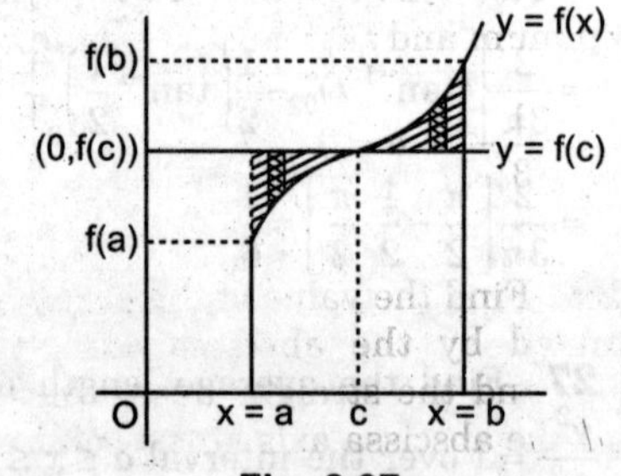

Fig. 3.37

$$A=[2c-(a+b)]f(c)+\int_c^b\{f(x)\}\,dx-\int_a^c\{f(x)\}\,dx$$

Differentiate w.r.t. c, $\dfrac{dA}{dc}=[2c-(a+b)]f'(c)+2f(c)+0-f(c)-\{f(c)-0\}$

For maxima and minima $\dfrac{dA}{dc}=0$

$$f'(c)[2c-(a+b)]=0 \qquad [\text{as } f'(c)\neq 0]$$

Hence, $\quad c=\dfrac{a+b}{2}$

Also, $\quad c<\dfrac{a+b}{2},\ \dfrac{dA}{dc}<0$ and $c>\dfrac{a+b}{2},\ \dfrac{dA}{dc}>0$

Hence, A is minimum when $c=\dfrac{a+b}{2}$

Illustration 28 If the area bounded by $f(x)=\dfrac{x^3}{3}-x^2+a$ and the straight lines $x=0$; $x=2$ and the x-axis is minimum, then find the value of 'a'.

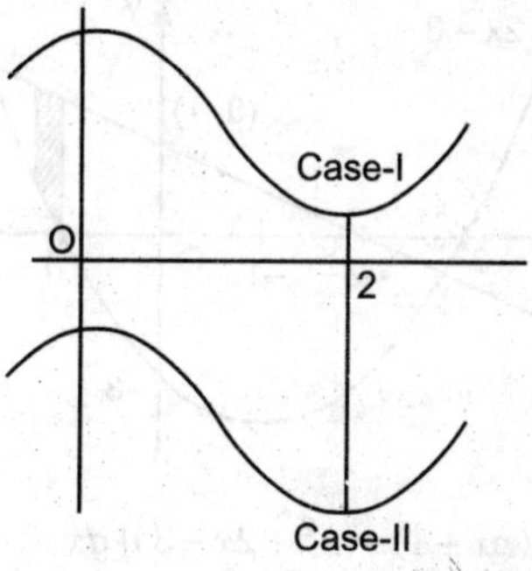

Solution. $f'(x)=x^2-2x=x(x-2)=0$ [note that $f(x)$ is monotonic in (0, 2)]

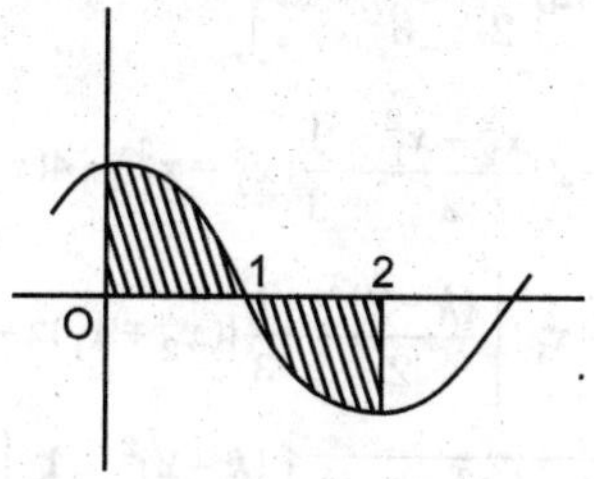

Thus, for the minimum and $f(x)$ must cross the x-axis are $\dfrac{0+2}{2}=1$.

Hence, $\quad f(1)=\dfrac{1}{3}-1+a=0 \Rightarrow a=\dfrac{2}{3}$

Illustration 29 Find the value of the parameter 'a' for which the area of the figure bounded by the abscissa axis, the graph of the function $y=x^3+3x^2+x+a$ and the straight lines, which are parallel to the axis of ordinates and cut the abscissa axis at the point of extremum of the function, is the least.

Solution. $f(x)=x^3+3x^2+x+a$

$$f'(x)=3x^2+6x+1=0 \Rightarrow x=-1\pm\frac{\sqrt{6}}{3}$$

Hence, $f(x)$ cuts the x-axis at $\frac{1}{2}\left[\left(-1+\frac{\sqrt{6}}{3}\right)+\left(-1-\frac{\sqrt{6}}{3}\right)\right]=-1$

$$f(-1)=-1+3-1+a=0 \Rightarrow a=-1$$

Illustration 30 If the area bounded by $y=x^2+2x-3$ and the line $y=kx+1$ is least, find k and also the least area.

Solution. x_1 and x_2 are the roots of the equation

$$x^2+2x-3=kx+1$$

$$x^2+(2-k)x-4=0$$

$$\left.\begin{aligned} x_1+x_2&=k-2\\ x_1x_2&=-4\end{aligned}\right\}$$

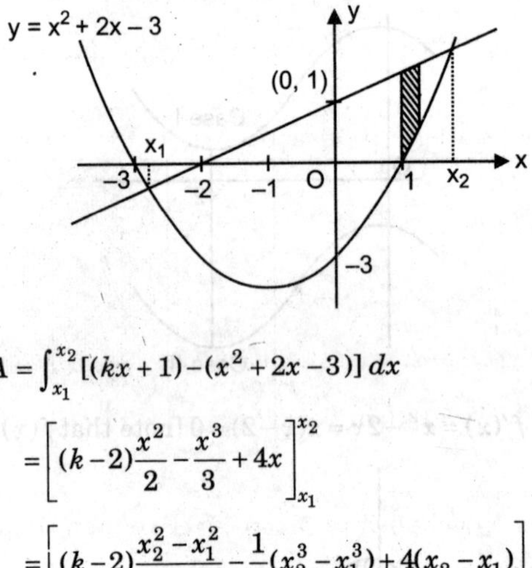

$$A=\int_{x_1}^{x_2}[(kx+1)-(x^2+2x-3)]\,dx$$

$$=\left[(k-2)\frac{x^2}{2}-\frac{x^3}{3}+4x\right]_{x_1}^{x_2}$$

$$=\left[(k-2)\frac{x_2^2-x_1^2}{2}-\frac{1}{3}(x_2^3-x_1^3)+4(x_2-x_1)\right]$$

$$=(x_2-x_1)\left[\frac{(k-2)^2}{2}-\frac{1}{3}\{(x_2+x_1)^2-x_1x_2\}+4\right]$$

$$=\sqrt{(x_2+x_1)^2-4x_1x_2}\left[\frac{(k-2)^2}{2}-\frac{1}{3}\{(k-2)^2+4\}+4\right]$$

$$=\frac{\sqrt{(k-2)^2+16}}{6}\left[\frac{1}{6}(k-2)^2+\frac{16}{3}\right]$$

$$=\frac{[(k-2)^2+16]^{3/2}}{6}$$

$\Rightarrow A$ is minimum, if $k=2$

Hence, $\quad A_{\min}=\dfrac{32}{3}$

Target Exercise 3.1

1. The area bounded by the curve $y = \frac{3}{|x|}$ and $y + |2 - x| = 2$, is

(a) $\frac{4 - \log 27}{3}$ (b) $2 - \log 3$

(c) $2 + \log 3$ (d) None of these

2. The area bounded by the curves $y = x^2 + 2$ and $y = 2|x| - \cos \pi x$ is equal to

(a) $\frac{2}{3}$ (b) $\frac{8}{3}$

(c) $\frac{4}{3}$ (d) $\frac{1}{3}$

3. If the area bounded by $f(x) = \sqrt{1 + x^2}$ $(x > 0)$ the line $y = x$, y-axis and $y = -x + a$ $(a > 1)$ is K, then area bounded by the graph of $f^{-1}(x)$, the line $y = x$ between the lines $y = -x + 1$ and $y = -x + a$, is

(a) $\frac{4K - 1}{4}$ (b) $\frac{2K - 1}{2}$

(c) K (d) None of these

4. The area bounded by the curve $y^2 = 4x$ and the circle $x^2 + y^2 - 2x - 3 = 0$ is equal to

(a) $2\pi + \frac{8}{3}$ (b) $4\pi + \frac{8}{3}$

(c) $\pi + \frac{8}{3}$ (d) $\pi - \frac{8}{3}$

5. A point P moves inside a triangle formed by $A\,(0, 0)$, $B\left(1, \frac{1}{\sqrt{3}}\right)$, $C\,(2, 0)$ such that $\min\{PA, PB, PC\} = 1$, then the area bounded by the curve traced by P is

(a) $3\sqrt{3} - \frac{3\pi}{2}$ (b) $\sqrt{3} + \frac{\pi}{2}$

(c) $\sqrt{3} - \frac{\pi}{2}$ (d) $3\sqrt{3} + \frac{3\pi}{2}$

Worked Examples

Type 1 : Subjective Type Questions

Example 1 Find the area common to the parabola $5x^2 - y = 0$ and $2x^2 - y + 9 = 0$.

Solution. Given curves are

$$y = 5x^2 \quad \text{...(i)}$$

and

$$y = 2x^2 + 9 \quad \text{...(ii)}$$

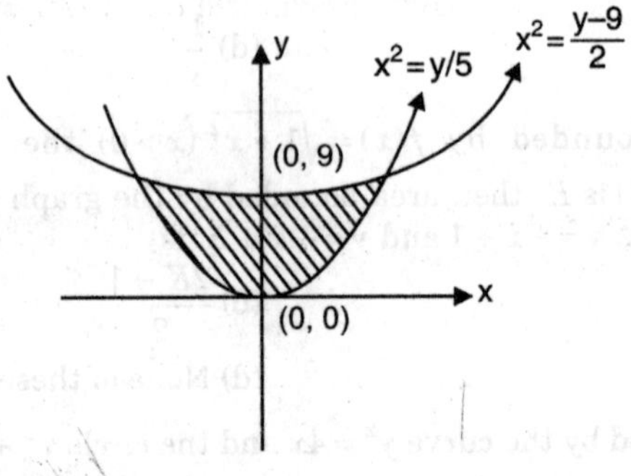

Points to Consider

In such examples, figure is the most essential thing. Without figure it just becomes difficult to judge whether y_1 to be subtracted from y_2 or otherwise.

Let us solve Eqs. (i) and (ii) simultaneously,

$$\therefore \quad 5x^2 = 2x^2 + 9 \;\therefore\; 3x^2 = 9;\; x^2 = 3$$

$$\therefore \quad x = -\sqrt{3} \quad \text{or} \quad x = \sqrt{3}$$

$\therefore$ In the usual notations, the required area is given by :

$$A = \int_{-\sqrt{3}}^{\sqrt{3}} (y_1 - y_2)\, dx$$

We have to find which curve is above and which is below w.r.t. x-axis in order to decide y_1 and y_2.

Take any point between $x = -\sqrt{3}$ and $x = \sqrt{3}$

Let us take $x = 0$, which lies between $x = -\sqrt{3}$ and $x = \sqrt{3}$

when $\quad x = 0$; from Eq. (i); $y = 0$

when $\quad x = 0$; from Eq. (ii); $y = 9$

Now, $9 > 0$

$\therefore$ Parabola Eq. (ii) is above parabola Eq. (i) between $x = -\sqrt{3}$ and $x = \sqrt{3}$

$\therefore$ y of the curve (ii) is to be taken as y_1 and y of the curve (i) is to be taken as y_2.

$$\therefore \quad A = \int_{-\sqrt{3}}^{\sqrt{3}} \{(2x^2 + 9) - 5x^2\}\, dx$$

$$= \int_{-\sqrt{3}}^{\sqrt{3}} (9 - 3x^2)\, dx = 2\int_0^{\sqrt{3}} (9 - 3x^2)\, dx$$

$$= 2\,[9x - x^3]_0^{\sqrt{3}} = 2\,[9\sqrt{3} - 3\sqrt{3}]$$

$\Rightarrow$ $\quad$ Area $= 12\sqrt{3}$ sq units

Example 2 Find the area enclosed by $y = x\,(x-1)(x-2)$ and x-axis.

Solution. The given curve is $y = x\,(x-1)(x-2)$. It passes through (0, 0), (1, 0) and (2, 0).
The sign scheme for $y = x\,(x-1)(x-2)$ is as shown in figure.

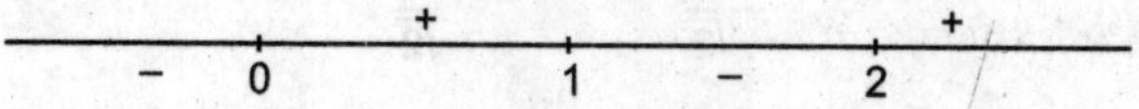

From the sign scheme it is clear that the curve is + ve when $0 < x < 1$ or $x > 2$, hence in these regions the curve lies above x-axis while in the rest regions the curve lies below x-axis.

Points to Consider

Sometimes the discussion of monotonicity of function helps us in sketching the polynomials. In the present case,

$$\frac{dy}{dx} = 3x^2 - 6x + 2, \quad \text{when} \quad \frac{dy}{dx} = 0;\ x = 1 \pm \frac{1}{\sqrt{3}}$$

Sign scheme for $\frac{dy}{dx}$ is

Thus, it is clear that the curve increases in $\left(-\infty, 1 - \frac{1}{\sqrt{3}}\right)$, decreases in $\left(1 - \frac{1}{\sqrt{3}}, 1 + \frac{1}{\sqrt{3}}\right)$ and again increases in $\left(1 + \frac{1}{\sqrt{3}}, \infty\right)$.
Therefore, the graph of the curve is as below :

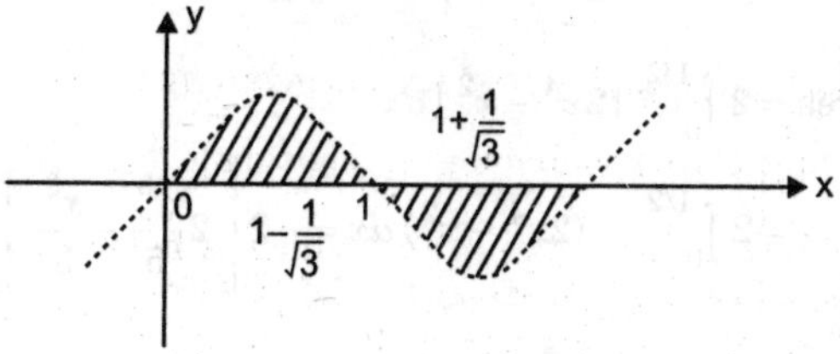

Hence, the required area $= \int_0^1 x\,(x-1)(x-2)\,dx + \left|\int_1^2 x\,(x-1)(x-2)\,dx\right|$

$$= \int_0^1 (x^3 - 3x^2 + 2x)\,dx + \left|\int_1^2 (x^3 - 3x^2 + 2x)\,dx\right|$$

$$= \left(\frac{x^4}{4} - x^3 + x^2\right)_0^1 + \left|\left(\frac{x^4}{4} - x^3 + x^2\right)_1^2\right|$$

$$= \left(\frac{1}{4} - 1 + 1\right) + \left|(4 - 8 + 4) - \left(\frac{1}{4} - 1 + 1\right)\right| = \frac{1}{4} + \frac{1}{4} = \frac{1}{2}$$

Example 3 Find the area between the curves $y = 2x^4 - x^2$, the x-axis and the ordinates of two minima of the curve.

Solution. The given curve is $y = 2x^4 - x^2$

when $\quad y = 0;\ x = 0, 0, \pm \dfrac{1}{\sqrt{2}}$

The sign scheme is as shown below :

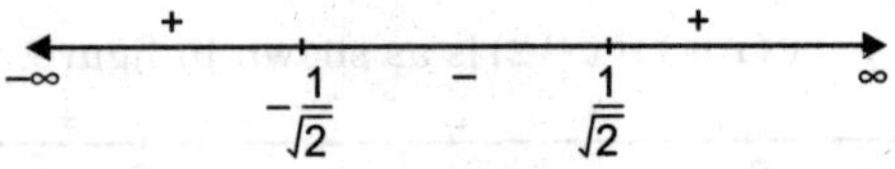

Therefore, it is clear that the curve cuts the x-axis at $x = -\dfrac{1}{\sqrt{2}}$, 0 and $\dfrac{1}{\sqrt{2}}$.

The curve is (− ve) in $\left(-\dfrac{1}{\sqrt{2}}, 0\right)$ and $\left(0, \dfrac{1}{\sqrt{2}}\right)$ while positive in the rest.

Now, $\qquad \dfrac{dy}{dx} = 8x^3 - 2x$

The sign scheme for $\dfrac{dy}{dx}$ is as below :

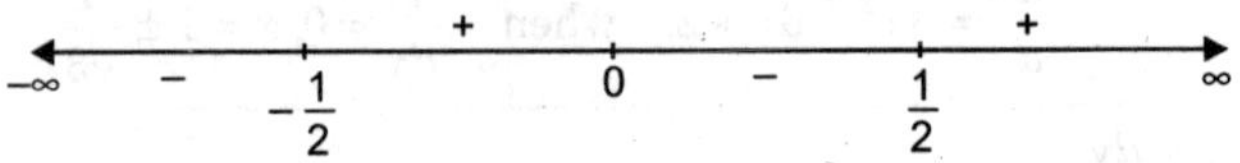

ie, The curve decreases in $(-\infty, -1/2)$ and $(0, 1/2)$ and increases in the rest of portions. Also, the function possess minimum at $x = -1/2$ and 1/2 while maximum at $x = 0$.

Therefore, the graph of the curve is as shown below :

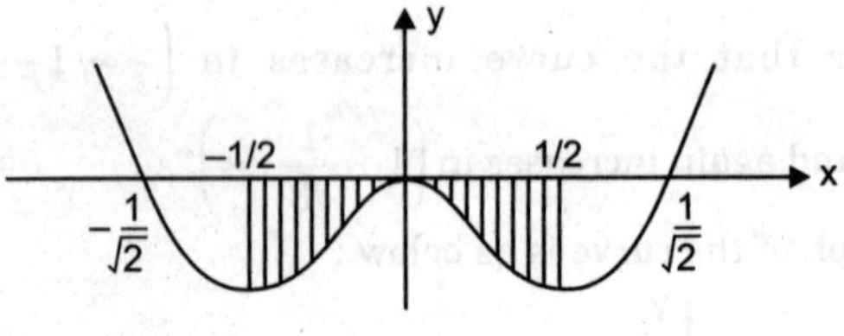

$\therefore$ The required area $= 2\displaystyle\int_0^{1/2} |2x^4 - x^2|\, dx$

$$= 2\int_0^{1/2} -(2x^4 - x^2)\, dx = -2\left[2\frac{x^5}{5} - \frac{x^3}{3}\right]_0^{1/2} = \frac{7}{120}$$

Example 4 Find the area of the region

$$\{(x, y): 0 \le y \le x^2 + 1, 0 \le y \le x + 1, 0 \le x \le 2\}$$

Solution. Let $R = \{(x, y): 0 \le y \le x^2 + 1, 0 \le y \le x + 1, 0 \le x \le 2\}$

$$= \{(x, y): 0 \le y \le x^2 + 1\} \cap \{(x, y): 0 \le y \le x + 1\} \cap \{(x, y): 0 \le x \le 2\}$$

$$= R_1 \cap R_2 \cap R_3$$

where $\qquad R_1 = \{(x, y): 0 \le y \le x^2 + 1\}$

$\qquad\qquad R_2 = \{(x, y): 0 \le y \le x + 1\}$

$\qquad\qquad R_3 = \{(x, y): 0 \le x \le 2\}$

Thus, the sketch of R_1, R_2 and R_3 are

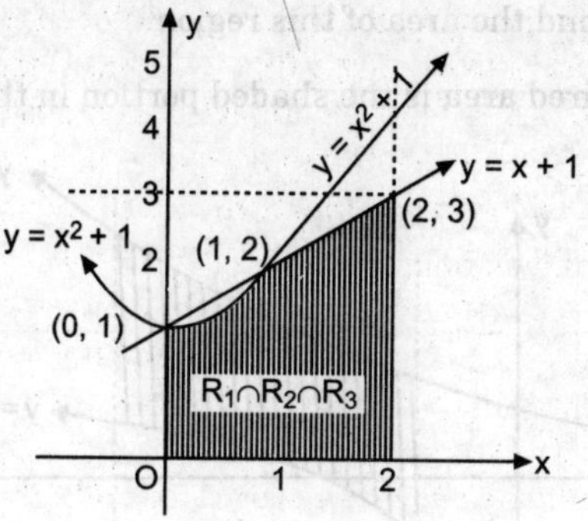

From above figure the required area

$$= \int_0^1 (x^2 + 1)\,dx + \int_1^2 (x + 1)\,dx$$

$$= \left(\frac{x^3}{3} + x\right)_0^1 + \left(\frac{x^2}{2} + x\right)_1^2 = \frac{23}{6} \text{ sq units}$$

Example 5 Find the area of the region bounded by the curves $y = x^2$, $y = |2 - x^2|$ and $y = 2$, which lies to the right of the line $x = 1$. **[IIT JEE 2002]**

Solution. The region bounded by given curves on the right side of $x = 1$ is shown as;

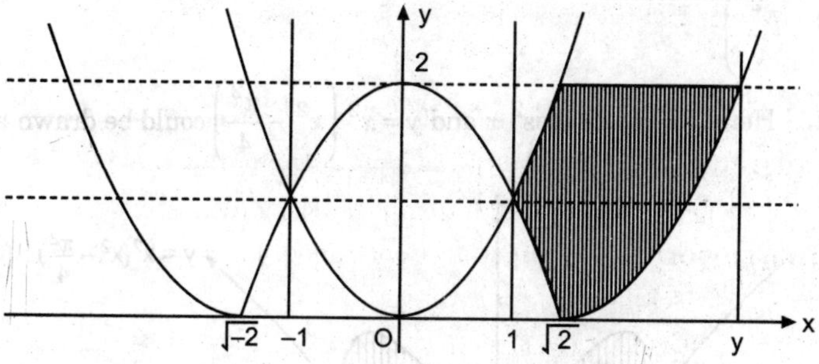

Thus, the required area $= \int_1^{\sqrt{2}} \{x^2 - (2 - x^2)\}\,dx + \int_{\sqrt{2}}^{2} \{4 - x^2\}\,dx$

$$= \int_1^{\sqrt{2}} (2x^2 - 2)\,dx + \int_{\sqrt{2}}^{2} (4 - x^2)\,dx$$

$$= \left(2\frac{x^3}{3} - 2x\right)_1^{\sqrt{2}} + \left(4x - \frac{x^3}{3}\right)_{\sqrt{2}}^{2}$$

$$= \left(\frac{4\sqrt{2}}{3} - 2\sqrt{2}\right) - \left(\frac{2}{3} - 2\right) + \left(8 - \frac{8}{3}\right) - \left(4\sqrt{2} - \frac{2\sqrt{2}}{3}\right)$$

$$= \left(-4\sqrt{2} + \frac{20}{3}\right) = \left(\frac{20 - 12\sqrt{2}}{3}\right) \text{ sq units}$$

Example 6 Sketch the curves and identify the region bounded by $x = 1/2, x = 2$, $y = \log_e x$ and $y = 2^x$. Find the area of this region. **[IIT JEE 1991]**

Solution. The required area is the shaded portion in the adjoining figure.

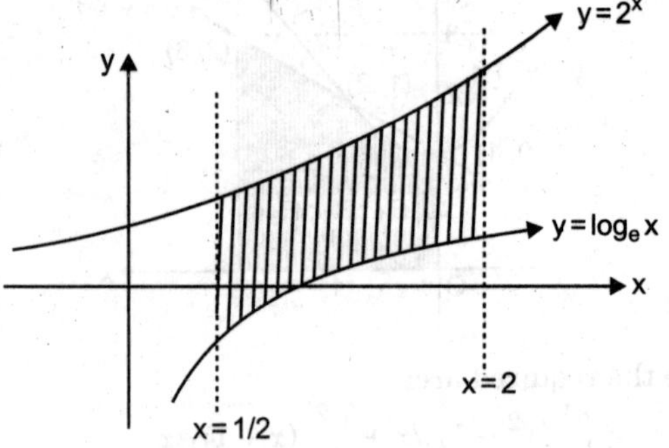

In the region $\frac{1}{2} \le x \le 2$ the curve $y = 2^x$ lies above as compared to $y = \log_e x$

Hence, the required area $= \int_{1/2}^{2} (2^x - \log x)\, dx = \left[\frac{2^x}{\log 2} - (x \log x - x) \right]_{1/2}^{2}$

$$= \frac{4 - \sqrt{2}}{\log 2} - \frac{5}{2} \log 2 + \frac{3}{2}$$

Example 7 Find the total area bounded by the curves $y = \cos x - \cos^2 x$ and $y = x^2 \left(x^2 - \frac{\pi^2}{4} \right)$.

Solution. Here, $y = \cos x - \cos^2 x$ and $y = x^2 \left(x^2 - \frac{\pi^2}{4} \right)$ could be drawn as in figure.

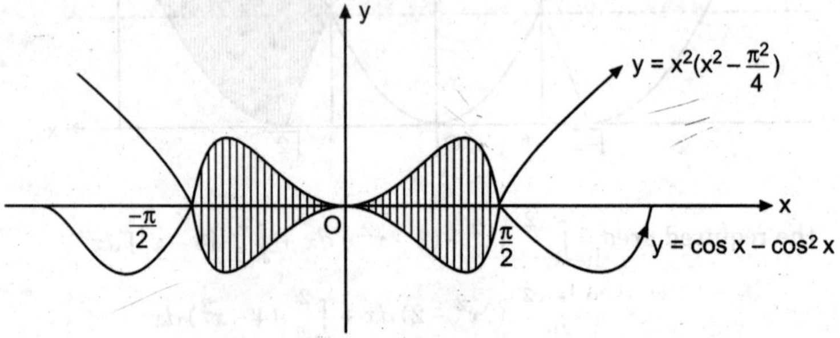

Thus, the area $= 2 \int_0^{\pi/2} \left[(\cos x - \cos^2 x) - \left\{ x^2 \left(x^2 - \frac{\pi^2}{4} \right) \right\} \right] dx$

$$= 2 \int_0^{\pi/2} \left(\cos x - \cos^2 x - x^4 + \frac{\pi^2}{4} x^2 \right) dx$$

$$= 2 \left[\sin x - \frac{x}{2} - \frac{\sin 2x}{4} - \frac{x^5}{5} + \frac{\pi^2 x^3}{12} \right]_0^{\pi/2} = \left(2 - \frac{\pi}{2} + \frac{\pi^5}{120} \right) \text{ sq units}$$

Example 8 Find the area of the region enclosed by the curve

$$5x^2 + 6xy + 2y^2 + 7x + 6y + 6 = 0.$$

Solution. Comparing $ax^2 + 2hxy + by^2 + 2gx + 2fy + c = 0$

We get $a = 5, b = 2, h = 3, g = 7/2, f = 3$ and $c = 6$

$\Rightarrow$ $$h^2 - ab = -1 < 0$$

So, above equation represents an ellipse

$\therefore$ $$2y^2 + 6(1 + x)y + (5x^2 + 7x + 6) = 0$$

$\Rightarrow$ $$y = \frac{-3(1 + x) \pm \sqrt{(3 - x)(x - 1)}}{2}$$

Clearly, the values of y are real for all $x \in [1, 3]$. Thus, the graph

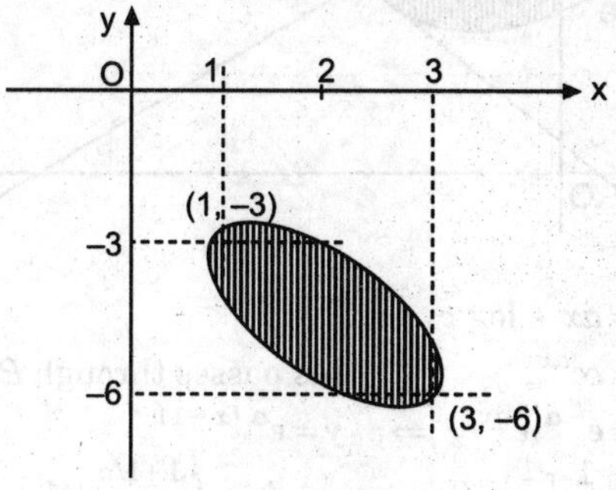

Thus, the required area

$$= \left| \int_1^3 \left(\frac{-3(1 + x) - \sqrt{(3 - x)(x - 1)}}{2} \right) - \left(\frac{-3(1 + x) + \sqrt{(3 - x)(x - 1)}}{2} \right) dx \right|$$

$$= \left| -\int_1^3 \sqrt{(3 - x)(x - 1)}\, dx \right| = \left| \int_1^3 \sqrt{1^2 - (x - 2)^2}\, dx \right|$$

$$= \left| -\left\{ \frac{1}{2}(x - 2)\sqrt{-x^2 + 4x - 3} + \frac{1}{2}\sin^{-1}\left(\frac{x - 2}{1} \right) \right\}_1^3 \right|$$

$$= \frac{\pi}{2} \text{ sq units}$$

Example 9 A curve $y = f(x)$ passes through the point $P(1, 1)$, the normal to the curve at P is $a(y - 1) + (x - 1) = 0$. If the slope of the tangent at any point on the curve is proportional to the ordinate of that point, determine the equation of the curve. Also obtain the area bounded by the y-axis, the curve and the normal to the curve at P.

Solution. Here, slope of the normal at $P(x, y)$

$\Rightarrow$ Slope of the line $a(y - 1) + (x - 1) = 0$ is $-\frac{1}{a}$

$\therefore$ Slope of the tangent at $P = a$

$\Rightarrow$ $$\left(\frac{dy}{dx} \right)_P = a \qquad \text{...(i)}$$

It is given that the slope of the tangent at any point on the curve $y = f(x)$ is proportional to the ordinate of the point.

$\therefore \qquad \dfrac{dy}{dx} \propto y \Rightarrow \dfrac{dy}{dx} = \lambda y$

$\Rightarrow \qquad \left(\dfrac{dy}{dx}\right)_{(1,1)} = \lambda \Rightarrow a = \lambda$

$\therefore \qquad \dfrac{dy}{dx} = ay \Rightarrow \dfrac{dy}{y} = a\,dx$

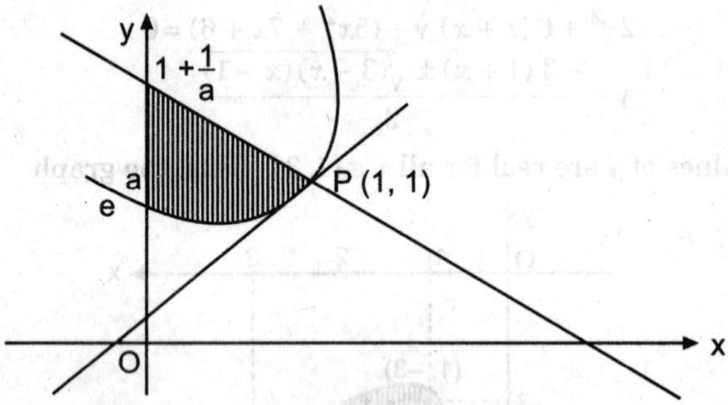

$\Rightarrow \qquad \log y = ax + \log c$

$\Rightarrow \qquad y = ce^{ax}$, this passes through $P(1, 1) \therefore c = e^{-a}$

$\Rightarrow$ Curve is, $\quad y = e^{-a}\,e^{ax} \Rightarrow y = e^{a(x-1)}$

$$\text{Required area} = \frac{1}{a}\int_{e^{-a}}^{1} (\log y + a)\,dy + \int_{1}^{1+1/a} \{(1+a) - ay\}\,dy$$

$$= \frac{1}{a}\{y(\log y - 1) + ay\}_{e^{-a}}^{1} + \left\{(1+a)\,y - \frac{a}{2}\,y^2\right\}_{a}^{1+\frac{1}{a}}$$

$$= \frac{1}{a}\{(-1+a) - e^{-a}(-a-1) - ae^{-a}\}$$

$$+ \left[(1+a)\left(1 + \frac{1}{a} - 1\right) - \frac{a}{2}\left\{\left(1 + \frac{1}{a}\right)^2 - 1^2\right\}\right]$$

$$= \frac{1}{a}[-1 + a + ae^{-a} + e^{-a} - ae^{-a}] + \left[\frac{1+a}{a} - \frac{a}{2}\left(2 + \frac{1}{a}\right)\frac{1}{a}\right]$$

$$= \frac{1}{a}\left(-1 + a + e^{-a}\right) + \left(\frac{1}{a} + 1 - 1 - \frac{1}{2a}\right)$$

$$= \left(1 + \frac{e^{-a}}{a} - \frac{1}{2a}\right) \text{ sq units}$$

Example 10 Sketch the region bounded by the curves $y = x^2$ and $y = \dfrac{2}{1+x^2}$. Find the area.

Solution. For intersection point, $x^2 = \dfrac{2}{1+x^2}$

ie, $\qquad x^4 + x^2 - 2 = 0$

ie, $\qquad (x^2 + 2)(x^2 - 1) = 0$

ie, $\qquad x = \pm 1$

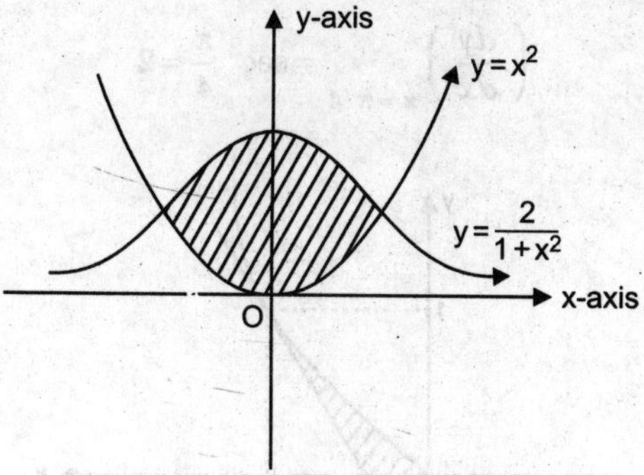

Hence, the required area $= 2\int_0^1 \left(\frac{2}{1+x^2} - x^2\right) dx$

$$= 2\left[2\tan^{-1} x - \frac{x^3}{3}\right]_0^1 = 2\left(\frac{\pi}{2} - \frac{1}{3}\right)$$

Example 11 Find the area enclosed between the curves

$$y = \log(x+e);\ x = \log_e\left(\frac{1}{y}\right) \text{ and } x\text{-axis.}$$

Solution. The given curves are $y = \log(x+e)$ and

$$x = \log_e\left(\frac{1}{y}\right) \Rightarrow \frac{1}{y} = e^x \Rightarrow y = e^{-x}$$

Using graph transformation we can sketch the curves.

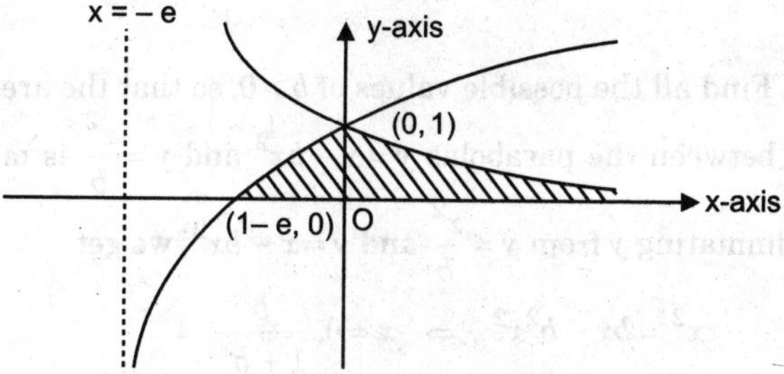

Hence, the required area $= \int_{1-e}^{0} \log(x+e)\, dx + \int_0^\infty e^{-x}\, dx$

$$= \int_1^e \log(t)\, dt + \int_0^\infty e^{-x}\, dx \qquad \text{(putting } x + e = t\text{)}$$

$$= [t\log t - t]_1^e - [e^{-x}]_0^\infty = 1 + 1 = 2$$

Example 12 Find the area of the region bounded by the curve $c : y = \tan x$, tangent drawn to c at $x = \pi/4$ and the x-axis.

Solution. The given curve is $y = \tan x$

$$\therefore \quad \frac{dy}{dx} = \sec^2 x$$

$$\left(\frac{dy}{dx}\right)_{x=\pi/4} = \sec^2\frac{\pi}{4} = 2$$

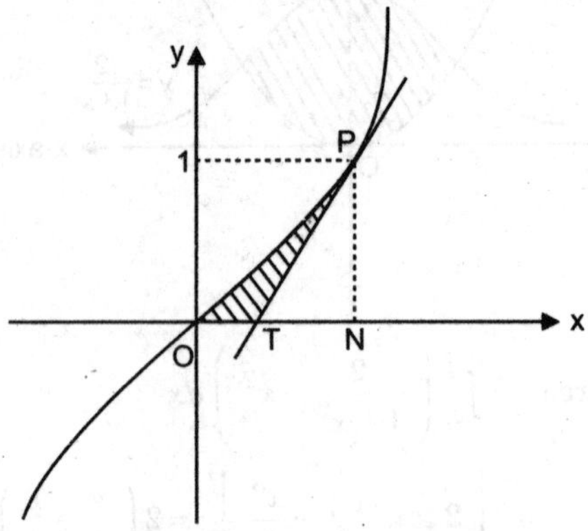

Also, at $x = \frac{\pi}{4}$; $y = 1$

$\therefore$ The equation of the tangent to the curve at the point $\left(\frac{\pi}{4}, 1\right)$ is $y - 1 = 2\left(x - \frac{\pi}{4}\right)$

when $\qquad y = 0;\ x = \frac{\pi}{4} - \frac{1}{2} = OT$

Now, the required area = Area of curvilinear ΔOPN – Area of ΔPTN

$$= \int_0^{\pi/4} (\tan x)\,dx - \frac{1}{2}\cdot NT\cdot PN$$

$$= [\log(\sec x)]_0^{\pi/4} - \frac{1}{2}\left(\frac{\pi}{4} - \frac{\pi}{4} + \frac{1}{2}\right)\cdot 1 = \frac{1}{2}\left(\log 2 - \frac{1}{2}\right)$$

Example 13 Find all the possible values of $b > 0$, so that the area of the bounded region enclosed between the parabolas $y = x - bx^2$ and $y = \frac{x^2}{b}$ is maximum.

Solution. Eliminating y from $y = \frac{x^2}{b}$ and $y = x - bx^2$, we get

$$x^2 = bx - b^2x^2 \quad \Rightarrow \quad x = 0, \frac{b}{1+b^2}$$

Thus, the area enclosed between the parabolas.

$$A = \int_0^{b/(1+b^2)} \left(x - bx^2 - \frac{x^2}{b}\right) dx$$

$$= \int_0^{b/1+b^2} \left\{x - x^2\left(\frac{1+b^2}{b}\right)\right\} dx$$

$$= \left(\frac{x^2}{2} - \frac{x^3}{3}\cdot\frac{1+b^2}{b}\right)_0^{b/1+b^2}$$

$$= \frac{1}{6}\cdot\frac{b^2}{(1+b^2)^2}$$

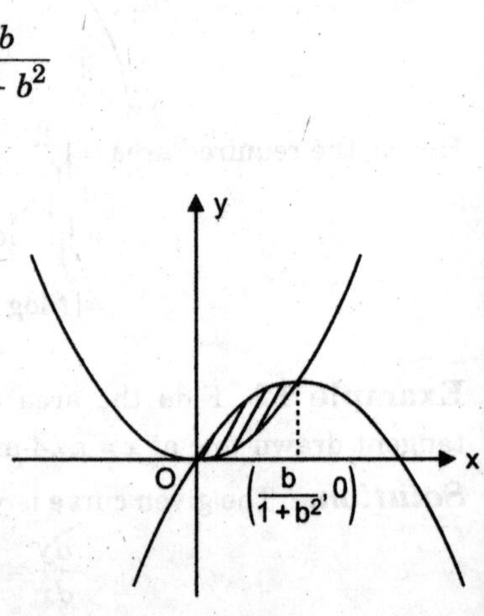

For maximum, value of A, $\frac{dA}{db} = 0$

But $$\frac{dA}{db} = \frac{1}{6} \cdot \frac{(1+b^2)^2 \cdot 2b - b^2 \cdot 2(1+b^2) \cdot 2b}{(1+b^2)^4} = \frac{1}{3} \cdot \frac{b(1-b^2)}{(1+b^2)^3}$$

Hence, $\frac{dA}{db} = 0$ gives $b = -1, 0, 1$ since $b > 0$

Therefore, we consider only $b = 1$

Sign scheme for $\frac{dA}{db}$ around $b = 1$ is as below

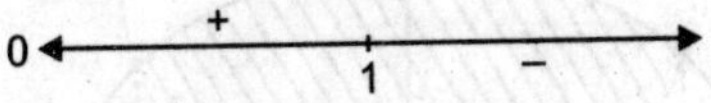

From sign scheme it is clear that A is maximum.

Example 14 Find the area given by $x + y \le 6$, $x^2 + y^2 \le 6y$ and $y^2 \le 8x$.

Solution. Let us consider the curves

$$P \equiv y^2 - 8x = 0 \qquad \text{...(i)}$$

$$C \equiv x^2 + y^2 = 6y$$

ie, $$x^2 + (y-3)^2 - 9 = 0 \qquad \text{...(ii)}$$

and $$S \equiv x + y - 6 = 0 \qquad \text{...(iii)}$$

The intersection points of the curves (ii) and (iii) are given by

$$(6-y)^2 + y^2 - 6y = 0$$

ie, $$y = 3, 6$$

Therefore, the points are (0, 6) and (3, 3). The intersection points of the curves (i) and (iii) are given by

$$y^2 = 8(6-y), \quad ie, \quad y = 4, -12$$

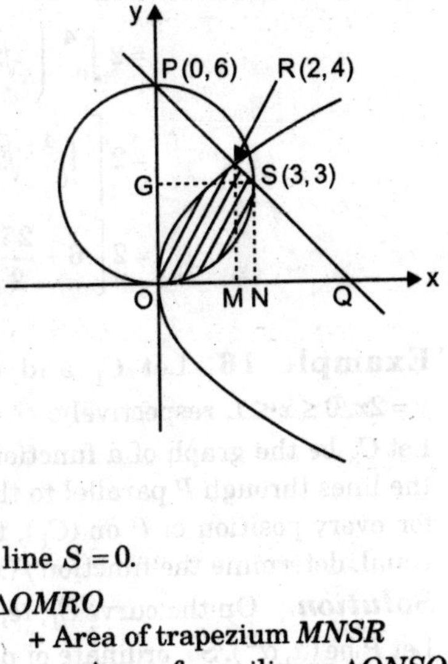

Therefore, the point of intersection in 1st quadrant is (2, 4).

Now, we know that

$C \le 0$ denotes the region, inside the circle $C = 0$.

$P \le 0$ denotes the region, inside the parabola $P = 0$.

$S \le 0$ denotes the region, which is (–ve) side of the line $S = 0$.

$\therefore$ The required area = Area of curvilinear $\Delta OMRO$

+ Area of trapezium $MNSR$

– Area of curvilinear $\Delta ONSO$

$$= \int_0^2 \sqrt{8x}\, dx + \frac{1}{2}(MR + NS) \cdot MN$$

– (Area of square $ONSG$ – Area of sector $OSGO$)

$$= \int_0^2 \sqrt{8x}\, dx + \frac{1}{2}(4+3) \cdot 1 - \left(3^2 - \frac{\pi \cdot 3^2}{4}\right)$$

$$= \frac{9\pi}{4} - \frac{1}{6}$$

Example 15 Find the area bounded by the curves $x^2 + y^2 = 25$, $4y = |4 - x^2|$ and $x = 0$ above the x-axis.

Solution. The 1st curve is a circle of radius 5 with centre at (0, 0).

The 2nd curve is $$y = \left|\frac{4 - x^2}{4}\right| = \left|1 - \frac{x^2}{4}\right|$$

Which can be traced easily by graph transformation.

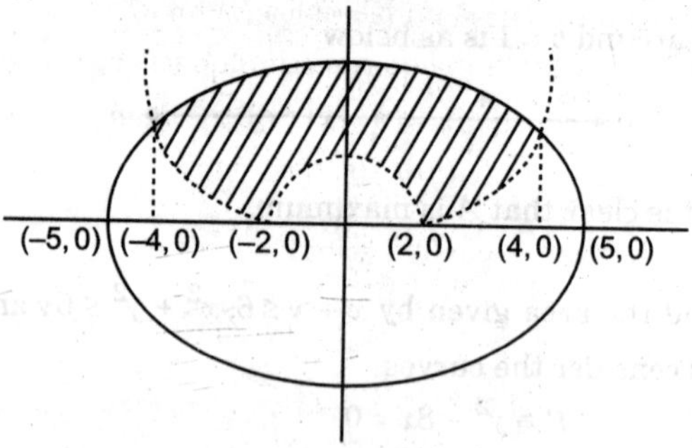

When the two curves intersect each other, then $x^2 + \left(1 - \frac{x^2}{4}\right)^2 = 25 \Rightarrow x = \pm 4$

Hence, the required area

$$= 2\int_0^4 \left(\sqrt{25 - x^2} - \left|1 - \frac{x^2}{4}\right|\right) dx$$

$$= 2\left[\int_0^4 \sqrt{25 - x^2}\, dx - \int_0^2 \left(1 - \frac{x^2}{4}\right) dx + \int_2^4 \left(1 - \frac{x^2}{4}\right) dx\right]$$

$$= 2\left[6 + \frac{25}{2}\sin^{-1}\left(\frac{4}{5}\right) - \frac{4}{3} - \frac{8}{3}\right] = 25\sin^{-1}\left(\frac{4}{5}\right) + 4$$

Example 16 Let C_1 and C_2 be the graphs of the function $y = x^2$ and $y = 2x$, $0 \le x \le 1$, respectively.

Let C_3 be the graph of a function $y = f(x)$; $0 \le x \le 1$, $f(0) = 0$. For a point P on C_1, let the lines through P parallel to the axes, meets C_2 and C_3 at Q and R respectively. If for every position of P on (C_1), the areas of the shaded region OPQ and ORP are equal, determine the function $f(x)$. **[IIT JEE 1998]**

Solution. On the curve C_1, *ie*, $y = x^2$.

Let P be (α, α^2). So, ordinate of point Q on C_2 is also α^2.

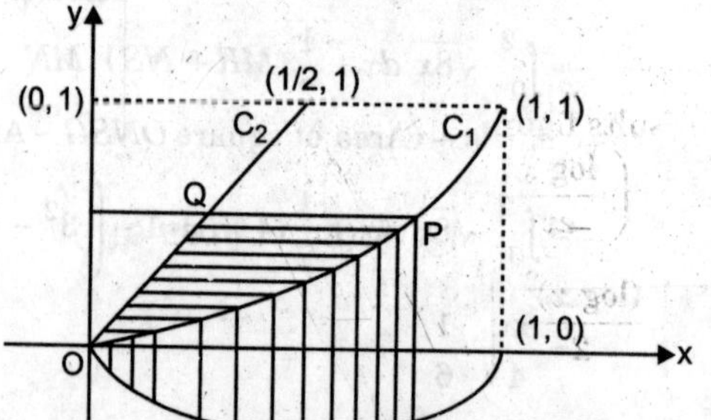

Now, on C_2 ($y = 2x$) the abscissae of Q is given by $x = \frac{y}{2} = \frac{\alpha^2}{2}$.

$\therefore$ Q is $\left(\frac{\alpha^2}{2}, \alpha\right)$ and R on C_3 is $\{\alpha, f(x)\}$.

Now, area of $\Delta OPQ = \int_0^{\alpha^2} (x_1 - x_2)\, dy = \int_0^{\alpha^2} \left(\sqrt{y} - \frac{y}{2}\right) dy = \frac{2}{3}\alpha^3 - \frac{\alpha^4}{4}$ ...(i)

Again, area of $\Delta ORP = \int_0^{\alpha} (y_1 - y_2)\, dx = \int_0^{\alpha^2} \{x^2 - f(x)\}\, dx$ ...(ii)

Thus, from Eqs. (i) and (ii), we get

$$\frac{2\alpha^3}{3} - \frac{\alpha^4}{4} = \int_0^{\alpha} \{x^2 - f(x)\}\, dx$$

Differentiating both the sides w.r.t. α, we get

$$2\alpha^2 - \alpha^3 = \alpha^2 - f(\alpha)$$

$\Rightarrow$ $f(\alpha) = \alpha^3 - \alpha^2 \quad \therefore \quad f(x) = x^3 - x^2$

Example 17 Find the area of the region bounded by the curves

$y = ex \log x$, and $y = \frac{\log x}{ex}$. **[IIT JEE 1990]**

Solution. Both the curves are defined for $x > 0$. Both are positive when $x > 1$ and negative when $0 < x < 1$.

We know that, $\lim_{x \to 0^+} \log x \to -\infty$

Therefore, $\lim_{x \to 0^+} \frac{\log x}{ex} \to -\infty$

Thus, y-axis is asymptote of second curve.

and $\lim_{x \to 0^+} ex \log x$ $[(0)(-\infty)$ form$]$

$= \lim_{x \to 0^+} \frac{e \log x}{1/x} \left(-\frac{\infty}{\infty} \text{ form}\right)$

$= \lim_{x \to 0^+} \frac{e\,(1/x)}{(-1/x^2)} = 0$ (using L'Hospital's rule)

Thus, the first curve starts from (0, 0) but does not include (0, 0).

Now, the given curves intersect therefore

$ex \log x = \frac{\log x}{ex}$

ie, $(e^2x^2 - 1) \log x = 0$

ie, $x = 1, \frac{1}{e}$ (since $x > 0$)

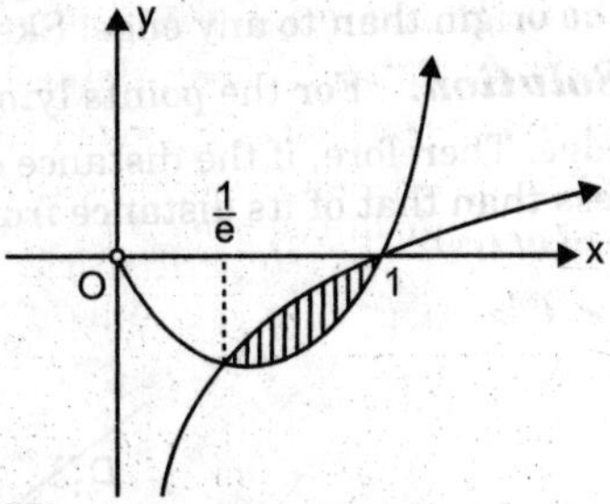

Therefore, using above results figure could be drawn as;

$\therefore$ The required area $= \int_{1/e}^{1} \left(\frac{\log x}{ex} - ex \log x\right) dx$

$$= \frac{1}{e}\left[\frac{(\log x)^2}{2}\right]_{1/e}^{1} - e\left[\frac{x^2}{4}(2 \log x - 1)\right]_{1/e}^{1} = \frac{e^2 - 5}{4e}$$

Example 18 Let A_n be the area bounded by the curve $y=(\tan x)^n$ and the lines $x=0$, $y=0$ and $x=\frac{\pi}{4}$. Prove that for $n>2$; $A_n + A_{n-2} = \frac{1}{n-1}$ and deduce that $\frac{1}{2n+2} < A_n < \frac{1}{2n-2}$.

Solution. **First part**

We have, therefore, $$A_n = \int_0^{\pi/4} (\tan x)^n \, dx$$

Hence $$A_{n-2} = \int_0^{\pi/4} (\tan x)^{n-2} \, dx$$

$\therefore$ $$A_n + A_{n-2} = \int_0^{\pi/4} (\tan x)^{n-2} (\tan^2 x + 1) \, dx$$
$$= \int_0^{\pi/4} (\tan x)^{n-2} \cdot \sec^2 x \, dx$$

Let $\tan x = t$, so that $\sec^2 x \, dx = dt$

$\therefore$ $$A_n + A_{n-2} = \int_0^1 t^{n-2} \, dt = \left(\frac{t^{n-1}}{n-1}\right)_0^1 = \frac{1}{n-1} \quad \text{...(i)}$$

Second part

Since, $\quad 0 \le x \le \pi/4 \;\therefore\; 0 \le \tan x \le 1$

$\Rightarrow \quad \tan^{n+2} x < \tan^n x < \tan^{n-2} x$

$\Rightarrow \quad \int_0^{\pi/4} \tan^{n+2} x \, dx < \int_0^{\pi/4} \tan^n x \, dx < \int_0^{\pi/4} \tan^{n-2} x \, dx$

$\Rightarrow \quad A_{n+2} < A_n < A_{n-2}$

$\Rightarrow \quad A_n + A_{n+2} < 2A_n < A_n + A_{n-2}$

$\therefore \quad \frac{1}{n+1} < 2A_n < \frac{1}{n-1}$ [using Eq. (i)]

$$\frac{1}{2(n+1)} < A_n < \frac{1}{2(n-1)}$$

Example 19 Consider a square with vertices at (1, 1) (–1, 1), (–1, –1) and (1, –1). Let S be the region consisting of all points inside the square which are nearer to the origin than to any edge. Sketch the region S and find its area. **[IIT JEE 1995]**

Solution. For the points lying in the ΔOAB the edge AB, *ie*, $x=1$ is the closest edge. Therefore, if the distance of a point $P(x, y)$ (lying in the ΔOAB) from origin is less than that of its distance from the edge $x=1$ it will fall in the region S.

$\therefore$ $$OP \le PQ$$

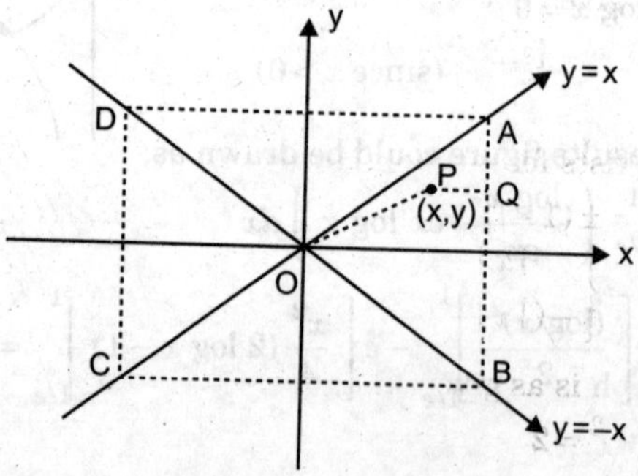

$\Rightarrow \qquad \sqrt{x^2 + y^2} \leq 1 - x$

$\Rightarrow \qquad x^2 + y^2 \leq x^2 - 2x + 1$

$\Rightarrow \qquad y^2 \leq 1 - 2x$

Similarly, for points lying in the ΔOAD the side $y = 1$ is the closest side and therefore the region S is determined by;

$$x^2 \leq 1 - 2y$$

Since, the edges are symmetric about the origin. Hence, by the above inequality and by symmetry, the required area will be the shaded portion in the figure given below :

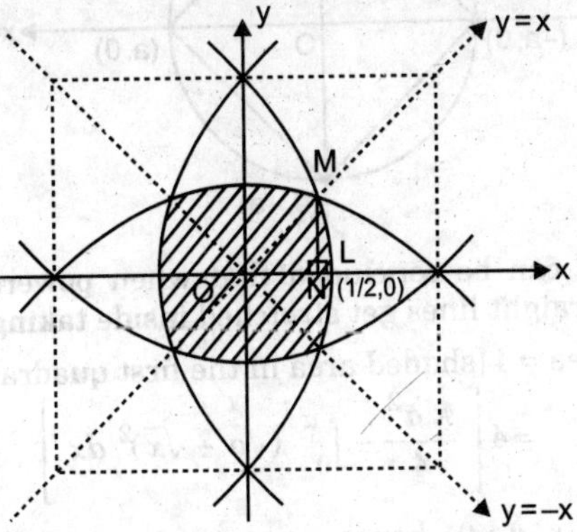

Now, when the curves $y^2 = 1 - 2x$ and $y = x$ intersect each other, then

$$x^2 = 1 - 2x \quad \Rightarrow \quad x^2 + 2x - 1 = 0$$

$\Rightarrow \qquad x = \sqrt{2} - 1, -\sqrt{2} - 1$

Hence, the intersection points in the first quadrant is $(\sqrt{2} - 1, \sqrt{2} - 1)$.

$\therefore$ The required area = 8 [Area of curvilinear ΔOLM]

$$= 8\left[\frac{1}{2}(\sqrt{2} - 1)(\sqrt{2} - 1) + \int_{\sqrt{2}-1}^{1/2} \sqrt{1 - 2x}\, dx\right]$$

$$= 8\left[\frac{1}{2}(3 - 2\sqrt{2}) + \left(\frac{2(1 - 2x)^{3/2}}{3(-2)}\right)_{\sqrt{2}-1}^{1/2}\right]$$

$$= \frac{4}{3}(4\sqrt{2} - 5)$$

Example 20 Find the area enclosed by $|x| + |y| = 1$.

Solution. From the given equation, we have

$|y| = 1 - |x| \,|\, 0$ [since $|y| \,|\, 0$]

$\Rightarrow \qquad -1 \leq x \leq 1$

Therefore, the curve exists for $x \in [-1, 1]$ only

and for $-1 \leq x \leq 1$; $\quad y = \pm(1 - |x|)$

ie, $\quad y = \begin{cases} |x| - 1 \\ -(|x| - 1) \end{cases}$

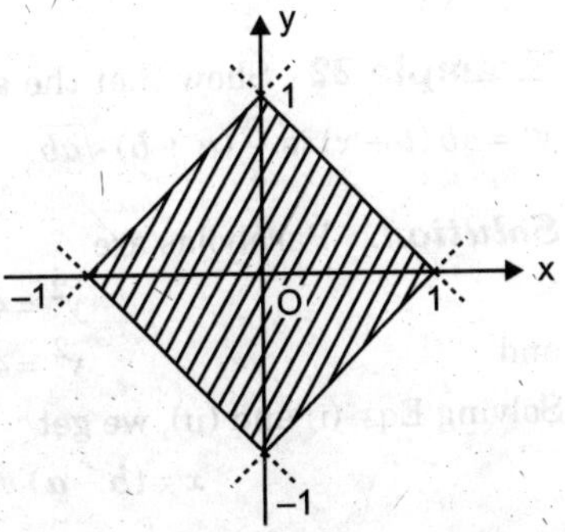

Thus, the required graph is as given in figure.

The area enclosed $= (\sqrt{2})^2 = 2$

Example 21 Sketch the region included between the curves $x^2 + y^2 = a^2$ and $\sqrt{|x|} + \sqrt{|y|} = \sqrt{a}$ $(a > 0)$ and find its area.

Solution. The graphs $|x| + |y| = a$ and $|x|^2 + |y|^2 = a^2$ are as shown in figure.

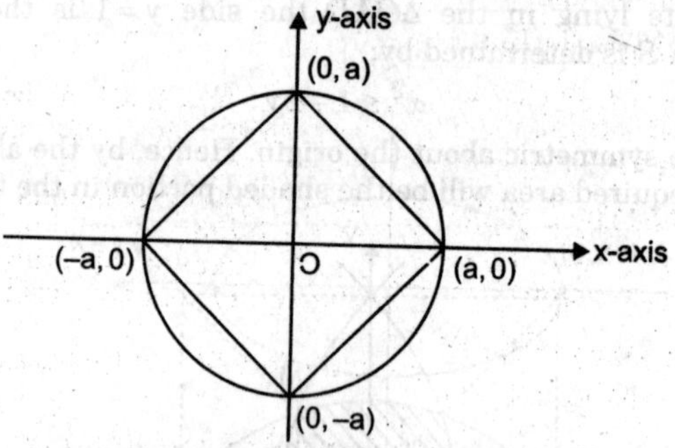

From the diagram it can be concluded that when powers of $|x|$ and $|y|$ both is reduced to half the straight lines get stretched inside taking the shape as above.

Thus, the required area = 4 [shaded area in the first quadrant]

$$= 4\left[\frac{\pi a^2}{4} - \int_0^a (\sqrt{a} - \sqrt{x})^2\, dx\right]$$

(Since in 1st quadrant $x, y > 0$), hence,

$$\sqrt{|x|} + \sqrt{|y|} = \sqrt{a} \Rightarrow \sqrt{x} + \sqrt{y} = \sqrt{a}$$

$$\Rightarrow \qquad y = (\sqrt{a} - \sqrt{x})^2$$

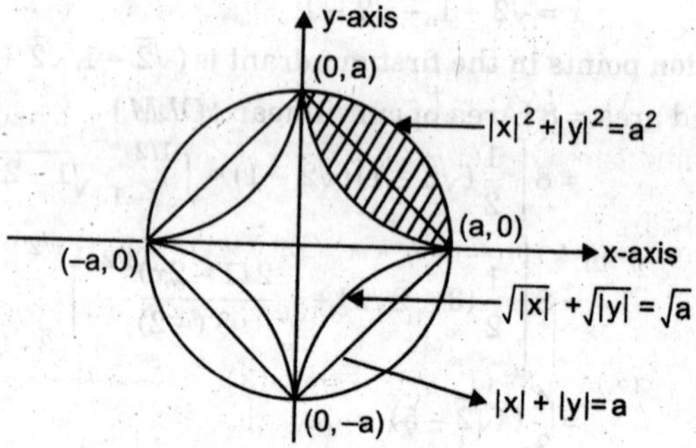

Hence, $\qquad$ the area $= \left(\pi - \dfrac{2}{3}\right) a^2$

Example 22 Show that the are included between the parabolas $y^2 = 4a(x + a)$, $y^2 = 4b(b - x)$ is $\frac{8}{3}(a + b)\sqrt{ab}$.

Solution. Parabolas are

$$y^2 = 4a(x + a) \qquad \text{...(i)}$$

and

$$y^2 = 4b(b - x) \qquad \text{...(ii)}$$

Solving Eqs. (i) and (ii), we get

$$x = (b - a) \quad \text{and} \quad y = \pm 2\sqrt{ab}$$

$\therefore$ P and Q are $(b-a, 2\sqrt{ab})$ and $(b-a, -2\sqrt{ab})$ respectively and the points A and A' are $(-a, 0)$ and $(b, 0)$ respectively.

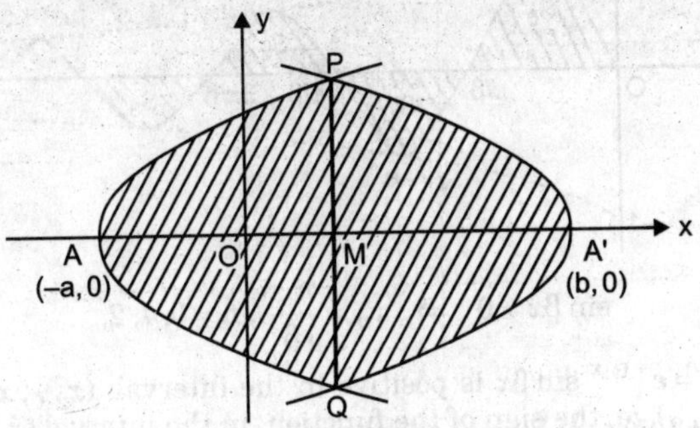

Now, required area = area $APA'QA = 2$ area $APA'A$

$$= 2[\text{area } APMA + \text{area } MPA'M]$$

$$= 2\left[\int_{-a}^{b-a} 2\sqrt{a(a+x)}\,dx + \int_{b-a}^{b} 2\sqrt{b(b-x)}\,dx\right]$$

$$= 4\sqrt{a}\int_{-a}^{b-a}\sqrt{a+x}\,dx + 4\sqrt{b}\int_{b-a}^{b}\sqrt{b-x}\,dx$$

$$= 4\sqrt{a}\left[\frac{2}{3}(a+x)^{3/2}\right]_{-a}^{b-a} + 4\sqrt{b}\left[-\frac{2}{3}(b-x)^{3/2}\right]_{b-a}^{b}$$

$$= \frac{8}{3}\sqrt{a}\,(b)^{3/2} + \frac{8\sqrt{b}}{3}(a)^{3/2} = \frac{8}{3}\sqrt{ab}\,(a+b) \text{ sq units}$$

Example 23 Determine the area of the figure bounded by two branches of the curve $(y-x)^2 = x^3$ and the straight line $x = 1$.

Solution. Curves are $(y-x)^2 = x^3$

$$y - x = \pm x\sqrt{x}$$

$$y = x + x\sqrt{x} \quad \text{...(i)}$$

$$y = x - x\sqrt{x} \quad \text{...(ii)}$$

and $\quad x = 1 \quad$ ...(iii)

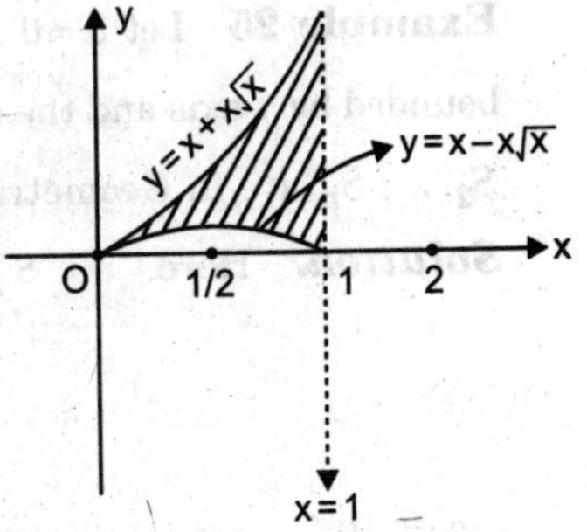

Which could be drawn as; shown in figure.
Hence, the required area

$$= \int_0^1 \{(x + x\sqrt{x}) - (x - x\sqrt{x})\}\,dx$$

$$= \int_0^1 (2x\sqrt{x})\,dx = 2\int_0^1 x^{3/2}\,dx = \frac{4}{5} \text{ sq unit}$$

Example 24 Prove that the areas $S_0, S_1, S_2, \ldots$ bounded by the x-axis and half-waves of the curve $y = e^{-\alpha x}\sin\beta x$, $x \,|\, 0$. form a geometric progression with the common ratio $g = e^{-\pi\alpha/\beta}$.

Solution. The curve $y = e^{-\alpha x}\sin\beta x$ intersects the positive x-axis at the points where $y = 0$.

$$\therefore \qquad e^{-\alpha x}\sin\beta x = 0$$

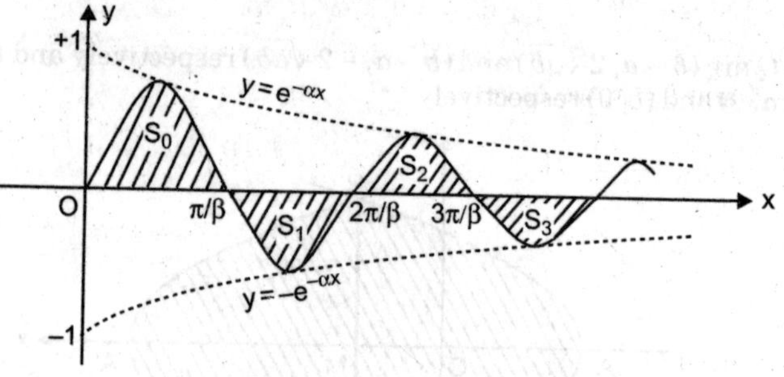

$$\Rightarrow \qquad \sin \beta x = 0 \quad \Rightarrow \quad x_n = \frac{n\pi}{\beta}, n = 0, 1, 2, \ldots$$

The function $y = e^{-\alpha x} \sin \beta x$ is positive in the interval (x_{2K}, x_{2K+1}) and negative in (x_{2K+1}, x_{2K+2}), *ie*, the sign of the function in the interval (x_n, x_{n+1}), therefore

$$S_n = \left| \int_{n\pi/\beta}^{(n+1)\pi/\beta} e^{-\alpha x} \sin \beta x \, dx \right|$$

$$= \left| \left\{ \frac{(-1)^{n+1} e^{-\alpha x}}{\alpha^2 + \beta^2} (\alpha \sin \beta x + \beta \cos \beta x) \right\}_{n\pi/\beta}^{(n+1)\pi/\beta} \right|$$

$$\therefore \qquad S_n = \frac{\beta e^{-n\pi\alpha/\beta}}{(\alpha^2 + \beta^2)} \{1 + e^{-\pi\alpha/\beta}\}$$

Hence,
$$g = \frac{S_{n+1}}{S_n} = \frac{\dfrac{\beta e^{-(n+1)\pi\alpha/\beta} \{1 + e^{-\pi\alpha/\beta}\}}{(\alpha^2 + \beta^2)}}{\dfrac{\beta e^{-n\pi\alpha/\beta} \{1 + e^{-\pi\alpha/\beta}\}}{\alpha^2 + \beta^2}} = e^{-\pi\alpha/\beta}$$

which completes the proof.

Example 25 Let $b \neq 0$ and for $j = 0, 1, 2, \ldots, n$. Let S_j be the area of the region bounded by y-axis and the curve $x \cdot e^{ay} = \sin by$, $\frac{j\pi}{b} \leq y \leq \frac{(j+1)\pi}{b}$. Show that S_0, S_1, $S_2, \ldots, S_n$ are in Geometric Progression. Also find their sum for $a = -1$ and $b = \pi$.

Solution. Here,
$$S_j = \left| \int_{j\pi/b}^{(j+1)\pi/b} x \, dy \right| = \left| \int_{j\pi/b}^{(j+1)\pi/b} e^{-ay} \sin by \, dy \right|$$

$$= \left| \left\{ \frac{e^{-ay}}{a^2 + b^2} (-a \sin by - b \cos by) \right\}_{j\pi/b}^{(j+1)\pi/b} \right|$$

$$= \left| \frac{e^{-a(j+1)\pi/b}}{a^2 + b^2} \times (-b)(-1)^{j+1} - \frac{e^{-aj\pi/b}}{a^2 + b^2} (-b)(-1)^j \right|$$

$$= |b| \left\{ \frac{e^{-a(j+1)\pi/b}}{a^2 + b^2} + \frac{e^{-aj\pi/b}}{a^2 + b^2} \right\}$$

$$= |b| \frac{e^{-aj\pi/b}}{a^2 + b^2} \{e^{-a\pi/b} + 1\}; \; \{j = 0, 1, 2, \ldots, n\}$$

Now,
$$\frac{S_{j+1}}{S_j} = |b| \frac{e^{-a(j+1)\pi/b}}{a^2 + b^2} \{e^{-a\pi/b} + 1\} \; |b| \frac{e^{-aj\pi/b}}{a^2 + b^2} \{e^{-a\pi/b} + 1\}$$

$$= e^{-a\pi/b} \quad \text{for all } j = 0, 1, 2, \ldots, n$$

Hence, $S_0, S_1, S_2, \ldots, S_n$ are in GP with common ratio $e^{-a\pi/b}$.
For $a = -1$ and $b = \pi$, we have

$$S_j = \frac{e^j \pi (e+1)}{\pi^2 + 1}; \quad j = 0, 1, 2, \ldots, n$$

$$\therefore \quad \sum_{j=0}^{n} S_j = \sum_{j=0}^{n} \frac{e^{j\pi}(e+1)}{\pi^2+1} = \frac{(e+1)\pi}{\pi+1}\left\{\frac{e^{n+1}-1}{e-1}\right\}$$

Example 26 For any real t, $x = 2 + \frac{e^t + e^{-t}}{2}$, $y = 2 + \frac{e^t - e^{-t}}{2}$ **is a point on the hyperbola $x^2 - y^2 - 4x + 4y - 1 = 0$. Find the area bounded by the hyperbola and the lines joining the centre to the points corresponding to t_1 and $-t_1$.**

Solution. The points $x = 2 + \frac{e^t + e^{-t}}{2}$, $y = 2 + \frac{e^t - e^{-t}}{2}$ is on the curve

$$(x-2)^2 - (y-2)^2 = 1 \quad \text{or} \quad x^2 - y^2 - 4x + 4y - 1 = 0$$

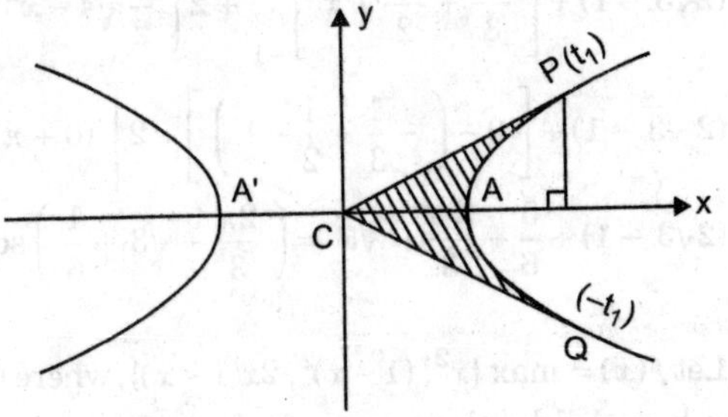

Put $\quad (x-2) = x, (y-2) = y \quad \therefore \quad x^2 - y^2 = 1$ **and** $x = \frac{e^t + e^{-t}}{2}$, $y = \frac{e^t - e^{-t}}{2}$

We have to find the area of the region bounded by the curve $x^2 - y^2 = 1$ and the lines joining the centre $x = 0$, $y = 0$ to the point (t_1) and $(-t_1)$.

$$\text{Required area} = 2\left[\text{area of } \Delta PCN - \int_1^{\frac{e^{t_1}+e^{-t_1}}{2}} y\, dx\right]$$

$$= 2\left[\frac{1}{2}\left(\frac{e^{t_1}+e^{-t_1}}{2}\right)\left(\frac{e^{t_1}-e^{-t_1}}{2}\right) - \int_1^{t_1} y \frac{dx}{dt}\, dt\right]$$

$$= 2\left[\left(\frac{e^{2t_1}-e^{-2t_1}}{8}\right) - \int_0^{t_1}\left(\frac{e^t - e^{-t}}{2}\right)^2 dt\right]$$

$$= \frac{e^{2t_1}-e^{-2t_1}}{4} - \frac{1}{2}\int_0^{t_1}(e^{2t}+e^{-2t}-2)\, dt$$

$$= \frac{e^{2t_1}-e^{-2t_1}}{4} - \frac{1}{2}\left[\frac{e^{2t_1}}{2} - \frac{e^{-2t_1}}{2} - 2t_1\right]_0^{t_1}$$

$$= \frac{e^{2t_1}-e^{-2t_1}}{4} - \frac{1}{2}\left[\frac{e^{2t_1}}{2} - \frac{e^{-2t_1}}{2} - 2t_1\right] = t_1$$

Example 27 Find the area enclosed by circle $x^2 + y^2 = 4$, parabola $y = x^2 + x + 1$, the curve $y = \left[\sin^2 \frac{x}{4} + \cos \frac{x}{4}\right]$ and x-axis (where [.] is the greatest integer function).

Solution. $\because y = \left[\sin^2 \frac{x}{4} + \cos \frac{x}{4}\right]$

$\because \quad 1 < \sin^2 \frac{x}{4} + \cos \frac{x}{4} < 2 \quad$ for $x \in (-2, 2]$

$\therefore \quad y = \left[\sin^2 \frac{x}{4} + \cos \frac{x}{4}\right] = 1$

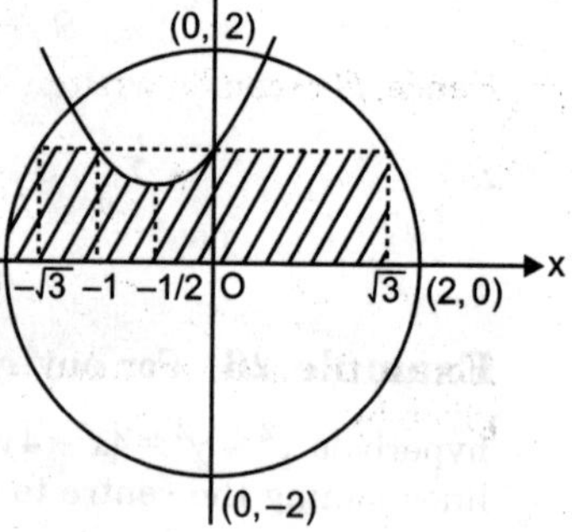

Now, we have to find out the area enclosed by the circle $x^2 + y^2 = 4$, parabola $\left(y - \frac{3}{4}\right) = \left(x + \frac{1}{2}\right)^2$, line $y = 1$ and x-axis.

Required area is shaded area in the figure.

Hence, required area

$$= \sqrt{3} \times 1 + (\sqrt{3} - 1) \times 1 + \int_{-1}^{0} (x^2 + x + 1)\, dx + 2\int_{\sqrt{3}}^{2} (\sqrt{4 - x^2})\, dx$$

$$= (2\sqrt{3} - 1) + \left[\frac{x^3}{3} + \frac{x^2}{2} + x\right]_{-1}^{0} + 2\left[\frac{x}{2}\sqrt{4 - x^2} + 2\sin^{-1}\left(\frac{x}{2}\right)\right]_{\sqrt{3}}^{2}$$

$$= (2\sqrt{3} - 1) + \left[0 - \left(-\frac{1}{3} + \frac{1}{2} - 1\right)\right] + 2\left[(0 + \pi) - \left(\frac{\sqrt{3}}{2} + \frac{2\pi}{3}\right)\right]$$

$$= (2\sqrt{3} - 1) + \frac{5}{6} + \frac{2\pi}{3} - \sqrt{3} = \left(\frac{2\pi}{3} + \sqrt{3} - \frac{1}{6}\right) \text{ sq units}$$

Example 28 Let $f(x) = \max\{x^2, (1 - x)^2, 2x(1 - x)\}$, where $0 \le x \le 1$. Determine the area of the region bounded by the curves $y = f(x)$, x-axis, $x = 0$ and $x = 1$.

[IIT JEE 1997]

Solution. We have, $f(x) = \max\{x^2, (1 - x^2), 2x(1 - x)\}$

Graphically it could be shown as;

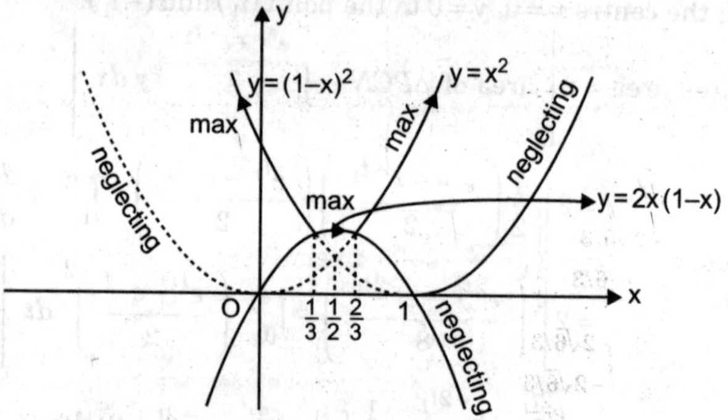

For figure it is clear that maximum graph (*ie*, above max graph is considered and others are neglected).

$\therefore \quad$ for $x \in \left[0, \frac{1}{3}\right], \quad x^2 \le 2x(1 - x) \le (1 - x)^2$

for $x \in \left[\frac{1}{3}, \frac{1}{2}\right], \quad x^2 \le (1 - x)^2 \le 2x(1 - x)$

for $x \in \left[\frac{1}{2}, \frac{2}{3}\right]$, $(1-x)^2 \le x^2 \le 2x(1-x)$

for $x \in \left[\frac{2}{3}, 1\right]$, $(1-x)^2 \le 2x(1-x) \le x^2$

Hence, $f(x)$ can be written as;

$$f(x) = \begin{cases} (1-x)^2, & \text{for } 0 \le x \le 1/3 \\ 2x(1-x), & \text{for } 1/3 \le x \le 2/3 \\ x^2, & \text{for } 2/3 \le x \le 1 \end{cases}$$

Hence, the area bounded by the curve $y = f(x)$; x-axis and the lines $x = 0$ and $x = 1$ is given by

$$= \int_0^{1/3} (1-x)^2\, dx + \int_{1/3}^{2/3} 2x(1-x)\, dx + \int_{2/3}^{1} (x^2)\, dx$$

$$= \frac{17}{27} \text{ sq units}$$

Example 29 Find the ratio in which the curve, $y = [-0.01x^4 - 0.02x^2]$ [where [·] denotes the greatest integer function) divides the ellipse $3x^2 + 4y^2 = 12$.

Solution. Here, $y = [-0.01x^4 - 0.02x^2]$

is $y = -1$, when $-2 < x < 2$

$y = -1$ cut the ellipse $3x^2 + 4y^2 = 12$

At $x^2 = \frac{8}{3}$ or $x = \pm \frac{2\sqrt{2}}{\sqrt{3}}$

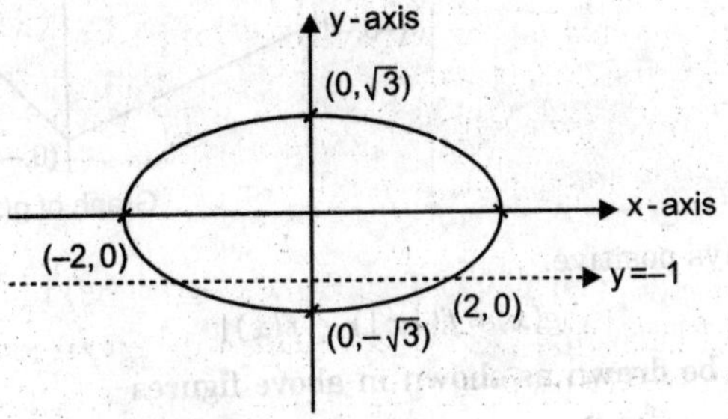

$$\text{Required area} = \int_{-2\sqrt{6}/3}^{2\sqrt{6}/3} \left\{\sqrt{\frac{12-3x^2}{4}} - 1\right\} dx$$

$$= \frac{\sqrt{3}}{2} \int_{-2\sqrt{6}/3}^{2\sqrt{6}/3} \sqrt{4-x^2}\, dx - \int_{-2\sqrt{6}/3}^{2\sqrt{6}/3} 1\, dx$$

$$= 2\sin^{-1}\sqrt{\frac{2}{3}} + \frac{2}{3} - \frac{4}{3}\sqrt{6}$$

Example 30 Let $f(x) = \begin{cases} -2, & -3 \le x \le 0 \\ x-2, & 0 < x \le 3 \end{cases}$

where $g(x) = \min\{f(|x|) + |f(x)|, f(|x|) - |f(x)|\}$. Find the area bounded by the curve $g(x)$ and the x-axis between the ordinates at $x = 3$ and $x = -3$.

Solution. Here, $f(|x|)=\begin{cases}-x-2, & -3\le x\le 0\\ x-2, & 0<x\le 3\end{cases}$

$$|f(x)|=\begin{cases}2, & -3\le x\le 0\\ -x+2, & 0<x\le 2\\ x-2, & 2<x\le 3\end{cases}$$

$\therefore$ $$f(|x|)-|f(x)|=\begin{cases}-x-4, & -3\le x\le 0\\ 2x-4, & 0<x\le 2\\ 0, & 2<x\le 3\end{cases}$$

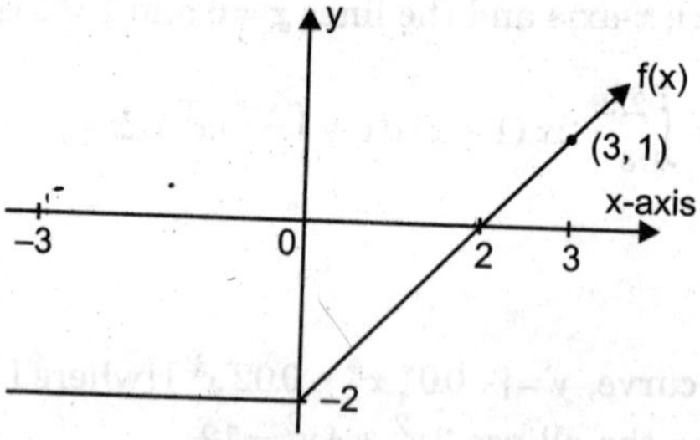

Graph of f(x)

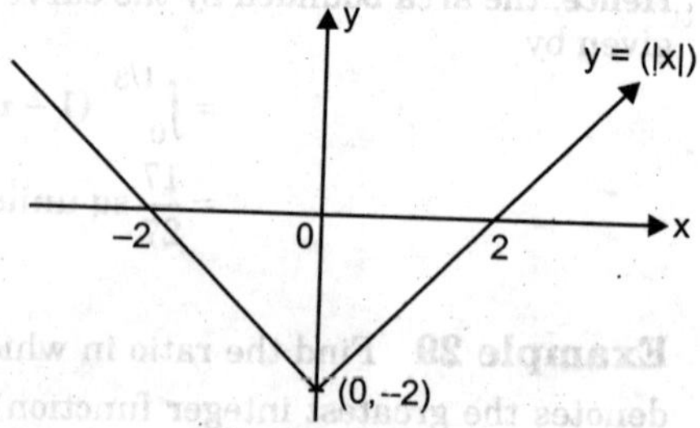

Graph of f(|x|)

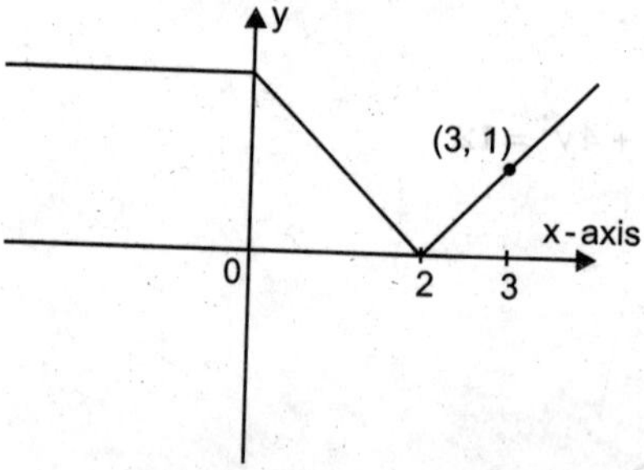

Graph of |f(x)|

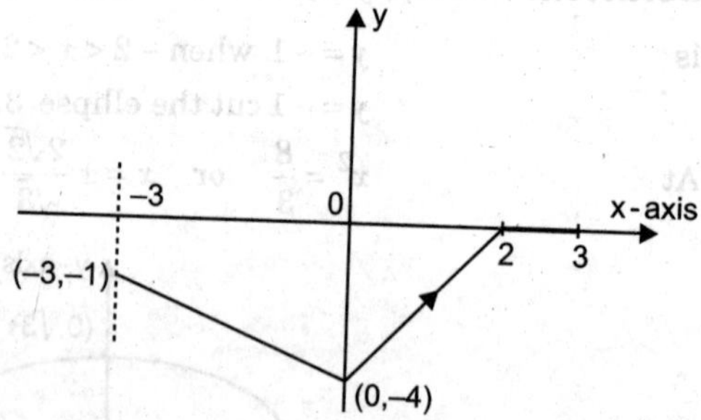

Graph of g(x)

Since, $|f(x)|$ is always positive.

$$g(x)=f(|x|)-|f(x)|$$

where graphs could be drawn as shown in above figures

From the graph the required area

$$=\frac{1}{2}(1+4)\times 3+\left(\frac{1}{2}\times 2\times 4\right)+0$$

$$=\frac{23}{2}\text{ sq units}$$

Example 31 Let $f(x)=\max\left\{\sin x, \cos x, \frac{1}{2}\right\}$, then determine the area of region bounded by the curves $y=f(x)$, x-axis, y-axis and $x=2\pi$.

Solution. We have,

$$f(x)=\max\left\{\sin x, \cos x, \frac{1}{2}\right\}$$

Graphically $f(x)$ could be drawn as;

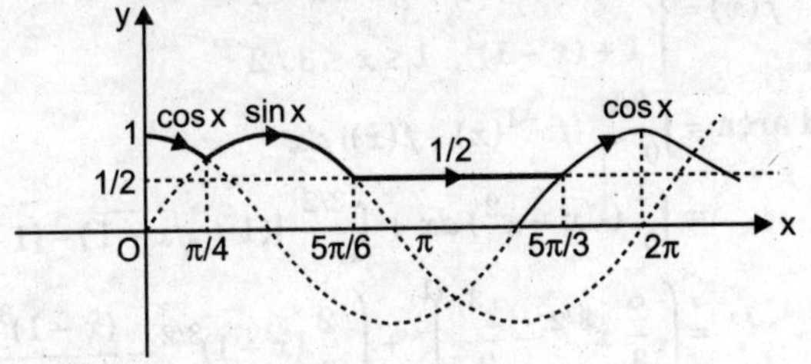

Here, the graph is plotted between 0 to 2π and between the points of intersection the maximum portion is included, thus the shaded part is required area;

	Interval	Value of $f(x)$
ie,	for $0 \le x < \pi/4$	$\cos x$
	for $\pi/4 \le x < 5\pi/6$	$\sin x$
	for $5\pi/6 \le x < 5\pi/3$	$1/2$
	for $5\pi/3 \le x < 2\pi$	$\cos x$

Hence, required area

$$I = \int_0^{\pi/4} \cos x\, dx + \int_{\pi/4}^{5\pi/6} \sin x\, dx + \int_{5\pi/6}^{5\pi/3} \frac{1}{2}\, dx + \int_{5\pi/3}^{2\pi} \cos x\, dx$$

$$= (\sin x)_0^{\pi/4} - (\cos x)_{\pi/4}^{5\pi/6} + \frac{1}{2}(x)_{5\pi/6}^{5\pi/3} + (\sin x)_{5\pi/3}^{2\pi}$$

$$= \left(\frac{1}{\sqrt{2}} - 0\right) - \left(-\frac{\sqrt{3}}{2} - \frac{1}{\sqrt{2}}\right) + \frac{1}{2}\left(\frac{5\pi}{3} - \frac{5\pi}{6}\right) + \left(0 + \frac{\sqrt{3}}{2}\right)$$

$$= \left(\frac{5\pi}{12} + \sqrt{2} + \sqrt{3}\right) \text{ sq units}$$

Example 32 Show that the function $f(x) = [x] + \{x\}^2$, (where [.] and { . } are respectively greatest integer and fractional part functions) is invertible. Find the area bounded by $y = f(x)$, $y = f^{-1}(x)$, $x = 0$ and $x = 3/2$.

Solution. We see that, f is continuous everywhere except possibly at integral points. Let n be an integer, then

$$f(n^-) = n - 1 + (1)^2 = n$$

$$f(n) = n$$

$$f(n^+) = n + (0)^2 = n$$

$\Rightarrow$ f is everywhere continuous; also $f(x) \to \pm\infty$ as $x \to \pm\infty$

$\Rightarrow$ Range of $f \in R$, thus onto.

and $f(x) = n + (x - n)^2$, for all $x \in [n, n+1)$, where $n \in I$.

$\therefore$ f is monotonically increasing.

$\Rightarrow$ f is one-one, thus invertible.

Obviously for $x \in [0, 1)$, $f(x) = x^2$ and $f^{-1}(x) = \sqrt{x}$

and for $x \in [1, 3/2]$, $f(x) = 1 + (x - 1)^2$

and $f^{-1}(x) = 1 + \sqrt{x - 1}$

$$\Rightarrow \qquad f(x)=\begin{cases} x^2, & 0\le x<1 \\ 1+(x-1)^2, & 1\le x\le 3/2 \end{cases}$$

$\therefore$ Required area $= \int_0^{3/2} \{f^{-1}(x)-f(x)\}\,dx$

$$= \int_0^1 (\sqrt{x}-x^2)\,dx + \int_1^{3/2} [(1+\sqrt{x-1})-\{1+(x-1)^2\}]\,dx$$

$$= \left(\frac{2}{3}x^{3/2}-\frac{x^3}{3}\right)_0^1 + \left(\frac{2}{3}(x-1)^{3/2}-\frac{(x-1)^3}{3}\right)_1^{3/2}$$

$$= \left(\frac{2}{3}-\frac{1}{3}\right)+\left\{\frac{2}{3}\left(\frac{1}{2}\right)^{3/2}-\frac{(1/2)^3}{3}\right\}=\frac{1}{3}+\frac{1}{3\sqrt{2}}-\frac{1}{6\sqrt{2}}$$

$$= \frac{2\sqrt{2}+2-1}{6\sqrt{2}}=\frac{1+2\sqrt{2}}{6\sqrt{2}} \text{ sq units}$$

Example 33 Let ABC be a triangle with vertices $A\equiv(6, 2(\sqrt{3}+1))$, $B\equiv(4, 2)$ and $C\equiv(8, 2)$. Let R be the region consisting of all those points P inside $\Delta\, ABC$ which satisfy

$$d(P, BC) \ge \max\{d(P, AB), d(P, AC)\}$$

where $d(P, L)$ denotes the distance of the point P from the line L. Sketch the region R and find its area.

Solution. It is easy to see that ABC is an equilateral triangle with side of length 4. BD and CE are angle bisectors of angle B and C respectively. Any point inside the $\Delta\, AEC$ is nearer to AC than BC and any point inside the $\Delta\, BDA$ is nearer to AB than BC. So any point inside the quadrilateral $AEGC$ will satisfy the given condition. Hence, shaded region is the required region, whose area is to be found, shown as in figure

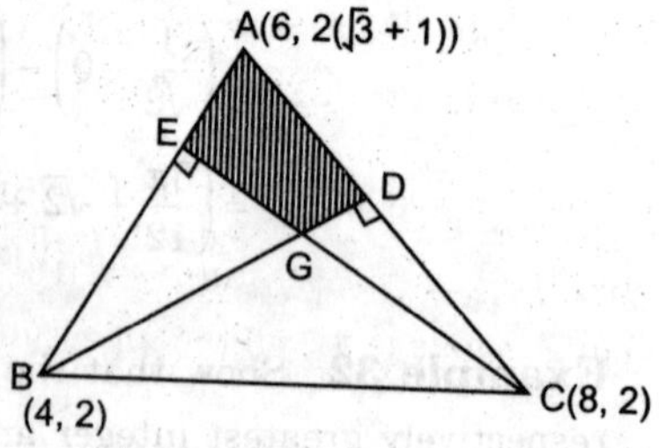

Thus, the required area $= 2$ area of $\Delta\, EAG = 2\times\frac{1}{2}AE\times EG$

$$= \frac{1}{2}AB\times\frac{1}{3}CE=\frac{1}{6}\times 4\times\sqrt{4^2-2^2}$$

$$= \frac{4\sqrt{3}}{3} \text{ sq units}$$

Example 34 Let $O(0, 0)$, $A(2, 0)$ and $B\left(1, \frac{1}{\sqrt{3}}\right)$ be the vertices of a triangle. Let R be the region consisting of all those points P inside $\Delta\, OAB$ which satisfy $d(P, OA)\le \min\{d(P, OB), d(P, AB)\}$, when '$d$' denotes the distance from the point to the corresponding line. Sketch the region R and find its area. **[IIT JEE 1997]**

Solution. Let the coordinate of P be (x, y).

Equation of line $OA \equiv y=0$

Equation of line $OB \equiv \sqrt{3}y=x$

Equation of line $AB = \sqrt{3}y=2-x$

$d(P, OA)=$ distance of P from line $OA = y$

$d(P, OB)$ = distance of P from line

$$OB = \frac{|\sqrt{3}y - x|}{2}$$

$d(P, AB)$ = distance of P from line

$$AB = \frac{|\sqrt{3}y + x - 2|}{2}$$

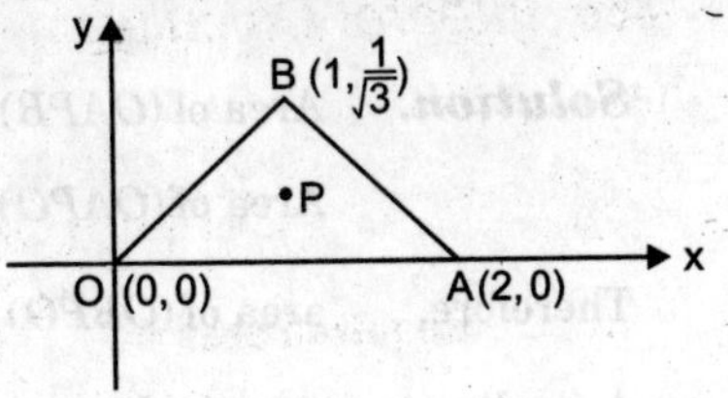

Given, $d(P, OA) \le \min\{d(P, OB), d(P, AB)\}$

$$y \le \min\left\{\frac{|\sqrt{3}y - x|}{2}, \frac{|\sqrt{3}y + x - 2|}{2}\right\}$$

$\Rightarrow$ $$y \le \frac{|\sqrt{3}y - x|}{2} \quad \text{...(i)}$$

and $$y \le \frac{|\sqrt{3}y + x - 2|}{2} \quad \text{...(ii)}$$

Case I If $$y \le \frac{|\sqrt{3}y - x|}{2}$$

$\therefore$ $y \le \dfrac{x - \sqrt{3}y}{2}$, *ie*, $x > \sqrt{3}y$ $\quad (\because \sqrt{3}y - x < 0)$

$\Rightarrow$ $(2 + \sqrt{3})\, y \le x$

$\Rightarrow$ $y \le (2 - \sqrt{3})\, x$

$\Rightarrow$ $y \le x \tan 15°$ ...(iii)

($\because$ $y = x \tan 15°$ is an acute angle bisector of $\angle AOB$.)

Case II If $$y \le \frac{|\sqrt{3}y + x - 2|}{2}$$

$\Rightarrow$ $2y \le 2 - x - \sqrt{3}y$ (*ie*, $\sqrt{3}y + x - 2 < 0$)

$\Rightarrow$ $(2 + \sqrt{3})\, y \le 2 - x$

$\Rightarrow$ $y \le -(2 - \sqrt{3})(x - 2)$

$\Rightarrow$ $y \le -(\tan 15°)(x - 2)$ ...(iv)

[$\because$ $y = (x - 2)\tan 15°$ is an acute angle bisector of CA]

From Eqs. (iii) and (iv), P moves inside the triangle as shown in figure.

As $\angle QOB = \angle OBQ = 15°$, ΔOQB is an isosceles triangle

$\Rightarrow$ $OC = AC = 1$ unit

Area of shaded region = area of ΔOQA.

$= \frac{1}{2}$(base) $\times$ (height)

$= \frac{1}{2}(2)(1 \tan 15°) = \tan 15°$

$= (2 - \sqrt{3})$ sq units

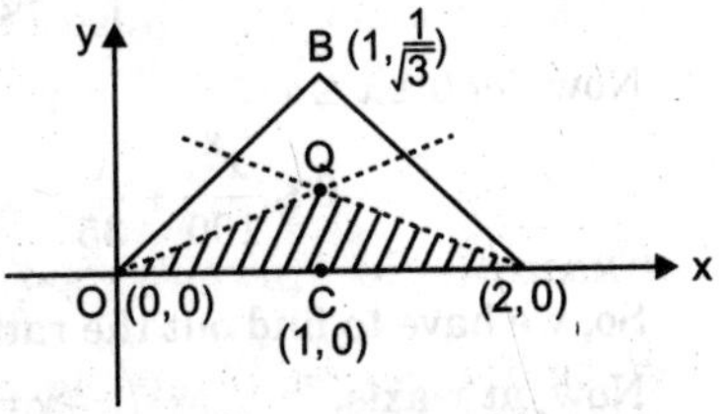

Example 35 A curve $y = f(x)$ passes through the origin and lies entirely in the first quadrant. Through any point $P(x, y)$ on the curve, lines are drawn parallel to the coordinate axes. If the curve divides the area formed by these lines and coordinate axes in $m : n$, find $f(x)$.

Solution. Area of $(OAPB) = xy$

Area of $(OAPO) = \int_0^x f(t)\,dt$

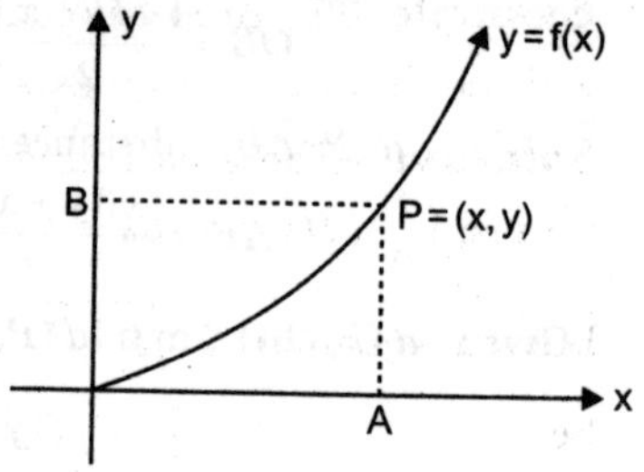

Therefore, area of $(OBPO) = xy - \int_0^x f(t)\,dt$

According to given condition;

$$\frac{xy - \int_0^x f(t)\,dt}{\int_0^x f(t)\,dt} = \frac{m}{n}$$

$$\Rightarrow \qquad n\,xy = (m+n)\int_0^x f(t)\,dt$$

Differentiating w.r.t. x, we get

$$n\left(x\frac{dy}{dx} + y\right) = (m+n)f(x) = (m+n)\,y, \text{ as } y = f(x)$$

$$\frac{m}{n}\cdot\frac{dx}{x} = \frac{dy}{y} \Rightarrow \frac{m}{n}\cdot(\log x) = \log y - \log c, \text{ where } c \text{ is a constant.}$$

$$\Rightarrow \qquad y = cx^{m/n}$$

Example 36 Find the ratio of the areas in which the curve $y = \left[\frac{x^3}{100} + \frac{x}{35}\right]$ divides the circle $x^2 + y^2 - 4x + 2y + 1 = 0$ (where [.] denotes the greatest integer function).

Solution. We have, $x^2 + y^2 - 4x + 2y + 1 = 0$

or $\qquad (x-2)^2 + (y+1)^2 = 4 \qquad$ ...(i)

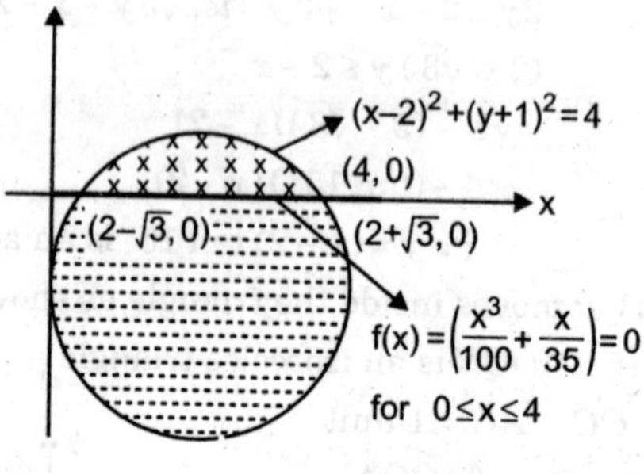

Now, for $0 \le x \le 4$

$$0 \le \frac{x^3}{100} + \frac{x}{35} < 1 \quad \Rightarrow \quad \left[\frac{x^3}{100} + \frac{x}{35}\right] = 0$$

So, we have to find out the ratio in which x-axis divides the circle (i).

Now, at x-axis, $\qquad y = 0$

So, $\qquad (x-2)^2 = 3$

So, it cuts the x-axis at $(2-\sqrt{3}, 0)$ and $(2+\sqrt{3}, 0)$.

Therefore, the required area $A = \int_{2-\sqrt{3}}^{2+\sqrt{3}} (\sqrt{4-(x-2)^2} - 1)\,dx = \frac{4\pi - 3\sqrt{3}}{3}$

The required ratio $= \frac{A}{4\pi - A} = \frac{4\pi - 3\sqrt{3}}{8\pi + 3\sqrt{3}}$

Example 37 Area bounded by the line $y = x$, curve $y = f(x)$, $(f(x) > x, \forall x > 1)$ and the lines $x = 1$, $x = t$ is $(t + \sqrt{1 + t^2}) - (1 + \sqrt{2})$ for all $t > 1$. Find $f(x)$.

Solution. The area bounded by $y = f(x)$ and $y = x$ between the lines $x = 1$ and $x = t$ is; $\int_{+1}^{t} (f(x) - x)\, dx$

But it is equal to $(t + \sqrt{1 + t^2}) - (1 + \sqrt{2})$

So, $$\int_1^t (f(x) - x)\, dx = (t + \sqrt{1 + t^2}) - (1 + \sqrt{2})$$

Differentiating both the sides w.r.t. t, we get

$$f(t) - t = 1 + \frac{t}{\sqrt{1 + t^2}} \Rightarrow f(t) = 1 + t + \frac{t}{\sqrt{1 + t^2}}$$

or $$f(x) = 1 + x + \frac{x}{\sqrt{1 + x^2}}$$

Example 38 The area bounded by the curve $y = f(x)$, x-axis and the ordinates $x = 1$ and $x = b$ is $(b - 1)\sin(3b + 4)$, find $f(x)$.

Solution. We know that the area bounded by the curve $y = f(x)$, x-axis and the ordinates $x = 1$ and $x = b$ is $\int_1^b f(x)\, dx$

$\therefore$ From the question; $\int_1^b f(x)\, dx = (b - 1)\sin(3b + 4)$

Differentiating w.r.t. b, we get

$$f(b) \cdot 1 = 3(b - 1)\cos(3b + 4) + \sin(3b + 4)$$

$$\Rightarrow \quad f(x) = 3(x - 1)\cos(3x + 4) + \sin(3x + 4)$$

Example 39 Find the area of region enclosed by the curve

$\frac{(x - y)^2}{a^2} + \frac{(x + y)^2}{b^2} = 2\ (a > b)$, the line $y = x$ and the positive x-axis.

Solution. The given curve $\frac{(x - y)^2}{a^2} + \frac{(x + y)^2}{b^2} = 2$ is an ellipse major and minor axes are $x - y = 0$ and $x + y = 0$ respectively.

The required area is shown with shaded region.

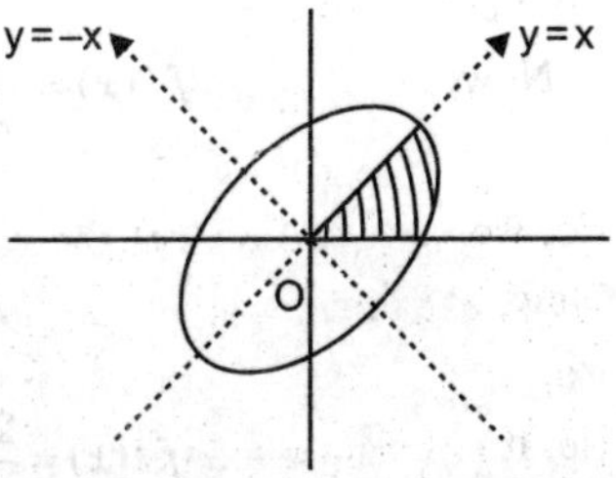

Instead of directly solving the problem we can solve equivalent problem with equivalent ellipse whose axes are $x = 0$ and $y = 0$.

The equivalent region is shown as $(OA'B'O)$ where the equation of ellipse is $\frac{x^2}{a^2} + \frac{y^2}{b^2} = 1$.

$\therefore$ Required area = area $(\Delta OA'A'' + A'A''B')$

where $$A' = \left(\frac{ab}{\sqrt{a^2 + b^2}}, \frac{ab}{\sqrt{a^2 + b^2}}\right)$$

Area of $\Delta OA'A'' = \frac{1}{2} \times \frac{ab}{\sqrt{a^2 + b^2}} \times \frac{ab}{\sqrt{a^2 + b^2}} = \frac{1}{2}\left(\frac{a^2b^2}{a^2 + b^2}\right)$ ...(i)

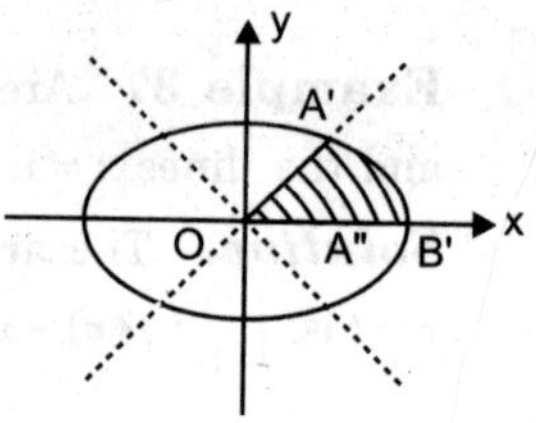

Area of $A'B'A'' = \int_{\frac{ab}{\sqrt{a^2+b^2}}}^{a} b\sqrt{\left(1-\frac{x^2}{a^2}\right)}\,dx$

$$= \frac{b}{a}\cdot\int_{\frac{ab}{\sqrt{a^2+b^2}}}^{a} \sqrt{a^2-x^2}\,dx$$

$$= \frac{b}{a}\left[\frac{x}{2}\sqrt{a^2-x^2}+\frac{a^2}{2}\sin^{-1}\frac{x}{a}\right]_{ab/\sqrt{a^2+b^2}}^{a}$$

$$= \frac{b}{a}\left[0+\frac{a^2}{2}\cdot\frac{\pi}{2}-\frac{ab}{2\sqrt{a^2+b^2}}\cdot\sqrt{\left(a^2-\frac{a^2b^2}{a^2+b^2}\right)}-\frac{a^2}{2}\sin^{-1}\frac{b}{\sqrt{a^2+b^2}}\right]$$

$$= \frac{b}{a}\left[\frac{\pi a^2}{4}-\frac{a^2}{2}\sin^{-1}\left(\frac{b}{\sqrt{a^2+b^2}}\right)-\frac{a^3b}{2(a^2+b^2)}\right]$$

$$= \frac{\pi ab}{4}-\frac{ab}{2}\sin^{-1}\left(\frac{b}{\sqrt{a^2+b^2}}\right)-\frac{a^2b^2}{2(a^2+b^2)} \quad \text{...(ii)}$$

Hence, the required area = Sum of Eqs. (i) and (ii)

$$= \frac{\pi ab}{4}-\frac{ab}{2}\sin^{-1}\left(\frac{b}{\sqrt{a^2+b^2}}\right)$$

Example 40 Let $f(x)$ be a function which satisfy the equation $f(xy)=f(x)+f(y)$ for all $x>0$, $y>0$ such that $f'(1)=2$. Find the area of the region bounded by the curves $y=f(x)$, $y=|x^3-6x^2+11x-6|$ and $x=0$.

Solution. Take $x=y=1 \Rightarrow f(1)=0$

Now, $$y=\frac{1}{x}$$

$\Rightarrow$ $$0=f\left(x\cdot\frac{1}{x}\right)=f(x)+f\left(\frac{1}{x}\right) \Rightarrow f\left(\frac{1}{x}\right)=-f(x)$$

$\therefore$ $$f\left(\frac{x}{y}\right)=f(x)+f\left(\frac{1}{y}\right)=f(x)-f(y) \quad \text{...(i)}$$

Now, $$f'(x)=\lim_{h\to 0}\frac{f(x+h)-f(x)}{h}=\lim_{h\to 0}\frac{f\left(\frac{x+h}{x}\right)}{h} \quad \text{[using Eq. (i)]}$$

$$= \lim_{h\to 0}\frac{f\left(1+\frac{h}{x}\right)-f(1)}{\frac{h}{x}\cdot x}=\frac{f'(1)}{x} \quad [\because f(1)=0]$$

$$f'(x)=\frac{2}{x}$$

$\Rightarrow$ $$f(x)=2\log x+c \quad \text{[since, } f(1)=0\Rightarrow c=0]$$

$\Rightarrow$ $$f(x)=2\log x$$

Thus, $f(x) = 2 \log x$ and $y = |x^3 - 6x^2 + 11x - 6|$ could be plotted as;

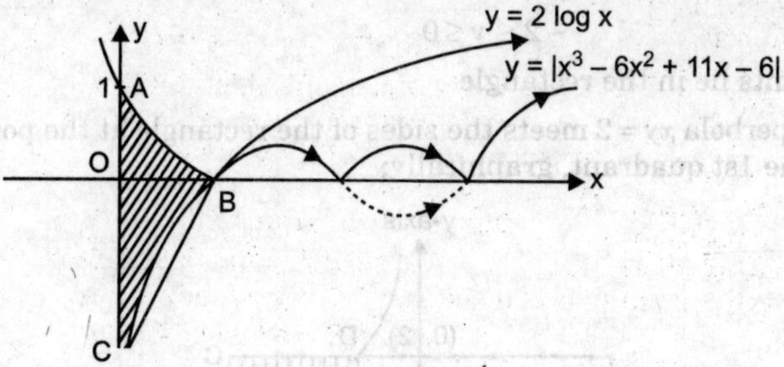

Hence, required area of $OABCO = \int_0^1 (x^3 - 6x^2 + 11x - 6)\, dx + \int_{-\infty}^{0} e^{y/2}\, dy$

$$= \left(\frac{x^4}{4} - \frac{6x^3}{3} + \frac{11x^2}{2} - 6x \right)_0^1 + 2(e^{y/2})_{-\infty}^{0}$$

$$= \left(\frac{1}{4} - 2 + \frac{11}{2} - 6 \right) - (0) + 2(e^0 - e^{-\infty})$$

$$= \frac{1}{4} + \frac{11}{2} - 8 + 2 = -\frac{1}{4} \text{ sq unit}$$

Example 41 Find the area of the region which contains all the points satisfying the condition $|x - 2y| + |x + 2y| \le 8$ and $xy \ge 2$.

Solution. The line $y = \pm \dfrac{x}{2}$ divide the xy plane in four parts :

Region I $\quad 2y - x \le 0$ and $2y + x \ge 0$

So that, $\quad |x - 2y| + |x + 2y| \le 8$

$\Rightarrow \quad (x - 2y) + (x + 2y) \le 8 \Rightarrow 0 \le x \le 4$

Region II $\quad 2y - x \,|\, 0$ and $2y + x \ge 0$

So that, $\quad |x - 2y| + |x + 2y| \le 8$

$\Rightarrow \quad -(x - 2y) + (x + 2y) \le 8 \Rightarrow 0 \le y \le 2$

Region III $\quad 2y + x \le 0, 2y - x \ge 0$

So that, $\quad |x - 2y| + |x + 2y| \le 8$

$\Rightarrow \quad -(x - 2y) - (x + 2y) \le 8 \Rightarrow -4 \le x \le 0$

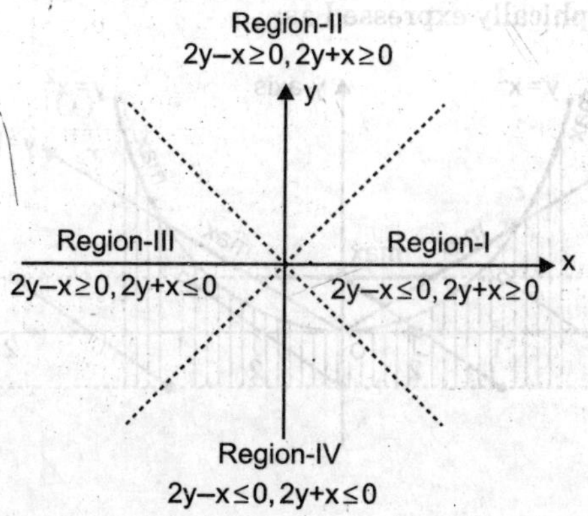

Region IV $\qquad 2y + x \le 0, 2y - x \le 0$

So that, $\qquad |x - 2y| + |x + 2y| \le 8$

$\Rightarrow \qquad -2 \le y \le 0$

Here, all points lie in the rectangle.

Also, the hyperbola $xy = 2$ meets the sides of the rectangle at the points (1, 2) and (4, 1/2) in the 1st quadrant, graphically;

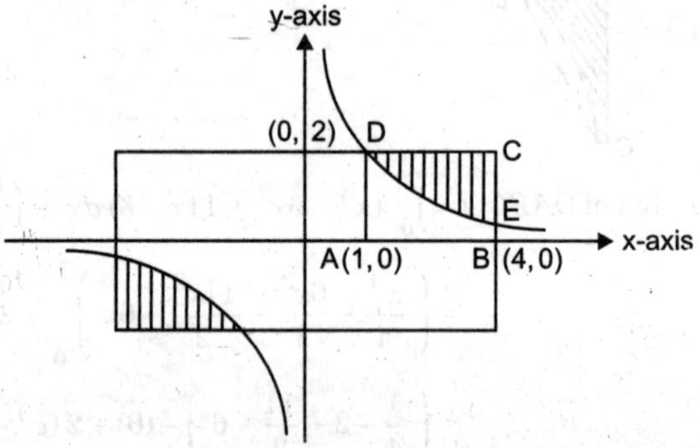

Hence, the required area = 2 (area of rectangle *ABCD* – area of *ABEDA*)

$$= 2\left(3 \times 2 - \int_1^4 \frac{2}{x}\,dx\right) = 2\,(6 - 2\log 4) \text{ sq units}$$

Example 42 Consider the function $f(x) = \begin{cases} x - [x] - \frac{1}{2}, & \text{if } x \notin I \\ 0, & \text{if } x \in I \end{cases}$

where [.] denotes the greatest integer function and I is the set of integers. If $g(x) = \max\{x^2, f(x), |x|\}, -2 \le x \le 2$, then find the area bounded by $g(x)$ when $-2 \le x \le 2$.

Solution. Here, $\quad f(x) = \begin{cases} x - [x] - \frac{1}{2}, & \text{if } x \notin I \\ 0, & \text{if } x \in I \end{cases}$

$$= \begin{cases} \{x\} - \frac{1}{2}, & \text{if } x \notin I \\ 0, & \text{if } x \in I \end{cases}$$

Thus, $\qquad g(x) = \max\{x^2, f(x), |x|\}$

Which could be graphically expressed as;

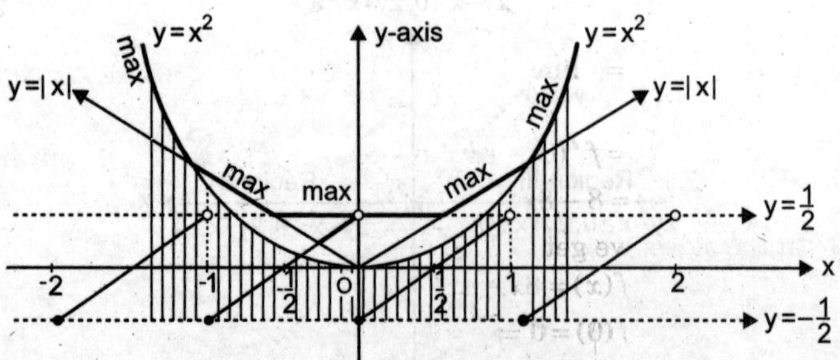

Clearly, $$g(x) = \begin{cases} x^2, & -2 \le x \le -1 \\ -x, & -1 \le x \le -1/4 \\ x + \frac{1}{2}, & -1/4 \le x \le 0 \\ x, & 0 \le x \le 1 \\ x^2, & 1 \le x \le 2 \end{cases}$$

Hence, the required area

$$= \int_{-2}^{2} g(x)\, dx$$

$$= \int_{-2}^{-1} (x^2)\, dx + \int_{-1}^{-1/4} (-x)\, dx + \int_{-1/4}^{0} \left(x + \frac{1}{2} \right) dx + \int_0^1 x\, dx + \int_1^2 x^2\, dx$$

$$= \left(\frac{x^3}{3} \right)_{-2}^{-1} - \left(\frac{x^2}{2} \right)_{-1}^{-1/4} + \left(\frac{x^2}{2} + \frac{1}{2} x \right)_{-1/4}^{0} + \left(\frac{x^2}{2} \right)_0^1 + \left(\frac{x^3}{3} \right)_1^2$$

$$= \frac{275}{48} \text{ sq units}$$

Example 43 Find the area of the region bounded by $y = f(x)$, $y = |g(x)|$ and the lines $x = 0$, $x = 2$. where 'f', 'g' are continuous functions satisfying

$$f(x + y) = f(x) + f(y) - 8xy, \ \forall\, x, y \in R$$

and $$g(x + y) = g(x) + g(y) + 3xy\,(x + y), \ \forall\, x, y \in R$$

Also, $f'(0) = 8$ and $g'(0) = -4$

Solution. Here, $f(x + y) = f(x) + f(y) - 8xy$

Replacing $x, y \to 0$, we get $f(0) = 0$

Now, $$f'(x) = \lim_{h \to 0} \frac{f(x + h) - f(x)}{h}$$

$$= \lim_{y \to 0} \frac{f(x + y) - f(x)}{y}$$

$$= \lim_{y \to 0} \frac{f(x) + f(y) - 8xy - f(x)}{y}$$

$$= \lim_{y \to 0} \left\{ \frac{f(y)}{y} - \frac{8xy}{y} \right\}$$

$$= \lim_{y \to 0} \left(\frac{f'(y)}{1} \right) - 8x \quad \text{(using L'Hospital's rule)}$$

$$= f'(0) - 8x = 8 - 8x \quad [\text{given } f'(0) = 8]$$

$\Rightarrow$ $f'(x) = 8 - 8x$

Integrating both the sides, we get

$$f(x) = 8x - 4x^2 + c$$

As $f(0) = 0 \Rightarrow c = 0$

$\Rightarrow$ $f(x) = 8x - 4x^2$ ...(i)

Also, $$g(x+y) = g(x) + g(y) + 3xy(x+y)$$

Replacing $x, y \Rightarrow 0$, we get $g(0) = 0$

Now, $$g'(x) = \lim_{y \to 0} \frac{g(x+y) - g(x)}{y}$$

$$= \lim_{y \to 0} \frac{g(x) + g(y) + 3x^2 y + 3xy^2 - g(x)}{y}$$

$$= \lim_{y \to 0} \left[\frac{g(y)}{y} + \frac{y(3x^2 + 3xy)}{y} \right]$$

$$= \frac{g'(0)}{1} + 3x^2 = -4 + 3x^2$$

$\therefore$ $$g'(x) = -4 + 3x^2$$

$\Rightarrow$ $$g(x) = x^3 - 4x \qquad \text{[as } g(0) = 0] \quad \ldots\text{(ii)}$$

Points where $f(x)$ and $g(x)$ meets, we have

$$f(x) = g(x) \quad \text{or} \quad 8x - 4x^2 = x^3 - 4x$$

$\Rightarrow$ $$x = 0, 2, -6$$

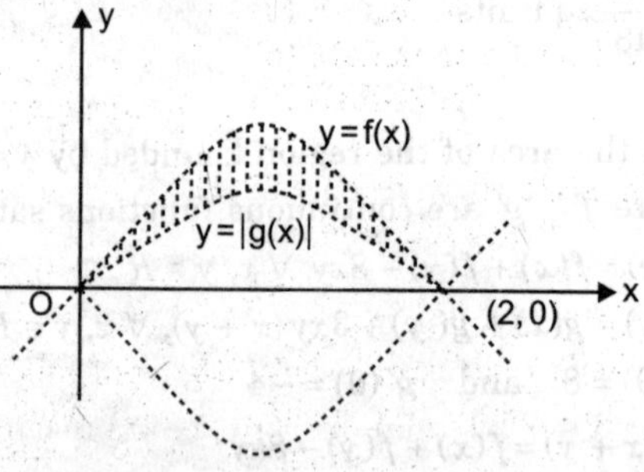

Now, $$|g(x)| = \begin{cases} x^3 - 4x, & x \in [-2, 0] \cup (2, \infty) \\ 4x - x^3, & x \in [-\infty, -2] \cup (0, 2) \end{cases}$$

Area bounded by $y = f(x)$ and $y = |g(x)|$ between $x = 0$ to $x = 2$ is

$$\int_0^2 \{(8x - 4x^2) - (4x - x^3)\}\, dx$$

$$= \int_0^2 (x^3 - 4x^2 + 4x)\, dx = \frac{4}{3} \text{ sq units}$$

Example 44 Find the area of the region bounded by the curve $y = x^2$ and $y = \sec^{-1}[-\sin^2 x]$, where [.] denotes the greatest integer function.

Solution. As we know, $[-\sin^2 x] = 0$ or -1

But $\sec^{-1}(0)$ is not defined.

$\Rightarrow$ $$\sec^{-1}[-\sin^2 x] = \sec^{-1}(-1) = \pi$$

Thus, to find area bounded between

$$y = x^2 \quad \text{and} \quad y = \pi$$

ie, when $x^2 = \pi$ or $(x = -\sqrt{\pi}$ to $x = \sqrt{\pi})$

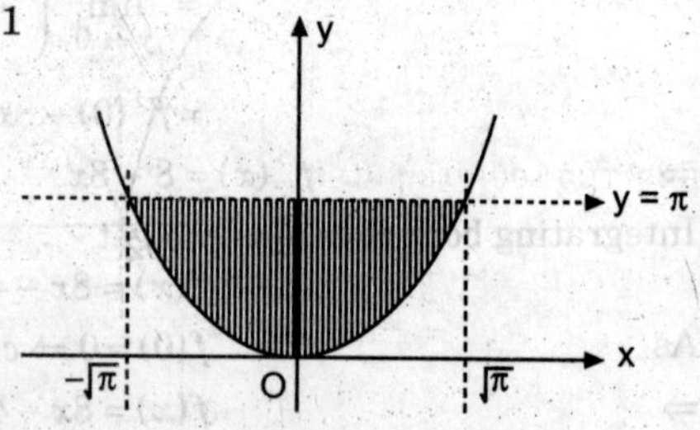

$$\therefore \qquad \text{Area} = \int_{-\sqrt{\pi}}^{\sqrt{\pi}} (\pi - x^2)\,dx = \left(\pi x - \frac{x^3}{3}\right)_{-\sqrt{\pi}}^{\sqrt{\pi}}$$

$$= \pi(\sqrt{\pi} + \sqrt{\pi}) - \frac{1}{3}(\pi\sqrt{\pi} + \pi\sqrt{\pi}) = \frac{4\pi}{3}\sqrt{\pi}$$

Example 45 Sketch the graph of $\cos^{-1}(4x^3 - 3x)$ and find the area enclosed between $y = 0$, $y = f(x)$ and $x \geq -1/2$.

Solution. Here, $\qquad f(x) = \cos^{-1}(4x^3 - 3x)$

Let $\qquad x = \cos\theta \quad \text{and} \quad 0 \leq \theta \leq \pi$

$$\Rightarrow \qquad f(x) = \cos^{-1}(4\cos^3\theta - 3\cos\theta)$$

$$= \cos^{-1}(\cos 3\theta);\ 0 \leq 3\theta \leq 3\pi$$

$$\Rightarrow \qquad f(x) = \begin{cases} 3\theta, & 0 \leq 3\theta \leq \pi \\ 2\pi - 3\theta, & \pi < 3\theta \leq 2\pi \\ 3\theta - 2\pi, & 2\pi < 3\theta \leq 3\pi \end{cases}$$

$$\Rightarrow \qquad f(x) = \begin{cases} 3\cos^{-1}x, & 1/2 \leq x \leq 1 \\ 2\pi - 3\cos^{-1}x, & -1/2 \leq x < 1/2 \\ 3\cos^{-1}x - 2\pi, & -1 \leq x < -1/2 \end{cases}$$

$$\therefore \qquad f'(x) = \begin{cases} -3/\sqrt{1 - x^2}, & 1/2 < x < 1 \\ 3/\sqrt{1 - x^2}, & -1/2 < x < 1/2 \\ -3/\sqrt{1 - x^2}, & -1 < x < -1/2 \end{cases}$$

and

$$f''(x) = \begin{cases} \dfrac{-3x}{(1 - x^2)^{3/2}}, & 1/2 < x < 1 \\ \dfrac{3x}{(1 - x^2)^{3/2}}, & -1/2 < x < 1/2 \\ \dfrac{-3x}{(1 - x^2)^{3/2}}, & -1 < x < -1/2 \end{cases}$$

Thus, the graph for $f(x) = \cos^{-1}(4x^3 - 3x)$

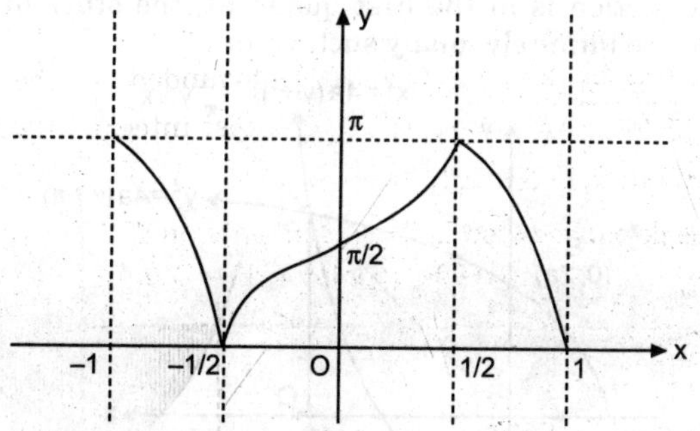

Thus, the required area, $\Delta = \int_{-1/2}^{1} f(x)\,dx$

$$= \int_{-1/2}^{1/2} (2\pi - 3\cos^{-1} x)\,dx + \int_{1/2}^{1} (3\cos^{-1} x)\,dx$$

$$= 2\pi - 3\int_{-1/2}^{1/2} \left(\frac{\pi}{2} - \sin^{-1} x\right) dx + 3\int_{1/2}^{1} (\cos^{-1} x)\,dx$$

$$= \frac{\pi}{2} + 3\int_{1/2}^{1} \cos^{-1} x\,dx \qquad \left[\text{as } \int_{-1/2}^{1/2} \sin^{-1} x\,dx = 0\right]$$

$$= \frac{\pi}{2} + 3\left\{(x\cos^{-1} x)_{1/2}^{1} + \int_{1/2}^{1} \frac{x}{\sqrt{1-x^2}}\,dx\right\}$$

On solving, we get, $\Delta = \frac{3\sqrt{3}}{2}$ sq units

Example 46 Consider two curves $y^2 = 4a(x - \lambda)$ and $x^2 = 4a(y - \lambda)$ where $a > 0$ and λ is a parameter show that :

(i) There is a single positive value of λ for which the two curves have exactly one point of intersection in the 1st quadrant find it.

(ii) There are infinitely many negative values of λ for which the two curves have exactly one point of intersection in the 1st quadrant.

(iii) If $\lambda = -a$, then find the area of the bounded by the two curves and the axes in the 1st quadrant.

Solution. The two curves are inverse of each other. Hence, the two curves always meet along the line $y = x$.

Consider, $y^2 = 4a(x - \lambda)$ and put $x = y \Rightarrow y^2 - 4ay + 4a\lambda = 0$

$$\Rightarrow \quad y = \frac{4a \pm 4\sqrt{a^2 - a\lambda}}{2} = 2a \pm \sqrt{a^2 - a\lambda}$$

Since, y is real $\Rightarrow a^2 - a\lambda \,|\, 0$ or $\lambda \le a$

(i) If $0 < \lambda < a$, then there are two distinct values of y and both $2(a + \sqrt{a^2 - a\lambda})$ and $2(a - \sqrt{a^2 - a\lambda})$ are positive, *ie*, both points lie in the first quadrant.

If $\lambda = a$, then $y = 2a$ only, *ie*, only one point of intersection $(2a, 2a)$.

Hence, there is exactly one point of intersection in 1st quadrant for $\lambda = a$. It is infact the points of tangency of the two curves.

(ii) If $\lambda < 0$, then $y = 2(a + \sqrt{a^2 - a\lambda}) > 0$ and $y = 2(a - \sqrt{a^2 - a\lambda}) < 0$. *ie*, the only point of intersection is in the first quadrant, the other in the 3rd quadrant. Hence, there are infinitely many such values.

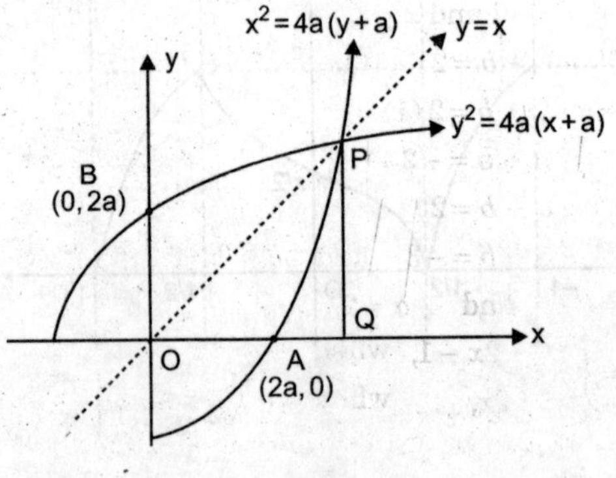

(iii) For $\lambda = -a$, we have $y^2 = 4a(x + a)$

$$x^2 = 4a(y + a)$$

The point of intersection in the 1st quadrant $P = ((2 + 2\sqrt{2})a, (2 + 2\sqrt{2})a)$

Required area = 2 (area of ΔOPQ − area $APQA$)

$$\text{Area of } \Delta OPQ = \frac{1}{2} \cdot (2 + 2\sqrt{2})^2 a^2 = 2(1 + \sqrt{2})^2 a^2$$

$$\text{Area of } APQA = \int_{2a}^{(2+2\sqrt{2})a} y\, dx = \int_{2a}^{(2+2\sqrt{2})a} \left(\frac{x^2}{4a} - a \right) dx$$

$$= \frac{1}{4a} \left[\frac{x^3}{3} \right]_{2a}^{(2+2\sqrt{2})a} - a(x)_{2a}^{(2+2\sqrt{2})a}$$

$$= \frac{1}{12a} [(2 + 2\sqrt{2})^3 \cdot a^3 - 8a^3] - 2\sqrt{2}\, a^2 = \left(\frac{4\sqrt{2} + 12}{3} \right) a^2$$

$$\text{Required area} = 2\left(2(1 + \sqrt{2})^2 + \frac{4\sqrt{2} + 12}{3} \right) a^2$$

$$= \frac{4}{3}(15 + 8\sqrt{2})\, a^2$$

Example 47 Let $f(x)$ be continuous function given by;

$$f(x) = \begin{cases} 2x, & |x| \le 1 \\ x^2 + ax + b, & |x| > 1 \end{cases}$$

Find the area of the region in the third quadrant bounded by the curves $x = -2y^2$ and $y = f(x)$ lying on the left of the line $8x + 1 = 0$. **[IIT JEE 1999]**

Solution. Given, a continuous function $f(x)$, given by

$$f(x) = \begin{cases} 2x, & |x| \le 1 \\ x^2 + ax + b, & |x| > 1 \end{cases}$$

ie,
$$f(x) = \begin{cases} x^2 + ax + b, & -\infty < x < -1 \\ 2x, & -1 \le x \le 1 \\ x^2 + ax + b, & 1 < x < \infty \end{cases}$$

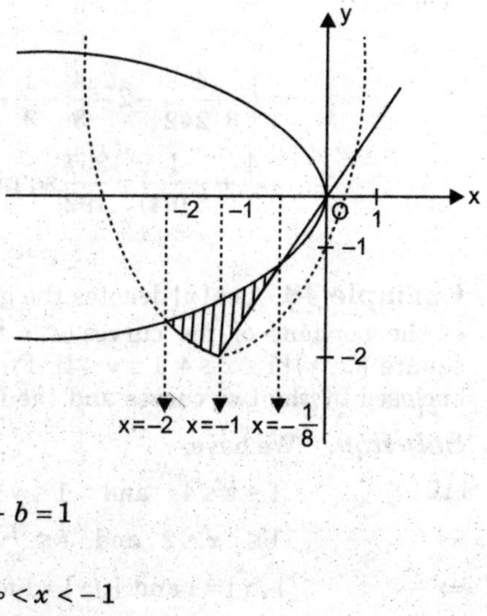

$\because$ f is continuous.

$\therefore$ It is continuous at $x = -1$ and $x = 1$.

$\therefore$ $(-1)^2 + a(-1) + b = 2(-1)$

and $(1)^2 + a(1) + b = 2(1)$

$\Rightarrow$ $1 - a + b = -2$

and $1 + a + b = 2$

$\Rightarrow$ $-a + b = -3$ and $a + b = 1$

$\Rightarrow$ $b = -1$ and $a = 2$

$$\therefore \quad f(x) = \begin{cases} x^2 + 2x - 1, & \text{when } -\infty < x < -1 \\ 2x, & \text{when } -1 \le x \le 1 \end{cases}$$

Now, $y = x^2 + 2x - 1 = (x+1)^2 - 2$

or $(y+2) = (x+1)^2$

We need the area of the region in third quadrant bounded by the curves $a = -2y^2$, $y = f(x)$ lying on the left of the line $8x + 1 = 0$.

$\Rightarrow$ $(y+2) = (x+1)^2$

Cuts the y-axis at $(0, -1)$ and the x-axis at $(-1-\sqrt{2}, 0)$ and $(-1+\sqrt{2}, 0)$.
When $x = 1$ and $y = 2$

Solving, $x = -2y^2$ and $(y+2) = (x+1)^2$

we get $(y+2) = (-2y^2+1)^2$

$\Rightarrow$ $4y^4 - 4y^2 + 1 - y - 2 = 0$

$\Rightarrow$ $4y^4 - 4y^2 - y - 1 = 0$

For $y = -1, 4y^4 - 4y^2 - y - 1 = 0$

$\therefore$ at $y = -1, x = -2$

$$\text{Required area} = \int_{-2}^{-1}\left\{-\sqrt{-\frac{x}{2}} - (x+1)^2 + 2\right\}dx + \int_{-1}^{-1/8}\left\{-\sqrt{-\frac{x}{2}} - 2x\right\}dx$$

$$= \left[\frac{(-x/2)^{3/2}}{\frac{3}{2}(1/2)} - \frac{(x+1)^3}{3} + 2x\right]_{-2}^{-1} + \left[\frac{(-x/2)^{3/2}}{\frac{3}{2}(1/2)} - \frac{2x^2}{2}\right]_{-1}^{-1/8}$$

$$= \left\{\frac{4}{3}\left(-\frac{x}{2}\right)^{3/2} - \frac{(x+1)^3}{3} + 2x\right\}_{-2}^{-1} + \left\{\frac{4}{3}\left(-\frac{x}{2}\right)^{3/2} - x^2\right\}_{-1}^{-1/8}$$

$$= \left\{\frac{4}{3}\left(\frac{1}{2}\right)^{3/2} - 0 - 2\right\} - \left\{\frac{4}{3}(1)^{3/2} - \frac{(-1)^3}{3} - 4\right\}$$

$$- \left\{\frac{4}{3}\left(\frac{1}{2}\right)^{3/2} - (-1)^2\right\} + \left\{\frac{4}{3}\left(\frac{1}{16}\right)^{3/2} - \frac{1}{64}\right\}$$

$$= \left\{\frac{4}{3\cdot 2\sqrt{2}} - 2 - \frac{4}{3} - \frac{1}{3} + 4 - \frac{4}{3\cdot 2\sqrt{2}} + 1 + \frac{4}{3\,(64)} - \frac{1}{64}\right\}$$

$$= \frac{4}{3} + \frac{1}{3\,(64)} = \frac{257}{192} \text{ sq units}$$

Example 48 Let $[x]$ denotes the greatest integer function. Draw a rough sketch of the portions of the curves $x^2 = 4[\sqrt{x}]\,y$ and $y^2 = 4[\sqrt{y}]\,x$ that lie within the square $\{(x, y) \mid 1 \le x \le 4, 1 \le y \le 4\}$. Find the area of the part of the square that is enclosed by the two curves and the line $x + y = 3$.

Solution. We have,

$1 \le x \le 4$ and $1 \le y \le 4$

$\Rightarrow$ $1 \le \sqrt{x} \le 2$ and $1 \le \sqrt{y} \le 2$

$\Rightarrow$ $[\sqrt{x}] = 1$ and $[\sqrt{y}] = 1$ for all (x, y) lying with in the square.

Thus, $x^2 = 4[\sqrt{x}]\,y$ and $y^2 = 4[\sqrt{y}]\,x$

$\Rightarrow$ $x^2 = 4y$ and $y^2 = 4x$ when $1 \le x, y \le 4$ which could be plotted as;

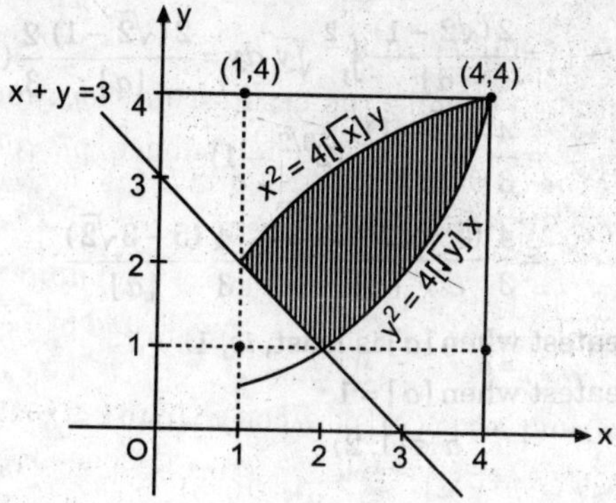

Thus, the required area

$$= \int_1^2 (2\sqrt{x} - 3 + x)\,dx + \int_2^4 \left(2\sqrt{x} - \frac{x^2}{4}\right) dx$$

$$= \left(\frac{4}{3}x^{3/2} - 3x + \frac{x^2}{2}\right)_1^2 + \left(\frac{4}{3}x^{3/2} - \frac{x^3}{12}\right)_2^4$$

$$= \left(\frac{8\sqrt{2}}{3} - 6 + 2\right) - \left(\frac{4}{3} - 3 + \frac{1}{2}\right) + \left(\frac{32}{3} - \frac{64}{12}\right) - \left(\frac{8\sqrt{2}}{3} - \frac{8}{12}\right)$$

$$= \frac{19}{6} \text{ sq units}$$

Example 49 Find all the values of the parameter a $(a \geq 1)$ for which the area of the figure bounded by pair of straight lines $y^2 - 3y + 2 = 0$ and the curves $y = [a]\,x^2$, $y = \frac{1}{2}[a]\,x^2$ is greatest, where [.] denotes the greatest integer function.

Solution. The curves $y = [a]\,x^2$ and $y = \frac{1}{2}[a]\,x^2$ represent parabolas which are symmetric about y-axis. The equation $y^2 - 3y + 2 = 0$ gives a pair of straight lines $y = 1$, $y = 2$ which are parallel to x-axis.
Thus, the area bounded is shown as;

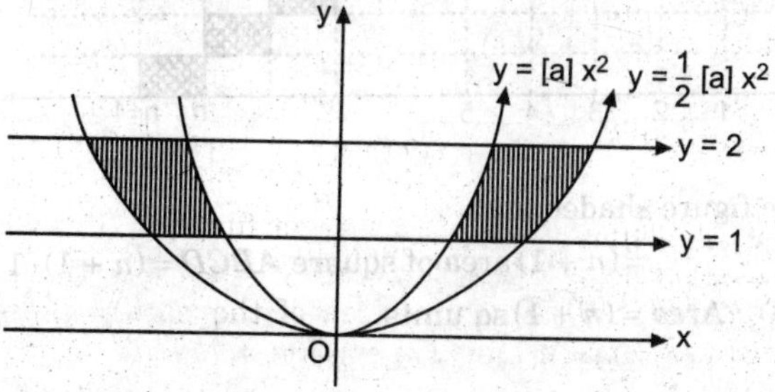

From above figure the required area,

$$= 2\int_{y=1}^{y=2} (x_2 - x_1)\,dy$$

$$= 2\int_1^2 \left(\sqrt{\frac{2y}{[a]}} - \sqrt{\frac{y}{[a]}}\right) dy$$

$$= \frac{2(\sqrt{2}-1)}{[a]} \int_1^2 \sqrt{y}\, dy = \frac{2(\sqrt{2}-1)}{[a]} \frac{2}{3} (y^{3/2})_1^2$$

$$= \frac{4}{3} \frac{(\sqrt{2}-1)}{[a]} \cdot (2^{3/2}-1)$$

$$= \frac{4}{3} \frac{(5-\sqrt{2}-2\sqrt{2})}{[a]} = \frac{4}{3} \frac{(5-3\sqrt{2})}{[a]}$$

$\therefore$ Area is the greatest when $[a]$ is least, *ie*, 1.

$\therefore$ Area is the greatest when $[a] = 1$

$\Rightarrow$ $a \in [1, 2)$

Example 50 Find the area in the first quadrant bounded by $[x] + [y] = n$ where $n \in N$ and $y = i$ (where $i \in N \ \forall \ i \le n+1$), where $[\cdot]$ denotes the greatest integer less than or equal to x.

Solution. As we know, $[x] + [y] = n \Rightarrow [x] = n - [y]$

when $y = 0, [x] = n \Rightarrow n \le x < n + 1$

when $y = 1, [x] = n - 1 \Rightarrow n - 1 \le x < n - 2$

when $y = 2, [x] = n - 2$

$\Rightarrow$ $n - 2 \le x < n - 3$. ..... and so on.

when $y = n$, $[x] = 0 \Rightarrow 0 \le x < 1$

which could be shown as;

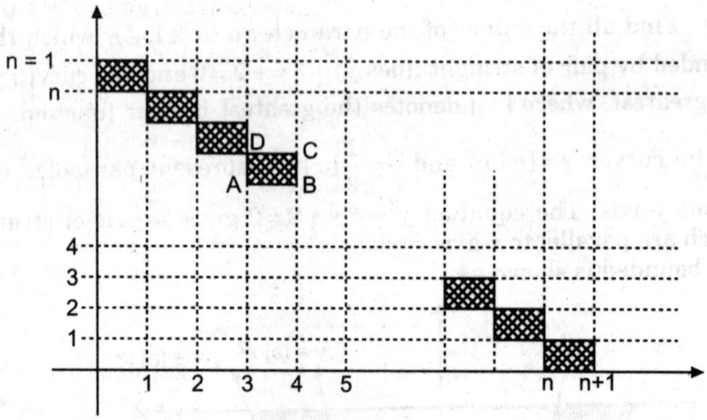

Thus from the above figure shaded area

$= (n+1)$ area of square $ABCD = (n+1) \cdot 1$

$\Rightarrow$ Area $= (n+1)$ sq units

Type 2 : Only One Correct Option

Example 51 If A denotes the area bounded by $f(x) = \left| \frac{\sin x + \cos x}{x} \right|$, x-axis, $x = \pi$ and $x = 3\pi$, then

(a) $1 < A < 2$ (b) $0 < A < 2$ (c) $2 < A < 3$ (d) None of these

Solution. $\frac{2\sqrt{2}}{2\pi} < \int_{\pi}^{2\pi} \frac{|\sin x + \cos x|}{x} dx < \frac{2\sqrt{2}}{2\pi} \left[\because \pi < x < 2\pi \Rightarrow \frac{1}{2\pi} < \frac{1}{x} < \frac{1}{\pi}\right]$ ...(i)

$\frac{2\sqrt{2}}{3\pi} < \int_{2\pi}^{3\pi} \frac{|\sin x + \cos x|}{x} dx < \frac{2\sqrt{2}}{2\pi}$ ...(ii)

On adding Eqs. (i) and (ii), we get

$$\frac{5\sqrt{2}}{3\pi} < A < \frac{3\sqrt{2}}{\pi} \Rightarrow 0.75 < A < 1.3$$

Hence, (b) is the correct answer.

Example 52 If $f(x) \geq 0, \forall x \in (0, 2)$ and $y = f(x)$ makes positive intercepts of 2 and 1 unit on x and y-axes respectively and encloses an area of 3/4 unit with axes, then $\int_0^2 xf'(x)\,dx$ is

(a) $\frac{3}{4}$ (b) 1 (c) $\frac{5}{4}$ (d) $-\frac{3}{4}$

Solution. $I = xf(x)|_0^2 - \int_0^2 f(x)\,dx = 0 - \frac{3}{4} = -\frac{3}{4}$

Hence, (d) is the correct answer.

Example 53 The area of the region included between the regions satisfying $\min(|x|, |y|) \geq 1$ and $x^2 + y^2 \leq 5$, is

(a) $\frac{5}{2}\left(\sin^{-1}\frac{2}{\sqrt{5}} - \sin^{-1}\frac{1}{\sqrt{5}}\right) - 4$ (b) $10\left(\sin^{-1}\frac{2}{\sqrt{5}} - \sin^{-1}\frac{1}{\sqrt{5}}\right) - 4$

(c) $\frac{2}{5}\left(\sin^{-1}\frac{2}{\sqrt{5}} - \sin^{-1}\frac{1}{\sqrt{5}}\right) - 4$ (d) $15\left(\sin^{-1}\frac{2}{\sqrt{5}} - \sin^{-1}\frac{1}{\sqrt{5}}\right) - 4$

Solution. Shaded region depicts $\min(|x|, |y|) \geq 1$

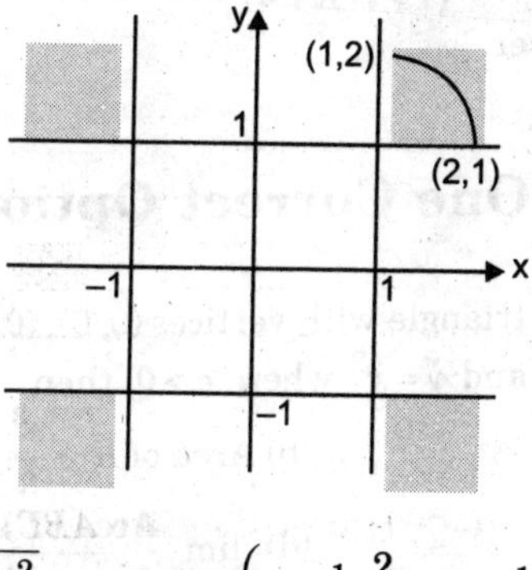

Required area $= 4\int_1^2 (\sqrt{5 - x^2} - 1)\,dx = 10\left(\sin^{-1}\frac{2}{\sqrt{5}} - \sin^{-1}\frac{1}{\sqrt{5}}\right) - 4$

Hence, (b) is the correct answer.

Example 54 The area of the region between the curves $y = \sqrt{\frac{1 + \sin x}{\cos x}}$ and $y = \sqrt{\frac{1 - \sin x}{\cos x}}$ and bounded by the lines $x = 0$ and $x = \frac{\pi}{4}$, is **[IIT JEE 2008]**

(a) $\int_0^{\sqrt{2}-1} \frac{t}{(1+t^2)\sqrt{1-t^2}}\,dt$ (b) $\int_0^{\sqrt{2}-1} \frac{4t}{(1+t^2)\sqrt{1-t^2}}\,dt$

(c) $\int_0^{\sqrt{2}+1} \frac{4t}{(1+t^2)\sqrt{1-t^2}}\,dt$ (d) $\int_0^{\sqrt{2}+1} \frac{t}{(1+t^2)\sqrt{1-t^2}}\,dt$

Solution. Required area $= \int_0^{\pi/4}\left(\sqrt{\frac{1+\sin x}{\cos x}} - \sqrt{\frac{1-\sin x}{\cos x}}\right)dx$

$$\left(\because \frac{1+\sin x}{\cos x} > \frac{1-\sin x}{\cos x} > 0\right)$$

$$= \int_0^{\pi/4}\left(\sqrt{\frac{1+\frac{2\tan\frac{x}{2}}{1+\tan^2\frac{x}{2}}}{\frac{1-\tan^2\frac{x}{2}}{1+\tan^2\frac{x}{2}}}} - \sqrt{\frac{1-\frac{2\tan\frac{x}{2}}{1+\tan^2\frac{x}{2}}}{\frac{1-\tan^2\frac{x}{2}}{1+\tan^2\frac{x}{2}}}}\right)dx$$

$$= \int_0^{\pi/4}\left(\sqrt{\frac{1+\tan\frac{x}{2}}{1-\tan\frac{x}{2}}} - \sqrt{\frac{1-\tan\frac{x}{2}}{1+\tan\frac{x}{2}}}\right)dx$$

$$= \int_0^{\pi/4} \frac{1+\tan\frac{x}{2} - 1 + \tan\frac{x}{2}}{\sqrt{1-\tan^2\frac{x}{2}}}\,dx = \int_0^{\pi/4} \frac{2\tan\frac{x}{2}}{\sqrt{1-\tan^2\frac{x}{2}}}\,dx$$

Put $\tan\frac{x}{2} = t$

$\therefore$ $$\text{Area} = \int_0^{\sqrt{2}-1} \frac{4t}{(1+t^2)\sqrt{1-t^2}}\,dt$$

Hence, (b) is the correct answer.

Type 3 : More Than One Correct Options

Example 55 Let T be the triangle with vertices $(0, 0)$, $(0, c^2)$ and (c, c^2) and let R be the region between $y = cx$ and $y = x^2$ where $c > 0$, then

(a) Area $(R) = \frac{c^3}{6}$ (b) Area of $R = \frac{c^3}{3}$

(c) $\lim\limits_{c \to 0^+} \frac{\text{Area } (T)}{\text{Area } (R)} = 3$ (d) $\lim\limits_{c \to 0^+} \frac{\text{Area } (T)}{\text{Area } (R)} = \frac{3}{2}$

Solution. Area $(T) = \frac{c \cdot c^2}{2} = \frac{c^3}{2}$

Area $(R) = \frac{c^3}{2} - \int_0^c x^2\,dx$

$= \frac{c^3}{2} - \frac{c^3}{3} = \frac{c^3}{6}$

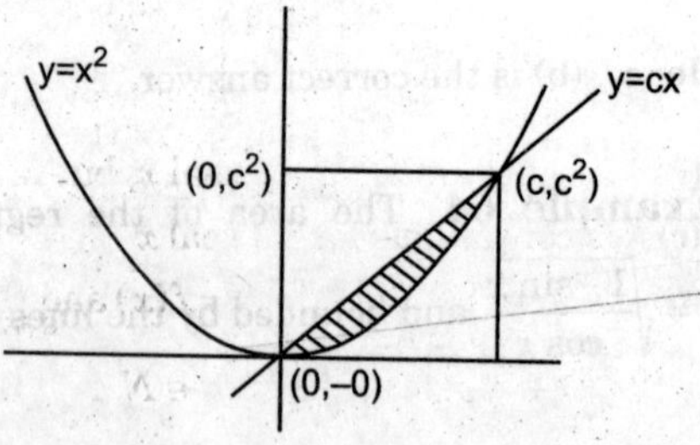

$$\therefore \quad \lim_{c \to 0^+} \frac{\text{Area}(T)}{\text{Area}(R)} = \lim_{c \to 0^+} \frac{c^3}{2} \cdot \frac{6}{c^3} = 3$$

Hence, (a) and (c) are the correct answers.

Example 56 Suppose f is defined from $R \to [-1, 1]$ as $f(x) = \dfrac{x^2 - 1}{x^2 + 1}$ where R is the set of real number. Then, the statement which does not hold is

(a) f is many-one onto

(b) f increases for $x > 0$ and decreases for $x < 0$

(c) minimum value is not attained even though f is bounded

(d) the area included by the curve $y = f(x)$ and the line $y = 1$ is π sq units

Solution. $y = f(x) = \dfrac{x^2 - 1}{x^2 + 1} = 1 - \dfrac{2}{x^2 + 1}$

$f'(x) = \dfrac{4x}{(x^2 + 1)^2}$ $x > 0, f$ is increasing and $x < 0$ f is decreasing.

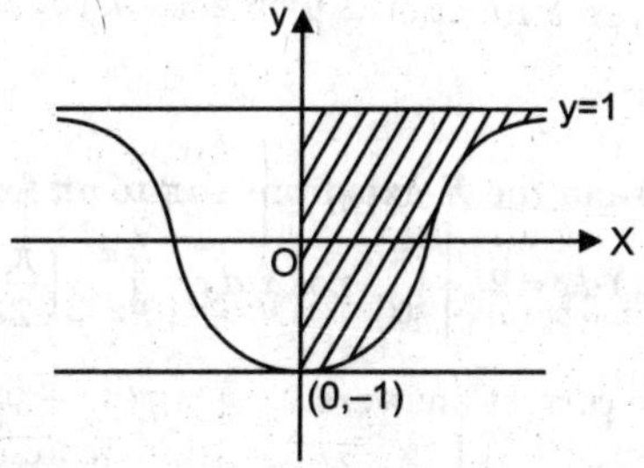

$\Rightarrow$ (b) is true; range is $[-1,1)$ $\Rightarrow$ into $\Rightarrow$ (a) is false; minimum value occurs at $x = 0$ and $f(0) = -1$ $\Rightarrow$ (c) is false.

$$A = 2\int_0^{\infty} \left(1 - \frac{x^2 - 1}{x^2 + 1}\right) dx = 4\int_0^{\infty} \frac{dx}{x^2 + 1}$$

$$= [4 \cdot \tan^{-1} x]_0^{\infty} = 4 \cdot \frac{\pi}{2} = 2\pi \Rightarrow \text{(d) is false.}$$

Hence, (a), (c) and (d) are the correct answers.

Example 57 Consider $f(x) = \begin{cases} \cos x, & 0 \le x < \dfrac{\pi}{2} \\ \left(\dfrac{\pi}{2} - x\right)^2, & \dfrac{\pi}{2} \le x < \pi \end{cases}$ such that f is periodic with period π, then

(a) the range of f is $\left[0, \dfrac{\pi^2}{4}\right)$

(b) f is continuous for all real x, but not differentiable for some real x

(c) f is continuous for all real x

(d) the area bounded by $y = f(x)$ and the x-axis from $x = -n\pi$ to $x = n\pi$ is $2n\left(1 + \dfrac{\pi^2}{24}\right)$ for a given $n \in N$

Solution. Given, $f(x)=\begin{cases}\cos x & , \quad 0 \le x < \dfrac{\pi}{2} \\ \left(\dfrac{\pi}{2}-x\right)^2, & \dfrac{\pi}{2} \le x < \pi\end{cases}$ and f is periodic with period π.

$\therefore$ Let us draw the graph of $y = f(x)$

From the graph, the range of the function is $\left[0, \dfrac{\pi^2}{4}\right)$.

It is discontinuous at $x = n\pi$, $n \in I$. It is not differentiable at $x = \dfrac{n\pi}{2}$, $n \in I$.

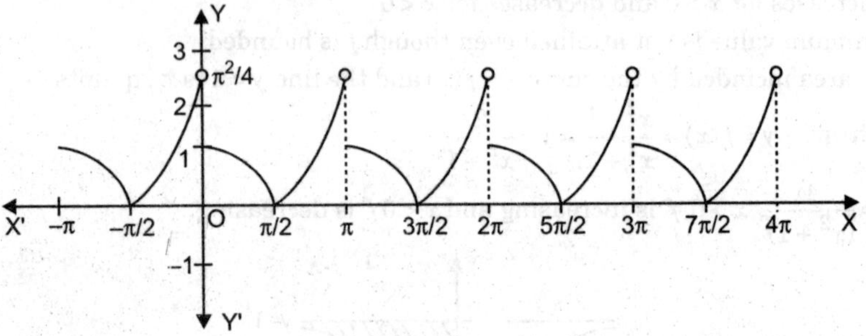

Area bounded by $y = f(x)$ and the X-axis from $-n\pi$ to $n\pi$ for $n \in N$

$$= 2n \int_0^{\pi} f(x)\,dx = 2n\left[\int_0^{\pi/2} \cos x\,dx + \int_{\pi/2}^{\pi}\left(\frac{\pi}{2}-x\right)^2 dx\right] = 2n\left(1+\frac{\pi^3}{24}\right)$$

Hence, (a) and (d) are the correct answers.

Example 58 Consider the functions $f(x)$ and $g(x)$, both defined from $R \to R$ and are defined as $f(x) = 2x - x^2$ and $g(x) = x^n$ where $n \in N$. If the area between $f(x)$ and $g(x)$ is 1/2, then n is a divisor of

(a) 12 (b) 15 (c) 20 (d) 30

Solution. Solving, $f(x) = 2x - x^2$ and $g(x) = x^n$ we have $2x - x^2 = x^n \Rightarrow x = 0$ and $x = 1$

$$A = \int_0^1 (2x - x^2 - x^n)\,dx$$

$$= \left[x^2 - \frac{x^3}{3} - \frac{x^{n+1}}{n+1}\right]_0^1$$

$$= 1 - \frac{1}{3} - \frac{1}{n+1} = \frac{2}{3} - \frac{1}{n+1}$$

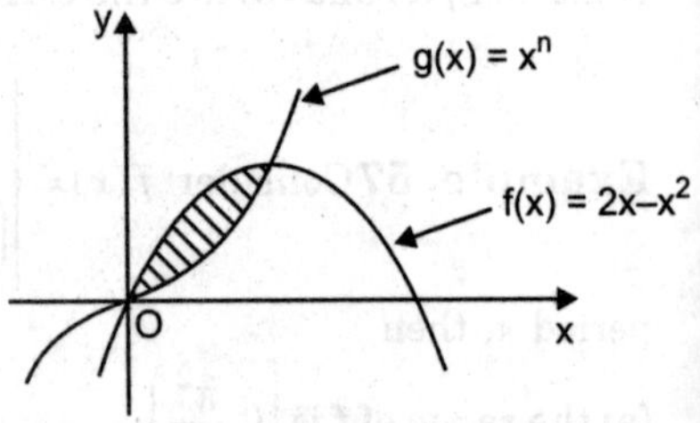

Since, $\dfrac{2}{3} - \dfrac{1}{n+1} = \dfrac{1}{2}$

$\Rightarrow \quad \dfrac{2}{3} - \dfrac{1}{2} = \dfrac{1}{n+1}$

$\Rightarrow \quad \dfrac{4-3}{6} = \dfrac{1}{n+1} \Rightarrow n+1 = 6 \Rightarrow n = 5$

Thus, n is a divisor of 15, 20, 30.

Hence, (b), (c) and (d) are the correct answers.

Example 59 Area of the region bounded by the curve $y = e^x$ and lines $x = 0$ and $y = e$ is **[IIT JEE 2009]**

(a) $e - 1$ (b) $\int_1^e \ln(e + 1 - y)\,dy$ (c) $e - \int_0^1 e^x\,dx$ (d) $\int_1^e \ln y\,dy$

Solution. Shaded area $= e - \left(\int_0^1 e^x\,dx\right) = 1$

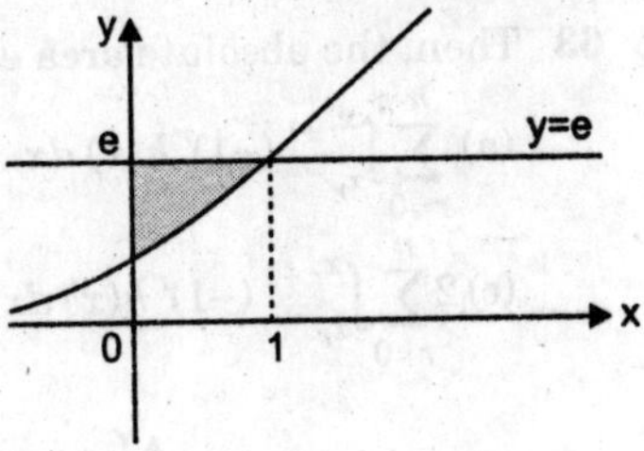

Also, $\int_1^e \ln(e + 1 - y)\,dy$

Put $e + 1 - y = t \Rightarrow -dy = dt$

$$= \int_0^1 \ln t(-dt) = \int_0^e \ln t\,dt$$

$$= \int_1^e \ln y\,dy = 1$$

Hence, (b), (c) and (d) are the correct answers.

Type 5 : Linked-Comprehension Based Questions

Passage I

(Q. Nos. 60 to 62)

Consider the function, $f(x) = x^3 - 8x^2 + 20x - 13$

60 Number of positive integers x for which $f(x)$ is a prime number, is

(a) 1 (b) 2 (c) 3 (d) 4

Solution. $f(x) = (x-1)(x^2 - 7x + 13)$ for $f(x)$ to be prime at least one of the factors must be prime.

Therefore, $x - 1 = 1 \Rightarrow x = 2$ or $x^2 - 7x + 13 = 1 \Rightarrow x^2 - 7x + 12 = 0$

$\Rightarrow \quad x = 3$ or $4 \quad \Rightarrow \quad x = 2, 3, 4$

Hence, (c) is the correct answer.

61 The function $f(x)$ defined for $R \to R$

(a) is one-one onto

(b) is many-one onto

(c) has 3 real roots

(d) is such that $f(x_1) \cdot f(x_2) < 0$ where x_1 and x_2 are the roots of $f'(x) = 0$

Solution. $f(x)$ is many-one as it increases and decreases, also range of $f(x) \in R$

$\Rightarrow$ many-one onto.

Hence, (b) is the correct answer.

62 Area enclosed by $y = f(x)$ and the coordinate axes is

(a) 65/12 (b) 13/12 (c) 71/12 (d) None of these

Solution. $A = \left|\int_0^1 f(x)\,dx\right| = -\int_0^1 (x^3 - 8x^2 + 20x - 13)\,dx = \dfrac{65}{12}$

Hence, (a) is the correct answer.

Passage II

(Q. Nos. 63 to 65)

Let $h(x) = f(x) - g(x)$, where $f(x) = \sin^4 \pi x$ and $g(x) = \ln x$. Let $x_0, x_1, x_2, \ldots, x_{n+1}$ be the roots of $f(x) = g(x)$ is increasing order.

63 Then, the absolute area enclosed by $y = f(x)$ and $y = g(x)$ is given by

(a) $\sum_{r=0}^{n} \int_{x_r}^{x_{r+1}} (-1)^r h(x)\, dx$ (b) $\sum_{r=0}^{n} \int_{x_r}^{x_{r+1}} (-1)^{r+1} h(x)\, dx$

(c) $2\sum_{r=0}^{n} \int_{x_r}^{x_{r+1}} (-1)^r h(x)\, dx$ (d) $\frac{1}{2}\sum_{r=0}^{n} \int_{x_r}^{x_{r+1}} (-1)^{r+1} h(x)\, dx$

Solution.

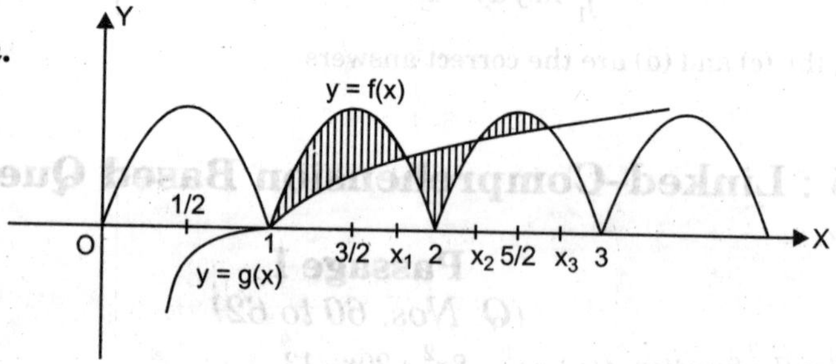

Hence, (a) is the correct answer.

64 In the above question the value of n is

(a) 1 (b) 2 (c) 3 (d) 4

Solution. $x_{n+1} = x_3 \Rightarrow n = 2$

Hence, (b) is the correct answer.

65 The whole area bounded by $y = f(x)$, $y = g(x)$ and $x = 0$ is

(a) $\frac{11}{8}$ (b) $\frac{8}{3}$ (c) 2 (d) $\frac{13}{3}$

Solution. Required area $= \int_0^1 \sin^4 \pi x\, dx - \int_0^1 \ln x\, dx = \frac{11}{8}$

Hence, (a) is the correct answer.

Passage III

(Q. Nos. 66 to 68)

Consider the function defined implicitly by the equation $y^3 - 3y + x = 0$ on various intervals in the real line. If $x \in (-\infty, -2) \cup (2, \infty)$, the equation implicitly defines a unique real-valued differentiable function $y = f(x)$. If $x \in (-2, 2)$, the equation implicitly defines a unique real-valued differentiable function $y = g(x)$ satisfying $g(0) = 0$.

[IIT JEE 2008]

66 If $f(-10\sqrt{2}) = 2\sqrt{2}$, then $f''(-10\sqrt{2})$ is equal to

(a) $\frac{4\sqrt{2}}{7^3 3^2}$ (b) $-\frac{4\sqrt{2}}{7^3 3^2}$ (c) $\frac{4\sqrt{2}}{7^3 3}$ (d) $-\frac{4\sqrt{2}}{7^3 3}$

Solution. $\therefore \quad y^3 - 3y + x = 0$

On differentiation $3y^2y' - 3y' - 1 = 0$

$$\Rightarrow \quad y' = \frac{1}{3(1-y^2)}$$

$$\Rightarrow \quad y'(-10\sqrt{2}) = \frac{1}{3\{1-(2\sqrt{2})^2\}} \quad \text{...(i)}$$

$$\Rightarrow \quad y'(-10\sqrt{2}) = \frac{1}{3\{1-(2\sqrt{2})^2\}} = \frac{1}{3\,(1-8)} = -\frac{1}{21}$$

Differentiating Eq. (i) again, we get

$$y'' = \frac{6yy'^2}{3(1-y^2)}$$

$$y''(-10\sqrt{2}) = \frac{6 \cdot 2\sqrt{2} \cdot \left(\frac{1}{21}\right)^2}{3(1-8)} = -\frac{4\sqrt{2}}{7^3 \cdot 3^2}$$

Hence, (b) is the correct answer.

67 The area of the region bounded by the curve $y = f(x)$, the x-axis and the line $x = a$ and $x = b$, where $-\infty < a < b < -2$, is

(a) $\int_a^b \frac{x}{3\,[\{f(x)\}^2 - 1]}\,dx + bf(b) - af(a)$ (b) $-\int_a^b \frac{x}{3\,[\{f(x)\}^2 - 1]}\,dx - bf(b) + af(a)$

(c) $\int_a^b \frac{x}{3\,[\{f(x)\}^2 - 1]}\,dx - bf(b) + af(a)$ (d) $-\int_a^b \frac{x}{3\,[\{f(x)\}^2 - 1]}\,dx - bf(b) + af(a)$

Solution. Required area $= \int_a^b f(x)\,dx = [xf(x)]_a^b - \int_a^b xf'(x)\,dx$

$$= bf(b) - af(a) + \int_a^b \frac{x}{3\,[\{f(x)\}^2 - 1]}\,dx$$

Hence, (a) is the correct answer.

68 $\int_{-1}^{1} g'(x)\,dx$ is equal to

(a) $2g(-1)$ (b) 0 (c) $-2g(1)$ (d) $2g(1)$

Solution. $I = \int_{-1}^{1} g'(x)\,dx = [g(x)]_{-1}^{1} = g(1) - g(-1)$

Since, $\quad y^3 - 3y + x = 0 \quad$ ...(i)

and $\quad y = g(x)$

Since, $\quad \{g(x)\}^3 - 3g(x) + x = 0 \quad$ [by Eq. (i)]

At $x = 1$, $\quad \{g(1)\}^3 - 3g(1) + 1 = 0 \quad$ ...(ii)

At $x = -1$, $\quad \{g(-1)\}^3 - 3g(-1) - 1 = 0 \quad$ ...(iii)

On adding Eqs. (i) and (ii), we get

$$\{g(1)\}^3 + \{g(-1)\}^3 - 3\{g(1) + g(-1)\} = 0$$

$$\{g(1) + g(-1)\}\{g(1)^2 + (g(-1)^2) - g(1)g(-1) - 3\} = 0$$

$$\Rightarrow \quad g(1) + g(-1) = 0,\ g(1) = -g(-1)$$

$$\Rightarrow \quad I = g(1) - g(-1) = g(1) - (-g(1)) = 2g(1).$$

Hence, (d) is the correct answer.

Proficiency Exercise 1

Type 1 : Only One Correct Option

1. Suppose $y = f(x)$ and $y = g(x)$ are two functions whose graphs intersect at the three points (0, 4), (2, 2) and (4, 0) with $f(x) > g(x)$ for $0 < x < 2$ and $f(x) < g(x)$ for $2 < x < 4$. If $\int_0^4 [f(x) - g(x)]\, dx = 10$ and $\int_2^4 [g(x) - f(x)]\, dx = 5$, the area between two curves for $0 < x < 2$, is

(a) 5 (b) 10 (c) 15 (d) 20

2. Let 'a' be a positive constant number. Consider two curves $C_1 : y = e^x$, $C_2 : y = e^{a-x}$. Let S be the area of the part surrounding by C_1, C_2 and the y-axis, then $\lim\limits_{a \to 0} \dfrac{S}{a^2}$ equals to

(a) 4 (b) 1/2 (c) 0 (d) 1/4

3. 3 points $O(0,0), P(a, a^2), Q(-b, b^2) (a > 0, b > 0)$ are on the parabola $y = x^2$. Let S_1 be the area bounded by the line PQ and the parabola and let S_2 be the area of the ΔOPQ, the minimum value of S_1/S_2 is

(a) 4/3 (b) 5/3 (c) 2 (d) 7/3

4. Area enclosed by the graph of the function $y = \ln^2 x - 1$ lying in the 4th quadrant is

(a) $\dfrac{2}{e}$ (b) $\dfrac{4}{e}$ (c) $2\left(e + \dfrac{1}{e}\right)$ (d) $4\left(e - \dfrac{1}{e}\right)$

5. The area bounded by $y = 2 - |2 - x|$ and $y = \dfrac{3}{|x|}$ is

(a) $\dfrac{4 + 3 \ln 3}{2}$ (b) $\dfrac{4 - 3 \ln 3}{2}$

(c) $\dfrac{3}{2} + \ln 3$ (d) $\dfrac{1}{2} + \ln 3$

6. Suppose $g(x) = 2x + 1$ and $h(x) = 4x^2 + 4x + 5$ and $h(x) = (fog)(x)$. The area enclosed by the graph of the function $y = f(x)$ and the pair of tangents drawn to it from the origin, is

(a) 8/3 (b) 16/3 (c) 32/3 (d) None of these

7. The area bounded by the curves $y = -\sqrt{-x}$ and $x = -\sqrt{-y}$ where $x, y \le 0$

(a) cannot be determined

(b) is 1/3

(c) is 2/3

(d) is same as that of the figure bounded by the curves $y = \sqrt{-x}$; $x \le 0$ and $x = \sqrt{-y}$; $y \le 0$

8. $y = f(x)$ is a function which satisfies

(i) $f(0) = 0$ (ii) $f''(x) = f'(x)$ and (iii) $f'(0) = 1$

then the area bounded by the graph of $y = f(x)$, the lines $x = 0, x - 1 = 0$ and $y + 1 = 0$, is

(a) e (b) $e - 2$ (c) $e - 1$ (d) $e + 1$

9. Area of the region enclosed between the curves $x = y^2 - 1$ and $x = |y|\sqrt{1 - y^2}$ is

(a) 1 (b) 4/3 (c) 2/3 (d) 2

10. The area bounded by the curve $y = xe^{-x}; xy = 0$ and $x = c$ where c is the x-coordinate of the curve's inflection point, is

(a) $1 - 3e^{-2}$ (b) $1 - 2e^{-2}$ (c) $1 - e^{-2}$ (d) 1

11. If $(a, 0); a > 0$ is the point where the curve $y = \sin 2x - \sqrt{3} \sin x$ cuts the x-axis first, A is the area bounded by this part of the curve, the origin and the positive x-axis, then

(a) $4A + 8 \cos a = 7$ (b) $4A + 8 \sin a = 7$ (c) $4A - 8 \sin a = 7$ (d) $4A - 8 \cos a = 7$

12. The curve $y = ax^2 + bx + c$ passes through the point (1, 2) and its tangent at origin is the line $y = x$. The area bounded by the curve, the ordinate of the curve at minima and the tangent line is

(a) $\frac{1}{24}$ (b) $\frac{1}{12}$ (c) $\frac{1}{8}$ (d) $\frac{1}{6}$

13. A function $y = f(x)$ satisfies the differential equation $\frac{dx}{dx} - y = \cos x - \sin x$, with initial condition that y is bounded when $x \to \infty$. The area enclosed by $y = f(x)$, $y = \cos x$ and the y-axis in the 1st quadrant

(a) $\sqrt{2} - 1$ (b) $\sqrt{2}$ (c) 1 (d) $1/\sqrt{2}$

14. If the area bounded between x-axis and the graph of $y = 6x - 3x^2$ between the ordinates $x = 1$ and $x = a$ is 19 sq units, then 'a' can take the value

(a) 4 or –2

(b) two values are in (2, 3) and one in (–1,0)

(c) two values one in (3, 4) and one in (–2, – 1)

(d) None of the above

15. Area bounded by $y = f^{-1}(x)$ and tangent and normal drawn to it at the points with abscissae π and 2π, where $f(x) = \sin x - x$ is

(a) $\frac{\pi^2}{2} - 1$ (b) $\frac{\pi^2}{2} - 2$ (c) $\frac{\pi^2}{2} - 4$ (d) $\frac{\pi^2}{2}$

16. If $f(x) = x - 1$ and $g(x) = |f(|x|) - 2|$, then the area bounded by $y = g(x)$ and the curve $x^2 - 4y + 8 = 0$ is equal to

(a) $\frac{4}{3}(4\sqrt{2} - 5)$ (b) $\frac{4}{3}(4\sqrt{2} - 3)$ (c) $\frac{8}{3}(4\sqrt{2} - 3)$ (d) $\frac{8}{3}(4\sqrt{2} - 5)$

17. Let $S = \left\{(x, y): \frac{y(3x - 1)}{x(3x - 2)} < 0\right\}$,

$S' = \{(x, y) \in A \times B: -1 \le A \le 1, -1 \le B \le 1\}$, then the area of the region enclosed by all points in $S \cap S'$ is

(a) 1 (b) 2 (c) 3 (d) 4

18. The area of the region bounded between the curves $y = e|\,|x| \ln |x|\,|$, $x^2 + y^2 - 2(|x| + |y|) + 1 \geq 0$ and x-axis where $|x| \leq 1$, if α is the x-coordinate of the point of intersection of curves in 1st quadrant, is

(a) $4\left[\int_0^{\alpha} ex \ln x\, dx + \int_{\alpha}^{1} (1 - \sqrt{1-(x-1)^2})\, dx\right]$

(b) $4\left[\int_0^{\alpha} ex \ln x\, dx + \int_{1}^{\alpha} (1 - \sqrt{1-(x-1)^2})\, dx\right]$

(c) $4\left[-\int_0^{\alpha} ex \ln x\, dx + \int_{\alpha}^{1} (1 - \sqrt{1-(x-1)^2})\, dx\right]$

(d) $2\left[\int_0^{\alpha} ex \ln x\, dx + \int_{1}^{\alpha} (1 - \sqrt{1-(x-1)^2})\, dx\right]$

19. A point P lying inside the curve $y = \sqrt{2ax - x^2}$ is moving such that its shortest distance from the curve at any position is greater than its distance from x-axis. The point P enclose a region whose area is equal to

(a) $\frac{\pi a^2}{2}$ (b) $\frac{a^2}{3}$ (c) $\frac{2a^2}{3}$ (d) $\left(\frac{3\pi - 4}{6}\right) a^2$

Type 2 : More Than One Correct Options

20. The triangle formed by the normal to the curve $f(x) = x^2 - ax + 2a$ at the point (2, 4) and the coordinate axes lies in second quadrant, if its area is 2 sq units, then a can be

(a) 2 (b) 17/4 (c) 5 (d) 19/4

21. Let f and g be continuous function on $a \leq x \leq b$ and set $p(x) = \max\{f(x), g(x)\}$ and $q(x) = \min\{f(x), g(x)\}$, then the area bounded by the curves $y = p(x)$, $y = q(x)$ and the ordinates $x = a$ and $x = b$ is given by

(a) $\int_a^b |f(x) - g(x)|\, dx$ (b) $\int_a^b |p(x) - q(x)|\, dx$

(c) $\int_a^b \{f(x) - g(x)\}\, dx$ (d) $\int_a^b \{p(x) - q(x)\}\, dx$

22. The area bounded by the parabola $y = x^2 - 7x + 10$ and x-axis equals

(a) area bounded by $y = -x^2 + 7x - 10$ and x-axis

(b) 1/6 sq units

(c) 5/6 sq units

(d) 9/2 sq units

23. Area bounded by the ellipse $\frac{x^2}{4} + \frac{y^2}{9} = 1$ is equal to

(a) 6π sq units

(b) 3π sq units

(c) 12π sq units

(d) area bounded by the ellipse $\frac{x^2}{9} + \frac{y^2}{4} = 1$

24. There is a curve in which the length of the perpendicular from the origin to tangent at any point is equal to abscissa of that point. Then

(a) $x^2 + y^2 = 2$ is one such curve

(b) $y^2 = 4x$ is one such curve

(c) $x^2 + y^2 = 2cx$ (c parameters) are such curves

(d) there are no such curves

Type 3 : Assertion and Reason

Directions
(Q. Nos. 25 to 29)

For the following questions, choose the correct answers from the codes (a), (b), (c) and (d) defined as follows :

(a) Statement I is true, Statement II is also true; Statement II is the correct explanation of Statement I.

(b) Statement I is true, Statement II is also true; Statement II is not the correct explanation of Statement I.

(c) Statement I is true, Statement II is false.

(d) Statement I is false, Statement II is true.

25. Statement I The area of the curve $y = \sin^2 x$ from 0 to π will be more than that of the curve $y = \sin x$ from 0 to π.

Statement II $x^2 > x$, if $x > 1$.

26. Statement I The area bounded by the curves $y = x^2 - 3$ and $y = kx + 2$ is least if $k = 0$.

Statement II The area bounded by the curves

$$y = x^2 - 3 \text{ and } y = kx + 2 \text{ is } \sqrt{k^2 + 20}.$$

27. Statement I The area of region bounded parabola

$$y^2 = 4x \text{ and } x^2 = 4y \text{ is } \frac{32}{3} \text{ sq units.}$$

Statement II The area of region bounded by parabola

$$y^2 = 4ax \text{ and } x^2 = 4by \text{ is } \frac{16}{3}ab.$$

28. Statement I The area by region $|x + y| + |x - y| \le 2$ is 8 sq units.

Statement II Area enclosed by region $|x + y| + |x - y| \le 2$ is symmetric about x-axis.

29. Statement I Area bounded by $y = x(x-1)$ and $y = x(1-x)$ is $\frac{1}{3}$.

Statement II Area bounded by $y = f(x)$ and $y = g(x)$ is $\left|\int_a^b (f(x) - g(x))\, dx\right|$ is true when $f(x)$ and $g(x)$ lies above x-axis. (Where a and b are intersection of $y = f(x)$ and $y = g(x)$).

Type 4 : Linked–Comprehension Based Questions

Passage I
(Q. Nos. 30 to 32)

Let $f(x) = \dfrac{ax^2 + bx + c}{x^2 + 1}$ *such that* $y = -2$ *is an asymptote of the curve* $y = f(x)$. *The curve* $y = f(x)$ *is symmetric about y-axis and its maximum values is 4.*

30. Maximum area of rectangle with two vertices on asymptote and two vertices on curve $y = f(x)$ is

(a) 12 (b) 8 (c) 6 (d) 4

31. If $\lambda_1, \lambda_2 > 0$ such that $\lambda_1 + \lambda_2 = 1$, then which of the following holds?

(a) $\lambda_1 f(x_1) + \lambda_2 f(x_2) > f(\lambda_1 x_1 + \lambda_2 x_2)$ for $x_1, x_2 \in \left(-\infty, -\frac{1}{\sqrt{3}}\right)$ or $x_1, x_2 \in \left(\frac{1}{\sqrt{3}}, \infty\right)$

(b) $\lambda_1 f(x_1) + \lambda_2 f(x_2) > f(\lambda_1 x_1 + \lambda_2 x_2)$ for $x_1, x_2 \in \left(-\frac{1}{\sqrt{3}}, \frac{1}{\sqrt{3}}\right)$

(c) $\lambda_1 f(x_1) + \lambda_2 f(x_2) < f(\lambda_1 x_1 + \lambda_2 x_2)$ for $x_1, x_2 \in \left(-\infty, -\frac{1}{\sqrt{3}}\right)$ or $x_1, x_2 \in \left(\frac{1}{\sqrt{3}}, \infty\right)$

(d) $\lambda_1 f(x_1) + \lambda_2 f(x_2) < f(\lambda_1 x_1 + \lambda_2 x_2)$ for $x_1, x_2 \in R$

32. Let $g(x) = \begin{cases} ke^{-kx} & : x < 0 \\ f(x) & : x \geq 0 \end{cases}$ the values of k such that $y = g(x)$ has a maxima at $x = 0$

(a) $(-\infty, 4]$ (b) $(-\infty, 4)$ (c) $(0, 4)$ (d) $(4, \infty)$

Passage II

(Q. Nos. 33 to 35)

Consider the function $f : (-\infty, \infty) \to (-\infty, \infty)$ *defined by* $f(x) = \frac{x^2 - ax + 1}{x^2 + ax + 1}$, $0 < a < 2$.

33. Which of the following is true?

(a) $(2 + a)^2 f''(1) + (2 - a)^2 f''(-1) = 0$

(b) $(2 - a)^2 f''(1) - (2 + a)^2 f''(-1) = 0$

(c) $f'(1) f'(-1) = (2 - a)^2$

(d) $f'(1) f'(-1) = -(2 + a)^2$

34. Which of the following is true?

(a) $f(x)$ is decreasing on $(-1, 1)$ and has a local minimum at $x = 1$

(b) $f(x)$ is increasing on $(-1, 1)$ and has a local maximum at $x = 1$

(c) $f(x)$ is increasing on $(-1, 1)$ but has neither a local maximum nor a local minimum at $x = 1$

(d) $f(x)$ is decreasing on $(-1, 1)$ but has neither a local maximum nor a local minimum at $x = 1$

35. Let $g(x) = \int_0^{e^x} \frac{f'(t)}{1 + t^2} dt$. Which of the following is true?

(a) $g'(x)$ is positive on $(-\infty, 0)$ and negative on $(0, \infty)$

(b) $g'(x)$ is negative on $(-\infty, 0)$ and positive on $(0, \infty)$

(c) $g'(x)$ change sign on both $(-\infty, 0)$ and $(0, \infty)$

(d) $g'(x)$ does not change sign on $(-\infty, \infty)$

Passage III

(Q. Nos. 36 to 38)

Computing areas with parametrically represented boundaries :

If the boundary of a figure is represented by parametric equations ie, $x = x(t)$, $y = y(t)$, *then the area of the figure is evaluated by one of the three formulas*

$$S = -\int_{\alpha}^{\beta} y(t)\cdot x'(t)\,dt$$

$$S = \int_{\alpha}^{\beta} x(t)\cdot y'(t)\,dt$$

$$S = \frac{1}{2}\int_{\alpha}^{\beta} (xy' - yx')\,dt$$

where α and β are the values of the parameter 't' corresponding, respectively to the beginning and the end of the traversal of the curve corresponding to increasing 't'.

36. The area enclosed by the asteroid $\left(\frac{x}{a}\right)^{2/3} + \left(\frac{y}{a}\right)^{2/3} = 1$ is

(a) $\frac{3}{4}a^2\pi$ (b) $\frac{3}{18}\pi a^2$ (c) $\frac{3}{8}\pi a^2$ (d) $\frac{3}{4}a\pi$

37. The area of the region bounded by an arc of the cycloid $x = a(t - \sin t)$, $y = a(1 - \cos t)$ and the x-axis is

(a) $6\pi a^2$ (b) $3\pi a^2$ (c) $4\pi a^2$ (d) None of these

38. Area of the loop described as $x = \frac{t}{3}(6 - t)$, $y = \frac{t^2}{8}(6 - t)$ is

(a) $\frac{27}{5}$ (b) $\frac{24}{5}$ (c) $\frac{27}{6}$ (d) $\frac{21}{5}$

Type 5 : Match the Columns

39. Match the statements of Column I with values of Column II.

Column I	Column II	
(A) The area bounded by the curve $y = x + \sin x$ and its inverse function between the ordinates $x = 0$ to $x = 2\pi$ is 4s. Then, the value of s is	(p)	2
(B) The area bounded by $y = xe^{\lvert x\rvert}$ and lies $\lvert x\rvert = 1$, $y = 0$ is	(q)	1
(C) The area bounded by the curves $y^2 = x^3$ and $\lvert y\rvert = 2x$ is	(r)	$\frac{16}{5}$
(D) The smaller are included between the curves $\sqrt{\lvert x\rvert} + \sqrt{\lvert y\rvert} = 1$ and $\lvert x\rvert + \lvert y\rvert = 1$ is	(s)	$\frac{1}{3}$

40. Match the following :

Column I	Column II	
(A) Area enclosed by $y = \lvert x\rvert$, $\lvert x\rvert = 1$ and $y = 0$ is	(p)	2
(B) Area enclosed by the curve $y = \sin x$, $x = 0$, $x = \pi$ and $y = 0$ is	(q)	4
(C) If the area of the region bounded by $x^2 \le y$ and $y \le x + 2$ is $\frac{k}{4}$, then k is equal to	(r)	27
(D) Area of the quadrilateral formed by tangents at the ends of latusrectum of ellipse of ellipse $\frac{x^2}{9} + \frac{y^2}{5} = 1$ is	(s)	18

Type 6 : Integer Answer Type Questions

41. Consider $f(x) = x^2 - 3x + 2$. The area bounded by $|y| = |f(|x|)|, x \geq 1$ is A, then find the value of $3A + 2$.

42. The value of $c + 2$ for which the area of the figure bounded by the curve $y = 8x^2 - x^5$; the straight lines $x = 1$ and $x = c$ and x-axis is equal to $\frac{16}{3}$, is ....

43. The area bounded by $y = 2 - |2 - x|$; $y = \frac{3}{|x|}$ is $\frac{k - 3 \ln 3}{2}$, then k is equal to ....

44. The area of the ΔABC, coordinates of whose vertices are $A(2, 0)$, $B(4, 5)$ and $C(6, 3)$ is ....

45. A point P moves in xy-plane in such a way that $[|x|] + [|y|] = 1$, where $[\cdot]$ denotes the greatest integer function. Area of the region representing all possible of the point P is equal to ....

46. Let $f : [0, 1] \rightarrow \left[0, \frac{1}{2}\right]$ be a function such that $f(x)$ is a polynomial of 2nd degree, satisfy the following condition :

(a) $f(0) = 0$ (b) has a maximum value of $\frac{1}{2}$ at $x = 1$.

If A be the area bounded by $y = f(x)$; $y = f^{-1}(x)$ and the line $2x + 2y - 3 = 0$ in 1st quadrant, then the value of $24A$ is equal to ....

Proficiency Exercise 2

1. Find the area bounded by the curve $y = \sin^{-1} x$ and the lines $x = 0, |y| = \frac{\pi}{2}$.

2. Find the area of the region bounded by the parabola $(y-2)^2 = (x-1)$, the tangent to it at the point where the ordinate is 3, and the x-axis.

3. The line $y = mx$ bisects the area enclosed by the lines $x = 0$, $y = 0$, $x = \frac{3}{2}$ and the curve $y = 1 + 4x - x^2$, find m.

4. Let f be a function satisfying the condition $f\left(\frac{x}{y}\right) = \frac{f(x)}{f(y)}$, $\forall\, x, y \in R;\ y \neq 0$, $f(y)\sqrt{0}$ and $f'(1) = 2$. Find the area enclosed by $y\,|\,f(x)$, the x-axis and the portion of $y^2 + x^2 \leq 2$ in the first quadrant.

5. Find the area of the closed figure bounded by $x = -1$, $y = 0$; $y = x^2 + x + 1$ and the tangent to the curve $y = x^2 + x + 1$ at $A\,(1, 3)$.

6. Find the area of the closed figure bounded by $y = 3 - |3 - x|$ and $y = \frac{6}{|x+1|}$.

7. Let f be a real valued function satisfying $f\left(\frac{x}{y}\right) = f(x) - f(y)$ and $\lim_{x \to 0} \frac{f(1+x)}{x} = 3$. Find the area bounded by the curve $y = f(x)$, the y-axis and the line $y = 3$, where $x, y \in R^+$.

8. Find a continuous function 'f', $(x^4 - 4x^2) \leq f(x) \leq (2x^2 - x^3)$ such that the area bounded by $y = f(x)$, $y = x^4 - 4x^2$, the y-axis and the line $x = t$, $(0 \leq t \leq 2)$ is k times the area bounded by $y = f(x)$, $y = 2x^2 - x^3$, y-axis and line $x = t\ (0 \leq t \leq 2)$.

9. Find out the area enclosed by the circle $x^2 + y^2 = 4$, parabola $y = x^2 + x + 1$, the curve $y = \left[\sin^2 \frac{x}{4} + \cos \frac{x}{4}\right]$ and x-axis (where [.] denotes the greatest integer function).

10. Let $f(t) = |t-1| - |t| + |t+1|\ \forall\, t \in R$ and $g(x) = \max\{f(t) : x+1 \leq t \leq x+2\}$; $\forall\, x \in R$. Find $g(x)$ and the area bounded by the curve $y = g(x)$, the x-axis and the lines $x = -3/2$ and $x = 5$.

11. Find the area of the region satisfying $|y| + \frac{1}{2} \geq e^{-|x|}$ and max. $\{|x|, |y|\} \leq 2$.

12. Let $f(x) =$ minimum $\{e^x, 3/2, 1 + e^{-x}\}$, $0 \leq x \leq 1$. Find the area bounded by $y = f(x)$, x-axis, y-axis and the line $x = 1$.

13. Find the area bounded by $y = f(x)$ and the curve $y = \dfrac{2}{1+x^2}$ where f is a continuous function satisfying the conditions $f(x) \cdot f(y) = f(xy)$, $\forall\, x, y \in R$ and $f'(1) = 2, f(1) = 1$.

14. Find out the area bounded by the curve $y = \int_{1/8}^{\sin^2 x} (\sin^{-1}\sqrt{t})\, dt + \int_{1/8}^{\cos^2 x} (\cos^{-1}\sqrt{t})\, dt$ $(0 \le x \le \pi/2)$ and the curve satisfying the differential equation $y(x + y^3)\, dx = x(y^3 - x)\, dy$ passing through $(4, -2)$.

15. Let T be an acute triangle. Inscribe a pair R, S of rectangles in T as shown :

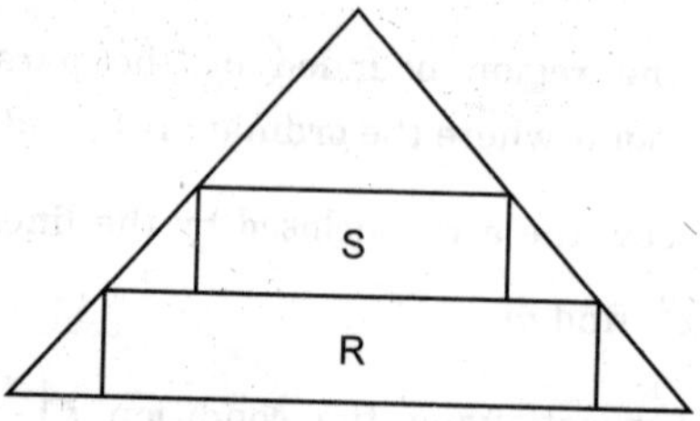

Let $A(x)$ denote the area of polygon X find the maximum value (or show that no maximum exists), of $\dfrac{A(R) + A(S)}{A(T)}$, where T ranges over all triangles and R, S over all rectangles as above.

16. Find the maximum area of the ellipse that can be inscribed in an isosceles triangles of area A and having one axis lying along the perpendicular from the vertex of the triangles to its base.

17. In the adjacent figure the graphs of two function $y = f(x)$ and $y = \sin x$ are given. $y = \sin x$ intersects, $y = f(x)$ at $A(a, f(a))$; $B(\pi, 0)$ and $C(2\pi, 0)$.

A_i $(i = 1, 2, 3)$ is the area bounded by the curves $y = f(x)$ and $y = \sin x$.

between $x = 0$ and $x = a$; $i = 1$

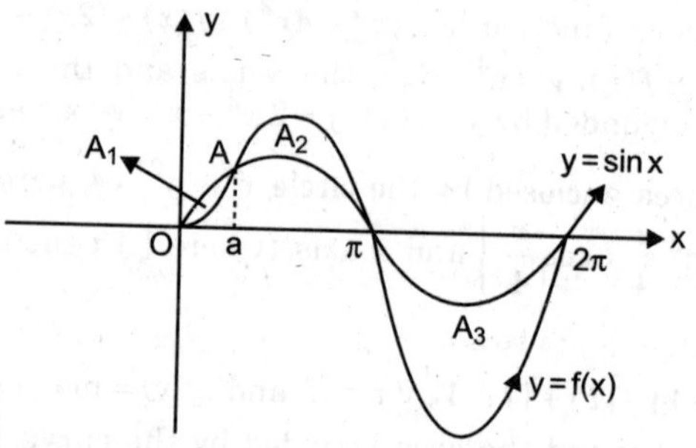

between $x = a$ and $x = \pi$; $i = 2$

between $x = \pi$ and $x = 2\pi$; $i = 3$.

If $A_1 = 1 - \sin a + (a - 1)\cos a$, determine the function $f(x)$. Hence, determine a and A_1. Also, calculate A_2 and A_3.

18. Find the area of the region bounded by the curve $y = 25^x + 16$ and the curve $y = b \,.\, 5^x + 4$, whose tangent at the point $x = 1$ make an angle $\tan^{-1}(40 \ln 5)$ with the x-axis.

19. If the circles of the maximum area inscribed in the region bounded by the curves $y = x^2 - x - 3$ and $y = 3 + 2x - x^2$, then the area of region $y - x^2 + 2x + 3 \le 0$, $y + x^2 - 2x - 3 \le 0$ and $s \le 0$.

20. Find limit of the ratio of the area of the triangle formed by the origin and intersection points of the parabola $y = 4x^2$ and the line $y = a^2$, to the area between the parabola and the line as a approaches to zero.

21. Find the area of curve enclosed by :

$$|x + y| + |x - y| \le 4, |x| \le 1, y \ge \sqrt{x^2 - 2x + 1}.$$

22. Calculate the area enclosed by the curve

$$4 \le x^2 + y^2 \le 2(|x| + |y|).$$

23. Find the area enclosed by the curve $[x] + [y] = 4$ in the 1st quadrant (where [.] denotes greatest integer function).

24. Sketch the region and find the area bounded by the curves

$$|y + x| \le 1, |y - x| \le 1 \text{ and } 2x^2 + 2y^2 = 1.$$

25. For what value of a the area of the figure bounded by the curve $y = 1/x$, $y = 1/(2x - 1)$, $x = 2$ and $x = a$ is equal to $\ln 4/\sqrt{5}$.

26. Find the area of the region bounded by the curve, $2^{|x|}|y| + 2^{|x|-1} \le 1$, with in the square formed by the lines $|x| \le 1/2$, $|y| \le 1/2$.

27. Find all the values of the parameter a $(a \le 1)$ for which the area of the figure bounded by the pair of straight lines $y^2 - 3y + 2 = 0$ and the curves $y = [a]x^2$, $y = \frac{1}{2}[a]x^2$ is the greatest, where [.] denotes greatest integer function.

28. Consider two curves $C_1 : y^2 = 4[\sqrt{y}]x$ and $C_2 : x^2 = 4[\sqrt{x}]y$, where [.] denotes the greatest integer function. Find the area of region enclosed by these two curves with in the square formed by the lines; $x = 1$, $y = 1$, $x = 4$, $y = 4$.

29. If $(x - y) f(x + y) - (x + y) f(x - y) = 4xy(x^2 - y^2)$, find the area common to both the curves $[f(y)]^{2/3} + [f(x)]^{1/3} = 0$ and $[f(y)]^{2/3} + [f(x)]^{2/3} = 12$.

30. If $f(x)$ is positive for all positive values of x, and $f'(x) < 0$, $f''(x) > 0 \;\forall\, x \in R^+$, prove that $\frac{1}{2} f(1) + \int_1^n f(x)\,dx < \sum_{r=1}^{n} f(r) < \int_1^n f(x)\,dx + f(1)$.

Answers

Target Exercise 3.1

1. (a) **2.** (b) **3.** (a) **4.** (a) **5.** (c)

Proficiency Exercise 1

1. (c) **2.** (d) **3.** (a) **4.** (b) **5.** (b) **6.** (b) **7.** (b)
8. (c) **9.** (d) **10.** (a) **11.** (a) **12.** (a) **13.** (a) **14.** (c)
15. (b) **16.** (a) **17.** (b) **18.** (d) **19.** (c) **20.** (b, c)
21. (a, b, d) **22.** (a, d) **23.** (a, d) **24.** (a, c) **25.** (d) **26.** (c) **27.** (d)
28. (b) **29.** (c) **30.** (c) **31.** (a) **32.** (b) **33.** (a) **34.** (a)
35. (b) **36.** (c) **37.** (b) **38.** (a)
39. (A) → (p), (B) → (p), (C) → (r), (D) → (s)
40. (A) → (p), (B) → (p), (C) → (r), (D) → (r)
41. (7) **42.** (1) **43.** (4) **44.** (7) **45.** (8) **46.** (5)

Proficiency Exercise 2

1. 2 sq units **2.** 9 sq units **3.** $m = 13/6$

4. $\left(\frac{3\pi}{4} - \frac{1}{6}\right)$ sq units **5.** $\frac{7}{6}$ sq units **6.** $\frac{13}{2} - 6\ln 2$

7. $3e$ sq units **8.** $f(x) = \frac{1}{k+1}\{x^4 - kx^3 + (2k-4)x^2\}$

9. $\left(\frac{2\pi}{3} + \sqrt{3} - \frac{1}{6}\right)$ sq units

10. $g(x) = \begin{cases} -x-1, & x \le -5/2 \\ 4+x, & -5/2 < x \le -2 \\ 2, & -2 < x \le -1 \\ 1-x, & -1 < x \le -1/2 \\ 1+x, & x > -1/2 \end{cases}$ and area $= \frac{101}{4}$ sq units

11. $(14 + 2\log 2)$ sq units **12.** $\left[2 + \log\left(\frac{4}{3\sqrt{3}}\right) - \frac{1}{e}\right]$ sq units

13. $\left(\pi - \frac{2}{3}\right)$ sq units **14.** $\frac{1}{8}\left(\frac{3\pi}{16}\right)^4$ sq units

15. Required maximum ratio $= \frac{2}{3}$ **16.** $(A)_{max} = \frac{\sqrt{3}\,\pi A}{9}$ sq units

17. $A_1 = 1 - \sin 1$, $A_2 = \pi - 1 - \sin 1$, $A_3 = 3\pi - 2$

18. $4\log_5\left(\frac{e^4}{27}\right)$ sq units **19.** $4\left(\frac{16}{3} - 4\pi\right)$ sq units

20. $\frac{3}{2}$ **21.** 2 sq units **22.** 8 sq units **23.** 5 sq units

24. $\left(2 - \frac{\pi}{2}\right)$ sq units **25.** $a = 8, \frac{16}{30}, \frac{12 - 2\sqrt{21}}{5}$

26. $\left[\frac{4}{\log 2}(1 - 2^{-1/2}) - 1\right]$ sq units **27.** $a \in [1, 2)$ **28.** $\frac{11}{3}$ sq units

29. $(2\pi - 3\sqrt{3})$ sq units

Solutions

(Proficiency Exercise 1)

Type 1 : Only One Correct Option

1. Given, $\int_0^4 f(x)\,dx - \int_0^4 g(x)\,dx = 10$

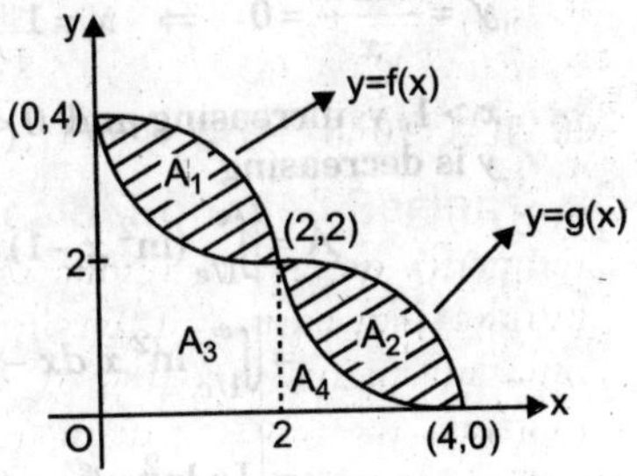

$$(A_1 + A_3 + A_4) - (A_2 + A_3 + A_4) = 10$$

$$A_1 - A_2 = 10 \quad \text{...(i)}$$

Again, $\int_2^4 g(x)\,dx - \int_2^4 f(x)\,dx = 5$

$$(A_2 + A_4) - A_4 = 5$$

$$A_2 = 5 \quad \text{...(ii)}$$

Adding Eqs. (i) and (ii),

$$A_1 = 15$$

2. Solving, $e^x = e^{a-x}$, we get

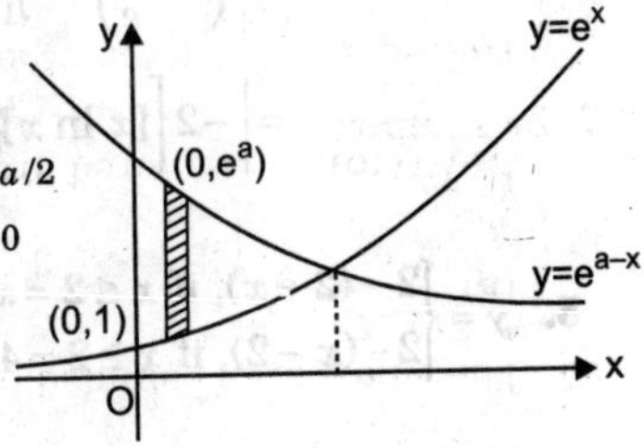

$$e^{2x} = e^a \Rightarrow x = \frac{a}{2}$$

$$S = \int_0^{a/2} (e^a \cdot e^{-x} - e^x)\,dx = \left[-(e^a \cdot e^{-x} + e^x)\right]_0^{a/2}$$

$$= (e^a + 1) - (e^{a/2} + e^{a/2})$$

$$= e^a - 2e^{a/2} + 1 = (e^{a/2} - 1)^2$$

$$\therefore \quad \frac{S}{a^2} = \left(\frac{e^{a/2} - 1}{a}\right)^2 = \frac{1}{4}\left(\frac{e^{a/2} - 1}{a/2}\right)^2$$

$$\therefore \quad \lim_{a \to 0} \frac{S}{a^2} = \frac{1}{4}$$

3. $m_{PQ} = \dfrac{a^2 - b^2}{a + b} = a - b$ equation of PQ

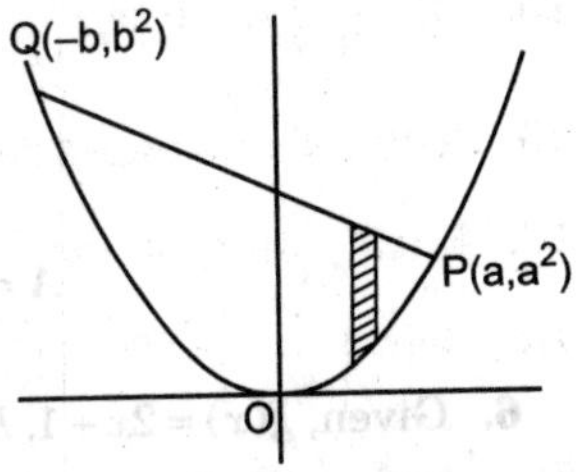

$$y - a^2 = \frac{a^2 - b^2}{a + b}(x - a)$$

or $\quad y - a^2 = (a - b)(x - a)$

$$y = a^2 + x(a - b) - a^2 + ab$$

$$y = (a - b)x + ab$$

$$\therefore \quad S_1 = \int_{-b}^{a} ((a - b)x + ab - x^2)\,dx$$

Which simplifies to $\dfrac{(a + b)^3}{6}$ ...(i)

Also, $\quad S_2 = \dfrac{1}{2}\begin{vmatrix} a & a^2 & 1 \\ -b & b^2 & 1 \\ 0 & 0 & 1 \end{vmatrix} = \dfrac{1}{2}[ab^2 + a^2 b] = \dfrac{1}{2}ab\,(a + b)$ ...(ii)

$$\therefore \quad \frac{S_1}{S_2} = \frac{(a+b)^3}{6} \cdot \frac{2}{ab(a+b)} = \frac{(a+b)^2}{3ab} = \frac{1}{3}\left[\frac{a}{b} + \frac{b}{a} + 2\right]$$

$$\therefore \quad \left.\frac{S_1}{S_2}\right|_{\min} = \frac{4}{3}$$

4. $y = \ln^2 x - 1$

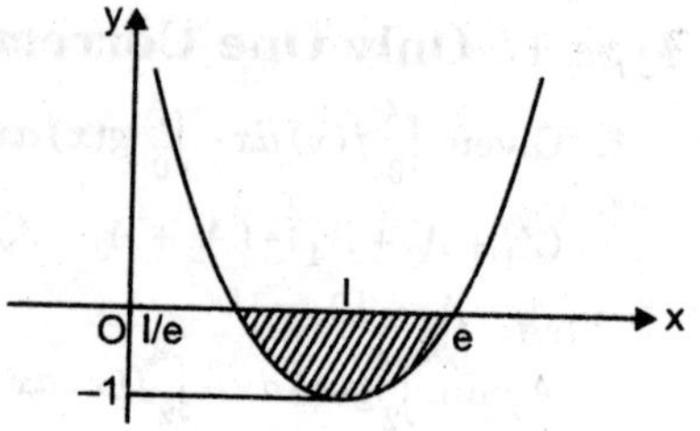

$$y' = \frac{2\ln x}{x} = 0 \quad \Rightarrow \quad x = 1$$

$x > 1$, y increasing and $0 < x < 1$, y is decreasing

$$A = \left| \int_{1/e}^{e} (\ln^2 x - 1)\, dx \right|$$

$$= \left| \int_{1/e}^{e} \ln^2 x\, dx - \int_{1/e}^{e} dx \right|$$

$$= \left| [x \ln^2 x]_{1/e}^{e} - 2\int_{1/e}^{e} \left(\frac{\ln x}{x}\right) \cdot x\, dx - \left(e - \frac{1}{e}\right) \right|$$

$$= \left| \left(e - \frac{1}{e}\right) - 2\int_{1/e}^{e} \left(\frac{\ln x}{x}\right) \cdot x\, dx - \left(e - \frac{1}{e}\right) \right|$$

$$= \left| -2\left[[x \ln x]_{1/e}^{e} - \int_{1/e}^{e} dx \right] \right| = \left| -2\left[\left(e + \frac{1}{e}\right) - \left(e - \frac{1}{e}\right) \right] \right| = \left| \frac{4}{e} \right| = \frac{4}{e}$$

5. $y = \begin{cases} 2 - (2 - x), & \text{if } x \le 2 = x \\ 2 - (x - 2), & \text{if } x \ge 2 = 4 - x \end{cases}$. Also, $y = \begin{cases} \dfrac{3}{x}, & \text{if } x > 0 \\ -\dfrac{3}{x}, & \text{if } x < 0 \end{cases}$

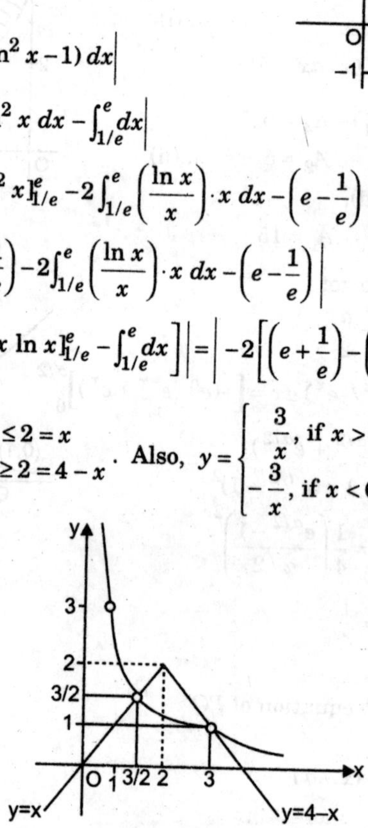

$$A = \int_{3/2}^{2} \left(x - \frac{3}{x}\right) dx + \int_{2}^{3} \left((4 - x) - \frac{3}{x}\right) dx$$

6. Given, $g(x) = 2x + 1$; $h(x) = (2x+1)^2 + 4$

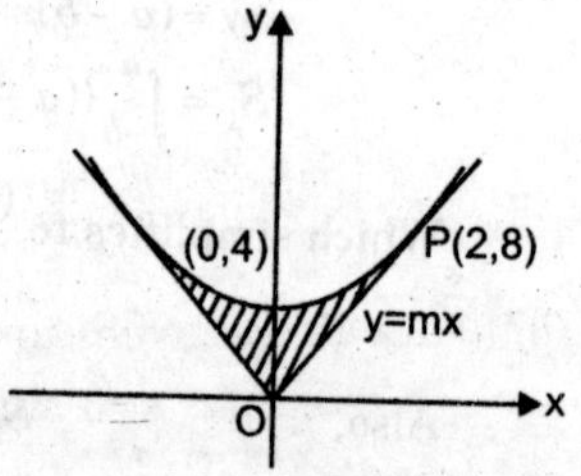

Now, $\quad h(x) = f[g(x)]$

$\quad (2x+1)^2 + 4 = f(2x+1)$

Let $\quad 2x + 1 = t \quad \Rightarrow \quad f(t) = t^2 + 4$

$\therefore \quad f(x) = x^2 + 4 \quad$ ...(i)

Solving, $\quad y = mx \quad$ and $\quad y = x^2 + 4$

$$x^2 - mx + 4 = 0$$

Put $D=0;\quad m^2=16 \Rightarrow m=\pm 4$

Tangents are $y=4x$ and $y=-4x$

$$A=2\int_0^2[(x^2+4)-4x]\,dx=2\int_0^2(x-2)^2\,dx=\left[\frac{2}{3}(x-2)^3\right]_0^2=\frac{16}{3}\text{ sq units}$$

7. $y=-\sqrt{-x}\quad \Rightarrow y^2=-x$ where x and y both negative

$x=-\sqrt{-y}\quad \Rightarrow x^2=-y$ where x and y both negative

Since, $\quad A=\dfrac{16\,ab}{3}$

Where, $\quad a=b=\dfrac{1}{4}$

$\therefore \quad A=\dfrac{1}{3}$

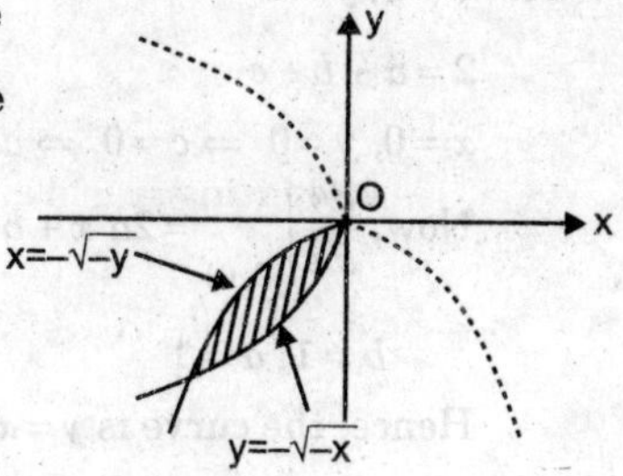

8. $\dfrac{f''(x)}{f'(x)}=1$

Integrating, $\ln f'(x)=x+C,\ f'(0)=1 \Rightarrow C=0$

$f'(x)=e^x$

$f(x)=e^x+k,\ f(0)=0 \Rightarrow k=-1$

$f(x)=e^x-1$

Area $=\int_0^1(e^x-1+1)\,dx=[e^x]_0^1=e-1$

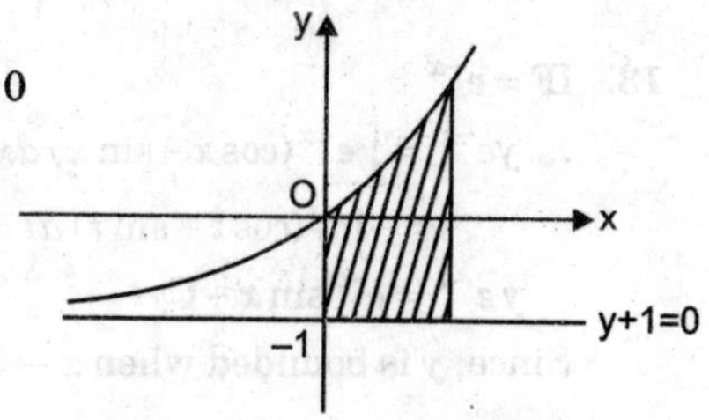

9. $A=2\int_0^1[y\sqrt{1-y^2}-(y^2-1)]\,dy=2$

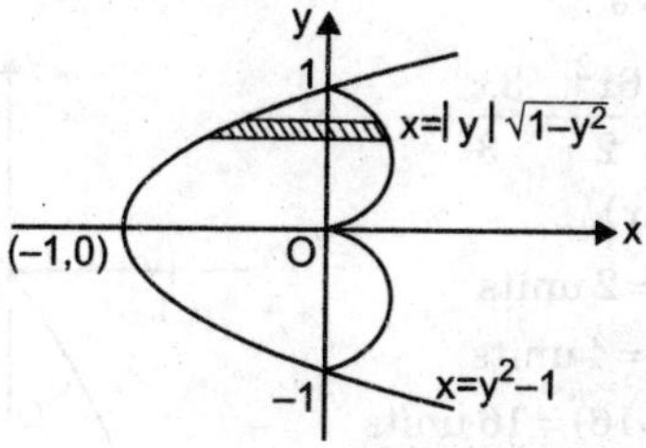

10. $y=xe^{-x}$

$y'=e^{-x}-x\,e^{-x}=(1-x)e^{-x}$ increasing for $x<1$

$y''=-e^{x}-[e^{-x}-xe^{-x}]=e^{-x}[-1-1+x]=(x-2)e^{-x}$

For point of inflection $y''=0 \quad\Rightarrow\quad x=2$

$$A=\int_0^2 xe^{-x}dx=[-xe^{-x}]_0^2+\int_0^2 e^{-x}$$

$$=(-2e^{-2})-(e^{-x})_0^2$$

$$=-2e^{-2}-(e^{-2}-1)=1-e^{-2}-2e^{-2}=1-3e^{-2}$$

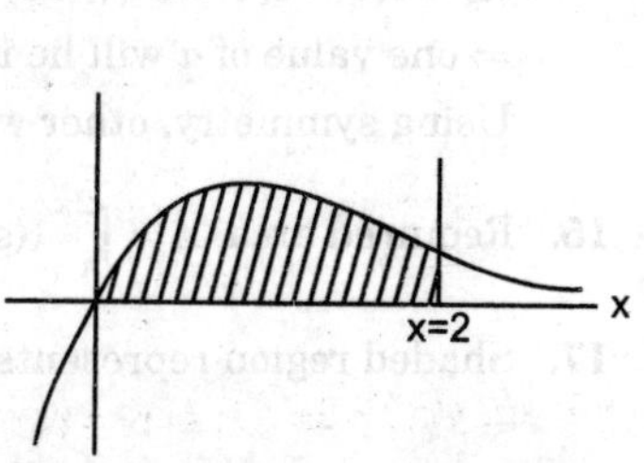

11. $(a,0)$ lies on the given curve

$\therefore \quad 0=\sin 2a-\sqrt{3}\sin a \quad\Rightarrow\quad \sin a=0$

or $\quad \cos a=\sqrt{3}/2$

$\Rightarrow \quad a=\dfrac{\pi}{6}$ (as $a>0$ and the first point of intersection with positive x-axis) and $\quad A=\int_0^{\pi/6}(\sin 2x-\sqrt{3}\sin x)\,dx=\left(-\frac{\cos 2x}{2}+\sqrt{3}\cos x\right)_0^{\pi/6}$

$$=\left(-\frac{1}{4}+\frac{3}{2}\right)-\left(-\frac{1}{2}+\sqrt{3}\right)=\frac{7}{4}-\sqrt{3}=\frac{7}{4}-2\cos a\Rightarrow 4A+8\cos a=7$$

12. $x=1;\ y=2$

$2=a+b+c \qquad$ ...(i)

$x=0,\ y=0\ \Rightarrow c=0\ \Rightarrow a+b=2$

Now, $\left.\frac{dy}{dx}\right|_{(0,0)}=2a\,x+b=1$

$\therefore\quad b=1,\ a=1$

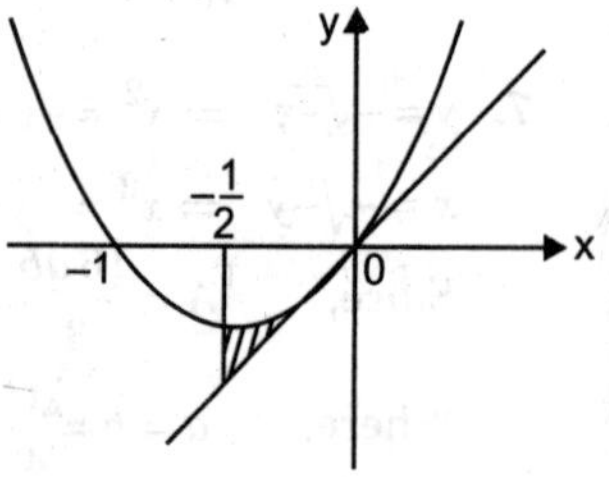

Hence, the curve is $y=x^2+x$

$$A=\int_{-\frac{1}{2}}^{0}(x^2+x-x)\,dx=\int_{-\frac{1}{2}}^{0}(x^2)\,dx=\frac{1}{24}\text{ sq units}$$

13. IF $=e^{-x}$

$\therefore\ ye^{-x}=\int e^{-x}(\cos x-\sin x)\,dx$ put $-x=t$

$=-\int e^{t}(\cos t+\sin t)\,dt=-e^{t}\sin t+C$

$ye^{-x}=e^{-x}\sin x+C$

Since, y is bounded when $x\to\infty\ \Rightarrow\ C=0$

$\therefore\qquad y=\sin x$

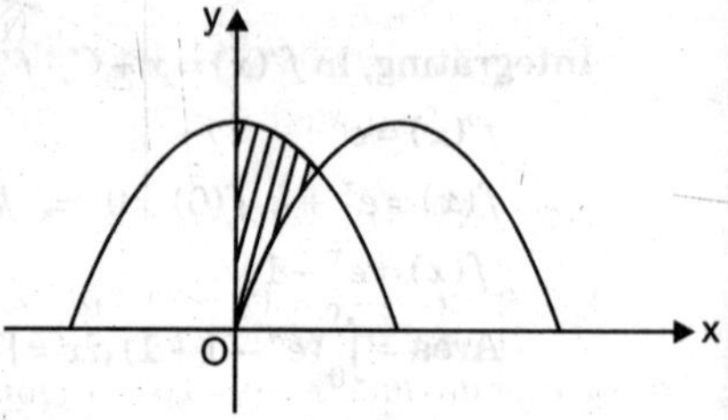

$$\text{Area}=\int_0^{\pi/4}(\cos x-\sin x)\,dx=\sqrt{2}-1$$

14. $I=\int(6x-3x^2)\,dx=\frac{6x^2}{2}-\frac{3x^3}{3}$

$=3x^2-x^3=x^2(3-x)$

$A_1=I(2)-I(1)=4-2=2$ units

$A_2=I(2)-I(3)=4-0=4$ units

$A_3=I(3)-I(4)=0-(-16)=16$ units

$\Rightarrow$ one value of a will lie in (3, 4).

Using symmetry, other will lie in (–2, –1).

15. Required area $A=\int_{\pi}^{2\pi}((\sin x-x)+2\pi)\,dx=\frac{\pi^2}{2}-2$ sq units

17. Shaded region represents $S\cap S'$ clearly area enclosed is 2 sq units.

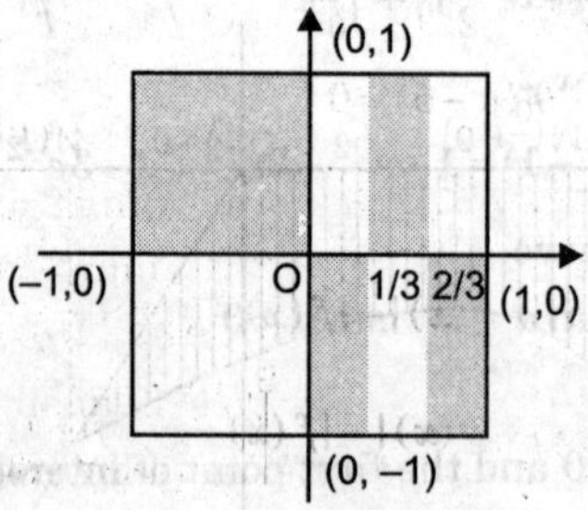

18. Required area is $2\left[\int_0^{\alpha} ex \ln x\, dx + \int_1^{\alpha} (1-\sqrt{1-(x-1)^2})\, dx\right]$

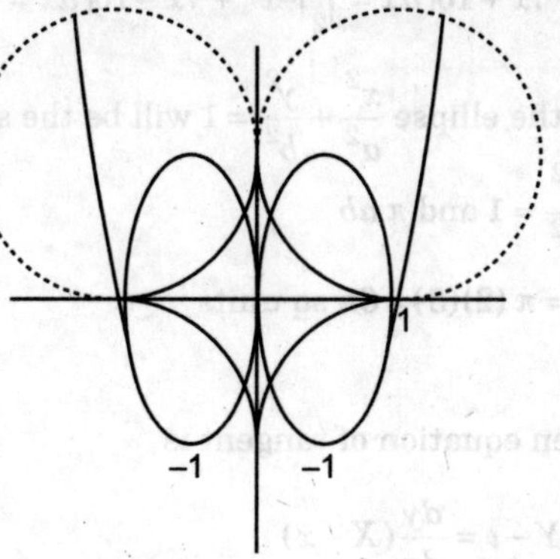

19. $y=\sqrt{2ax-x^2} \Rightarrow (x-a)^2+y^2=a^2$

Let $P(h, k)$ be a point, then $BP > PN$

For the bounded condition $BP = PN = k$

Now, $AP = a-k = \sqrt{(h-a)^2+k^2} \Rightarrow k = h - \dfrac{h^2}{2a}$

$\therefore$ Boundary of the region is $y = x - \dfrac{x^2}{2a}$

Required area $= 2\int_0^a \left(x - \dfrac{x^2}{2a}\right) dx = \dfrac{2a^2}{3}$

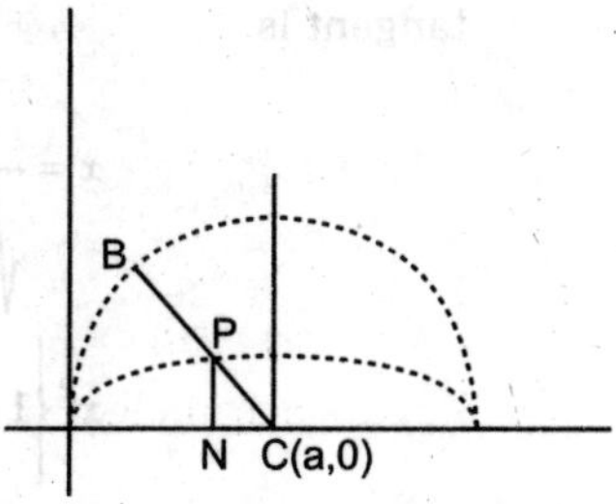

Type 2 : **More Than One Correct Options**

20. $f'(x) = 2x - a$

At (2, 4), $f'(x) = 4 - a$

Equation of normal at (2, 4) is $(y-4) = -\dfrac{1}{(4-a)}(x-2)$

Let point of intersection with x and y-axes be A and B respectively, then

$$A \equiv (-4a+18, 0)\ B \equiv \left(0, \frac{4a-18}{a-4}\right)$$

Since, $a > \dfrac{9}{2}$ as

$\therefore$ Area of triangle $= \dfrac{1}{2}(4a-18)\dfrac{(4a-18)}{(a-4)} = 2$

$\Rightarrow$ $(4a-17)(a-5) = 0$

$\Rightarrow$ $a = 5$ or $\dfrac{17}{4}$

21. $\text{Max}\{f(x), g(x)\} = \dfrac{1}{2}[|f(x)+g(x)| + |f(x)-g(x)|]$

$\text{Min}\{f(x), g(x)\} = \dfrac{1}{2}[|f(x)+g(x)| - |f(x)-g(x)|]$

22. Area bounded by parabola $y = x^2 - 7x + 10$ and x-axis is given by

$$\int_2^5 |x^2 - 7x + 10|\, dx = \int_2^5 |-x^2 + 7x - 10|\, dx = \frac{9}{2}$$

23. Area bounded by the ellipse $\frac{x^2}{a^2} + \frac{y^2}{b^2} = 1$ will be the same as the area bounded by the ellipse $\frac{x^2}{b^2} + \frac{y^2}{a^2} = 1$ and $\pi\, ab$

$\therefore$ Required area $= \pi\,(2)(3) = 6\pi$ sq units

24. $OP = x$

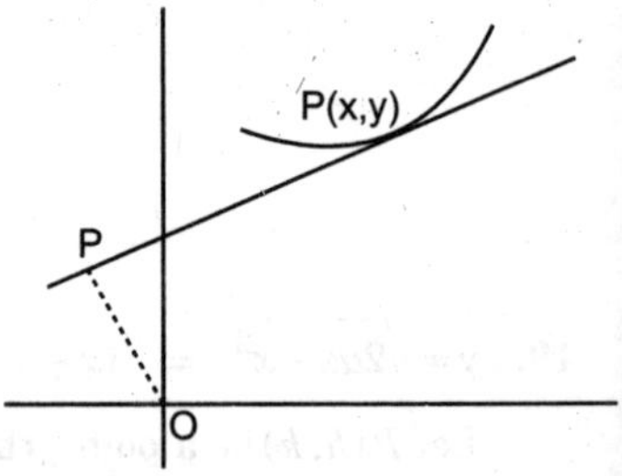

If slope be $\frac{dy}{dx}$, then equation of tangent is

$$Y - t = \frac{dy}{dx}(X - x)$$

Length of perpendicular from origin to this tangent is

$$x = \frac{Y - x\dfrac{dy}{dx}}{\sqrt{1 + \left(\dfrac{dy}{x}\right)^2}}$$

$$x^2\left\{1 + \left(\frac{dy}{dx}\right)^2\right\} = y^2 + x^2\left(\frac{dy}{dx}\right)^2 = 2xy\frac{dy}{dx}$$

$$\Rightarrow \quad \frac{dy}{dx} = -\frac{x^2 - y^2}{2xy} \qquad \text{(Homogeneous form)}$$

$$\frac{dy}{dx} = \frac{1-x}{y} = \frac{x - x^2}{xy} = \frac{\left[\dfrac{x^2 + y^2}{2}\right] - x^2}{xy} = \frac{y^2 - x^2}{2xy}$$

Since, option (a) is true (equation of circle)
If option (a) is true (b) can't be (It is parabola).
If option (a) is true, option (c) is also true where $c = 1$.

Type 3 : Assertion and Reason

25. $\therefore \quad \sin^2 x \le \sin x : \forall\, x \in (0, \pi)$

Therefore, area of $y = \sin^2 x$ will be lesser from area of $y = \sin x$.

Statement II is obviously true.

Hence, (d) is the correct answer.

26. Let the line $y = kx + 2$ cuts $y = x^2 - 3$ at $x = \alpha$ and $\alpha = \beta$, area bounded by the curves $= \int_\alpha^\beta (y_1 - y_2) = \int_\alpha^\beta \{(kx + 2) - (x^2 - 3)\}\, dx$

$$\Rightarrow \quad f(x) = \frac{(k^2 + 20)^{3/2}}{6}$$

Which clearly, shows the Statement II is false but $f(k)$ is least when $k = 0$.
Hence, (c) is the correct answer.

27. As of region bounded by parabola $y^2 = 4x$ and $x^2 = 4y$ is

$$\int_0^4 \left(2\sqrt{x} - \frac{x^2}{4}\right) dx = \left[\frac{4}{3}x^{3/2} - \frac{x^3}{12}\right]_0^4 = \frac{32}{3} - \frac{16}{3} = \frac{16}{3} \text{ sq units}$$

Hence, Statement I is false.

28. As the area enclosed by $|x| + |y| \le a$ is the area of square (*ie*, $2a^2$).

$\therefore$ Area enclosed by $|x + y| + |x - y| \le 2$ is area of square shown as

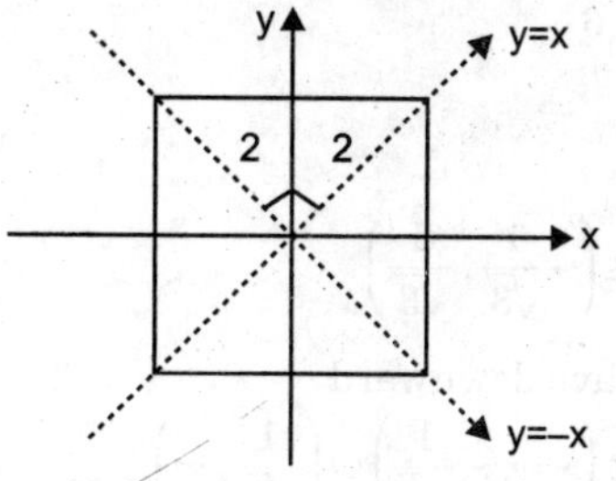

$\therefore$ $$\text{Area} = 4\left(\frac{1}{2} \times 2 \times 2\right) = 8 \text{ sq units}$$

Also the area enclosed by $|x + y| + |x - y| \le 2$ is symmetric about x-axis, y-axis, $y = x$ and $y = -x$.

$\therefore$ Both the Statements are true but Statement II is not the correct explanation of Statement I.

29. $\left|\int_a^b \{f(x) - g(x)\}\, dx\right|$ is true for all quadrants.

$\therefore$ Statement II is false.

The area bounded by $y = x(x - 1)$ and $y = x(1 - x)$.

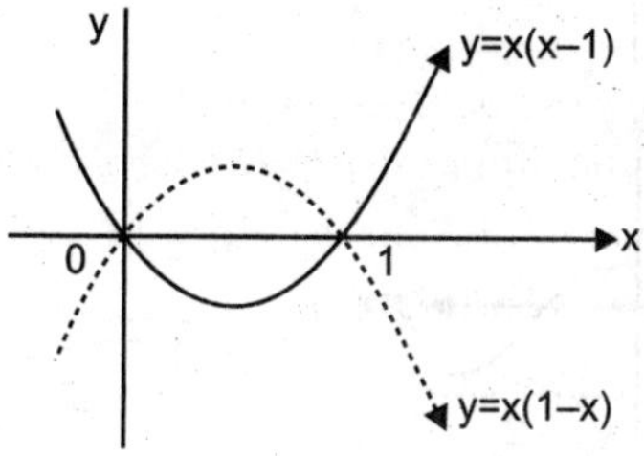

$\therefore$ $$\text{Area enclosed} = 2\int_0^1 x(1 - x)\, dx = 2\left\{\frac{x^2}{2} - \frac{x^3}{3}\right\}_0^1 = 2\left(\frac{1}{2} - \frac{1}{3}\right) = 2\left(\frac{3-2}{6}\right) = \frac{1}{3}$$

Type 4 : Linked-Comprehension Based Questions

30. Applying the given conditions, we get

$$f(x) = \frac{-2x^2 + 4}{x^2 + 1}$$

$$\text{Area} = 2t\left(\frac{-2t^2+4}{t^2+1}+2\right) = \frac{12t}{t^2+1}$$

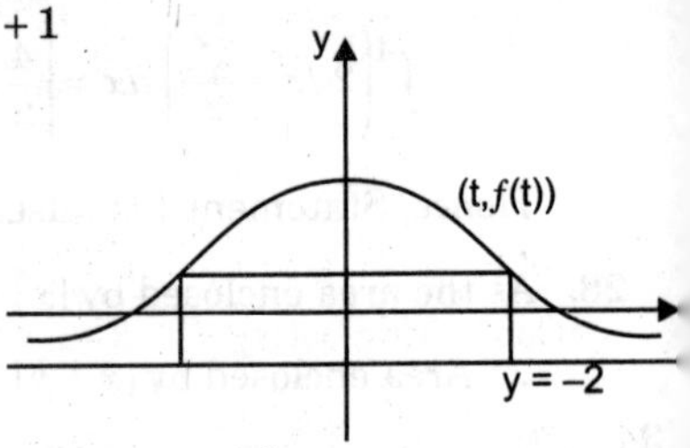

$$t > 0$$

$$\therefore \quad t + \frac{1}{t} \geq 2$$

$$\Rightarrow \quad \frac{1}{t+\frac{1}{t}} \leq \frac{1}{2} \Rightarrow \frac{12t}{t^2+1} \leq 6$$

∴ Maximum area = 6

31. $f''(x) = \dfrac{12(3x^2-1)}{(x^2+1)^3}$

$f''(x) < 0$ for $x \in \left(-\frac{1}{\sqrt{3}}, \frac{1}{\sqrt{3}}\right)$

$\Rightarrow$ $y = f(x)$ is concave downward

$f''(x) > 0$ for $x \in \left(-\infty, -\frac{1}{\sqrt{3}}\right) \cup \left(\frac{1}{\sqrt{3}}, \infty\right)$

$\Rightarrow$ $y = f(x)$ is concave upward

The result follows by considering points $(x_1, f(x_1)), (x_2, f(x_2))$ and point dividing these in the ratio $\lambda_2 : \lambda_1$.

32. It is clear from the graphs

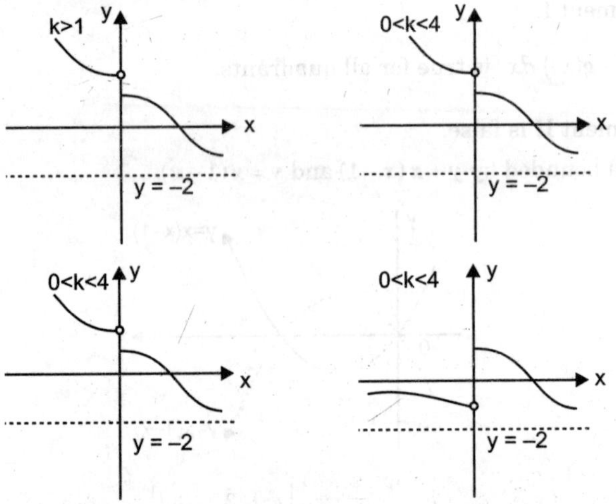

33. $f(x) = \dfrac{x^2 - ax + 1}{x^2 + ax + 1}$

For differentiation, better write $f(x)$ as $f(x) = 1 - \dfrac{2ax}{x^2+ax+1}$

Now, on differentiation $f(x) = \dfrac{2a(x^2-1)}{(x^2+ax+1)^2}$

$$f'(1) = 0 = f'(-1)$$

Then, the choices (b) and (d) are eliminated.

Again, for the differentiating Eq. (i) gives $(x^2+ax+1)^2 \cdot 2x-(x^2-1)$

$$f''(x)=2a\cdot\frac{2(x^2+ax+1)(2x+a)}{(x^2+ax+1)^4}$$

$$f''(-1)=-\frac{4a}{(2-a)^2},\ f''(1)=-\frac{4a}{(2+a)^2}$$

Combining both $f''(1)(2+a)^2+f''(-1)(2-a)^2=0$

34. $f'(x)=2a\dfrac{(x^2-1)}{(x^2+ax+1)^2}=2a\dfrac{(x-1)(x+1)}{(x^2+ax+1)^2}$

If is easily seen that $f(x)$ decreases on $(-1,1)$, and has a local minimum at $x=1$, because the derivatives changes its sign from negative to positive.

35. $\because$ $$g'(x)=\frac{f'(e^x)e^x}{1+e^{2x}}$$

Now, $$f'(e^x)=\frac{2a\cdot(e^{2x}-1)}{(e^{2x}+ae^x+1)^2}$$

It is seen from the above that $f'(e^x)$ and so $g'(x)$ is positive on $(0,\infty)$ and negative on $(-\infty,0)$.

36. Clearly, $x=a\sin^3 t,\ y=a\cos^3 t\ (0\le t\le 2\pi)$ $S=\int_0^{2\pi} a\sin^3 t\cdot a\cdot 3\cos^2 t(-\sin t)\,dt$

$$=-3a^2\times\int_0^{\pi/2}\sin^4 t\cos^2 t\,dt$$

$$=-3a^2\times\frac{\left\lceil\frac{5}{2}\right.\left\lceil\frac{3}{2}\right.}{2\left\lceil\frac{4+2+2}{2}\right.}$$

$$=-3a^2\,\frac{\frac{3}{2}\times\frac{1}{2}\sqrt{\pi}\times\frac{1}{2}\sqrt{\pi}}{2\times 3\times 2\times 1}$$

$$=-\frac{3}{8}\pi a^2=\frac{3}{8}\pi a^2$$ (absolute value)

37. $$S=-\int_0^{2\pi} a(1-\cos t)a(1-\cos t)\,dt$$

$$=-a^2\int_0^{2\pi}(1-2\cos t+\cos^2 t)\,dt$$

$$=-a^2\int_0^{2\pi}\left(1-2\cos t+\left(\frac{1+\cos 2t}{2}\right)\right)dt$$

$$=-\frac{a^2}{2}\int_0^{2\pi}[3-4\cos t+\cos 2t]_0^{2\pi}$$

$$=-3\pi a^2=3\pi a^2$$ (absolute value)

38. $x=\dfrac{1}{3}(6-t)=\dfrac{1}{3}(6t-t^2)$

$y=\dfrac{t^2}{8}(6-t)=\dfrac{1}{8}(6t^2-t^3)$

Type 5 : Match the Columns

39. (A) Required area = 4s

$$s = \int_0^\pi (x + \sin x)\, dx - \int_0^\pi x\, dx = \frac{\pi^2}{2} - \cos\pi + \cos 0 - \frac{\pi^2}{2} = 2 \text{ sq units}$$

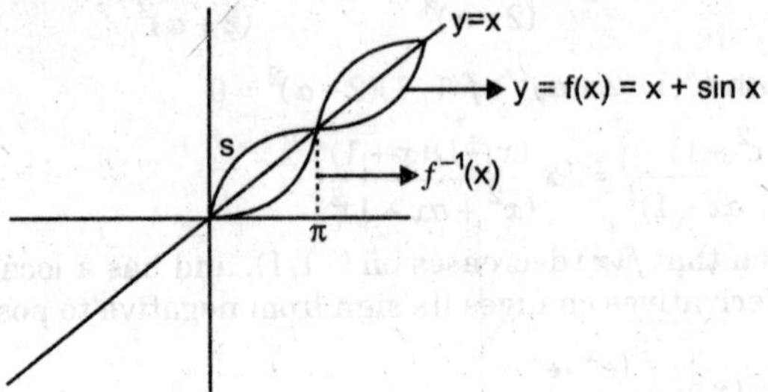

(B) Required area $= 2\int_0^1 xe^x\, dx = 2[xe^x - e^x]_0^1 = 2$

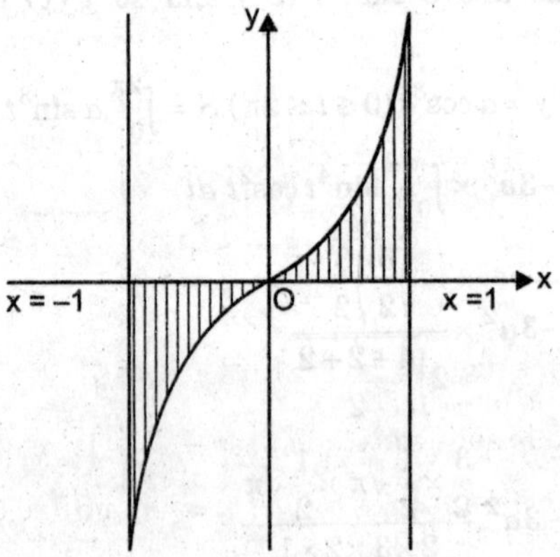

(C) $y^2 = x^3$ and $|y| = 2x$ both the curve are symmetric about y-axis

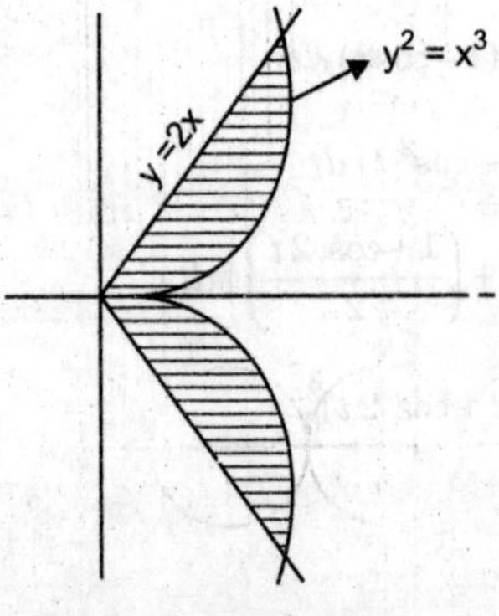

$$4x^2 = x^3 \Rightarrow x = 0, 4$$

Required area $= 2\int_0^4 (2x - x^{3/2})\, dx = \dfrac{16}{5}$

(D) $\sqrt{x}+\sqrt{|y|}=1$

Above curve is symmetric about x-axis

$$\sqrt{|y|}=1-\sqrt{x} \text{ and } \sqrt{x}=1-\sqrt{|y|}$$

$\Rightarrow$ for $x>0, y>0, \sqrt{y}=1-\sqrt{x}$

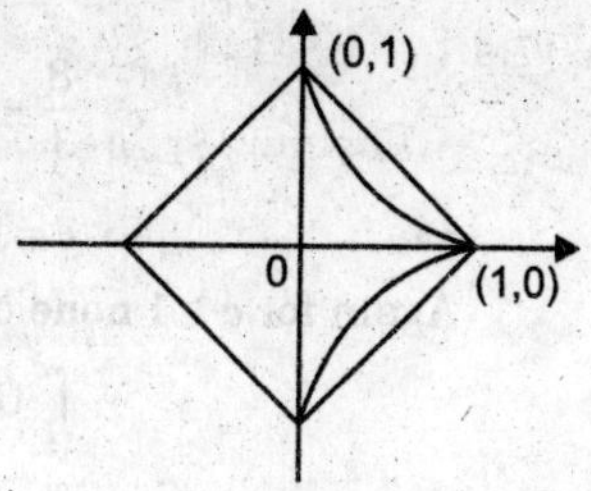

$$\frac{1}{2\sqrt{y}}\frac{dy}{dx}=-\frac{1}{2\sqrt{x}}$$

$$\frac{dy}{dx}=-\sqrt{\frac{x}{y}}$$

$$\frac{dy}{dx}<0$$

Function is decreasing required area $=\int_0^1(2\sqrt{x}-2x)=\frac{1}{3}$.

40. (A) The area = 2 unit

(B) Area enclosed $=\int_0^{\pi}\sin x\,dx=2$

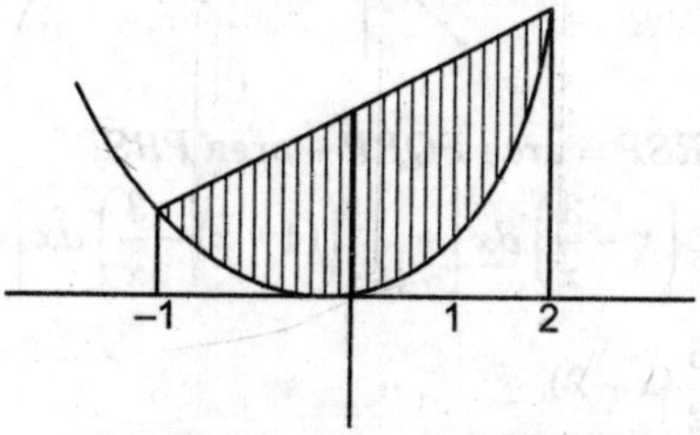

(C) The line $y=x+2$ intersects $y=x^2$ at $x=-1$ and $x=2$

The given region is shaded region area $=\frac{15}{2}-\int_{-1}^{2}x^2\,dx=\frac{9}{2}$

(D) Here, $a^2=9, b^2=5, b^2=a^2(1-e^2) \Rightarrow e^2=\frac{4}{9} \Rightarrow =\frac{2}{3}$

Equation of tangent at $\left(2,\frac{5}{3}\right)$ is $\frac{2x}{9}+\frac{y}{3}=1$

x-intercept $=\frac{9}{2}$, y-intercept $=3$

Area $=4\times\frac{9}{2}\times3\times\frac{1}{2}=27$ sq units

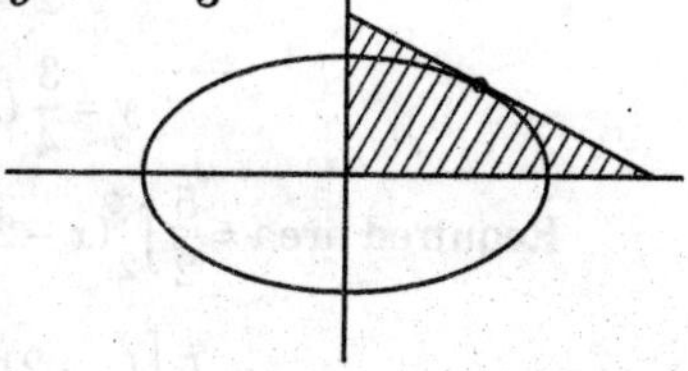

pe 6 : Integer Answer Type Questions

1.

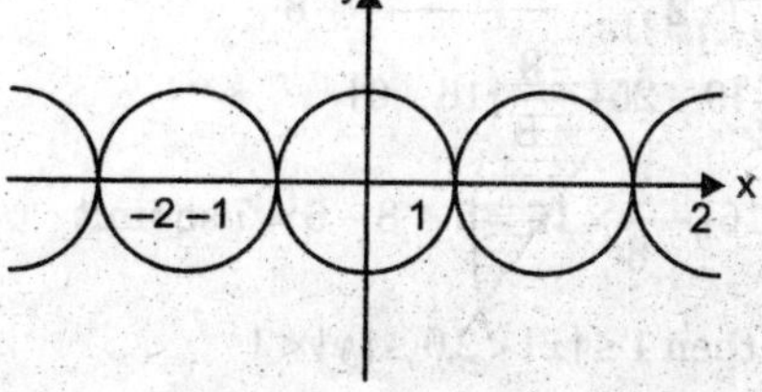

Area is given by $A=2\int_0^1(x^2-3x+2)\,dx=\frac{5}{3} \Rightarrow 3A+2=3\cdot\frac{5}{3}+2=7$

42. For $c < 1$; $\int_c^1 (8x^2 - x^5)\,dx = \frac{1}{3}$

$\Rightarrow \quad \frac{8}{3} - \frac{1}{6} - \frac{8c^3}{3} - \frac{c^6}{6} = \frac{16}{6} \Rightarrow c^3\left[-\frac{8}{3} + \frac{c^3}{6}\right] = \frac{16}{3} - \frac{8}{3} + \frac{1}{6} = \frac{17}{6}$

$\Rightarrow \quad c = -1$

Again for $c \geq 1$ none of the values of c satisfy the required condition that

$$\int_1^c (8x^2 - x^5)\,dx = \frac{16}{3} \Rightarrow c + 2 = 1$$

43. $y = \begin{cases} x; & x < 2 \\ 4 - x; & x \geq 2 \end{cases}$ and $y = \begin{cases} 3/x; & x > 0 \\ -3/x & x < 2 \end{cases}$

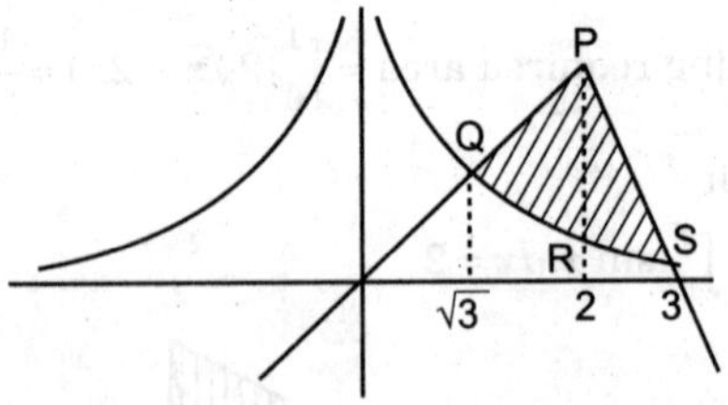

Required area $= PQRSP$ = area $PQRP$ + area $PRSP$

$$= \left|\int_{\sqrt{3}}^2 \left(x - \frac{3}{x}\right) dx\right| + \left|\int_2^3 \left((4 - x) - \frac{3}{x}\right) dx\right| = \frac{4 - 3\ln 3}{2} \text{ sq units}$$

44. Equation of AB $y = \frac{5}{2}(x - 2)$

Equation of BC $y - 5 = \frac{3-5}{6-4}(x - 4)$

$y = -x + 9$

Equation of CA $y - 3 = \frac{0-3}{2-6}(x - 6)$

$y = \frac{3}{4}(x - 2)$

Required area $= \frac{5}{2}\int_2^4 (x-2)\,dx + \int_4^6 -(x - 9)\,dx - \frac{3}{4}\int_2^6 (x-2)\,dx$

$= \frac{5}{2}\left[\frac{(x-2)^2}{2}\right]_2^4 - \left[\frac{(x-9)^2}{2}\right]_4^6 - \frac{3}{4}\left[\frac{(x-2)^2}{2}\right]_2^6$

$= \frac{5}{2}[2^2 - 0] - \frac{1}{2}[(-3)^2 - (-5)^2] - \frac{3}{8}[4^2 - 0]$

$= \frac{5}{4} \times 4 - \frac{1}{2}[9 - 25] - \frac{3}{8}[16 - 0]$

$= 5 - \frac{1}{2}[-16] - \frac{3}{8} \times 16 = 5 + 8 - 6 = 7$ sq unit

45. If $[|x|] = 1$ and $[|y|] = 0$, then $1 \leq |x| < 2, 0 \leq |y| < 1$

$\Rightarrow \quad x \in (-2, -1] \cup [1, 2),\ y \in (-1, 1)$

If $\quad [|x|] = 0, [|y|] = 1$

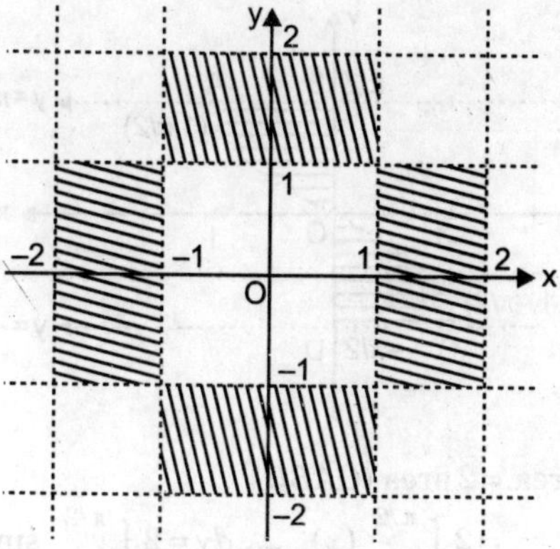

Then, $x \in (-1, 1),\ y \in (-2, -1] \cup [1, 2]$

Area of required region $= 4(2-1)(1-(-1)) = 8$ sq units

46. Clearly, $f(x) = \dfrac{2x - x^2}{2}$

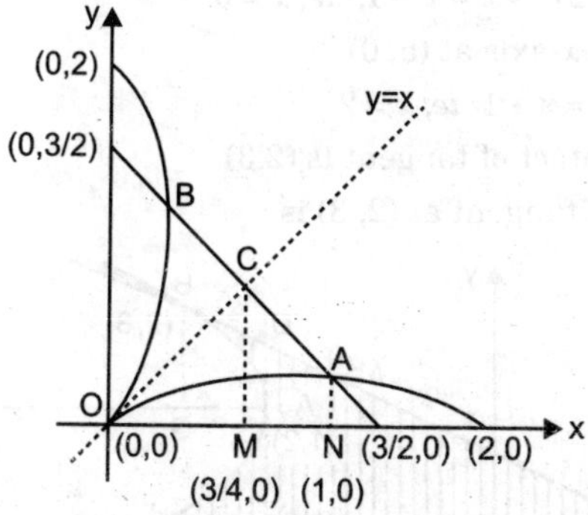

Since, $2x + 2y = 3$; passes through $A\left(1, \dfrac{1}{2}\right)$ and $B\left(\dfrac{1}{2}, 1\right)$ so bounded area A

$$= \text{area } OAB = 2\,[\text{area } OCM + \text{area } CMNA - \text{area } ONA]$$

$$= 2\left[\frac{1}{2}\times\frac{3}{4}\times\frac{3}{4} + \frac{1}{2}\left(\frac{3}{4}+\frac{1}{2}\right)\times\frac{1}{4} - \frac{1}{2}\int_0^1 (2x - x^2)\,dx\right] = \frac{5}{24}$$

$\Rightarrow\ 24A = 5$

(Proficiency Exercise 2)

1. The curve is $y = \sin^{-1} x$, *ie*, $x = \sin y$. This is standard curve lines $x = 0$, $y = \pi/2$, $y = -\pi/2$ are the y-axis and two lines parallel to the x-axis at distance $\dfrac{\pi}{2}$, one above and the other below the x-axis respectively.

Hence, the shaded part is the required area. By symmetry of the curve and the line,

$$\text{area } (OABO) = \text{area } (OCDO)$$

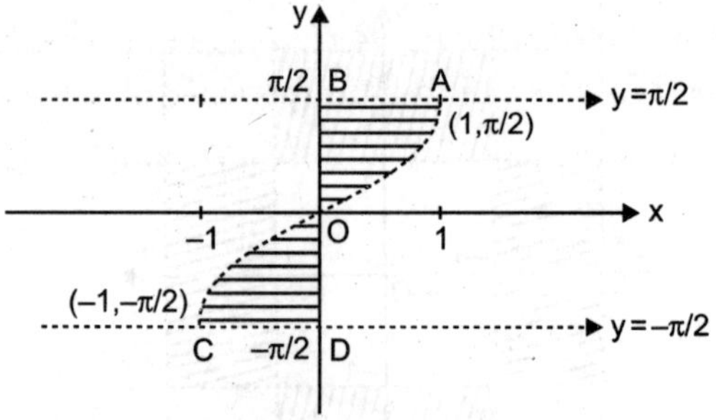

$$\therefore \quad \text{area} = 2 \text{ area } (OABO)$$

$$= 2\int_0^{\pi/2} (x)_{\text{curve}}\, dy = 2\int_0^{\pi/2} \sin y\, dy$$

($\because$ The equation of the curve is $x = \sin y$)

$$\Rightarrow \quad \text{area} = 2$$

2. The vertex of parabola is (1, 2) and the axis is $y - 2 = 0$.

When $y = 0, (0-2)^2 = x = x - 1,$ *ie*, $x = 5$

$\therefore$ The parabola cuts x-axis at (5, 0).

When $y = 3, (3-2)^2 = x - 1$ *ie*, $x = 2$

$\therefore$ The point of contact of tangent is (2,3).

The equation of tangent at (2, 3) is

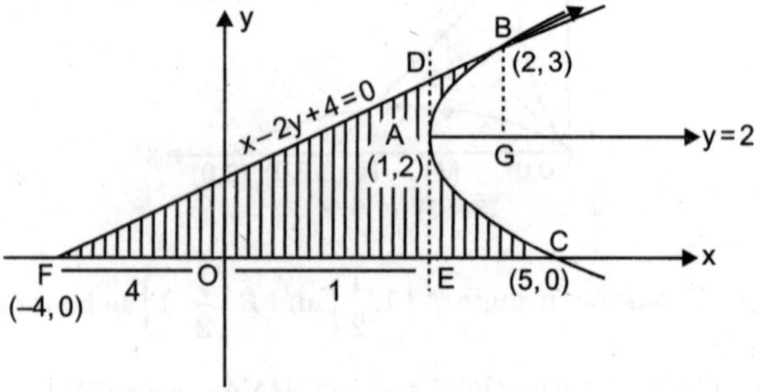

$$(y-3) = \left(\frac{dy}{dx}\right)_{(2,3)} (x-2) \quad \text{...(i)}$$

Differentiating $(y-2)^2 = (x-1)$ w.r.t. x.

$$2(y-2)\frac{dy}{dx} = 1$$

$$\therefore \quad \frac{dy}{dx} = \frac{1}{2(y-2)}$$

$$\therefore \quad \left(\frac{dy}{dx}\right)_{(2,3)} = \frac{1}{2}$$

$\therefore$ From Eq. (i) the equation of tangent at (2, 3) is

$$y - 3 = \frac{1}{2}(x-2)$$

or

$$x - 2y + 4 = 0 \quad \text{...(ii)}$$

The required area = area (ΔDEF) + area ($ADBA$) + area ($AECA$) ...(iii)

The tangent cuts the x-axis at $F = (-4, 0)$, obtained by solving Eq. (ii) with $y = 0$.

Also, the tangent cuts the line $x = 1$ (*ie*, DE) at $D = \left(1, \frac{5}{2}\right)$. So, $DE = \frac{5}{2}$

$$\therefore \quad \text{ar}(\Delta DEF) = \frac{1}{2} FE \cdot DE = \frac{1}{2} \times 5 \times \frac{5}{2} = \frac{25}{4}$$

$$\text{ar}(ADBA) = \text{ar}(ADBGA) - \text{ar}(ABGA)$$

$$= \int_1^2 (y-2)_{\text{Line}}\, dx - \int_1^2 (y-2)_{\text{Parabola}}\, dx$$

$$= \int_1^2 \left(\frac{x+4}{2} - 2\right) dx - \int_1^2 \sqrt{x-1}\, dx$$

{$\because$ for the tangent $y = \frac{x+4}{2}$, by Eq. (i) and for the parabola, $y - 2 = \sqrt{x-1}$}

$$= \left[\frac{1}{2} \cdot \frac{x^2}{2}\right]_1^2 - \left[\frac{(x-1)^{3/2}}{3/2}\right]_1^2$$

$$= \left(1 - \frac{1}{4}\right) - \frac{2}{3}(1 - 0) = \frac{1}{12}$$

$$\text{ar}(AECA) = \int_1^5 (y)_{\text{Parabola}}\, dx = \int_1^5 (2 - \sqrt{x-1})\, dx$$

{$\because$ for upper part of parabola $y = 2 + \sqrt{x-1}$ and for the lower part it is $y = 2 - \sqrt{x-1}$}.

$$= \left[2x - \frac{(x-1)^{3/2}}{3/2}\right]_1^5 = \frac{8}{3}$$

$\therefore$ By Eq. (iii); the required area $= \frac{25}{4} + \frac{1}{12} + \frac{8}{3} = 9$

3. Here, the curve is $x^2 - 4x = 1 - y$, *ie*, $(x-2)^2 = -(y-5)$

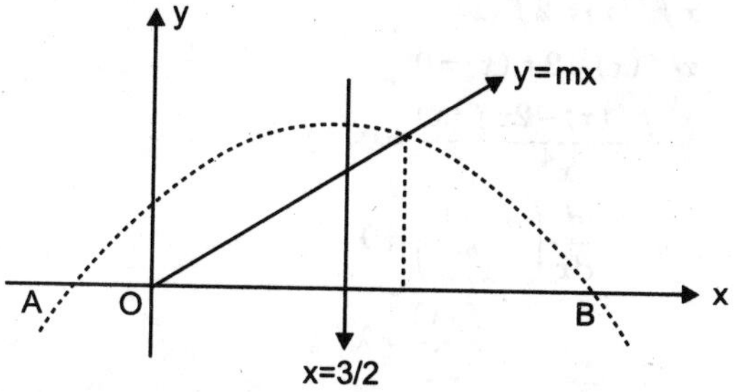

It is a parabola whose vertex is (2, 5), axis is $x - 2 = 0$ and $y \le 5$ at every point of the curve.

Also, when $x = 0$, $y = 1 + 4x - x^2$ gives $y = 1$

So, the parabola cuts the y-axis at (0, 1) when $y = 0$, $y = 1 + 4x - x^2$ gives;

$$1 + 4x - x^2 = 0 \quad ie, \quad x^2 - 4x - 1 = 0$$

$$\therefore \quad x = \frac{4 \pm \sqrt{20}}{2} = 2 \pm \sqrt{5}$$

$\therefore$ The parabola cuts the x-axis at $A(2-\sqrt{5}, 0)$ and $B(2+\sqrt{5}, 0)$.

The area enclosed by the lines $x = 0$, $y = 0$, $x = 3/2$ and the curve

$$= \int_0^{3/2} y\,dx = \int_0^{3/2} (1 + 4x - x^2)\,dx$$

$$= \left[x + 2x^2 - \frac{x^3}{3} \right]_0^{3/2}$$

$$= \left(\frac{3}{2} + 2 \cdot \frac{9}{4} - \frac{1}{3} \cdot \frac{27}{8} \right) = \frac{39}{8}$$

The area bounded by the lines ($y = mx$, $y = 0$ and $x = 3/2$)

$$= \int_0^{3/2} y\,dx = \int_0^{3/2} m\,x\,dx = \frac{9m}{8}$$

According to questions, $\frac{9}{8} m = \frac{1}{2} \cdot \frac{39}{8}$

$\Rightarrow$ $m = 13/6$

4. We have, $f\left(\frac{x}{y}\right) = \frac{f(x)}{f(y)}$. Putting $\frac{x}{y} = 1$, we get $f(1) = 1$.

Now, $$f'(x) = \lim_{h \to 0} \frac{f(x+h) - f(x)}{h}$$

$$= f(x) \cdot \lim_{h \to 0} \frac{\frac{f(x+h)}{f(x)} - 1}{h}$$

$$= f(x) \cdot \lim_{h \to 0} \frac{f\left(1 + \frac{h}{x}\right) - f(1)}{\frac{h}{x} \cdot x} \qquad \text{[as } f(1) = 1\text{]}$$

$$f'(x) = \frac{f(x)}{x} \cdot f'(1) = \frac{2f(x)}{x}$$

$\Rightarrow$ $x f'(x) = 2f(x)$

$\Rightarrow$ $xf'(x) - 2f(x) = 0$

$\Rightarrow$ $\frac{x^2 f'(x) - 2x f(x)}{x^4} = 0$

$\Rightarrow$ $\frac{d}{dx}\left(\frac{f(x)}{x^2}\right) = 0$

$\Rightarrow$ $\frac{f(x)}{x^2} = \lambda$ (constant

Now, $f(1) = 1 \Rightarrow \lambda = 1 \Rightarrow f(x) = x^2$

Now, required area $= \frac{\pi}{2} \cdot 2 - \int_0^1 (\sqrt{2 - x^2} - x^2)\,dx$

$$= \frac{\pi}{2} - \left(\frac{x}{2}\sqrt{2 - x^2} - \frac{2}{2}\sin^{-1}\left(\frac{x}{\sqrt{2}}\right) - \frac{x^3}{3} \right)_0^1$$

$$= \left(\frac{3\pi}{4} - \frac{1}{6} \right) \text{ sq units}$$

5. Given $y = x^2 + x + 1 = \left(x + \frac{1}{2}\right)^2 + \frac{3}{4}$

$\Rightarrow \quad y - 3/4 = \left(x + \frac{1}{2}\right)^2$

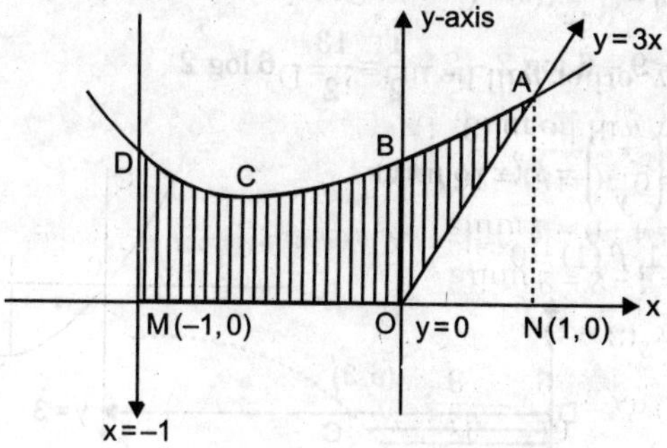

This is a parabola with vertex at $(-1/2, 3/4)$ and the curve is concave upwards

$$y = x^2 + x + 1 \quad \Rightarrow \quad \frac{dy}{dx} = 2x + 1$$

$$\Rightarrow \quad \left(\frac{dy}{dx}\right)_{(1,3)} = 3$$

Equation of tangent at $A = (1, 3)$ is $y = 3x$

Required (shaded) area = area $(ABCDMN)$ – area (ONA)

Now, area $(ABCDMN) = \int_{-1}^{1} (x^2 + x + 1)\, dx = 2\int_0^1 (x^2 + 1)\, dx = \frac{8}{3}$

and $\qquad$ area $(ONA) = \frac{1}{2} \times 1 \times 3 = \frac{3}{2}$

$\therefore \qquad$ Rquired area $= \frac{8}{3} - \frac{3}{2} = \frac{16 - 9}{6} = \frac{7}{6}$ sq units

6. First consider $y = 3 - |3 - x|$ for $x < 3$; $y = 3 - (3 - x) = x$

For $x > 3$; $y = 3 + (3 - x) = 6 - x$. Next, consider $y = \frac{6}{|x + 1|}$

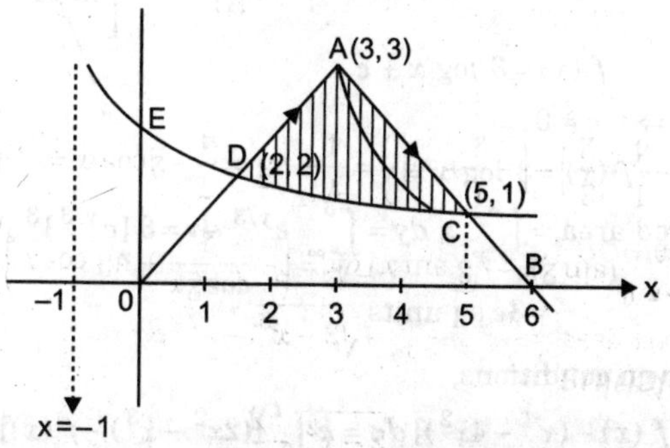

For $x < -1$; $\qquad y = \frac{6}{-(x + 1)}$

$\Rightarrow \qquad (x + 1)\, y = -6$

For $x > -1;\ y = \dfrac{6}{x+1} \Rightarrow (x+1)\,y = 6$

Required area = area of shaded portion

$$= \frac{1}{2} \times 6 \times 3 = \left(\int_2^5 \frac{6}{x+1}\,dx + \frac{1}{2} \times 2 \times 2 + \frac{1}{2} \times 1 \times 1 \right)$$

$$= 9 - 6 \log 2 - 2 - \frac{1}{2} = \frac{13}{2} - 6 \log 2$$

7. Given, $f\left(\dfrac{x}{y}\right) = f(x) - f(y)$ ...(i)

Putting, $x = y = 1$, $f(1) = 0$

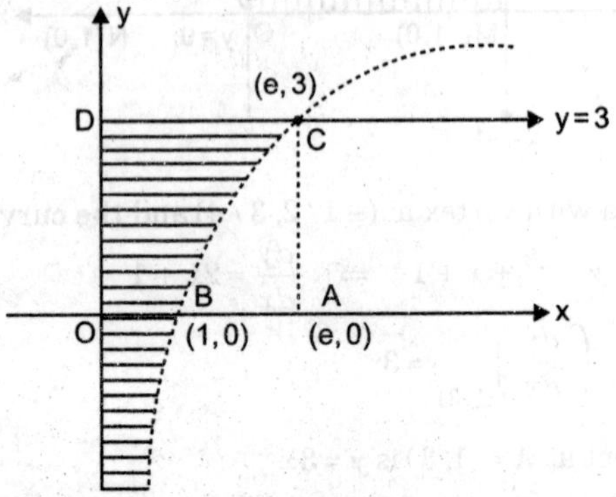

Now, $f'(x) = \lim_{h \to 0} \dfrac{f(x+h) - f(x)}{h}$

$$= \lim_{h \to 0} \frac{f\left(1 + \dfrac{h}{x}\right)}{h} \qquad \text{[from Eq. (i)]}$$

$$= \lim_{h \to 0} \frac{f\left(1 + \dfrac{h}{x}\right)}{\dfrac{h}{x} \cdot x} \Rightarrow f'(x) = \frac{3}{x}$$

$$\left[\text{since } \lim_{x \to 0} \frac{f(1+x)}{x} = 3\right]$$

$\Rightarrow \quad f(x) = 3 \log x + c$

Putting $x = 1 \Rightarrow c = 0$

$\Rightarrow \quad f(x) = 3 \log x = y$ (say)

$\therefore$ Required area $= \int_{-\infty}^{3} x\,dy = \int_{-\infty}^{3} e^{y/3}\,dy = 3\,[e^{y/3}]_{-\infty}^{3} = 3\,(e - 0)$

$= 3e$ sq units

8. According to given conditions,

$$\int_0^t [f(x) - (x^4 - 4x^2)]\,dx = k \int_0^t [(2x^2 - x^3) - f(x)]\,dx$$

Differentiable both the sides w.r.t. t, we get

$$f(t) - (t^4 - 4t^2) = k\,((2t^2 - t^3) - f(t))$$

or $\quad (1+k)f(t) = 2kt^2 - kt^3 + t^4 - 4t^2.$

$\Rightarrow \quad f(t) = \frac{1}{k+1}\{t^4 - kt^3 + (2k-4)t^2\}$

Hence, required f is given by; $f(x) = \frac{1}{k+1}[x^4 - kx^3 + (2k-4)x^2]$

10. $f(t) = |t-1| = |t| + |t+1|$

$$\Rightarrow \quad f(t) = \begin{cases} -t, & 1 \le -1 \\ 2+t, & -1 < t \le 0 \\ 2-t, & 0 < t \le 1 \\ t, & t > 1 \end{cases}$$

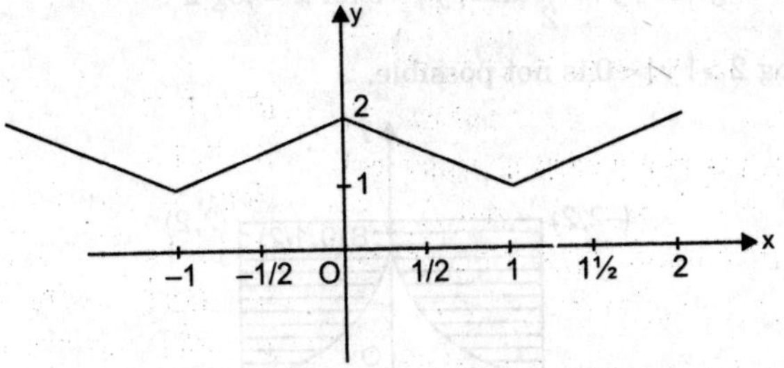

Case I $\quad x + 2 \le -\frac{1}{2}$

$\Rightarrow \quad x \le -\frac{5}{2}$

$$g(x) = \max\{f(t) : x+1 \le t \le x+2\}$$
$$= f(x+1) = -x-1,\ x \le -5/2$$

Case II $-\frac{1}{2} < x+2 \le 0 \Rightarrow -\frac{5}{2} < x \le -2$

$$g(x) = f(x+2) = 4+x,\ -\frac{5}{2} < x \le -2$$

Case III $0 < x+2 \le 1 \Rightarrow -2 < x \le -1$

$$g(x) = 2$$

Case IV $1 < x+2 \le 3/2 \Rightarrow -1 < x \le -1/2$

$$g(x) = f(x+1) = 1-x$$

Case V $x+2 > 3/2 \Rightarrow x > -1/2$

$$g(x) = f(x+2) = 2+x$$

Hence, $\quad g(x) = \begin{cases} -x-1, & x \le -5/2 \\ 4+x, & -5/2 < x \le -2 \\ 2, & -2 < x \le -1 \\ 1-x, & -1 < x \le -1/2 \\ 2+x, & x > -1/2 \end{cases}$

Now, required area $= \int_{-3/2}^{5} g(x)\,dx$

$$=\int_{-3/2}^{-1}(2)\,dx+\int_{-1}^{-1/2}(1-x)\,dx+\int_{-1/2}^{5}(2+x)\,dx$$

$$=2\left(-1+\frac{3}{2}\right)+\left(-\frac{1}{2}-\frac{1}{8}\right)-\left(-1-\frac{1}{2}\right)+2\left(5+\frac{1}{2}\right)+\frac{1}{2}\left(25-\frac{1}{4}\right)$$

$$=\frac{101}{4}\text{ sq units}$$

11. Consider the curve $|y|+\frac{1}{2}=e^{-|x|}\Rightarrow|y|=e^{-|x|}-\frac{1}{2}$. This curve is symmetrical about both the x and y-axes.

Also, for $x=0$ gives $|y|=\frac{1}{2}$ and $|y|=0$ for $x=\log 2$.

Also, $x>\log 2>|y|<0$ is not possible.

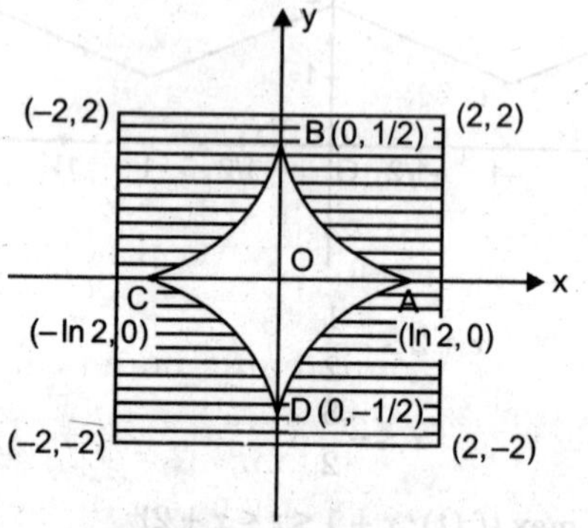

Also max $\{|x|,|y|\}\le 2$ is the interior of the square with vertices $(2, 2)$, $(2, -2)$, $(-2, 2)$ and $(-2, -2)$.

The region $|y|+\frac{1}{2}\ge e^{-|x|}$ and max $\{|x|,|y|\}\le 2$ is given above (as shaded portion).

$\therefore$ The required area $=4\,(4-\text{area } OABO)$

$$\text{Area } OABO=\int_0^{\log 2}\left(e^{-x}-\frac{1}{2}\right)dx=\frac{1}{2}(1-\log 2)$$

$$\therefore\quad \text{Required area}=4\left(\frac{7}{2}+\frac{\log 2}{2}\right)=(14+2\log 2)\text{ sq units}$$

12. It is easy to see that,

$$f(x)=\begin{cases} e^x, & 0\le x<\log(3/2)\\ 3/2, & \log(3/2)\le x<\log(2)\\ 1+e^x, & \log(2)\le x\le 1\end{cases}$$

Let A be the required area. Then,

$$A=\int_0^{(\log 3/2)}e^x\,dx+\int_{\log 3/2}^{\log 2}\frac{3}{2}\,dx+\int_{\log 2}^{1}(1+e^{-x})\,dx$$

$$=(e^x)_0^{\log 3/2}+\frac{3}{2}(x)_{\log 3/2}^{\log 2}+(x-e^{-x})_{\log 2}^{1}$$

$$= \left(\frac{3}{2} - 1\right) + \frac{3}{2}\left(\log 2 - \log\frac{3}{2}\right) + \left(1 - \frac{1}{e} - \log 2 + \frac{1}{2}\right)$$

$$= 2 + \log\left(\frac{4}{3\sqrt{3}}\right) - \frac{1}{e}$$

13. $$f'(x) = \lim_{h \to 0} \frac{f(x+h) - f(x)}{h}$$

$$= \lim_{h \to 0} \frac{f(x(1 + h/x)) - f(x)}{h}$$

$$= \frac{f(x)}{x} \lim_{h \to 0} \frac{f(1 + h/x) - f(1)}{h/x} = \frac{f(x)}{x} \cdot f'(1)$$

$\therefore$ $$f'(x) = \frac{2f(x)}{x} \text{ or } \frac{f'(x)}{f(x)} = \frac{2}{x}$$

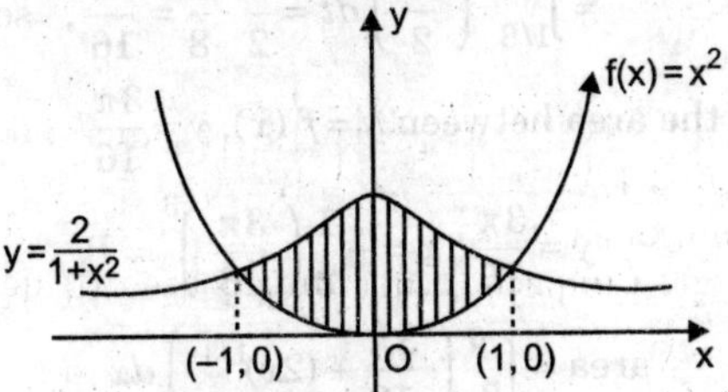

Integrating both the sides, we get $f(x) = Cx^2$, since $f(1) = 1 \Rightarrow C = 1$.

So, $$f(x) = x^2$$

Now, $$\frac{2}{1+x^2} = x^2 \quad \Rightarrow x^4 + x^2 - 2 = 0$$

$\Rightarrow$ $$x^2 = 1 \quad \Rightarrow \quad x = \pm 1$$

$$\text{Required area} = 2\left[\int_0^1 \left(\frac{2}{1+x^2} - x^2\right) dx\right]$$

$$= 2\left[2\tan^{-1} x - \frac{x^3}{3}\right]_0^1 = 2\left[\frac{\pi}{2} - \frac{1}{3}\right] = \left(\pi - \frac{2}{3}\right) \text{ sq units}$$

14. $y(x + y^3)\,dx = x(y^3 - x)\,dy$

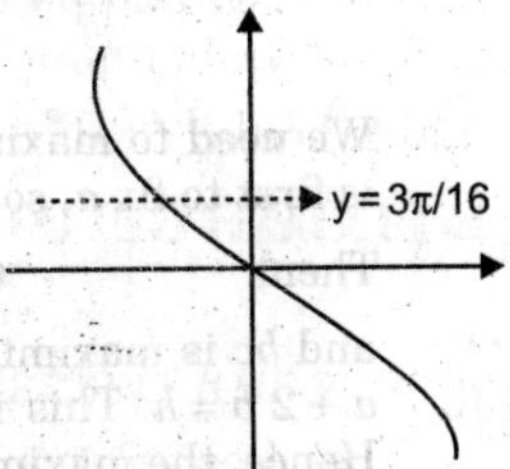

$\Rightarrow$ $$xy\,dx + y^2\,dx = xy^3\,dy - x^2\,dy$$

$\Rightarrow$ $$xd(xy) = x^2y^3\left(\frac{1}{x}\,dy - \frac{y}{x^2}\,dx\right)$$

$\Rightarrow$ $$\frac{d(xy)}{(xy)^2} = \frac{y}{x} \cdot d\left(\frac{y}{x}\right)$$

$\Rightarrow$ $$-\frac{1}{xy} = \frac{1}{2}\left(\frac{y}{x}\right)^2 + C$$

At, $x = 4$, $y = -2$

Hence, $$\frac{1}{8} = \frac{1}{2}\left(-\frac{1}{2}\right)^2 + C$$

$\Rightarrow \qquad C = 0$

Hence, $\qquad y^3 + 2x = 0$

So, $\qquad f(x) = (-2x)^{1/3}$

The second equation given is

$$y = \int_{1/8}^{\sin^2 x} \sin^{-1}\sqrt{t}\,dt + \int_{1/8}^{\cos^2 x} \cos^{-1}\sqrt{t}\,dt$$

$$\Rightarrow \qquad y' = x \cdot 2\sin x\cos x + x \cdot 2\cos x(-\sin x) = 0$$

Hence, y is constant.

Hence, put $\sin x = \cos x = \dfrac{1}{\sqrt{2}}$

Hence, $$y = \int_{1/8}^{1/2} (\sin^{-1}\sqrt{t} + \cos^{-1}\sqrt{t})\,dt$$

$$= \int_{1/8}^{1/2} \left(\frac{\pi}{2}\right) dt = \frac{\pi}{2} \cdot \frac{3}{8} = \frac{3\pi}{16}, \quad \text{so } g(x) = \frac{3\pi}{16}$$

So, we must find the area between $y = f(x)$, $y = \dfrac{3\pi}{16}$

At $$y = \frac{3\pi}{16};\ x = -\frac{1}{2}\left(\frac{3\pi}{16}\right)^3 = P \qquad \text{(say)}$$

Hence, $$\text{area} = \int_P^0 \left(\frac{3\pi}{16} + (2x)^{1/3}\right) dx$$

$$= \left(\frac{3\pi}{16}x + 2^{1/3} \cdot \frac{x^{4/3}}{4/3}\right)_P^0 = \frac{1}{8}\left(\frac{3\pi}{16}\right)^4$$

15. As in the figure

$$\frac{A(R) + A(S)}{A(T)} = \frac{ay + bz}{hx/2}$$

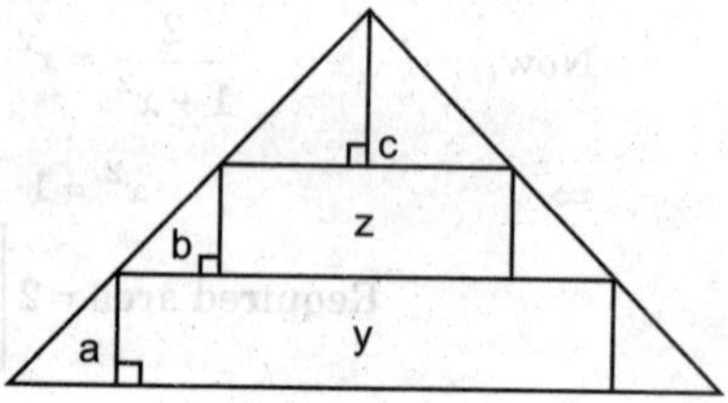

Where $h = a + b + c$, the altitude of T.

By similar triangles $\dfrac{x}{h} = \dfrac{y}{b+c} = \dfrac{z}{c}$,

So, $$\frac{A(R) + A(S)}{A(T)} = \frac{\dfrac{a(b+c)x}{h} + b \cdot \dfrac{cx}{h}}{hx/2}$$

$$= \frac{2}{h^2}(ab + ac + bc)$$

We need to maximise $(ab + bc + ca)$ subject to $a + b + c = h$. One way to do this is first to fix a, so $b + c = h - a$.

Then, $\qquad (ab + bc + ac) = a(h - a) + bc$

and bc is maximized when $b = c$. We now wish to maximize $2ab + b^2$ subject to $a + 2b = h$. This is a straight forward calculus problem giving $a = b = c = 1/3$. Hence, the maximum ratio is 2/3 (independent of T).

16. Consider a coordinate system with vertex P of the isosceles ΔPQR at $(a, 0)$ and Q and R at $(0, b)$ and $(0, -b)$ respectively.

$$A = \frac{1}{2} a \cdot 2b = ab \qquad \ldots\text{(i)}$$

Let the centre of ellipse be $(\alpha, 0)$ and the axes be of lengths, 2α and 2β.

So, the equation of ellipse is : $\dfrac{(x-\alpha)^2}{\alpha^2}+\dfrac{(y)^2}{\beta^2}=1$

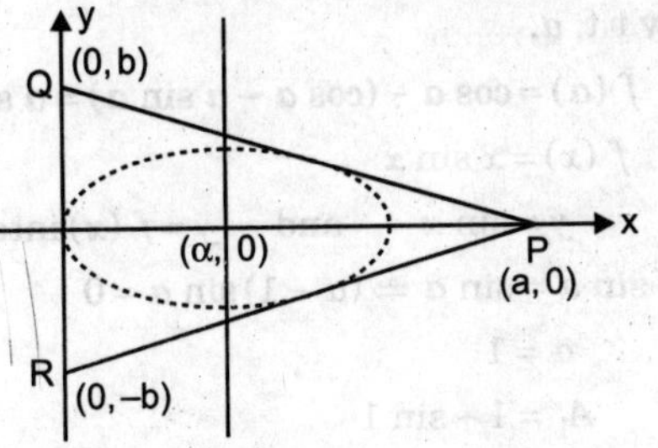

Now, the line PQ is tangent to the ellipse. To apply condition of tangency, let us take a new system $x'\,y'$ whose origin is at $(\alpha, 0)$.

Then, $x = x' + \alpha$ and $y = y'$.

So, the ellipse becomes $\dfrac{x'^2}{\alpha^2}+\dfrac{y'^2}{\beta^2}=1$

and the line PQ becomes $\dfrac{x'+\alpha}{a}+\dfrac{y'}{b}=1$

Which can be written as $y' = -\dfrac{b}{a}x' + b\left(1-\dfrac{\alpha}{a}\right)$

So, $$b^2\left(1-\frac{\alpha}{a}\right)^2=\alpha^2\left(-\frac{b}{a}\right)^2+\beta^2$$

$$\Rightarrow \quad \beta^2=b^2\left(1-\frac{2\alpha}{a}\right) \quad \text{...(ii)}$$

Now, area of ellipse $= \pi\alpha\beta \Rightarrow A^2 = \pi^2\alpha^2\beta^2$

Using Eq. (ii), $$A^2=\pi^2\alpha^2b^2\left(1-\frac{2\alpha}{a}\right)=f(x) \quad \text{(say)}$$

$$f'(x)=\pi^2b^2\left(2\alpha-\frac{6\alpha^2}{a}\right)$$

$$f'(\alpha)=0 \Rightarrow 2\alpha=\frac{6\alpha^2}{a} \Rightarrow \alpha=\frac{a}{3}\,(\alpha\neq 0)$$

Since, $$f''(\alpha)=\pi^2b^2\left(2-\frac{12\alpha}{a}\right)=-2\pi^2b^2=(-\text{ve}) \quad (\text{at } \alpha=a/3)$$

So, $$f\left(\frac{a}{3}\right)=\pi^2\cdot\frac{a^2}{9}\cdot b^2\left(1-\frac{2}{3}\right)=\frac{\pi^2A^2}{27} \quad \text{[using Eq. (i)]}$$

Hence, $$(A)_{\max}=\frac{\pi A}{3\sqrt{3}}=\frac{\sqrt{3}\pi A}{9}$$

17. $$A_1=\int_0^a\{\sin x-f(x)\}\,dx=-[\cos x]_0^a-\int_0^a f(x)\,dx$$

$$=-(\cos a-1)-\int_0^a f(x)\,dx=1-\sin a+(a-1)\cos a \quad \text{(given)}$$

$$\Rightarrow \quad -\cos a-\int_0^a f(x)\,dx=-\sin a+(a-1)\cos a$$

$$\Rightarrow -\int_0^a f(x)\,dx = -\sin a + a\cos a$$

$$\Rightarrow \int_0^a f(x)\,dx = \sin a - a\cos a$$

Differentiating w.r.t. a.

$$f(a) = \cos a - (\cos a - a\sin a) = a\sin a$$

$$\therefore \quad f(x) = x\sin x$$

Now, $y = \sin x$ and $y = f(x)$ intersects at

$$\Rightarrow a\sin a = \sin a \Rightarrow (a-1)\sin a = 0$$

$$\Rightarrow a = 1 \qquad \text{(as } \sin a = 0\text{)}$$

Hence,

$$A_1 = 1 - \sin 1$$

$$A_2 = \int_1^{\pi} (x\sin x - \sin x)\,dx = \pi - 1 - \sin 1$$

$$A_3 = \int_{\pi}^{2\pi} |x\sin x - \sin x|\,dx = 3\pi - 2$$

18. For $x = 1$, $y = b\cdot 5^x + 4 = 5b + 4$

and $\dfrac{dy}{dx} = b\cdot 5^x \log 5 \Rightarrow 5b\log 5 = 40\log 5$

$$\Rightarrow b = 8$$

The two curves intersects at points where

$$8\cdot 5^x + 4 = 25^x + 16$$

$$\Rightarrow 5^{2x} - 8\cdot 5 + 12 = 0$$

$$\Rightarrow x = \log_5 2,\ x = \log_5 6$$

Hence, the area of the given region;

$$= \int_{\log_5 2}^{\log_5 6} \{8\cdot 5^x + 4 - (25^x + 16)\}\,dx$$

$$= \int_{\log_5 2}^{\log_5 6} (8\cdot 5^x - 25^x - 12)\,dx$$

$$= \left[\frac{8\cdot 5^x}{\log_e 5} - 12x - \frac{25^x}{\log_e 25}\right]_{\log_5 2}^{\log_5 6}$$

$$= \frac{-5^{\log_5 36}}{\log_e 25} + \frac{8\cdot 5^{\log_5 6}}{\log_e 5} - 12(\log_5 6 - \log_5 2) + \frac{5^{\log_5 4}}{\log_e 25} - \frac{8\cdot 5^{\log_5 2}}{\log_e 5}$$

$$= \frac{-36}{2\log_e 5} + \frac{48}{\log_e 5} - 12[\log_5 3] + \frac{4}{2(\log_e 5)} - \frac{16}{(\log_e 5)}$$

$$= \frac{16}{\log_e 5} - 12\log_5 3 = 4\log_5 e^4 - 4\log_5 25$$

$$= 4\log_5\left[\frac{e^4}{27}\right]$$

19. By the symmetry of the figure circle(s) of maximum area will have the end point of diameter at the vertex of the two parabola.

$$\Rightarrow \quad \text{Radius of circle} = \frac{1}{2}\times AB = \frac{1}{2}\times 8 = 4 \text{ units}$$

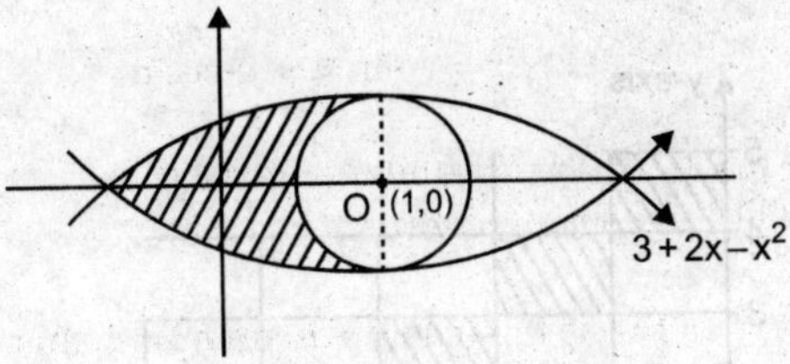

So, the area of shaded region $= 4 \times \left[\int_1^3 (3 + 2x - x^2)\, dx - \frac{1}{4} \text{ area of circle} \right.$

$$= 4\left(\frac{16}{3} - 4\pi\right)$$

20. Area of $\Delta AOB = 2 \times \left(\frac{1}{2} \times \frac{a}{2} \times a^2\right) = \frac{a^3}{2}$

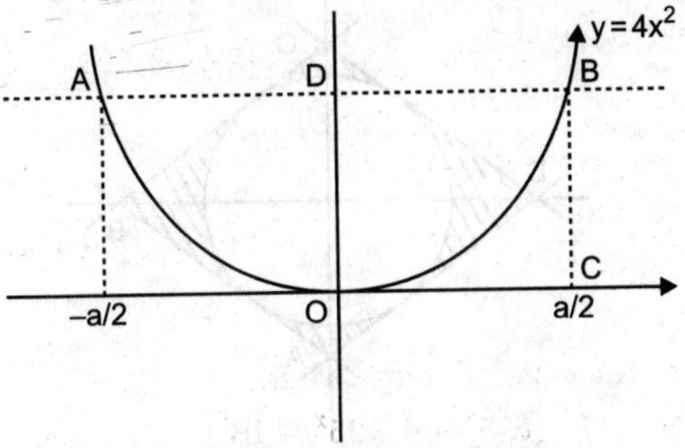

Area $=$ area $OCBD - \int_0^{a/2} 4x^2\, dx = \frac{a^3}{2} - \frac{a^3}{6} = \frac{a^3}{3}$

$$\lim_{x \to 0} \frac{\text{Area of triangle}}{\text{Area between the line and parabola}} = \lim_{x \to 0} \frac{a^3/2}{a^3/3} = \frac{3}{2}$$

21. Required area $= \frac{1}{2} \times 2 \times 2 = 2$ sq units

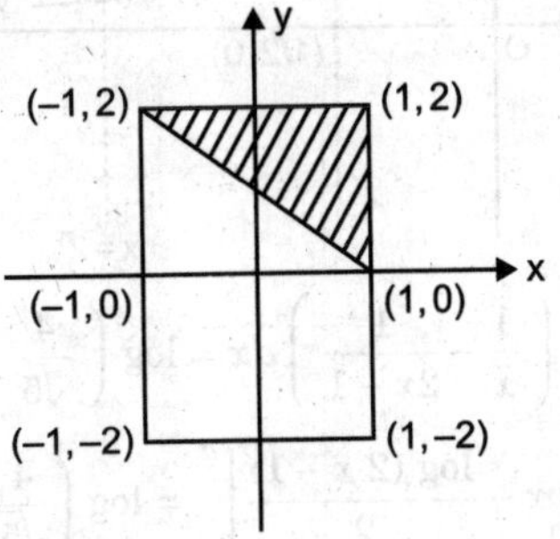

22. Required area $= 4 \times \left[\frac{\pi(\sqrt{2})^2}{2} - (\pi - 2)\right]$

$= 8$ sq units

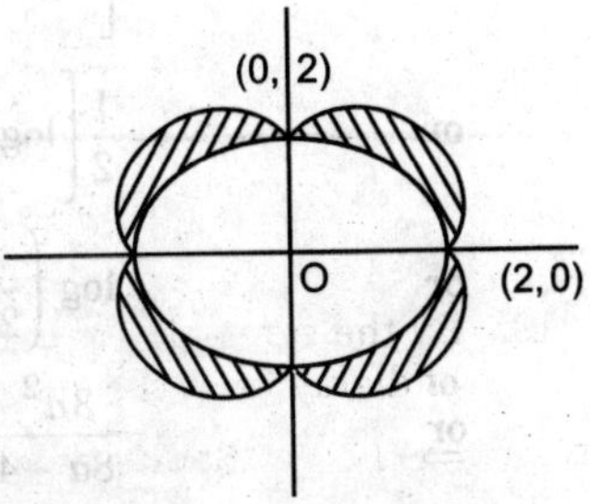

23. 5 sq units

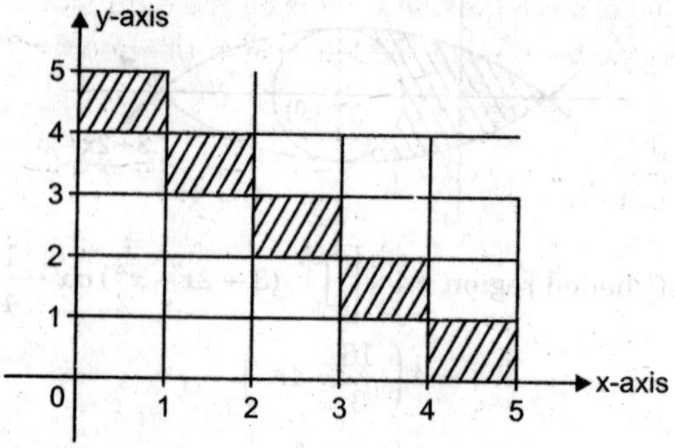

24. Area of the square $ABCD = 2$ sq units

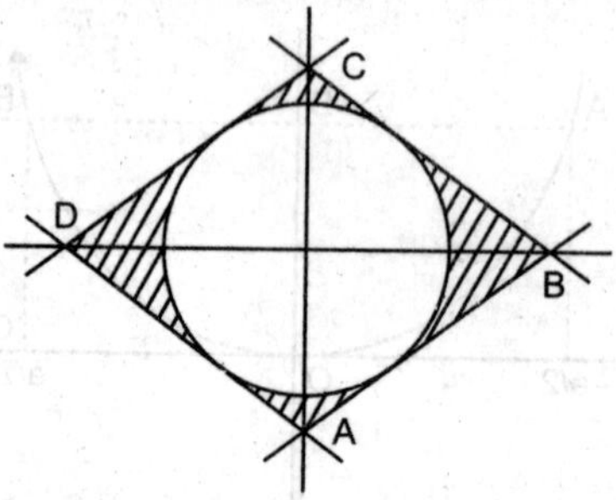

Area of the circle $= \pi \times \frac{1}{2} = \frac{\pi}{2}$ sq units. Required area $= \left(2 - \frac{\pi}{2}\right)$ sq units

25. ***Case I*** When the line $x = a$ is on the right side of the line $x = 2$.

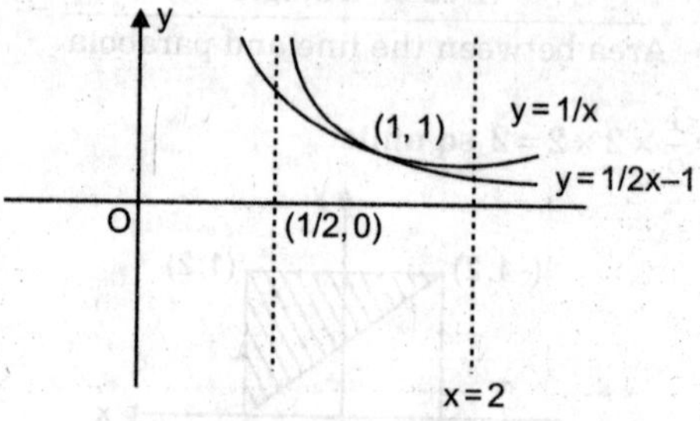

Then, $$\int_2^a \left(\frac{1}{x} - \frac{1}{2x-1}\right) dx = \log\left(\frac{4}{\sqrt{5}}\right)$$

or $$\left[\log x - \frac{\log(2x-1)}{2}\right]_2^a = \log\left(\frac{4}{\sqrt{5}}\right)$$

or $$\frac{1}{2}\left[\log \frac{x^2}{2x-1}\right]_2^a = \log\left(\frac{4}{\sqrt{5}}\right)$$

or $$\log\left(\frac{a^2}{2a-1}\right) - \log\left(\frac{4}{3}\right) = \log\left(\frac{16}{5}\right)$$

or $$\frac{3a^2}{8a-4} = \frac{16}{5} \Rightarrow a = 8, \frac{16}{30}$$

Acceptable value of $a = 8$ [as line $x = a$ is on the right side of $x = 2$]

Case II When line $x = a$ is on the left side of the line $x = 2$ and also left side of the point (1, 1).

Then, $$\int_a^1 \left(\frac{1}{2x-1} - \frac{1}{x}\right) dx + \int_1^2 \left(\frac{1}{x} - \frac{1}{2x-1}\right) dx = \left(\log \frac{4}{\sqrt{5}}\right)$$

$$\Rightarrow \quad \left[\frac{\log(2x-1)}{2} - \log(x)\right]_a^1 + \left(\log x - \frac{\log(2x+1)}{2}\right)_1^2 = \log \frac{4}{\sqrt{5}}$$

$$\Rightarrow \quad \frac{1}{2}\left(\frac{\log 2x-1}{x^2}\right)_a^1 + \frac{1}{2}\left[\frac{\log x^2}{2x-1}\right]_1^2 = \log\left(\frac{4}{\sqrt{5}}\right)$$

$$\Rightarrow \quad 0 - \frac{1}{2}\log\left(\frac{2a-1}{a^2}\right) + \frac{1}{2}\log\left(\frac{4}{3}\right) = \log\left(\frac{4}{\sqrt{5}}\right)$$

$$\Rightarrow \quad \frac{2a-1}{a^2} = \frac{5}{12}$$

$$\therefore \quad a = \frac{24 \pm 4\sqrt{21}}{10} = \frac{12 \pm 2\sqrt{21}}{5}$$

Acceptable value of $a = \dfrac{12 - 2\sqrt{21}}{5}$

From case I and case II $a = 8, \dfrac{12 - 2\sqrt{21}}{5}$

26. $$2^{|x|} \cdot |y| + 2^{|x|-1} \le 1 \qquad \ldots(i)$$

Clearly, this region is symmetrical about x and y-axes.

Let, $x < 0$, Eq. (i) gives,

$$2^x - y + 2^{x-1} \le 1 \quad \Rightarrow \quad y \le \frac{1 - 2^{x-1}}{2^x} = 2^{-x} - \frac{1}{2}$$

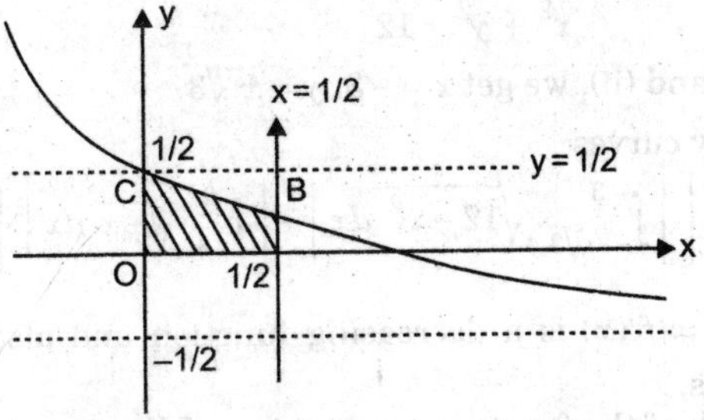

Clearly, bounded region in the first quadrant is $OABC$. The required area is 4 times the area of the region $OABC$.

Required area $= 4\int_0^{1/2} \left(2^{-x} - \frac{1}{2}\right) dx = 4\left(-\frac{2^{-x}}{\ln 2} - \frac{x}{2}\right)_0^{1/2} = \frac{4}{\ln 2}(1 - 2^{-1/2}) - 1$

27. Pair of lines $y^2 - 3y + 2 = 0$

$\Rightarrow$ Lines are $y = 2$, $y = 1$.

Let $[a] = b$

Now, curves are $y = bx^2$ and $y = \dfrac{b}{2}x^2$

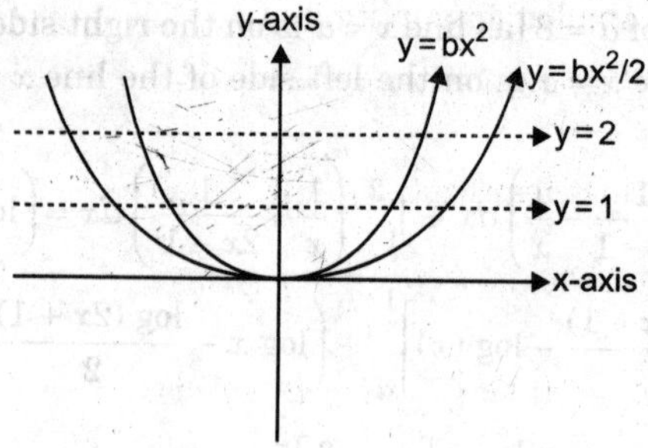

Bounded area $= 2\left[\int_1^2 (x_2 - x_1)\, dy\right] = 2\left[\int_1^2 \left(\sqrt{\frac{2y}{b}} - \sqrt{\frac{y}{b}}\right) dy\right]$

$$= 2\left[\sqrt{\frac{2}{b}} \cdot \frac{y^{3/2}}{3/2} - \frac{1}{\sqrt{b}} \cdot \frac{y^{3/2}}{3/2}\right]_1^2$$

$$= \frac{4}{3\sqrt{b}}((\sqrt{2}-1)\, y^{3/2})_1^2 = \frac{4}{3\sqrt{b}}(\sqrt{2}-1)(2\sqrt{2}-1)$$

Area will be maximum when $b = [a]$ is least.

As $a \geq 2 \Rightarrow [a]_{\text{least}} = 1 \Rightarrow 1 \leq a < 2$

28. $\frac{11}{3}$

29. Given, $(x-y)f(x+y) - (x+y)f(x-y) = 4xy(x^2 - y^2)$

$$= (x^2 - y^2)[(x+y)^2 - (x-y)^2] = (x-y)(x+y)^3 - (x+y)(x-y)^3$$

$\Rightarrow \quad f(x+y) = (x+y)^3 \to f(x) = x^3; f(y) = y^3.$

Now Eqs. of given curves;

$$y^2 + x = 0 \quad \ldots\text{(i)}$$

and

$$x^2 + y^2 = 12 \quad \ldots\text{(ii)}$$

Solving Eqs. (i) and (ii), we get $x = -3,\ y = \pm\sqrt{3}$

Area bounded by curves;

$$A = 2\left[\left|\int_{-2/\sqrt{3}}^{-3} \sqrt{12 - x^2}\, dx\right| + \left|\int_{-3}^{0} \sqrt{-x}\, dx\right|\right] = 2\pi - 3\sqrt{3}$$

30. Since, $f'(x) < 0 \Rightarrow f(x)$ is a decreasing function and also $f''(x) > 0 \Rightarrow f(x)$ is concave upwards.

Hence, the graph of the function $y = f(x)$ is as follows :

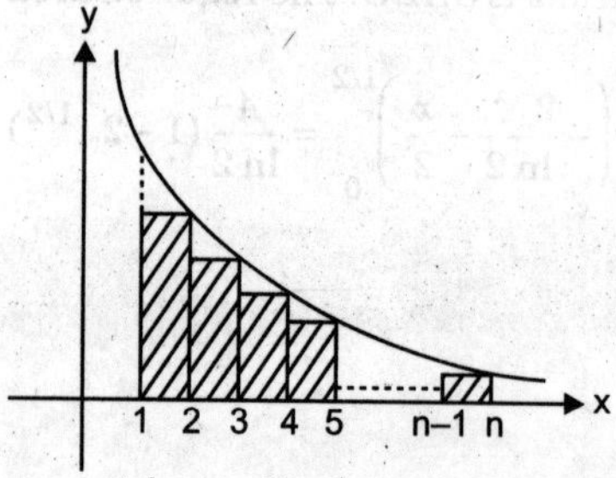

(i)

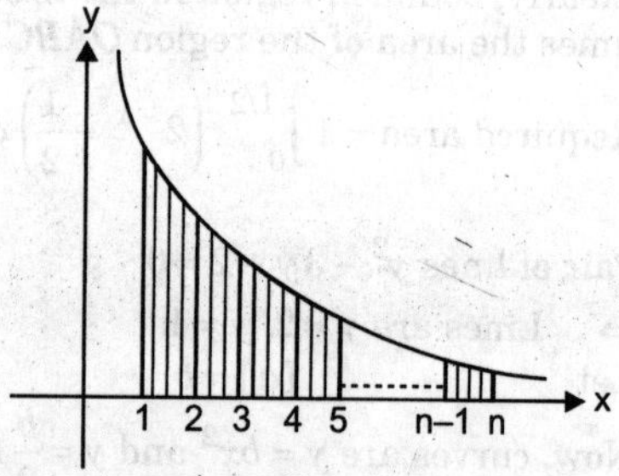

(ii)

Let S_L denotes the shaded area in figure (i).

$$\Rightarrow \qquad S_L = f(2) + f(3) + \ldots + f(n) = \sum_{r=1}^{n} f(r) - f(1)$$

From the figure (i) it is clear that, $S_L < \int_1^n f(x)\,dx$

$$\Rightarrow \qquad \sum_{r=1}^{n} f(r) - f(1) < \int_1^n f(x)\,dx$$

$$\Rightarrow \qquad \sum_{r=1}^{n} f(r) < \int_1^n f(x)\,dx + f(1) \qquad \ldots\text{(i)}$$

Let S_u denotes the area of the shaded region in figure (ii).

$$\Rightarrow \qquad S_u = \left(\frac{1}{2} f(1) + f(2)\right) + \frac{1}{2}(f(2) + f(3)) + \ldots + \frac{1}{2}(f(n-1) + f(n))$$

$$= [(f(1) + f(2) + f(3)) + \ldots + f(n-1) + f(n)] - \frac{1}{2}(f(1) + f(n))$$

$$= \sum_{r=1}^{n} f(r) - \frac{1}{2} f(1) - \frac{1}{2} f(n)$$

From figure (ii) it is clear that;

$$S_u > \int_1^n f(x)\,dx$$

$$\Rightarrow \qquad \sum_{r=1}^{n} f(r) - \frac{1}{2}(f(1) + f(n)) > \int_1^n f(x)\,dx$$

$$\Rightarrow \quad \sum_{r=1}^{n} f(r) > \int_1^n f(x)\,dx + \frac{1}{2}(f(1) + f(n)) > \int_1^n f(x)\,dx + 4 \cdot \frac{1}{2} f(1) \qquad \ldots\text{(ii)}$$

From Eqs. (i) and (ii), we get

$$\frac{1}{2} f(1) + \int_1^n f(x)\,dx < \sum_{r=1}^{n} f(r) < \int_1^n f(x)\,dx + f(1).$$

4

Differential Equations

Chapter in a Snapshot

- Introduction
- Solution of a Differential Equation
- Differential Equations of the First Order and First Degree
- Equations in Which the Variables are Separable
- Differential Equations Reducible to the Separable Variable Type
- Homogeneous Differential Equation
- Exact Differential Equations
- Equations Reducible to the Exact Form
- First Order Linear Differential Equations
- Bernoulli's Equation
- Orthogonal Trajectory
- Differential Equation of First Order and Higher Degrees
- Application of Differential Equations
- Application of First Order Differential Equations
- Falling Body Problems
- Electrical Circuits

Introduction

By a differential equation we mean an equation involving independent variable, dependent and the differential coefficients of the dependent variable, *ie*, it will be an equation in x, y and derivatives of y w.r.t. x.

eg,
$$\frac{dy}{dx} = \log x, \quad \frac{dy}{dx} + y = xe^x, \quad \frac{d^2y}{dx^2} + y = 0$$
$$\left(\frac{dy}{dx}\right)^2 = px + q, \quad \frac{d^2y}{dx^2} + 2\left(\frac{dy}{dx}\right) + 3y = 0, \ldots\ldots \text{etc.}$$

Here, y is some function of only one variable x.

(a) Order and Degree of a Differential Equation

The order of the highest differential coefficient appearing in the differential equation is called order of differential equation, while the exponent of the highest differential coefficient, when the differential equation is a polynomial in all the differential coefficients, is known as the degree of the differential equation.

Illustration 1 Find the order and degree (if defined) of the following differential equations :

(i) $y = 1 + \left(\frac{dy}{dx}\right) + \frac{1}{2!}\left(\frac{dy}{dx}\right)^2 + \frac{1}{3!}\left(\frac{dy}{dx}\right)^3 + \ldots$

(ii) $\left(\frac{d^3y}{dx^3}\right)^{2/3} = \frac{dy}{dx} + 2$

(iii) $\frac{d^2y}{dx^2} = x \ln\left(\frac{dy}{dx}\right)$

Solution. (i) The given differential equation can be rewritten as
$$y = e^{dy/dx} \quad \Rightarrow \quad \frac{dy}{dx} = \ln y.$$

Hence, its order is 1 and degree 1.

(ii) The given differential equation can be rewritten as
$$\left(\frac{d^3y}{dx^3}\right)^2 = \left(\frac{dy}{dx} + 2\right)^3.$$

Hence, its order is 3 and degree 2.

(iii) Its order is obviously 2.

Since, the given differential equation cannot be written as a polynomial in all the differential coefficients, the degree of the equation is not defined.

Illustration 2 Find the order and degree (if defined) of the following differential equations :

(i) $\sqrt{\frac{d^2y}{dx^2}} = \sqrt[3]{\frac{dy}{dx} + 3}$ (ii) $\frac{d^2y}{dx^2} = \sin\left(\frac{dy}{dx}\right)$ (iii) $\frac{dy}{dx} = \sqrt{3x + 5}$

Solution. (i) The given differential equation can be rewritten as

$$\left(\frac{d^2y}{dx^2}\right)^3 = \left(\frac{dy}{dx} + 3\right)^2$$

Hence, order is 2 and degree is 3.

(ii) The given differential equation has the order 2. Since, the given differential equation cannot be written as a polynomial in the differential coefficients, the degree of the equation is not defined.

(iii) Its order is obviously 1 and degree 1.

(b) Formation of Differential Equations

If an equation in independent and dependent variables involving some arbitrary constants is given, then a differential equation is obtained as follows :

(i) Differentiate the given equation w.r.t. the independent variable (say x) as many times as the number of arbitrary constants in it.

(ii) Eliminate the arbitrary constants.

(iii) The eliminant is the required differential equation. *ie,* If we have an equation $f(x, y, c_1, c_2, \ldots, c_n) = 0$

Containing n arbitrary constants $c_1, c_2, c_3, \ldots, c_n$, then by differentiating this n times, we shall get n-equations.

Now, among these n-equations and the given equation, in all $(n+1)$ equations, if the n arbitrary constants $c_1, c_2, c_3, \ldots, c_n$ are eliminated, we shall evidently get a differential equation of the nth order. For there being n differentiation, the resulting equation must contain a derivative of the nth order.

Illustration 3 Form the differential equations, if $y^2 = 4a(x + b)$, where a, b are arbitrary constants.

Solution. Differentiating $y^2 = 4a(x + b)$ w.r.t. x,

$$2y \cdot \frac{dy}{dx} = 4a$$

ie,
$$y\frac{dy}{dx} = 2a$$

Again, differentiating w.r.t. x, we get

$$y\frac{d^2y}{dx^2} + \left(\frac{dy}{dx}\right)^2 = 0$$

which is the required differential equation. Thus, the elimination of arbitrary leads to the formation of a differential equation.

Illustration 4 Find the differential equation whose solution represents the family $xy = ae^x + be^{-x}$.

Solution. $$xy = ae^x + be^{-x} \quad \ldots(i)$$

Differentiating Eq. (i) w.r.t. x, we get

$$x\frac{dy}{dx}+y=ae^{x}-be^{-x} \quad \text{...(ii)}$$

Differentiating Eq. (ii) w.r.t. x, we get

$$x\frac{d^2y}{dx^2}+\frac{dy}{dx}\cdot 1+\frac{dy}{dx}=ae^{x}+be^{-x} \quad \text{...(iii)}$$

Using Eqs. (i) and (iii), we get

$$x\frac{d^2y}{dx^2}+2\frac{dy}{dx}=xy$$

Which is the required differential equation.

Illustration 5 Find the differential equation whose solution represents the family $c\,(y+c)^2=x^3$.

Solution. Differentiating $c\,(y+c)^2=x^3$ ...(i)

We get $c\,[2\,(y+c)]\frac{dy}{dx}=3x^2$ but from Eq. (i), we have

$$\frac{2x^3}{(y+c)^2}(y+c)\frac{dy}{dx}=3x^2$$

$$\Rightarrow \quad \frac{2x^3}{y+c}\cdot\frac{dy}{dx}=3x^2$$

ie,

$$\frac{2x}{y+c}\cdot\frac{dy}{dx}=3 \Rightarrow \frac{2x}{3}\left[\frac{dy}{dx}\right]=y+c$$

$$\therefore \quad c=\frac{2x}{3}\left[\frac{dy}{dx}\right]-y$$

Substituting c in Eq. (i), we get

$$\left[\frac{2x}{3}\left(\frac{dy}{dx}\right)-y\right]\left[\frac{2x}{3}\frac{dy}{dx}\right]^2=x^3$$

Which is the required differential equation.

Illustration 6 Find the differential equation whose solution represents the family $y=ae^{3x}+be^{x}$.

Solution. $y=ae^{3x}+be^{x}$ ...(i)

Differentiating the given equation twice, we get

$$\frac{dy}{dx}=3ae^{3x}+be^{x} \quad \text{and} \quad \frac{d^2y}{dx^2}=9ae^{3x}+be^{x}$$

From the three equations by eliminating a and b, we obtain

$$\frac{d^2y}{dx^2}-\frac{4dy}{dx}+3y=0$$

Points to Consider

The order of the differential equation will be equal to number of independent parameters and is not equal to the number of all the parameters in the family of curves.

Illustration 7 Find the order of the family of curves

$$y = (c_1 + c_2)\, e^x + c_3\, e^{x + c_4}.$$

Solution. Here, the number of arbitrary parameters is 4 but the order of the corresponding differential equation will not be 4 as it can be rewritten as, $y = (c_1 + c_2 + c_3\, e^{c_4})\, e^x$, which is of the form $y = Ae^x$. Hence, the corresponding differential equation will be of order 1.

Illustration 8 The differential equation of all non-horizontal lines in a plane is given by

(a) $\dfrac{d^2y}{dx^2} = 0$ (b) $\dfrac{d^2x}{dy^2} = 0$

(c) $\dfrac{d^2y}{dx^2} = 0$ and $\dfrac{d^2x}{dy^2} = 0$ (d) All of these

Solution. The equation of the family of all non-horizontal lines in a plane is given by,

$$ax + by = 1 \qquad \text{(where } a \neq 0) \qquad \text{...(i)}$$

Differentiating w.r.t. y, we get

$$a\,\frac{dx}{dy} + b = 0 \qquad (\text{as } a \neq 0 \text{ and } b \in R)$$

Again, differentiating w.r.t. y, we get

$$a\,\frac{d^2x}{dy^2} = 0 \qquad (\text{as } a \neq 0 \text{ and } b \in R)$$

$$\Rightarrow \quad \frac{d^2x}{dy^2} = 0 \qquad (\text{as } a \neq 0)$$

$\therefore$ Differential equation of all non-horizontal lines in a plane is $\dfrac{d^2x}{dy^2} = 0.$

Hence, (b) is the correct answer.

Illustration 9 The differential equation of all non-vertical lines in a plane is given by

(a) $\dfrac{d^2y}{dx^2} = 0$ (b) $\dfrac{d^2x}{dy^2} = 0$

(c) $\dfrac{d^2x}{dy^2} = 0$ and $\dfrac{d^2y}{dx^2} = 0$ (d) All of these

Solution. The equation of the family of all non-vertical lines in a plane is given by $ax + by = 1$, where $b \neq 0$ and $a \in R$.
Differentiating both the sides w.r.t. x, we get

$$a + b\,\frac{dy}{dx} = 0 \qquad (\text{as } b \neq 0 \text{ and } a \in R)$$

Again, differentiating both the sides w.r.t. x, we get

$$b\,\frac{d^2y}{dx^2} = 0 \qquad (\text{as } b \neq 0 \text{ and } a \in R)$$

$$\Rightarrow \quad \frac{d^2y}{dx^2} = 0 \qquad (\text{as } b \neq 0)$$

∴ Differential equation of all non-vertical lines in a plane.

$$\Rightarrow \qquad \frac{d^2y}{dx^2} = 0$$

Hence, (a) is the correct answer.

Illustration 10 The differential equation of all straight lines which are at a constant distance p from the origin, is

(a) $(y + xy_1)^2 = p^2(1 + y_1^2)$ (b) $(y - xy_1^2) = p^2(1 + y_1)^2$

(c) $(y - xy_1)^2 = p^2(1 + y_1^2)$ (d) None of these

Solution. As, we know

$$x \cos\alpha + y \sin\alpha = p \qquad \text{...(i)}$$

Represents the family of straight lines which are at a constant distance p from origin.

Differentiating Eq. (i) w.r.t. x, we get

$$\cos\alpha + \sin\alpha \cdot \frac{dy}{dx} = 0$$

$$\Rightarrow \qquad \tan\alpha = -\frac{1}{y_1}$$

or

$$\sin\alpha = \frac{1}{\sqrt{1 + y_1^2}} \text{ and } \cos\alpha = -\frac{y_1}{\sqrt{1 + y_1^2}} \qquad \text{...(ii)}$$

From Eqs. (i) and (ii), we get

$$\frac{-xy_1}{\sqrt{1 + y_1^2}} + \frac{y}{\sqrt{1 + y_1^2}} = p$$

$\Rightarrow \quad (y - xy_1)^2 = p^2(1 + y_1^2)$, is required differential equations.

Hence, (c) is the correct answer.

Illustration 11 The differential equation of all circles of radius r, is given by

(a) $\{1 + (y_1)^2\}^2 = r^2 y_2^3$ (b) $\{1 + (y_1)^2\}^3 = r^2 y_2^3$

(c) $\{1 + (y_1)^2\}^3 = r^2 y_2^2$ (d) None of these

Solution. Equation of circle of radius r,

$$(x - a)^2 + (y - b)^2 = r^2 \qquad \text{...(i)}$$

(Here, a, b are two arbitrary constants)

Differentiating Eq. (i), we get

$$2(x - a) + 2(y - b)y_1 = 0 \qquad \text{...(ii)}$$

Again, differentiating Eq. (ii), we get

$$1 + (y - b)y_2 + y_1^2 = 0$$

$$\Rightarrow \qquad (y - b) = -\left(\frac{1 + y_1^2}{y_2}\right) \qquad \text{...(iii)}$$

Putting $(y - b)$ in Eq. (ii), we get

$$(x - a) = \frac{(1 + y_1^2)\,y_1}{y_2} \qquad \text{...(iv)}$$

From Eqs. (i), (iii) and (iv), we get

$$\frac{(1+y_1^2)^2 \cdot y_1^2}{y_2^2} + \frac{(1+y_1^2)^2}{y_2^2} = r^2$$

$$\Rightarrow \qquad (1+(y_1)^2)^3 = r^2 y_2^2$$

Hence, (c) is the correct answer.

Illustration 12 The differential equations of all circles touching the x-axis at origin is

(a) $(y^2 - x^2) = 2xy\left(\frac{dy}{dx}\right)$ (b) $(x^2 - y^2)\frac{dy}{dx} = 2xy$

(c) $(x^2 - y^2) = 2xy\left(\frac{dy}{dx}\right)$ (d) None of these

Solution. The equation of circle touches x-axis at origin.

$$\Rightarrow \qquad (x-0)^2 + (y-a)^2 = a^2$$

$$\text{or} \qquad x^2 + y^2 - 2ay = 0 \qquad \ldots\text{(i)}$$

Differentiating w.r.t. x, we get

$$2x + 2y\frac{dy}{dx} - 2a\frac{dy}{dx} = 0$$

$$\Rightarrow \qquad \frac{dy}{dx}(a-y) = x$$

$$\Rightarrow \qquad a = \frac{x + y\left(\frac{dy}{dx}\right)}{\left(\frac{dy}{dx}\right)} \qquad \ldots\text{(ii)}$$

From Eqs. (i) and (ii), we get

$$x^2 + y^2 - 2y\left[\frac{x + y\left(\frac{dy}{dx}\right)}{\left(\frac{dy}{dx}\right)}\right] = 0$$

$$\text{or} \qquad (x^2 - y^2)\frac{dy}{dx} = 2xy$$

Hence, (b) is the correct answer.

Illustration 13 The differential equation of all circles in the first quadrant which touch the coordinate axes is

(a) $(x-y)^2(1+(y')^2) = (x+yy')^2$ (b) $(x+y)^2(1+(y')^2) = (x+y')^2$

(c) $(x-y)^2(1+y') = (x+yy')^2$ (d) None of these

Solution. Equation of circles touching coordinate axes is

$$(x-a)^2 + (y-a)^2 = a^2 \qquad \ldots\text{(i)}$$

Differentiating, we get

$$2(x-a) + 2(y-a)y' = 0$$

$$\Rightarrow \qquad a = \frac{x + yy'}{1+y'}; \quad \text{where } y' = \frac{dy}{dx} \qquad \ldots\text{(ii)}$$

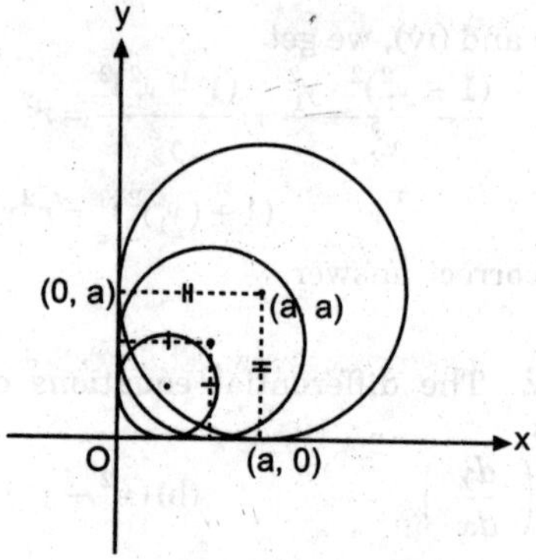

From Eqs. (i) and (ii),

$$\left(x - \frac{x + yy'}{1 + y'}\right)^2 + \left(y - \frac{x + yy'}{1 + y'}\right)^2 = \left(\frac{x + yy'}{1 + y'}\right)^2$$

$$\Rightarrow \quad \left(\frac{xy' - yy'}{1 + y'}\right)^2 + \left(\frac{y - x}{1 + y'}\right)^2 = \left(\frac{x + yy'}{1 + y'}\right)^2$$

$$\Rightarrow \quad (x - y)^2 (y')^2 + (y - x)^2 = (x + yy')^2$$

$$\Rightarrow \quad (x - y)^2 (1 + (y')^2) = (x + yy')^2$$

Hence, (a) is the correct answer.

Illustration 14 The differential equation satisfying the curve $\frac{x^2}{a^2 + \lambda} + \frac{y^2}{b^2 + \lambda} = 1$, where λ being arbitrary unknown, is

(a) $(x + yy_1)(xy_1 - y) = (a^2 - b^2) y_1$ (b) $(x + yy_1)(x - yy_1) = y_1$

(c) $(x - yy_1)(xy_1 + y) = (a^2 - b^2) y_1$ (d) None of these

Solution. Here, $\frac{x^2}{a^2 + \lambda} + \frac{y^2}{b^2 + \lambda} = 1$ ...(i)

Differentiating both the sides, we get

$$\frac{2x}{a^2 + \lambda} + \frac{2y}{b^2 + \lambda} \cdot \frac{dy}{dx} = 0$$

$$\Rightarrow \quad x(b^2 + \lambda) + y(a^2 + \lambda) y_1 = 0$$

$$\Rightarrow \quad \lambda = -\left(\frac{xb^2 + a^2 yy_1}{x + yy_1}\right)$$

$$\therefore \quad a^2 + \lambda = a^2 - \frac{xb^2 + a^2 yy_1}{x + yy_1}$$

$$\Rightarrow \quad a^2 + \lambda = \frac{(a^2 - b^2) x}{x + yy_1} \quad \text{...(ii)}$$

Also, $$b^2 + \lambda = \frac{(a^2 - b^2) yy_1}{x + yy_1} \quad \text{...(iii)}$$

From Eqs. (i), (ii) and (iii), we get

$$\frac{x^2 (x + yy_1)}{(a^2 - b^2) x} + \frac{y^2 (x + yy_1)}{(a^2 - b^2) yy_1} = 1$$

$$\Rightarrow \quad (x + yy_1)(xy_1 - y) = (a^2 - b^2) y_1$$

Hence, (a) is the correct answer.

Illustration 15 The differential equation of all conics whose centre lies at origin, is given by

(a) $(3xy_2 + x^2 y_3)(y - xy_1) = 3xy_2 (y - xy_1 - x^2 y_2)$
(b) $(3xy_1 + x^2 y_2)(y_1 - xy_3) = 3xy_1 (y - xy_2 - x^2 y_3)$
(c) $(3xy_2 + x^2 y_3)(y_1 - xy) = 3xy_1 (y - xy_1 - x^2 y_2)$
(d) None of the above

Solution. Equation of all conics whose centre lies at origin, is

$$ax^2 + 2hxy + by^2 = 1 \quad \text{...(i)}$$

Differentiating Eq. (i) w.r.t. x, we get

$$2ax + 2hxy_1 + 2hy + 2byy_1 = 0$$

$$\Rightarrow \quad ax + h (y + xy_1) + byy_1 = 0$$

Multiplying by x equation becomes,

$$ax^2 + h (xy + x^2 y_1) + bxyy_1 = 0 \quad \text{...(ii)}$$

Subtracting Eqs. (i) and (ii), we get

$$h (xy - x^2 y_1) + b (y^2 - xyy_1) = 1$$

$$\Rightarrow \quad (hx + by) y - xy_1 (hx + by) = 1$$

$$\Rightarrow \quad (hx + by)(y - xy_1) = 1$$

$$\Rightarrow \quad hx + by = \frac{1}{y - xy_1} \quad \text{...(iii)}$$

Again, differentiating w.r.t. x, we get

$$h + by_1 = -\frac{(y_1 - xy_2 - y_1)}{(y - xy_1)^2}$$

or

$$h + by_1 = \frac{xy_2}{(y - xy_1)^2} \quad \text{...(iv)}$$

From Eqs. (iii) and (iv), we get

$$b (y - xy_1) = \frac{y - xy_1 - x^2 y_2}{(y - xy_1)^2}$$

$$\Rightarrow \quad b = \frac{y - xy_1 - x^2 y_2}{(y - xy_1)^3} \quad \text{...(v)}$$

Again, differentiating both the sides w.r.t. x, we get

$$0 = \frac{y_1 - y_1 - 3xy_2 - x^2 y_3}{(y - xy_1)^3} + \frac{3 (y - xy_1 - x^2 y_2) xy_2}{(y - xy_1)^4}$$

$$\Rightarrow \quad (3xy_2 + x^2 y_3)(y - xy_1) = 3xy_2 (y - xy_1 - x^2 y_2)$$

Hence, (a) is the correct answer.

Target Exercise 4.1

1. The differential equation of all circles touching the y-axis at origin, is
   (a) $y^2 - x^2 = 2xy \frac{dy}{dx}$ (b) $y^2 - x^2 = 2xy \frac{dx}{dy}$
   (c) $x^2 - y^2 = 2xy \frac{dy}{dx}$ (d) $x^2 - y^2 = 2xy \frac{dx}{dy}$
2. The differential equation of all parabolas having their axes of symmetry coincident with the axes of x, is
   (a) $yy_2 + y_1^2 = y + y_1$ (b) $yy_2 + y_1^2 = 0$
   (c) $yy_2 + y_1^2 = y_1$ (d) None of these
3. The differential equation of all conics whose axes coincide with the coordinate axes, is
   (a) $xyy_2 + xy_1^2 - yy_1 = 0$ (b) $yy_2 + y_1^2 - yy_1 = 0$
   (c) $xyy_2 + (x - y) y_1 = 0$ (d) None of these
4. The differential equation having $y = (\sin^{-1} x)^2 + A (\cos^{-1} x) + B$, where A and B are arbitrary constant, is
   (a) $(1 - x^2) y_2 - xy_1 = 2$ (b) $(1 - x^2) y_2 + yy_1 = 0$
   (c) $(1 - x) y_2 + xy_1 = 0$ (d) None of these
5. The differential equation of circles passing through the points of intersection of unit circle with centre at the origin and the line bisecting the first quadrant, is
   (a) $y_1 (x^2 + y^2 - 1) + (x + yy_1) = 0$
   (b) $(y_1 - 1)(x^2 + y^2 - 1) + (x + yy_1) 2 (x - y) = 0$
   (c) $(x^2 + y^2 - 1) + yy_2 = 0$
   (d) None of the above

Solution of a Differential Equation

The solution of the differential equation is a relation between the variables of the equation not containing the derivatives, but satisfying the given differential equation (*ie*, from which the given differential equation can be derived).

Thus, the solution of $\frac{dy}{dx} = e^x$ could be obtained by simply integrating both the sides, *ie*, $y = e^x + C$ and that of, $\frac{dy}{dx} = px + q$ is $y = p\frac{x^2}{2} + qx + C$, where C is arbitrary constant.

(i) **A general solution or an integral** of a differential equation is a relation between the variables (not involving the derivatives) which contains the same number of the arbitrary constants as the order of the differential equation.

For example, a general solution of the differential equation $\frac{d^2x}{dt^2} = - 4x$ is $x = A \cos 2t + B \sin 2t$, where A and B are the arbitrary constants.

(ii) **Particular solution or particular integral** is that solution of the differential equation obtained from the general solution by assigning particular values to the arbitrary constant in the general solution.

For example, $x = 10\cot 2t + 5\sin 2t$ is a particular solution of differential equation $\dfrac{d^2x}{dt^2} = -4x$.

Differential Equations of the First Order and First Degree

In this section we shall discuss the differential equations which are of first order and first degree only. A differential equation of first order and first degree is of the form $\dfrac{dy}{dx} = f(x, y)$.

Points to Consider

All the differential equations, even of first order and first degree, cannot be solved. However, if they belong to any of the standard forms which we are going to discuss, in the subsequent articles they can be solved.

Equations in Which the Variables are Separable

The equation $\dfrac{dy}{dx} = f(x, y)$ is said to be variables separable form, if we can express it in the form $f(x)\,dx = g(y)\,dy$.

By integrating this, solution of the equation is obtained which is,

$$\int f(x)\,dx = \int g(y)\,dy + C$$

Illustration 16 Solve $\sec^2 x \tan y\,dx + \sec^2 y \tan x\,dy = 0$.

Solution. Dividing the given equation by $\tan x \tan y$, we get

$$\frac{\sec^2 x}{\tan x}dx + \frac{\sec^2 y}{\tan y}dy = 0$$

This is variable-separable type

Integrating, $$\int \frac{\sec^2 x}{\tan x}dx + \int \frac{\sec^2 y}{\tan y}dy = C'$$

$$\ln|\tan x| + \ln|\tan y| = \ln C;\ \text{where } C' = \ln C$$

or $$\ln|\tan x \cdot \tan y| = \ln C;\ (C > 0)$$

$\Rightarrow$ $$|\tan x \cdot \tan y| = C$$

This is the general solution.

Illustration 17 Solve $\dfrac{dy}{dx} = e^{x-y} + x^2 e^{-y}$.

Solution. Here, $$\frac{dy}{dx} = \frac{e^x}{e^y} + \frac{x^2}{e^y}$$

$\Rightarrow$ $$e^y\,dy = (x^2 + e^x)\,dx$$

This is variable-separable form,

∴ Integrating both the sides,

$$\int e^y\,dy = \int (x^2 + e^x)\,dx$$

$$\Rightarrow \qquad e^y = \frac{x^3}{3} + e^x + C$$

Which is the general solution of the given differential equation, where C is an arbitrary constant.

Illustration 18 Solve $\sqrt{1 + x^2 + y^2 + x^2y^2} + xy\dfrac{dy}{dx} = 0.$

Solution. The given differential equation can be written as

$$\sqrt{(1 + y^2)(1 + x^2)} = -xy\frac{dy}{dx}$$

$$\Rightarrow \qquad -\frac{\sqrt{1 + x^2}\,dx}{x} = \frac{y\,dy}{\sqrt{1 + y^2}}$$

This is the variable-separable form.

∴ Integrating both the sides, we get

$$-\int \frac{\sqrt{1 + x^2}}{x}\,dx = \int \frac{y}{\sqrt{1 + y^2}}\,dy$$

$$\Rightarrow \qquad -\left[\sqrt{1 + x^2} + \frac{1}{2}\log\left(\frac{\sqrt{1 + x^2} - 1}{\sqrt{1 + x^2} + 1}\right)\right] = \sqrt{1 + y^2} + C$$

This is the general solution to the given differential equation.

Illustration 19 Solve $y - x\dfrac{dy}{dx} = a\left(y^2 + \dfrac{dy}{dx}\right).$

Solution. Rewriting the given equation as

$$y - ay^2 = (x + a)\frac{dy}{dx}$$

$$\Rightarrow \qquad \frac{dy}{y(1 - ay)} = \frac{dx}{(x + a)}$$

This is the variable-separable form.

Integrating both the sides, we get

$$\int \frac{dy}{y(1 - ay)} = \int \frac{dx}{(x + a)}$$

$$\Rightarrow \qquad \int \left(\frac{1}{y} + \frac{a}{1 - ay}\right)dy = \int \frac{dx}{x + a}$$

$$\Rightarrow \qquad \ln y - \frac{a}{a}\ln(1 - ay) + \ln C = \ln(a + x)$$

$$\Rightarrow \qquad \ln\left(\frac{(a + x)(1 - ay)}{y}\right) = \ln(C)$$

or $\qquad Cy = (a + x)(1 - ay)$ is the general solution.

Illustration 20 Solve $e^{dy/dx} = x + 1$, given that when $x = 0$, $y = 3$.

Solution. This is an illustration of particular solution.

$$e^{(dy/dx)} = x + 1$$

$$\therefore \quad \frac{dy}{dx} = \ln(x+1)$$

$$\therefore \quad \int dy = \int \ln(x+1)\,dx \qquad \text{(integration by parts)}$$

ie,
$$y = x\ln(x+1) - \int \frac{x}{x+1}\,dx$$

ie,
$$y = x\ln(x+1) - x + \ln(x+1) + C$$

This is the general solution.

To find the particular solution, put $x = 0$, $y = 3$ in the general equation.

$$\therefore \quad 3 = 0 - 0 + 0 + C$$

$$\therefore \quad C = 3$$

$\therefore$ The required particular solution is,

$$y = (x+1)\ln(x+1) - x + 3$$

Differential Equations Reducible to the Separable Variable Type

Sometimes differential equation of the first order cannot be solved directly by variable separation but by some substitution we can reduce it to a differential equation with separable variable. "A differential equation of the form $\frac{dy}{dx} = f(ax + by + c)$ is solved by writing $ax + by + c = t$."

Illustration 21 Solve $\frac{dy}{dx} = \sin^2(x + 3y) + 5$.

Solution. Let $x + 3y = t$, so that $1 + \frac{3dy}{dx} = \frac{dt}{dx}$

The given differential equation becomes,

$$\frac{1}{3}\left(\frac{dt}{dx} - 1\right) = \sin^2(t) + 5$$

$$\frac{dt}{dx} = 3\sin^2 t + 16$$

$$\Rightarrow \quad \int \frac{dt}{3\sin^2 t + 16} = \int dx$$

$$\Rightarrow \quad \int \frac{\sec^2 t\,dt}{3\tan^2 t + 16\sec^2 t} = x + C$$

$$\Rightarrow \quad \int \frac{\sec^2 t\,dt}{19\tan^2 t + 16} = x + C$$

$$\Rightarrow \quad \int \frac{du}{19u^2 + 16} = x + C; \text{ where } \tan t = u \Rightarrow \sec^2 t\,dt = du$$

$$\Rightarrow \quad \frac{1}{19}\int \frac{du}{u^2+\frac{16}{19}} \Rightarrow \frac{1}{19}\,\frac{\sqrt{19}}{4}\tan^{-1}\left(\frac{\sqrt{19}u}{4}\right)+C$$

$$\Rightarrow \quad \frac{1}{4\sqrt{19}}\left\{\tan^{-1}\left(\frac{\sqrt{19}}{4}\tan(3y+x)\right)\right\}+C$$

Illustration 22 Solve $(x+y)^2\dfrac{dy}{dx}=a^2$.

Solution. $(x+y)^2\dfrac{dy}{dx}=a^2$ ...(i)

Put $x+y=t \Rightarrow 1+\dfrac{dy}{dx}=\dfrac{dt}{dx}$ or $\dfrac{dy}{dx}=\left(\dfrac{dt}{dx}-1\right)$

∴ Eq. (i) reduces to

$$t^2\left\{\frac{dt}{dx}-1\right\}=a^2$$

ie, $t^2\dfrac{dt}{dx}=a^2+t^2$, separating the variable and integrating.

$$\int dx=\int\frac{t^2}{a^2+t^2}\,dt=\int\left(1-\frac{a^2}{a^2+t^2}\right)dt$$

∴ $$x=t-a\tan^{-1}\left(\frac{t}{a}\right)+C$$

ie, $$x=x+y-a\tan^{-1}\left(\frac{x+y}{a}\right)+C$$

ie, $y=a\tan^{-1}\left(\dfrac{x+y}{a}\right)-C$, is the required general solution.

Illustration 23 Solve $(2x+3y-1)\,dx+(4x+6y-5)\,dy=0$.

Solution. $(2x+3y-1)\,dx+(4x+6y-5)\,dy=0$ ...(i)

Substitute $u=2x+3y-1$

∴ $\dfrac{du}{dx}=2+\dfrac{3dy}{dx}$ or $\dfrac{dy}{dx}=\dfrac{1}{3}\left(\dfrac{du}{dx}-2\right)$

∴ Eq. (i) reduces to

$$u+(2u-3)\frac{1}{3}\left(\frac{du}{dx}-2\right)=0$$

ie, $$u-\frac{2}{3}(2u-3)+\frac{1}{3}(2u-3)\frac{du}{dx}=0$$

ie, $$\frac{-u+6}{3}+\frac{1}{3}(2u-3)\frac{du}{dx}=0$$

Writing this in the variable-separable form

$$\left(\frac{2u-3}{u-6}\right)du=dx$$

∴ $$\int dx=\int\frac{2u-3}{u-6}\,du$$

$\therefore \quad x + C = \int \frac{2(u-6)+9}{(u-6)} du$

$\therefore \quad x + C = 2u + 9 \ln |u - 6|$

$\therefore \quad x + C = 2(2x + 3y - 1) + 9 \ln |2x + 3y - 7|$

$3x + 6y - 2 + 9 \ln |2x + 3y - 7| = C$ **is the general solution.**

Points to Consider

Sometimes conversion of cartesian coordinates into polar coordinates help us in separating the variables. See the following illustration.

Illustration 24 Solve $\frac{x\,dx + y\,dy}{x\,dy - y\,dx} = \sqrt{\frac{a^2 - x^2 - y^2}{x^2 + y^2}}$.

Solution. Let $x = r\cos\theta,\ y = r\sin\theta$

So that $\quad x^2 + y^2 = r^2 \quad$ ...(i)

and $\quad \tan\theta = \frac{y}{x} \quad$ ...(ii)

From Eq. (i), we have $\quad d(x^2 + y^2) = d(r^2)$

ie, $\quad x\,dx + y\,dy = r\,dr \quad$ ...(iii)

From Eq. (ii), we have $\quad d\left(\frac{y}{x}\right) = d(\tan\theta)$

ie, $\quad \frac{x\,dy - y\,dx}{x^2} = \sec^2\theta\,d\theta$

ie, $\quad x\,dy - y\,dx = x^2\sec^2\theta\,d\theta = r^2\cos^2\theta\sec^2\theta\,d\theta \quad$ ...(iv)

Using Eqs. (iii) and (iv) in the given equation, we get

$$\frac{r\,dr}{r^2\,d\theta} = \sqrt{\frac{a^2 - r^2}{r^2}} \quad ie, \quad \frac{dr}{\sqrt{a^2 - r^2}} = d\theta$$

ie, $\quad \sin^{-1}\left(\frac{r}{a}\right) = \theta + C$

or $\quad r = a\sin(\theta + C)$

or $\quad \sqrt{x^2 + y^2} = a\sin\{C + \tan^{-1}(y/x)\}$

It is advised to remember the results (iii) and (iv).

Homogeneous Differential Equation

(i) A differential equation $P(x, y)\,dx + Q(x, y)\,dy = 0 \quad$ ...(i)

is called homogeneous, if $P(x, y)$ and $Q(x, y)$ are homogeneous functions of the same degree in x and y. Eq. (i) may be reduced to the form $y' = f\left(\frac{y}{x}\right)$. By means of the substitution $y = xu$, where u is some unknown function, the differential equation is transformed to an equation with variable separables. We can also substitute $x = yu$.

(ii) If $\frac{dy}{dx} = f\left(\frac{a_1x + b_1y + c_1}{a_2x + b_2y + c_2}\right)$ ...(ii)

where $a_1b_2 - a_2b_1 \neq 0$, then substitutions $x = u + h$, $y = v + k$ transform the Eq. (ii) to a homogeneous type in the new variable u and v which can be solved by the method in (i). If $a_1b_2 - a_2b_1 = 0$, then substitute $u = a_1x + b_1y$ transforms the differential equation (ii) to an equation with variables separable.

Points to Consider

1. The function $P(x, y)$ is homogeneous function of degree n, if for any real number $t\ (\neq 0)$, we have $P(tx, ty) = t^n P(x, y)$.
   For instance, $P(x, y) = ax^{2/3} + hx^{1/3}y^{1/3} + by^{2/3}$ is a homogeneous function of degree 2/3 since.
   $$P(tx, ty) = a\,(tx)^{2/3} + h\,(tx)^{1/3}\cdot(ty)^{1/3} + b\,(ty)^{2/3}$$
   $$= t^{2/3}\,(ax^{2/3} + hx^{1/3}y^{1/3} + by^{2/3})$$
   $$= t^{2/3}\,P(x, y)$$
2. A differential equation of the form $\frac{dy}{dx} = f(x, y)$ is homogeneous, if $f(x, y)$ is a homogeneous function of degree zero *ie*, $f(tx, ty) = t^0 f(x, y) = f(x, y)$.

Illustration 25 Solve $y\,dx + (2\sqrt{xy} - x)\,dy = 0$.

Solution. $y\,dx + (2\sqrt{xy} - x)\,dy = 0$ ...(i)

This is homogeneous type. Substitute $y = ux$

$$\therefore \quad \frac{dy}{dx} = u + \frac{x\,du}{dx}$$

$\therefore$ Equation $ux\,dx + (2\sqrt{x^2u} - x)(u\,dx + x\,du) = 0$

ie, $\quad x\cdot\{udx + (2\sqrt{u} - 1)u\,dx + x\,du\,(2\sqrt{u} - 1)\} = 0$

ie, $\quad dx\,(2u^{3/2} - u + u) + x\,du\,(2\sqrt{u} - 1) = 0$

Separating the variables,

$$\frac{dx}{x} + \left(\frac{2\sqrt{u} - 1}{2u^{3/2}}\right)du = 0$$

Integrating both the sides

$$\ln|x| + \ln|u| + \frac{1}{\sqrt{u}} = C$$

or $\quad \ln|xu| + \frac{1}{\sqrt{u}} = C$ or $\ln|y| + \sqrt{\frac{x}{y}} = C \quad \left(\because u = \frac{y}{x}\right)$

Which is the general solution.

Illustration 26 Solve $(x^2 + y^2)\,dx - 2xy\,dy = 0$.

Solution. Here, $\frac{dy}{dx} = \frac{x^2 + y^2}{2xy} = \frac{1}{2}\left[\frac{x}{y} + \frac{y}{x}\right]$

With $y = ux$, $\dfrac{dy}{dx} = u + x\dfrac{du}{dx}$, so that the differential equation becomes

$$u + x\frac{du}{dx} = \frac{1}{2}\left(\frac{1}{u} + u\right) \Rightarrow \frac{x\,du}{dx} = \frac{1+u^2}{2u} - u$$

$$\Rightarrow \quad \frac{x\,du}{dx} = \frac{1-u^2}{2u} \Rightarrow \int \frac{2u}{1-u^2}\,du = \int \frac{dx}{x}$$

$$\Rightarrow \quad -\log|1-u^2| = \log|x| - \log|C|$$

$$\Rightarrow \quad x(1-u^2) = C$$

$$\Rightarrow \quad \frac{x^2-y^2}{x} = C \qquad \left(\because u = \frac{y}{x}\right)$$

Hence, $x^2 - y^2 = xC$, is the required solution.

Illustration 27 Solve $\dfrac{2\,dy}{dx} = \dfrac{y}{x} + \dfrac{y^2}{x^2}$.

Solution. The above equation is homogeneous so that we put $y = ux$.

$$\Rightarrow \quad 2\left[u + x\frac{du}{dx}\right] = u + u^2$$

$$\Rightarrow \quad 2u + 2x\frac{du}{dx} = u + u^2$$

$$\Rightarrow \quad 2x\frac{du}{dx} = u^2 - u$$

$$\Rightarrow \quad \frac{du}{u^2 - u} = \frac{dx}{2x}$$

$$\Rightarrow \quad \int \frac{du}{u(u-1)} = \frac{1}{2}\int \frac{dx}{x}$$

$$\Rightarrow \quad \int \frac{1}{u-1}\,du - \int \frac{1}{u}\,du = \frac{1}{2}\int \frac{dx}{x}$$

$$\Rightarrow \quad \log|u-1| - \log|u| = \frac{1}{2}\log|x| + \log|C|$$

$$\Rightarrow \quad \log\left|\frac{u-1}{u}\right| = \log|C\sqrt{x}|$$

$$\Rightarrow \quad \frac{u-1}{u} = C\sqrt{x} \Rightarrow \frac{y-x}{y} = C\sqrt{x}$$

$$\Rightarrow \quad y - x = C\sqrt{x}\cdot y$$

Which is the required solution.

Illustration 28 Solve $(1 + 2e^{x/y})\,dx + 2e^{x/y}(1 - x/y)\,dy = 0$.

Solution. The appearance of x/y in the equation suggests the substitution $x = vy$ or $dx = v\,dy + y\,dv$.

$\therefore$ The given equation is

$$(1 + 2e^v)(v\,dy + y\,dv) + 2e^v(1-v)\,dy = 0$$

ie,

$$y(1 + 2e^v)\,dv + (v + 2e^v)\,dy = 0$$

ie, $$\frac{1+2e^v}{v+2e^v}dv+\frac{dy}{y}=0$$

Integrating, $$\int\frac{1+2e^v}{v+2e^v}dv+\int\frac{dy}{y}=0$$

$$\log|v+2e^v|+\log|y|=\log|C|$$

$$\Rightarrow \quad (v+2e^v)\,y=C \qquad \left(v=\frac{x}{y}\right)$$

$$\Rightarrow \quad \left(\frac{x}{y}+2e^{x/y}\right)y=C$$

$$\Rightarrow \quad (x+2ye^{x/y})=C, \text{ is required solution.}$$

Illustration 29 Show that any equation of the form

$$y\,f(xy)\,dx+x\,g(xy)\,dy=0$$

can be converted to variable separable form by substituting $xy=v$.

Solution. Since, $xy=v$, $y=\frac{v}{x}$ and $d(x\,y)=d\,v$

ie, $$x\,dy+y\,dx=dv$$

and $$dy=d\left(\frac{v}{x}\right)=\frac{x\,dv-v\,dx}{x^2}$$

ie, $$x\,dy=dv-\frac{v}{x}dx$$

$$\therefore \quad \frac{v}{x}f(v)\,dx+g(v)\left\{dv-\frac{v}{x}dx\right\}=0$$

$$\therefore \quad \frac{v\{f(v)-g(v)\}}{x}dx+g(v)\,dv=0$$

ie, $$\frac{dx}{x}+\frac{g(v)\,dv}{v\{f(v)-g(v)\}}=0$$

Which is in variables separable form.

Illustration 30 Solve $(2x-y+4)\,dy+(x-2y+5)\,dx=0$.

Solution. $(2x-y+4)\,dy+(x-2y+5)\,dx=0$ ...(i)

The given equation is non-homogeneous and $a_1b_2-a_2b_1=-3\neq 0$

Solving $\quad 2x-y+4=0,\ x-2y+5=0$

We get $\quad x=-1,\ y=2$

Now, substitute $x=-1+u$, $y=2+v$ in Eq. (i)

$$\therefore \quad dx=du,\ dy=dv$$

$\therefore$ Eq. (i) reduce to $(2u-v)\,dv+(u-2v)\,du=0$ ...(ii)

This is homogeneous equation in u and v.

$\therefore$ Put $u=tv \quad \therefore \quad du=t\,dv+v\,dt$

$\therefore$ Eq. (ii) reduce to $(2tv-v)\,dv+(tv-2v)(t\,dv+v\,dt)=0$

ie, $$(2t-1)\,dv+(t-2)\,t\,dv+v(t-2)\,dt=0$$

ie, $$(t^2-2t+2t-1)\,dv+v(t-2)\,dt=0$$

Separating the variable v and t, $\frac{dv}{v} + \frac{t-2}{t^2-1}\, dt = 0$

$$\therefore \quad \int \frac{dv}{v} + \int \frac{t-2}{t^2-1}\, dt = C_1$$

ie,
$$\ln|v| + \int \left\{ \frac{-1}{2(t-1)} + \frac{3}{2(t+1)} \right\} dt = C_1$$

ie,
$$\ln|v| - \frac{1}{2}\ln|t-1| + \frac{3}{2}\ln|t+1| = \ln C, \text{ where } C_1 = \ln C\ (C > 0)$$

Put $\quad t = \frac{u}{v}$

ie,
$$\ln|v| - \frac{1}{2}\ln\left|\frac{u-v}{v}\right| + \frac{3}{2}\ln\left|\frac{u+v}{v}\right| = \ln C$$

or
$$\ln\left[\frac{|v|\cdot|v^{1/2}|\cdot(u+v)^{3/2}}{(u-v)^{1/2}\cdot v^{3/2}}\right] = \ln C$$

$$\therefore \quad (u+v)^{3/2} = C\,(u-v)^{1/2}$$

or
$$(u+v)^3 = A\,(u-v),\ A = C^2$$

ie,
$$(x+y-1)^3 = A\,(x-y+3) \text{ is the general solution.}$$

Illustration 31 Solve $\frac{dy}{dx} = \frac{1-3y-3x}{1+x+y}$.

Solution. In this example, $a_1b_2 - a_2b_1 = (1)(-3) - (1)(-3) = 0$ and the RHS is non-homogeneous.

Hence, substitute $x + y = t$

$$\therefore \quad 1 + \frac{dy}{dx} = \frac{dt}{dx}$$

$$\Rightarrow \quad \frac{dy}{dx} = \left(\frac{dt}{dx} - 1\right)$$

$\therefore$ The given differential equation

$$\Rightarrow \quad \left(\frac{dt}{dx} - 1\right) = \left(\frac{1-3t}{1+t}\right)$$

$$\Rightarrow \quad \frac{dt}{dx} = \frac{1-3t}{1+t} + 1 = \frac{2-2t}{1+t}$$

Separating the variables x and t

$$dx = \frac{1+t}{2(1-t)}\, dt = \frac{2-(1-t)}{2(1-t)}\, dt$$

ie,
$$dx = \frac{dt}{1-t} - \frac{dt}{2}$$

Integrating both the sides

$$x + C_1 = -\ln|1-t| - \frac{t}{2}$$

ie,
$$x + C_1 = -\ln|1-x-y| - \frac{(x+y)}{2}$$

or
$$3x + y + 2\ln|1-x-y| = C, \text{ where } C = -2C_1$$

Illustration 32 If $f(x)$ be a positive, continuous and differentiable on the interval (a, b). If $\lim_{x \to a^+} f(x) = 1$ and $\lim_{x \to b^-} f(x) = 3^{1/4}$.

Also, $f'(x) \geq f^3(x) + \dfrac{1}{f(x)}$, then

(a) $b - a \geq \pi/4$ (b) $b - a \leq \pi/4$ (c) $b - a \leq \pi/24$ (d) None of these

Solution. Since, $$f'(x) \geq f^3(x) + \frac{1}{f(x)}$$

$$\Rightarrow \quad f'(x) f(x) \geq 1 + f^4(x)$$

$$\Rightarrow \quad \frac{f(x) f'(x)}{1 + f^4(x)} \geq 1$$

On integrating w.r.t. 'x' from $x = a$ to $x = b$.

$$\frac{1}{2}(\tan^{-1}(f^2(x)))_a^b \geq (b - a)$$

or $$(b - a) \leq \frac{1}{2}\left\{\lim_{x \to b^-}(\tan^{-1}(f^2(x))) - \lim_{x \to a^+}(\tan^{-1}(f^2(x)))\right\}$$

or $$(b - a) \leq \pi/24$$

Hence, (c) is the correct answer.

Illustration 33 The solution of the differential equation $\dfrac{dy}{dx} = \dfrac{\sin y + x}{\sin 2y - x \cos y}$ is

(a) $\sin^2 y = x \sin y + \dfrac{x^2}{2} + C$ (b) $\sin^2 y = x \sin y - \dfrac{x^2}{2} + C$

(c) $\sin^2 y = x + \sin y + \dfrac{x^2}{2} + C$ (d) $\sin^2 y = x - \sin y + \dfrac{x^2}{2} + C$

Solution. Here, $$\frac{dy}{dx} = \frac{\sin y + x}{\sin 2y - x \cos y}$$

$$\Rightarrow \quad \cos y \frac{dy}{dx} = \frac{\sin y + x}{2 \sin y - x}, \text{ put } \sin y = t$$

$$\Rightarrow \quad \frac{dt}{dx} = \frac{t + x}{2t - x}, \text{ put } t = vx$$

$$\frac{x\,dv}{dx} + v = \frac{vx + x}{2vx - x} = \frac{v + 1}{2v - 1}$$

$$\therefore \quad x\frac{dv}{dx} = \frac{v + 1}{2v - 1} - v = \frac{v + 1 - 2v^2 + v}{2v - 1}$$

or $$\frac{2v^2 - v}{-2v^2 + 2v + 1}\,dv = \frac{dx}{x}$$

On solving, we get

$$\sin^2 y = x \sin y + \frac{x^2}{2} + C$$

Hence, (a) is the correct answer.

Illustration 34 The equation of curve passing through (1, 0) and satisfying $\left(y\frac{dy}{dx}+2x\right)^2=(y^2+2x^2)\left(1+\left(\frac{dy}{dx}\right)^2\right)$, is given by

(a) $\sqrt{2}\,x^{\pm\frac{1}{\sqrt{2}}}=\frac{y+\sqrt{y^2+2x^2}}{x}$ (b) $\sqrt{2}x^{\pm\sqrt{2}}=\frac{y+\sqrt{y^2+\sqrt{2}\,x^2}}{x}$

(c) $\sqrt{2}\,y^{\pm\frac{1}{\sqrt{2}}}=\frac{y+\sqrt{x^2+2y^2}}{x}$ (d) None of these

Solution. The given differential equation can be written as

$$y^2\left(\frac{dy}{dx}\right)^2+4x^2+4xy\cdot\frac{dy}{dx}=(y^2+2x^2)\left(1+\left(\frac{dy}{dx}\right)^2\right)$$

$$\Rightarrow \qquad \frac{dy}{dx}=\frac{y}{x}\pm\sqrt{\frac{1}{2}\left(\frac{y}{x}\right)^2+1} \qquad \text{...(i)}$$

Let $\quad y=vx \Rightarrow v+x\frac{dv}{dx}=\frac{dy}{dx}$

$\therefore$ Eq. (i) becomes

$$v+x\frac{dv}{dx}=v\pm\sqrt{\frac{1}{2}v^2+1}$$

or

$$\int\frac{dv}{\sqrt{\frac{1}{2}v^2+1}}=\int\frac{dx}{x}$$

$$\Rightarrow \qquad \sqrt{2}\log|v+\sqrt{v^2+2}|=\log|xC|$$

$$\Rightarrow \qquad \sqrt{2}\log\left|\frac{y+\sqrt{y^2+2x^2}}{x}\right|=\log|xC|, \text{ putting } x=1 \text{ and } y=0$$

$$\Rightarrow \qquad C=(\sqrt{2})^{\sqrt{2}}$$

$\therefore$ Curves are given by $\frac{y+\sqrt{y^2+2x^2}}{x}=\sqrt{2}\,x^{\pm\frac{1}{\sqrt{2}}}$

Hence, (a) is the correct answer.

Target Exercise 4.2

1. The solution of $\frac{dy}{dx}=\frac{(x+y)^2}{(x+2)(y-2)}$, is given by

(a) $(x+2)^4\left(1+\frac{2y}{x}\right)=ke^{2y/x}$ (b) $(x+2)^4\left(1+\frac{2(y-2)}{(x+2)}\right)=k\,e^{\frac{2(y-2)}{(x+2)}}$

(c) $(x+2)^3\left(1+2\frac{(y-2)}{x+2}\right)=ke^{\frac{2(y-2)}{x+2}}$ (d) None of these

2. If $(y^3-2x^2y)\,dx+(2xy^2-x^3)\,dy=0$, then the value of $xy\sqrt{y^2-x^2}$, is

(a) y^2+x (b) xy^2

(c) any constant (d) None of these

3. The solution of $dy/dx = \cos(x+y) + \sin(x+y)$, is given by

(a) $\log\left|1+\tan\left(\frac{x+y}{2}\right)\right| = x + C$ (b) $\log|1+\tan(x+y)| = x + C$

(c) $\log|1-\tan(x+y)| = x + C$ (d) None of these

4. The solution of $\frac{dy}{dx} = (x+y-1) + \frac{x+y}{\log(x+y)}$, is given by

(a) $\{1+\log(x+y)\} - \log\{1+\log(x+y)\} = x + C$
(b) $\{1-\log(x+y)\} - \log\{1-\log(x+y)\} = x + C$
(c) $\{1+\log(x+y)\}^2 - \log\{1+\log(x+y)\} = x + C$
(d) None of the above

5. The solution of $(2x^2+3y^2-7)\,x\,dx - (3x^2+2y^2-8)\,y\,dy = 0$, is given by

(a) $(x^2+y^2-1) = (x^2+y^2-3)^5 C$ (b) $(x^2+y^2-1)^2 = (x^2+y^2-3)^5 C$
(c) $(x^2+y^2-3) = (x^2+y^2-1)^5 C$ (d) None of these

6. The solution of $\frac{dy}{dx} = \frac{(x-1)^2 + (y-2)^2 \tan^{-1}\left(\frac{y-2}{x-1}\right)}{(xy-2x-y+2)\tan^{-1}\left(\frac{y-2}{x-1}\right)}$, is equal to

(a) $\{(x-1)^2+(y-1)^2\}\tan^{-1}\left(\frac{y-2}{x-1}\right) - 2(x-1)(y-2) = 2(x-1)^2 \log C(x-1)$

(b) $\{(x-1)^2+(y-1)^2\} - 2(x-1)(y-2)\tan^{-1}\left(\frac{y-2}{x-1}\right) = 2(x-1)^2 \log C$

(c) $\{(x-1)^2+(y-1)^2\}\tan^{-1}\left(\frac{y-2}{x-1}\right) - 2(x-1)(y-2) = \log C(x-1)$

(d) None of the above

7. The solution of $\frac{dy}{dx} = \left(\frac{x+2y-3}{2x+y+3}\right)^2$, is

(a) $(x+3)^3 - (y-3)^3 = C(x-y+6)^4$ (b) $(x+3)^3 - (y-3)^3 = C$
(c) $(x+3)^4 + (y-3)^4 = C$ (d) None of these

8. The solution of $\frac{dy}{dx} = \frac{-\cos x\,(3\cos y - 7\sin x - 3)}{\sin y\,(3\sin x - 7\cos y + 7)}$, is

(a) $(\cos y - \sin x - 1)^2 (\sin x + \cos y - 1)^5 = C$
(b) $(\cos y - \sin x - 1)^2 (\sin x + \cos y - 1)^3 = C$
(c) $(\cos y - \sin x - 1)^2 (\sin x + \cos y - 1)^7 = C$
(d) None of the above

Exact Differential Equations

A differential equation of the form $M(x, y)\,dx + N(x, y)\,dy = 0$ is said to be exact (or total) if its left hand expression is the exact differential of some function $u(x, y)$.

ie, $$du = M \cdot dx + N \cdot dy$$

Hence, its solution is $u(x, y) = c$ (where c is an arbitrary constant). But then there is a question that how do we confirm whether the above mentioned equation is exact. The answer to this question is the following theorem.

Theorem The necessary and sufficient condition for the differential equation $M\,dx + N\,dy = 0$ to be exact is $\dfrac{\partial M}{\partial y} = \dfrac{\partial N}{\partial x}$.

Solution. The solution of $M\,dx + N\,dy = 0$ is,

$$\int_{y-\text{constant}} M\,dx + \int (\text{terms of } N \text{ not containing } x)\,dy = C$$

provided $\dfrac{\partial M}{\partial y} = \dfrac{\partial N}{\partial x}$

Illustration 35 Solve $(x^2 - ay)\,dx + (y^2 - ax)\,dy = 0$.

Solution. Here, we have $M = x^2 - ay$ and $N = y^2 - ax$

$$\therefore \quad \frac{\partial M}{\partial y} = -a = \frac{\partial N}{\partial x}$$

Thus, the equation is exact.

$\therefore$ The solution is, $\displaystyle\int_{y-\text{constant}} (x^2 - ay)\,dx + \int y^2\,dy = C$

$$\Rightarrow \quad \frac{x^3}{3} - axy + \frac{y^3}{3} = C$$

Illustration 36 Solve $(2x \log y)\,dx + \left(\dfrac{x^2}{y} + 3y^2\right)dy = 0$.

Solution. Here, we have $M = 2x \log y$ and $N = \dfrac{x^2}{y} + 3y^2$

$\therefore$ $\dfrac{\partial M}{\partial y} = \dfrac{2x}{y}$ and $\dfrac{\partial N}{\partial x} = \dfrac{2x}{y}$ and hence the equation is exact.

$\therefore$ The solution is, $\displaystyle\int_{y-\text{constant}} (2x \log y)\,dx + \int 3y^2\,dy = C$

$$\Rightarrow \quad x^2 \log y + y^3 = C$$

Equations Reducible to the Exact Form

Sometimes a differential equation of the form $M\,dx + N\,dy = 0$ which is not exact can be reduced to an exact form by multiplying by a suitable function $f(x, y)$ which is not identically zero. This function $f(x, y)$ which then multiplied to a non-exact differential equation makes it exact is known as integrating factor.

One can find integrating factors by inspection but for that some experience and practice is required.

For finding the integrating factors by inspection, the following identities must be remembered.

1. $x\,dy + y\,dx = d\,(xy)$
2. $x\,dx + y\,dy = \dfrac{1}{2}\,d\,(x^2 + y^2)$

3. $\dfrac{x\,dy - y\,dx}{x^2} = d\left(\dfrac{y}{x}\right)$

4. $\dfrac{y\,dx - x\,dy}{y^2} = d\left(\dfrac{x}{y}\right)$

5. $\dfrac{x\,dy - y\,dx}{xy} = \dfrac{dy}{y} - \dfrac{dx}{x} = d\left[\log\left(\dfrac{y}{x}\right)\right]$

6. $\dfrac{y\,dx - x\,dy}{xy} = d\left[\log\left(\dfrac{x}{y}\right)\right]$

7. $\dfrac{x\,dy - y\,dx}{x^2 + y^2} = \dfrac{\dfrac{x\,dy - y\,dx}{x^2}}{1 + \dfrac{y^2}{x^2}} = \dfrac{d\left(\dfrac{y}{x}\right)}{1 + \left(\dfrac{y}{x}\right)^2} = d\left[\tan^{-1}\left(\dfrac{y}{x}\right)\right]$

8. $\dfrac{dx + dy}{x + y} = d\,[\ln (x + y)]$

9. $d(\ln (xy)) = \dfrac{x\,dy + y\,dx}{xy}$

10. $d\left(\dfrac{1}{2}\ln (x^2 + y^2)\right) = \dfrac{x\,dx + y\,dy}{x^2 + y^2}$

11. $d\left(-\dfrac{1}{xy}\right) = \dfrac{x\,dy + y\,dx}{x^2\,y^2}$

12. $d\left(\dfrac{e^y}{x}\right) = \dfrac{x\,e^y\,dy - e^y\,dx}{x^2}$

13. $d\left(\dfrac{e^x}{y}\right) = \dfrac{ye^x dx - e^x dy}{y^2}$

14. $d(x^m\,y^n) = x^{m-1}\,y^{n-1}\,(my\,dx + nx\,dy)$

15. $\dfrac{d\,[f(x, y)]^{1-n}}{1 - n} = \dfrac{f\,'(x, y)}{(f(x, y))^n}$

Points to Consider

Sometimes transformation to the polar coordinates facilitates separation of variables. It is convenient to remember the following differentials.

If $x = r\cos\theta$, $y = r\sin\theta$

1. $x\,dx + y\,dy = r\,dr$ **2.** $x\,dy - y\,dx = r^2\,d\theta$ **3.** $dx^2 + dy^2 = dr^2 + r^2\,d\theta^2$

Illustration 37 Solve $(x^2 - ay)\,dx + (y^2 - ax)\,dy = 0$.

Solution. The given differential equation is

$$x^2dx + y^2dy - a\,(y\,dx + x\,dy) = 0$$

$$\Rightarrow \quad d\left(\frac{x^3}{3}\right) + d\left(\frac{y^3}{3}\right) - ad\,(xy) = 0$$

Integrating, we get $\frac{x^3}{3} + \frac{y^3}{3} - a\,xy = k$

$$\Rightarrow \quad x^3 + y^3 - 3axy = 3k = C$$

Illustration 38 Solve $(2x \log y)\,dx + \left(\frac{x^2}{y} + 3y^2\right) dy = 0.$

Solution. The given differential equation is

$$(\log y)\,2x\,dx + \frac{x^2}{y}\,dy + 3y^2\,dy = 0$$

$$\Rightarrow \quad (\log y)\,d\,(x^2) + x^2\,d\,(\log y) + d\,(y^3) = 0$$

$$\Rightarrow \quad d\,(x^2 \log y) + d\,(y^3) = 0$$

$$\Rightarrow \quad x^2 \log y + y^3 = C \quad \text{(integrating both the sides)}$$

Illustration 39 Solve $x\,dx + y\,dy = x\,dy - y\,dx.$

Solution. The given equation can be written as

$$\frac{1}{2}\,d\,(x^2 + y^2) = x^2\,d\left(\frac{y}{x}\right)$$

$$\Rightarrow \quad \frac{d\,(x^2 + y^2)}{x^2 + y^2} = \frac{2x^2\,d\left(\frac{y}{x}\right)}{x^2 + y^2}$$

$$\Rightarrow \quad \frac{d\,(x^2 + y^2)}{x^2 + y^2} = \frac{2d\,(y/x)}{1 + (y/x)^2}$$

$$\Rightarrow \quad \log\,(x^2 + y^2) = 2\tan^{-1}\left(\frac{y}{x}\right) + C$$

Illustration 40 Solve $\frac{y + \sin x \cos^2(xy)}{\cos^2(xy)}\,dx + \left(\frac{x}{\cos^2(xy)} + \sin y\right) dy = 0.$

Solution. The given differential equation can be written as

$$\frac{y\,dx + x\,dy}{\cos^2(xy)} + \sin x\,dx + \sin y\,dy = 0$$

$$\Rightarrow \quad \sec^2(xy)\,d\,(xy) + \sin x\,dx + \sin y\,dy = 0$$

$$d\,(\tan(xy)) + d\,(-\cos x) + d\,(-\cos y) = 0$$

$$\Rightarrow \quad \tan(xy) - \cos x - \cos y = C$$

Illustration 41 Solve $\dfrac{x + y\dfrac{dy}{dx}}{y - x\dfrac{dy}{dx}} = x^2 + 2y^2 + \dfrac{y^4}{x^2}.$

Solution. The given equation can be written as

$$\frac{x\,dx + y\,dy}{(x^2 + y^2)^2} = \frac{y\,dx - x\,dy}{y^2} \cdot \frac{y^2}{x^2}$$

$$\Rightarrow \quad \int \frac{d(x^2+y^2)}{(x^2+y^2)^2} = 2\int \frac{1}{x^2/y^2}\, d\left(\frac{x}{y}\right)$$

Integrating both the sides, we get

$$-\frac{1}{(x^2+y^2)} = -\frac{1}{(x/y)} + C \Rightarrow \frac{y}{x} - \frac{1}{x^2+y^2} = C$$

Illustration 42 The solution of

$$e^{x\frac{(y^2-1)}{y}}\{xy^2\, dy + y^3 dx\} + \{ydx - xdy\} = 0, \text{ is}$$

(a) $e^{xy} + e^{x/y} + C = 0$ (b) $e^{xy} - e^{x/y} + C = 0$

(c) $e^{xy} + e^{y/x} + C = 0$ (d) $e^{xy} - e^{y/x} + C = 0$

Solution. Here, $e^{x\frac{(y^2-1)}{y}} \cdot y^2\{xdy + ydx\} + \{ydx - xdy\} = 0$

$$\Rightarrow \quad e^{xy} \cdot y^2 \cdot \{xdy + ydx\} + e^{x/y}\{ydx - xdy\} = 0$$

or $$e^{xy} \cdot \{xdy + ydx\} + e^{x/y}\frac{\{ydx - xdy\}}{y^2} = 0$$

or $$e^{xy} \cdot d(xy) + e^{x/y} \cdot d\left(\frac{x}{y}\right) = 0$$

or $$d(e^{xy}) + d(e^{x/y}) = 0$$

Integrating both the sides, we get

$$e^{xy} + e^{x/y} + C = 0$$

Hence, (a) is the correct answer.

Illustration 43 The solution of $x^2 dy - y^2 dx + xy^2(x - y)\, dy = 0$, is

(a) $\log\left|\frac{x-y}{xy}\right| = \frac{y^2}{2} + C$ (b) $\log\left|\frac{xy}{x-y}\right| = \frac{x^2}{2} + C$

(c) $\log\left|\frac{x-y}{xy}\right| = \frac{x^2}{2} + C$ (d) $\log\left|\frac{x-y}{xy}\right| = x + C$

Solution. Here, $x^2y^2\left(\frac{dy}{y^2} - \frac{dx}{x^2}\right) + x^2y^3\left(\frac{1}{y} - \frac{1}{x}\right) dy = 0$

$$\Rightarrow \quad -d\left(\frac{1}{y} - \frac{1}{x}\right) + y\left(\frac{1}{y} - \frac{1}{x}\right) dy = 0$$

$$\Rightarrow \quad -\frac{d\left(\frac{1}{y} - \frac{1}{x}\right)}{\left(\frac{1}{y} - \frac{1}{x}\right)} + ydy = 0$$

or $$\frac{d\left(\frac{1}{y} - \frac{1}{x}\right)}{\left(\frac{1}{y} - \frac{1}{x}\right)} = d\left(\frac{y^2}{2}\right)$$

Integrating both the sides, we get

$$\log\left|\frac{1}{x}-\frac{1}{y}\right|=\frac{y^2}{2}+C$$

Hence, (a) is the correct answer.

Illustration 44 The solution of the differential equation

$$ydx - xdy + xy^2\,dx = 0, \text{ is}$$

(a) $\frac{x}{y}+x^2=\lambda$ (b) $\frac{x}{y}+\frac{x^2}{2}=\lambda$ (c) $\frac{x}{2y^2}+\frac{x^2}{4}=\lambda$ (d) None of these

Solution. Given equation is,

$$ydx - xdy + xy^2\,dx = 0$$

Which could be converted into exact form

ie, $$\frac{ydx - xdy}{y^2} + xdx = 0$$

$$\Rightarrow \quad d\left(\frac{x}{y}\right)+d\left(\frac{x^2}{2}\right)=0$$

Integrating both the sides, we get

$$\frac{x}{y}+\frac{x^2}{2}=\text{constant}$$

or $$\frac{x}{y}+\frac{x^2}{2}=\lambda$$

Hence, (b) is the correct answer.

Illustration 45 The solution of differential equation

$$xdy\,(y^2e^{xy}+e^{x/y}) = ydx\,(e^{x/y}-y^2e^{xy}), \text{ is}$$

(a) $xy=\log(e^x+\lambda)$ (b) $x^2/y=\log(e^{x/y}+\lambda)$

(c) $xy=\log(e^{x/y}+\lambda)$ (d) $xy^2=\log(e^{x/y}+\lambda)$

Solution. The given equation is

$$(xy^2e^{xy})\,dy + (xe^{x/y})\,dy = (ye^{x/y})\,dx - (y^3e^{xy})\,dx$$

$$\Rightarrow \quad y^2e^{xy}\,(xdy + ydx) = e^{x/y}\,(ydx - xdy)$$

$$\Rightarrow \quad e^{xy}\,(d\,(xy)) = e^{x/y}\cdot\left(\frac{ydx - xdy}{y^2}\right)$$

$$\Rightarrow \quad e^{xy}\,(d\,(xy)) = e^{x/y}\cdot d\left(\frac{x}{y}\right)$$

$$\Rightarrow \quad d\,(e^{xy}) = d\,(e^{x/y})$$

Integrating both the sides, we get

$$e^{xy} = e^{x/y}+\lambda$$

$$\Rightarrow \quad xy = \log\,(e^{x/y}+\lambda)$$

Hence, (c) is the correct answer.

Illustration 46 The solution of the differential equation

$$(y + x\sqrt{xy}\,(x + y))\,dx + (y\sqrt{xy}\,(x + y) - x)\,dy = 0, \text{ is}$$

(a) $\dfrac{x^2 + y^2}{2} + 2\tan^{-1}\sqrt{\dfrac{x}{2y}} = C$ (b) $\dfrac{x^2 + y^2}{2} + 2\tan^{-1}\sqrt{\dfrac{x}{y}} = C$

(c) $\dfrac{x^2 + y^2}{\sqrt{2}} + 2\tan^{-1}\sqrt{\dfrac{x}{y}} = C$ (d) None of these

Solution. The given equation can be written as

$$(ydx - xdy) + x\sqrt{xy}\,(x + y)\,dx + y\sqrt{xy}\,(x + y)\,dy = 0$$

$$\Rightarrow \quad (ydx - xdy) + (x + y)\sqrt{xy}\,(xdx + ydy) = 0$$

$$\Rightarrow \quad \frac{ydx - xdy}{y^2} + \left(\frac{x}{y} + 1\right)\cdot\sqrt{\frac{x}{y}}\,d\left(\frac{x^2 + y^2}{2}\right) = 0$$

$$\Rightarrow \quad d\left(\frac{x}{y}\right) + d\left(\frac{x^2 + y^2}{2}\right)\left(\frac{x}{y} + 1\right)\sqrt{\frac{x}{y}} = 0$$

$$\Rightarrow \quad d\left(\frac{x^2 + y^2}{2}\right) + \frac{d\left(\frac{x}{y}\right)}{\left(\frac{x}{y} + 1\right)\cdot\sqrt{\frac{x}{y}}} = 0$$

or
$$d\left(\frac{x^2 + y^2}{2}\right) + 2d\left(\tan^{-1}\left(\sqrt{\frac{x}{y}}\right)\right) = 0$$

Integrating both the sides, we get

$$\Rightarrow \quad \frac{x^2 + y^2}{2} + 2\tan^{-1}\sqrt{\frac{x}{y}} = C$$

Hence, (b) is the correct answer.

Target Exercise 4.3

1. The solution of $xdy + ydx + 2x^3dx = 0$, is

   (a) $xy + x^4 = C$ (b) $xy + \frac{1}{2}x^4 = C$

   (c) $\dfrac{x^2}{y} + \dfrac{x^4}{4} = C$ (d) None of these

2. The solution of, $ydx - xdy + (1 + x^2)\,dx + x^2 \sin y\, dy = 0$, is given by

   (a) $x + 1 - y^2 + \cos y + C = 0$ (b) $y + 1 - x^2 + x\cos y + C = 0$

   (c) $\dfrac{x}{y} + \dfrac{1}{y} - y + \cos y + C = 0$ (d) $\dfrac{y}{x} + \dfrac{1}{x} - x + \cos y + C = 0$

3. The solution of $(1 + x\sqrt{x^2 + y^2})\,dx + (-1 + \sqrt{x^2 + y^2})\,ydy = 0$, is

   (a) $2x - y^2 + \frac{2}{3}(x^2 + y^2)^{3/2} = C$ (b) $2x - y + \frac{2}{3}(x^2 + y^2)^{3/2} = C$

   (c) $2y - x^2 + \frac{2}{3}(x^2 + y^2)^{3/2} = C$ (d) None of these

4. The solution of, $\frac{xdy}{x^2+y^2} = \left(\frac{y}{x^2+y^2} - 1\right) dx$, is given by

(a) $\tan^{-1}\left(\frac{x}{y}\right) + x = C$ (b) $\tan^{-1}\left(\frac{y}{x}\right) + x = C$

(c) $\tan^{-1}\left(\frac{y}{x}\right) + xy = C$ (d) $\tan^{-1}\left(\frac{y}{x}\right) + x^2 = C$

5. The solution of $ye^{x/y}\,dx = (xe^{x/y} + y^2 \sin y)\,dy$, is given by

(a) $e^{x/y} = -\cos y + C$ (b) $e^{x/y} + 2\cos y = C$

(c) $e^{x/y} = x\cos y + C$ (d) $e^{x/y} = 2\cos y\, e^{x/y} + C$

6. The solution of $x \sin\left(\frac{y}{x}\right) dy = \left\{ y \sin\left(\frac{y}{x}\right) - x \right\} dx$, is given by

(a) $\log x - \cos\left(\frac{y}{x}\right) = \log C$ (b) $\log x - \sin\left(\frac{y}{x}\right) = C$

(c) $\log\left(\frac{x}{y}\right) - \cos\left(\frac{y}{x}\right) = \log C$ (d) None of these

7. The solution of $\frac{xdx + ydy}{xdy - ydx} = \sqrt{\frac{a^2 - x^2 - y^2}{x^2 + y^2}}$, is given by

(a) $\sin^{-1}(\sqrt{x^2+y^2}) = a\tan^{-1}\left(\frac{y}{x}\right) + C$

(b) $\sin^{-1}(\sqrt{x^2+y^2}) = \frac{1}{a}\tan^{-1}\left(\frac{y}{x}\right) + C$

(c) $\sin^{-1}\left(\frac{\sqrt{x^2+y^2}}{a}\right) = \tan^{-1}\left(\frac{y}{x}\right) + C$

(d) None of the above

8. The solution of $(1 + e^{x/y})\,dx + e^{x/y}\left(1 - \frac{x}{y}\right) dy = 0$, is given by

(a) $x - ye^{x/y} = C$ (b) $x + ye^{x/y} = C$

(c) $y - \frac{x}{y}e^{x/y} = C$ (d) None of these

9. The solution of $\frac{x + y\,dy/dx}{y - x\,dy/dx} = \frac{x\sin^2(x^2+y^2)}{y^3}$, is given by

(a) $-\cot(x^2+y^2) = \left(\frac{x}{y}\right)^2 + C$ (b) $\tan(x^2+y^2) = x^2y^2 + C$

(c) $\cot(x^2+y^2) = \frac{x}{y} + C$ (d) None of these

10. The solution of $\frac{dy}{dx} + \frac{y}{x} = \frac{1}{(1 + \log x + \log y)^2}$, is given by

(a) $xy(1 + \log(xy)) = C$ (b) $xy^2(1 + \log(xy)) = C$

(c) $xy(1 + \log(xy))^2 = C$ (d) $xy(1 + (\log xy)^2) = C$

First Order Linear Differential Equations

A differential equation is said to be linear if an unknown variable and its derivative occur only in the first degree.

An equation of the form

$$\frac{dy}{dx} + P(x) \cdot y = Q(x).$$

Where $P(x)$ and $Q(x)$ are functions of x only or constant is called a linear equation of the first order.

To get the general solution of the above equation we proceeds as follows. By multiplying both the sides of the above equation by $e^{\int Pdx}$, we get

$$e^{\int Pdx} \cdot \frac{dy}{dx} + yP \cdot e^{\int Pdx} = Q\, e^{\int Pdx}$$

ie, $$e^{\int Pdx} \cdot \frac{dy}{dx} + y \frac{d}{dx}(e^{\int Pdx}) = Q\, e^{\int Pdx}$$

ie, $$\frac{d}{dx}(y\, e^{\int Pdx}) = Q \cdot e^{\int Pdx}$$

∴ Integrating, we get $$y\, e^{\int Pdx} = \int Q\, e^{\int Pdx}\, dx + C$$

Points to Consider

Here, the term $e^{\int Pdx}$ which converts the left hand expression of the equation into a perfect differential is called an Integrating factor. In short it is written as IF. Thus, we remember the solution of the above equation as $y\,(\text{IF}) = \int Q\,(\text{IF})\, dx + C$.

Illustration 47 Solve $\frac{dy}{dx} + 2y = \cos x$.

Solution. It is a linear equation of the form

$$\frac{dy}{dx} + Py = Q(x)$$

where $P = 2$ and $Q = \cos x$

Then $$\text{IF} = e^{\int Pdx} = e^{\int 2dx} = e^{2x}$$

Hence, the general solution is $y\,(\text{IF}) = \int Q\,(\text{IF})\, dx$

ie, $$y \cdot e^{2x} = \int e^{2x} \cos x\, dx + C$$

$$y \cdot e^{2x} = \frac{e^{2x}}{5}[2 \cos x + \sin x] + C$$

Illustration 48 Solve $\frac{dy}{dx} + \frac{y}{x} = \log x$.

Solution. It is a linear differential equation of the form

$$\frac{dy}{dx} + Py = Q(x)$$

Here, $P = \frac{1}{x}, Q = \log x$

Then $\text{IF} = e^{\int P dx} = e^{\int 1/x\, dx} = e^{\log x} = x$

Hence, the general solution is

$$y\,(\text{IF}) = \int Q\,(\text{IF})\, dx + C$$

ie, $$yx = \int (\log x)\, x\, dx + C$$

ie, $$yx = (\log x) \cdot \frac{x^2}{2} - \int \frac{1}{x} \cdot \frac{x^2}{2} + C$$

ie, $$yx = \frac{x^2}{2} (\log x) - \frac{x^2}{4} + C$$

Illustration 49 Solve $\frac{dy}{dx} = \frac{y}{2y \ln y + y - x}$.

Solution. The equation can be written as

$$\frac{dx}{dy} = \frac{2y \ln y + y - x}{y} = (2 \ln y + 1) - \frac{x}{y}$$

ie, $$\frac{dx}{dy} + \frac{1}{y} \cdot x = (2 \ln y + 1)$$

In this equation it is clear that $P = \frac{1}{y}$ and $Q = (2 \ln y + 1)$.

Which are function of y only because equation contains derivatives of x with respect to y.

$$\text{IF} = e^{\int P dy} = e^{\int 1/y\, dy} = e^{\ln y} = y$$

$\therefore$ The solution is; $x\,(\text{IF}) = \int (2 \ln y + 1)(\text{IF})\, dy$

ie, $$xy = \int (2 \ln y + 1) \cdot y\, dy = y^2 \ln y + C$$

ie, $$x = y \ln y + \frac{C}{y}$$

Points to Consider

In some cases a linear differential equation may be of the form $\frac{dx}{dy} + P_1 x = Q_1$, where P_1 and Q_1 are function of y alone. In such a case the integrating factor is $e^{\int P_1 dy}$.

Illustration 50 Solve $\cos^2 x \frac{dy}{dx} - y \tan 2x = \cos^4 x$, where $|x| < \frac{\pi}{4}$ and $y\left(\frac{\pi}{6}\right) = \frac{3\sqrt{3}}{8}$.

Solution. The given equation can be written as

$$\frac{dy}{dx} - \sec^2 x \cdot \tan 2x \cdot y = \cos^2 x$$

$$\text{IF} = e^{\int -\tan 2x \cdot \sec^2 x\, dx} = e^{\int + \frac{2\tan x}{\tan^2 x - 1}\sec^2 x\, dx}$$

$$= e^{\int \frac{dt}{t}}, \text{ where } t = \tan^2 x - 1$$

$$= e^{\ln|t|} = |t| = |\tan^2 x - 1|$$

It is given that $|x| < \dfrac{\pi}{4}$ and for this region $\tan^2 x < 1$.

$\therefore$ $$\text{IF} = (1 - \tan^2 x)$$

$\therefore$ The solution is

$$y(1 - \tan^2 x) = \int \cos^2 x\,(1 - \tan^2 x)\, dx = \int (\cos^2 x - \sin^2 x)\, dx$$

$$= \int (\cos 2x)\, dx = \frac{\sin 2x}{2} + C$$

Now, when $x = \dfrac{\pi}{6}$, $y = \dfrac{3\sqrt{3}}{8}$

$\therefore$ $$\frac{3\sqrt{3}}{8}\left(1 - \frac{1}{3}\right) = \frac{1}{2} \cdot \frac{\sqrt{3}}{2} + C \Rightarrow C = 0$$

$\therefore$ $$y = \frac{\sin 2x}{2(1 - \tan^2 x)}$$

Points to Consider

If interval x is not given, then conventionally we write $\int \frac{1}{x}\, dx = \log(x) + C$, otherwise $\log|x|$.

Illustration 51 Solve $\dfrac{dy}{dx} + y\,\phi'(x) = \phi(x) \cdot \phi'(x)$, where $\phi(x)$ is a given function.

Solution. Here, $P = \phi'(x)$ and $Q = \phi(x)\,\phi'(x)$

$$\text{IF} = e^{\int \phi'(x)\, dx} = e^{\phi(x)}$$

$\therefore$ The solution is

$$y\, e^{\phi(x)} = \int \phi(x) \cdot \phi'(x) \cdot e^{\phi(x)}\, dx = \int t \cdot e^t\, dt, \quad \text{where } \phi(x) = t$$

$$y\, e^{\phi(x)} = e^t (t - 1) + C$$

ie, $$y e^{\phi(x)} = \{\phi(x) - 1\}\, e^{\phi(x)} + C$$

Bernoulli's Equation

Sometimes a differential equation is not linear but it can be converted into a linear differential equation by a suitable substitution.

An equation of the form

$$\frac{dy}{dx} + Py = Q\,y^n. \qquad (n \neq 0, 1)$$

Where P and Q are functions of x only, is known as Bernoulli's equation (for $n = 0$ the equation is linear.)

It is easy to reduce the above equation into linear form as below :
Dividing both the sides by y^n, we get

$$y^{-n}\frac{dy}{dx}+P\,y^{1-n}=Q$$

Putting $y^{1-n}=z$ and hence, $(1-n)\,y^{-n}\dfrac{dy}{dx}=\dfrac{dz}{dx}$ the equation becomes $\dfrac{dz}{dx}+(1-n)\,Pz=(1-n)\,Q$ which is linear in z.

Here, $\qquad \text{IF}=e^{\int(1-n)\,P\,dx}$

$\therefore$ The solution is, $z\,e^{\int(1-n)\,Pdx}=\int\{(1-n)\cdot Q\cdot e^{\int(1-n)\,Pdx}\}\,dx$

Illustration 52 Solve $(y\log x-1)\,y\,dx=x\,dy$.

Solution. The given differential equation can be written as

$$x\frac{dy}{dx}+y=y^2\log x \qquad \text{...(i)}$$

Dividing by xy^2, hence

$$\frac{1}{y^2}\cdot\frac{dy}{dx}+\frac{1}{xy}=\frac{1}{x}\log x$$

Let $\qquad \dfrac{1}{y}=v\Rightarrow-\dfrac{1}{y^2}\cdot\dfrac{dy}{dx}=\dfrac{dv}{dx}$

So that $\qquad \dfrac{dv}{dx}-\dfrac{1}{x}v=-\dfrac{1}{x}\log x$

Which is the standard linear differential equations, with

$$P=-\frac{1}{x},\ Q=-\frac{1}{x}\log x$$

$$\text{IF}=e^{\int-1/x\,dx}=e^{-\ln x}=e^{\ln x^{-1}}=\frac{1}{x}$$

The solution is given by

$$v\cdot\frac{1}{x}=\int\frac{1}{x}\left(-\frac{1}{x}\log x\right)dx=-\int\frac{\log x}{x^2}\,dx$$

$$=\frac{\log x}{x}-\int\frac{1}{x}\cdot\frac{1}{x}\,dx=\frac{\log x}{x}+\frac{1}{x}+C$$

$\Rightarrow \qquad v=1+\log x+Cx=\log(ex)+Cx$

or $\qquad \dfrac{1}{y}=\log(ex)+Cx \quad$ or $\quad y\{\log(ex)+Cx\}=1$

Illustration 53 Solve $\dfrac{dy}{dx}+xy=xy^2$.

Solution. Dividing by y^2, we get

$$y^{-2}\frac{dy}{dx}+\frac{1}{y}\cdot x=x$$

Let $\qquad \dfrac{1}{y}=z$

So that $$-\frac{1}{y^2}\cdot\frac{dy}{dx}=\frac{dz}{dx}$$

$\therefore$ The given equation reduces to

$$\frac{dz}{dx}-xz=-x$$

$$\text{IF}=e^{\int -x\,dx}=e^{-x^2/2}$$

$\therefore$ The solution is

$$z\,e^{-x^2/2}=\int -x\cdot e^{-x^2/2}\,dx=e^{-x^2/2}+C$$

ie, $$\frac{1}{y}=1+Ce^{x^2/2}$$

Illustration 54 Solve $\dfrac{dy}{dx}=\dfrac{y\,\phi'(x)-y^2}{\phi(x)}$, where $\phi(x)$ is a given function.

Solution. The equation can be written as

$$\frac{dy}{dx}-\frac{\phi'(x)}{\phi(x)}y=-\frac{y^2}{\phi(x)}$$

ie, $$-y^{-2}\frac{dy}{dx}+\frac{\phi'(x)}{\phi(x)}\cdot\frac{1}{y}=\frac{1}{\phi(x)}$$

Let $\dfrac{1}{y}=z$. So that, $$-\frac{1}{y^2}\frac{dy}{dx}=\frac{dz}{dx}$$

$\therefore$ $$\frac{dz}{dx}+\frac{\phi'(x)}{\phi(x)}\cdot z=\frac{1}{\phi(x)}$$

$$\text{IF}=e^{\int\frac{\phi'(x)}{\phi(x)}dx}=e^{\ln\phi(x)}=\phi(x)$$

$\therefore$ The solution is $$z\cdot\phi(x)=\int\frac{1}{\phi(x)}\cdot\phi(x)\,dx=x+C$$

ie, $$\frac{\phi(x)}{y}=x+C\quad ie,\ \frac{\phi(x)}{x+C}=y$$

Points to Consider

Another type of equation which is reducible to the linear form is

$$f'(y)\frac{dy}{dx}+P(x)\cdot f(y)=Q(x)$$

An equation of this type can be easily reduced the linear form by taking $z=f(y)$.

Illustration 55 Solve $\sec^2 y\dfrac{dy}{dx}+2x\tan y=x^3$.

Solution. Let $\tan y=z$ so that $\sec^2 y\cdot\dfrac{dy}{dx}=\dfrac{dz}{dx}$

Thus, the given equation reduces to

$$\frac{dz}{dx}+2x\cdot z=x^3$$

$$\text{IF} = e^{\int 2x\,dx} = e^{x^2}$$

$\therefore$ The solution is, $\quad z \cdot e^{x^2} = \int x^3 \cdot e^{x^2}\, dx$

ie, $\quad \tan y \cdot e^{x^2} = \frac{1}{2}\int x^2 e^{x^2} \cdot (2x)\, dx = \frac{1}{2}\int t \cdot e^t\, dt, \qquad$ where $t = x^2$

$$\tan y \cdot e^{x^2} = \frac{1}{2}(t \cdot e^t - e^t) + C$$

ie, $\quad \tan y = Ce^{-x^2} + \frac{e^{-x^2}}{2} \cdot e^{x^2}(x^2 - 1)$

$$\tan y = Ce^{-x^2} + \frac{1}{2}(x^2 - 1)$$

Illustration 56 Solve $\frac{dy}{dx} + x(x + y) = x^3(x + y)^3 - 1.$

Solution. The given equation can be written as

$$\left(\frac{dy}{dx} + 1\right) + x(x + y) = x^3(x + y)^3$$

ie, $\quad \frac{d(x + y)}{dx} + x(x + y) = x^3(x + y)^3$

ie, $\quad (x + y)^{-3} \cdot \frac{d(x + y)}{dx} + x(x + y)^{-2} = x^3$

Let $(x + y)^{-2} = z$ so that $-2(x + y)^{-3}\frac{d(x + y)}{dx} = \frac{dz}{dx}$

The given equation reduces to

$$-\frac{1}{2}\frac{dz}{dx} + xz = x^3 \quad ie, \quad \frac{dz}{dx} - 2xz = -2x^3$$

$$\text{IF} = e^{\int -2x\,dx} = e^{-x^2}$$

$\therefore$ The solution is

$$z \cdot e^{-x^2} = \int -2x^3 \cdot e^{-x^2}\, dx = (x^2 + 1)e^{-x^2} + C$$

$$\frac{1}{(x + y)^2} = Ce^{x^2} + x^2 + 1$$

Illustration 57 Solve $\sin y \cdot \frac{dy}{dx} = \cos y(1 - x\cos y).$

Solution. The given differential equation is

$$\sin y\frac{dy}{dx} = \cos y(1 - x\cos y)$$

or $\quad \sin y\frac{dy}{dx} - \cos y = -x\cos^2 y$

Dividing by $\cos^2 y$, we get

$$\tan y \cdot \sec y \cdot \frac{dy}{dx} - \sec y = -x$$

Let $\quad \sec y = v \Rightarrow \sec y \tan y \cdot \frac{dy}{dx} = \frac{dv}{dx}$

So that $\quad \frac{dv}{dx} - v = -x$

Which is linear differential equation with $P = -1, Q = -x$

$$\text{IF} = e^{\int P dx} = e^{\int -1\, dx} = e^{-x}$$

The solution is given by

$$v \cdot e^{-x} = \int -x \cdot e^{-x}\, dx = xe^{-x} + e^{-x} + C = e^{-x}(x+1) + C$$

or
$$v = (1+x) + Ce^{x}$$

or
$$\sec y = (1+x) + Ce^{x}$$

Orthogonal Trajectory

Any curve, which cuts every member of a given family of curves at right angles, is called an orthogonal trajectory of the family. For example, each straight line passing through the origin, *ie*, $y = kx$ is an orthogonal trajectory of the family of the circles $x^2 + y^2 = a^2$.

Procedure for Finding the Orthogonal Trajectory

(i) Let $f(x, y, c) = 0$ be the equation of the given family of curves, where c is an arbitrary parameter.

(ii) Differentiate $f = 0$; w.r.t. 'x' and eliminate 'c', *ie*, form a differential equation.

(iii) Substitute $-\frac{dx}{dy}$ for $\frac{dy}{dx}$ in the above differential equation. This will give the differential equation of the orthogonal trajectories.

(iv) By solving this differential equation, we get the required orthogonal trajectories.

Illustration 58 Find the orthogonal trajectories of the hyperbola $xy = C$.

Solution. The equation of the given family of curves is $xy = c$ ...(i)

Differentiating Eq. (i) w.r.t. x, we get

$$\frac{x\, dy}{dx} + y = 0 \quad \text{...(ii)}$$

Substitute $-\frac{dx}{dy}$ for $\frac{dy}{dx}$ in Eq. (ii), we get

$$-\frac{x\, dx}{dy} + y = 0 \quad \text{...(iii)}$$

This is the differential equation for the orthogonal trajectory of given family of hyperbola.

Eq. (iii) can be rewritten as $x\, dx = y\, dy$, which on integration gives

$$x^2 - y^2 = C.$$

This is the family of required orthogonal trajectories.

Illustration 59 Find the orthogonal trajectories of the curves $y = cx^2$.

Solution. Here, $y = cx^2$ ...(i)

Differentiating w.r.t. x, we get

$$\frac{dy}{dx} = 2cx \qquad \text{...(ii)}$$

Eliminating c from Eqs. (i) and (ii),

$$y = \left(\frac{1}{2x} \frac{dy}{dx} \right) x^2$$

$$\Rightarrow \qquad 2y = x \frac{dy}{dx} \qquad \text{...(iii)}$$

This is the differential equation of the family of curves given in Eq. (i).

Now, to obtain orthogonal trajectory replace $\frac{dy}{dx}$ by $-\frac{dx}{dy}$ in Eq. (iii).

$$\Rightarrow \qquad 2y = -x \frac{dx}{dy}$$

or $\qquad 2y\,dy = -x\,dx$

Integrating both the sides, we get

$$y^2 = -\frac{x^2}{2} + C_1$$

$\Rightarrow \qquad x^2 + 2y^2 = C_1$, is the required family of orthogonal trajectory.

Illustration 60 Find the equation of all possible curves that will cut each member of the family of circles $x^2 + y^2 - 2cx = 0$ at right angle.

Solution. Here, $x^2 + y^2 - 2cx = 0$ ...(i)

Differentiating w.r.t. x, we get

$$2x + 2yy_1 - 2c = 0$$

$$\Rightarrow \qquad c = x + yy_1 \qquad \text{...(ii)}$$

From Eqs. (i) and (ii), we eliminate c

$$\Rightarrow \qquad x^2 + y^2 - 2(x + yy_1)x = 0$$

or

$$-x^2 + y^2 - 2xy\frac{dy}{dx} = 0$$

This is the differential equation representing the given family of circles. To find differential equation of the orthogonal trajectories, we replace $\frac{dy}{dx}$ by $-\frac{dx}{dy}$.

$$\Rightarrow \qquad y^2 - x^2 = -2xy\frac{dx}{dy}$$

$$\Rightarrow \qquad y^2\,dy = x^2\,dy - 2xy\,dx$$

$$\Rightarrow \qquad -dy = \frac{y\,d(x^2) - x^2\,dy}{y^2} \Rightarrow -dy = d\left(\frac{x^2}{y}\right)$$

Integrating both the sides, we get

$$-y = \frac{x^2}{y} + C \Rightarrow x^2 + y^2 + Cy = 0$$

Represents family of orthogonal trajectory.

Illustration 61 Find the orthogonal trajectory of the circles

$$x^2 + y^2 - ay = 0.$$

Solution. Here, $$x^2 + y^2 - ay = 0 \quad \text{...(i)}$$

Differentiating, we get

$$2x + 2yy_1 - ay_1 = 0$$

$$\Rightarrow \quad a = \frac{2(x + yy_1)}{y_1} \quad \text{...(ii)}$$

Substituting '*a*' in Eq. (i), we get

$$x^2 + y^2 - \frac{2(x + yy_1)}{y_1} y = 0$$

$$\Rightarrow \quad (x^2 - y^2) y_1 - 2xy = 0$$

This is the differential equation of the family of circles given in Eq. (i).

$\therefore$ The differential representing the orthogonal trajectory is obtained by replacing $\frac{dy}{dx}$ by $-\frac{dy}{dx}$.

ie, $$-(x^2 - y^2)\frac{dx}{dy} - 2xy = 0$$

$$\Rightarrow \quad 2xy\,dy - y^2\,dx = -x^2\,dx$$

$$\Rightarrow \quad \frac{x\,d(y^2) - y^2\,dx}{x^2} = -dx$$

$$\Rightarrow \quad d\left(\frac{y^2}{x}\right) = -dx$$

Integrating both the sides, we get

$y^2 + x^2 = Cx$, is required family of orthogonal trajectories.

Target Exercise 4.4

1. The solution of $(1 + x^2)\frac{dy}{dx} + y = e^{\tan^{-1} x}$, is given by
   (a) $2ye^{\tan^{-1} x} = e^{2\tan^{-1} x} + C$
   (b) $ye^{\tan^{-1} y} = e^{2\tan^{-1} x} + C$
   (c) $2ye^{\tan^{-1} y} = e^{2\tan^{-1} y} + C$
   (d) None of these
2. The solution of $\frac{dy}{dx} + \frac{x}{1 - x^2} y = x\sqrt{y}$, is given by
   (a) $3\sqrt{y} + (1 - x^2) = C(1 - x^2)^{1/4}$
   (b) $\frac{3}{2}\sqrt{y} + (1 - x^2) = C(1 - x^2)^{3/2}$
   (c) $3\sqrt{y} - (1 - x^2) = C(1 - x^2)^{3/2}$
   (d) None of these
3. The solution of $\frac{dy}{dx} + x \sin 2y = x^3 \cos^2 y$, is
   (a) $e^{x^2} = (x^2 - 1)e^{x^2} \tan y + C$
   (b) $e^{x^2} \tan y = \frac{1}{2}(x^2 - 1)e^{x^2} + C$
   (c) $e^{x^2} \tan y = (x^2 - 1) \tan y + C$
   (d) None of these

4. The solution of $3x(1-x^2)y^2\, dy/dx + (2x^2-1)y^3 = ax^3$ is

(a) $y^3 = ax + C\sqrt{1-x^2}$ (b) $y^3 = ax + Cx\sqrt{1-x^2}$

(c) $y^2 = ax + C\sqrt{1-x^2}$ (d) None of these

5. The solution of $\frac{dy}{dx} + \frac{y}{x}\log y = \frac{y}{x^2}(\log y)^2$, is

(a) $x = \frac{1}{2x}\log y + C$ (b) $x^2 + \log y = C$

(c) $\frac{1}{x\log y} = \frac{1}{2x^2} + C$ (d) None of these

Differential Equation of First Order and Higher Degrees

A differential equation of first order is of the form $f(x, y, P)$ where $P = dy/dx$. If in the equation degree of P is greater than one, then the equation is of first order and higher degree.

The differential equation of first order and higher degree can be written in the form

$$P^n + F_1(x, y)P^{n-1} + \ldots + F_{n-1}(x, y)P + F_n(x, y) = 0$$

The differential equations of these category can be solved by one or more of the following methods :

(i) Equations solvable for P.
(ii) Equations solvable for y.
(iii) Equations solvable for x.
(iv) Clairaut's equations.

Now, we shall discuss these cases.

(i) Equations Solvable For *P*

If the equation

$$P^n + F_1(x, y)P^{n-1} + \ldots + F_{n-1}(x, y)P + F_n(x, y) = 0,$$

is solvable for P, then LHS expression can be resolved into n linear factors and hence can be put in the form

$$(P - f_1(x, y))(P - f_2(x, y))\ldots(P - f_n(x, y)) = 0.$$

Equating each of these factors to zero, we get n differential equations of the first order and first degree.

$$\frac{dy}{dx} = f_1(x, y), \frac{dy}{dx} = f_2(x, y), \ldots, \frac{dy}{dx} = f_n(x, y)$$

Let the solutions of these obtained equations are

$$\phi_1(x, y, c_1) = 0, \phi_2(x, y, c_2) = 0, \ldots, \phi_n(x, y, c_n) = 0 \text{ respectively.}$$

Hence, the general solution is given by

$$\phi_1(x, y, c), \phi_2(x, y, c), \ldots, \phi_n(x, y, c) = 0$$

Here, the arbitrary constant $c_1, c_2, \ldots, c_n$ are replaced by a single arbitrary constant c because every first order equation has only one arbitrary constant in its solution.

Illustration 62 Solve $(p-x)(p-e^x)(p-1/y)=0$; where $p=\frac{dy}{dx}$.

Solution. The component linear equations are

$$p=x,\ p=e^x,\ p=\frac{1}{y}$$

If $\frac{dy}{dx}=x$, then $dy=x\,dx \Rightarrow y=\frac{x^2}{2}+C_1$

If $\frac{dy}{dx}=e^x$, then $dy=e^x\,dx \Rightarrow y=e^x+C_2$

If $\frac{dy}{dx}=\frac{1}{y}$, then $y\,dy=dx \Rightarrow \frac{y^2}{2}=x+C_3$

$\therefore$ The required solution is

$$\left(y-\frac{x^2}{2}+C\right)(y-e^x+C)\left(\frac{y^2}{2}-x+C\right)=0$$

Illustration 63 Solve $x^2p^2+xyp-6y^2=0$.

Solution. The given equation is

$$x^2p^2+xyp-6y^2=0$$

Solving as a quadratic in p, we get

$$p=\frac{(-xy\pm\sqrt{x^2y^2+24x^2y^2})}{2x^2}=\frac{2y}{x},\ -\frac{3y}{x}$$

If $p=\frac{2y}{x}$, then $\frac{dy}{dx}=\frac{2y}{x}\Rightarrow\frac{dy}{y}=\frac{2dx}{x}$

$$\Rightarrow \qquad \log\left|\frac{y}{x^2}\right|=k\Rightarrow y=C_1x^2$$

If $p=-\frac{3y}{x}$, then $\frac{dy}{dx}=-\frac{3y}{x}\Rightarrow\frac{dy}{y}=-\frac{3dx}{x}$

$$\Rightarrow \qquad x^3y=C_2$$

$\therefore$ The required solution is $(y-Cx^2)(x^3y-C)=0$.

Illustration 64 Solve $xy^2(p^2+2)=2py^3+x^3$.

Solution. The given equation can be written as

$$(xy^2p^2-x^3)+2(xy^2-py^3)=0$$

$$\Rightarrow \qquad x(y^2p^2-x^2)+2y^2(x-py)=0$$

$$\Rightarrow \qquad (py-x)\{x(py+x)-2y^2\}=0$$

If $py-x=0$, then $y\,dy-x\,dx=0\Rightarrow y^2-x^2=C_1$

If $xyp+x^2-2y^2=0$, then $2y\frac{dy}{dx}-\frac{4y^2}{x}=-2x$

$$\Rightarrow \quad \frac{dt}{dx} - \frac{4}{x} t = -2x, \qquad \text{where } t = y^2$$

$$\text{IF} = e^{-\int \frac{4}{x} dx} = e^{-4 \ln x} = \frac{1}{x^4}$$

Its solution is $t\left(\frac{1}{x^4}\right) = \int -2x \cdot \frac{1}{x^4} dx$

ie, $$\frac{t}{x^4} = \frac{1}{x^2} + C_2 \quad ie, \quad y^2 = x^2 + C_2 x^4$$

Hence, the required solution is

$$(y^2 - x^2 - C)(y^2 - x^2 - Cx^4) = 0$$

(ii) Equations Solvable For *y*

Equation that comes under this category, can be expressed in the form

$y = g(x, p)$ (*ie*, an explicit function y in term of x and p) ...(i)

Differentiating Eq. (i) w.r.t. x, we get

$$\frac{dy}{dx} = p = F\left(x, p, \frac{dp}{dx}\right)$$

Which is a differential equation of the first order containing x and p. Let us suppose that its solution is $\phi(x, p, c) = 0$ ...(ii)

Then, the solution is obtained by eliminating p between $y = g(x, p)$ and $\phi(x, p, c) = 0$. However, if eliminating of p is difficult express x and y as a function of the parameter p.

Illustration 65 Solve $xp^2 - 2yp + ax = 0$.

Solution. The given equation can be written as,

$$y = \frac{xp}{2} + \frac{ax}{2p} \qquad \text{...(i)}$$

$$\frac{dy}{dx} = \frac{p}{2} + \frac{x}{2} \cdot \frac{dp}{dx} + \frac{a}{2p} - \frac{ax}{2p^2} \cdot \frac{dp}{dx}$$

$$\Rightarrow \quad p(p^2 - a) = x(p^2 - a) \cdot \frac{dp}{dx}$$

$$\Rightarrow \quad \frac{dp}{dx} = \frac{p}{x} \Rightarrow p = Cx$$

(The equation $p^2 - a = 0$ gives us singular solution in which we are not interested).

The substitute p in Eq. (i), we get the required solution

$$2y = Cx^2 + \frac{a}{C}$$

Illustration 66 Solve $y = 2px - p^2$.

Solution. Differentiating the given equation w.r.t. x, we get

$$\frac{dy}{dx} = 2p + 2x\frac{dp}{dx} - 2p\frac{dp}{dx}$$

or $$2(x-p)\frac{dp}{dx}+p=0$$

or $$\frac{dx}{dp}+\frac{2}{p}x=2$$

It is a linear equation in x and p.

$$\text{IF}=e^{\int \frac{2}{p}dx}=e^{2\log p}=p^2$$

$\therefore$ The solution is $$xp^2=\int p^2\cdot 2\,dp=\frac{2}{3}p^3+C$$

Thus, the solution of the given equation is

$$x=\frac{2}{3}p+Cp^{-2}\text{, where } p \text{ is parameter.}$$

(iii) Equations Solvable For x

This type of equation can be put in the form

$$x=g(y,p) \quad \text{...(i)}$$

Differentiating w.r.t. y, we get

$$\frac{1}{p}=\frac{dx}{dy}=G\left(x,p,\frac{dp}{dx}\right)$$

which is a differential equation of 1st order containing y and p and its solution is

$$\phi(y,p,c)=0$$

Then, the solution is obtained by eliminating p between $x=g(y,p)$ and $\phi(y,p,c)=0$. However, if eliminating of p is difficult express x and y as a function of the parameter p.

Illustration 67 Solve $y=2px+y^2p^3$.

Solution. Solving for x, we get

$$x=\frac{y}{2p}-\frac{y^2p^2}{2} \quad \text{...(i)}$$

Differentiating Eq. (i) w.r.t. y, we get

$$\frac{dx}{dy}=\frac{1}{2p}-\frac{y}{2p^2}\cdot\frac{dp}{dy}-yp^2-y^2p\cdot\frac{dp}{dy}$$

or $$\frac{1}{p}-\frac{1}{2p}+yp^2=-y\left(\frac{1}{2p^2}+yp\right)\cdot\frac{dp}{dy}$$

or $$(1+2yp^3)\,p=-y(1+2p^3y)\cdot\frac{dp}{dy}$$

or $$\frac{dp}{p}+\frac{dy}{y}=0\Rightarrow py=C\Rightarrow p=\frac{C}{y}$$

Substituting this in the Eq. (i), we get

$$x=\frac{y^2}{2C}-\frac{C^2}{2}\quad\Rightarrow\quad y^2=2Cx+C^2$$

(iv) Clairaut's Equation

The differential equation $y = px + f(p)$ is known as Clairaut's equation. The solution of equation of this type is given by $y = cx + f(c)$.

Which is obtained by replacing p by c in the given equation.

Points to Consider

Some equations can be reduced to Clairaut's form by suitable substitution.

Illustration 68 Solve $y = px + \dfrac{p}{\sqrt{1+p^2}}$.

Solution. Its solution is, $y = cx + \dfrac{c}{\sqrt{1+c^2}}$

Illustration 69 Solve $\sqrt{1+p^2} = \tan(px - y)$.

Solution. The given equation is

$$\sqrt{1+p^2} = \tan(px - y) \quad \text{or} \quad px - y = \tan^{-1}(\sqrt{1+p^2})$$

or $$y = px - \tan^{-1}(\sqrt{1+p^2})$$

Its solution is, $$y = cx - \tan^{-1}(\sqrt{1+c^2})$$

Illustration 70 Solve $y^2 \log y = pxy + p^2$.

Solution. Let $\log y = t$. Then $\dfrac{1}{y}\dfrac{dy}{dx} = \dfrac{dt}{dx}$

So, if $\dfrac{dt}{dx} = p$, then $\dfrac{p}{y} = p$

Substituting these in the given equation, we have

$$y^2 t = y \cdot p\,xy + p^2 y^2 \quad \text{or} \quad t = px + p^2$$

Which is in Clairaut's form.

Thus, the required solution is

$$t = cx + c^2$$

or $$\log y = cx + c^2$$ (c being an arbitrary constant.)

Application of Differential Equations

Differential Equation of First Order But not of First Degree

1. The most general form of a first order and higher degree differential equation is $p^n + P_1 p^{n-1} + P_2 p^{n-2} + \ldots\ldots + P_n = 0$ where $P_1, P_2, \ldots\ldots, P_n$ are function of x, y and $p = dy/dx$. If a 1st order any degree equation can be resolved into differential equation (involving p) of first degree and 1st order, in such case we say that the equation is solvable for p.

Let their solution be $g_1(x, y, c_1) \times g_2(x, y, c_2) \times \ldots \times g_n(x, y, c_n) = 0$, (where c_1, c_2, ..., c_n, are arbitrary constant) we take $c_1 = c_2 = \ldots\ldots = c_n = c$ because the differential equation of 1st order 1st degree contain only one arbitrary constant. So solution is

$$g_1(x, y, c) \times g_2(x, y, c) \times \ldots \times g_n(x, y, c) = 0$$

2. The most general form of a first order and higher degree differential equation is $p^n + P_1\, p^{n-1} + P_2\, p^{n-2} + \ldots + P_n = 0$, where P_1, P_2, ..., P_n are function of x, y and $p = dy/dx$. If differential equation is expressible in the form $y = f(x, p)$, then

Step 1 Differentiate w.r.t. x, we get p or $\frac{dy}{dx} = f\left(x, p, \frac{dp}{dx}\right)$.

Step 2 Solving this we obtain $\phi(x, p, c) = 0$.

Step 3 The solution of differential equation is obtained by eliminating p.

Application of First Order Differential Equations

Growth and Decay Problems

Let $N(t)$ denotes the amount of substance (or population) that is either growing or decaying. If we assume that dN/dt, the time rate of change of this amount of substance, is proportional to the amount of substance present then

$$\frac{dN}{dt} = kN \quad \text{or} \quad \frac{dN}{dt} - kN = 0 \qquad \ldots\text{(i)}$$

Where k is the constant of proportionality. We are assuming that $N(t)$ is a differentiable, hence continuous, function of time.

Illustration 71 The population of a certain country is known to increase at a rate proportional to the number of people presently living in the country. If after two years the population has doubled and after three year the population is 20000, estimate the number of people initially living in the country.

Solution. Let N denotes the number of people living in the country at any time t, and let N_0 denote the number of people initially living in the country. Then, from Eq. (i)

$$\frac{dN}{dt} - kN = 0$$

Which has the solution $N = Ce^{kt}$ ...(i

At $t = 0$, $N = N_0$; hence, it follows from Eq. (i) that $N_0 = Ce^{k(0)}$ or that $C = N_0$.

Thus, $$N = N_0 e^{kt} \qquad \ldots\text{(ii)}$$

At $t = 2$, $N = 2N_0$. Substituting these values into Eq. (ii), we have

$$2N_0 = N_0 e^{2k} \text{ from which } k = \frac{1}{2}\ln 2 = 0.347$$

Substituting this value into Eq. (i) gives

$$N = N_0 e^{(0.347)t} \quad \text{...(iii)}$$

At $t = 3$, $N = 20000$. Substituting these values into Eq. (iii), we obtain

$$20000 = N_0 e^{(0.347)(3)}$$

Illustration 72 A certain radioactive material is known to decay at a rate proportional to the amount present. If initially there is 50 mg of the material present and after two hours it is observed that the material has lost 10% of its original mass, find (a) and expression for the mass of the material remaining at any time t, (b) the mass of the material after four hours, and (c) the time at which the material has decayed to one half of its initial mass.

Solution. (a) Let N denotes the amount of material present at time t. Then, from Eq. (i)

$$\frac{dN}{dt} - kN = 0$$

This differential equation is separable and linear, its solution is

$$N = Ce^{kt} \quad \text{...(i)}$$

At $t = 0$, we are given that $N = 50$. Therefore, from Eq. (i), $50 = Ce^{k(0)}$ or $C = 50$.

Thus, $$N = 50\,e^{kt} \quad \text{...(ii)}$$

At $t = 2$, 10% of the original mass of 50 mg or 5 mg has decayed. Hence, at $t = 2$, $N = 50 - 5 = 45$. Substituting these values into Eq. (ii) and solving for k, we have

$$45 = 50e^{2k} \quad \text{or} \quad k = \frac{1}{2}\ln\frac{45}{50} = -0.053$$

Substituting this value into Eq. (ii), we obtain the amount of mass present at any time t as

$$N = 50e^{-0.053\,t} \quad \text{...(iii)}$$

Where t is measured in hours.

(b) We require N at $t = 4$. Substituting $t = 4$ into Eq. (iii) and then solving for N, we find $N = 50e^{(-0.053)(4)}$

(c) We require t when $N = 50/2 = 25$. Substituting $N = 25$ into Eq. (iii) and solving for t, we find

$$25 = 50e^{-0.053t} \quad \text{or} \quad -0.053\,t = \ln\frac{1}{2} \quad \text{or} \quad t = 13 \text{ h}$$

Illustration 73 Five mice in a stable population of 500 are intentionally infected with a contagious disease to test a theory of epidemic spread that postulates the rate of change in the infected population is proportional to the product of the number of mice who have the disease with the number that are disease free.

Assuming the theory is correct, how long will it take half the population to contract the disease?

Solution. Let $N(t)$ denotes the number of mice with the disease at time t. We are given that $N(0) = 5$, and it follows that $500 - N(t)$ is the number of mice without the disease at time t. The theory predicts that

$$\frac{dN}{dt} = kN(500 - N) \qquad ...(i)$$

Where k is a constant of proportionality. This equation is different from Eq. (i) because the rate of change is no longer proportional to just the number of mice who have the disease. Eq. (i) has the differential form

$$\frac{dN}{N(500 - N)} - kdt = 0 \qquad ...(ii)$$

Which is separable. Using partial fraction decomposition, we have

$$\frac{1}{N(500 - N)} = \frac{1/500}{N} + \frac{1/500}{500 - N}$$

Hence, Eq. (ii) may be rewritten as $\frac{1}{500}\left(\frac{1}{N} + \frac{1}{500 - N}\right) dN - kdt = 0$

It solution is $\frac{1}{500}\int\left(\frac{1}{N} + \frac{1}{500 - N}\right) dN - \int kdt = C$

or

$$\frac{1}{500}(\ln|N| - \ln|500 - N|) - kt = C$$

Which may be rewritten as $\ln\left|\frac{N}{500 - N}\right| = 500(C + kt)$

$$\frac{N}{500 - N} = e^{500(C + kt)} \qquad ...(iii)$$

But $e^{500(C+kt)} = e^{500}\ e^{kt}$. Setting $C_1 = e^{500C}$, we can write Eq. (iii) as

$$\frac{N}{500 - N} = C_1 e^{500kt} \qquad ...(iv)$$

At $t = 0$, $N = 5$. Substituting these values into Eq. (iv), we find

$$\frac{4}{495} = C_1\, e^{500k(0)} = C_1$$

So, $C_1 = 1/99$ and Eq. (iv) becomes $\frac{N}{500 - N} = \frac{1}{99}e^{500kt}$ ...(v)

We could solve Eq. (v) for N, but this is not necessary. We seek a value of when $N = 250$, one half the population. Substituting $N = 250$ into Eq. (v) and solving for t, we obtain

$$1 = \frac{1}{99}e^{500kt}; \qquad \ln 99 = 500kt$$

or $t = 0.0091/k$ time units. Without additional information, we cannot obtain a numerical value for the constant of proportionality k or be more definitive about t.

Temperature Problems

Newton's law of cooling, which is equally applicable to heating, states that the time rate of change of the temperature of a body is proportional to the temperature difference between the body and its surrounding medium. Let T denote the temperature of the body and let T_m denote the temperature of the surrounding medium. Then, the time rate of change temperature of the body is $\frac{dT}{dt}$ and

Newton's law of cooling can be formulated as

$$\frac{dT}{dt} = -k(T - T_m) \quad \text{or} \quad \text{as} \frac{dT}{dt} + kT = kT_m \qquad \text{...(ii)}$$

where k is a positive constant of proportionality. Once k is chosen positive, the minus sign is required in Newton's law to make $\frac{dT}{dt}$ negative in a cooling process, when T is greater than T_m and positive in a heating process, when T is less than T_m.

Illustration 74 A metal bar at a temperature of 100°F is placed in a room at a constant temperature of 0°F. If after 20 min the temperature of the bar is 50°F, find (a) the time it will take the bar to reach a temperature of 25°F and (b) the temperature of the bar after 10 min.

Solution. Use Eq. (ii) with $T_m = 0$; the medium here is the room which is being held at a constant temperature of 0°F. Thus, we have

$$\frac{dT}{dt} = KT = 0$$

whose solution is $\quad T = Ce^{-kt}$ ...(i)

Since, $T = 100$ at $t = 0$ (the temperature of the bar is initially 100°F), it follows Eq. (i) that $100 = Ce^{-k(0)}$ or $100 = C$. Substituting this value into Eq. (i), we obtain

$$t = 100\,e^{-kt} \qquad \text{...(ii)}$$

At $t = 20$, we are given that $T = 50$, hence from Eq. (ii),

$$50 = 100e^{-20k}$$

From which $k = -\frac{1}{20}\ln\frac{50}{100} = -\frac{1}{20(-0.693)} = 0.035$

Substituting this value into Eq. (ii), we obtain the temperature of the bar at any time t as

$$T = 100e^{-0.035t} \qquad \text{...(iii)}$$

(a) We require t when $T = 25$. Substituting $T = 25$ into Eq. (iii), we have

$$25 = 100e^{-0.035t} \quad \text{or} \quad -0.035t = \ln\frac{1}{4}$$

Solving, we find that $t = 39.6$ min

(b) We require T when $t = 10$. Substituting $t = 10$ into Eq. (iii) and then solving for T, we find that

$$T = 100e^{(-0.035)(10)} = 100\,(0.705) = 70.5°\text{F}$$

It should be noted that, since Newton's law is valid only for small temperature differences, the above calculations represent only a first approximation to the physical situation.

Illustration 75 A body at a temperature of 50°F is placed outdoors where the temperature is 100°F. If after 5 min the temperature of the body is 60°F, find (a) how long it will take the body to reach a temperature of 75°F and (b) the temperature of the body after 20 min.

Solution. Using Eq. (ii) with $T_m = 100$ (the surrounding medium is the outside air), we have

$$\frac{dT}{dt} + kT = 100k$$

This differential equation is linear $T = Ce^{-kt} + 100$ ...(i)

Since, $T = 50$ when $t = 0$, it follows from Eq. (i) that $50 = Ce^{-k(0)} + 100$ or $C = -50$. Substituting this value into Eq. (i), we obtain

$$t = -50e^{-kt} + 100 \quad \text{...(ii)}$$

At $t = 5$, we are given that $T = 60$; hence, from Eq. (ii), $60 = -50e^{-5k} + 100$. Solving for k, we obtain

$$-40 = -50e^{-5k}$$

or $$k = -\frac{1}{5}\ln\frac{40}{50} = -\frac{1}{5}(-0.223) = 0.045$$

Substituting this value into Eq. (ii), we obtain the temperature of the body at any time t as

$$T = -50e^{-0.045t} + 100 \quad \text{...(iii)}$$

(a) We require t when $T = 75$. Substituting $T = 75$ into Eq. (iii), we have

$$75 = -50e^{-0.045t} + 100$$

or $$e^{-0.045t} = \frac{1}{2}$$

Solving for t, we find $\quad -0.045t = \ln\frac{1}{2} \quad$ or $\quad t = 15.4$ min

(b) We require T when $t = 20$. Substituting $t = 20$ into Eq. (iii) and then solving for T, we find

$$T = -50e^{(-0.045)(20)} + 100$$

Illustration 76 A body at an temperature is placed in a room which is held at a constant temperature of 30°F. If after 10 min the temperature of the body is 0°F and after 20 min the temperature of the body is 15°F, find the unknown initial temperature.

Solution. From Eq. (ii), $\frac{dT}{dt} + kT = 30k$

Solving, we obtain $\quad T = Ce^{-kt} + 30$ ...(i)

At $t = 10$, we are given that $T = 0$, Hence, form Eq. (i)

$$0 = Ce^{-10k} + 30 \text{ or } Ce^{-100} = -30 \quad \text{...(ii)}$$

At $t = 20$, we are given that $T = 15$. Hence, from Eq. (i) again,

$$15 = Ce^{-20k} + 30 \text{ or } Ce^{-20} = -15 \quad \text{...(iii)}$$

Solving Eqs. (ii) and (iii) for k and C, we find

$$k = \frac{1}{10}\ln 2 = 0.069$$

and $$C = -30\,e^{10k} = -30(2) = -60$$

Substituting these values into Eq. (i), we have for the temperature of the body at any time t

$$T = -60e^{-0.069t} + 30 \quad \text{...(iv)}$$

Since, we require T at the initial time $t = 0$, it follows from Eq. (iv) that

$$T_0 = -60e^{(-0.069)(0)} + 30 = -60 + 30 = -30° \text{ F}$$

Dilution Problems

Consider a tank which intially holds V_0 gal of brine that contains a lb of salt. Another brine solution, containing b lb of salt per gallon, is poured into the tank at the rate of e gal/min while, simultaneously, the well-stirred solution leaves the tank at the rate of f gal/min. The problem is to find the amount of salt in the tank at any time t.

Let Q denote the amount of salt in the tank at any time. The time rate of change of Q, dQ/dt, equals the rate at which salt enters the tank minus the rate at which salt leaves the tank. Salt enters the tank at the rate of be lb/min. To determine the rate at which salt leaves the tank, we first calculate the volume of brine in the tank at any time t, which is the initial volume V_0 plus the volume of brine added et minus the volume of brine removed ft. Thus, the volume of brine at any time is

$$V_0 + et - ft \qquad \text{...(i)}$$

The concentration of salt in the tank at any time is $Q/(V_0 + et - ft)$, from which it follows that salt leaves the tank at the rate of $f\left(\frac{Q}{V_0 + et - ft}\right)$ lb/min

Thus,
$$\frac{dQ}{dt} = be - f\left(\frac{Q}{V_0 + et - ft}\right)$$

or
$$\frac{dQ}{dt} + \frac{f}{V_0 + (e - f)t} Q = be \qquad \text{...(ii)}$$

Illustration 77 A tank initially holds 100 gal of a brine solution containing 20 lb of salt. At $t = 0$, fresh water is poured into the tank at the rate of 5 gal/min, while the well-stirred mixture leaves the tank at the same rate. Find amount of salt in the tank at any time t.

Solution. Here, $V_0 = 100$, $a = 20$, $b = 0$ and $e = f = 5$, Eq. (iii)

$$\frac{dQ}{dt} + \frac{1}{20} Q = 0$$

The solution of this linear equation is $Q = Ce^{-t/20}$ ...(i)

At $t = 0$, we are given that $Q = a = 20$. Substituting these values into Eq. (i), we find that $C = 20$, so that Eq. (i) can be rewritten as $Q = 20e^{-t/20}$. Note that as $t \to \infty$, $Q \to 0$ as it should, since only fresh water is being added.

Illustration 78 A tank initially holds 100 gal of a brine solution containing 1 lb of salt. At $t = 0$ another brine solution containing 1 lb of salt per gallon is poured into the tank at the rate of 3 gal/min, while the well-stirred mixture leaves the tank at the same rate. Find (a) the amount of salt in the tank at any time t and (b) the time at which the mixture in the tank contains 2 lb of salt.

Solution. (a) Here, $V_0 = 100$, $a = 1$, $b = 1$ and $e = f = 3$

Hence, Eq. (iii)
$$\frac{dQ}{dt} + 0.03Q = 3$$

The solution to this linear differential equation is

$$Q = Ce^{-0.03t} + 100 \qquad \text{...(i)}$$

At $t = 0, Q = a = 1$.

Substituting these values into Eq. (i), we find $1 = Ce^0 + 100$,

or $$C = -99$$

Then, Eq. (i) can be rewritten as

$$Q = -99e^{-0.03t} + 100 \qquad \text{...(ii)}$$

(b) We require t when $Q = 2$. Substituting $Q = 2$ into Eq. (ii), we obtain

$$2 = -99e^{-0.03t} + 100 \text{ or } e^{-0.03t} = \frac{98}{99}$$

From which $t = -\dfrac{1}{0.03} \ln \dfrac{98}{99} = 0.338$ min

Illustration 79 A 50 gal tank initially contains 10 gal of fresh water. At $t = 0$, a brine solution containing 1 lb of salt per gallon is poured into the tank at the rate of 4 gal/min, while the well-stirred mixture leaves the tank at the rate of 2 gal/min. Find (a) the amount of time required for overflow to occure and (b) the amount of salt in the tank at the moment of overflow.

Solution. (a) Here, $a = 0$, $b = 1$, $e = 4$, $f = 2$ and $V_0 = 10$. The volume of brine in the tank at any time t is given in Eq. (i) as $V_0 + et - ft = 10 + 2t$. We require t when $10 + 2t = 50$, hence, $t = 20$ min.

(b) For the equation $\dfrac{dQ}{dt} + \dfrac{2}{10 + 2t} Q = 4$

This is a linear equation; its solution is

$$Q = \frac{40t + 4t^2 + C}{10 + 2t} \qquad \text{...(i)}$$

At $t = 0$, $Q = a = 0$. Substituting these values into Eq. (i), we find that $C = 0$

We require Q at the moment of overflow, which from part (a) is $t = 20$.

Thus, $$Q = \frac{40(20) + 4(40)^2}{10 + 2(20)} = 48 \text{ lb}$$

Falling Body Problems

Consider a vertically falling body of mass m that is being influenced only by gravity g and an air resistance that is proportional to the velocity of the body. Assume that both gravity and mass remain constant and, for convenience, choose the downward direction as the positive direction.

Newton's second law of motion The net force acting on a body is equal to the time rate of change of the momentum of the body; or for constant mass

$$F = m \frac{dv}{dt} \qquad \text{...(iii)}$$

where F is the net force on the body and v is the velocity of the body, both at time t.

For the problem at hand, there are two forces acting, on the body : (i) the force due to gravity given by the weight w of the body, which equals mg and (ii) the force due to air resistance given by $-kv$, where $k \geq 0$ is a constant of proportionality. The minus sign is required because this force opposes the velocity; that is, it acts the upwards, or negative direction. The net force F on the body is, therefore $F = mg - kv$, substituting this result into Eq. (iii)

$$mg - kv = m\frac{dv}{dt}$$

or
$$\frac{dv}{dt} + \frac{k}{m}v = g \qquad \text{...(i)}$$

As the equation of motion for the body. If air resistance is negligible or non-existent, then $k = 0$ and Eq. (i) simplifies to

$$\frac{dv}{dt} = g \qquad \text{...(ii)}$$

When $k > 0$, the limiting velocity v_1 is defined by

$$v_1 = \frac{mg}{k} \qquad \text{...(iii)}$$

Caution Eqs. (i), (ii) and (iii) are valid only if the given conditions are satisfied. These equations are not valid, if for example, air resistance is not proportional to velocity but to the velocity squared or if the upward direction is taken to be the positive direction.

Electrical Circuits

The basic equation governing the amount of current I (in amperes) in a simple RL circuit consisting of a resistance R (in ohms), an inductor L (in henries) and an electromotive force abbreviated emf) E (in volts) is

$$\frac{DI}{dt} + \frac{R}{L}I = \frac{E}{R} \qquad \text{...(i)}$$

For an RC circuit consisting of a resistance, a capacitance C (in farads), an emf, and no inductance, the equation governing the amount of electrical charge q (in coulombus) on the capacitor is

$$\frac{dq}{dt} + \frac{1}{RC}q = \frac{E}{R} \qquad \text{...(ii)}$$

The relationship between q and I is

$$I = \frac{dq}{dt} \qquad \text{...(iii)}$$

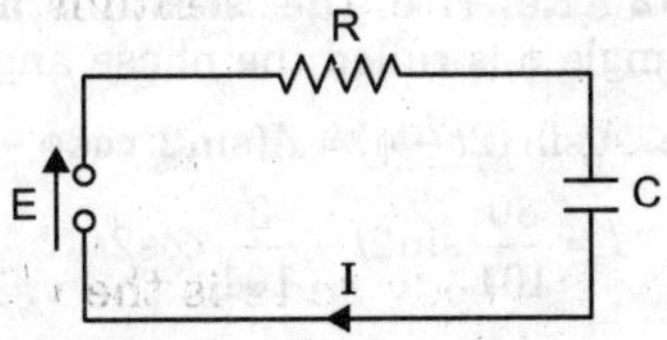

Illustration 80 An *RL* circuit has an emf of 5 volt, a resistance of 50 ohm, an inductance of 1 henry, and no initial circuit. Find the current in the circuit at any time *t*.

Solution. Here, $E = 5$, $R = 50$ and $L = 1$; becomes $\frac{dI}{dt} + 50I = 5$

This equation is linear, its solution is $I = Ce^{-50t} + \frac{1}{10}$

At $t = 0$, $I = 0$; thus $0 = Ce^{-50(0)} + \frac{1}{10}$ or $C = -\frac{1}{10}$.

The current at any time *t* is, then

$$I = -\frac{1}{10}e^{-50t} + \frac{1}{10} \quad \text{...(i)}$$

The quantity $-\frac{1}{10}e^{-50t}$ in Eq. (i) is called the transient current, since this quantity goes to zero (dies out) at $t \to \infty$. The quantity $\frac{1}{10}$ in Eq. (i) is called the steady-state current. As $t \to \infty$, the current I approaches the value of the steady-state current.

Illustration 81 An *RL* circuit has an emf given (in volts) by $3\sin 2t$, a resistance of 10 ohm, an inductance of 0.5 henry and an initial current of 6 amperes. Find the current in the circuit at any time *t*.

Solution. Here, $E = 3\sin 2t$, $R = 10$ and $L = 0.5$; becomes $\frac{dI}{dt} + 20I = 6\sin 2t$

This equation is linear, with solution $\int d(Ie^{20t}) = \int 6e^{20t} \sin 2t \, dt$

Carrying out the integrations (the second integral requires two integration by parts), we obtain

$$I = Ce^{-20t} + \frac{30}{101}\sin 2t - \frac{3}{101}\cos 2t$$

At $t = 0$, $I = 6$; hence $6 = Ce^{-20(0)} + \frac{30}{101}\sin 2(0) - \frac{3}{101}\sin 2(0)$

or
$$6 = C - \frac{3}{101}$$

Where $C = 609/101$. The current at any time *t* is

$$I = \frac{609}{101}e^{-20t} + \frac{30}{101}\sin 2t - \frac{3}{101}\cos 2t$$

the current is the sum of a transient current, here $(690/101)e^{-20t}$ and a steady-state current,

$$\frac{30}{101}\sin 2t - \frac{3}{101}\cos 2t$$

Illustration 82 Rewrite the steady-state current in the form $A\sin(2t - \phi)$. The angle ϕ is called the phase angle.

Solution. Since, $A\sin(2t - \phi) = A(\sin 2t\cos\phi - \cos 2t\sin\phi)$, we require

$$I = \frac{30}{101}\sin 2t - \frac{3}{101}\cos 2t$$
$$= A\cos\phi \sin 2t - A\sin\phi \cos 2t$$

Thus, $\quad A\cos\phi = \dfrac{30}{101}$ and $A\sin\phi = \dfrac{3}{101}$

It now follows that

$$\left(\frac{30}{101}\right)^2 + \left(\frac{3}{101}\right)^2 = A^2\cos^2\phi + A^2\sin^2\phi$$

$$= A^2(\cos^2\phi + \sin^2\phi) = A^2$$

and
$$\tan\phi = \frac{A\sin\phi}{A\cos\phi} = \frac{\left(\frac{3}{101}\right)}{\left(\frac{30}{101}\right)} = \frac{1}{10}$$

Consequently, I has the required form, if $A = \sqrt{\dfrac{909}{(101)^2}} = \dfrac{3}{\sqrt{101}}$

and
$$\phi = \arctan\frac{1}{10} = 0.0997 \text{ rad}$$

Illustration 83 An *RC* circuit has an emf given (in volts) by $400\cos 2t$, a resistance of 100 ohm, and a capacitance of 10^{-2} farad. Initially there is no charge on the capacitor. Find the current in the circuit at any time *t*.

Solution. We first find the charge *q* and then the current.

Here, $E = 400\cos 2t$, $R = 100$ and $C = 10^{-2}$;

becomes
$$\frac{dq}{dt} + q = 4\cos 2t$$

This equations is linear and its solution is (two integrations by parts are required)

$$q = Ce^{-t} + \frac{8}{5}\sin 2t + \frac{4}{5}\cos 2t$$

At $t = 0$, $q = 0$; hence, $0 = Ce^{-(0)} + \dfrac{8}{5}\sin 2(0) + \dfrac{4}{5}\cos 2(0)$

or
$$C = -\frac{4}{5}$$

Thus,
$$q = -\frac{4}{5}e^{-t} + \frac{8}{5}\sin 2t + \frac{4}{5}\cos 2t$$

and we obtain
$$I = \frac{dq}{dt} = \frac{4}{5}e^{-t} + \frac{16}{5}\cos 2t - \frac{8}{5}\sin 2t$$

Illustration 84 A body of mass 2 slugs is dropped with no initial velocity and encounters an air resistance that is proportional to the square of its velocity. Find an expression for the velocity of the body at any time *t*.

Solution. The force due to air resistance is $-kv^2$, so that Newton's second law of motion becomes

$$m\frac{dv}{dt} = mg - kv^2 \quad \text{or} \quad 2\frac{dv}{dt} = 64 - kv^2$$

Rewriting this equation in differential form, we have

$$\frac{2}{64 - kv^2}\,dv - dt = 0 \qquad \text{...(i)}$$

Which is separable, by partial fractions, $\dfrac{2}{64 - kv^2} = \dfrac{2}{(8 - \sqrt{kv})(8 + \sqrt{kv})}$

Hence, Eq. (i) can be rewritten as $\dfrac{1}{8}\left(\dfrac{1}{8 - \sqrt{kv}} + \dfrac{1}{8 + \sqrt{kv}}\right) dv - dt = 0$

This last equation has as its solution $\dfrac{1}{8}\displaystyle\int\left(\dfrac{1}{8 - \sqrt{kv}} + \dfrac{1}{8 + \sqrt{kv}}\right) dv - \int dt = C$

Which can be rewritten as $\ln\left|\dfrac{8 + \sqrt{kv}}{8 - \sqrt{kv}}\right| = 8\sqrt{kt} + 8\sqrt{k}C$

At $t = 0$, we are given that $v = 0$. This implies $C_1 = 1$ and the velocity is given by $\dfrac{8 + \sqrt{kv}}{8 - \sqrt{kv}} = e^{8\sqrt{kt}}$

Geometrical Applications

Let $P(x_1, y_1)$ be any point on the curve $y = f(x)$, then slope of the tangent at $P(= \tan \psi) = \left(\dfrac{dy}{dx}\right)_{(x_1, y_1)}$ and hence we find the following facts.

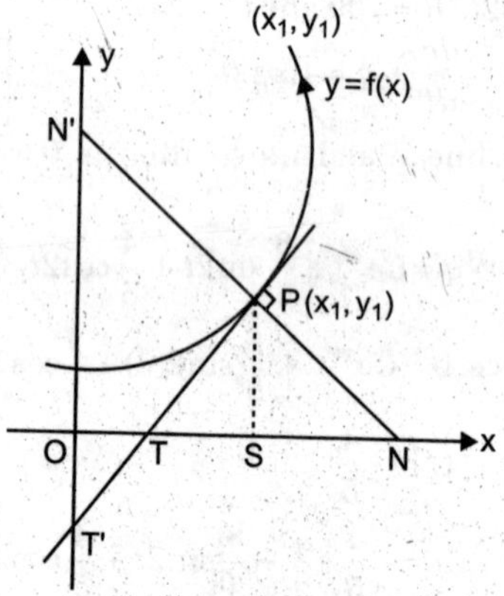

Fig. 4.2

(i) The equation of the tangent at P is, $y - y_1 = \dfrac{dy}{dx}(x - x_1)$ when it cuts x-axis, $y = 0$.

$\therefore$ x-intercept of the tangent $= x_1 - y_1\left(\dfrac{dx}{dy}\right)$

y-intercept of the tangent $= y_1 - x_1\dfrac{dy}{dx}$

(ii) The equation of normal at P is, $y - y_1 = -\dfrac{1}{(dy/dx)}(x - x_1)$

x and y-intercepts of normal are; $x_1 + y_1\dfrac{dy}{dx}$ and $y_1 + x_1\dfrac{dx}{dy}$

(iii) Length of tangent $= PT = |y_1| \sqrt{1+(dx/dy)^2_{(x_1, y_1)}}$

(iv) Length of normal $= PN = |y_1| \sqrt{1+(dy/dx)^2_{(x_1, y_1)}}$

(v) Length of subtangent $= ST = \left| y_1 \left(\frac{dx}{dy}\right)_{(x_1, y_1)} \right|$

(vi) Length of subnormal $= SN = \left| y_1 \left(\frac{dy}{dx}\right)_{(x_1, y_1)} \right|$

(vii) Length of radius vector $= \sqrt{x_1^2 + y_1^2}$

Illustration 85 Find the curve for which the area of the triangle formed by the x-axis tangent drawn at any point on the curve and radius vector of the point of tangency is constant equal to a^2.

Solution. Tangent drawn at any point (x, y) is

$$Y - y = \frac{dy}{dx}(X - x)$$

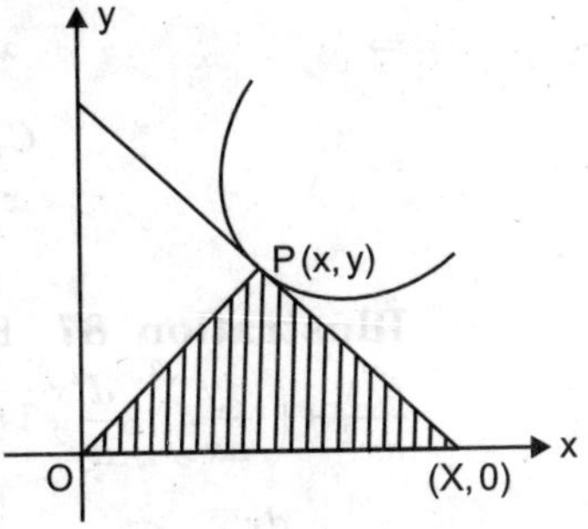

When $Y = 0$, $X = x - y\frac{dx}{dy}$

$$\text{Area of } \Delta = 2a^2 \text{ (given)}$$

ie, $$\left|\frac{1}{2} \cdot X \cdot Y\right| = 2a^2$$

ie, $$\left|xy - y^2\frac{dx}{dy}\right| = 2a^2$$

ie, $$xy - y^2\frac{dx}{dy} = \pm 2a^2$$

ie, $$\frac{dx}{dy} - \frac{x}{y} = \pm\frac{2a^2}{y^2}$$

$$\text{IF} = e^{\int -\frac{1}{y}dy} = \frac{1}{y}$$

∴ The solution is

$$x \cdot \frac{1}{y} = \int \pm \frac{2a^2}{y^2} \cdot \frac{1}{y}\, dy$$

$$\frac{x}{y} = \pm\frac{2a^2 \cdot y^{-2}}{-2} + C, \; ie, \; x = Cy + \frac{a^2}{y}$$

Illustration 86 Find the curve for which the intercept cut off by any tangent on y-axis is proportional to the square of the ordinate of the point of tangency.

Solution. The equation of tangent at any point (x, y) is

$$Y - y = \frac{dy}{dx}(X - x)$$

When $X = 0$; $Y = y - x\dfrac{dy}{dx}$ = y-intercept

It is given $Y \propto y^2$.

ie, $Y = ky^2$ (k being constant of proportionality)

ie, $y - x\dfrac{dy}{dx} = ky^2$ ie, $\dfrac{dy}{dx} - \dfrac{1}{x}y = -\dfrac{ky^2}{x}$

ie, $-y^{-2}\dfrac{dy}{dx} + \dfrac{1}{x}\cdot\dfrac{1}{y} = \dfrac{k}{x}$

Let $\dfrac{1}{y} = z$ so that, $-\dfrac{1}{y^2}\cdot\dfrac{dy}{dx} = \dfrac{dz}{dx}$ $\therefore$ $\dfrac{dz}{dx} + \dfrac{z}{x} = \dfrac{k}{x}$

$$\text{IF} = e^{\int \frac{1}{x}dx} = e^{\log x} = x$$

$\therefore$ The solution is $zx = \displaystyle\int \frac{kx}{x}\,dx = kx + C$

$\Rightarrow$ $\dfrac{x}{y} = kx + C$

$\Rightarrow$ $x = kxy + Cy$

$\Rightarrow$ $x - Cy = kxy \Rightarrow \dfrac{1}{ky} + \dfrac{-C}{k}\cdot\dfrac{1}{x} = 1$

$\Rightarrow$ $\dfrac{C_1}{x} + \dfrac{C_2}{y} = 1$ $\left(\text{where } \dfrac{1}{k} = C_2 \text{ and } -\dfrac{C}{k} = C_1\right)$

Illustration 87 For any differential function $y = f(x)$, the value of $\dfrac{d^2y}{dx^2} + \left(\dfrac{dy}{dx}\right)^3\cdot\dfrac{d^2x}{dy^2}$ is equal to

(a) $2y\dfrac{dy}{dx}$ (b) $y^2\dfrac{dy}{dx}$ (c) $y\dfrac{dy}{dx} + \left(\dfrac{d^2x}{dy^2}\right)^2$ (d) None of these

Solution. We know, $\dfrac{dy}{dx} = \left(\dfrac{dx}{dy}\right)^{-1}$, differentiating both the sides

or $$\frac{d^2y}{dx^2} = -\left(\frac{dx}{dy}\right)^{-2}\cdot\frac{d}{dy}\left(\frac{dx}{dy}\right)\cdot\frac{dy}{dx}$$

$$= -\left(\frac{dx}{dy}\right)^{-2}\cdot\frac{d^2x}{dy^2}\cdot\frac{dy}{dx}$$

$\Rightarrow$ $$\frac{d^2y}{dx^2} = -\frac{d^2x}{dy^2}\cdot\left(\frac{dy}{dx}\right)^3$$

or $$\frac{d^2y}{dx^2} + \left(\frac{dy}{dx}\right)^3\cdot\frac{d^2x}{dy^2} = 0$$

Hence, (d) is the correct answer.

Illustration 88 The solution of $y = x\dfrac{dy}{dx} + \dfrac{dy}{dx} - \left(\dfrac{dy}{dx}\right)^2$, is

(a) $y = (x-1)^2$ (b) $4y = (x+1)^2$ (c) $(y-1)^2 = 4x$ (d) None of these

Solution. The given equation can be written as

$$y = xp + p - p^2; \text{ where } p = \frac{dy}{dx} \quad \text{...(i)}$$

Differentiating both the sides w.r.t. x, we get

$$p = p + \frac{xdp}{dx} + \frac{dp}{dx} - 2p\frac{dp}{dx}$$

$$\therefore \quad \frac{dp}{dx}(x + 1 - 2p) = 0$$

$$\therefore \text{ Either} \quad \frac{dp}{dx} = 0 \quad ie, \; p = C \quad \text{...(ii)}$$

$$\text{or} \quad x + 1 - 2p = 0 \quad ie, \; p = \frac{1}{2}(x + 1) \quad \text{...(iii)}$$

Eliminating p between Eqs. (i) and (ii), we get

$$y = Cx + C - C^2$$

As the complete solution and eliminating p between Eqs. (i) and (iii)

$$y = \frac{1}{2}(x + 1)x + \frac{1}{2}(x + 1) - \frac{1}{4}(x + 1)^2$$

ie, $4y = (x + 1)^2$ as the singular solution.

Hence, (b) is the correct answer.

Illustration 89 The solution of $\left(\frac{dy}{dx}\right)^2 + (2x + y)\frac{dy}{dx} + 2xy = 0$, is

(a) $(y + x^2 - C_1)(x + \log y + y^2 + C_2) = 0$
(b) $(y + x^2 - C_1)(x - \log y - C_2) = 0$
(c) $(y + x^2 - C_1)(x + \log y - C_2) = 0$
(d) None of the above

Solution. The given equation can be written as

$$p^2 + (2x + y)p + 2xy = 0, \text{ where } p = \frac{dy}{dx}$$

ie, $(p + 2x)(p + y) = 0$

$\therefore$ $p + 2x = 0$, otherwise $p + y = 0$

$\Rightarrow$ $\frac{dy}{dx} + 2x = 0$ or $\frac{dy}{dx} + y = 0$

$\Rightarrow$ $\int dy + 2\int x\,dx = C_1$ or $\int \frac{dy}{y} + \int dx = C_2$

$\Rightarrow$ $y + x^2 = C_1$ or $\log y + x = C_2$

$\therefore$ $(y + x^2 - C_1) = 0$ or $(x + \log y - C_2) = 0$

$\Rightarrow$ $(y + x^2 - C_1)(x + \log y - C_2) = 0$ is the required solution.

Hence, (c) is the correct answer.

Illustration 90 A curve $y = f(x)$ passes through the origin. Through any point (x, y) on the curve, lines are drawn parallel to the coordinate axes. If the curve divides the area formed by these lines and coordinate axes in the ratio $m : n$. Then the equation of curve is

(a) $y = Cx^{m/n}$ (b) $my^2 = Cx^{m/n}$ (c) $y^3 = Cx^{m/n}$ (d) None of these

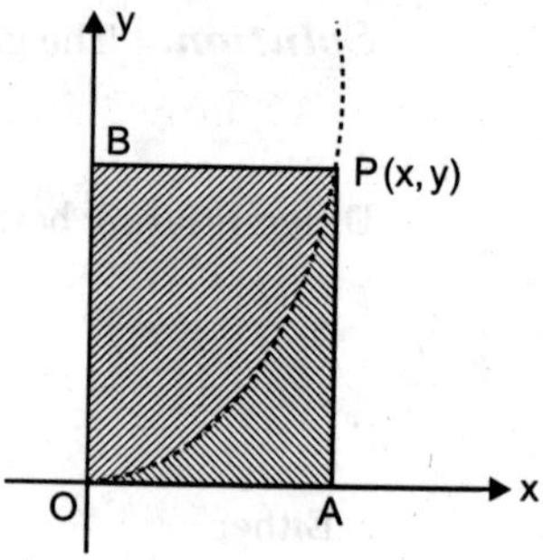

Solution. $\dfrac{\text{Area of } OBPO}{\text{Area of } OPAO} = \dfrac{m}{n}$

$$\Rightarrow \quad \frac{xy - \int_0^x y\,dx}{\int_0^x y\,dx} = \frac{m}{n}$$

$$\Rightarrow \quad nxy = (m+n)\int_0^x y\,dx$$

Differentiating w.r.t. x, we get

$$n\left(x\frac{dy}{dx} + y\right) = (m+n)\,y$$

$$\Rightarrow \quad nx\frac{dy}{dx} = my$$

$$\Rightarrow \quad \frac{m}{n}\cdot\frac{dx}{x} = \frac{dy}{y}, \text{ integrating both the sides}$$

$$y = Cx^{m/n}$$

Hence, (a) is the correct answer.

Illustration 91 The equation of the curve passing through the points $(3a, a)(a > 0)$ in the form $x = f(y)$ which satisfy the differential equation;

$$\frac{a^2}{xy}\cdot\frac{dx}{dy} = \frac{x}{y} + \frac{y}{x} - 2, \text{ is}$$

(a) $x = y + a\left(\dfrac{1+e^{y-k}}{1-2e^{y-k}}\right)$ (b) $x = y + a\left(\dfrac{1+e^{y-k}}{1-e^{y-k}}\right)$

(c) $y = x + a\left(\dfrac{1+e^{y-k}}{1-e^{y-k}}\right)$ (d) None of these

Solution. Here, $\dfrac{a^2}{xy}\cdot\dfrac{dx}{dy} = \dfrac{x}{y} + \dfrac{y}{x} - 2$

$$\Rightarrow \quad a^2 = \frac{dy}{dx}(x^2 + y^2 - 2xy)$$

$$\Rightarrow \quad (x-y)^2\cdot\frac{dy}{dx} = a^2, \text{ put } x - y = v \;\therefore\; 1 - \frac{dy}{dx} = \frac{dv}{dx}$$

$$\Rightarrow \quad v^2\left(1 - \frac{dv}{dx}\right) = a^2$$

$$\Rightarrow \quad v^2 - a^2 = v^2\frac{dv}{dx} \Rightarrow \frac{v^2\,dv}{v^2 - a^2} = dx$$

$$\Rightarrow \quad \left(1 + \frac{a^2}{v^2 - a^2}\right)dv = dx$$

Integrating both the sides, we get

$$v + \frac{a}{2}\log\left|\frac{v-a}{v+a}\right| = x + C$$

$$\Rightarrow \quad (x-y) + \frac{a}{2}\log\left(\frac{x-y-a}{x-y+a}\right) = x + C$$

$$\Rightarrow \quad y + C = \frac{a}{2} \log\left(\frac{x - y - a}{x - y + a}\right) \quad \text{...(i)}$$

It passes through $(3a, a)$

$$\Rightarrow \quad a + C = \frac{a}{2} \log\left(\frac{1}{3}\right)$$

$$\Rightarrow \quad C = -a + \frac{a}{2} \log\left(\frac{1}{3}\right)$$

$$\Rightarrow \quad C = -a\left(\frac{2 + \log 3}{2}\right)$$

$$\therefore \quad y = \frac{a}{2}(2 + \log 3) + \log\left(\frac{x - y - a}{x - y + a}\right)$$

$$\Rightarrow \quad \frac{x - y - a}{x - y + a} = e^{y - k}, \text{ where } k = \frac{a}{2}(2 + \log 3)$$

$$\therefore \quad \frac{x - y}{a} = \frac{1 + e^{y - k}}{1 - e^{y - k}}$$

$$\Rightarrow \quad x = y + a\left(\frac{1 + e^{y - k}}{1 - e^{y - k}}\right), \text{ where } k = \frac{a}{2}(2 + \log 3)$$

Which is required equation of curve.
Hence, (b) is the correct answer.

Illustration 92 The family of curves, the subtangent at any point of which is the arithmetic mean of the coordinates of the point of tangency, is given by

(a) $(x - y)^2 = Cy$ (b) $(y - x)^2 = Cx$ (c) $(x - y)^2 = Cxy$ (d) None of these

Solution. Let the family of curves be $y = f(x)$

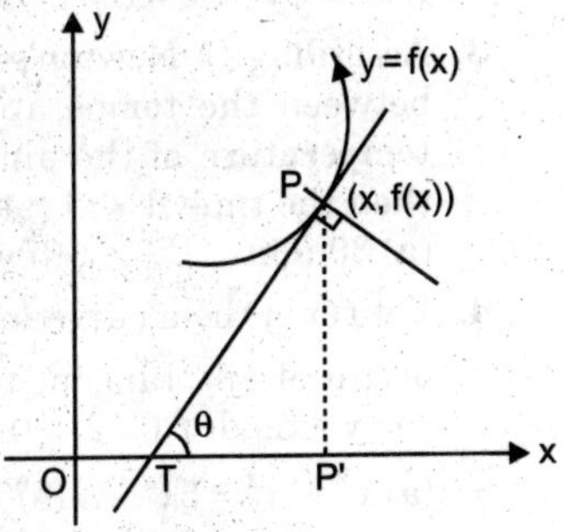

$$\tan \theta = \frac{l(PP')}{l(TP')}$$

$$\tan \theta = \frac{l(PP')}{l(TP')}$$

$$\therefore \quad l(\text{subtangent}) = \frac{f(x)}{f'(x)}$$

$$\therefore \quad \frac{y}{y'} = \frac{x + y}{2} \text{ (given)}$$

$$\therefore \quad y' = \frac{2y}{x + y}$$

$$\Rightarrow \quad \frac{dy}{dx} = \frac{2y}{x + y} \quad \therefore \quad \frac{dx}{dy} = \frac{x + y}{2xy} \quad \text{...(i)}$$

It is a homogeneous differential equation.

$\therefore$ Put $x = vy$

Differentiating w.r.t. y, we get

$$\frac{dx}{dy} = v + y\frac{dv}{dy} \quad \text{...(ii)}$$

In Eq. (i) replacing $\frac{dx}{dy}$ by Eq. (ii), we get

$$v + y\frac{dv}{dy} = \frac{vy + y}{2y} = \frac{1+v}{2}$$

$$\Rightarrow \quad y\frac{dv}{dy} = \frac{1+v}{2} - v = \frac{1+v-2v}{2} = \frac{1-v}{2}$$

$$\Rightarrow \quad \frac{2}{1-v}\,dv = \frac{dy}{y}$$

Integrating, $\dfrac{2\log|1-v|}{-1} = \log|y| + \log C_1 \; (C_1 > 0)$

$\therefore \quad -2\log|y-x| + 2\log|y| = \log|y| + \log C_1$

$\Rightarrow \quad \log|y-x|^2 = \log|y| - \log C_1$

$\Rightarrow \quad \log|y-x|^2 = \log|y| + \log C$, where $\log C = -\log C_1$

$\Rightarrow \quad \log|y-x|^2 = \log|yC|$

$\Rightarrow \quad (x-y)^2 = Cy$, is the required equation of family of curves.

Hence, (a) is the correct answer.

Target Exercise 4.5

1. The equation of curve for which the normal at every point passes through a fixed point, is
   (a) a circle (b) an ellipse (c) a hyperbola (d) None of these
2. If the tangent at any point P of a curve meets the axis of x in T. Then the curve for which $OP = PT$, O being the origin is
   (a) $x = Cy^2$ (b) $x = Cy^2$ or $x = C/y^2$
   (c) $x = Cy$ or $x = C/y$ (d) None of these
3. According to Newton's law, the rate of cooling is proportional to the difference between the temperature of the body and the temperature of the air. If the temperature of the air is 20°C and body cools for 20 min from 100°C to 60°C, then the time it will take for it temperature to drop to 30°C, is
   (a) 30 min (b) 40 min (c) 60 min (d) 80 min
4. Let $f(x, y)$ be a curve in the x-y plane having the property that distance from the origin of any tangent to the curve is equal to distance of point of contact from the y-axis. If $f(1, 2) = 0$, then all such possible curves are
   (a) $x^2 + y^2 = 5x$ (b) $x^2 - y^2 = 5x$ (c) $x^2y^2 = 5x$ (d) All of these
5. Given the curves $y = f(x)$ passing through the point $(0, 1)$ and $y = \int_{-\infty}^{x} f(t)$ passing through the point $\left(0, \frac{1}{2}\right)$. The tangents drawn to both the curves at the points with equal abscissae intersect on the x-axis. Then the curve $y = f(x)$, is
   (a) $f(x) = x^2 + x + 1$ (b) $f(x) = \frac{x^2}{e^x}$ (c) $f(x) = e^{2x}$ (d) $f(x) = x - e^x$

Worked Examples

Type 1 : Subjective Type Questions

Example 1 For a certain curve $y = f(x)$ satisfying $\frac{d^2y}{dx^2} = 6x - 4$; $f(x)$ has a local minimum value 5 when $x = 1$. Find the equation of the curve and also the global maximum and global minimum values of $f(x)$ given that $0 \le x \le 2$.

Solution. Integrating, $\frac{d^2y}{dx^2} = 6x - 4$, we get $\frac{dy}{dx} = 3x^2 - 4x + C$

when $x = 1$, $\frac{dy}{dx} = 0$. So that $C = 1$

Hence, $$\frac{dy}{dx} = 3x^2 - 4x + 1 \quad \text{...(i)}$$

Integrating, we get

$$y = x^3 - 2x^2 + x + C_1, \text{ when } x = 1,\ y = 5, \text{ so that } C_1 = 5$$

Thus, we have $$y = x^3 - 2x^2 + x + 5$$

Form Eq. (i), we get the critical points $x = \frac{1}{3}, x = 1$

At the critical point $x = \frac{1}{3}$, $\frac{d^2y}{dx^2}$ is (–ve).

Therefore, at $x = \frac{1}{3}$, y has a local maximum.

At $x = 1$, $\frac{d^2y}{dx^2}$ is (+ ve).

Therefore, at $x = 1$, y has a local minimum.

Also, $$f(1) = 5$$
$$f\left(\frac{1}{3}\right) = \frac{139}{27}$$
$$f(0) = 5, f(2) = 7$$

Hence, the global maximum value = 7, the global minimum value = 5.

Example 2 If $\phi(x)$ is a differentiable real-valued function satisfying $\phi'(x) + 2\phi(x) \le 1$, prove that $\phi(x) - \frac{1}{2}$ is a non-increasing function of x.

Solution. $$\phi'(x) + 2\phi(x) \le 1$$

$$\Rightarrow \quad e^{2x}\phi'(x) + 2\phi(x)e^{2x} \le e^{2x}$$

$$\Rightarrow \quad \frac{d}{dx}\left(e^{2x}\phi(x) - \frac{1}{2}e^{2x}\right) \le 0$$

$\Rightarrow e^{2x}\left(\phi(x) - \frac{1}{2}\right)$ is a non-increasing function of x.

$\Rightarrow \phi(x) - \frac{1}{2}$ is a non-increasing function of x.

Example 3 Determine all curve for which the ratios of the length of the segment intercepted by any tangent on the y-axis to the length of the radius vector is a constant.

Solution. Let $y = f(x)$ be the equation of the required curve.

Given that $\dfrac{\left| y - x \dfrac{dy}{dx} \right|}{\sqrt{x^2 + y^2}} = k$ (a constant)

$$\Rightarrow \quad \frac{dy}{dx} = \frac{y}{x} \pm k \sqrt{1 + \left(\frac{y}{x}\right)^2}$$

Let $y = vx$, then $v + x \dfrac{dv}{dx} = v \pm k \sqrt{1 + v^2}$

$$\Rightarrow \quad \frac{dv}{\sqrt{1 + v^2}} = \pm k \frac{dx}{x}, \text{ integrating we get}$$

$$\Rightarrow \quad \log \left| v + \sqrt{1 + v^2} \right| = \pm k \ln x + C$$

$$\Rightarrow \quad \log \left| \frac{xy + \sqrt{x^2 + y^2}}{x} \right| = \pm k \ln x + C$$

Which are the equations of the required curves.

Example 4 Let $u(x)$ and $v(x)$ satisfy the differential equations $\dfrac{du}{dx} + p(x) u = f(x)$ and $\dfrac{dv}{dx} + p(x) v = g(x)$, where $p(x)$, $f(x)$ and $g(x)$ are continuous functions. If $u(x_1) > v(x_1)$ for some x_1 and $f(x) > g(x)$, for all $x > x_1$, prove that any point (x, y), where $x > x_1$, does not satisfy the equation $y = u(x)$ and $y = v(x)$. **[IIT JEE 1997]**

Solution. Given that

$$\frac{du}{dx} + p(x) \cdot u = f(x) \text{ and } \frac{dv}{dx} + p(x) \cdot v = g(x)$$

Subtracting, we get

$$\frac{d(u - v)}{dx} + p(x) \cdot (u - v) = f(x) - g(x)$$

Multiplying by $e^{\int p(x)\, dx}$, we get

$$e^{\int p(x)\, dx} \cdot \frac{d(u - v)}{dx} + (u - v) \cdot p(x) \cdot e^{\int p(x)\, dx} = \{f(x) - g(x)\} \cdot e^{\int p(x)\, dx}$$

ie, $$\frac{d}{dx} \{(u - v) e^{\int p dx}\} = \{f(x) - g(x)\} e^{\int p(x)\, dx}$$

Since, exponential function takes only positive values and

$f(x) > g(x)$ for all $x > x_1$, RHS is + ve; $x > x_1$

$$\therefore \quad \frac{d}{dx} \{(u - v) \cdot e^{\int p(x)\, dx}\} = 0$$

ie, $(u - v) \cdot e^{\int p(x)\, dx}$ is increasing function.

Hence, if $e^{\int p(x)\, dx} = \phi(x)$, then for $x > x_1$

We have, $$\{u(x) - v(x)\} \phi(x) > \{u(x_1) - v(x_1)\} \phi(x_1)$$

ie, $$u(x) - v(x) > \frac{\{u(x_1) - v(x_1)\} \cdot \phi(x_1)}{\phi(x)} > 0 \qquad [\because u(x_1) > v(x_1)]$$

Thus, $u(x) > v(x), \forall x > x_1$

ie, $u(x) \neq v(x), \forall x > x_1$

Hence, no point (x, y) such that $x > x_1$ can satisfy the equations $y = u(x)$ and $y = v(x)$.

Example 5 A normal is drawn at a point $P(x, y)$ of a curve. It meets the x-axis at Q. If PQ is of constant length k, then show that the differential equation describing such curves is, $y\frac{dy}{dx} = \pm\sqrt{k^2 - y^2}$ and the equation of such a curve passing through $(0, k)$. **[IIT JEE 1994]**

Solution. Let $y = f(x)$ be the curve such that the normal at $P(x, y)$ to this curve meets x-axis at Q. Then,

$$PQ = \text{length of the normal at } P$$
$$= y\sqrt{1 + \left(\frac{dy}{dx}\right)^2}$$

But $PQ = k$

$\therefore$ $$y\sqrt{1 + \left(\frac{dy}{dx}\right)^2} = k$$

$\Rightarrow$ $$y^2 + y^2\left(\frac{dy}{dx}\right)^2 = k^2$$

or $$y\frac{dy}{dx} = \pm\sqrt{k^2 - y^2}$$

$\Rightarrow$ $$\frac{y\,dy}{\sqrt{k^2 - y^2}} = \pm\,dx$$

Integrating both the sides, we get

$$-\sqrt{k^2 - y^2} = \pm\,x + C, \text{ since it passes through } (0, k) \rightarrow C = 0.$$

$\therefore$ $$-\sqrt{k^2 - y^2} = \pm\,x$$

or $$k^2 - y^2 = x^2$$

$\Rightarrow$ $x^2 + y^2 = k^2$, is required equation of the curve.

Example 6 A curve passing through the point (1, 1) has the property that the perpendicular distance of the normal at any point P on the curve from the origin is equal to the distance of P from x-axis. Determine the equation of the curve. **[IIT JEE 1999]**

Solution. Let $P(x, y)$ be any point on the curve $y = f(x)$. Then, the equation of the normal at P is,

$$Y - y = -\frac{1}{(dy/dx)}(X - x)$$

or $$X + Y\frac{dy}{dx} - \left(y\frac{dy}{dx} + x\right) = 0 \qquad \text{...(i)}$$

It is given that distance of Eq. (i) from origin = Distance from x-axis (*ie*, y)

ie,
$$\left|\frac{0-\left(y\frac{dy}{dx}+x\right)}{\sqrt{1+\left(\frac{dy}{dx}\right)^2}}\right| = y$$

$$\Rightarrow \quad \left(y\frac{dy}{dx}+x\right)^2 = y^2\left[1+\left(\frac{dy}{dx}\right)^2\right]$$

$$\Rightarrow \quad x^2 + 2xy\frac{dy}{dx} = y^2$$

or
$$\frac{dy}{dx} = \frac{y^2 - x^2}{2xy}$$

which is homogeneous differential equation and we can solve by homogeneous or by total differential.

Here, using total differential,

$$2xy\,dy - y^2\,dx = -x^2\,dx$$

$$\Rightarrow \quad \frac{x\,d(y^2) - y^2\,dx}{x^2} = -dx$$

$$\Rightarrow \quad d\left(\frac{y^2}{x}\right) = -dx$$

Integrating both the sides, we get

$$\Rightarrow \quad \frac{y^2}{x} = -x + C \quad \text{...(ii)}$$

It passes through (1, 1) $\Rightarrow C = 2$

$$\therefore \quad \frac{y^2}{x} = -x + 2$$

or
$$y^2 = -x^2 + 2x$$

$\Rightarrow$ $x^2 + y^2 - 2x = 0$, is required equation of curve.

Example 7 A country has a food deficit of 10%. Its population grows continuously at a rate of 3% per year. Its annual food production every year is 4% more than that of the last year. Assuming that the average food requirement per person remains constant, prove that the country will become self-sufficient in food after n years, where n is the smallest integer bigger than or equal to $\dfrac{\log_e 10 - \log_e 9}{(\log_e 1.04) - 0.03}$.

[IIT JEE 2000]

Solution. Let P_0 be the initial population, Q_0 be its initial food production.

Let P be the population of the country in year t and Q be its food production in year t.

$$\Rightarrow \quad \frac{dP}{dt} = \frac{3P}{100} \quad \text{or} \quad \frac{dP}{P} = \frac{3}{100}\,dt$$

Integrating, we get

$$\log P = \frac{3}{100}t + C$$

At $t = 0$, we have $P = P_0$

$$\Rightarrow \quad C = \log P_0$$

$$\Rightarrow \qquad P = P_0\, e^{0.03\,t} \qquad \text{...(i)}$$

It is given that the annual food production every year is 4% more than that of last year.

$$\Rightarrow \qquad Q = Q_0 \left(1 + \frac{4}{100}\right)^t$$

Let the average consumption per person be k units.

$$\Rightarrow \qquad Q_0 = kP_0 \left(\frac{90}{100}\right) = 0.9\, kP_0$$

$$\therefore \qquad Q = 0.9\, kP_0\, (1.04)^t \qquad \text{...(ii)}$$

This gives quantity of food available in year t. The population in year t is,

$$P = P_0\, e^{0.03\,t} \qquad \text{[from Eq. (i)]}$$

$\therefore$ Consumption in year, $\quad t = kP_0 e^{0.03\,t}$ ...(iii)

The country will be self sufficient, if

$$Q \ge P$$

$$\Rightarrow \qquad 0.9k\, P_0\, (1.04)^t \ge kP_0 e^{0.03\,t}$$

$$\Rightarrow \qquad \frac{9}{10}(1.04)^t \ge e^{0.03\,t}$$

$$\Rightarrow \qquad (1.04)^t\, e^{-0.03\,t} \ge \frac{10}{9}$$

$$\Rightarrow \qquad t \log(1.04) - 0.03\,t \ge \log\left(\frac{10}{9}\right)$$

$$\Rightarrow \qquad t\,\{\log(1.04) - 0.03\} \ge \log\left(\frac{10}{9}\right)$$

$$\Rightarrow \qquad t \ge \frac{\log 10 - \log 9}{\log(1.04) - 0.03}$$

Thus, the least number of year in which country becomes self sufficient.

$$\Rightarrow \qquad t = \frac{\log 10 - \log 9}{\log(1.04) - 0.03}$$

Example 8 A right circular cone with radius R and height H contains a liquid which evaporates at a rate proportional to its surface area in contact with air (proportionality constant $= k > 0$), find the time after which the cone is empty.

[IIT JEE 2003]

Solution. Let the semi-vertical angle of the cone be θ and let the height of the liquid at time 't' be 'h' from the vertex V and radius of the liquid cone be r. Let V be the volume at time t. Then,

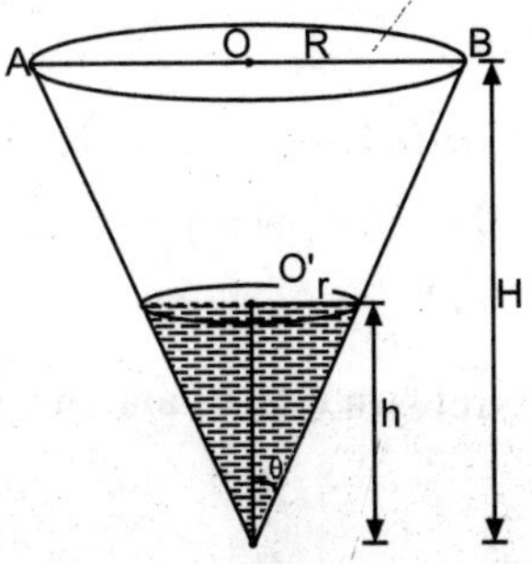

$$V = \frac{1}{3}\pi r^2 h$$

$$\Rightarrow \qquad V = \frac{1}{3}\pi r^3 \cot\theta \qquad \left(\because \tan\theta = \frac{r}{h}\right)$$

Let S be the surface area of the liquid in contact with air at time t.

Then, $$S = \pi r^2$$

$$\Rightarrow \quad \frac{dV}{dt} \propto S$$

$$\Rightarrow \quad \frac{dV}{dt} = -kS, \; k \text{ is constant of proportionality.}$$

$$\Rightarrow \quad \frac{d}{dt}\left(\frac{1}{3}\pi r^3 \cot\theta\right) = -k\pi r^2$$

$$\Rightarrow \quad \pi r^2 \frac{dr}{dt}\cot\theta = -k\pi r^2 \Rightarrow \cot\theta\, dr = -k\, dt$$

On integrating, we get $\cot\theta \int_R^0 dr = -k\int_0^T dt$

$$\Rightarrow \quad R\cot\theta = +kT, \text{ where } T \text{ is required time.}$$

$$\Rightarrow \quad T = H/k \qquad (\text{as } \tan\theta = R/H)$$

Example 9 Solve the equation $x\int_0^x y(t)\,dt = (x+1)\int_0^x t\,y(t)\,dt, \; x > 0.$

Solution. Differentiating the equation w.r.t. x, we get

$$xy(x) + 1\cdot\int_0^x y(t)\,dt = (x+1)\,xy(x) + 1\cdot\int_0^x t\,y(t)\,dt$$

ie,
$$\int_0^x y(t)\,dt = x^2\,y(x) + \int_0^x t\,y(t)\,dt$$

Again, differentiating w.r.t. x, we get

$$y(x) = x^2\,y'(x) + 2xy(x) + xy(x)$$

ie,
$$(1-3x)\,y(x) = \frac{x^2\,dy(x)}{dx}$$

ie,
$$\frac{(1-3x)\,dx}{x^2} = \frac{dy(x)}{y(x)}$$

Integrating, we get $y = \frac{C}{x^3}e^{-1/x}$

Example 10 If (y_1, y_2) are two solutions of the differential equation

$$\frac{dy}{dx} + P(x)\cdot y = Q(x)$$

Then prove that $y = y_1 + C\,(y_1 - y_2)$ is the general solution of the equation where C is any constant. For what relation between the constant α, β will the linear combination $\alpha y_1 + \beta y_2$ also be a solution.

Solution. As y_1, y_2 are the solutions of the differential equation;

$$\frac{dy}{dx} + P(x)\cdot y = Q(x) \qquad \text{...(i)}$$

$$\therefore \quad \frac{dy_1}{dx} + P(x)\cdot y_1 = Q(x) \qquad \text{...(ii)}$$

and
$$\frac{dy_2}{dx} + P(x)\cdot y_2 = Q(x) \qquad \text{...(iii)}$$

From Eqs. (i) and (ii),
$$\left(\frac{dy}{dx} - \frac{dy_1}{dx}\right) + P(x)(y - y_1) = 0$$

$$\therefore \quad \frac{d}{dx}(y - y_1) + P(x)\cdot(y - y_1) = 0 \qquad \text{...(iv)}$$

From Eqs. (ii) and (iii),
$$\frac{d}{dx}(y_1 - y_2) + P(x)(y_1 - y_2) = 0 \qquad \text{...(v)}$$

From Eqs. (iv) and (v), $$\frac{\frac{d}{dx}(y+y_1)}{\frac{d}{dx}(y_1-y_2)}=\frac{y-y_1}{y_1-y_2}$$

$$\Rightarrow \quad \frac{\frac{d}{dx}(y-y_1)}{y-y_1}=\frac{\frac{d}{dx}(y_1-y_2)}{y_1-y_2}$$

Integrating both the sides, we get

$$\log(y-y_1)=\log(y_1-y_2)$$

$$\therefore \quad y=y_1+C(y_1-y_2)$$

Now, $y=\alpha y_1+\beta y_2$ will be a solutions, if

$$\frac{d}{dx}(\alpha y_1+\beta y_2)+P(x)(\alpha y_1+\beta y_2)=Q(x)$$

or $$\alpha\left(\frac{dy_1}{dx}+P(x)\,y_1\right)+\beta\left(\frac{dy_2}{dx}+P(x)\,y_2\right)=Q(x)$$

or $$\alpha\,Q(x)+\beta\,Q(x)=Q(x)$$ [using Eqs. (ii) and (iii)]

$$\therefore \quad (\alpha+\beta)\,Q(x)=Q(x)$$

Hence, $$\alpha+\beta=1$$

Example 11 Find a pair of curves such that
(a) the tangents drawn at points with equal abscissae intersect on the y-axis.
(b) the normal drawn at points with equal abscissae intersect on x-axis.
(c) one curve passes through (1, 1) and other passes through (2, 3).

Solution. Let the curve be $y=f_1(x)$ and $y=f_2(x)$ equation of tangents with equal abscissa, x are

$$(y-f_1(x))=f'_1(x)(X-x)$$

and $$Y-f_2(x)=f'_2(x)(X-x)$$

These tangent intersect at y-axis,

$$\Rightarrow \quad -x\,f'_1(x)+f_1(x)=-x\,f'_2(x)+f_2(x)$$

$$\Rightarrow \quad f_1(x)-f_2(x)=x\,(f'_1(x)-f'_2(x))$$

Integrating both the sides, we get

$$\Rightarrow \quad \ln|f_1(x)-f_2(x)|=\ln|x|+C$$

$$\Rightarrow \quad f_1(x)-f_2(x)=\pm C_1\,x \qquad \text{...(i)}$$

Now, equations of normal with equal abscissa x, are

$$y-f_1(x)=-\frac{1}{f'_1(x)}(X-x)$$

and $$(y-f_2(x))=-\frac{1}{f'_2(x)}(X-x)$$

As these normal intersect on the x-axis,

$$x+f_1(x)\cdot f'_1(x)=x+f_2(x)\cdot f'_2(x)$$

$$\Rightarrow \quad f_1(x)\cdot f_1'(x)=f_2(x)\cdot f_2'(x)$$

Integrating

$$\Rightarrow \quad f'^2_1(x)-f_2^2(x)=C_2$$

$$\Rightarrow \qquad f_1(x)+f_2(x)=\frac{C_2}{f_1(x)-f_2(x)}=\pm\frac{C_2}{C_1\,x}=\pm\frac{\lambda_2}{x} \quad \text{[using Eq. (i)]} \quad \text{...(ii)}$$

From Eqs. (i) and (ii), we get

$$2f_1(x)=\pm\left(\frac{\lambda_2}{x}+C_1\,x\right),\ 2f_2(x)=\pm\left(\frac{\lambda_2}{x}-C_1\,x\right)$$

We have, $\quad f_1(1)=1 \quad$ and $\quad f_2(2)=3$

$$\Rightarrow \qquad f_1(x)=\frac{2}{x}-x \quad \text{and} \quad f_2(x)=\frac{2}{x}+x$$

Example 12 Given two curves $y=f(x)$ passing through (0, 1) and $y=\int_{-\infty}^{x} f(t)\,dt$ passing through $(0, 1/n)$. The tangents drawn to both the curves at the points with equal abscissae intersect on the x-axis find the curve $y=f(x)$.

Solution. Equation of the tangent to the curve; $y=f(x)$ is

$$(Y-y)=f'(x)(X-x)$$

Equation of tangent to the curve $g(x)=y_1=\int_{-\infty}^{x} f(t)\,dt$ is

$$(Y-y_1)=g'(x)(X-x)=f(x)(X-x)$$

Given that tangent with equal abscissas intersect on the x-axis.

$$\Rightarrow \qquad x-\frac{y}{f'(x)}=x-\frac{y_1}{f(x)}$$

$$\Rightarrow \qquad \frac{f(x)}{f'(x)}=\frac{y_1}{f(x)} \qquad [\because y=f(x)]$$

$$\Rightarrow \qquad \frac{f(x)}{y_1}=\frac{f'(x)}{f(x)} \ \Rightarrow\ \frac{g'(x)}{g(x)}=\frac{f'(x)}{f(x)}$$

$$\Rightarrow \qquad \frac{g'(x)}{g(x)}=k \ \Rightarrow\ g(x)=Ce^{kx}$$

$$\Rightarrow \qquad g'(x)=k\,Ce^{kx} \ \Rightarrow\ f(x)=k\,Ce^{kx}$$

$y=f(x)$ passes through $(0, 1)\Rightarrow kC=1$

$y_1=g(x)$ passes through $(0, 1/n)\Rightarrow C=\frac{1}{n}\Rightarrow k=n$

$$\Rightarrow \qquad f(x)=e^{nx}$$

Example 13 A normal is drawn at a point $P(x, y)$ of a curve. It meets the x-axis and the y-axis in point A and B, respectively, such that $\frac{1}{OA}+\frac{1}{OB}=1$, where O is the origin, find the equation of such a curve passing through (5, 4).

Solution. The equation of the normal at (x, y) is

$$(X-x)+(Y-y)\frac{dy}{dx}=0$$

$$\Rightarrow \qquad \frac{X}{x+y\dfrac{dy}{dx}}+\frac{Y}{\dfrac{(x+y\,dy/dx)}{dy/dx}}=1 \ \Rightarrow\ OA=x+y\frac{dy}{dx},\ OB=\frac{\left(x+y\dfrac{dy}{dx}\right)}{\dfrac{dy}{dx}}$$

Also, $\frac{1}{OA}+\frac{1}{OB}=1$

$\Rightarrow \quad 1+\frac{dy}{dx}=x+y\frac{dy}{dx}$

$\Rightarrow \quad (y-1)\frac{dy}{dx}+(x-1)=0$

Integrating, we get

$$(y-1)^2+(x-1)^2=C$$

Since, the curve passes through (5, 4), $C=25$.

Hence, the curve is $(x-1)^2+(y-1)^2=25$.

Example 14 A line is drawn from a point $P(x, y)$ on curve $y=f(x)$, making an angle with the x-axis which is supplementary to the one made by the tangent to the curve at $P(x, y)$. The line meets the x-axis at A. Another line perpendicular to the first, is drawn from $P(x, y)$ meeting the y-axis at B. If $OA=OB$, where O is origin, find all curve which passes through (1, 1).

Solution. The equation of the line through $P(x, y)$ making an angle with the x-axis which is supplementary to the angle made by the tangent at $P(x, y)$ is

$$Y-y=-\frac{dy}{dx}(X-x) \qquad \text{...(i)}$$

where it meets the x-axis.

$$Y=0, \quad X=x+\frac{y}{dy/dx} \Rightarrow OA=x+y\frac{dx}{dy} \qquad \text{...(ii)}$$

The line through $P(x, y)$ and perpendicular to Eq. (i) is

$$Y-y=\frac{dx}{dy}(X-x)$$

where it meets the y-axis.

$$X=0, Y=y-x\frac{dx}{dy} \Rightarrow OB=y-x\frac{dx}{dy} \qquad \text{...(iii)}$$

Since, $OA=OB$

$\Rightarrow \quad x+y\frac{dx}{dy}=y-x\frac{dx}{dy}$

or $\quad (y-x)=(y+x)\frac{dx}{dy}$

or $\quad \frac{dy}{dx}=\frac{y+x}{y-x}$, put $y=vx$

$\Rightarrow \quad x\frac{dv}{dx}=\frac{1+2v-v^2}{v-1}$

$\Rightarrow \quad \frac{(1-v)\,dv}{1+2v-v^2}+\frac{dx}{x}=0$

$\Rightarrow \quad \log(1+2v-v^2)+\log x=C_1$

$\Rightarrow \quad x^2+2xy-y^2=C \qquad$ where $C_1=\log\sqrt{C}$

Since, curve passes through $(1, 1) \rightarrow C=2$

$\therefore$ Required curve, $\quad x^2-y^2+2xy=2$

Example 15 The tangent and a normal to a curve at any point P meet the x and y axes at A, B, C and D respectively. Find the equation of the curve passing through (1, 0) if the centre of circle through O, C, P and B lies on the line $y = x$ (where O is origin).

Solution. Let $P(x, y)$ be a point on the curve.

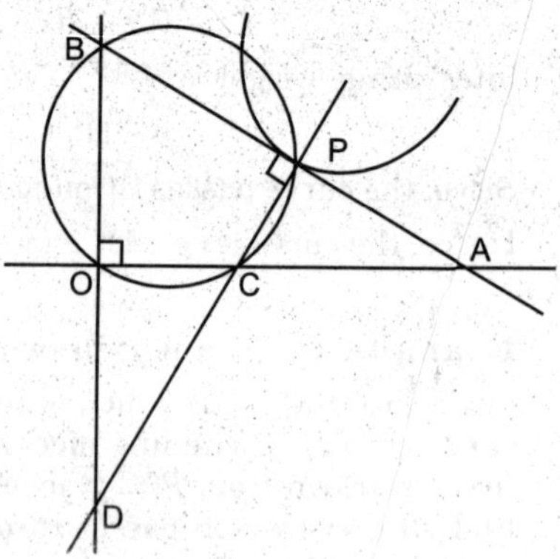

$$\Rightarrow \qquad C \equiv \left(x + y\frac{dy}{dx}, 0 \right)$$

$$B \equiv \left(0, y - x\frac{dy}{dx} \right)$$

Circle passing through O, C, P and B has its centre at mid-point of BC.

Let the centre of the circle be (α, β).

$$\Rightarrow \qquad 2\alpha = x + y\frac{dy}{dx}$$

and $$2\beta = y - x\frac{dy}{dx}$$

and since $\beta = \alpha$, $$y - x\frac{dy}{dx} = x + y\frac{dy}{dx}$$

$$\Rightarrow \qquad \frac{dy}{dx} = \frac{y - x}{y + x}$$

Let $y = vx \Rightarrow x\dfrac{dv}{dx} = -\dfrac{(1 + v^2)}{1 + v} \Rightarrow \dfrac{1 + v}{v^2 + 1}dv = -\dfrac{dx}{x}$

Integrating both the sides, we get

$$\int \frac{1 + v}{v^2 + 1}dv = -\int \frac{dx}{x}$$

$$\Rightarrow \qquad \frac{1}{2}\int \frac{2v}{v^2 + 1}dv + \int \frac{dv}{v^2 + 1} = -\int \frac{dx}{x}$$

$$\Rightarrow \qquad \frac{1}{2}\log|v^2 + 1| + \tan^{-1}|v| = -\log x + C$$

$$\Rightarrow \qquad \log\{(\sqrt{v^2 + 1})x\} + \tan^{-1} v = C$$

$$\Rightarrow \qquad \log\sqrt{x^2 + y^2} + \tan^{-1}\frac{y}{x} = C$$

As $x = 1$ and $y = 0$, $\log 1 + \tan^{-1} 0 = C \Rightarrow C = 0$

$\therefore$ Required curve, $(\log\sqrt{x^2 + y^2}) + \tan^{-1}\left(\dfrac{y}{x}\right) = 0$

Type 2 : Only One Correct Option

Example 16 Spherical rain drop evaporates at a rate proportional to its surface area. The differential equation corresponding to the rate of change of the radius of the rain drop, if the constant of proportionality is $K > 0$, is

(a) $\dfrac{dr}{dt} + K = 0$ (b) $\dfrac{dr}{dt} - K = 0$ (c) $\dfrac{dr}{dt} = Kr$ (d) None of these

Solution. $\dfrac{dV}{dt} = -k4\pi r^2$ ...(i)

But $V = \dfrac{4}{3}\pi r^3 \Rightarrow \dfrac{dV}{dt} = 4\pi r^2 \dfrac{dr}{dt}$ ...(ii)

Therefore, $\dfrac{dr}{dt} = -K$

Hence, (a) is the correct answer.

Example 17 A function $y = f(x)$ satisfies the differential equation $f(x)\cdot \sin 2x - \cos x + (1+\sin^2 x) f'(x) = 0$ with initial condition $y(0) = 0$. The value of $f(\pi/6)$ is equal to

(a) 1/5 (b) 3/5 (c) 4/5 (d) 2/5

Solution. $y \sin 2x - \cos x + (1+\sin^2 x)\dfrac{dy}{dx} = 0$ where $y = f(x)$

$$\frac{dy}{dx} + \left(\frac{\sin 2x}{1+\sin^2 x}\right) y = \frac{\cos x}{1+\sin^2 x}$$

$$\text{IF} = e^{\int \frac{\sin 2x}{1+\sin^2 x}dx} = e^{\int \frac{dt}{t}} = e^{\ln(1+\sin^2 x)}$$

$$= 1+\sin^2 x \qquad \text{(by putting } 1+\sin^2 x = t\text{)}$$

$$y(1+\sin^2 x) = \int \cos x \, dx$$

$$y(1+\sin^2 x) = \sin x + C;\ y(0) = 0 \Rightarrow C = 0$$

Therefore, $y = \dfrac{\sin x}{1+\sin^2 x};\ y\left(\dfrac{\pi}{6}\right) = \dfrac{2}{5}$

Hence, (d) is the correct answer.

Example 18 The general solution of the differential equation $\dfrac{dy}{dx} = \dfrac{1-x}{y}$ is a family of curves which looks most like which of the following?

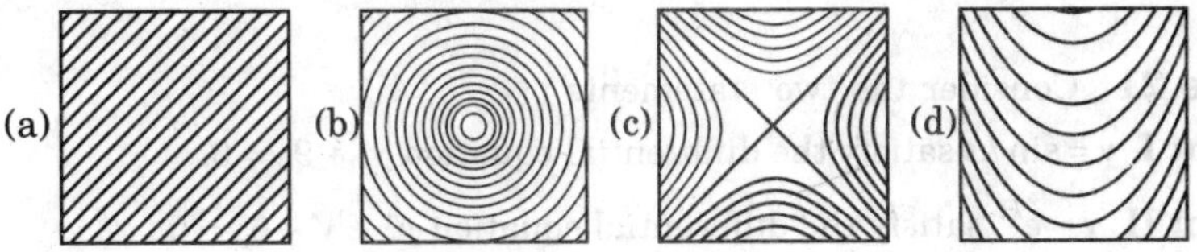

Solution. $\int y\,dy = \int (1-x)\,dx$

$$\frac{y^2}{2} = x - \frac{x^2}{2} + C$$

$$x^2 + y^2 - 2x = C$$

$$(x-1)^2 + y^2 = C + 1 = C$$

Hence, (b) is the correct answer.

Points to Consider

Family of concentric circles with (1, 0) as the centre and variable radius.

Example 19 Water is drained from a vertical cylindrical tank by opening a valve at the base of the tank. It is known that the rate at which the water level drops is proportional to the square root of water depth y, where the constant of proportionality $k > 0$ depends on the acceleration due to gravity and the geometry of the hole. If t is measured in minutes and $k = \frac{1}{15}$, then the time to drain the tank, if the water is 4 m deep to start with is

(a) 30 min (b) 45 min (c) 60 min (d) 80 min

Solution. $\frac{dy}{dt} = -k\sqrt{y}$; when $t = 0$; $y = 4$

$$\int_4^0 \frac{dy}{\sqrt{y}} = -k\int_0^t dt$$

$$[2\sqrt{y}]_4^0 = -kt = -\frac{t}{15}$$

$$0 - 4 = -\frac{t}{15}$$

$\Rightarrow$ $$t = 60 \text{ min}$$

Hence, (c) is the correct answer.

Example 20 Number of straight lines which satisfy the differential equation $\frac{dy}{dx} + x\left(\frac{dy}{dx}\right)^2 - y = 0$ is

(a) 1 (b) 2 (c) 3 (d) 4

Solution. $y = kx + b$; $\frac{dy}{dx} = k$

$\Rightarrow$ $$kx + b \equiv k + xk^2$$

$\Rightarrow$ $$k = k^2 \text{ and } b = k$$

$$k = 0 \quad \text{or} \quad k = 1$$

Hence, (b) is the correct answer.

Example 21 Consider the two statements :

Statement I $y = \sin kt$ satisfy the differential equation $y'' + 9y = 0$.

Statement II $y = e^{kt}$ satisfy the differential equation $y'' + y' - 6y = 0$.

The value of k for which both the statements are correct is

(a) –3 (b) 0 (c) 2 (d) 3

Solution. **Statement I** $y = \sin kt$, $y' = k\cos kt$; $y'' = -k^2 \sin kt$

$\therefore$ $$-k^2 \sin kt + 9\sin kt = 0$$

$$\sin kt\,[9 - k^2] = 0$$

$\Rightarrow$ $$k = 0, k = 3, k = -3$$

Statement II $y = e^{kt}$, $y' = ke^{kt}$; $y'' = k^2e^{kt}$

$\therefore$ $$k^2e^{kt} + ke^{kt} - 6e^{kt} = 0$$

$$e^{kt}[k^2 + k - 6] = 0$$

$$(k+3)(k-2)=0$$
$$k=-3 \text{ or } 2$$

Common value is $k=-3$.

Hence, (a) is the correct answer.

Example 22 If $y=\dfrac{x}{\ln|cx|}$ (where c is an arbitrary constant) is the general solution of the differential equation $\dfrac{dy}{dx}=\dfrac{y}{x}+\phi\left(\dfrac{x}{y}\right)$, then the function $\phi\left(\dfrac{x}{y}\right)$ is

(a) $\dfrac{x^2}{y^2}$ (b) $-\dfrac{x^2}{y^2}$ (c) $\dfrac{y^2}{x^2}$ (d) $-\dfrac{y^2}{x^2}$

Solution. $\ln c+\ln|x|=\dfrac{x}{y}$

Differentiating w.r.t. x, $\dfrac{1}{x}=\dfrac{y-xy_1}{y^2}$

$$\frac{y^2}{x}=y-x\frac{dy}{dx}$$

$$\frac{dy}{dx}=\frac{y}{x}-\frac{y^2}{x^2} \Rightarrow \phi\left(\frac{x}{y}\right)=-\frac{y^2}{x^2}$$

Hence, (d) is the correct answer.

Example 23 If $\int_a^x ty(t)\,dt=x^2+y(x)$, then y as a function of x is

(a) $y=2-(2+a^2)e^{\frac{x^2-a^2}{2}}$ (b) $y=1-(2+a^2)e^{\frac{x^2-a^2}{2}}$

(c) $y=2-(1+a^2)e^{\frac{x^2-a^2}{2}}$ (d) None of these

Solution. Differentiating both the sides, we get

$$xy(x)=2x-y'(x)$$

Hence, $\dfrac{dy}{dx}-xy=-2x$ $\left[y'(x)=\dfrac{dy}{dx};\ y(x)=y\right]$

$$\text{IF}=e^{\int -x\,dx}=e^{\frac{-x^2}{2}}$$

$$ye^{\frac{-x^2}{2}}=\int -2xe^{\frac{-x^2}{2}}\,dx$$

Let $e^{\frac{-x^2}{2}}=t$

$\Rightarrow$ $-xe^{\frac{-x^2}{2}}\,dx=dt$

$$I=\int 2dt$$

$$ye^{\frac{-x^2}{2}}=2e^{\frac{-x^2}{2}}+C$$

$$y=2+Ce^{\frac{x^2}{2}}$$

If $x=a \Rightarrow a^2+y=0 \Rightarrow y=-a^2$ (from the given equation)

Hence, $$-a^2 = 2 + Ce^{\frac{a^2}{2}}; Ce^{\frac{a^2}{2}} = -(2 + a^2)$$

$$C = -(2 + a^2)e^{-\frac{a^2}{2}}; y = 2 - (2 + a^2)e^{\frac{x^2 - a^2}{2}}$$

Hence, (a) is the correct answer.

Example 24 The differential equation $\frac{dy}{dx} = \frac{\sqrt{1 - y^2}}{y}$ determines a family of circles with **[IIT JEE 2007]**

(a) variable radii and a fixed centre at (0, 1)

(b) variable radii and fixed centre at (0, –1)

(c) fixed radius 1 and variable centres along the x-axis

(d) fixed radius 1 and variable centres along the y-axis

Solution. $\because$ $$\frac{dy}{dx} = \frac{\sqrt{1 - y^2}}{y}$$

$$\Rightarrow \int \frac{y}{\sqrt{1 - y^2}} dy = \int dx \Rightarrow -\frac{1}{2} \cdot 2\sqrt{1 - y^2} = x + C$$

$$\Rightarrow -\sqrt{1 - y^2} = x + C \Rightarrow 1 - y^2 = (x + C)^2$$

$$\Rightarrow (x + C)^2 + y^2 = 1$$

Therefore, the differential equation represents a circle of fixed radius 1 and variable centres along the x-axis.
Hence, (c) is the correct answer.

Type 3 : More Than One Correct Options

Example 25 A curve $y = f(x)$ has the property that the perpendicular distance of the origin from the normal at any point P of the curve is equal to the distance of the point P from the x-axis. Then the differential equation of the curve

(a) is homogeneous

(b) can be converted into linear differential equation with some suitable substitution

(c) is the family of circles touching the x-axis at the origin

(d) the family of circles touching the y-axis at the origin

Solution. Equation of normal

$$Y - y = -\frac{1}{m}(X - x)$$

$$-my + my = X - x$$

$$X + my - (x + my) = 0$$

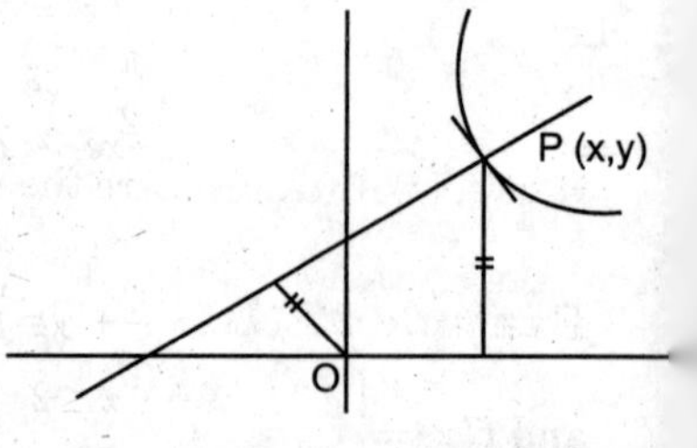

Perpendicular from (0, 0) $= \left|\frac{x + my}{\sqrt{1 + m^2}}\right| = y$

$$x^2 + 2xym = y^2$$

$$\frac{dy}{dx}=\frac{y^2-x^2}{2xy} \Rightarrow \text{homogeneous}$$

Also, $$x\cdot 2y\cdot\frac{dy}{dx}-x^2=y^2$$

Put $y^2=t$; $$2y\frac{dy}{dx}=\frac{dt}{dx};\ x\cdot\frac{dt}{dx}+x^2=t$$

$$\frac{dt}{dx}-\frac{1}{x}t=-x \text{ which is linear differential equation.}$$

Hence, (a), (b) and (d) are the correct answers.

Example 26 A differentiable function satisfies $f(x)=\int_0^x\{f(t)\cos t-\cos(t-x)\}\,dt.$

Which is of the following hold good?

(a) $f(x)$ has a minimum value $1-e$

(b) $f(x)$ has a maximum value $1-e^{-1}$

(c) $f''\left(\frac{\pi}{2}\right)=e$

(d) $f'(0)=1$

Solution. $$f(x)=\int_0^x\{f(t)\cos t-\cos(t-x)\}\,dx$$

$$=\int_0^x f(t)\cos t\,dt-\int_0^x\cos(-t)\,dt \qquad \left[\int_0^a f(x)\,dx=\int_0^a f(a-x)\,dx\right]$$

$$f(x)=\int_0^x f(t)\cos t\,dt-\sin x$$

Differentiating both the sides, we get

$$f'(x)=f(x)\cos x-\cos x$$

Let $$f(x)=y;\ f'(x)=\frac{dy}{dx}$$

$$\frac{dy}{dx}-y\cos x=-\cos x \qquad \text{(L.D.E.)}$$

$$\text{IF}=e^{-\int\cos x\,dx}=e^{-\sin x}$$

Therefore, $y\cdot e^{-\sin x}=-\int e^{-\sin x}\cos x\,dx;\ y\cdot e^{-\sin x}=C+e^{-\sin x};\ y=Ce^{\sin x}+1$

If $x=0$; $y=0$ (from the given relation)

$\Rightarrow$ $C=-1$

Therefore, $f(x)=1-e^{\sin x}$

Now, minimum value $=1-e$ (when $x=\pi/2$)

Maximum value $=1-e^{-1}$ (when $x=-\pi/2$)

$$f'(x)=-e^{\sin x}\cos x$$

Therefore, $$f'(0)=-1$$

$$f''(x)=-[\cos^2 x\cdot e^{\sin x}-e^{\sin x}\cdot\sin x]$$

$$f''\left(\frac{\pi}{2}\right)=e$$

Hence, (a), (b) and (c) are the correct answers.

Example 27 Let $\frac{dy}{dx}+y=f(x)$ where y is a continuous function of x with $y(0)=1$ and $f(x)=\begin{cases}e^{-x}, & \text{if } 0\le x\le 2\\ e^{-2}, & \text{if } x>2\end{cases}$. Which is of the following hold(s) good?

(a) $y(1)=2e^{-1}$ (b) $y'(1)=-e^{-1}$ (c) $y(3)=-2e^{-3}$ (d) $y'(3)=-2e^{-3}$

Solution. $$\frac{dy}{dx} + y = f(x)$$

$$\text{IF} = e^x$$

$$ye^x = \int e^x f(x)\, dx + C$$

Now, if $0 \le x \le 2$, then $ye^x = \int e^x e^{-x}\, dx + C$

$\Rightarrow$ $ye^x = x + C$

$x = 0,\ y(0) = 1,\ C = 1$

$\therefore$ $$ye^x = x + 1 \quad \text{...(i)}$$

$$y = \frac{x+1}{e^x};\ y(1) = \frac{2}{e}$$

$$y' = \frac{e^x - (x+1)e^x}{e^{2x}}$$

$$y'(1) = \frac{e - 2e}{e^2} = \frac{-e}{e^2} = -\frac{1}{e}$$

If $x > 2$, $$ye^x = \int e^{x-2}\, dx$$

$$ye^x = e^{x-2} + C$$

$$y = e^{-2} + Ce^{-x}$$

As y is continuous.

$\therefore$ $$\lim_{x \to 2} \frac{x+1}{e^x} = \lim_{x \to 2} (e^{-2} + Ce^{-x})$$

$$3e^{-2} = e^{-2} + Ce^{-2} \Rightarrow C = 2$$

$\therefore$ for $x > 2$

$$y = e^{-2} + 2e^{-x}$$

Hence, $$y(3) = 2e^{-3} + e^{-2} = e^{-2}(2e^{-1} + 1)$$

$$y' = -2e^{-x}$$

$$y'(3) = -2e^{-3}$$

Hence, (a), (b) and (d) are the correct answers.

Example 28 A curve $y = f(x)$ passes through (1, 1) and tangent at $P(x, y)$ cuts the x-axis and y-axis at A and B respectively such that $BP : AP = 3 : 1$, then **[IIT JEE 2006]**

(a) equation of curve is $xy' - 3y = 0$

(b) normal at (1, 1) is $x + 3y = 4$

(c) curve passes through $\left(2, \frac{1}{8}\right)$

(d) equation of curve is $xy' + 3y = 0$

Solution. Equation of the tangent to the curve $y = f(x)$ at (x, y) is

$$Y - y = \frac{dy}{dx}(X - x)$$

$\because$ Thus, cuts the x-axis at A and y-axis at B.

$\therefore$ $$A\left(\frac{x\frac{dy}{dx} - y}{\frac{dy}{dx}}, 0\right) \text{ and } B\left(0, -x\frac{dy}{dx} + y\right)$$

$\because$ $$BP : PA = 3 : 1$$

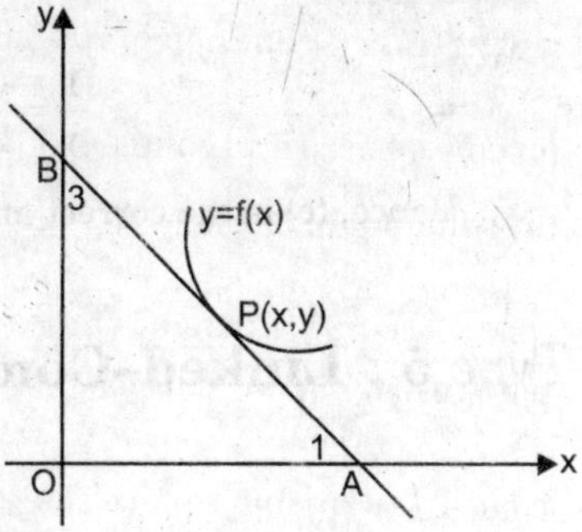

$$\Rightarrow \quad x = \frac{\dfrac{3\left(x\dfrac{dy}{dx} - y\right)}{(dy/dx)} + 1 \times 0}{4}$$

$$\Rightarrow \quad x\frac{dy}{dx} + 3y = 0$$

$$\Rightarrow \quad \int \frac{dy}{y} = \int -3\frac{dx}{x}$$

$$\Rightarrow \quad \log y = -3\log x + \log C$$

$$\Rightarrow \quad y = \frac{C}{x^3}$$

$\because$ Curve passes through (1, 1) $\therefore C = 1$

$\therefore$ Curve is $x^3 y = 1$ which also passes through $\left(2, \frac{1}{8}\right)$.

Hence, (c) and (d) are the correct answers.

Type 4 : **Assertion and Reason**

29 Let a solution $y = y(x)$ of the differential equation $x\sqrt{x^2-1}\,dy - y\sqrt{y^2-1}\,dx = 0$ satisfy $y(2) = \frac{2}{\sqrt{3}}$. **[IIT JEE 2008]**

Statement I $y(x) = \sec\left(\sec^{-1} x - \frac{\pi}{6}\right)$.

Statement II $y(x)$ is given by $\frac{1}{y} = \frac{2\sqrt{3}}{x} - \sqrt{1 - \frac{1}{x^2}}$.

(a) Statement I is true, Statement II is also true; Statement II is the correct explanation of Statement I.
(b) Statement I is true, Statement II is also true; Statement II is not the correct explanation of Statement I.
(c) Statement I is true, Statement II is false.
(d) Statement I is false, Statement II is true.

Solution. $\because \; x\sqrt{x^2-1}\,dy - y\sqrt{y^2-1}\,dx = 0$

Which can be rewritten as $\dfrac{dx}{x\sqrt{x^2-1}} = \dfrac{dy}{y\sqrt{y^2-1}}$

Integration yields, $\displaystyle\int \frac{dx}{x\sqrt{x^2-1}} = \int \frac{dy}{y\sqrt{y^2-1}}$

$$\Rightarrow \quad \sec^{-1} x = \sec^{-1} y + C$$

$$\Rightarrow \quad \sec^{-1}(2) = \sec^{-1}(2/\sqrt{3}) + C$$

$$\Rightarrow \quad \frac{\pi}{3} = \frac{\pi}{6} + C \Rightarrow C = \frac{\pi}{6}$$

Thus, $$\sec^{-1} x = \sec^{-1} y + \frac{\pi}{6}$$

$$\Rightarrow \quad y = \sec\left(\sec^{-1} x - \frac{\pi}{6}\right) = \frac{1}{\cos\left(\cos^{-1}\dfrac{1}{x} - \dfrac{\pi}{6}\right)} = \frac{1}{\dfrac{1}{x}\cdot\dfrac{\sqrt{3}}{2} + \sqrt{1 - \dfrac{1}{x^2}}\cdot\dfrac{1}{2}}$$

$$\Rightarrow \qquad \frac{1}{y} = \frac{\sqrt{3}}{2x} + \frac{\sqrt{1 - \frac{1}{x^2}}}{2}$$

Hence, (c) is the correct answer.

Type 5 : **Linked–Comprehension Based Questions**

Passage
(Q. Nos. 30 to 32)

A curve $y = f(x)$ *satisfies the differential equation* $(1 + x^2)\frac{dy}{dx} + 2yx = 4x^2$ *and passes through the origin.*

30 The function $y = f(x)$

(a) is strictly increasing, $\forall\, x \in R$
(b) is such that it has a minima but no maxima
(c) is such that it has a maxima but no minima
(d) has no inflection point

31 The area enclosed by $y = f^{-1}(x)$, the x-axis and the ordinate at $x = 2/3$ is

(a) $2 \ln 2$ (b) $\frac{4}{3} \ln 2$ (c) $\frac{2}{3} \ln 2$ (d) $\frac{1}{3} \ln 2$

32 For the function $y = f(x)$ which one of the following does not hold good?

(a) $f(x)$ is a rational function
(b) $f(x)$ has the same domain and same rage
(c) $f(x)$ is a transcendental function
(d) $y = f(x)$ is a bijective mapping

Solution. (Q. Nos. 30 to 32)

$$\frac{dy}{dx} + \left(\frac{2x}{1 + x^2}\right) y = \frac{4x^2}{1 + x^2}$$

$$\text{IF} = e^{\int \frac{2x}{1 + x^2} dx} = e^{\ln (1 + x^2)} = (1 + x^2)$$

$$\therefore \qquad y(1 + x^2) = \int 4x^2\, dx = \frac{4x^3}{3} + C$$

Passing through (0, 0) $\Rightarrow C = 0$

$$\therefore \qquad y = \frac{4x^3}{3(1 + x^2)}$$

$$\frac{dy}{dx} = \frac{4}{3}\left[\frac{(1 + x^2)3x^2 - x^3 \cdot 2x}{(1 + x^2)^2}\right] = \frac{4}{3}\left[\frac{3x^2 + x^4}{(1 + x^2)^2}\right] = \frac{4x^2(3 + x^2)}{3(1 + x^2)^2}$$

Hence, $\frac{dy}{dx} > 0, \forall\, x \neq 0$; $\frac{dy}{dx} = 0$ at $x = 0$ and it does not change sign $\Rightarrow x = 0$ is the point of inflection

$y = f(x)$ is increasing for all $x \in R$.

$x \to \infty;\ y \to \infty;\ x \to -\infty;\ y \to -\infty$

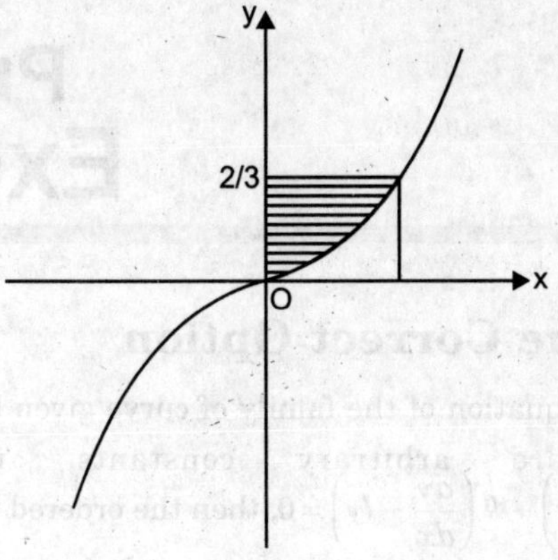

Area enclosed by $y = f^{-1}(x)$, x-axis and ordinate at $x = \frac{2}{3}$

$$A = \frac{2}{3} - \frac{4}{3}\int_0^1 \frac{x^3}{1+x^2}\,dx$$

Put $\quad 1 + x^2 = t \Rightarrow 2x\,dx = dt$

$$A = \frac{2}{3} - \frac{2}{3}\int_1^2 \frac{(t-1)}{t}\,dt = \frac{2}{3} - \frac{2}{3}\int_1^2 \left(1 - \frac{1}{t}\right) dt$$

$$= \frac{2}{3} - \frac{2}{3}[t - \ln t]_1^2 = \frac{2}{3} - \frac{2}{3}[(2 - \ln 2) - 1]$$

$$= \frac{2}{3} - \frac{2}{3}[1 - \ln 2] = \frac{2}{3}\ln 2$$

Ans. 30. (a) **31.** (c) **32.** (c)

Proficiency Exercise 1

Type 1 : **Only One Correct Option**

1. If the differential equation of the family of curve given by $y = Ax + Be^{2x}$, where A and B are arbitrary constants, is of the form $(1-2x)\frac{d}{dx}\left(\frac{dy}{dx} + ly\right) + k\left(\frac{dy}{dx} + ly\right) = 0$, then the ordered pair (k, l) is

(a) $(2, -2)$ (b) $(-2, 2)$ (c) $(2, 2)$ (d) $(-2, -2)$

2. A curve passes through the point $\left(1, \frac{\pi}{4}\right)$ and its slope at any point is given by $\frac{y}{x} - \cos^2\left(\frac{y}{x}\right)$. Then, the curve has the equation

(a) $y = x\tan^{-1}\left(\ln\frac{e}{x}\right)$ (b) $y = x\tan^{-1}(\ln 2)$

(c) $y = \frac{1}{x}\tan^{-1}\left(\ln\frac{e}{x}\right)$ (d) None of these

3. The x-intercept of the tangent to a curve is equal to the ordinate of the point of contact. The equation of the curve through the point (1, 1) is

(a) $ye^{\frac{x}{y}} = e$ (b) $xe^{\frac{x}{y}} = e$ (c) $xe^{\frac{y}{x}} = e$ (d) $ye^{\frac{y}{x}} = e$

4. A function $y = f(x)$ satisfies the condition $f'(x)\sin x + f(x)\cos x = 1$, $f(x)$ being bounded when $x \to 0$. If $I = \int_0^{\pi/2} f(x)\,dx$, then

(a) $\frac{\pi}{2} < I < \frac{\pi^2}{4}$ (b) $\frac{\pi}{4} < I < \frac{\pi^2}{2}$ (c) $1 < I < \frac{\pi}{2}$ (d) $0 < I < 1$

5. A curve is such that the area of the region bounded by the coordinate axes, the curve and the ordinate of any point on it is equal to the cube of that ordinate. The curve represents

(a) a pair of straight lines (b) a circle

(c) a parabola (d) an ellipse

6. The value of the constant 'm' and 'c' for which $y = mx + c$ is a solution of the differential equation $D^2y - 3Dy - 4y = -4x$.

(a) is $m = -1$; $c = 3/4$ (b) is $m = 1$; $c = -3/4$

(c) no such real m, c (d) is $m = 1$; $c = 3/4$

7. The real value of m for which the substitution, $y = u^m$ will transform the differential equation, $2x^4y\frac{dy}{dx} + y^4 = 4x^6$ into a homogeneous equation is

(a) $m = 0$ (b) $m = 1$ (c) $m = 3/2$ (d) No value of m

8. The solution of the differential equation, $x^2 \dfrac{dy}{dx} \cdot \cos\dfrac{1}{x} - y\sin\dfrac{1}{x} = -1$, where $y \to -1$ as $x \to \infty$ is

(a) $y = \sin\dfrac{1}{x} - \cos\dfrac{1}{x}$ (b) $y = \dfrac{x+1}{x\sin\dfrac{1}{x}}$

(c) $y = \sin\dfrac{1}{x} + \cos\dfrac{1}{x}$ (d) $y = \dfrac{x+1}{x\cos\dfrac{1}{x}}$

9. A wet porous substance in the open air loses its moisture at a rate proportional to the moisture content. If a sheet hung in the wind loses half its moisture during the first hour, then the time when it would have lost 99.9% of its moisture is (weather conditions remaining same)

(a) more than 100 h (b) more than 10 h

(c) approximately 10 h (d) approximately 9 h

10. A curve C passes through origin and has the property that at each point (x, y) on it, the normal line at that point passes through (1, 0). The equation of a common tangent to the curve C and the parabola $y^2 = 4x$ is

(a) $x = 0$ (b) $y = 0$ (c) $y = x + 1$ (d) $x + y + 1 = 0$

11. A function $y = f(x)$ satisfies $(x+1) \cdot f'(x) - 2(x^2 + x) f(x) = \dfrac{e^{x^2}}{(x+1)}$, $\forall\, x > -1$. If $f(0) = 5$, then $f(x)$ is

(a) $\left(\dfrac{3x+5}{x+1}\right) \cdot e^{x^2}$ (b) $\left(\dfrac{6x+5}{x+1}\right) \cdot e^{x^2}$ (c) $\left(\dfrac{6x+5}{(x+1)^2}\right) \cdot e^{x^2}$ (d) $\left(\dfrac{5-6x}{x+1}\right) \cdot e^{x^2}$

12. The curve, with the property that the projection of the ordinate on the normal is constant and has a length equal to 'a', is

(a) $x + a\ln(\sqrt{y^2 - a^2} + y) = C$ (b) $x + \sqrt{a^2 - y^2} = C$

(c) $(y - a)^2 = Cx$ (d) $ay = \tan^{-1}(x + C)$

13. The differential equation corresponding to the family of curves $y = e^x(ax + b)$ is

(a) $\dfrac{d^2y}{dx^2} + 2\dfrac{dy}{dx} - y = 0$ (b) $\dfrac{d^2y}{dx^2} - 2\dfrac{dy}{dx} + y = 0$

(c) $\dfrac{d^2y}{dx^2} + 2\dfrac{dy}{dx} + y = 0$ (d) $\dfrac{d^2y}{dx^2} - 2\dfrac{dy}{dx} - y = 0$

14. The equation to the orthogonal trajectories of the system of parabolas $y = ax^2$ is

(a) $\dfrac{x^2}{2} + y^2 = C$ (b) $x^2 + \dfrac{y^2}{2} = C$ (c) $\dfrac{x^2}{2} - y^2 = C$ (d) $x^2 - \dfrac{y^2}{2} = C$

15. A function $f(x)$ satisfying $\int_0^1 f(tx)\,dt = n\,f(x)$, where $x > 0$, is

(a) $f(x) = C \cdot x^{\frac{1-n}{n}}$ (b) $f(x) = C \cdot x^{\frac{n}{n-1}}$

(c) $f(x) = C \cdot x^{\frac{1}{n}}$ (d) $f(x) = C \cdot x^{(1-n)}$

16. The substitution $y = z^\alpha$ transforms the differential equation $(x^2y^2 - 1)\,dy + 2xy^3 dx = 0$ into a homogeneous differential equation for

(a) $\alpha = -1$ (b) 0 (c) $\alpha = 1$ (d) No value of α

17. A curve passing through (2, 3) and satisfying the differential equation $\int_0^x t y(t)\, dt = x^2 y(x), (x > 0)$ is

(a) $x^2 + y^2 = 13$ (b) $y^2 = \frac{9}{2}x$ (c) $\frac{x^2}{8} + \frac{y^2}{18} = 1$ (d) $xy = 6$

18. Which one of the following curves represents the solution of the initial value problem $Dy = 100 - y$, where $y(0) = 50$?

(a) 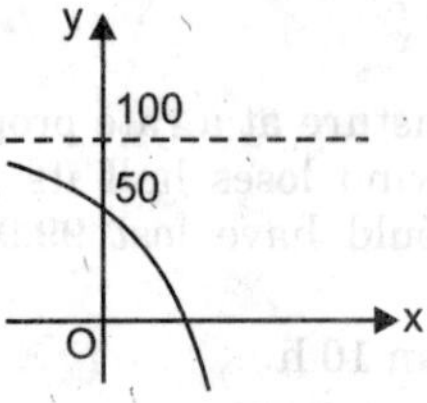

(b) 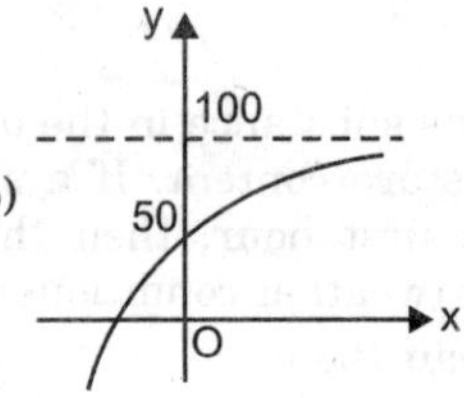

(c) 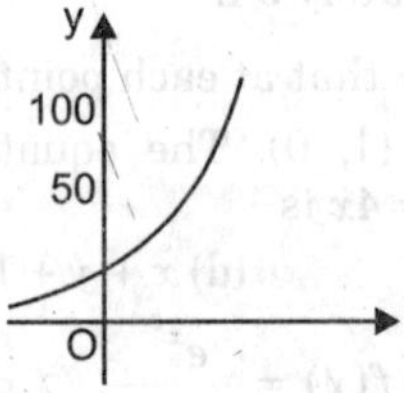

(d) 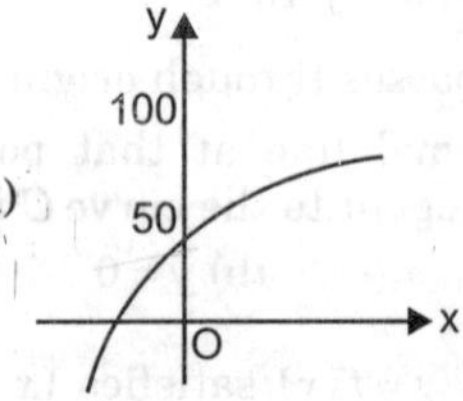

Type 2 : More Than One Correct Options

19. The differential equation $x\frac{dy}{dx} + \frac{3}{\frac{dy}{dx}} = y^2$

(a) is of order 1 (b) is of degree 2 (c) is linear (d) is non-linear

20. The function $f(x)$ satisfying the equation $f^2(x) + 4f'(x)\, f(x) + [f'(x)]^2 = 0$

(a) $f(x) = C \cdot e^{(2-\sqrt{3})x}$ (b) $f(x) = C \cdot e^{(2+\sqrt{3})x}$

(c) $f(x) = C \cdot e^{(\sqrt{3}-2)x}$ (d) $f(x) = C \cdot e^{-(2+\sqrt{3})x}$

where C is an arbitrary constant.

21. Which of the following pair(s) is/are orthogonal?

(a) $16x^2 + y^2 = C$ and $y^{16} = kx$ (b) $y = x + Ce^{-x}$ and $x + 2 = y + ke^{-y}$

(c) $y = Cx^2$ and $x^2 + 2y^2 = k$ (d) $x^2 - y^2 = C$ and $xy = k$

22. Family of curves whose tangent at a point with its intersection with the curve $xy = c^2$ form an angle of $\frac{\pi}{4}$ is

(a) $y^2 - 2xy - x^2 = k$ (b) $y^2 + 2xy - x^2 = k$

(c) $y = x - 2c\tan^{-1}\left(\frac{x}{c}\right) + k$ (d) $y = c\ln\left|\frac{c+x}{c-x}\right| - x + k$

23. The general solution of the differential equation, $x\left(\frac{dy}{dx}\right) = y \cdot \log\left(\frac{y}{x}\right)$ is

(a) $y = xe^{1-Cx}$ (b) $y = xe^{1+Cx}$ (c) $y = ex \cdot e^{Cx}$ (d) $y = xe^{Cx}$

where C is an arbitrary constant.

24. Which of the following equation(s) is/are linear?

(a) $\frac{dy}{dx}+\frac{y}{x}=\ln x$ (b) $y\left(\frac{dy}{dx}\right)+4x=0$ (c) $dx+dy=0$ (d) $\frac{d^2y}{dx^2}=\cos x$

25. The equation of the curve passing through (3, 4) and satisfying the differential equation, $y\left(\frac{dy}{dx}\right)^2+(x-y)\frac{dy}{dx}-x=0$ can be

(a) $x-y+1=0$ (b) $x^2+y^2=25$

(c) $x^2+y^2-5x-10=0$ (d) $x+y-7=0$

26. Identify the statement(s) which is/are true?

(a) $f(x, y)=e^{y/x}+\tan\frac{y}{x}$ is homogeneous of degree zero.

(b) $x\cdot\log\frac{y}{x}\,dx+\frac{y^2}{x}\sin^{-1}\frac{y}{x}\,dy=0$ is homogeneous differential equation.

(c) $f(x, y)=x^2+\sin x\cdot\cos y$ is not homogeneous.

(d) $(x^2+y^2)\,dx-(xy^2-y^3)\,dy=0$ is a homogeneous differential equation.

27. The graph of the function $y=f(x)$ passing through the point $(0,-1)$ and satisfying the differential equation $\frac{dy}{dx}+y\cos x=\cos x$ is such that

(a) it is a constant function

(b) it is periodic

(c) it is neither an even nor an odd function

(d) it is continuous and differentiable for all x

28. A function $y=f(x)$ satisfying the differential equation

$$\frac{dy}{dx}\cdot\sin x-y\cos x+\frac{\sin^2 x}{x^2}=0$$

is such that, $y\to 0$ as $x\to\infty$, then the statement which is correct?

(a) $\lim_{x\to 0} f(x)=1$ (b) $\int_0^{\pi/2} f(x)\,dx$ is less than $\frac{\pi}{2}$

(c) $\int_0^{\pi/2} f(x)\,dx$ is greater than unity (d) $f(x)$ is an odd function

29. Identify the statement(s) which is/are true?

(a) The order of differential equation $\sqrt{1+\frac{d^2y}{dx^2}}=x$ is 1.

(b) Solution of the differential equation $xdy-ydx=\sqrt{x^2+y^2}\,dx$ is $y+\sqrt{x^2+y^2}=Cx^2$.

(c) $\frac{d^2y}{dx^2}=2\left(\frac{dy}{dx}-y\right)$ is differential equation of family of curves $y=e^x(A\cos x+B\sin x)$.

(d) The solution of differential equation $(1+y^2)+(x-2e^{\tan^{-1}y})\frac{dy}{dx}=0$ is $xe^{\tan^{-1}y}=e^{2\tan^{-1}y}+k$.

30. Let $y=(A+Bx)e^{3x}$ is a solution of the differential equation

$$\frac{d^2y}{dx^2}+m\frac{dy}{dx}+ny=0,\ m, n\in I,\text{ then}$$

(a) $m=-6$ (b) $n=-6$ (c) $m=9$ (d) $n=9$

Type 3 : Assertion and Reason

Directions
(Q. Nos. 31 to 40)

For the following questions, choose the correct answers from the codes (a), (b), (c) and (d) defined as follows :

(a) Statement I is true, Statement II is true and Statement II is the correct explanation for Statement I.

(b) Statement I is true, Statement II is true and Statement II is not the correct explanation for Statement I.

(c) Statement I is true, Statement II is false.

(d) Statement I is false, Statement II is true.

31. A curve C has the property that its initial ordinate of any tangent drawn is less than the abscissa of the point of tangency by unity.

Statement I Differential equation satisfying the curve is linear.

Statement II Degree of differential equation is one.

32. **Statement I** Differential equation corresponding to all lines, $ax + by + c = 0$ has the order 3.

Statement II General solution of a differential equation of nth order contains n independent arbitrary constants.

33. **Statement I** Integral curves denoted by the first order linear differential equation $\frac{dy}{dx} - \frac{1}{x} y = -x$ are family of parabolas passing through the origin.

Statement II Every differential equation geometrically represents a family of curve having some common property.

34. **Statement I** The solution of $(y\,dx - x\,dy)\cot\left(\frac{x}{y}\right) = ny^2\,dx$ is $\sin\left(\frac{x}{y}\right) = Ce^{nx}$

Statement II Such type of differential equations can only be solved by the substitution $x = vy$.

35. **Statement I** The order of the differential equation whose general solution is $y = c_1 \cos 2x + c_2 \sin^2 x + c_4 e^{2x} + c_5 e^{2x + c_6}$ is 3.

Statement II Total number of arbitrary parameters in the given general solution in the Statement I is 6.

36. Consider the differential equation

$$(x^2 + 1)\cdot\frac{d^2y}{dx^2} = 2x\cdot\frac{dy}{dx}$$

Statement I For any member of this family $y \to \infty$ as $x \to \infty$.

Statement II Any solution of this differential equation is a polynomial of odd degree with positive coefficient of maximum power.

37. **Statement I** Order of differential equation of family of parallel whose axis is parallel to y-axis and latusrectum is fixed is 2.

Statement II Order of first equation is same as actual number of arbitrary constant present in differential equation.

38. Statement I The differential equation of all non-vertical lines in a plane is $\frac{d^2x}{dy^2} = 0$.

Statement II The general equation of all non-vertical lines in a plane is $ax + by = 1$, where $b \neq 0$.

39. Statement I The order of differential equation of all conics whose centre lies at origin is, 2.

Statement II The order of differential equation is same as number of arbitrary unknowns in the given curve.

40. Statement I $y = a \sin x + b \cos x$ is general solution of $y'' + y = 0$.

Statement II $y = a \sin x + b \cos x$ is a trigonometric function.

Type 4 : Linked–Comprehension Based Questions

Passage I

(Q. Nos. 41 to 43)

Let $y = f(x)$ *satisfies the equation* $f(x) = (e^{-x} + e^{x}) \cos x - 2x - \int_0^x (x - t) f'(t) dt.$

41. y satisfies the differential equation

(a) $\frac{dy}{dx} + y = e^x (\cos x - \sin x) - e^{-x} (\cos x + \sin x)$

(b) $\frac{dy}{dx} - y = e^x (\cos x - \sin x) + e^{-x} (\cos x + \sin x)$

(c) $\frac{dy}{dx} + y = e^x (\cos x + \sin x) - e^{-x} (\cos x - \sin x)$

(d) $\frac{dy}{dx} - y = e^x (\cos x - \sin x) + e^{-x} (\cos x - \sin x)$

42. The value of $f'(0) + f''(0)$ equals to

(a) -1 (b) 2

(c) 1 (d) 0

43. $f(x)$ as a function of x equals to

(a) $e^{-x} (\cos x - \sin x) + \frac{e^x}{5} (3 \cos x + \sin x) + \frac{2}{5} e^{-x}$

(b) $e^{-x} (\cos x + \sin x) + \frac{e^x}{5} (3 \cos x - \sin x) - \frac{2}{5} e^{-x}$

(c) $e^{-x} (\cos x - \sin x) + \frac{e^x}{5} (3 \cos x - \sin x) + \frac{2}{5} e^{-x}$

(d) $e^{-x} (\cos x + \sin x) + \frac{e^x}{5} (3 \cos x - \sin x) - \frac{2}{5} e^{-x}$

Passage II

(Q. Nos. 44 to 46)

For certain curves $y = f(x)$ satisfying $\frac{d^2y}{dx^2} = 6x - 4$, $f(x)$ has local minimum value 5 when $x = 1$.

44. Number of critical point for $y = f(x)$ for $x \in [0, 2]$

(a) 0 (b) 1 (c) 2 (d) 3

45. Global minimum value of $y = f(x)$ for $x \in [0, 2]$ is

(a) 5 (b) 7 (c) 8 (d) 9

46. Global maximum value of $y = f(x)$ for $x \in [0, 2]$ is

(a) 5 (b) 7 (c) 8 (d) 9

Passage III

(Q. Nos. 47 to 49)

If any differential equation in the form

$$f(f_1(x, y))\, d(f_1(x, y)) + \phi(f_2(x, y))\, d(f_2(x, y)) + \dots = 0$$

then each term can be integrated separately.

For example, $\int \sin xy\, d(xy) + \int \left(\frac{x}{y}\right) d\left(\frac{x}{y}\right) = -\cos xy + \frac{1}{2}\left(\frac{x}{y}\right)^2 + C$

47. The solution of the differential equation $x dy - y\, dx = \sqrt{x^2 - y^2}\, dx$ is

(a) $Cx = e^{\sin^{-1}\frac{y}{x}}$ (b) $xe^{\sin^{-1}\frac{y}{x}} = C$ (c) $x + e^{\sin^{-1}\frac{y}{x}} = C$ (d) None of these

48. The solution of the differential equation $(xy^4 + y)\, dx - x dy = 0$ is

(a) $\frac{x^3}{4} + \frac{1}{2}\left(\frac{x}{y}\right)^2 = C$ (b) $\frac{x^4}{4} + \frac{1}{3}\left(\frac{x}{y}\right)^3 = C$ (c) $\frac{x^4}{4} - \frac{1}{2}\left(\frac{x}{y}\right)^3 = C$ (d) $\frac{x^3}{4} - \frac{1}{2}\left(\frac{x}{y}\right)^2 = C$

49. Solution of differential equation

$(2x \cos y + y^2 \cos x)\, dx + (2y \sin x - x^2 \sin y)\, dy = 0$ is

(a) $x^2 \cos y + y^2 \sin x = C$ (b) $x \cos y - y \sin x = C$

(c) $x^2 \cos^2 y + y^2 \sin^2 x = C$ (d) None of these

Passage IV

(Q. Nos. 50 to 52)

Differential equation $\frac{dy}{dx} = f(x)\, g(x)$ can be solved by separating variable $\frac{dy}{g(y)} = f(x)\, dx$.

50. The equation of the curve to the point (1, 0) which satisfies the differential equation $(1 + y^2)\, dx - xy\, dy = 0$ is

(a) $x^2 + y^2 = 1$ (b) $x^2 - y^2 = 1$ (c) $x^2 + y^2 = 2$ (d) $x^2 - y^2 = 2$

51. Solution of the differential equation $\frac{dy}{dx} + \frac{1+y^2}{\sqrt{1-x^2}} = 0$ is

(a) $\tan^{-1} y + \sin^{-1} x = C$ (b) $\tan^{-1} x + \sin^{-1} y = C$

(c) $\tan^{-1} y \cdot \sin^{-1} x = C$ (d) $\tan^{-1} y - \sin^{-1} x = C$

52. If $\frac{dy}{dx} = 1 + x + y + xy$ and $y(-1) = 0$, then y is equal to

(a) $e^{\frac{(1-x)^2}{2}}$ (b) $e^{\frac{(1+x)^2}{2}} - 1$

(c) $\ln(1+x) - 1$ (d) $1 + x$

Type 5 : Match the Columns

53. Match the following :

Column I	Column II
(A) $\frac{xdx + ydy}{xdy - ydx} = \sqrt{\frac{a^2 - x^2 - y^2}{x^2 + y^2}}$	(p) $y = \frac{C}{x^3} e^{1/x}$
(B) Solution of $\cos^2 x \frac{dy}{dx} - \tan 2x \cdot y = \cos^4 x$, where $\lvert x \rvert < \frac{\pi}{4}$ and $y\left(\frac{\pi}{6}\right) = \frac{3\sqrt{3}}{8}$	(q) $\sqrt{x^2 + y^2} = a \sin\left\{C + \tan^{-1}\left(\frac{y}{x}\right)\right\}$
(C) The equation of all possible curves that will cut each member of the family of circles $x^2 + y^2 - 2cx = 0$ at right angle	(r) $x^2 + y^2 + Cy = 0$
(D) Solution of the equation $x\int_0^x y(t)\,dt = (x+1)\int_0^x ty(t)\,dt, x > 0$ is	(s) $y = \frac{\sin 2x}{2(1 - \tan^2 x)}$

54. Match the following :

Column I	Column II
(A) Circular plate is expanded by heat from radius 5 cm to 5.06 cm. Approximate increase in area is	(p) 4
(B) Side of cube increasing by 1%, then percentage increase in volume is	(q) 0.6π
(C) If the rate of decrease of $\frac{x^2}{2} - 2x + 5$ is twice the rate of decrease of x, then x is equal to	(r) 3
(D) Rate of increase in area of equilateral triangle of side 15 cm, when each side is increasing at the rate of 0.1 cm/s; is	(s) $\frac{3\sqrt{3}}{4}$

55. Match the following :

Column I	Column II
(A) The differential equation of the family of curves $y = e^x (A\cos x + B\sin x)$, where A, B are arbitrary constants, has the degree n and order m. Then, the values of n and m are, respectively	(p) 2,1
(B) The degree and order of the differential equation of the family of all parabolas whose axis is the x-axis, are respectively	(q) 1,1
(C) The order and degree of the differential equations of the family of circles touching the x-axis at the origin, are respectively	(r) 2,2
(D) The degree and order of the differential equation of the family of ellipse having the same foci, are respectively.	(s) 1,2

Type 6 : Integer Answer Type Questions

56. Find the constant of integration by the general solution of the differential equation $(2x^2y - 2y^4)\,dx + (2x^3 + 3xy^3)\,dy = 0$ if curve passes through (1, 1).

57. A tank initially contains 50 gallons of fresh water. Brine contains 2 pounds per gallon of salt, flows into the tank at the rate of 2 gallons per minutes and the mixture kept uniform by stirring, runs out at the same rate. If it will take for the quantity of salt in the tank to increase from 40 to 80 pounds (in seconds) is 206λ, then find λ.

58. If $f : R - \{-1\} \rightarrow$ and f is differentiable function which satisfies :

$f\{x + f(y) + xf(y)\} = y + f(x) + yf(x) \; \forall \; x, y \in R - \{-1\}$,

then find the value of $2010\,[1 + f(2009)]$.

59. If $\phi(x)$ is a differential real-valued function satisfying $\phi'(x) + 2\phi(x) \le 1$, then the value of $2\phi(x)$ is always less than or equal to ....

60. The degree of the differential equation satisfied by the curves $\sqrt{1+x} - a\sqrt{1+y} = 1$, is ....

Proficiency Exercise 2

1. Solve the following differential equations :

(i) $x\dfrac{dy}{dx} = y - x\cos^2\left(\dfrac{y}{x}\right)$

(ii) $(x\sqrt{x^2+y^2} - y)\,dx + (y\sqrt{x^2+y^2} - x)\,dy = 0$

(iii) $y^4dx + 2xy^3dy = \dfrac{y\,dx - x\,dy}{x^3y^3}$

(iv) $\dfrac{dy}{dx} + y f'(x) - f(x)\cdot f'(x) = 0,\ y \ne f(x)$

(v) $(x^2-1)\dfrac{dy}{dx}\sin y - 2x\cos y = 2x - 2x^3$

(vi) $\cos\left(\dfrac{x}{y}\right)(y\,dx - x\,dy) = xy^3\,(x\,dy + y\,dx)$

(vii) $\dfrac{dy}{dx} = \dfrac{(x-1)^2 + (y-2)^2\tan^{-1}\left(\dfrac{y-2}{x-1}\right)}{(xy-2x-y+2)\tan^{-1}\left(\dfrac{y-2}{x-1}\right)}$

(viii) $\dfrac{dy}{dx} = \dfrac{(x-1)^2 + (y+3)^2\, e^{\frac{y+3}{x-1}}}{(xy+3x-y-3)\,e^{\frac{y+3}{x-1}}}$

(ix) $\dfrac{dy}{dx} = \dfrac{y^2\cos x + 2y\sin 2x + 2\cos x\cdot\sin^2 x}{2\sin^2 x}$

(x) $\dfrac{dy}{dx} = \sqrt{\dfrac{1}{2} + \left\{\displaystyle\int_{\cos^4\theta}^{-\sin^4\theta} \dfrac{\sqrt{f(\phi)}\,d\phi}{\sqrt{f(\cos(2\theta-\phi))} + \sqrt{f(\phi)}}\right\}}$

Where $\theta = x + y$ and $\dfrac{\pi}{2} < x + y < \dfrac{3\pi}{4}$.

2. A curve C has the property that if the tangent drawn at any point P on C meets the coordinate axes at A and B, then P is the mid-point of AB. The curve passes through the point (1, 1). Determine the equation of the curve. **[IIT JEE 1998]**

3. Find the curve possessing the property that the intercept made by the tangent at any point of the curve on the y-axis is equal to square of the abscissa of the point of tangency.

4. Find the family of curves whose tangent form an angle $\pi/4$ with the hyperbola $xy = c$.

5. A and B are two separate reservoirs of water capacity of reservoir A is double the capacity of reservoir B. Both the reservoirs are filled completely with water their inlets are closed and then the water is released simultaneously from both the reservoirs. The rate of flow of water out of each reservoir any instant of time is proportional to the quantity of water in the reservoir at that time. One hour after the water is released, the quantity of the water in the reservoir A is $1\frac{1}{2}$ times the quantity of the water in reservoir B. After how many hours do both the reservoirs have the same quantity of water. **[IIT JEE 1997]**

6. Find the time required for a cylindrical tank of radius r and height H to empty through a round hole of area 'a' at the bottom. The flow through a hole is according to the law $U(t) = u\sqrt{2gh(t)}$ where $v(t)$ and $h(t)$ are respectively the velocity of flow through the hole and the height of the water level above the hole at time t and g is the acceleration due to gravity.

7. The hemispherical tank of radius 2 m is initially full of water and has an outlet of 12 cm^2 cross-sectional area at the bottom. The outlet is opened at some instant. The flow through the outlet is according to the law $v(t) = 0\cdot 6\sqrt{2gh(t)}$, where $v(t)$ and $h(t)$ are respectively velocity of the flow through the outlet and the height of water level above the outlet at time t and g is the acceleration due to gravity. Find the time it takes to empty the tank. **[IIT JEE 2001]**

8. The tangent at a point P of a curve meets the y-axis at A, and the line parallel to y-axis through P meets the x-axis at B. It area of ΔOAB is constant (O being the origin), then form the differential equation of the curve and show that the curve is a hyperbola.

9. Find the value of k such that the family of parabolas $y = cx^2 + k$ is the orthogonal trajectory of the family of ellipses $x^2 + 2y^2 - y = c$.

10. A curve passing through (1, 0) is such that the ratio of the square of the intercept cut by any tangent on the y-axis to the subnormal is equal to the ratio of the product of the coordinates of the point of tangency to the product of square of the slope of the tangent and the subtangent at the same point. Determine all such possible curves.

11. Consider a curve $y = f(x)$ in xy-plane. The curve passes through (0, 0) and has the property that a segment of tangent drawn at any point. $P(x, f(x))$ and the line $y = 3$ gets bisected by the line $x + y = 1$. Determine the equation of the curve.

12. A curve passes through (2, 1) and is such that the square of the ordinate is twice the rectangle contained by the abscissa and the intercept of the normal. Find its equation.

13. A normal at $P(x, y)$ on a curve meets the x-axis at Q and N is the foot of the ordinate at P. If $NQ = \dfrac{x(1+y^2)}{(1+x^2)}$ find the equation of the curve, given that it passes through the point (3, 1).

14. Consider the curved mirror $y = f(x)$ passing through (8, 6) having the property that all light rays emerging from origin, after getting reflected from the mirror becomes parallel to x-axis, find the equation of curve.

15. Let $f : R^+ \to R$ satisfies the functional equation $f(xy) = e^{xy-x-y}(e^y f(x) + e^x f(y))\ \forall\ x, y \in R^+$. If $f'(1) = e$, determine $f(x)$.

16. Let $y = f(x)$ be curve passing through $(1, \sqrt{3})$ such that tangent at any point P on the curve lying in the first quadrant has positive slope and the tangent and the normal at the point P cut the x-axis at A and B respectively so that the mid-point of AB is origin. Find the differential equation of the curve and hence determine $f(x)$.

17. Determine all the curves for which the ratio of the length of the segment intercepted by any tangent on the y-axis to the length of the radius vector is constant.

18. A point $p(x, y)$ moves on the curve $x^{2/3} + y^{2/3} = a^{2/3}$; $a > 0$ for each position (x, y) of p, perpendiculars are drawn from origin upon the tangent and normal at p, the length (absolute value) of them being $p_1(x)$ and $p_2(x)$ respectively. Prove that $\frac{dp_1}{dx} \cdot \frac{dp_2}{dx} \leq 0$.

19. If y_1 and y_2 are the solution of differential equation $dy/dx + Py = Q$, where P and Q are function of x alone and $y_2 = y_1 z$, then prove that $z = 1 + c \cdot e^{-\int \frac{Q}{y_1} dx}$, where c is an arbitrary constant.

20. Let $f(x, y) = 0$ be a curve passing through the point $(0, 2)$ such that all the rays coming from a fixed point source when reflected from it are parallel. Form the differential equation and determine all such possible curves.

21. Let $y = f(x)$ be a differentiable function $\forall\ x \in R$ and satisfies :

$$f(x) = x + \int_0^1 x^2 z f(z)\, dz + \int_0^1 x z^2 f(z)\, dz.$$

Determine the function.

22. If $f : R - \{-1\} \to R$ and f is differentiable function satisfies :

$$f((x) + f(y) + x f(y)) = y + f(x) + y f(x)\ \forall\ x, y \in R - \{-1\}$$

Find $f(x)$.

23. Two points move in the same straight line starting at the same moment in the same direction. The first moves with constant velocity U and the second with constant acceleration f(its initial velocity being zero). Show that the greatest distance between the points before the second catches first is $\frac{U^2}{2f}$.

24. An inverted cone of height H and radius R is pointed at bottom. It is filled with a volatile liquid completely. If the rate of evaporation is directly proportional to the surface area of liquid in contact with air. Find the time in which whole liquid evaporates?

25. A tank contains 100 litres of an aqueous solution containing 10 kg of salt water is entering in the tank at the rate of 3 litres per minute, and mixture is flowing out at 2 litre per minute, the concentric is maintained uniform by stirring. How much salt will the tank contain at the end of one hour?

Answers

Target Exercise 4.1

1. (a) 2. (b) 3. (a) 4. (a) 5. (b)

Target Exercise 4.2

1. (b) 2. (c) 3. (a) 4. (a) 5. (c) 6. (a) 7. (a)
8. (a)

Target Exercise 4.3

1. (b) 2. (d) 3. (a) 4. (b) 5. (a) 6. (a) 7. (c)
8. (b) 9. (a) 10. (d)

Target Exercise 4.4

1. (a) 2. (a) 3. (b) 4. (b) 5. (c)

Target Exercise 4.5

1.(a) 2. (c) 3. (c) 4. (a) 5. (c)

Proficiency Exercise 1

1. (a) 2. (a) 3. (a) 4. (a) 5. (c) 6. (b) 7. (c)
8. (a) 9. (c) 10. (a) 11. (b) 12. (a) 13. (b) 14. (a)
15. (a) 16. (a) 17. (d) 18. (b) 19. (a,b,d) 20. (c,d)
21. (a,b,c,d) 22. (a,b,c,d) 23. (a,b,c) 24. (a,c,d) 25. (a,b) 26. (a,b,c)
27. (a,b,d) 28. (a,b,c) 29. (b,c,d) 30. (a,d) 31. (b) 32. (d) 33. (d)
34. (c) 35. (c) 36. (a) 37. (b) 38. (d) 39. (d) 40. (b)
41. (a) 42. (d) 43. (c) 44. (c) 45. (a) 46. (b) 47. (a)
48. (b) 49. (a) 50. (b) 51. (a) 52. (b)
53. (A) → (q), (B) → (s), (C) → (r), (D) → (p)
54. (A) → (q), (B) → (r), (C) → (p), (D) → (s)
55. (A) → (s), (B) → (s), (C) → (q), (D) → (p)
56. (1) 57. (8) 58. (1) 59. (1) 60. (1)

Proficiency Exercise 2

1. (i) $\tan\frac{y}{x} = \log\frac{K}{x}$ (ii) $(x^2 + y^2)^{3/2} = 3xy + C$

(iii) $\frac{x^3y^6}{3} = \log\left(\frac{x}{y}\right) + C$ (iv) $y = f(x) - 1 + Ce^{-f(x)}$

(v) $(x^2 - 1)\cos y = \frac{x^4}{2} - x^2 + C$ (vi) $\sin\left(\frac{x}{y}\right) = \frac{(xy)^2}{2} + C$

(vii) $\{(x-1)^2 + (y-1)^2\}\tan^{-1}\left(\frac{y-2}{x-1}\right) = (x-1)(y-2) + 2(x-1)^2\log C(x-1)$

(viii) $(y - x + 4)e^{\frac{x+3}{x-1}} = (x-1)\log|(x-1)| + C(x-1)$

(ix) $\frac{2}{\sqrt{5}} \tan^{-1}\left(\frac{y}{\sin x} + \frac{1}{4}\right) = \log(\sin x) + C$

(x) $\frac{2}{\sqrt{3}} \tan^{-1}\left(\frac{2\tan\theta - 1}{\sqrt{3}}\right) = x + C$

2. $xy = 1$

3. $y = -x^2 + Cx$

4. $\frac{x+y}{\sqrt{c}} + \log\left|\frac{x-\sqrt{c}}{x+\sqrt{c}}\right| = K$

5. $t_0 = \log_{3/4}\left(\frac{1}{2}\right)$

6. $t = \frac{\pi r^2}{\mu a}\sqrt{\frac{2H}{g}}$

7. $t = \frac{7\pi \times 10^5}{135\sqrt{g}}$

8. Required curve is, $Cx^2 - xy + K = 0$

9. $k = 1/4$

10. $x = e^{\pm 2\sqrt{y/x}}$

11. $(y-3)^2 = 9(1 - x - y)$

12. $4x^2 + 2y^2 = 9x$

13. $5(1 + y^2) = (1 + x^2)$

14. $y^2 = 4(1 + x)$ or $y^2 = 36(9 - x)$

15. $f(x) = e^x \log x$

16. $x + f(x) f'(x) = \sqrt{x^2 + f^2(x)}$ and $f^2(x) = 1 + 2x$

17. $(y + \sqrt{x^2 + y^2})x^{k-1} = C_1$

20. $y^2 = 4(x + 1)$ or $y^2 = -4(x - 1)$

21. $f(x) = \frac{20x}{119}(4 + 9x)$

22. $f(x) = -\left(\frac{x}{1+x}\right)$

24. $T = \frac{H}{k}$

25. 3.9 kg

Solutions

(*Proficiency Exercise* 1)

Type 1 : Only One Correct Option

1.
$$y \cdot e^{-2x} = Ax\, e^{-2x} + B$$
$$e^{-2x} \cdot y_1 - 2ye^{-2x} = A(e^{-2x} - 2x\, e^{-2x})$$

Cancelling e^{-2x} throughout

$$y_1 - 2y = A(1 - 2x) \quad \text{...(i)}$$

Differentiating again $y_2 - 2y_1 = -2A$

$\Rightarrow$
$$A = \frac{2y_1 - y_2}{2}$$

On substituting A in Eq. (i)

$$2(y_2 - 2y) = (2y_1 - y_2)(1 - 2x)$$
$$2y_1 - 4y = 2y_1(1 - 2x) - (1 - 2x)y_2$$
$$(1 - 2x)\frac{d}{dx}\left(\frac{dy}{dx} - 2y\right) + 2\left(\frac{dy}{dx} - 2y\right) = 0$$

Hence, $k = 2$ and $l = -2$

$\Rightarrow$ Ordered pair $(k, l) \equiv (2, -2)$

2. $\dfrac{dy}{dx} = \dfrac{y}{x} - \cos^2 \dfrac{y}{x}$

$y = vx$

$$v + x\frac{dv}{dx} = v - \cos^2 v$$

$$\int \frac{dv}{\cos^2 v} + \int \frac{dx}{x} = C$$

$$\tan v + \ln x = C$$

$$\tan\frac{y}{x} + \ln x = C$$

If $x = 1,\ y = \dfrac{\pi}{4} \Rightarrow C = 1$

$$\tan\frac{y}{x} = 1 - \ln x = \ln\frac{e}{x}$$

$$y = x\tan^{-1}\left(\ln\frac{e}{x}\right)$$

3. $Y - y = m(X - x)$

For X-intercept $Y = 0$

$$X = x - \frac{y}{m}$$

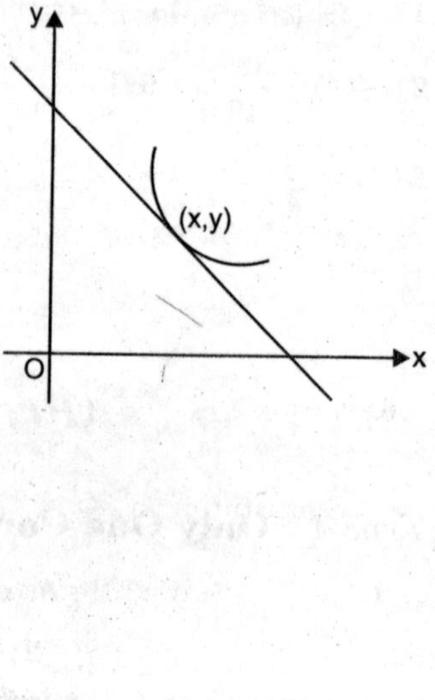

Therefore, $x - \dfrac{y}{m} = y$

or $\dfrac{dy}{dx} = \dfrac{y}{x - y}$

Put $y = vx$

$$v + x\frac{dv}{dx} = \frac{v}{1 - v}$$

$$x\frac{dv}{dx} = \frac{v}{1 - v} - v = \frac{v - v + v^2}{1 - v}$$

$$\int \frac{1 - v}{v^2}\,dv = \int \frac{dx}{x}$$

$$-\frac{1}{v} - \ln v = \ln x + C$$

$$-\frac{x}{y} - \ln\frac{y}{x} = \ln x + C$$

$$-\frac{x}{y} = \ln y + C$$

$x = 1,\ y = 1 \Rightarrow C = -1$

$$1 - \frac{x}{y} = \ln y$$

$$y = e \cdot e^{-x/y}$$

$$e^{-x/y} = \frac{e}{y}$$

$$ye^{x/y} = e$$

4. $\sin x \dfrac{dy}{dx} + y\cos x = 1$

$$\frac{dy}{dx} + y\cot x = \text{cosec}\, x$$

$$\text{IF} = e^{\int \cot x\, dx} = e^{\ln(\sin x)} = \sin x$$

$$y\sin x = \int \text{cosec}\, x \cdot \sin x\, dx$$

$$y\sin x = x + C$$

If $x = 0$, y is finite $\therefore\ C = 0$

$$y = x(\text{cosec}\, x) = \frac{x}{\sin x}$$

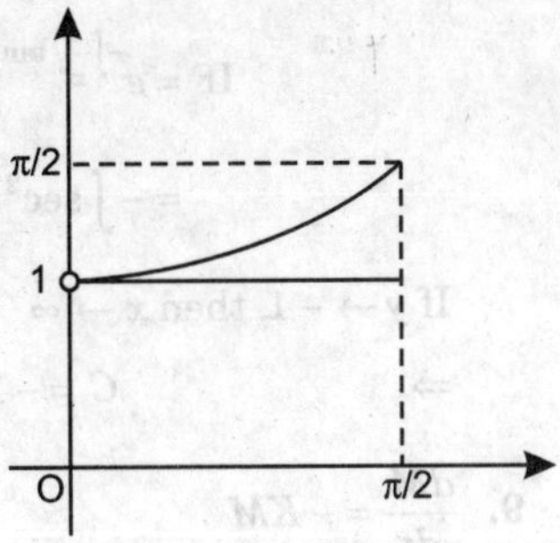

Now, $I < \dfrac{\pi^2}{4}$ and $I > \dfrac{\pi}{2}$

Hence, $\dfrac{\pi}{2} < I < \dfrac{\pi^2}{4}$.

5. $\int_0^x f(x)\, dx = y^3$

Differentiating $f(x) = 3y^2 \cdot \dfrac{dy}{dx}$

$$y = 3y^2 \frac{dy}{dx}$$

$\Rightarrow \quad y = 0 \quad$ (rejected)

or $\quad 3y\, dy = dx$

$$\frac{3y^2}{2} = x + C \Rightarrow \text{Parabola}$$

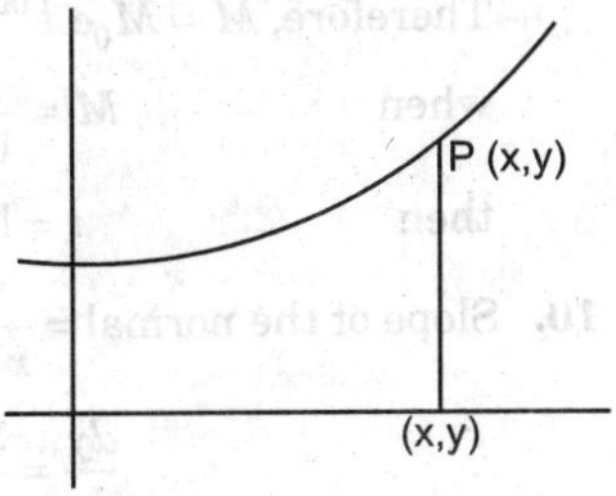

6. $y = mx + c;\ \dfrac{dy}{dx} = m;\ \dfrac{d^2y}{dx^2} = 0$

Substituting in $\dfrac{d^2y}{dx^2} - 3\dfrac{dy}{dx} - 4y = -4x$

$$0 - 3m - 4(mx + c) = -4x$$

$$-3m - 4c - 4mx = -4x$$

$$-(3m + 4c) = 4x(m - 1) \quad \ldots\text{(i)}$$

Eq. (i) is true for all real x, if $m = +1$ and $c = -3/4$.

7. $y = u^m \Rightarrow \dfrac{dy}{dx} = mu^{m-1}\dfrac{dy}{dx}$

Since, $2x^4 \cdot u^m \cdot mu^{m-1} \cdot \dfrac{du}{dx} + u^{4m} = 4x^6$

$$\frac{du}{dx} = \frac{4x^6 - u^{4m}}{2m\, x^4\, u^{2m-1}}$$

$\Rightarrow \quad 4m = 6 \Rightarrow m = \dfrac{3}{2}$

and $\quad 2m - 1 = 2 \Rightarrow m = \dfrac{3}{2}$

8. $\frac{dy}{dx} - \frac{y}{x^2}\tan\frac{1}{x} = -\sec\frac{1}{x}\cdot\frac{1}{x^2}$

$$\text{IF} = e^{-\int\frac{1}{x^2}\tan\frac{1}{x}dx} = \sec\frac{1}{x} \Rightarrow y\cdot\sec\frac{1}{x}$$

$$= -\int \sec^2\left(\frac{1}{x}\right)\frac{1}{x^2}\,dx = \tan\frac{1}{x} + C$$

If $y \to -1$, then $x \to \infty$

$\Rightarrow$ $$C = -1 \Rightarrow y = \sin\frac{1}{x} - \cos\frac{1}{x}$$

9. $\frac{dM}{dt} = -KM$

$$M = Ce^{-kt}$$

when $t = 0$; $M = M_0$

$\Rightarrow$ $$C = M_0 \Rightarrow M = M_0e^{-kt}$$

when $t = 1$, $M = \frac{M_0}{2}$

$\Rightarrow$ $$k = \ln 2$$

Therefore, $M = M_0e^{-t\ln 2}$

when $$M = \frac{M_0}{1000},$$

then $$t = \log_2 1000$$

10. Slope of the normal $= \frac{y}{x-1}$

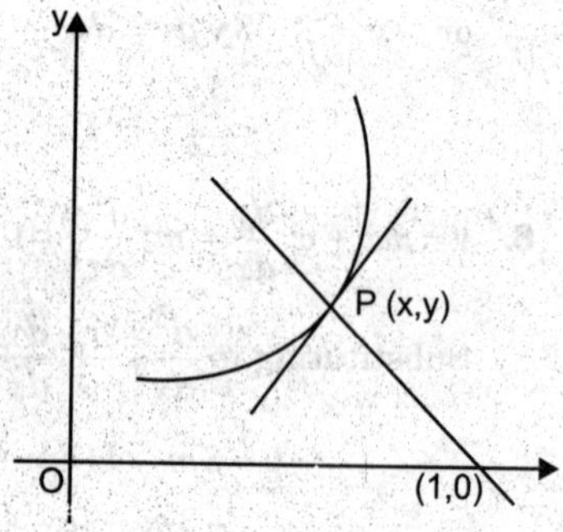

$\therefore$ $$\frac{dy}{dx} = \frac{1-x}{y}$$

$$\frac{y^2}{2} = x - \frac{x^2}{2} + C \quad \text{...(ii)}$$

Eq. (ii) passes through (0, 0). Thus, $C = 0$

$$x^2 + y^2 - 2x = 0$$

Now, tangent to $y^2 = 4x$

$$y = mx + \frac{1}{m} \quad \text{...(iii)}$$

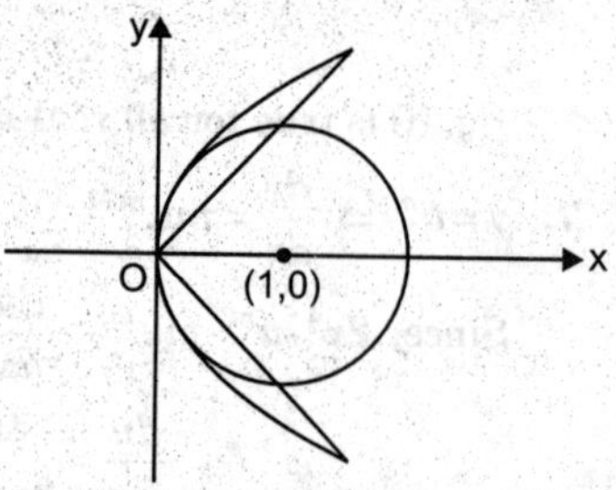

If it touches the circle

$$x^2 + y^2 - 2x = 0$$

Then, $$\left|\frac{m + (1/m)}{\sqrt{1+m^2}}\right| = 1$$

$\Rightarrow$ $$1 + m^2 = m^2 \Rightarrow m \to \infty$$

Hence, tangent is y-axis *ie*, $x = 0$.

11. $f'(x) - \frac{2x(x+1)}{x+1}f(x) = \frac{e^{x^2}}{(x+1)^2}$

$$\text{IF} = e^{\int -2x\,dx} = e^{-x^2}$$

$$\therefore \qquad f(x)\cdot e^{-x^2} = \int \frac{dx}{(x+1)^2}$$

$$\Rightarrow \qquad f(x)\cdot e^{-x^2} = -\frac{1}{x+1} + C$$

At $x = 0, f(0) = 5 \Rightarrow C = 6$

$$\therefore \qquad f(x) = \left(\frac{6x+5}{x+1}\right)\cdot e^{x^2}$$

12. Ordinate $= PM$. Let $P \equiv (x, y)$

Projection of ordinate on normal $= PN$

$$\therefore \qquad PN = PN\cos\theta = a \qquad \text{(given)}$$

$$\therefore \qquad \frac{y}{\sqrt{1+\tan^2\theta}} = a$$

$$\Rightarrow \qquad y = a\sqrt{1+(y_1)^2}$$

$$\Rightarrow \qquad \frac{dy}{dx} = \frac{\sqrt{y^2-a^2}}{a}$$

$$\Rightarrow \qquad \int \frac{a\,dy}{\sqrt{y^2-a^2}} = \int dx$$

$$\Rightarrow \qquad a\ln|y+\sqrt{y^2-a^2}| = x + C$$

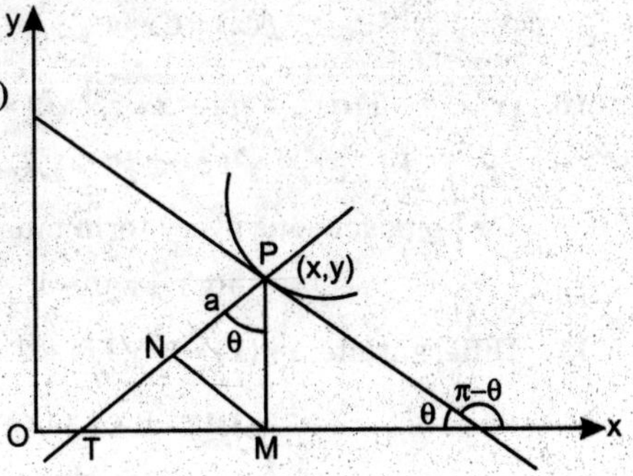

13. $$y = e^x(ax+b) \qquad \text{...(i)}$$

Differentiating w.r.t. x, we get

$$\frac{dy}{dx} = e^x(ax+b) + e^x\cdot a \quad \text{or} \quad \frac{dy}{dx} = y + ae^x \qquad \text{...(ii)}$$

Again, differentiating both the sides

$$\frac{d^2y}{dx^2} = \frac{dy}{dx} + ae^x \qquad \text{...(iii)}$$

From Eq. (iii) – Eq. (ii)

$$\frac{d^2y}{dx^2} - \frac{2dy}{dx} + y = 0$$

is required differential equation.

14. $\frac{dy}{dx} = 2ax = 2x\cdot\frac{y}{x^2}; \ \frac{dy}{dx} = \frac{2y}{x};$

Now, $$m\frac{dy}{dx} = -1 \ \Rightarrow\ m = -\frac{x}{2y} \ \Rightarrow\ \frac{dy}{dx} = -\frac{x}{2y}$$

$$y^2 = -\frac{x^2}{2} + C$$

15. $\int_0^1 f(tx)\,dt = n\cdot f(x)$

Put, $x = y \ \Rightarrow\ dt = \frac{1}{x}dy$

$$\therefore \qquad \frac{1}{x}\int_0^x f(y)\,dy = nf(x)$$

$$\therefore \quad \int_0^x f(y)\,dy = x \cdot n \cdot f(x)$$

Differentiating, $f(x) = n\,[f(x) + xf'(x)]$

$$f(x)(1-n) = nx\,f'(x)$$

$$\therefore \quad \frac{f'(x)}{f(x)} = \frac{1-n}{nx}$$

Integrating, $\ln f(x) = \left(\frac{1-n}{n}\right) \ln Cx = \ln (Cx)^{\frac{1-n}{n}}$;

$$\therefore \quad f(x) = Cx^{\frac{1-n}{n}}$$

16. $(x^2 z^{2\alpha} - 1)\,\alpha z^{\alpha-1}\,dz + 2x\,z^{3\alpha}\,dx = 0$

or $\quad \alpha(x^2\,z^{3\alpha-1} - z^{\alpha-1})\,dz + 2x\,z^{3\alpha}\,dx = 0$

for homogeneous every term must be of the same degree,

$$3\alpha + 1 = \alpha - 1 \Rightarrow \alpha = -1$$

17. Differentiate, $xy(x) = x^2 y'(x) + 2\,xy(x)$

or $\quad xy(x) + x^2\,y'(x) = 0$

$$x\frac{dy}{dx} + y = 0$$

$$\ln y + \ln x = \ln C$$

$$xy = C$$

18. $\int \frac{dy}{100-y} = \int dx$

$$-\ln(100-y) = x + C$$

$$\ln(100-y) = -x + C$$

$$x = 0,\ y = 50$$

Since, $\quad C = \ln 50$

$$x = \ln 50 - \ln(100-y)$$

$$\ln\frac{50}{100-y} = x \Rightarrow \frac{50}{100-y} = e^x$$

$$\Rightarrow \quad 100 - y = 50e^{-x} \Rightarrow y = 100 - 50e^{-x}$$

Type 2 : **More Than One Correct Options**

19. Here, $x\left(\frac{dy}{dx}\right)^2 - 3y^2\left(\frac{dy}{dx}\right) + 3 = 0$

has order 1, degree 2 and non-linear.

20. Here, $(f'(x))^2 + 4f'(x)\cdot f(x) + (f(x))^2 = 0$

$$\Rightarrow \quad \left(\frac{f'(x)}{f(x)}\right)^2 + 4\left(\frac{f'(x)}{f(x)}\right) + 1 = 0$$

$$\Rightarrow \quad \frac{f'(x)}{f(x)} = \frac{-4 \pm \sqrt{16-4}}{2} \Rightarrow \frac{f'(x)}{f(x)} = -2 \pm \sqrt{3}$$

Integrating both the sides

$$\log |f(x)| = (-2 \pm \sqrt{3})x + C_1$$

$$\Rightarrow \quad f(x) = e^{(-2 \pm \sqrt{3})x} \cdot C$$

21. (a) $32x + 2y\dfrac{dy}{dx} = 0$

$$\Rightarrow \quad \frac{dy}{dx} = m_1 = -\frac{16x}{y} \text{ and } 16y^{15}\frac{dy}{dx} = k$$

$$\Rightarrow \quad \frac{dy}{dx} = m_2 = \frac{k}{16y^{15}}$$

$$m_1 m_2 = -\frac{16x}{y} \cdot \frac{k}{16y^{15}} = -\frac{x}{y^{16}} \cdot k = -\frac{x}{y^{16}} \cdot \frac{y^{16}}{x} = -1$$

(b) $\dfrac{dy}{dx} = 1 - ce^{-x} = 1 - (y - x) = -(y - x - 1)$ [using $ce^{-x} = y - x$]

and $\dfrac{dy}{dx} - k \cdot \dfrac{dy}{dx} e^{-y} = 1$

$$\frac{dy}{dx}[1 - ke^{-y}] = 1$$

or $[1 - (x + 2 - y)]\dfrac{dy}{dx} = 1$ [using $ke^{-y} = x - y + 2$]

$$\frac{dy}{dx} = m_2 = \frac{1}{y - x - 1}$$

$$\Rightarrow \quad m_1 m_2 = -1$$

(c) $\dfrac{dy}{dx} = 2cx = 2x \cdot \dfrac{y}{x^2} = \dfrac{2y}{x} = m_1$

Also, $2x + 4y\dfrac{dy}{dx} = 0 \Rightarrow \dfrac{dy}{dx} = -\dfrac{x}{2y} = m_2$

Hence, $m_1 m_2 = -1$

(d) $x^2 - y^2 = c$

$$2x - 2y\frac{dy}{dx} = 0 \Rightarrow \frac{dy}{dx} = \frac{x}{y} = m_1$$

$$xy = k$$

$$x\frac{dy}{dx} + y = 0 \Rightarrow \frac{dy}{dx} = -\frac{y}{x} = m_2$$

$\therefore \quad m_1 m_2 = -1$

Hence, (a), (b), (c) and (d) are all correct.

22. Let $m = \dfrac{dy}{dx}$ be the slope of tangent (x, y) to the required curve.

m_1 = slope of the tangent at $xy = c^2 = -\dfrac{c^2}{x^2}$ or $-\dfrac{y}{x}$

Hence, $$\frac{m + \dfrac{c^2}{x^2}}{1 - \dfrac{c^2}{x^2}m} = \pm 1 \quad \text{or} \quad \frac{m + \dfrac{y}{x}}{1 - \dfrac{y}{x}m} = \pm 1$$

23. $x\dfrac{dy}{dx} = y\log\left(\dfrac{y}{x}\right)$

$\dfrac{dy}{dx} = \dfrac{y}{x}\log\left(\dfrac{y}{x}\right)$, put $y = vx$

$$\Rightarrow \quad \frac{dy}{dx} = v + x\frac{dv}{dx}$$

$$\therefore \quad v + x\frac{dv}{dx} = v\log v \Rightarrow x\frac{dv}{dx} = v(\log v - 1)$$

$$\Rightarrow \quad \int\frac{dv}{v(\log v - 1)} = \int\frac{dx}{x} \Rightarrow \int\frac{dt}{t-1} = \int\frac{dx}{x}, \text{ let } \log v = t$$

$$\Rightarrow \quad \log(t-1) = \log(x) + \log C$$

$$\Rightarrow \quad \log(t-1) = \log(xC)$$

$$\Rightarrow \quad t = 1 + xC \Rightarrow \log\frac{y}{x} = 1 + xC$$

$$\Rightarrow \quad y = x\cdot e^{1+Cx} \quad \text{or} \quad y = x\cdot e^{1-Cx}$$

$$\text{and} \quad \log\frac{y}{x} = \log e + Cx \Rightarrow y = ex\cdot e^{Cx}$$

24. Clearly, (a) and (c) are of the form $\dfrac{dy}{dx} + Py = Q$, which is linear in y.

Also, (d) is $\dfrac{d^2y}{dx^2} = \cos x$, on integrating $\dfrac{dy}{dx} = \sin x + C$

which is also linear in y.

25. $\dfrac{dy}{dx} = \dfrac{(y-x) \pm \sqrt{(x-y)^2 + 4xy}}{2y}$

$$\Rightarrow \quad \frac{dy}{dx} = 1 \quad \text{or} \quad \frac{dy}{dx} = -\frac{x}{y}$$

$$\Rightarrow \quad x - y + 1 = 0 \quad \text{and} \quad x^2 + y^2 = 25$$

26. (a) $f(x, tx) = e^t + \tan^{-1}(t)$, independent of x.

$\Rightarrow$ Homogeneous differential equation.

(b) $\log\left(\dfrac{y}{x}\right)dx + \dfrac{y^2}{x^2}\sin^{-1}\dfrac{y}{x}\cdot dy = 0$

$$\Rightarrow \quad \frac{dy}{dx} = \frac{\log\left(\frac{y}{x}\right)}{\frac{y^2}{x^2}\sin^{-1}\left(\frac{y}{x}\right)}$$

$$f(x, y) = \frac{\log\left(\frac{y}{x}\right)}{\frac{y^2}{x^2}\sin^{-1}\left(\frac{y}{x}\right)}$$

$\therefore f(x, tx) = \dfrac{\log(t)}{t^2\sin^{-1}(t)}$, independent of x

$\Rightarrow$ Homogeneous differential equation.

(c) $f(x, y) = x^2 + \sin x \cdot \cos y$

$f(x, tx) = x^2 + \sin x \cdot \cos(tx)$, not independent of x.

$\Rightarrow$ Not homogeneous differential equation.

(d) $f(x, y) = \dfrac{x^2 + y^2}{xy^2 - y^3}$

$\Rightarrow f(x, tx)$ is not independent of x.

$\Rightarrow$ Not homogeneous differential equation.

27. Integrating Factor $= e^{\int \cos x\, dx} = e^{-\sin x}$

or $\quad y \cdot e^{-\sin x} = \int e^{+\sin x} \cos x = -e^{+\sin x} + C$

$y = 1$

28. $\dfrac{dy}{dx} - y \cot x = -\dfrac{\sin x}{x^2}$

Integrating Factor $= e^{\int -\cot x\, dx} = e^{-\log \sin x} = \dfrac{1}{\sin x}$

$\therefore$ Required solution is $y \cdot \dfrac{1}{\sin x} = \int -\dfrac{\sin x}{x^2} \cdot \dfrac{1}{\sin x}\, dx + C$

$$y \cdot \frac{1}{\sin x} = \frac{1}{x} + C$$

As $x \to \infty$, $y \to 0 \Rightarrow C = 0$

$\therefore \quad y = \dfrac{\sin x}{x} \Rightarrow \lim_{x \to 0} f(x) = 1 \quad$ ...(i)

$$I = \int_0^{\pi/2} \frac{\sin x}{x}\, dx$$

Since, $\dfrac{\sin x}{x}$ is decreasing.

$$x > 0 \Rightarrow f(x) < f(0)$$

$\Rightarrow \quad \int_0^{\pi/2} f(x) < \dfrac{\pi}{2}$ and $x < \dfrac{\pi}{2}$

$\Rightarrow \quad f(x) > \int \left(\dfrac{\pi}{2}\right)$

$$\int_0^{\pi/2} f(x)\, dx > 1$$

29. (a) Order of the differential equation is 2.

(b) $\quad \dfrac{xdy - ydx}{\sqrt{x^2 + y^2}} = dx$

$$\frac{\dfrac{xdy - ydx}{x^2}}{\sqrt{\left(1 + \dfrac{y^2}{x^2}\right)}} = \frac{dx}{x}$$

$\therefore \quad \ln \left| \dfrac{y}{x} + \sqrt{1 + \dfrac{y^2}{x^2}} \right| = \ln |Cx|$

$\therefore \quad \dfrac{y}{x} + \dfrac{\sqrt{x^2 + y^2}}{x} = Cx \quad ie, \quad y + \sqrt{x^2 + y^2} = Cx^2$

(c) $y = e^x (A\cos x + B\sin x)$

$$\frac{dy}{dx} = e^x (A\cos x + B\sin x) + e^x(-A\sin x + B\cos x)$$

$$= y + e^x (-A\sin x + B\cos x)$$

$$\therefore \quad \frac{d^2y}{dx^2} = \frac{dy}{dx} + e^x (-A\sin x + B\cos x) + e^x (-A\cos x - B\sin x)$$

$$\therefore \quad \frac{d^2y}{dx^2} = \frac{dy}{dx} + e^x (-A\sin x + B\cos x) - y$$

$$\frac{dy}{dx} + \frac{dy}{dx} - y - y = 2\left(\frac{dy}{dx} - y\right)$$

(d) $(1 + y^2)\dfrac{dy}{dx} + x = 2e^{\tan^{-1} y}$

$$\Rightarrow \quad \frac{dx}{dy} + \frac{1}{1+y^2}x = \frac{e^{\tan^{-1} y}}{1+y^2}$$

$$\text{IF} = e^{\int \frac{1}{1+y^2}} = e^{\tan^{-1} y} \quad \Rightarrow \quad xe^{\tan^{-1} y} = 2\int e^{\tan^{-1} y} \cdot \frac{e^{\tan^1 y}}{1+y^2}\,dy$$

$$\Rightarrow \quad xe^{\tan^{-1} y} = e^{2\tan^{-1} y} + k$$

30. $\dfrac{dy}{dx} = 3\,(A + Bx)\,e^{3x} = Be^{3x}$

$$\Rightarrow \quad \frac{dy}{dx} + my = (3 + m)(A + Bx)\,e^{3x} + Be^{3x}$$

$$\Rightarrow \quad \frac{d^2y}{dx^2} + m\frac{dy}{dx} + ny = (9 + 3m + n)(A + Bx)\,e^{3x} + B\,(6 + m)\,e^{3x} = 0$$

$\Rightarrow \quad 3m + n + 9 = 0$ and $m + 6 = 0$

$\Rightarrow \quad m = -6$ and $n = 9$

Type 3 : Assertion and Reason

31. Equation of tangent

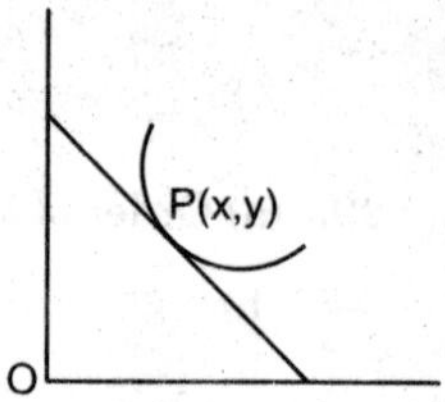

$$Y - y = m\,(X - x)$$

Put $X = 0$, $\quad Y = y - mx$

Since, initial ordinate is

$$y - mx = x - 1$$

$$\Rightarrow \quad mx - y = 1 - x$$

$$\frac{dy}{dx} - \frac{1}{x}y = \frac{1-x}{x}$$ which is a linear differential equation.

Hence, Statement I is correct and its degree is 1.

$\Rightarrow$ Statement II is also correct. Since, every 1st degree differential equation need not be linear, hence Statement II is not the correct explanation of Statement I.

32. **Statement I** The order of differential equation is 2.

$\therefore$ Statement I is false.

33. Integral curves are $y = cx - x^2$
The differential equation does not represent all the parabolas passing through origin but it represents all parabolas through origin with axis of symmetry parallel to y-axis and coefficient of x^2 as -1, hence Statement I is false.
Statement II is universally true.

34. $$\frac{ydx - xdy}{y^2}\cdot\cot\frac{x}{y} = x\,dx$$

or $$\int \cot\frac{x}{y}\cdot d\left(\frac{x}{y}\right) = \int x\,dx$$

or $$\int \cot t\,dx = nx + C$$

$$\ln(\sin t) = nx + C;\ \sin\frac{x}{y} = Ce^{nx}$$

35. $y = c_1\cos 2x + c_2\sin^2 x + c_3\cos^2 x + c_4e^{2x} + c_5$

$$e^{2x+c_6} = c_1\cos 2x + c_2\left[\frac{1-\cos 2x}{2}\right] + c_3\left[\frac{\cos 2x + 1}{2}\right] + c_4e^{2x} + c_5e^{2x}\cdot e^{c_6}$$

$$= \left(c_1 - \frac{c_2}{2} + \frac{c_3}{2}\right)\cos 2x + \left(\frac{c_2}{2} + \frac{c_3}{2}\right) + (c_4 + c'_5)\,e^{2x} = \lambda_1\cos 2x + \lambda_2 e^{2x} + \boldsymbol{\lambda_3}$$

$\Rightarrow$ Total number of independent parameters in the given general solution is **3.**

36. The given differential equation is

$$\frac{d\left(\frac{dy}{dx}\right)}{\frac{dy}{dx}} = \frac{2x}{x^2+1}\,dx$$

$$\ln\left(\frac{dy}{dx}\right) = \ln(x^2+1) + \ln C,\ C > 0$$

$$\Rightarrow \quad \frac{dy}{dx} = C\,(x^2+1)$$

$$\Rightarrow \quad y = C\left(\frac{x^3}{3} + x\right) + C',\ C' \in R$$

Obviously $y \to \infty$, as $x \to \infty$; as $C > 0$

37. $(x-h)^2 = 4b\,(y-k)$ here b is constant and h, k are parameters.

38. The general equation of all non-vertical lines in a plane is $ax + by = 1$, **where** $b \neq 0$.

Now, $$ax + by = 1$$

$$\Rightarrow \quad a + b\frac{dy}{dx} = 0 \quad \text{(differentiating w.r.t. } x\text{)}$$

$$\Rightarrow \quad b\frac{d^2y}{dx^2} = 0 \quad \text{(differentiating w.r.t. } x\text{)}$$

$$\Rightarrow \quad \frac{d^2y}{dx^2} = 0 \quad (\text{as } b \neq 0)$$

Hence, the differential equation is $\dfrac{d^2y}{dx^2} = 0.$

39. The equation $ax^2 + 2hxy + by^2 = 1$ represents the family of all conics whose centre lies at the origin for different values of a, h, b.

$\therefore$ Order is 3.

Thus, Statement I is false and Statement II is true.

Hence, option (d) is the correct answer.

40. $y = a \sin x + b \cos x$

$$\frac{dy}{dx} = a \cos x - b \sin x \Rightarrow \frac{d^2y}{dx^2} = -a \cos x - b \sin x = -y$$

$$\Rightarrow \quad \frac{d^2y}{dx^2} + y = 0$$

But Statement II is not the correct explanation of the Statement I.

Type 4 : **Linked–Comprehension Based Questions**

Solutions (Q. Nos. 41 to 43)

$f(0) = 2$

41. $f(x) = (e^x + e^{-x}) \cos x - 2x - \left[x\int_0^x f'(t)\,dt - \int_0^x \underset{\text{I}}{t}\, \underset{\text{II}}{f'(t)}\,dt\right]$

$f(x) = (e^x + e^{-x}) \cos x - 2x - [x(f(x) - f(0)) - \{t \cdot f(t)|_0^x - \int_0^x f(t)\,dt\}]$

$f(x) = (e^x + e^{-x}) \cos x - 2x - xf(x) + 2x + \left[x\, f(x) - \int_0^x f(t)\,dt\right]$

$f(x) = (e^x + \quad ^{x}) \cos x - \int_0^x f(t)\,dt$ ...(i)

On differentiating Eq. (i)

$f'(x) + f(x) = \cos x\,(e^x - e^{-x}) - (e^x + e^{-x}) \sin x$ ...(ii)

Hence, $\dfrac{dy}{dx} + y = e^x (\cos x - \sin x) - e^x (\cos x + \sin x)$

42. $f'(0) + f(0) = 0 - 2 \cdot 0 = 0$

43. IF of Eq. (i) is e^x

$$y \cdot e^x = \int e^{2x}(\cos x - \sin x)\,dx - \int (\cos x + \sin x)\,dx$$

$$y \cdot e^x = \int e^{2x}(\cos x - \sin x)\,dx - (\sin x - \cos x) + C$$

Let $\quad I = \int e^{2x}(\cos x - \sin x)\,dx = e^{2x}(A \cos x + B \sin x)$

Solving, $\quad A = 3/5$ and $B = -1/5$ and $C = 2/5$

$$\therefore \quad y = e^x \left(\frac{3}{5} \cos x - \frac{1}{5} \sin x\right) - (\sin x - \cos x)\, e^{-x} + \frac{2}{5} e^{-x}$$

Solutions (Q. Nos. 44 to 46)

Integrating, $\dfrac{d^2y}{dx^2} = 6x - 4$, we get $\dfrac{dy}{dx} = 3x^2 - 4x + A$

When $x = 1$, $\dfrac{dy}{dx} = 0$ so that $A = 1$.

Hence, $$\frac{dy}{dx} = 3x^2 - 4x + 1 \quad \ldots\text{(i)}$$

Integrating, we get $y = x^3 - 2x^2 + x + B$.

When $x = 1$, $y = 5$, so that $B = 5$

Thus, we have $y = x^3 - 2x^2 + x + 5$

From Eq. (i), we get the critical points $x = \frac{1}{3}, x = 1$

At the critical point $x = \frac{1}{3}$, $\frac{d^2y}{dx^2}$ is negative.

Therefore, at $x = \frac{1}{3}$, y has a local maximum.

At $x = 1$, $\frac{d^2y}{dx^2}$ is positive.

Therefore, at $x = 1$, y has a local minimum.

Also, $f(1) = 5$, $f\left(\frac{1}{3}\right) = \frac{157}{27}$, $f(0) = 5$, $f(2) = 7$

Hence, the global maximum value = 7.

and the global minimum value = 5.

47. $xdy - ydx = \sqrt{x^2 - y^2}\, dx$

$$\Rightarrow \quad d\left(\frac{y}{x}\right) = \frac{\sqrt{1-(y/x)^2}}{x}\, dx \quad \Rightarrow \int \frac{d(y/x)}{\sqrt{1-(y/x)^2}} = \int \frac{dx}{x}$$

$$\Rightarrow \sin^{-1}\frac{y}{x} = \ln x = \ln C \text{ or } Cx = e^{\sin^{-1} y/x}$$

48. $$(xy^4 + y)\, dx - x\, dy = 0$$

$$\Rightarrow \quad xy^4 dx + y\, dx - x\, dy = 0$$

$$\Rightarrow \quad \int x^3\, dx - \int \left(\frac{x}{y}\right)^2 d\left(\frac{x}{y}\right) = 0$$

$$\Rightarrow \quad \frac{x^4}{4} + \frac{1}{3}\left(\frac{x}{y}\right)^3 = C$$

49. $2x \cos y\, dx - x^2 \sin y\, dy + y^2 \cos x\, dx + 2y \sin x\, dy = 0$

$$\Rightarrow \quad \int d(x^2 \cos y) + \int d(y^2 \sin x) = 0$$

$$\Rightarrow \quad x^2 \cos y + y^2 \sin x = C$$

50. $$\frac{dx}{x} = \frac{ydy}{1+y^2}$$

$$\Rightarrow \quad \ln x = \frac{1}{2} \cdot \ln(1+y^2) + C$$

From the given condition, $C = 0$

$$\therefore \quad x^2 - y^2 = 1$$

51. $$\frac{dy}{dx} + \frac{1+y^2}{\sqrt{1-x^2}} = 0$$

$$\Rightarrow \quad \frac{dy}{1+y^2} + \frac{dx}{\sqrt{1-x^2}} = 0$$

$$\Rightarrow \quad \tan^{-1} y + \sin^{-1} x = C$$

52. $\frac{dy}{dx} = (1+x)\cdot(1+y)$ gives $y = e^{\frac{(1+x)^2}{2}} - 1$

Type 5 : Match the Columns

53. (A) Let $x = r\cos\theta$, $y = r\sin\theta$

$$\therefore \quad x^2 + y^2 = r^2(\sin^2\theta + \cos^2\theta) = r^2 \quad \text{...(i)}$$

$$\text{and} \quad \tan\theta = \frac{y}{x} \quad \text{...(ii)}$$

$$\therefore \quad d(x^2+y^2) = d(r^2)$$

From Eq. (i),

$$x\,dx + y\,dy = r dr \quad \text{...(iii)}$$

From Eq. (ii),

$$d\left(\frac{y}{x}\right) = d(\tan\theta)$$

$$\frac{xdy - ydx}{x^2} = \sec^2\theta\, d\theta$$

$$\therefore \quad x\,dy - y\,dx = x^2\sec^2\theta\, d\theta$$

$$= r^2\cos^2\theta\sec^2\theta\, d\theta = r^2 d\theta \quad \text{...(iv)}$$

From Eqs. (iii) and (iv)

$$\frac{rdr}{r^2 d\theta} = \sqrt{\frac{a^2 - r^2}{r^2}}$$

$$\text{or} \quad \frac{dr}{\sqrt{a^2 - r^2}} = d\theta$$

$$\sin^{-1}\left(\frac{r}{a}\right) = \theta + C$$

$$\Rightarrow \quad r = a\sin(\theta + C)$$

$$\Rightarrow \quad \sqrt{x^2 + y^2} = a\sin\left\{C + \tan^{-1}\left(\frac{y}{x}\right)\right\}$$

(B) $\frac{dy}{dx} - \sec^2 x \tan 2xy = \cos^2 x$

$$\text{IF} = e^{\int \tan 2x \sec^2 x\, dx} = e^{\int \frac{2\tan x}{\tan^2 x - 1} \times \sec^2 x\, dx}$$

$$= e^{\int \frac{dt}{t}}, \text{ where } t = \tan^2 x - 1$$

$$= e^{\ln|t|} = |t| = |\tan^2 x - 1|$$

Given that, $|x| < \frac{\pi}{4}$

$\therefore \quad \tan^2 x < 1$

$\therefore \quad IF = 1 - \tan^2 x$

Solution is $y(1 - \tan^2 x) = \int \cos^2 x (1 - \tan^2 x)\, dx = \int (\cos^2 x - \sin^2 x)\, dx$

$$= \int \cos 2x \, dx = \frac{\sin 2x}{2} + C$$

when $x = \frac{\pi}{6}, y = \frac{3\sqrt{3}}{8}$

$$\therefore \quad \frac{3\sqrt{3}}{8}\left(1 - \frac{1}{3}\right) = \frac{1}{2} \times \frac{\sqrt{3}}{2} + C$$

$$\therefore \quad C = 0 \quad \Rightarrow \quad y = \frac{\sin 2x}{2(1 - \tan^2 x)}$$

(C) $$x^2 + y^2 - 2cx = 0 \quad \text{...(i)}$$

$$\therefore \quad 2x + 2y\frac{dy}{dx} - 2c = 0$$

or $$C = x + \frac{ydy}{dx} \quad \text{...(ii)}$$

From Eqs. (i) and (ii), $x^2 + y^2 - 2\left(x + \frac{ydy}{dx}\right)x = 0$

or $-x^2 + y^2 - 2xy\frac{dy}{dx} = 0$ is the differential equation representing the given family of circles. To find the orthogonal trajectories.

$$y^2 - x^2 = -2xy\frac{dx}{dy}$$

or $$y^2\, dy = x^2\, dy - 2xy\, dx$$

or $$-dy = \frac{yd(x^2) - x^2\, dy}{y^2}$$

or $$-dy = d\left(\frac{x^2}{y}\right) \quad \text{or} \quad -y = \frac{x^2}{y} + C$$

$\Rightarrow$ $$x^2 + y^2 + Cy = 0$$

$\Rightarrow$ Orthogonal trajectory.

(D) Differentiating the equation w.r.t. x, we get

$$xy(x) + 1\int_0^x y(t)\, dt = (x+1)\, xy(x) + 1\int_0^x ty(t)\, dt$$

Again, differentiating, w.r.t. x, we get

$$y(x) = x^2 y'(x) + 2xy(x) + xy(x)$$

or $$(1 - 3x)\, y(x) = \frac{x^2 dy(x)}{dx}$$

or $$\frac{(1 - 3x)\, dx}{x^2} = \frac{dy(x)}{y(x)}$$

or $$y = \frac{C}{x^3} e^{-1/x}$$

54. (A) $r = 5$ cm

$$\delta r = 0.06$$

$$A = \pi r^2$$

$$\delta A = 2\pi r\, \delta r = 10\pi \times 0.06 = 0.6\pi$$

(B) $v = x^3$

$$\delta v = 3x^2\, \delta x$$

$$\frac{\delta v}{v} \times 100 = 3\frac{\delta x}{x} \times 100 = 3 \times 1 = 3$$

(C) $(x-2)\dfrac{dx}{dt} = 2\dfrac{dx}{dt}$

$$x = 4$$

(D) $$A = \frac{\sqrt{3}}{4}x^2$$

$$\frac{dA}{dt} = \frac{\sqrt{3}}{2}x\frac{dx}{dt} = \frac{\sqrt{3}}{2}\cdot 15 \cdot \frac{1}{10} = \frac{3\sqrt{3}}{4}$$

55. (A) $\dfrac{dy}{dx} = e^x(A\cos x + B\sin x) + e^x(-A\sin x + B\cos x)$

$$= y + e^x(-A\sin x + B\cos x)$$

$$\therefore \frac{d^2y}{dx^2} = \frac{dy}{dx} + e^x(-A\sin x + B\cos x) + e^x(-A\cos x - B\sin x)$$

$$= \frac{dy}{dx} + \left(\frac{dy}{dx} - y\right) - y = 2\left(\frac{dy}{dx} - y\right)$$

So, degree = 1 and order = 2

(B) The equation of the family is $y^2 = 4a(x-b)$ where a, b are arbitrary constants.

$$\therefore \quad 2y\frac{dy}{dx} = 4a \quad \text{or} \quad \left(\frac{dy}{dx}\right)^2 + y\frac{d^2y}{dx^2} = 0$$

So, degree = 1 and order = 2

(C) The equation of the family is

$$(x-c)^2 + y^2 = c^2$$

or $$x^2 + y^2 - 2cx = 0$$

or $$\frac{x^2 + y^2}{x} = 2c$$

$$\Rightarrow \quad \frac{\left(2x + 2y\dfrac{dy}{dx}\right)x - (x^2 + y^2)\cdot 1}{x^2} = 0$$

So, degree = 1 and order = 1

(D) The equation of the family is $\dfrac{x^2}{a^2 + \lambda} + \dfrac{y^2}{b^2 + \lambda} = 1$ because they have the same foci $(\pm\sqrt{a^2 - b^2}, 0)$.

On differentiating, $$\frac{2x}{d^2 + \lambda} + \frac{2y}{b^2 + \lambda}\cdot\frac{dy}{dx} = 0$$

or $$\frac{x}{a^2+\lambda}+\frac{yp}{b^2+\lambda}=0 \qquad \left(\text{Let } p=\frac{dy}{dx}\right)$$

or $$x(b^2+\lambda)+yp(a^2+\lambda)=0$$

$\Rightarrow$ $$\lambda=\frac{-b^2x-a^2yp}{x+yp}$$

$\therefore$ The differential equation is $$\frac{x^2}{a^2+\dfrac{-b^2x-a^2yp}{x+yp}}+\frac{y^2}{b^2+\dfrac{-b^2x-a^2yp}{x+yp}}=1$$

or $$\frac{x(x+yp)}{a^2-b^2}+\frac{y(x+yp)}{(b^2-a^2)p}=1$$

So, order = 1 and degree = 2

Type 6 : Integer Answer Type Questions

56. $2x^2y\,dx-2y^4\,dx+2x^3dy+3xy^3\,dy=0$

Dividing throughout by x^3y, we get

$$2\frac{dx}{x}-\frac{2y^3}{x^3}dx+2\cdot\frac{dy}{y}+\frac{3y^2}{x^2}dy=0$$

$\Rightarrow$ $$2\frac{dx}{x}+2\frac{dy}{y}+\frac{3y^2x^4\,dy-2y^3x^3dx}{x^6}=0$$

$\Rightarrow$ $$2\frac{dx}{x}+2\frac{dy}{y}+\frac{3y^2x^2\,dy-2y^3x\,dx}{x^4}=0$$

$\Rightarrow$ $$2d(\ln x)+2d(\ln y)+d\left(\frac{y^3}{x^2}\right)=0$$

$\Rightarrow$ $$2\ln|x|+2\ln|y|+\frac{y^3}{x^2}=C$$

when $x=1$, $y=1$

So, $C=1$

57. Let the salt content at time 't' be u lb, so that its rate of change is du/dt.

$$=2\text{ gal}\times 2\text{ lb}=4\text{ lb/min}$$

If c be the concentration of the brine at time t, the rate at which the salt content decreases due to the out flow = 2 gal × c lb/min = $2c$ lb/min

$\therefore$ $$\frac{du}{dt}=4-2c \qquad \text{...(i)}$$

Also, since there is no increase in the volume of the liquid, the concentrations

$$c=\frac{u}{50}$$

$\therefore$ Eq. (i) becomes $$\frac{du}{dt}=4-\frac{2u}{50}$$

Separating the variables and integrating, we have

$$\int dt=25\int\frac{du}{100-u}$$

or $$t = -25 \ln(100 - u) + K \quad \text{...(ii)}$$

Initially when $t = 0, u = 0$

$$0 = -25 \ln 100 + K \quad \text{...(iii)}$$

Eliminating 'K' from Eqs. (ii) and (iii), we get $t = 25 \ln\left(\dfrac{100}{100 - u}\right)$

Taking $t = t_1$ when $u = 40$ and $t = t_2$ when $u = 80$, we have

$$t_1 = 25 \ln\left(\frac{100}{60}\right) \quad \text{and} \quad t_2 = 25 \ln\left(\frac{100}{20}\right)$$

$\therefore$ The required time $(t_2 - t_1)$

$$= 25\left(\ln 5 - \ln \frac{5}{3}\right) = 25 \ln 3 = 25 \times 1.0986 = 26 \text{ min } 28 \text{ s}$$

$$= 1648 \text{ s} = 206 \times 8 = 206 \times \lambda$$

$\therefore$ $$\lambda = 8$$

58. $f(x + f(y) + xf(y)) = y + f(x) + yf(x)$ ...(i)

Differentiating w.r.t. x and y is constant

$$f'(x + f(y) + xf(y))(1 + f(y)) = f'(x) + yf'(x) \quad \text{...(ii)}$$

From Eq. (i) again differentiating w.r.t. y as x is constant

$$f'(x + f(y) + xf(y))(1 + x)f'(y) = 1 + f(x) \quad \text{...(iii)}$$

From Eqs. (ii) and (iii)

$$\frac{1 + f(y)}{(1 + x)f'(y)} = \frac{(1 + y)f'(x)}{1 + f(x)}$$

$$\frac{(1 + y)f'(y)}{1 + f(y)} = \frac{1 + f(x)}{(1 + x)f'(x)} = \lambda$$

$\therefore$ $$f'(y) = \frac{1 + f(y)}{1 + y}\lambda \quad \text{and} \quad f'(x) = \frac{1 + f(x)}{\lambda(1 + x)}$$

$\therefore$ $$\lambda = \frac{1}{\lambda} \Rightarrow \lambda = \pm 1 \quad \therefore \quad \frac{f'(x)}{1 + f(x)} = \pm \frac{1}{1 + x}$$

Integrating both the sides $f(x) = C(1 + x)^{\pm 1} - 1$

From Eq. (i) put $x = y = 0$

$$f(f(0)) = f(0)$$

From Eq. (ii), $f(0) = C - 1$

So, $$f(C - 1) = C - 1$$

$\therefore$ $$C = 0, 1 \quad \text{(taking +ve sign)}$$

So, $f(x) = -1$ and $f(x) = (1 + x) - 1 = x$ and $C = 1$

$\therefore$ $$f(x) = (1 + x)^{-1} - 1 = \frac{-x}{1 + x}$$

$$1 + f(x) = \frac{1}{1 + x}$$

$\therefore$ $$1 + f(2009) = \frac{1}{2010}$$

$\therefore$ $$(2010)(1 + f(2009)) = 1$$

59. $\phi'(x) + 2\phi(x) \le 1$

$$e^{2x}\,\phi'(x) + 2\phi(x)\,e^{2x} \le e^{2x}$$

$$\frac{d}{dx}\left(e^{2x}\,\phi(x) - \frac{1}{2}e^{2x}\right) \le 0$$

$\therefore$ $\left(e^{2x}\,\phi(x) - \frac{1}{2}\right)$ is a non-increasing function of x.

$\Rightarrow$ $\phi(x) - \frac{1}{2}$ is a non-increasing function of x.

$\therefore$ $2\phi(x)$ is always less than or equal to 1.

60. $\frac{1}{2}(1+x)^{-1/2} - \frac{a}{2}(1+y)^{-1/2}\frac{dy}{dx} = 0$

$$\Rightarrow \quad \frac{1}{\sqrt{1+x}} = \frac{a}{\sqrt{1+y}}\cdot\frac{dy}{dx} \quad \Rightarrow \quad a = \frac{\sqrt{1+y}}{\sqrt{1+x}}\cdot\frac{1}{dy/dx}$$

Putting this value in the given equation,

$$\frac{dy}{dx}\sqrt{1+x} - \frac{1+y}{\sqrt{1+x}} = \frac{dy}{dx}$$

$$\Rightarrow \quad (1+x)\frac{dy}{dx} = 1 + y + \sqrt{1+x}\,\frac{dy}{dx}$$

$$\Rightarrow \quad (1 + x - \sqrt{1+x})\frac{dy}{dx} = 1 + y$$

$\therefore$ Degree of the given equation is 1.

(*Proficiency Exercise 2*)

1. (i) $\tan\frac{y}{x} = \ln\frac{k}{x}$

(ii) $(x^2 + y^2)^{3/2} - 3xy = C$

(iii) $\frac{(xy^2)^3}{3} = \ln\left(\frac{x}{y}\right) + C$

(iv) $y = f(x) - 1 + Ce^{-f(x)}$

(v) $(x^2 - 1)\cos y = \frac{x^4}{2} - x^2 + C$

(vi) $\sin\left(\frac{x}{y}\right) = (xy)^2/2 + C$

(vii) $[(x-1)^2 + (y-1)^2]\tan^{-1}\left(\frac{y-2}{x-1}\right) = (x-1)(y-2) + 2(x-1)^2\log C(x-1)$

(viii) $(y - x + 4)\,e^{\frac{x+3}{x-1}} = (x-1)\log|(x-1)| + C(x-1)$

(ix) $2\tan^{-1}\left(\frac{y}{\sin x} + 1\right) = \ln\sin x + C$

(x) $\frac{2}{\sqrt{3}}\tan^{-1}\left(\frac{2\tan\theta - 1}{\sqrt{3}}\right) = x + C$

2. Let the curve C is $y = f(x)$. If $P(x, y)$ be a point on $y = f(x)$, then the equation of tangent at P is $Y - y = \dfrac{dy}{dx}(X - x)$.

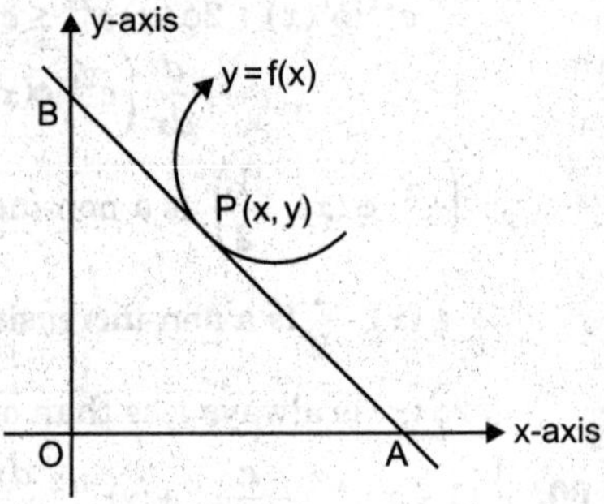

Putting $Y = 0$, x-intercept $= x - y\dfrac{dx}{dy}$

Putting $X = 0$, y-intercept $= y - x\dfrac{dy}{dx}$

The tangent cuts coordinate axes at A and B and P is the mid-point of AB.

$$\therefore \quad x = \frac{0 + x - y\dfrac{dx}{dy}}{2}$$

and
$$y = \frac{0 + y - x\dfrac{dy}{dx}}{2}$$

In each case, we get $x\dfrac{dy}{dx} = -y$

$$\Rightarrow \quad \frac{dy}{y} + \frac{dx}{x} = 0 \quad \Rightarrow \quad \ln|xy| = K$$

$$\Rightarrow \quad xy = \pm e^K = C$$

Since, the curve passes through (1, 1).

$$\therefore \quad 1 \times 1 = C$$

Hence, required curve is $xy = 1$.

3. The equation of the tangent at any point (x, y) is given by

$$Y - y = \frac{dy}{dx}(X - x)$$

When $X = 0$, $Y = y - x\dfrac{dy}{dx} = y$-intercept

It is given that $Y = x^2$

ie,
$$y - x\frac{dy}{dx} = x^2$$

ie,
$$\frac{dy}{dx} - \frac{y}{x} = -x$$

$$\text{IF} = e^{\int -\frac{1}{x}dx} = e^{-\log x} = \frac{1}{x}$$

$\therefore$ The solution is $y \cdot \dfrac{1}{x} = \displaystyle\int -x \cdot \frac{1}{x}\,dx = -x + C$

ie,
$$y = -x^2 + Cx$$

4. The equation of hyperbola is $xy = c$

or
$$y = \frac{c}{x} \quad \therefore \quad \frac{dy}{dx} = -\frac{c}{x^2}$$

Given that the angle which the required family of curve makes with the hyperbola is $\dfrac{\pi}{4}$.

$$\therefore \qquad \tan\left(\frac{\pi}{4}\right) = \left|\frac{dy/dx - (-c/x^2)}{1 + (dy/dx)\cdot\left(-\frac{c}{x^2}\right)}\right|$$

$\frac{dy}{dx}$ being the slope of the unknown curve

$$\Rightarrow \qquad \frac{dy}{dx} + \frac{c}{x^2} = \pm\left(1 - \frac{c}{x^2}\cdot\frac{dy}{dx}\right)$$

Taking +ve sign; $1 - \frac{c}{x^2} = \left(1 + \frac{c}{x^2}\right)\frac{dy}{dx}$

$$\Rightarrow \qquad dy = \frac{x^2 - c}{x^2 + c}\,dx = \left(1 - \frac{2c}{x^2 + c}\right)dx$$

$$\Rightarrow \qquad y = x - 2\sqrt{c}\,\tan^{-1}\left(\frac{x}{\sqrt{c}}\right) + K$$

Taking – ve sign; $\frac{c}{x^2} + 1 = \frac{dy}{dx}\left(\frac{c}{x^2} - 1\right)$

$$\Rightarrow \qquad dy = \frac{c + x^2}{c - x^2}\,dx = -\left(1 + \frac{2c}{x^2 - c}\right)dx$$

$$\Rightarrow \qquad x + y = -\frac{2c}{2\sqrt{c}}\log\left|\frac{x - \sqrt{c}}{x + \sqrt{c}}\right| + K$$

$$\Rightarrow \qquad \frac{x + y}{\sqrt{c}} + \log\left|\frac{x - \sqrt{c}}{x + \sqrt{c}}\right| = K$$

5. $\frac{dv}{dt} \propto v$ for each reservoir.

For reservoir A; $\frac{dv_A}{dt} \int - v_A$

$$\frac{dv_A}{dt} = -k_1 v_A \qquad (k_1 \text{ is +ve proportional constant})$$

$$\Rightarrow \qquad \int_{v_A}^{v_A'} \frac{dv_A}{v_A} = -k_1 \int_0^t dt$$

$$\Rightarrow \qquad \left(\log\frac{v_A'}{v_A}\right) = -k_1 t \Rightarrow v_A' = v_A \cdot e^{-k_1 t} \qquad \text{...(i)}$$

Similarly, for B, $\quad v_B' = v_B \cdot e^{-k_2 t}$ ...(ii)

Dividing Eq. (i) by Eq. (ii), we get

$$\frac{v_A'}{v_B'} = \frac{v_A}{v_B}\cdot e^{-(k_1 - k_2)t} \qquad \text{...(iii)}$$

It is given that at $t = 0, v_A = 2v_B$ and at $t = 3/2, v_A' = 1\frac{1}{2}v_B'$

Thus, $\qquad \frac{3}{2} = 2\cdot e^{-(k_1 - k_2)t}$

$$\Rightarrow \qquad e^{-(k_1 - k_2)} = 3/4$$

Now, let at $t = t_0$ both the reservoirs have some quantity of water.

Then, $v_A' = v_B'$

Hence, from Eq. (iii), $2 \cdot e^{-(k_2 - k_1) t_0} = 1$

$$\Rightarrow \quad \frac{1}{2} = \{e^{-(k_1 - k_2)}\}^{t_0} = \left(\frac{3}{4}\right)^{t_0}$$

$$\Rightarrow \quad t_0 = \log_{3/4}\left(\frac{1}{2}\right)$$

6. Let in time dt the decrease in water level in the tank is dh, then amount of water flown out in time $dt = \pi r^2 \cdot dh$

Now, through the hole the amount of water flown = (Volume of cylinder of cross sectional area 'a' and length vdt).

$$= a \cdot v \cdot dt = a\,\mu\,\sqrt{2gh}\,dt$$

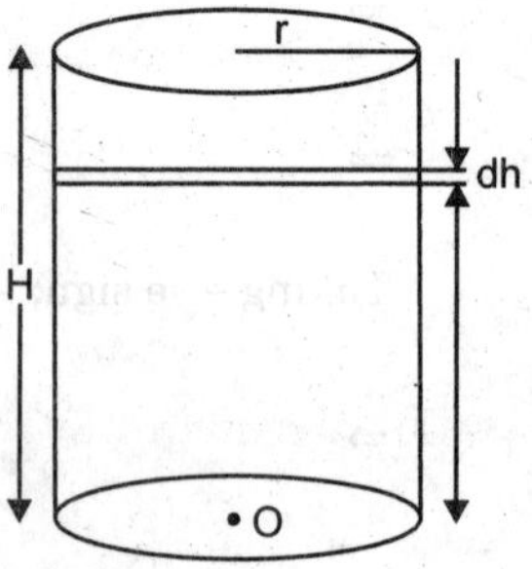

Hence, $\mu a \sqrt{2gh}\,dt = -\pi r^2 dh$

$$\Rightarrow \quad dt = \frac{-\pi r^2}{\mu a \sqrt{2g}} \cdot h^{-1/2}\,dh$$

Now, when $t = 0$, $h = H$ and when $t = t$, $h = 0$

Thus, $$\int_0^t dt = \frac{-\pi r^2}{\mu a \sqrt{2g}} \int_H^0 h^{-1/2}\,dh$$

$$\Rightarrow \quad t = \frac{-\pi r^2}{\mu a \sqrt{2g}} \cdot 2(\sqrt{h})_H^0$$

$$\Rightarrow \quad t = \frac{\pi r^2}{\mu a}\sqrt{\frac{2H}{g}}$$

7. Let at time t the depth of water is h; the radius of water surface is r.

Then, $r^2 = R^2 - (R - h)^2$

$$\Rightarrow \quad r^2 = 2Rh - h^2$$

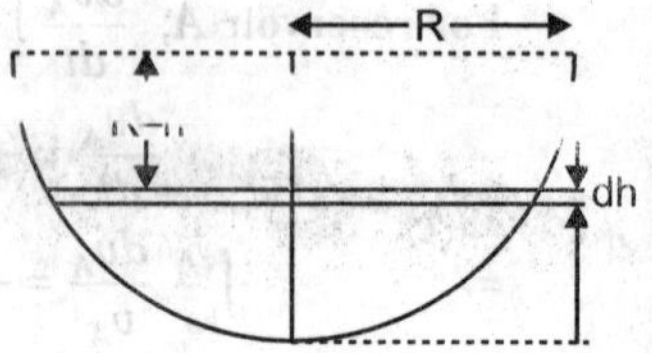

Now, if in time dt the decrease in water level is dh, then

$$-\pi r^2\,dh = 0.6\sqrt{2gh} \cdot a\,dt$$

(a is cross-sectional area of the outlet)

$$\Rightarrow \quad \frac{-\pi}{(0.6)\,a\sqrt{2g}}(2Rh - h^2)\frac{dh}{\sqrt{h}} = dt$$

$$\Rightarrow \quad \frac{\pi}{(0.6)\,a\sqrt{2g}}\int_R^0 (h^{3/2} - 2R\,h^{1/2})\,dh = \int_0^t dt$$

$$\Rightarrow \quad \frac{\pi}{(0.6)\,a\sqrt{2g}}\left[\frac{2}{5}h^{5/2} - \frac{4}{3}Rh^{3/2}\right]_R^0 = t$$

$$\Rightarrow \quad t = \frac{\pi}{(0.6)\,a\sqrt{2g}}\left[0 - R^{5/2}\left(\frac{2}{5} - \frac{4}{3}\right)\right] = \frac{7\pi \times 10^5}{135\sqrt{g}}$$

8. Tangent at point $P(x, y)$ to the curve is given by

$$Y - y = \frac{dy}{dx}(X - x)$$

At the y-axis, $n = 3$

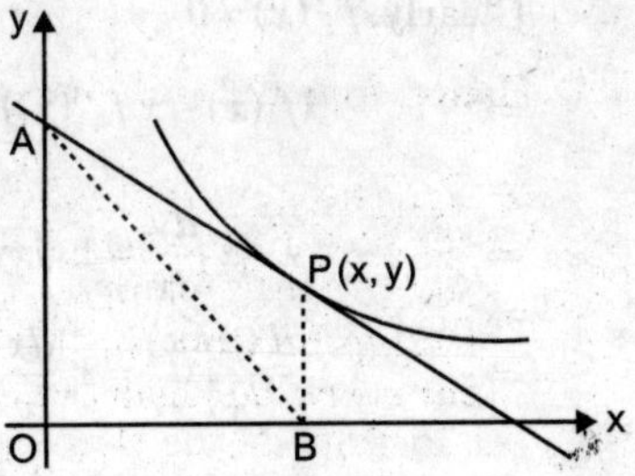

$\Rightarrow \qquad A = \left(0, y - x\dfrac{dy}{dx}\right)$

Clearly, $\qquad B = (x, 0)$

Given that, $\qquad \dfrac{1}{2} \cdot x\left(y - x\dfrac{dy}{dx}\right) = K$

$\Rightarrow \qquad x^2 \dfrac{dy}{dx} - xy = -2K$

or $\qquad \dfrac{dy}{dx} - \dfrac{y}{x} = -\dfrac{2K}{x^2}$

$$\text{IF} = e^{\int -\frac{1}{x}dx} = e^{-\log x} = +1/x$$

$\therefore$ The solution is given by

$$y \cdot \frac{1}{x} = \int \frac{-2K}{x^3}\, dx + C \quad \text{or} \quad \frac{y}{x} = \frac{K}{x^2} + C$$

$\Rightarrow \qquad Cx^2 - xy + K = 0$

(Which is clearly a hyperbola as $\Delta \neq 0$ and $H^2 > AB$)

9. Differentiating both the sides of $x^2 + 2y^2 - y = c$ w.r.t. x, we get

$$2x + 4y\frac{dy}{dx} - \frac{dy}{dx} = 0$$

or $2x + (4y-1)\dfrac{dy}{dx} = 0$; is the differential equation of the given family of curves.

Replacing $\dfrac{dy}{dx}$ by $-\dfrac{dx}{dt}$ to obtain the differential equation of orthogonal trajectories, we get

$$2x + \frac{(1-4y)}{dy/dx} = 0 \Rightarrow \frac{dy}{dx} = \frac{4y-1}{2x}$$

$\Rightarrow \qquad \displaystyle\int \frac{dy}{4y-1} = \int \frac{dx}{2x} \Rightarrow \frac{1}{4}\ln|4y-1| = \frac{1}{2}\ln|x| + \ln|C|$

$\Rightarrow \qquad \ln|4y-1| = 2\ln|x| + \ln|C|$

$\Rightarrow \qquad y = \dfrac{1}{4}Cx^2 + \dfrac{1}{4}$

Which is required orthogonal trajectory.

But it is given that $y = Cx^2 + k$ is the orthogonal trajectory.

Comparing both, we get $k = 1/4$.

10. Let $y = f(x)$ be the curve. Equations of tangent at $P(x, y)$

$$Y - y = \frac{dy}{dx}(X - x) = f'(x)(X - x)$$

Intercept made on y-axis is $f(x) - xf'(x)$.

Subnormal is $f(x) \cdot f'(x)$ and subtangent is $\dfrac{f(x)}{f'(x)}$.

As per the given condition

$$\frac{(f(x) - xf'(x))^2}{f(x) \cdot f'(x)} = \frac{xf(x) \cdot f'(x)}{(f'(x))^2 \cdot f(x)}$$

Clearly, $f'(x) \neq 0$

$$\Rightarrow \quad (f(x) - x f'(x))^2 = x f(x) \text{ or } \left(y - x \frac{dy}{dx} \right)^2 = xy$$

$$\Rightarrow \quad y - x \frac{dy}{dx} = \pm \sqrt{xy} \Rightarrow \frac{x\,dy - y\,dx}{x^2} = \pm \frac{\sqrt{xy}}{x^2} dx$$

$$\Rightarrow \quad \frac{d(y/x)}{\sqrt{y/x}} = \pm \frac{dx}{x} \Rightarrow 2\sqrt{y/x} = \pm \ln|x| + C$$

Curves passes through (1, 0) $\Rightarrow C = 0$

$$\Rightarrow \quad 2\sqrt{y/x} = \ln|x| \text{ or } x = e^{\pm 2\sqrt{y/x}}$$

11. Equation of tangent at 'p' is

$$(Y - y) = f'(x)(X - x)$$

Let us assume that this tangent meets the line $y = 3$ at T

$$\Rightarrow \quad T \equiv (X, 3)$$

Where $(3 - y) = f'(x)(X - x) \Rightarrow X = x + \frac{3 - y}{f'(x)}$

Let $Q(h, k)$ is the mid-point of PT

$$\Rightarrow \quad 2h = 2x + \frac{3 - y}{f'(x)} \text{ and } 2K = 3 + y$$

This point must lie on the line $x + y = 1$

$$\Rightarrow \quad 2h + 2K = 2$$

$$\Rightarrow \quad 2x + \frac{3 - y}{f'(x)} + 3 + y = 2$$

$$\Rightarrow \quad 2x + \frac{3 - y}{f'(x)} + 1 + y = 0$$

$$\Rightarrow \quad f'(x) = \frac{y - 3}{2x + y + 1} \Rightarrow \frac{dy}{dx} = \frac{y - 3}{2(x - 2) + y - 3}$$

Let $x_1 = x + 2$, $y_1 = y - 3$

$$\Rightarrow \quad \frac{dy}{dx} = \frac{dy_1}{dx_1} = \frac{y_1}{2x_1 + y_1}$$

$$\Rightarrow \quad \frac{dx_1}{dy_1} = \frac{2x_1 + y_1}{y_1} = \frac{2x_1}{y_1} + 1$$

$$\Rightarrow \quad \frac{dx_1}{dy_1} - \frac{2x_1}{y_1} = 1$$

Multiplying by $e^{\int -\frac{2}{y_1} dy_1} = \frac{1}{y_1^2}$ on both the sides, we get

$$\frac{1}{y_1^2} \cdot \frac{dx_1}{dy_1} - \frac{2x_1}{y_1^3} = \frac{1}{y_1^2}$$

$$\Rightarrow \quad d\left(\frac{x_1}{y_1^2} \right) = \frac{dy_1}{y_1^2} \Rightarrow d\left(\frac{x_1}{y_1} \right) = \frac{dy_1}{y_1^2}$$

$$\Rightarrow \quad x_1 = y_1^2\left(-\frac{1}{y_1} + C\right) = y_1 + Cy_1^2$$

$$\Rightarrow \quad (x+2) = -(y-3) + C(y-3)^2$$

$$\Rightarrow \quad C(y-3)^2 = x + y - 1$$

As the curve passes through (0, 0), we get $9C = -1$

$\Rightarrow$ Required curve is $(y-3)^2 = 9(1 - x - y)$.

12. Let P be (x, y) and PN be the perpendicular from P to x-axis. According to the given problem $y^2 = 2x \cdot OG$

or $$y^2 = 2x\,[ON + NG]$$

$$\Rightarrow \quad y^2 = 2x\left[x + y\frac{dy}{dx}\right] = 2x^2 + 2xy\cdot\left(\frac{dy}{dx}\right)$$

$$\Rightarrow \quad 2xy\left(\frac{dy}{dx}\right) = y^2 - 2x^2$$

Which is a homogeneous differential equation. Therefore, by putting $y = vx$, the above equation reduces to

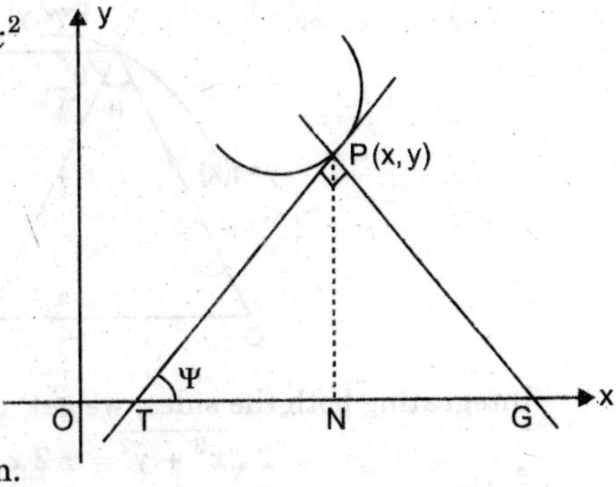

$$2v\,x^2\left(v + x\frac{dv}{dx}\right) = v^2x^2 - 2x^2$$

$$\Rightarrow \quad 2v\left(v + x\frac{dv}{dx}\right) = v^2 - 2$$

$$\Rightarrow \quad 2v\,x\left(\frac{dv}{dx}\right) = -v^2 - 2$$

$$\Rightarrow \quad \frac{2v}{v^2+2}\,dv + \frac{dx}{x} = 0$$

Integration $\ln(v^2+2) + \ln x = \ln C$

or $\quad y^2 + 2x^2 = Cx$ is the solution.

If it passes through $(2, 1) \Rightarrow C = 9/2$

$\therefore$ Required solution $y^2 + 2x^2 = \frac{9}{2}x$

$$\Rightarrow \quad 4x^2 + 2y^2 = 9x$$

13. In ΔPNQ, $\quad \tan\psi = \frac{NQ}{y}$; $NQ = y\cdot\frac{dy}{dx}$

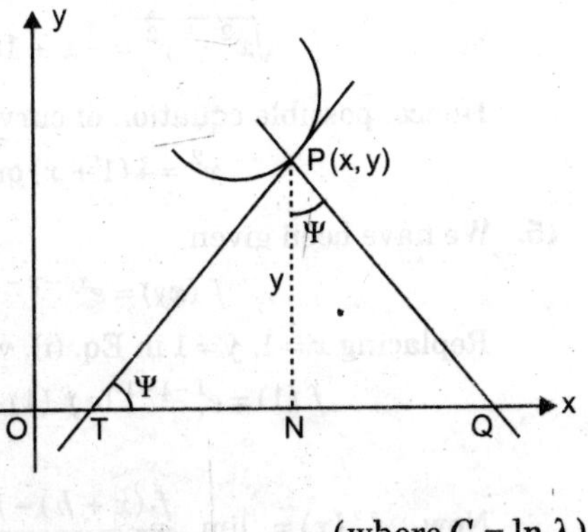

Given that, $$y\frac{dy}{dx} = \frac{x(1+y^2)}{(1+x^2)}$$

$$\Rightarrow \quad \frac{y\,dy}{1+y^2} = \frac{x\,dx}{1+x^2}$$

Integrating $\int\frac{y\,dy}{1+y^2} = \int\frac{x\,dx}{1+x^2}$

$$\Rightarrow \quad \frac{1}{2}\ln(1+y^2) = \frac{1}{2}\ln(1+x^2) + C$$

$$\Rightarrow \quad \ln\sqrt{\frac{1+y^2}{1+x^2}} = \ln\lambda \qquad \text{(where } C = \ln\lambda\text{)}$$

$$\Rightarrow \quad 1+y^2=\lambda^2(1+x^2)$$

Given that the curve passes through the point (3, 1).

$$1+1=\lambda^2(1+9) \Rightarrow \lambda^2=\frac{1}{5}$$

Required equation is $5(1+y^2)=(1+x^2)$

14. Given that, $(y-f(x))=-\dfrac{1}{f'(x)}(X-x)$

$$\Rightarrow \quad N \equiv (x+f(x)\cdot f'(x), 0) \quad \text{or} \quad \left(x+y\frac{dy}{dx}, 0\right)$$

Since, $ON=OP$, $\sqrt{x^2+y^2}=\left|x+y\dfrac{dy}{dx}\right|$

$$\Rightarrow \quad \frac{2x\,dx+2y\,dy}{\sqrt{x^2+y^2}}=\pm 2\,dx$$

$$\Rightarrow \quad \frac{d(x^2+y^2)}{\sqrt{x^2+y^2}}=\pm 2\,dx$$

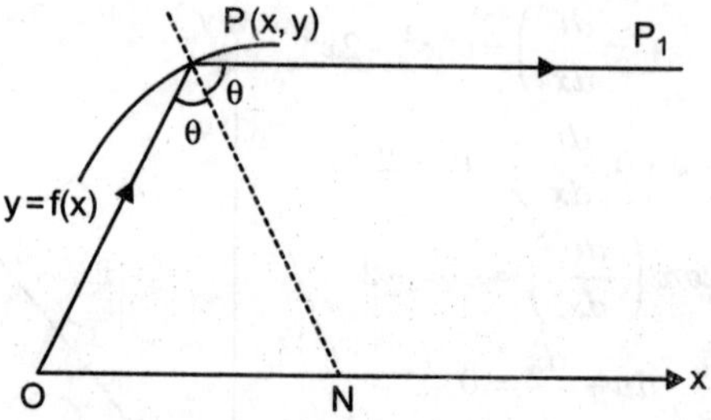

Integrating both the sides, we get

$$2\sqrt{x^2+y^2}=\pm 2x+2K$$

$$\Rightarrow \quad \sqrt{x^2+y^2}=\pm x+K$$

Curve passes through (8, 6).

$$\Rightarrow \quad 10=\pm 8+K \Rightarrow K=2, 18$$

Thus, curve is $\sqrt{x^2+y^2}=x+2$, *ie*, $y^2=4+4x$

or $\quad \sqrt{x^2+y^2}=-x+18$, *ie*, $y^2=324-36x$

Hence, possible equation of curves are

$$y^2=4(1+x) \text{ or } y^2=36(9-x)$$

15. We have been given,

$$f(xy)=e^{xy-x-y}\{e^y f(x)+e^x f(y)\}, \forall\, x, y \in R^+ \quad \text{...(i)}$$

Replacing $x=1$, $y=1$ in Eq. (i), we get

$$f(1)=e^{1-1-1}\{e\,f(1)+e\,f(1)\} \Rightarrow f(1)=0$$

Now, $f'(x)=\lim\limits_{h\to 0}\dfrac{f(x+h)-f(x)}{h}=\lim\limits_{h\to 0}\dfrac{f\left(x\left(1+\frac{h}{x}\right)\right)-f(x)}{h}$

$$= \lim_{h \to 0} \frac{e^{x+h-x-1-\frac{h}{x}} (e^{1+h/x} \cdot f(x) + e^x \cdot f(1+h/x)) - f(x)}{h}$$

$$= \lim_{h \to 0} \frac{e^h f(x) + e^{h-1-\frac{h}{x}+x} \cdot f\left(1+\frac{h}{x}\right) - f(x)}{h}$$

$$= \lim_{h \to 0} \frac{f(x)(e^h - 1) + e^{h-1-\frac{h}{x}+x} \cdot \left(f\left(1+\frac{h}{x}\right) - f(1)\right)}{h}$$

$$= \lim_{h \to 0} \frac{f(x)(e^h - 1)}{h} + \lim_{h \to 0} \frac{e^{h-1-\frac{h}{x}+x} \left(f\left(1+\frac{h}{x}\right) - f(1)\right)}{\frac{h}{x} \cdot x}$$

$$= f(x) + \frac{e^{x-1} \cdot f'(1)}{x} = f(x) + \frac{e^x}{x}$$

$$\Rightarrow \quad f'(x) - f(x) = \frac{e^x}{x}$$

$$\Rightarrow \quad \frac{e^x f'(x) - f(x) e^x}{e^{2x}} = \frac{1}{x}$$

$$\Rightarrow \quad \frac{d}{dx}\left(\frac{f(x)}{e^x}\right) = \frac{1}{x}$$

Integrating both the sides, we get

$$\frac{f(x)}{e^x} = \ln|x| + C$$

Since, $f(1) = 0 \Rightarrow C = 0$

Thus, $\quad f(x) = e^x \cdot \ln(x)$

16. Let $P(x, f(x))$ be a point lying on the curve in first quadrant. Equation of normal and tangent at P are

$$(Y - f(x)) = -\frac{1}{f'(x)}(X - x) \text{ and } (Y - f(x)) = f'(x)(X - x) \text{ respectively.}$$

$$\Rightarrow \quad A \equiv \left(x - \frac{f(x)}{f'(x)}, 0\right), B \equiv (x + f(x) f'(x), 0)$$

Since, the mid-point of segment AB is origin.

$$\Rightarrow \quad 2x - \frac{f(x)}{f'(x)} + f(x) \cdot f'(x) = 0$$

$$\Rightarrow \quad f(x)(f'(x))^2 + 2x f'(x) - f(x) = 0$$

$$\Rightarrow \quad f'(x) = \frac{-2x \pm \sqrt{4x^2 + 4f^2(x)}}{2f(x)} = \frac{-x + \sqrt{x^2 + f^2(x)}}{f(x)}$$

Negative sign been neglected as $f'(x) > 0$

Thus, we have, $x + f(x) \cdot f'(x) = \sqrt{x^2 + f^2(x)}$

$$\Rightarrow \quad \frac{d}{2dx}(x^2 + f^2(x)) = \sqrt{x^2 + f^2(x)} \Rightarrow \frac{d(x^2 + f^2(x))}{2\sqrt{x^2 + f^2(x)}} = dx$$

Integrating both the sides, we get

$$\sqrt{x^2 + f^2(x)} = x + \lambda$$

It passes through $(1, \sqrt{3})$.

Hence, $\lambda + 1 = 2 \Rightarrow \lambda = 1$

$\therefore$ Curve is $x^2 + f^2(x) = (x+1)^2$

or $y^2 = 1 + 2x$

17. Let the curve be $y = f(x)$

The equation of the tangent at any point (x, y) is $Y - y = \frac{dy}{dx}(X - x)$

Its intercept on the y-axis is given by $(X = 0)$

$$Y = y - x\frac{dy}{dx} = k\sqrt{x^2 + y^2} \qquad \text{(given)}$$

So that $x\frac{dy}{dx} - y + k\sqrt{x^2 + y^2} = 0$ is the differential equation governing the curve this can be written as

$$\frac{dy}{dx} = \frac{y - k\sqrt{x^2 + y^2}}{x}$$

Let $y = vx$, so that $\frac{dy}{dx} = v + x\frac{dv}{dx}$

The differential equation becomes

$$v + x\frac{dv}{dx} = v - k\sqrt{1 + v^2}$$

$$\Rightarrow \quad \frac{dv}{\sqrt{1 + v^2}} = -k\frac{dx}{x}$$

Let $v = \tan\theta \Rightarrow dv = \sec^2\theta \, d\theta$

Hence, $\sec\theta \, d\theta = -k\frac{dx}{x}$

$$\Rightarrow \quad \log|\sec\theta + \tan\theta| = -k\ln|x| + C$$

$$\Rightarrow \quad \log|v + \sqrt{1 + v^2}| + k\log|x| = C$$

$$\Rightarrow \quad \log|y + \sqrt{x^2 + y^2}| + (k-1)\log|x| = C$$

$$\Rightarrow \quad (y + \sqrt{x^2 + y^2}) \cdot x^{k-1} = C_1$$

18. Any point (x, y) on the curve can be represented by using parameter θ, as $(a\cos^3\theta, a\sin^3\theta)$

$$\therefore \quad \frac{dy}{dx} = \frac{dy/d\theta}{dx/d\theta} = \frac{3a\sin^2\theta\cos\theta}{-3a\cos^2\theta\sin\theta} = -\tan\theta$$

$\therefore$ Equation of the tangent at P is

$$(y - a\sin^3\theta) = -\tan\theta(x - a\cos^3\theta)$$

or

$$(\tan\theta)x + y = a\sin^3\theta + a\cos^3\theta \cdot \tan\theta \qquad \text{...(i)}$$

$$\therefore \quad p_1 = \left| \frac{a \sin^3 \theta + a \cos^3 \theta \tan \theta}{\sec \theta} \right| = \frac{1}{2} | a \sin 2\theta |$$

Also, equation of the normal at P is

$$(y - a \sin^3 \theta) = \cot \theta \, (x - a \cos^3 \theta)$$

or $$x \cot \theta - y = a \cos^3 \theta \cot \theta - a \sin^3 \theta$$

$$\therefore \quad p_2 = \left| \frac{a \cos^3 \theta \cot \theta - a \sin^3 \theta}{\operatorname{cosec} \theta} \right| = | a \cos 2\theta |$$

$\Rightarrow$ $4p_1^2 + p_2^2 = a^2 =$ constant $\forall$ point $P(x, y)$

$\Rightarrow$ If p_1 increases, p_2 decreases and conversely.

$\Rightarrow$ $\frac{dp_1}{dx}$ and $\frac{dp_2}{dx}$ are of opposite signs.

$$\Rightarrow \quad \frac{dp_1}{dx} \cdot \frac{dp_2}{dx} \le 0$$

19. We have been given, $\frac{dy_1}{dx} + Py_1 = Q, \; \frac{dy_2}{dx} + Py_2 = Q$

Now, $$y_2 = y_1 z \Rightarrow \frac{dy_2}{dx} = y_1 \frac{dz}{dx} + z \frac{dy_1}{dx}$$

$$\Rightarrow \quad y_1 \frac{dz}{dx} + z \frac{dy_1}{dx} + Py_1 z = Q$$

$$\Rightarrow \quad y_1 \cdot \frac{dz}{dx} + z \left(\frac{dy_1}{dx} + py_1 \right) = Q$$

$$\Rightarrow \quad y_1 \cdot \frac{dz}{dx} + z Q = Q$$

$$\Rightarrow \quad y_1 \frac{dz}{dx} = Q(1 - z)$$

$$\Rightarrow \quad \frac{dz}{1 - z} = \frac{Q}{y_1} dx$$

$\Rightarrow$ $\ln | z - 1 | = -\int \frac{Q}{y_1} dx + \lambda$, '$\lambda$' being constant of integration

$$\Rightarrow \quad z = 1 + c e^{\int -\frac{Q}{y_1} dx}$$

20. Let the fixed point source be situated at the origin and the x-axis be parallel to the reflected ray. Where P is the point (x, y) on the curve. Let α be the angle which the tangent at P makes with the x-axis.

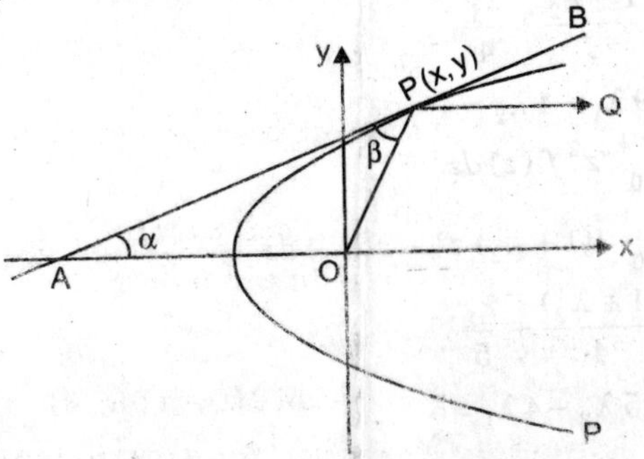

The angle of incidence = The angle of reflection

$\Rightarrow \quad \beta = \angle OPA = \angle BPQ = \angle OAP = \alpha$

$\Rightarrow \quad \beta = \alpha$

$\angle XOP = \alpha + \beta = 2\alpha$

$\tan(\angle XOP) = \frac{y}{x} \Rightarrow \tan 2\alpha = \frac{y}{x}$

$\Rightarrow \quad \frac{2\tan\alpha}{1-\tan^2\alpha} = \frac{y}{x} \Rightarrow \frac{2m}{1-m^2} = \frac{y}{x}$ (where $m = \tan\alpha$)

$\Rightarrow \quad 2x = \frac{y}{m} - ym$ ...(i)

Differentiating w.r.t. y, we get

$$\frac{2dx}{dy} = \frac{1}{m} - \frac{y}{m^2}\frac{dm}{dy} - m - y\frac{dm}{dy}$$

$\Rightarrow \quad \frac{2}{m} = \frac{1}{m} - \frac{y}{m^2}\frac{dm}{dy} - m - y\frac{dm}{dy}$

$\Rightarrow \quad \left(\frac{1}{m} + m\right) + y\left(\frac{1}{m^2} + 1\right)\frac{dm}{dy} = 0 \Rightarrow \left(\frac{1}{m} + m\right)\left(1 + \frac{y}{m}\frac{dm}{dy}\right) = 0$

$\Rightarrow \quad \frac{dm}{m} = -\frac{dy}{y} \Rightarrow \log m = \log\frac{C}{y}$ $\left(\text{since } \frac{m+1}{m} \neq 0\right)$

$\Rightarrow \quad m = \frac{C}{y}$ ...(ii)

Thus, eliminating m from Eqs. (i) and (ii),

$$2x = \frac{y}{C/y} - y\frac{C}{m} \Rightarrow 2x = \frac{y^2}{C} - C$$

$\Rightarrow \quad y^2 = 2Cx + C^2$, this is passing through (0, 2)

$\Rightarrow \quad C^2 = 4 \Rightarrow C = \pm 2$

Hence, $\quad y^2 = \pm 4x + 4$

$y^2 = 4(x+1)$ and $y^2 = -4(x-1)$

21. We have, $f(x) = x + x^2\int_0^1 z f(z)\,dz + x\int_0^1 z^2 f(z)\,dz$

$f(x) = x + x^2\lambda_1 + x\lambda_2$ (say)

Now, $\quad \lambda_1 = \int_0^1 z f(z)\,dz = \int_0^1 ((1+\lambda_2)z + z^2\lambda_1)\,z\,dz$

$= \frac{1+\lambda_2}{3} + \frac{\lambda_1}{4}$

$\Rightarrow \quad 9\lambda_1 - 4\lambda_2 = 4$ ...(i)

Also, $\quad \lambda_2 = \int_0^1 z^2 f(z)\,dz$

$= \int_0^1 ((1+\lambda_2)z^3 + z^4\lambda_1)\,dz$

$= \frac{(1+\lambda_2)}{4} + \frac{\lambda_1}{5}$

$\Rightarrow \quad 15\lambda_2 - 4\lambda_1 = 5$ ...(ii)

From Eqs. (i) and (ii),

$$\lambda_1 = \frac{80}{119} \quad \text{and} \quad \lambda_2 = \frac{61}{119}$$

$$\Rightarrow \qquad f(x) = x + \frac{80}{119}x^2 + \frac{61}{119}x = \frac{20x}{119}(4 + 9x)$$

22. $f(x + f(y) + x f(y)) = y + f(x) + y f(x)$

Keep y constant and differentiating the expression w.r.t. 'x', we get

$$f'(x + f(y) + x(y))(1 + f(y)) = f'(x)(1 + y) \qquad \text{...(i)}$$

Similarly, differentiate the expression w.r.t. 'y' and keep x constant, we get

$$f'(x + f(y) + x f(y))(f'(y)(1 + x)) = (1 + f(x)) \qquad \text{...(ii)}$$

Dividing Eq. (i) by Eq. (ii), we get

$$\frac{1 + f(y)}{f'(y)(1 + x)} = \frac{f'(x)(1 + y)}{(1 + f(x))}$$

$$\Rightarrow \qquad \frac{1 + f(y)}{f'(y)(1 + y)} = \frac{f'(x)(1 + x)}{(1 + f(x))} = C$$

$$\Rightarrow \qquad f'(y) = \frac{1}{C}\left[\frac{1 + f(y)}{1 + y}\right] \text{ and } f'(x) = C\frac{1 + f(x)}{(1 + x)}$$

$$\Rightarrow \qquad C = \frac{1}{C} \Rightarrow C = \pm 1$$

$$\therefore \qquad \frac{f'(x)}{1 + f(x)} = \pm\frac{1}{1 + x}$$

$$\Rightarrow \qquad f(x) = \lambda_1 (1 + x)^{\pm 1} - 1$$

Replacing $x, y \to 0$ in the Eq. (i), we get

$$f(f(0)) = f(0)$$

Now, $\quad f(x) = \lambda_1 (1 + x)^{\pm 1} - 1 \Rightarrow f(0) = \lambda_1 - 1$

and $\quad f(f(0)) = f(\lambda_1 - 1) = \lambda_1 \lambda_1^{\pm 1} - 1$

Since, $\quad f(f(0)) = f(0),\ \lambda_1 \lambda_1^{\pm 1} - 1 = \lambda_1 - 1$

$$\Rightarrow \qquad \lambda_1^{1 \pm 1} = \lambda_1$$

By taking +ve sign, we get $\lambda_1 = 0, 1$

$$\Rightarrow \qquad f(x) = -1 \text{ or } f(x) = x$$

By taking – ve sign, we get $\lambda_1 = 1$

$$\Rightarrow \qquad f(x) = \frac{1}{1 + x} - 1 = -\frac{x}{1 + x}$$